MW01008109

Distillation Troubleshooting

Distillation Troubleshooting

Henry Z. Kister

Fluor Corporation

A JOHN WILEY & SONS, INC., PUBLICATION

DISCLAIMER

The author and contributors to "Distillation Troubleshooting" do not represent, warrant, or otherwise guarantee, expressly or impliedly, that following the ideas, information, and recommendations outlined in this book will improve tower design, operation, downtime, troubleshooting, or the suitability, accuracy, reliability or completeness of the information or case histories contained herein. The users of the ideas, the information, and the recommendations contained in this book apply them at their own election and at their own risk. The author and the contributors to this book each expressly disclaims liability for any loss, damage or injury suffered or incurred as a result of or related to anyone using or relying on any of the ideas or recommendations in this book. The information and recommended practices included in this book are not intended to replace individual company standards or sound judgment in any circumstances. The information and recommendations in this book are offered as lessons from the past to be considered for the development of individual company standards and procedures.

Published by John Wiley & Sons, Inc., Hoboken, New Jersey.
Published simultaneously in Canada.

Library of Congress Cataloging-in-Publication Data:

Kister, Henry Z.
Distillation troubleshooting / Henry Z. Kister.
p. cm.
Includes bibliographical references.
ISBN-13 978-0-471-46744-1 (Cloth)
ISBN-10 0-471-46744-8 (Cloth)
1. Distillation apparatus—Maintenance and repair. I. Title.
TP159.D9K57 2005
660′.28425—dc22 2004016490

Printed in the United States of America

15 14 13 12

To my son, Abraham and my wife, Susana, who have been my love, inspiration, and the lighthouses illuminating my path,

and to my life-long mentor, Dr. Walter Stupin – it is easy to rise when carried on the shoulders of giants.

Contents

Preface xxiii

Acknowledgments xxvii

How to Use this Book xxix

Abbreviations xxxi

1. Troubleshooting Distillation Simulations 1

2. Where Fractionation Goes Wrong 25

3. Energy Savings and Thermal Effects 61

4. Tower Sizing and Material Selection Affect Performance 73

5. Feed Entry Pitfalls in Tray Towers 97

6. Packed-Tower Liquid Distributors: Number 6 on the Top 10 Malfunctions 111

7. Vapor Maldistribution in Trays and Packings 133

8. Tower Base Level and Reboiler Return: Number 2 on the Top 10 Malfunctions 145

9. Chimney Tray Malfunctions: Part of Number 7 on the Top 10 Malfunctions 163

10. Draw-Off Malfunctions (Non–Chimney Tray) Part of Number 7 on the Top 10 Malfunctions 179

11. Tower Assembly Mishaps: Number 5 on the Top 10 Malfunctions 193

12. Difficulties During Start-Up, Shutdown, Commissioning, and Abnormal Operation: Number 4 on the Top 10 Malfunctions 215

13. Water-Induced Pressure Surges: Part of Number 3 on the Top 10 Malfunctions 225

14. Explosions, Fires, and Chemical Releases: Number 10 on the Top 10 Malfunctions 233

15. Undesired Reactions in Towers 237

16. Foaming 241

17. The Tower as a Filter: Part A. Causes of Plugging—Number 1 on the Top 10 Malfunctions 253

18. The Tower as a Filter: Part B. Location of Plugging—Number 1 on the Top 10 Malfunctions 257

19. Coking: Number 1 on the Top 10 Malfunctions 271

20. Leaks 281

21. Relief and Failure 287

22. Tray, Packing, and Tower Damage: Part of Number 3 on the Top 10 Malfunctions 291

23. Reboilers That Did Not Work: Number 9 on the Top 10 Malfunctions 315

24. Condensers That Did Not Work 335

25. Misleading Measurements: Number 8 on the Top 10 Malfunctions 347

26. Control System Assembly Difficulties 357

27. Where Do Temperature and Composition Controls Go Wrong? 373

28. Misbehaved Pressure, Condenser, Reboiler, and Preheater Controls 377

29. Miscellaneous Control Problems 395

DISTILLATION TROUBLESHOOTING DATABASE OF PUBLISHED CASE HISTORIES

1. Troubleshooting Distillation Simulations 398

1.1 VLE 398
1.1.1 Close-Boiling Systems 398
1.1.2 Nonideal Systems 399
1.1.3 Nonideality Predicted in Ideal System 400
1.1.4 Nonideal VLE Extrapolated to Pure Products 400
1.1.5 Nonideal VLE Extrapolated to Different Pressures 401
1.1.6 Incorrect Accounting for Association Gives Wild Predictions 401
1.1.7 Poor Characterization of Petroleum Fractions 402
1.2 Chemistry, Process Sequence 402
1.3 Does Your Distillation Simulation Reflect the Real World? 404
1.3.1 General 404
1.3.2 With Second Liquid Phase 406
1.3.3 Refinery Vacuum Tower Wash Sections 406
1.3.4 Modeling Tower Feed 406
1.3.5 Simulation/Plant Data Mismatch Can Be Due to an Unexpected Internal Leak 406
1.3.6 Simulation/Plant Data Mismatch Can Be Due to Liquid Entrainment in Vapor Draw 407
1.3.7 Bug in Simulation 407
1.4 Graphical Techniques to Troubleshoot Simulations 407
1.4.1 McCabe–Thiele and Hengstebeck Diagrams 407
1.4.2 Multicomponent Composition Profiles 407
1.4.3 Residue Curve Maps 407
1.5 How Good Is Your Efficiency Estimate? 407
1.6 Simulator Hydraulic Predictions: To Trust or Not to Trust 409
1.6.1 Do Your Vapor and Liquid Loadings Correctly Reflect Subcool, Superheat, and Pumparounds? 409
1.6.2 How Good Are the Simulation Hydraulic Prediction Correlations? 409

2. Where Fractionation Goes Wrong **410**

2.1 Insufficient Reflux or Stages; Pinches 410
2.2 No Stripping in Stripper 412
2.3 Unique Features of Multicomponent Distillation 412
2.4 Accumulation and Hiccups 413
2.4.1 Intermediate Component, No Hiccups 413
2.4.2 Intermediate Component, with Hiccups 414
2.4.3 Lights Accumulation 416
2.4.4 Accumulation between Feed and Top or Feed and Bottom 417
2.4.5 Accumulation by Recycling 418
2.4.6 Hydrates, Freeze–Ups 418
2.5 Two Liquid Phases 419
2.6 Azeotropic and Extractive Distillation 421
2.6.1 Problems Unique to Azeotroping 421
2.6.2 Problems Unique to Extractive Distillation 423

3. Energy Savings and Thermal Effects **424**

3.1 Energy-Saving Designs and Operation 424
3.1.1 Excess Preheat and Precool 424
3.1.2 Side-Reboiler Problems 424
3.1.3 Bypassing a Feed around the Tower 424
3.1.4 Reducing Recycle 425
3.1.5 Heat Integration Imbalances 426
3.2 Subcooling: How It Impacts Towers 428
3.2.1 Additional Internal Condensation and Reflux 428
3.2.2 Less Loadings above Feed 429
3.2.3 Trapping Lights and Quenching 429
3.2.4 Others 430
3.3 Superheat: How It Impacts Towers 430

4. Tower Sizing and Material Selection Affect Performance **431**

4.1 Undersizing Trays and Downcomers 431
4.2 Oversizing Trays 431
4.3 Tray Details Can Bottleneck Towers 433
4.4 Low Liquid Loads Can Be Troublesome 434
4.4.1 Loss of Downcomer Seal 434
4.4.2 Tray Dryout 435
4.5 Special Bubble-Cap Tray Problems 436
4.6 Misting 437
4.7 Undersizing Packings 437
4.8 Systems Where Packings Perform Different from Expectations 437

4.9 Packed Bed Too Long 438
4.10 Packing Supports Can Bottleneck Towers 439
4.11 Packing Hold-downs Are Sometimes Troublesome 440
4.12 Internals Unique to Packed Towers 440
4.13 Empty (Spray) Sections 440

5. Feed Entry Pitfalls in Tray Towers 441

5.1 Does the Feed Enter the Correct Tray? 441
5.2 Feed Pipes Obstructing Downcomer Entrance 441
5.3 Feed Flash Can Choke Downcomers 441
5.4 Subcooled Feeds, Refluxes Are Not Always Trouble Free 442
5.5 Liquid and Unsuitable Distributors Do Not Work with Flashing Feeds 442
5.6 Flashing Feeds Require More Space 443
5.7 Uneven or Restrictive Liquid Split to Multipass Trays at Feeds and Pass Transitions 443
5.8 Oversized Feed Pipes 444
5.9 Plugged Distributor Holes 444
5.10 Low ΔP Trays Require Decent Distribution 445

6. Packed-Tower Liquid Distributors: Number 6 on the Top 10 Malfunctions 446

6.1 Better Quality Distributors Improve Performance 446
6.1.1 Original Distributor Orifice or Unspecified 446
6.1.2 Original Distributor Weir Type 447
6.1.3 Original Distributor Spray Type 447
6.2 Plugged Distributors Do Not Distribute Well 448
6.2.1 Pan/Trough Orifice Distributors 448
6.2.2 Pipe Orifice Distributors 449
6.2.3 Spray Distributors 450
6.3 Overflow in Gravity Distributors: Death to Distribution 451
6.4 Feed Pipe Entry and Predistributor Problems 454
6.5 Poor Flashing Feed Entry Bottleneck Towers 455
6.6 Oversized Weep Holes Generate Undesirable Distribution 456
6.7 Damaged Distributors Do Not Distribute Well 457
6.7.1 Broken Flanges or Missing Spray Nozzles 457
6.7.2 Others 457
6.8 Hole Pattern and Liquid Heads Determine Irrigation Quality 458
6.9 Gravity Distributors Are Meant to Be Level 459
6.10 Hold-Down Can Interfere with Distribution 460
6.11 Liquid Mixing Is Needed in Large-Diameter Distributors 460
6.12 Notched Distributors Have Unique Problems 461
6.13 Others 461

7. Vapor Maldistribution in Trays and Packings 462

7.1 Vapor Feed/Reboiler Return Maldistributes Vapor to Packing Above 462
7.1.1 Chemical/Gas Plant Packed Towers 462
7.1.2 Packed Refinery Main Fractionators 463
7.2 Experiences with Vapor Inlet Distribution Baffles 465
7.3 Packing Vapor Maldistribution at Intermediate Feeds and Chimney Trays 465
7.4 Vapor Maldistribution Is Detrimental in Tray Towers 466
7.4.1 Vapor Cross-Flow Channeling 466
7.4.2 Multipass Trays 467
7.4.3 Others 467

8. Tower Base Level and Reboiler Return: Number 2 on the Top 10 Malfunctions 468

8.1 Causes of High Base Level 468
8.1.1 Faulty Level Measurement or Level Control 468
8.1.2 Operation 469
8.1.3 Excess Reboiler Pressure Drop 470
8.1.4 Undersized Bottom Draw Nozzle or Bottom Line 470
8.1.5 Others 470
8.2 High Base Level Causes Premature Tower Flood (No Tray/Packing Damage) 470
8.3 High Base Liquid Level Causes Tray/Packing Damage 471
8.4 Impingement by the Reboiler Return Inlet 472
8.4.1 On Liquid Level 472
8.4.2 On Instruments 473
8.4.3 On Tower Wall 473
8.4.4 Opposing Reboiler Return Lines 474
8.4.5 On Trays 474
8.4.6 On Seal Pan Overflow 474
8.5 Undersized Bottom Feed Line 475
8.6 Low Base Liquid Level 475
8.7 Issues with Tower Base Baffles 476
8.8 Vortexing 476

9. Chimney Tray Malfunctions: Part of Number 7 on the Top 10 Malfunctions 477

9.1 Leakage 477
9.2 Problem with Liquid Removal, Downcomers, or Overflows 478
9.3 Thermal Expansion Causing Warping, Out-of-Levelness 479
9.4 Chimneys Impeding Liquid Flow to Outlet 480

9.5 Vapor from Chimneys Interfering with Incoming Liquid 480
9.6 Level Measurement Problems 481
9.7 Coking, Fouling, Freezing 482
9.8 Other Chimney Tray Issues 482

10. Drawoff Malfunctions (Non–Chimney Tray): Part of Number 7 on the Top 10 Malfunctions 484

10.1 Vapor Chokes Liquid Draw Lines 484
10.1.1 Insufficient Degassing 484
10.1.2 Excess Line Pressure Drop 485
10.1.3 Vortexing 486
10.2 Leak at Draw Tray Starves Draw 486
10.3 Draw Pans and Draw Lines Plug Up 488
10.4 Draw Tray Damage Affects Draw Rates 488
10.5 Undersized Side-Stripper Overhead Lines Restrict Draw Rates 488
10.6 Degassed Draw Pan Liquid Initiates Downcomer Backup Flood 489
10.7 Other Problems with Tower Liquid Draws 489
10.8 Liquid Entrainment in Vapor Side Draws 490
10.9 Reflux Drum Malfunctions 490
10.9.1 Reflux Drum Level Problems 490
10.9.2 Undersized or Plugged Product Lines 490
10.9.3 Two Liquid Phases 490

11. Tower Assembly Mishaps: Number 5 on the Top 10 Malfunctions 491

11.1 Incorrect Tray Assembly 491
11.2 Downcomer Clearance and Inlet Weir Malinstallation 491
11.3 Flow Passage Obstruction and Internals Misorientation at Tray Tower Feeds and Draws 492
11.4 Leaking Trays and Accumulator Trays 493
11.5 Bolts, Nuts, Clamps 493
11.6 Manways/Hatchways Left Unbolted 493
11.7 Materials of Construction Inferior to Those Specified 494
11.8 Debris Left in Tower or Piping 494
11.9 Packing Assembly Mishaps 495
11.9.1 Random 495
11.9.2 Structured 496
11.9.3 Grid 496
11.10 Fabrication and Installation Mishaps in Packing Distributors 496
11.11 Parts Not Fitting through Manholes 498
11.12 Auxiliary Heat Exchanger Fabrication and Assembly Mishaps 498
11.13 Auxiliary Piping Assembly Mishaps 498

12. Difficulties during Start-Up, Shutdown, Commissioning, and Abnormal Operation: Number 4 on the Top 10 Malfunctions 499

12.1 Blinding/Unblinding Lines 499
12.2 Backflow 500
12.3 Dead-Pocket Accumulation and Release of Trapped Materials 501
12.4 Purging 501
12.5 Pressuring and Depressuring 502
12.6 Washing 502
12.7 On-Line Washes 504
12.8 Steam and Water Operations 506
12.9 Overheating 506
12.10 Cooling 507
12.11 Overchilling 507
12.12 Water Removal 508
12.12.1 Draining at Low Points 508
12.12.2 Oil Circulation 508
12.12.3 Condensation of Steam Purges 508
12.12.4 Dehydration by Other Procedures 508
12.13 Start-Up and Initial Operation 509
12.13.1 Total-Reflux Operation 509
12.13.2 Adding Components That Smooth Start-Up 509
12.13.3 Siphoning 509
12.13.4 Pressure Control at Start-Up 510
12.14 Confined Space and Manhole Hazards 510

13. Water-Induced Pressure Surges: Part of Number 3 on the Top 10 Malfunctions 512

13.1 Water in Feed and Slop 512
13.2 Accumulated Water in Transfer Line to Tower and in Heater Passes 513
13.3 Water Accumulation in Dead Pockets 513
13.4 Water Pockets in Pump or Spare Pump Lines 514
13.5 Undrained Stripping Steam Lines 515
13.6 Condensed Steam or Refluxed Water Reaching Hot Section 516
13.7 Oil Entering Water-Filled Region 517

14. Explosions, Fires, and Chemical Releases: Number 10 on the Top 10 Malfunctions 518

14.1 Explosions Due to Decomposition Reactions 518
14.1.1 Ethylene Oxide Towers 518
14.1.2 Peroxide Towers 519
14.1.3 Nitro Compound Towers 520
14.1.4 Other Unstable-Chemical Towers 521

14.2 Explosions Due to Violent Reactions 523
14.3 Explosions and Fires Due to Line Fracture 524
14.3.1 C_3–C_4 Hydrocarbons 524
14.3.2 Overchilling 525
14.3.3 Water Freeze 526
14.3.4 Other 527
14.4 Explosions Due to Trapped Hydrocarbon or Chemical Release 527
14.5 Explosions Induced by Commissioning Operations 528
14.6 Packing Fires 529
14.6.1 Initiated by Hot Work Above Steel Packing 529
14.6.2 Pyrophoric Deposits Played a Major Role, Steel Packing 530
14.6.3 Tower Manholes Opened While Packing Hot, Steel Packing 532
14.6.4 Others, Steel Packing Fires 532
14.6.5 Titanium, Zinconium Packing Fires 533
14.7 Fires Due to Opening Tower before Cooling or Combustible Removal 533
14.8 Fires Caused by Backflow 534
14.9 Fires by Other Causes 535
14.10 Chemical Releases by Backflow 536
14.11 Trapped Chemicals Released 536
14.12 Relief, Venting, Draining, Blowdown to Atmosphere 537

15. Undesired Reactions in Towers 539

15.1 Excessive Bottom Temperature/Pressure 539
15.2 Hot Spots 539
15.3 Concentration or Entry of Reactive Chemical 539
15.4 Chemicals from Commissioning 540
15.5 Catalyst Fines, Rust, Tower Materials Promote Reaction 540
15.6 Long Residence Times 541
15.7 Inhibitor Problems 541
15.8 Air Leaks Promote Tower Reactions 542
15.9 Impurity in Product Causes Reaction Downstream 542

16. Foaming 543

16.1 What Causes or Promotes Foaming? 543
16.1.1 Solids, Corrosion Products 543
16.1.2 Corrosion and Fouling Inhibitors, Additives, and Impurities 544
16.1.3 Hydrocarbon Condensation into Aqueous Solutions 545
16.1.4 Wrong Filter Elements 546
16.1.5 Rapid Pressure Reduction 546
16.1.6 Proximity to Solution Plait Point 546

16.2 What Are Foams Sensitive To? 546
16.2.1 Feedstock 546
16.2.2 Temperature 547
16.2.3 Pressure 547
16.3 Laboratory Tests 547
16.3.1 Sample Shake, Air Bubbling 547
16.3.2 Oldershaw Column 547
16.3.3 Foam Test Apparatus 548
16.3.4 At Plant Conditions 548
16.4 Antifoam Injection 548
16.4.1 Effective Only at the Correct Quantity/Concentration 548
16.4.2 Some Antifoams Are More Effective Than Others 549
16.4.3 Batch Injection Often Works, But Continuous Can Be Better 549
16.4.4 Correct Dispersal Is Important, Too 550
16.4.5 Antifoam Is Sometimes Adsorbed on Carbon Beds 550
16.4.6 Other Successful Antifoam Experiences 550
16.4.7 Sometimes Antifoam Is Less Effective 551
16.5 System Cleanup Mitigates Foaming 551
16.5.1 Improving Filtration 551
16.5.2 Carbon Beds Mitigate Foaming But Can Adsorb Antifoam 553
16.5.3 Removing Hydrocarbons from Aqueous Solvents 553
16.5.4 Changing Absorber Solvent 553
16.5.5 Other Contaminant Removal Techniques 554
16.6 Hardware Changes Can Debottleneck Foaming Towers 555
16.6.1 Larger Downcomers 555
16.6.2 Smaller Downcomer Backup (Lower Pressure Drop, Larger Clearances) 556
16.6.3 More Tray Spacing 556
16.6.4 Removing Top Two Trays Does Not Help 556
16.6.5 Trays Versus Packings 556
16.6.6 Larger Packings, High-Open-Area Distributors Help 557
16.6.7 Increased Agitation 557
16.6.8 Larger Tower 557
16.6.9 Reducing Base Level 557

17. The Tower as a Filter: Part A. Causes of Plugging—Number 1 on the Top 10 Malfunctions **558**

17.1 Piping Scale/Corrosion Products 558
17.2 Salting Out/Precipitation 559
17.3 Polymer/Reaction Products 560
17.4 Solids/Entrainment in the Feed 561
17.5 Oil Leak 561

17.6 Poor Shutdown Wash/Flush 562
17.7 Entrainment or Drying at Low Liquid Rates 562
17.8 Others 562

18. The Tower as a Filter: Part B. Locations of Plugging—Number 1 on the Top 10 Malfunctions 563

18.1 Trays 563
18.2 Downcomers 564
18.3 Packings 565
18.4 How Packings and Trays Compare on Plugging Resistance 565
18.4.1 Trays versus Trays 565
18.4.2 Trays versus Packings 566
18.4.3 Packings versus Packings 567
18.5 Limited Zone Only 567
18.6 Draw, Exchanger, and Vent Lines 569
18.7 Feed and Inlet Lines 570
18.8 Instrument Lines 570

19. Coking: Part of Number 1 on Tower Top 10 Malfunctions 571

19.1 Insufficient Wash Flow Rate, Refinery Vacuum Towers 571
19.2 Other Causes, Refinery Vacuum Towers 572
19.3 Slurry Section, FCC Fractionators 573
19.4 Other Refinery Fractionators 574
19.5 Nonrefinery Fractionators 574

20. Leaks 575

20.1 Pump, Compressor 575
20.2 Heat Exchanger 575
20.2.1 Reboiler Tube 575
20.2.2 Condenser Tube 576
20.2.3 Auxiliary Heat Exchanger (Preheater, Pumparound) 576
20.3 Chemicals to/from Other Equipment 577
20.3.1 Leaking from Tower 577
20.3.2 Leaking into Tower 577
20.3.3 Product to Product 578
20.4 Atmospheric 578
20.4.1 Chemicals to Atmosphere 578
20.4.2 Air into Tower 579

21. Relief and Failure 580

21.1 Relief Requirements 580
21.2 Controls That Affect Relief Requirements and Frequency 580
21.3 Relief Causes Tower Damage, Shifts Deposits 581

21.4 Overpressure Due to Component Entry 581
21.5 Relief Protection Absent or Inadequate 582
21.6 Line Ruptures 583
21.7 All Indication Lost When Instrument Tap Plugged 584
21.8 Trips Not Activating or Incorrectly Set 584
21.9 Pump Failure 585
21.10 Loss of Vacuum 585
21.11 Power Loss 585

22. Tray, Packing, and Tower Damage: Part of Number 3 on the Top 10 Malfunctions 586

22.1 Vacuum 586
22.2 Insufficient Uplift Resistance 587
22.3 Uplift Due to Poor Tightening during Assembly 587
22.4 Uplift Due to Rapid Upward Gas Surge 589
22.5 Valves Popping Out 590
22.6 Downward Force on Trays 590
22.7 Trays below Feed Bent Up, above Bent Down and Vice Versa 591
22.8 Downcomers Compressed, Bowed, Fallen 592
22.9 Uplift of Cartridge Trays 593
22.10 Flow-Induced Vibrations 593
22.11 Compressor Surge 594
22.12 Packing Carryover 595
22.13 Melting, Breakage of Plastic Packing 595
22.14 Damage to Ceramic Packing 595
22.15 Damage to Other Packings 595

23. Reboilers That Did Not Work: Number 9 on the Top 10 Malfunctions 596

23.1 Circulating Thermosiphon Reboilers 596
23.1.1 Excess Circulation 596
23.1.2 Insufficient Circulation 596
23.1.3 Insufficient ΔT, Pinching 596
23.1.4 Surging 596
23.1.5 Velocities Too Low in Vertical Thermosiphons 597
23.1.6 Problems Unique to Horizontal Thermosiphons 597
23.2 Once-Through Thermosiphon Reboilers 597
23.2.1 Leaking Draw Tray or Draw Pan 597
23.2.2 No Vaporization/Thermosiphon 598
23.2.3 Slug Flow in Outlet Line 599
23.3 Forced-Circulation Reboilers 599
23.4 Kettle Reboilers 599
23.4.1 Excess ΔP in Circuit 599
23.4.2 Poor Liquid Spread 601
23.4.3 Liquid Level above Overflow Baffle 602

23.5 Internal Reboilers 602
23.6 Kettle and Thermosiphon Reboilers in Series 603
23.7 Side Reboilers 603
23.7.1 Inability to Start 603
23.7.2 Liquid Draw and Vapor Return Problems 603
23.7.3 Hydrates 603
23.7.4 Pinching 604
23.7.5 Control Issues 604
23.8 All Reboilers, Boiling Side 604
23.8.1 Debris/Deposits in Reboiler Lines 604
23.8.2 Undersizing 604
23.8.3 Film Boiling 604
23.9 All Reboilers, Condensing Side 605
23.9.1 Non condensables in Heating Medium 605
23.9.2 Loss of Condensate Seal 605
23.9.3 Condensate Draining Problems 606
23.9.4 Vapor/Steam Supply Bottleneck 606

24. Condensers That Did Not Work 607

24.1 Inerts Blanketing 607
24.1.1 Inadequate Venting 607
24.1.2 Excess Lights in Feed 608
24.2 Inadequate Condensate Removal 608
24.2.1 Undersized Condensate Lines 608
24.2.2 Exchanger Design 609
24.3 Unexpected Condensation Heat Curve 609
24.4 Problems with Condenser Hardware 610
24.5 Maldistribution between Parallel Condensers 611
24.6 Flooding/Entrainment in Partial Condensers 611
24.7 Interaction with Vacuum and Recompression Equipment 612
24.8 Others 612

25. Misleading Measurements: Number 8 on the Top 10 Malfunctions 613

25.1 Incorrect Readings 613
25.2 Meter or Taps Fouled or Plugged 614
25.3 Missing Meter 615
25.4 Incorrect Meter Location 615
25.5 Problems with Meter and Meter Tubing Installation 616
25.5.1 Incorrect Meter Installation 616
25.5.2 Instrument Tubing Problems 616
25.6 Incorrect Meter Calibration, Meter Factor 617
25.7 Level Instrument Fooled 617
25.7.1 By Froth or Foam 617
25.7.2 By Oil Accumulation above Aqueous Level 618
25.7.3 By Lights 619

25.7.4 By Radioactivity (Nucleonic Meter) 619
25.7.5 Interface-Level Metering Problems 619
25.8 Meter Readings Ignored 619
25.9 Electric Storm Causes Signal Failure 619

26. Control System Assembly Difficulties 620

26.1 No Material Balance Control 620
26.2 Controlling Two Temperatures/Compositions Simultaneously Produces Interaction 621
26.3 Problems with the Common Control Schemes, No Side Draws 622
26.3.1 Boil-Up on TC/AC, Reflux on FC 622
26.3.2 Boil-Up on FC, Reflux on TC/AC 623
26.3.3 Boil-Up on FC, Reflux on LC 624
26.3.4 Boil-Up on LC, Bottoms on TC/AC 625
26.3.5 Reflux on Base LC, Bottoms on TC/AC 626
26.4 Problems with Side-Draw Controls 626
26.4.1 Small Reflux below Liquid Draw Should Not Be on Level or Difference Control 626
26.4.2 Incomplete Material Balance Control with Liquid Draw 628
26.4.3 Steam Spikes with Liquid Draw 628
26.4.4 Internal Vapor Control makes or Breaks Vapor Draw Control 628
26.4.5 Others 628

27. Where Do Temperature and Composition Controls Go Wrong? 629

27.1 Temperature Control 629
27.1.1 No Good Temperature Control Tray 629
27.1.2 Best Control Tray 630
27.1.3 Fooling by Nonkeys 630
27.1.4 Averaging (Including Double Differential) 631
27.1.5 Azeotropic Distillation 631
27.1.6 Extractive Distillation 631
27.1.7 Other 632
27.2 Pressure-Compensated Temperature Controls 632
27.2.1 ΔT Control 632
27.2.2 Other Pressure Compensation 633
27.3 Analyzer Control 633
27.3.1 Obtaining a Valid Analysis for Control 633
27.3.2 Long Lags and High Off-Line Times 633
27.3.3 Intermittent Analysis 634
27.3.4 Handling Feed Fluctuations 635
27.3.5 Analyzer–Temperature Control Cascade 635
27.3.6 Analyzer On Next Tower 635

28. Misbehaved Pressure, Condenser, Reboiler, and Preheater Controls 636

28.1 Pressure Controls by Vapor Flow Variations 636
28.2 Flooded Condenser Pressure Controls 637
28.2.1 Valve in the Condensate, Unflooded Drum 637
28.2.2 Flooded Drum 637
28.2.3 Hot-Vapor Bypass 637
28.2.4 Valve in the Vapor to the Condenser 639
28.3 Coolant Throttling Pressure Controls 640
28.3.1 Cooling-Water Throttling 640
28.3.2 Manipulating Airflow 640
28.3.3 Steam Generator Overhead Condenser 640
28.3.4 Controlling Cooling-Water Supply Temperature 640
28.4 Pressure Control Signal 641
28.4.1 From Tower or from Reflux Drum? 641
28.4.2 Controlling Pressure via Condensate Temperature 641
28.5 Throttling Steam/Vapor to Reboiler or Preheater 641
28.6 Throttling Condensate from Reboiler 642
28.7 Preheater Controls 643

29. Miscellaneous Control Problems 644

29.1 Interaction with the Process 644
29.2 ΔP Control 644
29.3 Flood Controls and Indicators 644
29.4 Batch Distillation Control 645
29.5 Problems in the Control Engineer's Domain 645
29.6 Advanced Controls Problems 646
29.6.1 Updating Multivariable Controls 646
29.6.2 Advanced Controls Fooled by Bad Measurements 646
29.6.3 Issues with Model Inaccuracies 647
29.6.4 Effect of Power Dips 647
29.6.5 Experiences with Composition Predictors in Multivariable Controls 647

References 649

Index 669

About the Author 713

Preface

"To every problem, there's always an easy solution—neat, plausible, and wrong."

—Mencken's Maxim

The last half-century has seen tremendous progress in distillation technology. The introduction of high-speed computers revolutionized the design, control, and operation of distillation towers. Invention and innovation in tower internals enhanced tower capacity and efficiency beyond previously conceived limits. Gamma scans and laser-guided pyrometers have provided troubleshooters with tools of which, not-so-long-ago, they would only dream. With all these advances, one would expect the failure rate in distillation towers to be on the decline, maybe heading towards extinction as we enter the 21st century. Our recent survey of distillation failures (255) brought disappointing news: Distillation failures are not on the path to extinction. Instead, the tower failure rate is on the rise and accelerating.

Our survey further showed that the rise is not because distillation is moving into new, unchartered frontiers. By far, the bulk of the failures have been repetitions of previous ones. In some cases, the literature describes 10–20 repetitions of the same failure. And for every case that is reported, there are tens, maybe hundreds, that are not.

In the late 1980s, I increased tray hole areas in one distillation tower in an attempt to gain capacity. Due to vapor cross flow channeling, a mechanism unknown at the time, the debottleneck went sour and we lost 5% capacity. Half a year of extensive troubleshooting, gamma scans, and tests taught us what went wrong and how to regain the lost capacity. We published extensively on the phenomenon and how to avoid. A decade later, I returned to investigate why another debottleneck (this time by others) went sour at the same unit. The tower I previously struggled with was replaced by a larger one, but the next tower in the sequence (almost the same hydraulics as the first) was debottlenecked . . . by increasing tray hole areas!

It dawned on me how short a memory the process industries have. People move on, the lessons get forgotten, and the same mistakes are repeated. It took only one decade to forget. Indeed, people moved on: only one person (beside me) that experienced the 1980s debottleneck was involved in the 1990s efforts. This person actually questioned the debottleneck proposal, but was overruled by those who did not believe it will happen again.

Likewise, many experiences are repeatedly reported in the literature. Over the last two decades, there has been about one published case history per year of a tower flooding prematurely due to liquid level rising above the reboiler return nozzle, or of a kettle reboiler bottleneck due to an incorrectly compiled force balance. One would think that had we learned from the first case, all the repetitions could have been avoided. And again, for every case that is reported, there are tens, maybe hundreds that are not.

Why are we failing to learn from past lessons? Mergers and cost-cuts have retired many of the experienced troubleshooters and thinly spread the others. The literature offers little to bridge the experience gap. In the era of information explosion, databases, and computerized searches, finding the appropriate information in due time has become like finding a needle in an evergrowing haystack. To locate a useful reference, one needs to click away a huge volume of wayward leads. Further, cost-cutting measures led to library closures and to curtailed circulation and availability of some prime sources of information, such as, AIChE meeting papers.

The purpose of this book is pick the needles out of the haystack. The book collects lessons from past experiences and puts them in the hands of troubleshooters in a usable form. The book is made up of two parts: the first is a collection of "war stories," with the detailed problems and solutions. The second part is a database mega-table which presents summaries of all the "war stories" I managed to find in the literature. The summaries include some key distillation-related morals. For each of these, the literature reference is described fully, so readers can seek more details. Many of the case histories could be described under more than one heading, so extensive cross references have been included.

If an incident that happened in your plant is described, you may notice that some details could have changed. Sometimes, this was done to make it more difficult for people to tell where the incident occurred. At other times, this was done to simplify the story without affecting the key lessons. Sometimes, the incident was written up several years after it occurred, and memories of some details faded away. Sometimes, and this is the most likely reason, the case history did not happen in your plant at all. Another plant had a similar incident.

The case histories and lessons drawn are described to the best of my and the contributors' knowledge and in good faith, but do not always correctly reflect the problems and solutions. Many times I thought I knew the answer, possibly even solved the problem, only to be humbled by new light or another experience later. The experiences and lessons in the book are not meant to be followed blindly. They are meant to be taken as stories told in good faith, and to the best of knowledge and understanding of the author or contributor. We welcome any comments that either affirm or challenge our perception and understanding.

If you picked the book, you expressed interest in learning from past experiences. This learning is an essential major step along the path traveled by a good troubleshooter or designer. Should you select this path, be prepared for many sleepless nights in the plant, endless worries as to whether you have the right answer, tests that will shatter your favorite theories, and many humbling experiences. Yet, you will share the glory when your fix or design solves a problem where others failed. You will enjoy harnessing the forces of nature into a beneficial purpose. Last but not least, you will experience the electric excitement of the "moments of insight," when all the facts you have been struggling with for months suddenly fall together into a simple explanation. I hope this book helps to get you there.

HENRY Z. KISTER

March 2006

Acknowledgments

Many of the case histories reported in this book have been invaluable contributions from colleagues and friends who kindly and enthusiastically supported this book. Many of the contributors elected to remain anonymous. Kind thanks are due to all contributors. Special thanks are due to those who contributed multiple case histories, and to those whose names do not appear in print. To those behind-the-scenes friends, I extends special appreciation and gratitude.

Writing this book required breaking away from some of the everyday work demands. Special thanks are due to Fluor Corporation, particularly to my supervisors, Walter Stupin and Paul Walker, for their backing, support and encouragement of this book-writing effort, going to great lengths to make it happen.

Recognition is due to my mentors who, over the years, encouraged my work, immensely contributed to my achievements, and taught me much about distillation and engineering: To my life-long mentor, Walter Stupin, who mentored and encouraged my work, throughout my career at C F Braun and later at Fluor, being a ceaseless source of inspiration behind my books and technical achievements; Paul Walker, Fluor, whose warm encouragement and support have been the perfect motivators for professional excellence and achievement; Professor Ian Doig, University of NSW, who inspired me over the years, showed me the practical side of distillation, and guided me over a crisis early in my career; Reno Zack, who enthusiastically encouraged and inspired my achievements throughout my career at C F Braun; Dick Harris and Trevor Whalley, who taught me about practical distillation and encouraged my work and professional pursuits at ICI Australia; and Jack Hull, Tak Yanagi, and Jim Gosnell, who were sources of teaching and inspiration at C F Braun. The list could go on, and I express special thanks to all that encouraged, inspired, and contributed to my work over the years. Much of my mentors' teachings found their way into the following pages.

Special thanks are due to family members and close friends who have helped, supported and encouraged my work—my mother, Dr. Helen Kister, my father, Dr. John Kister, and Isabel Wu—your help and inspiration illuminated my path over the years.

Last but not least, special thanks are due to Mireille Grey and Stan Okimoto at Fluor, who flawlessly and tirelessly converted my handwritten scrawl into a typed manuscript, putting up with my endless changes and reformats.

H.Z.K.

How to Use this Book

The use of this book as a story book or bedtime reading is quite straight forward and needs no guidance. Simply select the short stories of specific interest and read them.

More challenging is the use of this book to look for experiences that could have relevance to a given troubleshooting endeavor. Here the database mega-Table in the second part of the book is the key. Find the appropriate subject matter via the table of contents or index, and then explore the various summaries, including those in the cross-references. The database mega-Table also lists any case histories that are described in full in this book. Such case histories will be prefixed "DT" (acronym for Distillation Troubleshooting). For instance, if the mega-Table lists DT2.4, it means that the full experience is reported as case history 2.4 in this book.

The database as well as many of the case histories list only some of the key lessons drawn. The lessons listed are not comprehensive, and omit nondistillation morals (such as the needs for more staffing or better training). The reader is encouraged to review the original reference for additional valuable lessons.

For quick reference, the acronyms used in Distillation Troubleshooting are listed up front, and the literature references are listed alphabetically.

Some of the case histories use English units, others use metric units. The units used often reflect the unit system used in doing the work. The conversions are straightforward and can readily be performed by using the conversion tables in Perry's Handbook (393) or other handbooks.

The author will be pleased to hear any comments, experiences or challenges any readers may wish to share for possible inclusion in a future edition. Also, the author is sure that despite his intensive literature search, he missed several invaluable references, and would be very grateful to receive copies of such references. Feedback on any errors, as well as rebuttal to any of the experiences described, is also greatly appreciated and will help improve future editions. Please write, fax or e-mail to Henry Z. Kister, Fluor, 3 Polaris Way, Aliso Viejo, CA 92698, phone 1-949-349-4679; fax 1-949-349-2898; e-mail *henry.kister@fluor.com.*

Abbreviations

AC	Analyzer control
AGO	Atmospheric gas oil
aMDEA	Activated MDEA
AMS	Alpha-methyl styrene
APC	adaptive process control
AR	on-line analyzer
ASTM	American Society for Testing and Materials
atm	atmospheres, atmospheric
B	Bottoms
barg	bars, gauge
BFW	Boiler feed water
BMD	2-bromomethyl-1, 3-dioxolane
BPD	Barrels per day
BPH	Barrels per hour
BSD	bottom side draw
BTEX	Benzene, toluene, ethylbenzene, xylene
BTX	Benzene, toluene, xylene
$C_1, C_2, C_3 \ldots$	Number of carbon atoms in compound
CAT	computed axial tomography
Cat	Catalytic
C-factor	Vapor capacity factor, defined by equation 2 in Case Study 1.14
CFD	computational fluid dynamics
CHP	cumene hydroperoxide
CO_2	Carbon dioxide
Co.	Company
CS	Carbon steel
CT	Chimney Tray
CTC	Carbon tetrachloride
CW	Cooling water
CWR	Cooling water return
CWS	Cooling water supply
D	Distillate
D86	ASTM atmospheric distillation test of petroleum fraction

DAA	diacetone alcohol
DC_1	Demethanizer
DC_2	Deethanizer
DC_3	Depropanizer
DC_4	Debutanizer
DC_5	Depentanizer
DCM	Dichloromethane
DCS	Distributed control system
DEA	Diethanol amine
DFNB	2, 4-difluoronitrobenzene
DIB	Deisobutanizer
DMAC	dimethylacetamide
DMC	Dynamic matrix control
DMF	Dimethylformamide
DMSO	Dimethyl sulphoxide
DO	Decant oil
dP	Same as ΔP
DQI	Distribution quality index
DRD	distillation region diagram
dT	Same as ΔT
DT	Distillation troubleshooting (this book)
EB	Energy balance; ethylbenzene
ED	extractive distillation
EDC	Ethylene dichloride
EG	Ethylene glycol
EGEE	Ethylene glycol monoethyl ether
EO	ethylene oxide
EOR	End of run
ETFE	Ethylene tetrafluoroethylene, a type of teflon
FC	Flow control
FCC	Fluid catalytic cracker
FI	Flow indicator
fph	feet per hour
FR	Flow recorder
FS	Flow Switch
ft	Feet
gal	gallons
GC	Gas chromatographs
GC-MS	Gas chromatography–mass spectrometry
gpm	gallons per minute
GS	A process of concentrating deutrium by dual-temperature isotope exchange between water and hydrogen sulfide with no catalyst
h	hours
H_2	Hydrogen
H_2O	Water

H_2S	Hydrogen sulfide
HA	Hydroxyl amine
HAZOP	Hazard and operability study
HC	Hydrocarbon
HCGO	Heavy coker gas oil
HCl	Hydrogen chloride
HCN	Hydrogen cyanide
HCO	Heavy cycle oil
HD	Heavy diesel
HETP	Height equivalent of a theoretical plate
HF	Hydrogen fluoride
Hg	Mercury
HK	Heavy key
HN	Heavy naphtha
HP	High pressure
HR	High reflux
HSS	Heat-stable salts
HV	hand valve
HVGO	Heavy vacuum gas oil
IBP	Initial boiling point
ICO	intermediate cycle oil
ID	Internal diameter
IK	Intermediate key
in.	inch
IPA	Isopropyl alcohol
IPE	Isopropyl ether
IR	Infrared
IVC	Internal vapor control
kPa	Kilopascals
kPag	Kilopascals gage
lb	pounds
LC	Level control
LCGO	Light coker gas oil
LCO	Light cycle oil
LD	Light diesel
LI	Level indicator
LK	Light key
LL	Liquid–liquid
LMTD	Log mean temperature difference
LP	Low pressure
LPB	Loss Prevention Bullletin
LPG	Liquefied petroleum gas; refers to C_3 and C_4 hydrocarbons
LR	Low reflux
LT	Level transmitter
L/V	Liquid-to-vapor molar ratio

LVGO Light vacuum gas oil
m meters
MB Material balance
MDEA Methyl diethanol amine
MEA Monoethanol amine
MEK Methyl ethyl ketone
MF Main fractionator
min Minutes or minimum
MISO Multiple inputs, single output
mm millimeters
MNT Mononitrotoluene
MOC Management of change
MP Medium Pressure
MPC Model predictive control
mpy mils per year, refers to a measure of conosion rates. 1 mil is 1/1000 inch
MSDS Material safety data sheets
MTS Refers to a proprietary liquid distributor marketed by Sulzer under license from Dow Chemical
MV Manual valve
MVC Multivariable control, or more volitle component
N_2 nitrogen
NC Normally closed
NGL Natural gas liquids
NNF Normally no flow
NO Normally open
NPSH Net positive suction head
NRTL Nonrandom two liquid; refers to a popular VLE prediction method
NRU Nitrogen rejection unit
O_2 oxygen
ORS Oxide redistillation still
OSHA Occupational Safety and Health Administration
PA Pumparound
P&ID Process and instrumentation diagram
PC Pressure control
PCV Pressure control valve
PI Pressure indicator
PR Peng–Robinson; refers to a popular VLE prediction method
psi pounds per square inch
psia psi absolute
psig psi gauge
PSV Pressure safety valve
PT Pressure transmitter
PVC Polyvinyl chloride
PVDF Polyvynilidene fluoride
R22 Freon 22

R/D	Reflux-to-distillate molar ratio
Ref.	Reference
Refrig	Refrigeration
RO	Restriction orifice
RVP	Reid vapor pressure
s	seconds
SBE	Di-*Sec*-butyl ether
sec.	secondary
SG	specific gravity
SPA	Slurry pumparound
SRK	Soave, Redlich, and Kwong; refers to a popular VLE method
SS	Stainless steel
STM	Steam
T/A	Turnaround
TBP	True boiling point
TC	Temperature control
TCE	Trichloroethylene
TDC	Temperature difference controller
TEA	Triethanol amine
TEG	Triethylene glycol
TI	Temperature indicator
Ti	Titanium
TRC	temperature recorder/controller
UNIQAC	Unified quasi-chemical; refers to a popular VLE prediction method
VAM	Vinyl acetate monomer
V/B	Stripping ratio, i.e., molar ratio of stripping section vapor flow rate to tower bottom flow rate
VCFC	Vapor cross-flow channeling
VCM	Vinyl chloride monomer
VGO	Vacuum gas oil
VLE	Vapor–liquid equilibrium
VLLE	Vapor–liquid–liquid equilibrium
VOC	Volatile organic carbon
vol	Volume
w.g.	water gage
wt	by weight
ΔP	Pressure difference
ΔT	Temperature difference

Chapter 1

Troubleshooting Distillation Simulations

It may appear inappropriate to start a distillation troubleshooting book with a malfunction that did not even make it to the top 10 distillation malfunctions of the last half century. Simulations were in the 12th spot (255). Countering this argument is that simulation malfunctions were identified as the fastest growing area of distillation malfunctions, with the number reported in the last decade about triple that of the four preceding decades (252). If one compiled a distillation malfunction list over the last decade only, simulation issues would have been in the equal 6th spot. Simulations have been more troublesome in chemical than in refinery towers, probably due to the difficulty in simulating chemical nonidealities. The subject was discussed in detail in another paper (247).

The three major issues that affect simulation validity are using good vapor–liquid equilibrium (VLE) predictions, obtaining a good match between the simulation and plant data, and applying graphical techniques to troubleshoot the simulation (255). Case histories involving these issues account for about two-thirds of the cases reported in the literature. Add to this ensuring correct chemistry and correct tray efficiency, these items account for 85% of the cases reported in the literature.

A review of the VLE case studies (247) revealed major issues with VLE predictions for close-boiling components, either a pair of chemicals [e.g., hydrocarbons (HCs)] of similar vapor pressures or a nonideal pair close to an azeotrope. Correctly estimating nonidealities has been another VLE troublespot. A third troublespot is characterization of heavy components in crude oil distillation, which impacts simulation of refinery vacuum towers. Very few case histories were reported with other systems. VLE prediction for reasonably ideal, relatively high volatility systems (e.g., ethane–propane or methanol–ethanol) is not frequently troublesome.

The major problem in simulation validation appears to be obtaining a reliable, consistent set of plant data. Getting correct numbers out of flowmeters and laboratory analyses appears to be a major headache requiring extensive checks and rechecks. Compiling mass, component, and energy balances is essential for catching a

---

Distillation Troubleshooting. By Henry Z. Kister

misleading flowmeter or composition. One specific area of frequent mismatches between simulation and plant data is where there are two liquid phases. Here comparison of measured to simulated temperature profiles is invaluable for finding the second liquid phase. Another specific area of frequent mismatches is refinery vacuum towers. Here the difficult measurement is the liquid entrainment from the flash zone into the wash bed, which is often established by a component balance on metals or asphaltenes.

The key graphical techniques for troubleshooting simulations are the McCabe–Thiele and Hengstebeck diagrams, multicomponent distillation composition profiles, and in azeotropic systems residue curve maps. These techniques permit visualization and insight into what the simulation is doing. These diagrams are not drawn from scratch; they are plots of the composition profiles obtained by the simulation using the format of one of these procedures. The book by Stichlmair and Fair (472) is loaded with excellent examples of graphical techniques shedding light on tower operation.

In chemical towers, reactions such as decomposition, polymerization, and hydrolysis are often unaccounted for by a simulation. Also, the chemistry of a process is not always well understood. One of the best tools for getting a good simulation in these situations is to run the chemicals through a miniplant, as recommended by Ruffert (417).

In established processes, such as separation of benzene from toluene or ethanol from water, estimating efficiency is quite trouble free in conventional trays and packings. Problems are experienced in a first-of-a-kind process or when a new mass transfer device is introduced and is on the steep segment of its learning curve.

Incorrect representation of the feed entry is troublesome if the first product leaves just above or below or if some chemicals react in the vapor and not in the liquid. A typical example is feed to a refinery vacuum tower, where the first major product exits the tower between 0.5 and 2 stages above the feed.

The presentation of liquid and vapor rates in the simulation output is not always user friendly, especially near the entry of subcooled reflux and feeds, often concealing higher vapor and liquid loads. This sometimes precipitates underestimates of the vapor and liquid loads in the tower.

Misleading hydraulic predictions from simulators is a major troublespot. Most troublesome have been hydraulic predictions for packed towers, which tend to be optimistic, using both the simulator methods and many of the vendor methods in the simulator (247, 254). Simulation predictions of both tray and packing efficiencies as well as downcomer capacities have also been troublesome. Further discussion is in Ref. 247.

CASE STUDY 1.1 METHANOL IN C_3 SPLITTER OVERHEAD?

Installation Olefins plant C_3 splitter, separating propylene overhead from propane at pressures of 220–240 psig, several towers.

Background Methanol is often present in the C_3 splitter feed in small concentrations, usually originating from dosing upstream equipment to remove hydrates. Hydrates are loose compounds of water and HCs that behave like ice, and methanol is used like antifreeze. The atmospheric boiling points of propylene, propane, and methanol are −54, −44, and 148°F, respectively. The C_3 splitters are large towers, usually containing between 100 and 300 trays and operating at high reflux, so they have lots of separation capability.

Problem Despite the large boiling point difference (about 200°F) and the large tower separation capability, some methanol found its way to the overhead product in all these towers. Very often there was a tight specification on methanol in the tower overhead.

Cause Methanol is a polar component, which is repelled by the nonpolar HCs. This repulsion is characterized by a high activity coefficient. With the small concentration of methanol in the all-HC tray liquid, the repulsion is maximized; that is, the activity coefficient of methanol reaches its maximum (infinite dilution) value. This high activity coefficient highly increases its volatility, to the point that it almost counterbalances the much higher vapor pressure of propylene. The methanol and propylene therefore become very difficult to separate.

Simulation All C_3 splitter simulations that the author worked with have used equations of state, and these were unable to correctly predict the high activity coefficient of the methanol. They therefore incorrectly predicted that all the methanol would end up in the bottom and none would reach the tower top product.

Solution In most cases, the methanol was injected upstream for a short period only, and the off-specification propylene product was tolerated, often blended in storage. In one case, the methanol content of the propylene was lowered by allowing some propylene out of the C_3 splitter bottom at the expense of lower recovery.

Related Experience A very similar experience occurred in a gas plant depropanizer separating propane from butane and heavier HCs. Here the methanol ended in the propane product.

Other Related Experiences Several refinery debutanizers that separated C_3 and C_4 [liquefied petroleum gases (LPGs)] from C_5 and heavier HCs (naphtha) contained small concentrations of high-boiling sulfur compounds. Despite their high boiling points (well within the naphtha range), these high boilers ended in the overhead LPG product. Sulfur compounds are polar and are therefore repelled by the HC tray liquid. The repulsion (characterized by their infinite dilution activity coefficient) made these compounds volatile enough to go up with the LPG. Again, tower simulations that were based on equations of state incorrectly predicted that these compounds would end up in the naphtha.

In one refinery and one petrochemical debutanizer, mercury compounds with boiling points in the gasoline range were found in the LPG, probably reaching it by a similar mechanism.

CASE STUDY 1.2 WATER IN DEBUTANIZER: QUO VADIS?

Installation A debutanizer separating C_4 HCs from HCs in the C_5–C_8 range. Feed to the tower was partially vaporized in an upstream feed-bottom interchanger. The feed contained a small amount of water. Water has a low solubility in the HCs and distilled up. The reflux drum was equipped with a boot designed to gravity-separate water from the reflux.

Problem When the feed contained a higher concentration of water or the reflux boot was inadvertently overfilled, water was seen in the tower bottoms.

Cause The tower feed often contained caustic. Caustic deposits were found in the tower at shutdown. Sampling the water in the tower bottom showed a high pH. Analysis showed that the water in the bottom was actually concentrated caustic solution.

Prevention Good coalescing of water and closely watching the interface level in the reflux drum boot kept water out of the feed and reflux. Maximizing feed preheat kept water in the vapor.

CASE STUDY 1.3 BEWARE OF HIGH HYDROCARBON VOLATILITIES IN WASTEWATER SYSTEMS

Benzene was present in small concentration, of the order of ppm, in a refinery wastewater sewer system. Due to the high repulsion between the water and benzene molecules, benzene has a high activity coefficient, making it very volatile in the wastewater.

Poor ventilation, typical of sewer systems, did not allow the benzene to disperse, and it concentrated in the vapor space above the wastewater. The lower explosive limit of benzene in air is quite low, about a few percent, and it is believed that the benzene concentration exceeded it at least in some locations in the sewer system.

The sewer system had one vent pipe discharging at ground level without a gooseneck. A worker was doing hot work near the top of that pipe. Sparks are believed to have fallen into the pipe, igniting the explosive mixture. The pipe blew up into the worker's face, killing him.

Morals

- Beware of high volatilities of HCs and organics in a wastewater system.
- Avoid venting sewer systems at ground level.

CASE STUDY 1.4 A HYDROCARBON VLLE METHOD USED FOR AQUEOUS FEED EQUILIBRIUM

Contributed by W. Randall Hollowell, CITGO, Lake Charles, Louisiana

Installation Feed for a methanol–water separation tower was the water–methanol phase from a three-phase gas–oil–aqueous separator. Gas from the separator was moderately high in H_2S and in CO_2. Tower preliminary design used a total overhead condenser to produce 95% methanol. Methanol product was cooled and stored at atmospheric pressure. Off gas from storage was not considered a problem because the calculated impurities in the methanol product were predominantly water.

Problem Tower feed had been calculated with a standard gas-processing vapor–liquid–liquid equilibrium (VLLE) method (Peng–Robinson equation of state). A consultant noted that the VLLE method applied only to aqueous phases that behaved like pure water and only to gas-phase components that had low solubility in the aqueous phase.

The large methanol content of the aqueous phase invalidated these feed composition calculations. Every gas component was far more soluble in the tower feed than estimated. The preliminary tower design would have produced a methanol product with such a high H_2S vapor pressure that it could not be safely stored in the atmospheric tank.

Better Approach Gas solubility in a mixed, non-HC solvent (methanol and water) is a Henry's constant type of relationship for which process simulation packages often do not have the methods and/or parameters required.

Addition of a pasteurization section to the top of a tower is a common fix for removing light impurities from the distillate product. After condensing most of the overhead vapor, a small overhead vent gas stream is purged out of the tower to remove light ends. Most or all of the overhead liquid is refluxed to minimize loss of desired product in the purges. The pasteurization section typically contains 3–10 trays or a short packed bed, used to separate light ends from the distillate product. The distillate product is taken as a liquid side draw below the pasteurization trays. The side draw may be stripped to further reduce light ends. The vent gas may be refrigerated and solvent washed or otherwise treated to reduce loss of desired product.

Solution An accurate, specific correlation (outside of the process simulation package) was used to calculate H_2S and CO_2 concentration in the methanol–water tower feed. Solubility of HC components was roughly estimated because they were at relatively low concentrations in the tower feed. A high-performance coalescer was used to minimize liquid HC droplets in the tower feed.

A pasteurization section was added to the top of the tower. The overhead vent gas purge stream was designed to remove most of the H_2S, CO_2, and light HCs. Downstream recovery of methanol from the vent gas and stripping of the methanol product side draw were considered but found to be uneconomical.

Moral Poor simulation and design result from poor selection of VLE and VLLE methods. Computer output does not include a warning when the selected VLE method produces garbage.

CASE STUDY 1.5 MODELING TERNARY MIXTURE USING BINARY INTERACTION PARAMETERS

Contributed by Stanislaw K. Wasylkiewicz, Aspen Technology, Inc., Calgary, Alberta, Canada

This case study describes a frequently encountered modeling problem during simulation of heterogeneous azeotropic distillation. Phase diagrams are invaluable for troubleshooting this type of simulation problems.

Distillation Simulation A sequence of distillation columns for separation of a mixture containing water and several organic alcohols was set up in a simulator. Since some of the alcohols are not fully miscible with water, a nonrandom two-liquid (NRTL) model was selected to model VLLE in the system. At atmospheric pressure, the vapor phase was treated as an ideal gas.

Problem Simulation of the sequence of distillation columns never converged, giving many warnings about flash failures.

Investigation For the three key components (methanol, water, and *n*-butanol) a phase diagram was created (508) (Fig. 1.1*a*). As expected, the water–methanol and methanol–*n*-butanol edges are homogeneous and the water–*n*-butanol edge contained an immiscibility gap. Surprisingly, the three-liquid region and three two-liquid regions covered almost the entire composition space. Since water and methanol, as well as butanol and methanol, are fully miscible, the diagram should have been dominated by a single-liquid region. Just looking at the phase diagram one can conclude that the model is not correct.

Analysis Binary interaction parameters for activity models used for VLLE calculations are published for thousands of components [see, e.g., DECHEMA (158) series]. They are regressed based on various experimental data and usually fit the experimental points quite well. NRTL, UNIQUAC, and Wilson models extend these binary data to multicomponent systems without requiring additional ternary, quaternary, and so on, interaction parameters. That is why these models are so popular for modeling VLE for strongly nonideal azeotropic mixtures. This extension, however, is not always performed correctly by the model.

For the ternary mixture methanol–water–*n*-butanol, the binary interaction parameters have been taken from DECHEMA (158). Some of them are recommended values. All of them describe all the binary pairs very well. But what they predict when combined together can be seen in Figure 1.1*a*. Notice that to create this VLLE diagram an extremely robust flash calculation with stability test is essential. Without a reliable global stability test, flash calculation can easily fail at some points in this component space or give unstable solutions (526).

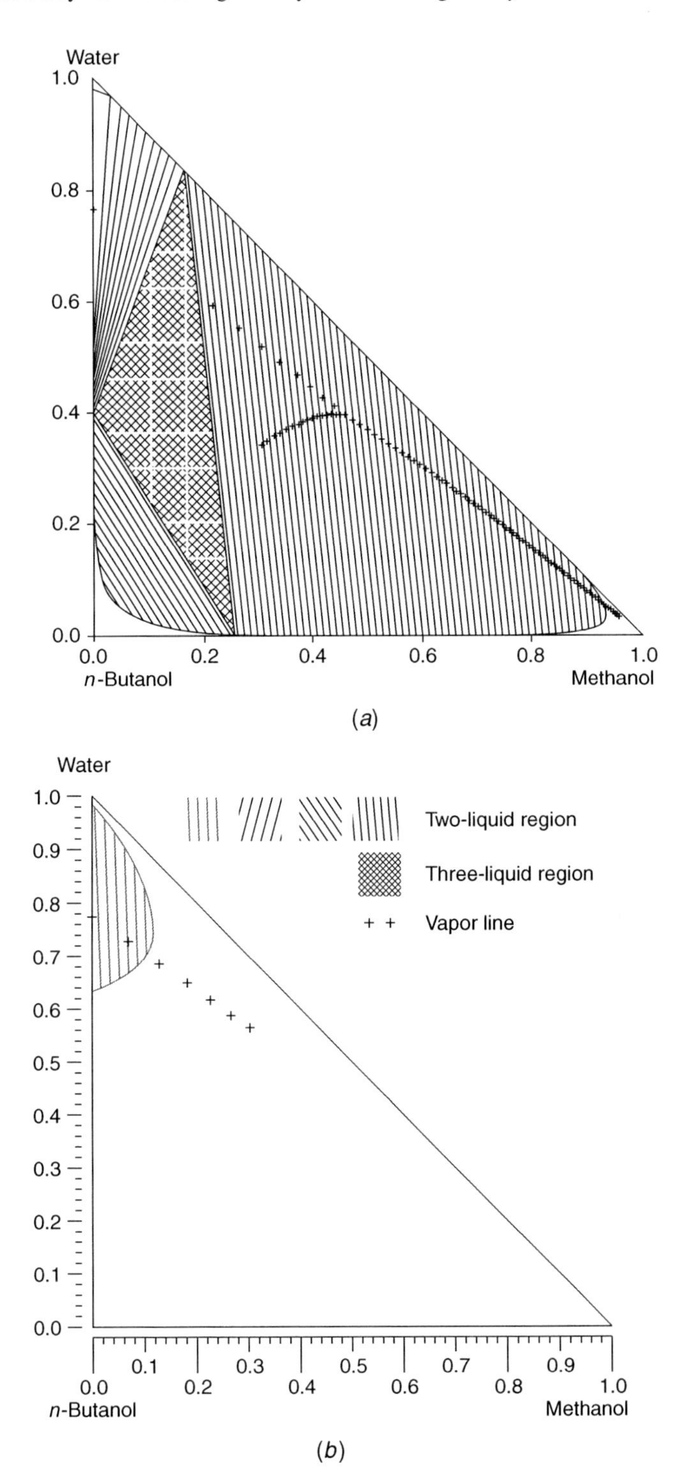

Figure 1.1 Phase diagram for nonideal system methanol–water–*n*-butanol, based on extension of good binary data using NRTL model: (*a*) incorrect extension; (*b*) correct extension.

Solution Another set of binary interaction parameters was carefully selected and a new phase diagram was recreated (34). The VLLE changed dramatically (Fig. 1.1*b*). There is no more three-liquid phase region and only one two-liquid phase region covers only a small part of the composition space. After proper selection of interaction parameters of the thermodynamic model, the sequence of distillation columns converged quickly without any problems.

Morals

- To simulate multicomponent, nonideal distillation, the behavior of the mixture must be carefully verified, starting from binary mixtures, then ternary subsystems, and so on.
- Since there may be many pairs of binary interaction parameters of an activity thermodynamic model that describe behavior of a binary mixture equally well, it is recommended to select one with the lowest absolute values. It is our experience that such values extrapolate better to multicomponent mixtures.
- To correctly create a multicomponent, nonideal VLLE model, an extremely robust VLLE calculation routine with a reliable global stability test is a must [even if liquid–liquid (LL) split is not expected].
- Because of their visualization capabilities, VLLE phase diagrams are invaluable (for ternary and quaternary mixtures) for verification of thermodynamic models used in distillation simulations.

CASE STUDY 1.6 VERY LOW CONCENTRATIONS REQUIRE EXTRA CARE IN VLE SELECTION

Contributed by W. Randall Hollowell, CITGO, Lake Charles, Louisiana

Problem Bottoms from a tower recovering methanol from a methanol–water mixture contained 6 ppm methanol, exceeding the maximum specification of 4 ppm required for discharging to the ocean.

Investigation A consultant pointed out that unusual hydrogen-bonding behavior had been reported at very low concentration of methanol in water. He recommended use of the UNIQUAC equation.

Wilson's equation is generally the method of choice for alcohol–water mixtures when there is no unusual behavior. The more complex NRTL equation is the usual choice for systems that cannot be handled by Wilson's equation. The UNIQUAC equation often applies to systems with chemicallike interactions (i.e., hydrogen bonding, which behaves like weak chemical bonding) that neither Wilson's nor the NRTL equations can represent.

Solution Schedule constraints precluded independently developing UNIQUAC parameters. Various process simulation packages were checked for methanol–water VLE with Wilson's, NRTL, and UNIQUAC equations. All of the equations in all of the packages gave essentially the same VLE, except that UNIQUAC in one major simulator gave lower methanol relative volatilities (by as much as 15%) at very low methanol concentrations. This package executed much slower than the other alternatives. The only methanol concentration predictions that were in line with the field data came from this UNIQAC equation.

Postmortem Exceptions to the typical choices of chemical VLE methods are often not reflected in process simulation packages. For this case, the same data base was probably used by all of the process simulation packages for the regression of UNIQUAC parameters. Predicting VLE for high-purity mixture often requires extrapolation of activity coefficients. Only one method and one simulation package did a good extrapolation to the low-methanol end. Cross checking of VLE equations and packages is a useful way to identify potential problems.

CASE STUDY 1.7 DIAGRAMS TROUBLESHOOT ACETIC ACID DEHYDRATION SIMULATION

Contributed by Stanislaw K. Wasylkiewicz, Aspen Technology, Inc., Calgary, Alberta, Canada

This case study describes a typical thermodynamic modeling problem in distillation simulation and an application of residue curve maps for troubleshooting and proper model selection. The problem described here happened far too many times for many of our clients.

Dehydration of Acetic Acid At atmospheric pressure, there is no azeotrope in the binary mixture of water and acetic acid. However, there is a tangent pinch close to pure water. This makes this binary separation very expensive if only a small concentration of acetic acid in water is allowed (high reflux, many rectifying stages). The difficult separation caused by the tangent pinch can be avoided by adding an entrainer that forms a new heterogeneous azeotrope, moving the distillation profile away from the binary pinch toward the minimum-boiling heterogeneous azeotrope. A decanter can then be used to obtain required distillate purity in far fewer stages than in the original binary distillation (525).

Distillation Simulation A column with top decanter was set up in a simulator to remove water from a mixture containing mostly water and acetic acid. *N*-Butyl acetate was selected as an entrainer. The vapor phase was treated as an ideal gas [*Idel* (227) option]. For the liquid phase, the NRTL model was selected.

Problem Even with an extreme reflux and a large number of stages, the simulation never achieved the required high-purity water in the bottom product of the column.

Troubleshooting For the three key components (water, acetic acid, and the entrainer) a distillation region diagram (DRD) was created (227) to examine the three-component space for multiple liquid regions, azeotropes, and distillation boundaries, as shown in Figure 1.2*a*.

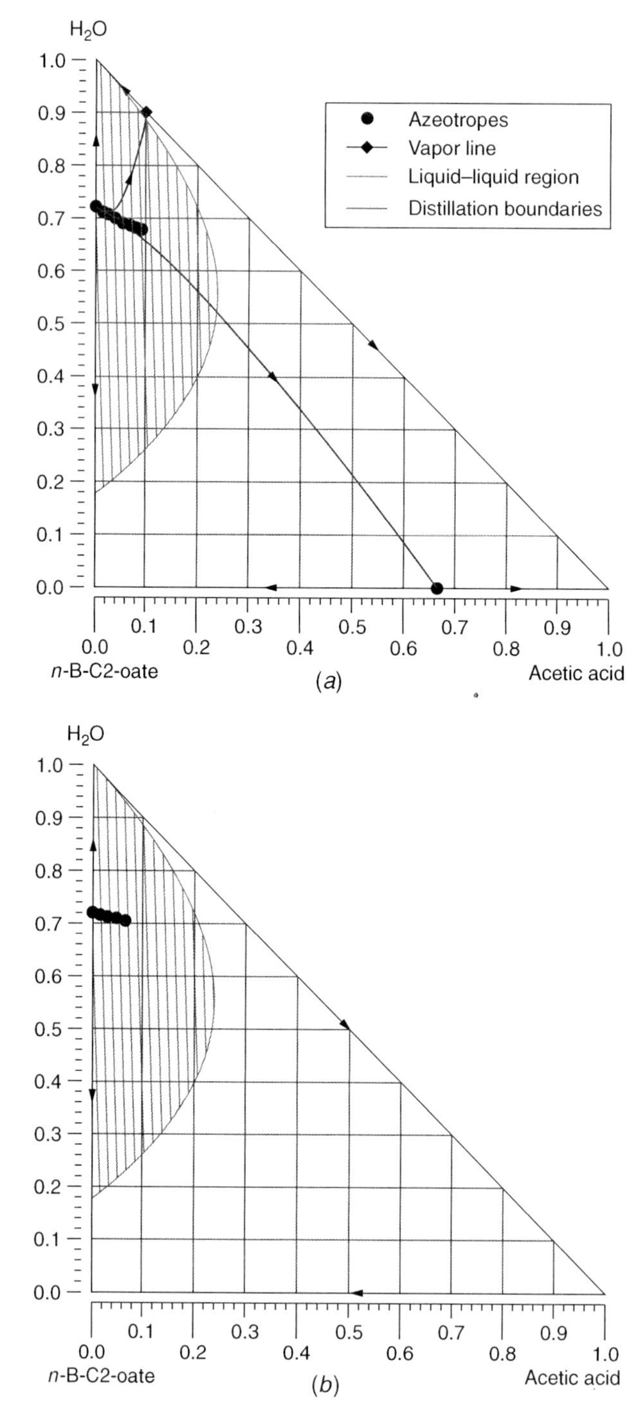

Figure 1.2 Phase diagram for dehydration of acetic acid using *n*-butyl acetate (*n*-B-C2-oate) entrainer at 1 atm: (*a*) with ideal vapor phase, incorrect; (*b*) accounting for dimerization, correct.

Analysis By examining the DRD, one can easily conclude that there is something wrong with the model. We know that there is no binary acetic acid–water azeotrope at 1 atm. The model (ideal vapor phase) is not capable of describing the system properly. It is well known that carboxylic acids associate in the vapor phase and this has to be taken into account, for example, by vapor dimerization model (158) [*Dimer* option (227)].

Solution Instead of *Idel*, the *Dimer* option was selected (227). The DRD for the system changed tremendously (see Fig. 1.2*b*). There are no more binary azeotropes between acetic acid and water or *n*-butyl acetate. After proper selection of the thermodynamic model, the distillation column converged quickly to the required high-purity water specifications in the bottoms.

Morals

- It is important to select the proper thermodynamic model and carefully verify the behavior of the mixture.
- Because of their visualization capabilities, DRDs are extremely useful for evaluating thermodynamic models for ternary and quaternary mixtures.

CASE STUDY 1.8 EVERYTHING VAPORIZED IN A CRUDE VACUUM TOWER SIMULATION

Contributed by W. Randall Hollowell, CITGO, Lake Charles, Louisiana

Problem Atmospheric crude tower bottom was heated, then entered a typical, fuel-type vacuum tower. A hand-drawn curve estimated the atmospheric crude tower bottom composition from assay distillation data for a light crude oil. The simulation estimated that all of the vacuum tower feed vaporized in the flash zone. This was a preposterous result inconsistent with plant data.

Investigation The heaviest assay cuts fell progressively lower than those from another assay of the same crude oil. The heaviest cut was at 850°F atmospheric cut point, compared to the other assay at 1000°F. The assay data were extrapolated on a linear scale to 100% at 1150°F atmospheric boiling point.

The high-boiling part of crude assay data must be carefully assessed. The last several assay points are often poor, particularly when coming from laboratories that cut back on quality control for increased productivity. Crude oils have very high boiling point material. Even light crude oils have material boiling above 1500°F. Extrapolation should be done with percent distilled on a probability-type scale, particularly for light crudes where the slope increases very rapidly on a linear scale.

Solution A new boiling point curve was developed. Another assay was used up to 1000°F cut point, thus reducing the needed extrapolation range. Extrapolation and smoothing of assay data were based upon a probability scale for percent distilled.

A 95% point (whole crude oil basis) of 1400°F was estimated by this extrapolation. Simulation based upon the new boiling point curve was in reasonable agreement with plant data.

Moral Crude oil high-boiling-point data are often poor and must be extrapolated. Experience, following good procedures, and cross checks with plant data are essential for reliable results.

CASE STUDY 1.9 CRUDE VACUUM TOWER SIMULATION UNDERESTIMATES RESIDUE YIELD

Contributed by W. Randall Hollowell, CITGO, Lake Charles, Louisiana

Problem Process simulation estimated much lower vacuum residue yields than obtained from plant towers and from pilot unit runs. Vacuum tower feed boiling point curves were based upon high-temperature gas chromatography (GC) analyses.

Investigation Vacuum tower feed boiling point curves from the GC fell well below curves estimated from assays. The GC analyses assumed that all of the feed oil vaporized in the test and was analyzed.

The highest boiling part of crude oil is too heavy to vaporize in a GC test. Thus the reported GC results did not include the highest boiling part (that above about 1250°F boiling point) of the feed. Simulations based upon this GC data estimated much higher vaporization than actual because they were missing the heaviest part of the feed.

Solution The GC method was modified to include a standard that allowed estimation of how much oil remained in the GC column and was not measured. New GC data and extrapolations of assay data indicated that 10–15% of the feed oil was not vaporized and thus had not been measured by the earlier GC method.

With this improved GC data, simulations agreed well with most of the pilot data. The agreement between simulation and plant data was much better than before but was still not good. This may have been due to poor plant data. Specifically, measured flash zone pressures were often bad.

Moral The analyses used for process simulations must be thoroughly understood.

CASE STUDY 1.10 MISLED BY ANALYSIS

Contributed by Geert Hangx and Marleen Horsels, DSM Research, Geleen, The Netherlands

Problem After a product change in a multipurpose plant, a light-boiling by-product could not be removed to the proper level in the (batch) distillation. The concentration of the light-boiling component in the final product was 0.5%. It should have been (and was in previous runs) 200 ppm.

Investigation The feed was analyzed by GC per normal procedure. The concentration levels of different components looked good. No significant deviation was found. Then some changes in the distillation were performed, such as

- increasing the "lights fraction" in the batch distillation,
- increasing the reflux ratio during the lights fraction, and
- decreasing the vapor load during the lights fraction.

These changes yielded no significant improvement.

The off-specification product was redistilled. The purity was improved, but still the specification could not be met. The GC analysis was checked (recalibrated) again. Everything was OK.

As all of the above-mentioned actions did not improve the product quality, it seemed that something was wrong with the column. After long discussions it was decided to open the handhole at the top of the column and to have a closer look at the feed distributor. Nothing suspicious was found.

Then it was decided to have a closer look at the analysis again. A gas chromatography–mass spectrometry (GC-MS) analysis was performed. This method showed that the impurity was not the light-boiling component as presumed. This component was a remainder from the previous run in the multipurpose plant. Having a boiling point much closer to the end product, this component could not be separated in the column.

Moral It is a good idea to check the analysis with GC-MS before shutting down a column.

CASE STUDY 1.11 INCORRECT FEED CHARACTERIZATION LEADS TO IMPOSSIBLE PRODUCT SPECIFICATIONS

Contributed by Chris Wallsgrove

Installation A new, entirely conventional depentanizer, recovering a C_5 distillate stream from a $C_5/C_6/C_7$ raffinate mixture from a catalytic reformer/aromatics extraction unit, with some light pyrolysis gasoline feed from an adjacent naphtha-cracking ethylene plant. The column had 30 valve trays, a steam-heated reboiler, and a condenser on cooling water.

Problem The C_5 distillate was guaranteed by the process licensor to contain a maximum of 0.5% wt. C_6's. Laboratory testing by the on-site laboratory as well as an impartial third-party laboratory consistently showed about 1.0% of C_6's in the distillate. Increasing reflux ratio or other operation adjustments did not improve distillate purity.

Troubleshooting The tower was shut down after about 6 weeks of operation to inspect the trays. No damage was found and the trays were reported to be "cleaner than new."

The design simulation was rerun with a variety of options: correlations, convergence criteria, and plant analysis data. The laboratory methods, which were established American Society for Testing and Materials (ASTM) test methods, were reviewed. It became apparent that the feed contained some low-boiling components, such as certain methyl-cyclo C_5's which were analyzed (correctly) as C_6's but whose boiling points are in the C_5 range. Since these components would end up in the distillate, it was thermodynamically impossible to achieve the specified performance.

Solution The higher impurity level could be lived with without excessive economic penalty and was accepted.

Moral Correct characterization of feed components is essential even for an "ideal" hydrocarbon mixture.

CASE STUDY 1.12 CAN YOU NAME THE KEY COMPONENTS?

Henry Z. Kister, reference 254. Reproduced with permission.

Installation A stabilizer separating C_3 and lighter HCs from nC_4 and heavier operated at its capacity limit. It was to be debottlenecked for a 25% increase in capacity. In addition, it was required to handle several different feedstocks at high throughputs. Due to the tight requirements, thorough tests were conducted and formed the basis for a simulation, which was used for the debottlenecking. We have seen very few tests as extensive and thorough as the stabilizer tests. Two tests were conducted: a high-reflux (HR) test and a low-reflux (LR) test.

Simulation Versus Measurement With two seemingly minor and insignificant exceptions, all reliable measurements compared extremely well with simulated values. In most tests, the accuracy and reliability of the data would have made it difficult to judge whether the exceptions were real or reflected a minor test data problem. In this case, however, consistency checks verified that the exceptions were real. The high accuracy and reliability of the test data made even small discrepancies visible and significant.

The discrepancies occurred in the HR test, while the LR test showed no discrepancy. This was strange because the stabilizer was extremely steady and smooth during the HR test. Any data problems should have occurred in the LR test or in both tests, but not in the HR test alone.

The two exceptions were interlinked. For the HR test, the simulation predicted three times the measured C_5 concentration in the stabilizer overhead, which would lead to a warmer rectifying section. Indeed, the second exception was simulated rectifying section temperatures 2–5°F warmer than measured.

What Does the Stabilizer Do? At first glance, this question appears stupid. But it turned out to be the key for understanding the test versus the simulation discrepancy.

There was a tight specification on the content of C_3 in the stabilizer bottoms. An excessive amount of C_3 would lead to excessive Reid vapor pressure (RVP) in the bottom, which was undesirable. For similar reasons, it was desirable to minimize iC_4 in the stabilizer bottom, although there was *no* set specification. In the bottoms, nC_4 and heavier were desirable components and were to be maximized. Any C_5 and heavier, and even nC_4, ending up in the overhead product incurred an economic penalty because the bottoms were far more valuable than the overheads. There were no set specifications for any of these components.

With the above in mind, what is the stabilizer actually doing? Which pair is the key components? Initially, we thought it was iC_4/nC_4—but could it have been C_3/iC_4, nC_4/C_5, C_3/C_5, or maybe some other pair? Computer simulations do not answer such questions; Hengstebeck diagrams (211, described in detail in Ref. 251) do.

Hengstebeck diagrams (Fig. 1.3) were prepared from the compositions calculated by the simulation. The HR and LR tests each require one Hengstebeck diagram for each choice of key components: C_3/iC_4, iC_4/nC_4, and nC_4/iC_5. A Hengstebeck diagram for the iC_4/nC_4 separation was included in a more detailed description of the case (254) and showed that this pair behaved the same as the C_3/iC_4 pair.

Figure 1.3*a* shows that in the HR test, below the feed, the stabilizer effectively separated C_3 and lighter from iC_4 and heavier. The diagram also shows that a limited degree of separation of these components occurred in the top two stages of the rectifying section, but pinching occurred below these. Overall, very little separation of C_3 and lighter from iC_4 and heavier occurred in the rectifying section. The stabilizer essentially behaved as a stripper for separating C_3 and lighter from iC_4 and heavier.

Figure 1.3*b* shows that in the HR test, above the feed, the stabilizer effectively separated nC_4 and lighter from iC_5 and heavier. It also illustrates that some separation of these components took place in the bottom five stages of the stripping section, but pinching occurred above these.

Together, Figures 1.3*a* and *b* underscore that the stripping section of the stabilizer separated C_3 and lighter from iC_4 and heavier and, per Ref. 254, also iC_4 and lighter from nC_4 and heavier. The rectifying section of the stabilizer separated iC_5 and heavier from nC_4 and lighter.

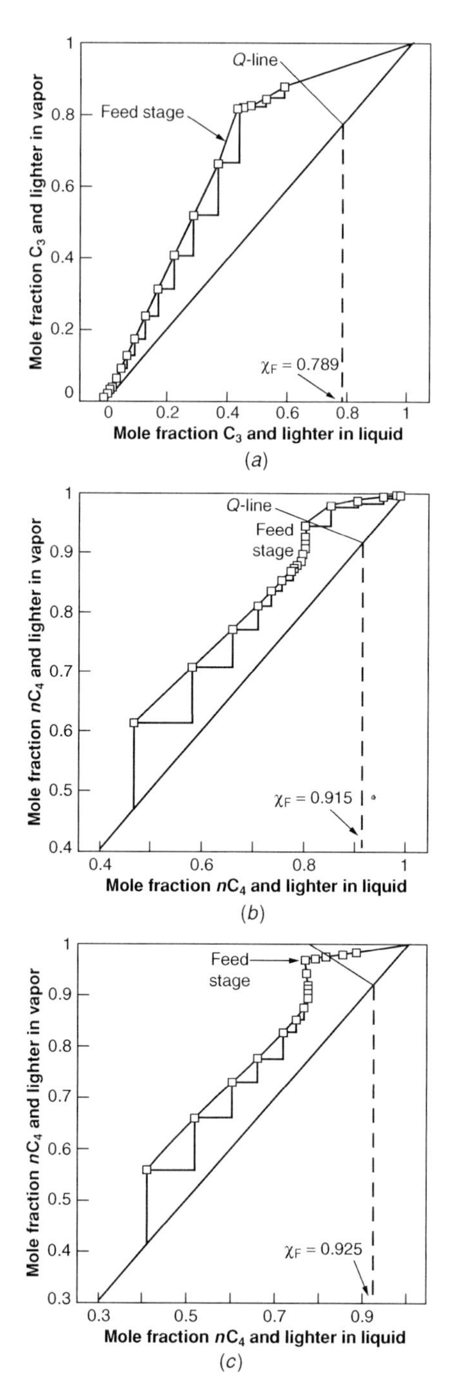

Figure 1.3 Hengstebeck diagrams for stabilizer tests: (*a*) C_3–iC_4 separation, HR test; (*b*) nC_4–iC_5 separation, HR test; (*c*) nC_4–iC_5 separation, LR test. (From Ref. 254. Reproduced with permission.)

In the LR test, the Hengstebeck diagrams for the C_3/iC_4 and iC_4/nC_4 separation were similar to those for the HR test (Fig. 1.3*a*). In this test, too, the stabilizer stripping section effectively separated C_3 and lighter from iC_4 and heavier and iC_4 and lighter from nC_4 and heavier.

Figure 1.3*c* indicates that in the LR test, above the feed, the separation of nC_4 and lighter from iC_5 and heavier was pinched. This is different from the HR test, where the rectifying section effectively separated nC_4 and lighter from iC_5 and heavier. The diagram also shows that, as in the HR test, the nC_4/iC_5 separation was pinched in the stripping section.

Overall, the stabilizer behavior in the LR test resembled that of HR test, with the exception that the rectifying section, which separated nC_4 from iC_5 in the HR test, was pinched and did little of this separation in the LR test.

Why the Differences Between Measurement and Simulation? There were two conceivable explanations to the high C_5 concentration in the HR test simulation:

1. *Inaccuracies in VLE data.* Detailed checks of the VLE confirmed that the values used were very good and superior to those predicted by the commercial simulator program, but not perfect. Two relevant inaccuracies were a high C_3/iC_4 volatility prediction for the stripping section and a low C_4/C_5 volatility prediction for the rectifying section.
2. *Efficiency differences between different binary pairs.* This explanation was unlikely because the simulation would suggest a considerably higher efficiency for the higher volatility pair, nC_4/iC_5, than for the lower volatility pair, iC_4/nC_4. In contrast, test data (52, 379, 381) show that lower volatility pairs have a higher efficiency.

It was therefore concluded that VLE inaccuracy is the most likely explanation.

One unanswered question is why the differences between measurement and simulation were observed only in the HR test and not in the LR test. Again, the Hengstebeck diagrams provided the answer.

For the HR test, the Hengstebeck diagram (Fig. 1.3*b*) shows that the rectifying section rectifies C_5 from the nC_4 and lighter. Any error in the relative volatility of the nC_4/iC_5 and nC_4/nC_5 pairs is magnified at each separation stage. The final result is a large difference between measured and simulated top-product compositions.

For the LR test, the Hengstebeck diagram (Fig. 1.3*c*) shows very little separation of nC_4 from C_5 in the rectifying section. Because of the pinch, an error in the relative volatility of the nC_4/iC_5 and nC_4/nC_5 pairs is not magnified in each separation stage. Such an error, therefore, has little effect on the separation and the temperature profile. For this reason, the LR test simulation gave a good match to measured data.

Would the Inaccuracy Affect the Debottlenecking Predictions? The simulation predicted higher C_5 in the top product, giving a conservative forecast of the stabilizer performance under test conditions. The remaining question is whether the simulation will continue to give conservative predictions under different process conditions. The question of extrapolating test data into different process conditions is addressed rigorously on pp. 400–405 of Ref. 251. In fact, the analysis in Ref. 251 was part of the stabilizer-debottlenecking assignment. The conclusion reached was that when test data are simulated with too low a volatility the simulation compensates by using a greater number of stages (and, hence, higher efficiencies) to match the measured separation. In this case (e.g., the nC_4/C_5 pair in the stabilizer), the simulation will continue to give conservative predictions when extrapolated into different process conditions.

The converse occurs when test data are simulated with too high a relative volatility. The simulation compensates by using a smaller number of stages to match the measured separation. In this case (e.g., the C_3/iC_4 pair in the stabilizer), extrapolation to other process conditions will be optimistic, sometimes grossly so.

Based on the above, it was concluded that the simulation was a reliable basis for debottlenecking for the base case (similar feedstock to that used in one of the tests) and for alternative feedstocks that are not widely different from the base case. However, for those cases of feedstock variations where feed composition varied widely from the base case, the simulation could not be used with confidence until the inaccuracy in the C_3/iC_4 relative volatility was mitigated.

Postmortem The column was successfully debottlenecked. The same simulation (modified to account for the debottlenecking hardware modifications) was found to give superb predictions of the post-revamp performance.

CASE STUDY 1.13 LOCAL EQUILIBRIUM FOR CONDENSERS IN SERIES

Contributed by W. Randall Hollowell, CITGO, Lake Charles, Louisiana

This is my all-time favorite fractionation simulation problem. The entire refinery capacity was sometimes limited by the gas rate, which was calculated to be zero.

Installation An atmospheric crude distillation tower had an extremely broad boiling range overhead vapor with significant ethane, high propane, through full-range kerosene. There were three long, double split-flow condensers in series. The shells were flange to flange and located directly above the overhead accumulator.

Problem Simulation predicted a zero off-gas rate at peak summer temperatures. But actual off-gas rates were substantial, even in winter. Summer crude charge rate was sometimes reduced to avoid flaring of gas in excess of compressor capacity. There was a strong economic incentive to increase butane spiking of crude, but this was not done due to concerns that the gas rate would increase.

Component Balances Earlier calculations had failed to obtain an adequate material balance of the lightest components in the overhead. The naphtha GC analyses were found to be poor. Procedures were corrected by the laboratory, and good material balance closures were obtained.

Simulations predicted that all of the exiting vapor off gas should have been absorbed into the naphtha stream at the operating temperature and pressure. The naphtha had much lower light-ends concentrations than predicted: 30% of the predicted for propane, 50% for butanes, and 75% for pentanes concentration. These low concentrations in the naphtha provided the vapor off-gas flow.

With many sets of data, each giving good material balance closure, it was obvious that the vapor exiting the overhead accumulator was not in equilibrium with liquid exiting the accumulator. Condensers fouled severely on the tube side, but this did not explain the large deviations from equilibrium.

Theory Conventional process simulation assumes what can be called the "universal VLE model." This model assumes that VLE is universal, that is, holds at every location, between the total vapor flow and the total liquid flow. In shell-side condensation, the liquid and vapor are usually close to equilibrium locally when the liquid condenses on the tube surface. But after the liquid drops off the tube (and to the bottom of the shell), there is not enough vapor–liquid mixing to maintain equilibrium with the downstream vapor. Thus there is usually "local VLE" at the tube surface, but not universal VLE for the system. This local equilibrium is responsible for the phenomena of subcooled refluxes coexisting with uncondensed vapor. Condensers designed for total condensation have frequently been partial condensers because of local VLE.

Deviations from universal equilibrium can be large for condensers in series with broad boiling range mixtures. Deviations are particularly high for mixtures with high light-ends content and for arrangements where the liquid stays largely separated from the downstream vapor. This case study represents an extreme example of these deviations.

For the overhead accumulator, universal VLE requires that the operating pressure and the exiting liquid bubble point pressure be equal. But bubble point pressure was half of the operating pressure. If the entire exiting vapor flow had been absorbed into the naphtha stream, the bubble point pressure would still have been less than the operating pressure.

Solution A model was developed to more closely represent the condensation steps. Liquid condensed in each shell was assumed to be in equilibrium with the gas leaving that shell. After the liquid left the shell in which it condensed, it was assumed to have zero mass transfer with the gas phase but to be cooled to the local operating temperature. This model had only one-third of the total liquid (the one-third that condensed in the last shell) in equilibrium with the off gas. The other two-thirds of the liquid was much heavier and caused the overall liquid bubble point pressure to be about half that of the liquid that condensed in the last shell. The actual system was more complex than the above model, in particular:

- The liquid condensed in each shell was heavier than the calculated liquid in equilibrium with the exiting vapor.
- Liquid condensed in an upstream shell experienced a moderate amount of mixing (and thus mass transfer) with downstream vapor.

The above two effects are in opposite directions and largely cancel each other for this case study (perhaps because of the double-split arrangement and three shells in series). This cancellation of errors caused the model to adequately match actual liquid composition and actual vapor rate leaving the overhead accumulator.

Morals For broad boiling range mixtures, condensers (particularly condensers in series) have less capacity than estimated by conventional simulation with universal VLE. This is a failure in simulation and design rather than an equipment failure.

A simulation based upon good operating data can often be used to adequately model the effect of local equilibrium. Good heat and material balances and confidence in them are necessary to step away from universal VLE assumptions and obtain realistic simulations.

Process designers have compensated for their lack of understanding by using large design margins for condensers, by specifying off-gas compressors for zero calculated gas rates, and by greatly oversizing off-gas compressors. These practices can still result in lack of capacity for installations such as in this case study. Even very rough estimates of local equilibrium effects can be far better than conventional calculations for series condensation.

For a single shell and moderate deviations from universal VLE, a reasonable subcooling delta temperature can sometimes be used for simulation and design. In extreme cases, calculations for zones in each shell may be necessary to give good simulation or design. For this case study, the zone method would probably have been required if the condenser paths had been many times longer than in a double split-flow configuration.

CASE STUDY 1.14 SIMULATOR HYDRAULIC PREDICTIONS: TO TRUST OR NOT TO TRUST?

Henry Z. Kister, reference 254. Reproduced with permission.

In this case study, a simulator hydraulic calculation led a plant to expect a capacity gain almost twice as high as the tower revamp actually achieved.

History A refinery vacuum tower was debottlenecked for a 30% capacity gain by replacing 2-in. Pall rings in the wash and heavy vacuum gas oil (HVGO) sections with 3-in. modern proprietary random packings. Only about 15–20% capacity gain was achieved. It was theorized that above this throughput vapor maldistribution set in and caused the tower to lose separation. The refinery sought improvements to vapor distribution in an effort to gain the missing 10–15%.

Troubleshooting A vacuum manometer pressure survey showed that at the point where the tower lost separation the pressure drop was 0.65 in. H_2O/ft packing. Based on air/water measurements, many suppliers' packages take the capacity limit (or flood point) to occur at a pressure drop of 1.5–2 in. H_2O/ft packing. Work by Strigle (473), Rukovena and Koshy (418), and Kister and Gill (257, 259) demonstrated that such numbers are grossly optimistic for modern, high-capacity random and structured packings. Using published flood data, Kister and Gill (257, 259) showed that, for random and structured packings, the flood pressure drop is given by

$$\Delta P_{\text{flood}} = 0.115 F_{\text{P}}^{0.7} \tag{1}$$

where ΔP_{flood} is the flood pressure drop (in. H_2O/ft packing) and F_{P} is the packing factor (ft^{-1}). This equation was shown to give a good fit to experimental data (many of which were generated by suppliers) and was later endorsed by Strigle (473) with a slight change of coefficient. For the high-capacity packing in the vacuum tower, the packing factor was 12. Equation 1 predicts that ΔP_{flood} was 0.65 in. H_2O/ft packing, which coincided with the limit observed by the refinery.

For hydraulic calculations, gas velocity usually is expressed as a C-factor (C_{S}), (ft/s), given by

$$C_{\text{S}} = U_{\text{S}} \left(\frac{\rho_{\text{G}}}{\rho_{\text{L}} - \rho_{\text{G}}} \right)^{0.5} \tag{2}$$

where U_{S} is the gas superficial velocity based on tower cross-sectional area (ft/s), ρ is the density (lb/ft^3), and the subscripts G and L denote gas and liquid, respectively. The C-factor essentially is a density-corrected superficial velocity. The fundamental relevance of the C-factor is discussed elsewhere (251).

Based on a flood pressure drop of 0.65 in. H_2O/ft packing derived from Equation 1, the maximum efficient capacity of the new 3-in. random packing calculated by the Kister and Gill method (251) was at a C-factor of 0.38 ft/s. This is about 17% higher than the maximum efficient capacity for the previous 2-in. Pall rings, just as the refinery observed.

According to the supplier's published hand correlation, which we believe was similar to the one in the computer simulation, the maximum efficient capacity of the packing was at a C-factor of 0.43 ft/s, which is 13% higher than observed. This high C-factor matched a pressure drop of between 1 and 1.5 in. H_2O/ft packing, well above the value where the packing reached a capacity limit.

Epilogue Based on the hydraulic calculation in the computer simulation, the refinery expected that changing the 2-in. Pall rings to the 3-in. high-capacity random packing would increase capacity by 30%. In real life, just over half of the capacity increase materialized. The half that did not materialize is attributed to the optimistic prediction from the simulation.

CASE STUDY 1.15 PACKING HYDRAULIC PREDICTIONS: TO TRUST OR NOT TO TRUST

Background This case presents a number of experiences which were very similar to Case Study 1.14. In each one of these, vendor and simulator predictions for a packed tower were optimistic. In each one of these, the Kister and Gill equation (257, 259) gave excellent prediction for the maximum capacity. The Kister and Gill equation is

$$\Delta P_{\text{flood}} = 0.115 F_{\text{P}}^{0.7} \tag{1}$$

where ΔP_{flood} is the flood pressure drop (in. H_2O/ft packing) and F_P is the packing factor (ft^{-1}).

Tower A This was a chemical tower, equipped with wire-mesh structured packing with a packing factor of 21. The tower ran completely smoothly until reaching a pressure drop of 1 in. H_2O/ft packing, then would rapidly lose efficiency. This compares to a flood pressure drop of 0.97 in. H_2O/ft packing from Equation 1. Simulation prediction (both vendor and general options) predicted a much higher capacity.

Tower B This was a chemical tower equipped with random packing with a packing factor of 18. This column would rapidly lose efficiency when the pressure drop increased above 0.67 in. H_2O/ft packing. This compares to a flood pressure drop of 0.87 from Equation 1. The measurement was slightly lower than the prediction because the vapor load varied through the packings, so much of the bed operated at lower pressure drop. Simulation prediction (both vendor and general options) predicted a much higher capacity. Similar to Case Study 1.14, the plant initially theorized that the shortfall in capacity was due to vapor maldistribution.

Tower C This was a chemical absorber equipped with random packing with a packing factor of 18. The highest pressure drop at which operation was stable was 0.8 in. H_2O/ft packing. Above this, the pressure drop would rapidly rise. This compares to a flood pressure drop of 0.87 from Equation 1. Simulation predictions (both vendor and general options) were of a 20% higher capacity.

Tower D Random packing installed in a chemical tower fell short of achieving design capacity. The vendor method predicted flooding at a pressure drop of 1.5 in. H_2O/ft packing. With a packing factor of 18, Equation 1 predicted that the packing would flood much earlier at a pressure drop of 0.8 in. H_2O/ft packing. The packing flooded at exactly that pressure drop.

CASE STUDY 1.16 DO GOOD CORRELATIONS MAKE THE SIMULATION HYDRAULIC CALCULATIONS RELIABLE?

Henry Z. Kister, reference 254. Reproduced with permission.

What follows is an actual letter circulated by an engineer working for a reputable company. The names of the correlations cited, as well as a few sentences, were changed to protect those involved.

> *We have had a problem recently with the prediction of flooding in packed towers using the Smith correlation for packed tower capacity in the Evertrue Simulator. We used this for sizing a packed tower at 400 psia. The program predicted a percentage flood of 56 percent using the Smith correlation. The vendor predicted 106 percent of flood, and 123 percent of the packing useful capacity.*
>
> *The Evertrue calculation is based on an article by Smith in Quality Chemical Engineering magazine. Smith's method, in turn, depends on an earlier correlation by Jones, also published in an article in Quality Chemical Engineering.*
>
> *These correlations are neither well developed nor tested. Neither of these articles (Smith's and Jones') have undergone very close scrutiny, nor are the correlations from well-known textbooks or journals that have a tradition of peer review. One of the failings is the use of the correlation at high pressure with hydrocarbon systems. Smith's correction factor for high pressures produces numbers that are unreasonably high. There is no indication that this factor is supported either by correlation or by theory. In addition to the lack of credibility of Smith's values, the correlation of Jones, used as the basis of the Smith method, appears inaccurate for the high-pressure systems.*
>
> *For these reasons, I would not recommend use of the Evertrue Smith correlation, regardless of the system pressure, for predicting whether or not a packed tower will work. Instead, the 1960 correlation included on Evertrue should be used. This correlation is based on well-known methods, and can be found in "Perry's Handbook." It predicts the tower would be at 96 percent of flood, compared to the 106 percent predicted by the vendor, which is much closer than the Smith correlation.*
>
> *In either case, calculations must be verified by the packing vendor. I recommend that the vendor verifies the results even for estimates.*

What Really Happened In our experience, both the Smith and the Jones correlations are excellent. The correlation that leaves a lot to be desired for modern packing calculations is the 1960 one. Nevertheless, the letter's author appeared to have reached the converse conclusion.

It is a sad fact of life that correlation authors always examine their correlations for good statistical fit but seldom properly explore and clearly define their correlation limitations. On page 39 of Ref. 259, Kister and Gill remark: "An excellent fit to experimental data is insufficient to render a packing pressure-drop correlation suitable for design. In addition, the correlation's limitations must be fully explored."

In contrast to the letter writer's comment, the problem is more acute in articles that are peer reviewed. These contain correlations based on fundamental models that are inherently complex. This complexity makes it very difficult to properly identify the limitations. A peer review offers little help unless the reviewer spends several days checking the calculations. This rarely happens.

The Smith correlation works very well for vacuum and atmospheric pressures, perhaps up to 50 psia. It was never intended to apply to 400 psia. Unfortunately, Smith's article only contained a hint of the pressure limitation but no firm statements to that effect. It, therefore, went into the Evertrue simulator without a warning flag above 50 psia. In this case, the 1960 correlation was found to work well. This appears to be a case of two wrongs making a right.

Epilogue There are many correlations in the published and proprietary literature for which the limitations are neither well explored nor well defined. Limitations unflagged in the original articles remain unflagged in the simulator version.

Despite the letter writer's wrong conclusion, his bottom line is broadly valid. A simulator correlation cannot be trusted, even when the correlation is good, unless the correlation's limitations are known and included in the simulation. An independent verification, say, by a supplier or an independent method, is a good idea. When *Distillation Design* (251) was compiled, special effort was made to talk to authors of good correlations, with the objective of exploring their limitations and filling in the missing blanks. For instance, the pressure ranges for the application of Smith's correlation were listed in *Distillation Design* almost two years before the above letter was written.

Chapter 2

Where Fractionation Goes Wrong

Fractioination issues featured very low on the distillation malfunctions list for the last half century (255). Only two issues rated a mention, intermediate-component accumulation and two liquid phases. Neither of these made it to the top 20 distillation malfunctions. This contrasts the author's experience. Intermediate-component accumulation is experienced frequently enough to justify a place in the top 20, maybe even close to the 10th spot. The large number of cases of intermediate-component accumulation reported in this book will testify to that. In many cases, the accumulation led to periodic flooding in the tower. Other problems induced by the accumulation include corrosion, product losses, product contamination, and inability to draw a product stream.

A second liquid phase, either present where undesirable or absent where desired, was troublesome in several case histories, most from chemical towers. In many cases, issues in the overhead decanter or its piping induced an undesirable phase either into the reflux or into the product. Presence or absence of a second liquid phase caused not only separation issues and production bottlenecks but in some cases also violent reactions, damage, and explosions.

Other fractionation issues include insufficient reflux, insufficient stages, insufficient stripping, and excessive bottom temperatures. Although basic to fractionation, it is amazing how often it is overlooked. Unique multicomponent issues include absorption effects in wide-boiling mixtures and location of side draws. Azeotropic and extractive distillation have their own unique challenges.

CASE STUDY 2.1 NO REFLUX, NO SEPARATION

Contributed by Ron F. Olsson, Celanese Corp.

The feed to a 55-tray tower came in 10 trays below the top. The tower was separating an alcohol as the distillate from a glycol as a bottom product. A simulation detected that the losses of glycol in the distillate were excessive. The glycol losses were

---

Distillation Troubleshooting. By Henry Z. Kister

estimated to cost about $250,000 per year. Further investigation revealed that the reflux had been eliminated. Apparently, the reflux rate was cut out during the 1970s, when energy savings were most important. Over the years, this mode of operation became the norm. Further, corrosion of the trays reduced their efficiency, causing the separation to deteriorate.

The glycol losses were drastically cut once the reflux was reintroduced.

CASE STUDY 2.2 HEAVIER FEEDSTOCK IMPEDES STRIPPING

Contributed by Dmitry Kiselev and Oleg Karpilovskiy, Koch-Glitsch, Moscow, Russia

Installation A diesel hydrotreating unit was revamped to a dewaxing process. Due to increase in production of wild naphtha and gases, the diesel stabilizer was revamped also. The revamp design proposed to use a fired heater reboiler to provide the desired diesel flash point. The refinery did not have enough time to revamp the heater, so the unit started with stripping steam injection under the bottom tray, instead of the heater reboiler circuit.

Problem After several months of operation, the refinery decided to complete the unit revamp. The heater reboiling circuit was made operational while the steam line was disconnected. The result was surprising: The flash point of diesel decreased by 20°C, even though design specifications of the reboiling circuit (flow rate and heater outlet temperature) were achieved.

Investigation The first suspicion was that tray damage occurred during start-up, but checking of this required a shutdown or gamma scans, which were expensive options in that location. A complete set of process data was collected instead and a tower simulation prepared. The feed composition was surprisingly much heavier than design. The ASTM D86 50% percent point increased from 265 to 320°C. The reflux rate was half the design value. The simulation showed almost no vapor in the stripping part of the column. The heater outlet temperature could not be increased beyond 330°C to generate additional vapor due to vibration of the 75-m-long heater outlet line. The simulation showed that heater outlet temperatures even as high as 350–360°C would have been insufficient for achieving the diesel flash point specification.

The reason for poor operation was the new feed composition. The reason for the heavier feedstock was a revamp of the atmospheric tower of the crude oil distillation unit that took place at the same time as the last stage of the revamp of the hydrotreating unit. The crude tower revamp added a diesel draw in order to send light diesel directly to product blending and to dewax the heavy diesel only.

Solution During the next turnaround, the stripping steam line was reconnected. Simultaneous use of stripping steam and reboiling allowed the tower to achieve the product specification.

CASE STUDY 2.3 POOR H_2S REMOVAL FROM NAPHTHA HYDROTREATER STRIPPER

Contributed by Mark Pilling, Sulzer Chemtech, Tulsa, Oklahoma

Installation Naphtha hydrotreater stripper, stripping H_2S from naphtha.

Problem Tower had been operating fine for extended period. At a later time, it could no longer meet H_2S specification for bottom product.

Troubleshooting The tower was operated at the same bottom temperature as it always had been, but the reflux rate was much lower than normal. Investigations revealed that the feed to the unit had become considerably heavier. For this heavier feed, the operating bottom temperature was too low to provide sufficient stripping for H_2S removal.

Solution Bottom operating temperature and reflux ratio were raised to ensure proper H_2S removal.

Morals Tower operation needs to vary to accommodate changing feedstocks. Operators need to be trained to recognize the critical operating set points.

CASE STUDY 2.4 HEAVIES ACCUMULATION INTERRUPTS BOIL-UP

Contributed by Ron F. Olsson, Celanese Corp.

Figure 2.1 shows a system that recovered product from residues. The system removed product continuously as an overhead product from column C1. The heavy residues were periodically removed from drum D1.

Occasionally, the temperature of drum D1 would rise to the point where the reboiler could no longer boil it. The plant would then dump the drum content out of the bottom (route B). When the drum contents were dumped, lots of lights were lost in the dump. When the reboil ceased, liquid from column C1 dumped and much of it ended in the D1 drum dump.

The problem was fixed by removing bottom residue streams continuously from both points A and B. It took some trial and error to correctly set the bottom rate.

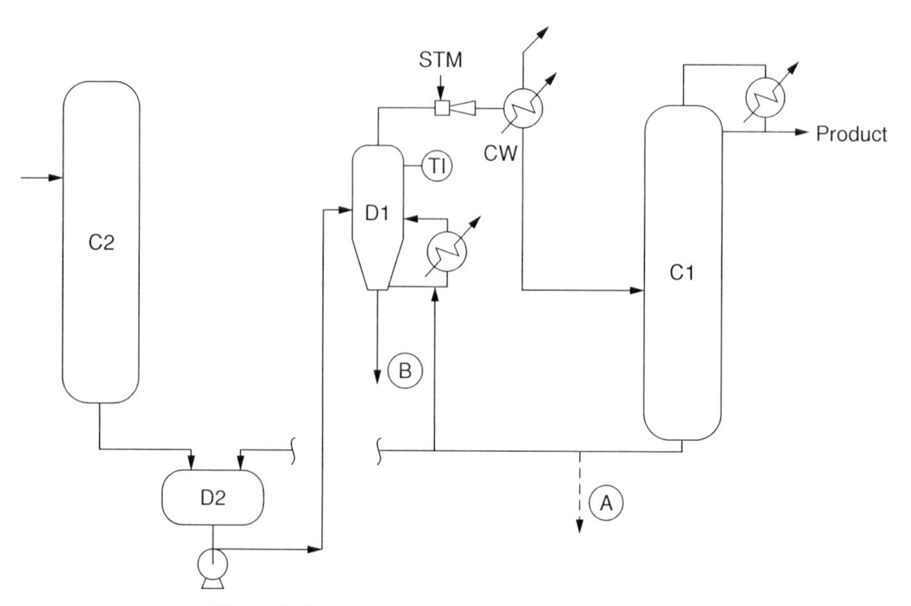

Figure 2.1 System that recovers product from residue.

CASE STUDY 2.5 INTERREBOILER DRIVES TOWER TO A PINCH

Henry Z. Kister, references 254, 276. Reproduced with permission.

A composition pinch occurs when, due to an insufficient driving force, the change in composition on each successive distillation stage diminishes and approaches zero. In the stripping section, an insufficient driving force usually coincides with an excessively low stripping ratio (V/B). Increasing the stripping ratio can reinstate composition changes, but at the expense of higher vapor and liquid hydraulic load. These higher loads cannot be tolerated when the tower nears a capacity bottleneck.

In this case, a clever debottleneck scheme looked great on the simulation. Yet pinches and mislocated feeds, readily visible on McCabe–Thiele diagrams, remain hidden on computer screens. It took a McCabe–Thiele diagram (341, described in detail in Ref. 251) to see that the scheme would drive the tower too close to a pinch and would be risky. Fortunately, the McCabe–Thiele diagram was prepared before the scheme was implemented.

Background An olefins plant was being debottlenecked for a 15% increase in throughput. The C_2 splitter (Fig. 2.2*a*) was a major bottleneck. The tower contained 95 trays in the rectifying section and 45 trays in a smaller diameter stripping section. The vapor feed entered close to its dew point.

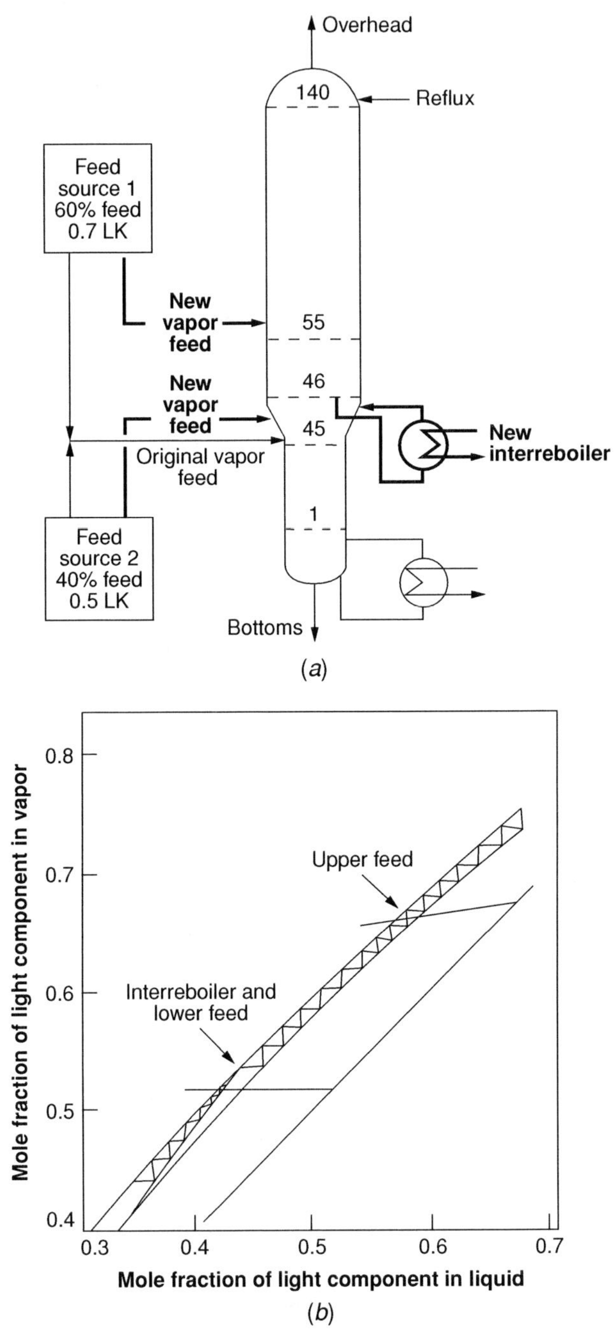

Figure 2.2 Proposed C_2 splitter debottleneck: (*a*) proposed changes, adding an interreboiler and splitting the feeds; (*b*) McCabe–Thiele diagram that clearly warned of imminent pinch. (From Ref. 254, 276. Reproduced with permission.)

Hydraulic calculations showed that the rectifying section would be barely capable of handling the increased throughput. The stripping section was undersized for the higher throughput and would require an expensive retray with specialty high-capacity trays.

Alternative Scheme A clever alternative scheme was conceived with a potential of slashing the revamp costs as well as saving energy. There was scope to have the feed enter as two separate streams. One contained 70% ethylene, made up 60% of the feed, and was to enter on tray 55 (Fig. 2.2*a*). The second contained 50% ethylene, made up 40% of the feed, and was to enter on tray 45. To unload the bottom section, an interreboiler was to be added at tray 46, supplying about 10% of the total column heat duty. Since the interreboiler was to convert 10% of the liquid into vapor, the vapor and liquid traffic in the narrow-diameter section below would diminish by 10%. This unloading was enough to accommodate the post-debottleneck throughput.

In principle, the interreboiler was to unload the narrow-diameter section that bottlenecked the tower. Splitting the feed was to assist in expanding the stripping section from 45 to 55 trays without adversely affecting separation in the rectifying section. The extra stripping trays were needed to accommodate for the lower V/B (stripping ratio) generated below the interreboiler.

A computer simulation showed that the scheme would work well. There were no convergence problems, nor was there anything about the simulation that may indicate a potential problem. The scheme received the go-ahead.

Just prior to going into the final design, a McCabe–Thiele diagram was constructed to explore hidden traps (Fig. 2.2*b*). The pinch just below the interreboiler was glaring.

Postmortem The interreboiler caused the V/B for the section below to diminish almost to the minimum stripping. Although hydraulically the interreboiler would have fulfilled its function, the column may not have achieved the design separation due to the pinch. Alternatively, to overcome the pinch, the operator would have needed to raise both the reflux and reboil and would have possibly encountered a hydraulic bottleneck.

CASE STUDY 2.6 TEMPERATURE MULTIPLICITY IN MULTICOMPONENT DISTILLATION

Henry Z. Kister and Tom C. Hower, reference 263. Reproduced with permission.

Installation A lean-oil still in an absorption–refrigeration gas plant (Fig. 2.3*a*). This still was the last step of purification of the absorption oil before the oil was returned to the plant absorber to absorb heavy components from natural gas. Feed to the still was the absorption oil, containing the absorbed gasolines and some LPG. Lighter components were removed from the oil upstream of the still. Lean oil left

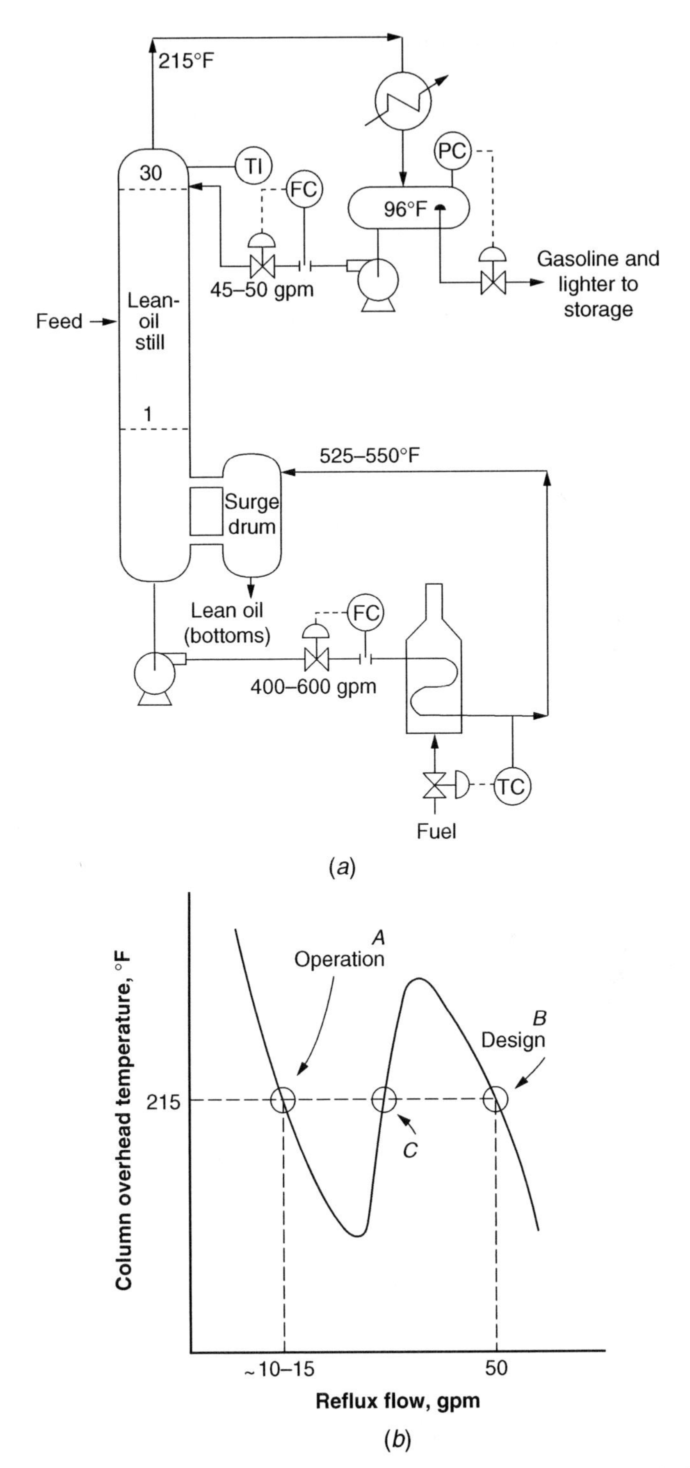

Figure 2.3 Multicomponent still that showed temperature multiciplity: (*a*) lean-oil still; (*b*) variation of lean-oil still top temperature with reflux. (From Ref. 263. Reproduced with permission. Copyright © (1987) AIChE. All rights reserved.)

as the still bottom product, while gasoline and LPG were the top product. The still operated at 210 psig. Note the large temperature difference between the bottom and top of the still.

The main objective of the still was to keep gasoline out of the column bottom. This was achieved by the furnace outlet temperature control. The reflux rate was trimmed by manually adjusting the flow controller set point, so as to give a reasonably constant column overhead temperature. The reflux drum was flooded, and liquid level in the condenser was used to control column pressure.

Problem The plant absorber appeared to be malfunctioning. It did not absorb all the heavy ends out of the gas.

Investigation The initial boiling point (IBP) of the lean oil leaving the still was low, which indicated the presence of a substantial quantity of gasoline in the still bottom. This suggested that the still was malfunctioning. The still showed no signs of flooding. The control temperature, the overhead temperature, and the reflux rate appeared to be at their design values. The composition of the top product was not analyzed.

Solution The problem was caused by insufficient reflux rate. The low reflux rate was unnoticed because of an incorrectly sized orifice plate in the still reflux line. When the orifice plate was replaced and the correct reflux flow set, the plant observed a large permanent increase in fuel usage and a large drop in the apparent quantity of absorption oil, indicating that the gasoline was being stripped off the bottom. Following this, the plant absorber started functioning normally and absorbing heavy ends out of the gas.

Analysis The problem was particularly difficult to detect because of the unusual behavior of the overhead and bottom temperatures. Normally, when a substantial amount of light impurity is present in the bottom, one would expect the bottom temperature to drop; when a substantial amount of heavy impurity is present in the column overhead, one would expect this temperature to rise. Over-reboiling can bring the bottom temperature back up, but in such a case, one would expect the top temperature to rise further above design.

The above considerations are generally valid for binary distillation and often, but not always, for multicomponent separations. This case is an example of a multicomponent distillation where the above considerations do not apply.

In general terms, at the low reflux rates the column was operated as a gasoline–LPG separator instead of an absorption oil–gasoline/LPG separator. This lowered temperatures throughout. However, the column was over-reboiled; this returned the bottom and top temperatures to their design values. This is explained in detail below.

At the low reflux ratio, a substantial fraction of the gasoline reached the bottom. This would have caused a lower temperature at the base of the column, but the control system increased the reboil rate (i.e., over-reboiled) to keep the bottom temperature up at design. Because of the low reflux ratio, however, the over-reboil action boiled over a significant fraction of absorption oil and perhaps most of the gasoline. The column probably fractionated out most, but not all, of the absorption oil. The mixture arriving at the top tray therefore contained the LPGs, some gasoline, and a small quantity of absorption oil. The presence of the absorption oil acted to increase the top tray temperature; the absence of gasoline that was lost to the bottom acted to decrease it. By varying the reflux rate as in normal operation, one could keep the top temperature at its design value.

Variations of the column overhead temperature are shown in Figure 2.3*b*. Under all these conditions, bottom temperature was controlled at 525–550°F. The column overhead initially operated at point *A* at the low reflux conditions. At the correct reflux rate, the overhead temperature operated at point *B*. Note the existence of point *C* on this curve, at which an increase in reflux rate causes an increase in overhead temperature. This operating condition (point *C*) has actually been observed in this type of column.

CASE STUDY 2.7 COMPOSITION PROFILES ARE KEY TO MULTICOMPONENT DISTILLATION

Contributed by Frank Wetherill (retired), C. F. Braun, Inc., Alhambra, California

Installation A product column in a specialty chemical plant producing a heavy, water-soluble glycol product. The process is similar to that described in Case Study 15.1. The column separated glycol product from high-boiling residues. The column is shown in Figure 2.4*a*.

Problem Although water was removed from the column feed and water-forming reactions were suppressed by lowering the base temperature in a manner similar to that described in Case Study 15.1, a very small quantity of water (about 0.1%) was still present in the product. It was economical to remove even that amount of water from the product.

Investigation This amount of water was very small and could have originated either in the column feed or from water-forming condensation reactions at the column base. Tackling this problem at the source would have been difficult.

It was realized that the product was very hygroscopic. Therefore, it was suspected that after the product was condensed and subcooled in the overhead condenser it reabsorbed water from the inerts stream.

Solution It appeared beneficial to withdraw the product upstream of the point where it was being subcooled. A suitable point was the top tray of the column. The column was modified to withdraw product from this tray, as shown in Figure 2.4*b*. This eliminated the water problem.

Postmortem The relative volatility for glycol–water separation was large (the atmospheric boiling point of the glycol was greater than 400°F). Any liquid water present in the reflux stream therefore easily vaporized on the top tray.

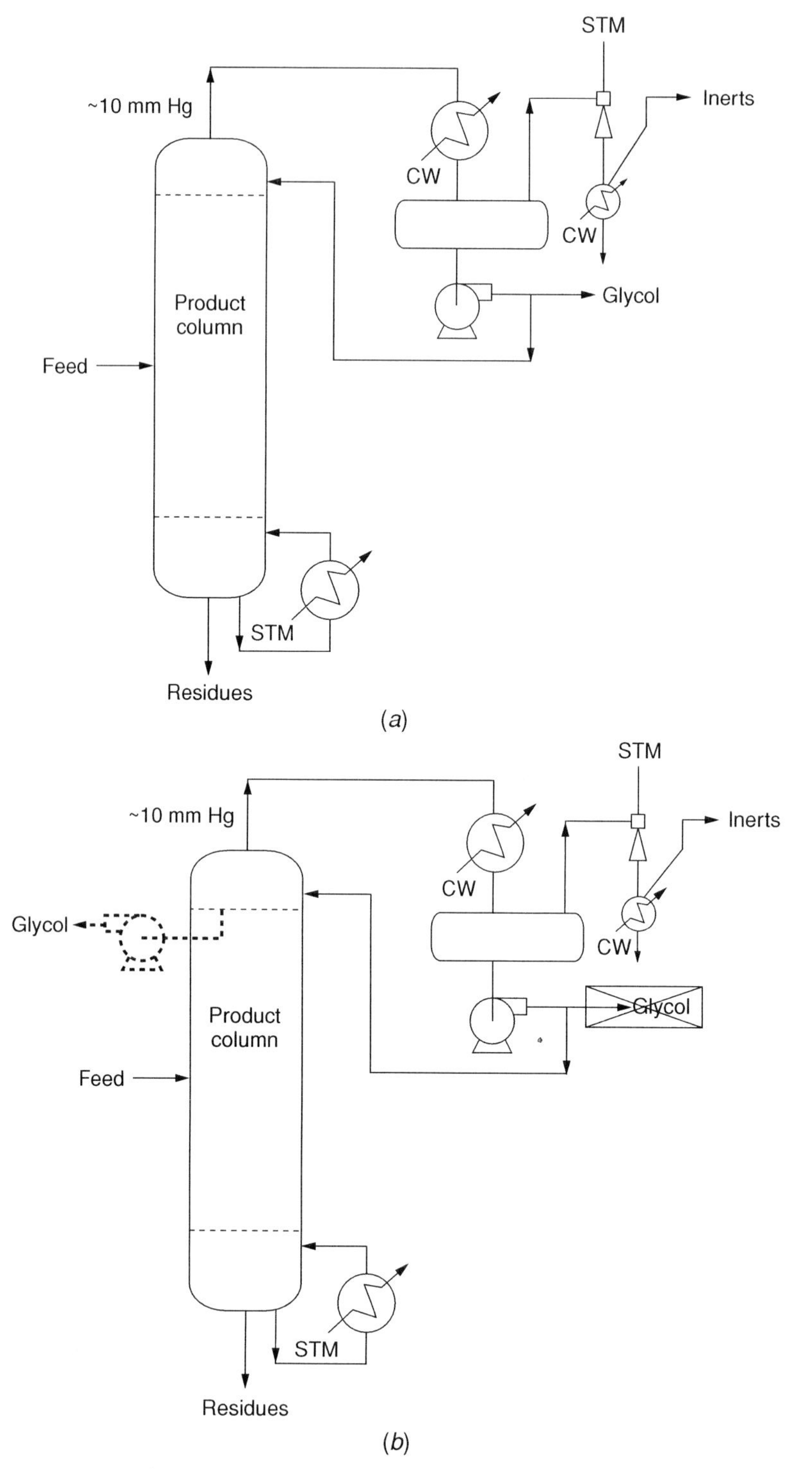

Figure 2.4 Glycol product column: (*a*) initial; (*b*) modified.

It may appear that withdrawing water from the top tray, instead of from the reflux drum, would have enriched the product with the heavier impurity because the condenser stage was no longer available for the product–residue separation. This enrichment, however, was minimal, because even before the modification the condenser behaved as a total condenser from the product–residue separation viewpoint (product was withdrawn as liquid) and had therefore contributed little to the product–residue separation.

Another Plant A glycol/residue tower in a completely different plant and operated by a different company experienced a somewhat similar problem. In that case, the amount of water was small. Instead of escaping in the inerts route, the water was condensed and refluxed back into the tower. Over a period of time, water built up in the overhead loop and adversely affected product purity. The problem was solved by periodically running the reflux drum liquid to a flash tank.

CASE STUDY 2.8 COMPOSITION PROFILE PLOT TROUBLESHOOTS MULTICOMPONENT SEPARATION

Henry Z. Kister, Rusty Rhoad, and Kimberly A. Hoyt, references 254, 273.

Engineers seldom bother plotting composition profiles in multicomponent distillation. Like the McCabe–Thiele and Hengstebeck diagrams, column composition profiles (generated from the compositions calculated by the simulation; Refs. 243 and 251 have detailed examples) are a superb analytical and troubleshooting tool that provides visualization that simulations do not. Undetected abnormalities often reveal themselves as a column malfunction after start-up. This case shows how a composition profile identified a very unforgiving column design.

Background A chemical vacuum tower containing structured packing (Fig. 24.1a and 2.5) separated a heavy key (HK) component from an intermediate key (IK) component in its lower section. There was a specification of 0.3% maximum IK in the bottom and 1.0% maximum HK in the vapor side product. Feed to the column contained many other components that were lighter or heavier than the keys.

Problem While the bottom product was on specification, the vapor side product contained about 10% HK, which was several times higher that the design.

Troubleshooting Initial suspicion was a malfunction of the structured packing or the distributors. The design height equivalent of a theoretical plate (HETP) was on the low side, but not grossly so. The lower bed was simulated by eight stages; six or seven would have been a closer estimate. The distributor design was found to be good, and the distributor was successfully water tested and debugged at the manufacturer's shop before being installed in the tower. The VLE data were examined. While not perfect, the volatility estimate was quite reasonable.

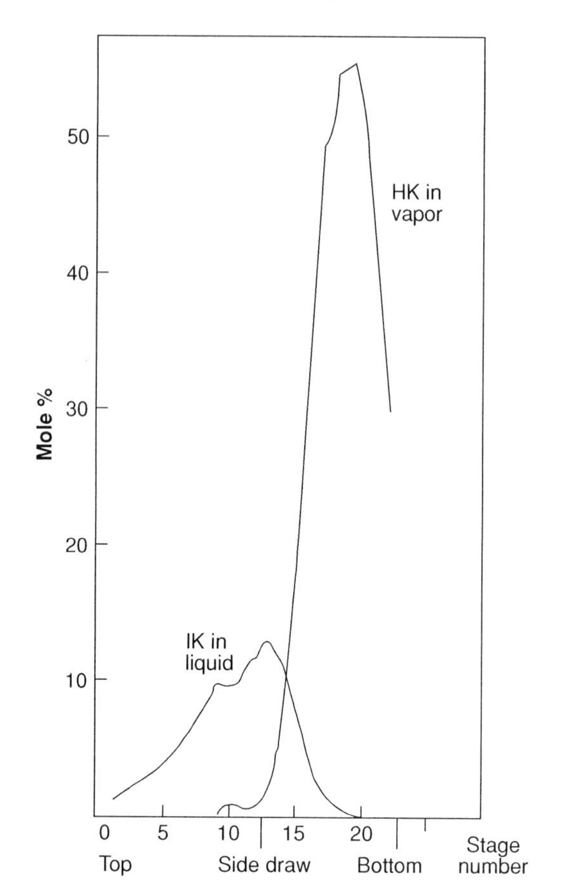

Figure 2.5 Composition profile pinpoints sensitivity of heavy key. (From Ref. 254, 273. Reproduced with permission. Copyright © (1996) AIChE. All rights reserved.)

Next, plugging of the packing or distributors was suspected. Extensive field tests, described in detail in Case Study 4.9, were performed and showed that the tower operated well below flood and that both the pressure drop and flood point were well inline with predictions. Gamma scans verified that distribution below the flood point was quite reasonable and there were no signs of plugging.

Likely Cause During the troubleshooting, the design simulation was revisited and the composition profiles plotted. The profiles plot the concentrations of each component in the liquid (one plot) and the vapor (a second plot) against the theoretical stage number. Figure 2.5 is a condensed version, singling out the IK in the liquid and the HK in the vapor. These were the prime actors in the current problem. The diagram shows an extremely steep peak for the HK in the vapor. Stage 18 vapor contains 55% HK. By the time the vapor draw-off is reached on stage 13 (five stages up), the HK concentration is supposed to drop to 1%. On stage 14, the HK concentration is about 7%, and on stage 15, it is 18%. Figure 2.5 therefore depicts a very unforgiving composition profile.

Achieving the design separation depends upon the lower packed bed successfully developing eight theoretical stages. Should this number fall a stage or two short, the concentration of HK in the vapor side draw would skyrocket, with product going severely off specification. Sources that can make the number of stages fall short of expectation by one or two were (and usually are) abundant. These include a slightly optimistic design HETP, inaccuracies in VLE, differences between design and actual feed compositions, relatively small scale fouling or maldistribution, and even disturbances to the feed and the heating and cooling media.

CASE STUDY 2.9 WATER ACCUMULATION CAUSES CORROSION IN CHLORINATED HYROCARBON TOWER

Installation A tower separating HCl and HC gases from chlorinated HCs (Fig. 2.6). There was a very small amount of water (~3 ppm) in the feed.

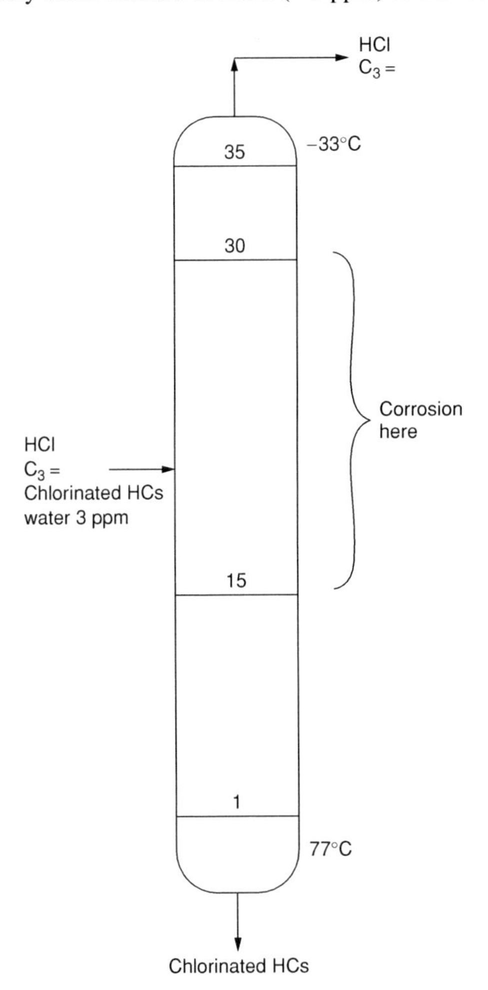

Figure 2.6 Water accumulation in chlorinated hydrocarbon column.

Problem There was severe corrosion on trays 15–30. There was no corrosion at the top 5 trays and the bottom 15 trays. The column run length was less than a month; afterward it needed shutting down to replace the trays.

Cause Top temperature was too cold, and bottom temperature too hot, to allow water to escape. In the bottom section, repulsion of water by the chlorinated HCs increased its volatility. As a result, the water became trapped in the tower and concentrated near the feed. The accumulation could be predicted using a NRTL or UNIQAC model, but not using ideal solution or equation-of-state models.

Solution The problem was resolved by replacing trays 15–30 by trays fabricated from Hastelloy C.

Related Experience A decomposition reaction took place near the bottom of one chemical tower, yielding a corrosive compound. The boiling point of that compound was well below the tower overhead temperature. It therefore accumulated and corroded trays in the middle of the tower.

CASE STUDY 2.10 HICCUPS IN A REBOILED DEETHANIZER ABSORBER

Installation A refinery reboiled deethanizer absorber (Fig. 2.7). The top section of the tower used a naphtha stream to absorb C_3 and C_4 HCs from a gas stream that went to fuel gas.

Feed to the tower contained a small amount of water. Free water was removed in the feed drum upstream of the tower, but the separation was not perfect. In addition, the small quantity of water dissolved in the HC feed would not be removed in the feed drum.

Bottoms from the tower went to a debutanizer that operated much hotter than the deethanizer. The debutanizer recovered the C_3 and C_4 HCs in the top product, leaving gasoline as the bottom product.

Problem Plant economics favored maximizing recovery of C_3 and C_4 in the deethanizer bottoms. To achieve this, the control temperature in the stripping section was lowered. The system worked well for 2–3 days following the change. Then the debutanizer pressure suddenly shot up, and a large slug of water was observed to fill the boot of the debutanizer reflux drum. A few minutes later the water disappeared. Two to three days later the process repeated. The possibility of steam or water leaks was investigated, but none were found.

Solution The tower was returned to its previous mode of operation with the higher deethanizer control temperature.

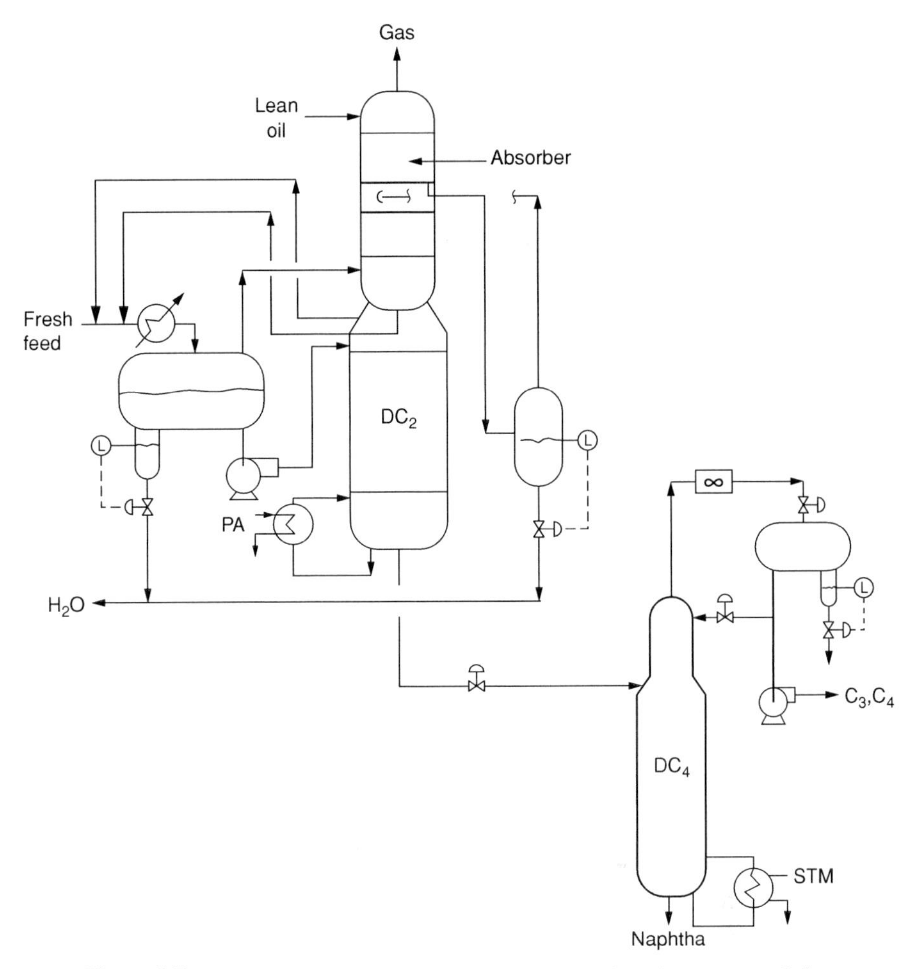

Figure 2.7 Reboiled deethanizer absorber system that experienced water accumulation.

Postmortem The symptom described, that is, periodic sudden slugging of water (hiccups), is typical of accumulation of a component in the tower. Water accumulation is a common experience in reboiled deethanizer absorbers because their top temperatures are often too cold, while their bottom temperatures are too hot, to allow the water to escape out of the tower at the same rate at which it comes in. In most cases, this accumulation leads to corrosion in the accumulation areas (see Case Study 2.11), but the accumulation usually does not go far enough to lead to hiccups.

The design of the current system recognized water accumulation as a potential problem and provided a special water draw to circumvent it. This was successful when the tower operated close to the design temperature profile. However, when the control tray temperature was lowered, the water accumulation zone descended further down the column, where no water draw was available.

Related Experiences Another reboiled deethanizer stripper experienced periodic flooding whenever a coalescer that removed water from the tower feed (upstream of the drum in Fig. 2.7) malfunctioned. The water accumulated in the stripper and initiated flooding there. Overcome by slumping the tower and allowing water escape in the bottoms.

CASE STUDY 2.11 WATER ACCUMULATION IN REBOILED DEETHANIZER ABSORBER

Installation A refinery reboiled deethanizer absorber, similar to that in Case Study 2.10 and Figure 2.7, except that the absorber and stripper were two separate towers, not mounted one above the other. All the water draw-offs were in the upper part of the absorber. There were no water draws in the stripper and in the lower part of the absorber.

Experience As in Case Study 2.10, plant economics historically favored maximizing C_3 and C_4 in the deethanizer bottoms. After three decades of operation, the absorber bottom temperature was 30°F colder than initially. The bottom trays of the absorber, and most trays in the stripper, experienced severe corrosion and required frequent repair and replacement.

Cause The cause was water accumulation, as explained in Case Study 2.10. Once water accumulated, it formed a second liquid phase, which contributed to the large observed temperature drop. The water phase dissolved acidic components, and the weak acid corroded the tower carbon steel (CS) trays. The plant lived with the problem, replacing and repairing trays at turnarounds.

Related Experiences Another refinery reboiled deethanizer experienced a similar corrosion problem, especially when running the feeds colder. A second deethanizer cured a corrosion problem in the stripper by improving coaleseing of the feed. Two other reboiled deethanizers containing stainless steel (SS) internals (one packed the other trayed) experienced no significant corrosion. At least one of these had water in the feed.

CASE STUDY 2.12 WATER ACCUMULATION AND HICCUPS IN A REFLUXED GAS PLANT DEETHANIZER

Henry Z. Kister and Tom C. Hower, reference 263. Reproduced with permission.

Installation A gas plant refluxed deethanizer (Fig. 2.8*a*). Feed to the column was rich absorption oil saturated with absorbed gas components (C_1 to gasoline). Reflux was condensed using C_3 refrigeration and entered the column at −30°F.

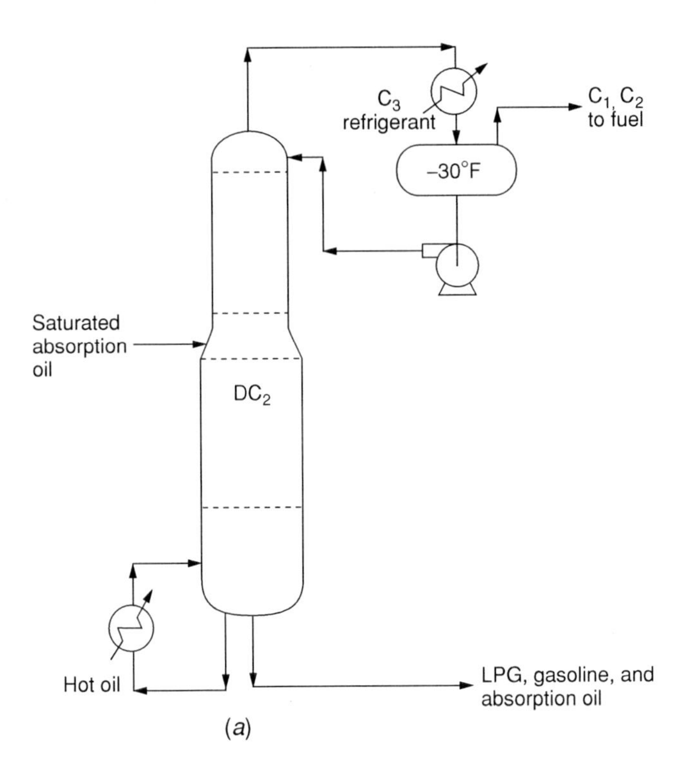

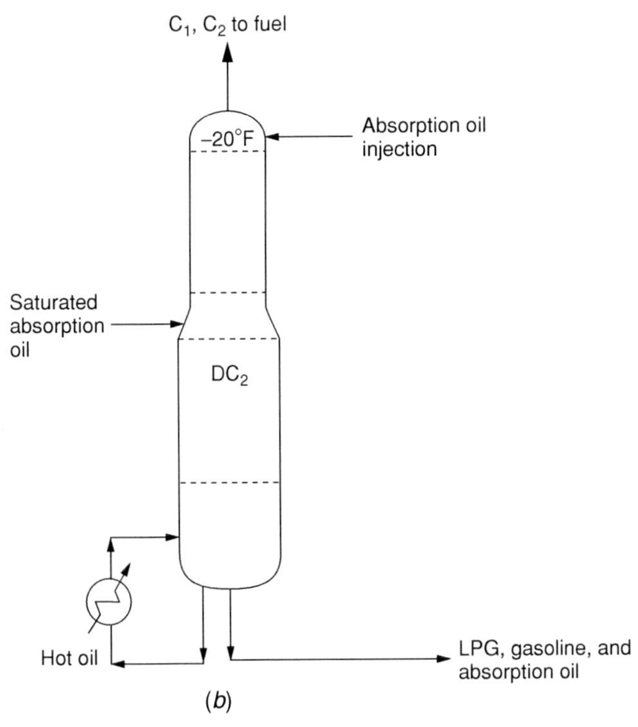

Figure 2.8 Gas plant deethanizer that experienced water hiccups: (*a*) initial, refluxed; (*b*) modified, with absorption oil injection. (From Ref. 263. Reproduced with permission.)

Problem At unpredictable time intervals, a slug of water would empty out from either the top or the bottom of the column. Emptying out from the top appeared to occur by massive carryover of fluids out of the top. Some absorption oil, water, and gasoline were found in equipment downstream of the reflux drum following emptying out from the top. Emptying out of the bottom appeared to take place by a massive slug of fluids. This slug caused a large increase in feed to a downstream depropanizer, resulting in a major upset in the column train downstream.

Cause Trace quantities of water, absorbed in the absorption oil, would enter the column. Top temperature was too cold, while bottom temperature was too hot, to permit the water to leave, so it accumulated in the column. The accumulation continued to a point where a water slug would empty out.

Cure Refluxing the column was discontinued. Instead presaturated absorption oil (i.e., absorption oil that was previously contacted with C_1 and C_2 to eliminate heat-of-absorption effects) was injected onto the top tray, so that tray temperature could be raised to −20°F (Fig. 2.8*b*). This was sufficient to ensure that the water vapor left with the top product. This eliminated the water slug problem. Absorption oil losses in the column overhead stream were negligible. The only unfavorable effect of this modification was to increase the absorption oil circulation throughout the plant by 1–2%. An alternative solution would have been to install water draw-off trays in the column; however, this solution would have been less economical and would also have suffered from the difficulty in predicting the location of the main points of water accumulation.

Related Experience Overhead vapor from a similar refluxed gas plant deethanizer contained water. This water was condensed, then refluxed back into the tower. Over a period of time, the water built up to a concentration high enough to generate hydrates in the overhead refrigerated condenser. Problem was mitigated by adding a boot that separated and removed the water from the reflux drum.

CASE STUDY 2.13 HICCUPS IN A COKER DEBUTANIZER

Installation Feeds to a debutanizer were hydrogenated coker naphtha and straight-run naphtha. The tower operated at about 200 psig. The feed was preheated.

Problem Once every 4 hours or so the tower would experience a hiccup, in which it emptied itself out from either the top or the bottom. Once the tower hiccupped, it took 20 minutes to restabilize. On occasions when the preheater was fouled, the problem became more frequent and more severe.

Cause Components intermediate between light key (LK) and (HK) at times accumulate in columns. This occurs when top temperature is too cold while bottom temperature is too hot to let enough of these components out. The symptom of this phenomenon is cyclic flooding, as described above, often at a period of every few hours or every few days.

It appears that the component accumulating here came in with the coker naphtha. This is because the refinery had another debutanizer that was fed only with straight-run naphtha, and it did not experience hiccups. Also, problem aggravation by preheater fouling suggests that the component was trapped just below the feed.

Solution The top of the tower was operated warm to allow enough of the trapped component to escape. This mitigated the hiccups at the expense of losing C_5 in the overhead, which was an economic loss. Typically, the overhead product contained 8–10% C_5.

Related Experience In one fluid catalytic cracker (FCC) debutanizer, a tray temperature about 10 trays above the bottom manipulated reboiler steam. Periodically, the boil-up rate significantly rose over a period of time without any changes to the control temperature set point. If the operators did nothing, the boil-up kept rising. The corrective action was to cut back the boil-up for about an hour so that bottom temperature went down 20–30°F. After this, normal boil-up was reestablished and remained steady. It appears like an intermediate component irregularly accumulated in the tower. The frequency of accumulation was as high as several times a week or as low as once every several weeks. Reducing the bottom temperature allowed the component to escape out of the bottom.

CASE STUDY 2.14 HICCUPS IN A SOLVENT RECOVERY COLUMN

Installation Feed to the column in Figure 2.9 was typically 75% water (H_2O), 8% ethanol (EtOH), 8% *n*-propanol (*n*-PrOH), 8% other alcohols and acetates, and 1% of ethylene glycol monoethyl ether (EGEE). The tower separated a solvent–water azeotrope from water. The azeotrope, which contained 15–20% water, went to dehydration. The water was sewered.

Problems The column experienced two instability problems:

1. The column would run steadily for 2–3 hours. With the control tray temperature operating at its normal 185°F, temperatures below the control tray would slowly begin to creep down. Then suddenly the column would appear to flood, and temperatures would drop throughout. The operators tackled this by reducing feed to about 40% of the normal rate, even more. This would allow the column to stabilize, regain a good temperature profile, and return to normal. The column would then run steadily for another 2–3 hours and the above reoccurred.

   Raising steam to tackle this problem was tried but was found less effective than cutting the feed. It also created the potential problem of releasing liquid via the flame arrestor.

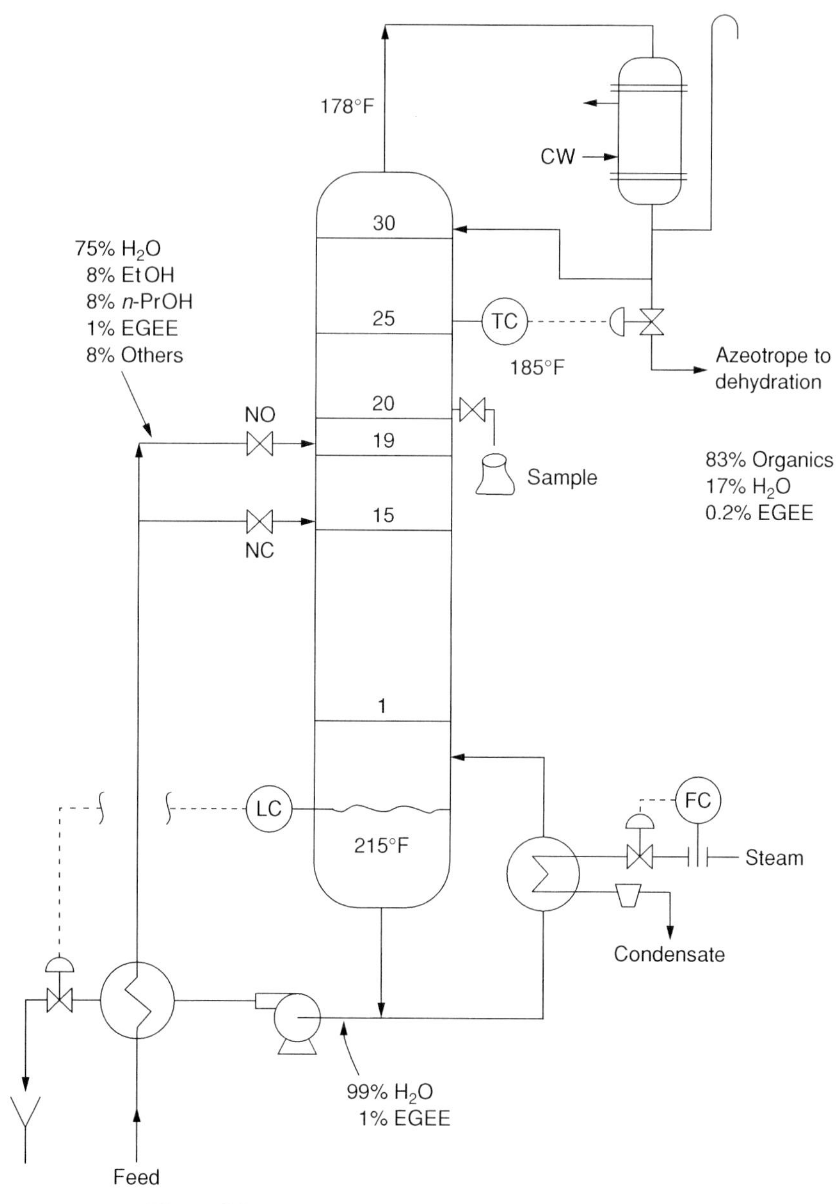

Figure 2.9 Solvent recovery column that experienced hiccups.

2. *The converse problem.* The column would run steadily for 2–3 hours. Then the bottom temperature, usually 215°F, would rise. Initially it would rise slowly, then it would jump up to 230°F within 5 minutes or so. Bottom pressure would jump up from 3 psig to 5–7 psig. Then the rest of the column temperatures would jump up. The overhead temperature would rise from 176 to 190°F. The

problem occurred regardless of whether the top-temperature controller ran on automatic or manual. Opening the distillate control valve only made the column hotter.

To tackle this problem, the operator would drastically cut back steam flow, reduce top-product rate, and divert the bottom to a rerun tank. When a good temperature profile was reestablished, normal operation was resumed and sustained for another 2–3 hours.

Testing A drain valve was found coming out of a downcomer from tray 20. A sample connection was added and a sample was removed. It was found to contain 38% water, 5% EtOH, and as much as 44% *n*-PrOH and 12% EGEE. It looks like *n*-PrOH and EGEE were building up near the center of the tower.

The feed was then switched to tray 15 and the test repeated. This time, the sample contained 49% H_2O, 3% EtOH, 17% *n*-PrOH, and as much as 28% EGEE.

Proposed Mechanism The two problems appear related and were caused by concentration of *n*-PrOH and EGEE in the tower. The atmospheric boiling points of these components are 208 and 275°F, respectively. In a water-rich mixture, an activity coefficient effect makes *n*-PrOH far more volatile (volatility of about 12), so it easily distills upward. The same activity coefficient effect makes the higher boiler EGEE much more volatile than water in a water-rich solution, so it too distills upward. The bottom temperature was too hot to permit these components to escape out of the bottom in sufficient quantity.

In the upper part of the tower, water was in small concentration, and EtOH becomes a major component. Both *n*-PrOH and EGEE are far less volatile than EtOH. So the top temperature was too cold and did not allow sufficient *n*-PrOH and/or EGEE to escape in the overhead. These components therefore had nowhere to go. They accumulated in the tower.

Problem 1 occurred when the dominant accumulation was that of *n*-PrOH. It azeotroped with water, and the concentration of this azeotrope in regions usually occupied by water caused temperatures to drop. To clear, the feed rate was reduced, and the accumulated *n*-PrOH–H_2O azeotrope was batch distilled over the top.

Problem 2 occurred when the EGEE accumulation predominated. The EGEE–H_2O azeotrope boils at much the same temperature as water. The accumulation would raise pressure drop, so the bottom temperatures rose. To clear, the operator cut back steam and distillate, allowing the EGEE (together with other components) to escape in the bottom.

Solution Control tray temperature was raised to 190°F to avoid accumulation and allow enough *n*-PrOH and EGEE to leave in the overhead.

Initially, there was a concern that raising this temperature would increase the concentration of water in the distillate and overload the dehydration system downstream. To minimize this effect, the feed point was lowered to tray 15. This change added rectifying stages and was demonstrated by test to minimize the amount of water in the overhead.

The net result was a marginal increase in the amount of water in the overhead. This marginal increase was easily handled by the dehydration system downstream. The accumulation problem was eliminated. Stable operation, requiring little operator intervention, had been established.

Related Experience A tower separating ethanol from water experienced a similar accumulation problem of EGEE. The cure was drawing the EGEE as a side-draw.

CASE STUDY 2.15 THREE-PHASE DISTILLATION CALCULATIONS AND TRAPPED COMPONENTS

Contributed by W. Randall Hollowell, CITGO, Lake Charles, Louisiana

Installation The water–methanol phase from a three-phase gas–oil–aqueous separator was fractionated for recovery of methanol.

Problem The tower experienced periodic upsets with large amounts of water and oil puking into the methanol overhead product.

Cause Some of the oil phase from the separator was entrained in the methanol–water phase feeding the tower, in addition to the oil dissolved in the methanol–water phase. Oil components rapidly became less volatile as they went up the tower, where methanol concentrations were higher. Oil components rapidly became more volatile as they went down the tower, where methanol concentrations were lower. The lightest oil components could escape with tower off gas and/or side-draw methanol product because they had enough volatility, even in high methanol concentrations. The heaviest oil components could go out the bottom of the tower as a separate liquid oil phase. Midrange oil components would accumulate in the tower.

Buildup of the trapped components caused the "puking," or periodic flooding. Since the latent heats of vaporization of the oil components were much lower than those of the methanol–water mix, their accumulation increased the gas and liquid traffic in the accumulation zone. This was a major contributor to the flooding.

Simulation Simulations tried on a number of different packages all failed to converge. Accumulation of components is a non-steady-state condition. Process simulation packages could not converge steady-state simulations with a small amount of a third phase.

When a simulation cannot solve a system, it is often possible to define related systems that can be solved with solutions that provide insight for analyzing the actual system. This may involve trial and error and careful consideration of how well the solutions extend to the actual system.

Simulator component library data were often limited for the C_{10} + HCs, and they were not corrected until much later. A similar molecule could usually be found with good library data, allowing a fair selection of components for simulation.

Three-phase tower solutions were finally obtained by using larger than actual oil rates in the feed. Small oil rates caused convergence failure by causing many trays to alternate between two- and three-phase conditions. Excess oil in the feed gave a continuous, three-phase condition for most of these trays, leaving only a few trays that alternated between two and three phases.

Numerous feed oil compositions were tried to obtain estimations of accumulation of components after various run times. Convergence was obtained by the usual techniques, "sneaking up" on any substantial changes, 16-hour days, strong coffee, and long, loud strings of invectives.

The simulations confirmed that this was a classic example of trapping components.

Related Experience In another methanol–water tower, the feed contained a small percentage of xylenes. These accumulated in the tower, causing puking, or periodic flooding. The problem was alleviated by raising the top temperature and allowing the xylenes to escape with the methanol at the expense of reduced methanol purity.

Another Related Experience In another methanol–water tower, periodic puking was caused by accumulation of isopropanol. There was a side draw for withdrawing the isopropanol, but when opened, only water came out. Above the side draw was a bed of packing and no other side draw.

One More In this small methanol–water tower, the puking was caused by accumulation of heavier alcohols and oxygenates. Following a puking episode, this tower returned to normal without operator intervention.

CASE STUDY 2.16 HICCUPS IN AN AMMONIA STRIPPER

Installation Ammonia stripper tower with a rectifying section, separating an ammonia-rich overhead stream from water bottoms.

Problem Small quantities of methanol (0.2–0.3% of the feed) accumulated, giving hiccups every second or third day. The top of the tower was too cold, while the bottom was too hot, to allow the methanol to escape at a sufficient rate.

Solution Tower overhead temperature was raised, which allowed the methanol to escape in the overheads. The increase in water content of the overhead stream was small due to the unique nature of the ammonia–water temperature–composition relationship. In ammonia-rich mixtures, large changes in equilibrium temperature are equivalent to only small changes in water content.

CASE STUDY 2.17 EXCESS PREHEAT LEADS TO HICCUPS

Henry Z. Kister and Tom C. Hower, reference 263.

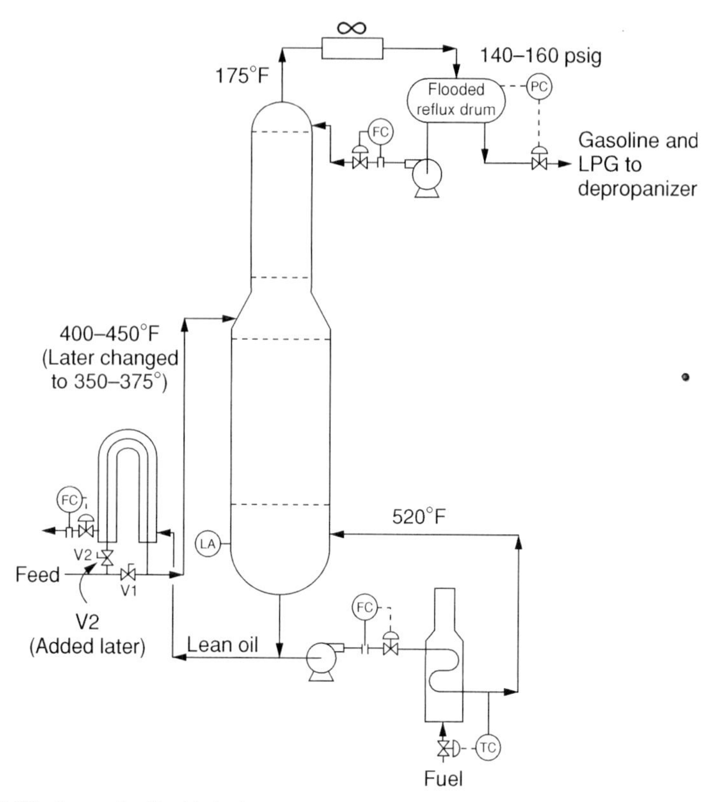

Figure 2.10 Lean-oil still with feed-bottom interchanger. (From Ref. 263. Reproduced with permission.)

Installation Lean-oil still in an absorption–refrigeration gas plant (Fig. 2.10). Feed to the still was rich absorption oil, containing gasoline and some LPG absorbed from the natural gas. Lean oil left as the still bottom product, while gasoline and LPG were the top product. The still was reboiled by a fired heater, and reflux was supplied by an air condenser. Top temperature was about 175°F, bottom temperature 520°F, and column pressure about 140–160 psig. To reduce the duty of the fired reboiler, the still feed was preheated by exchanging heat with the still bottom stream. The preheater had a valved bypass on the feed side (valve V1 in Fig. 2.10). The preheater performed better than expected and preheated the feed to 400–450°F.

Problem Most of the time, the column operated normally. However, every 4–12 hours or so, the column would suddenly unload, or hiccup, and empty itself out either through the top or into the bottom surge section. It was difficult to predict through which end the column would unload. Unloading occurred with no prior warning. Each occurrence of unloading lasted less than 1 minute and was accompanied by a rapid rise in column differential pressure. The column then stabilized itself automatically over a 15–20-minute period. When the column unloaded out of the top, the bottom temperature rose and cut back the heater fuel supply. When the column unloaded out of the bottom, the heater fuel consumption greatly increased, and even this could not supply sufficient heat to maintain the bottom temperature.

Analysis There appeared to be some vague connection but no direct cause-and-effect relationship between the column feed composition and the occurrence of the problem. An increase in the heavies content of the feed appeared to aggravate the problem. The still feed composition varied widely, depending on the gas wells feeding the plant. In addition, the raw gas lines were pigged about every 6 hours to clear condensate which settled in the lines, and this temporarily increased the heavies content of the feed to the still.

The plant experimented with several variables in an attempt to overcome the problem. Different bottom temperatures, varying rates of circulation through the reboiler, changing reflux rate, opening and closing the feed bypass valve were all tried, but no improvement was observed. Plant rates also appeared to have little effect; the problem occurred at plant rates as low as 40% of design.

Theory The feed temperature of about 400–450°F was sufficiently high to boil a significant fraction of the heavier components of the gas contained in the rich oil. These heavy components would normally end up in the column bottom and add up to the lean oil. Column temperature at the top was too cold to let these components escape with the LPG and gasoline, while feed tray temperature was too hot to allow them to flow toward the bottom of the column. These components therefore accumulated in the top section until the column would flood and empty itself, by either massive entrainment or postflood dumping.

Solution To eliminate the accumulation, it was desired to lower the still feed temperature. Opening the feed bottom interchanger bypass did not drop the temperature significantly because of the good performance of the exchanger. A valve V2 was installed in the feed line to the exchanger and was throttled to force more feed via the bypass. Adjustment of both valves enabled feed temperature to be reduced to 350–375°F. Once the feed temperature was lowered, the column no longer emptied itself.

CASE STUDY 2.18 RECYCLING CAUSES WATER TRAPPING

Tom C. Hower and Henry Z. Kister, reference 224. Reprinted with permission from *Hydrocarbon Processing*, by Gulf Publishing Co.

This case describes the trapping and freezing of water in an absorber–deethanizer system following a modification in which the deethanizer overhead was chilled and recycled to the absorber.

Installation A natural gas plant using an absorption–regeneration process for recovering heavy HCs from natural gas (Fig. 2.11). Natural gas was first dehydrated

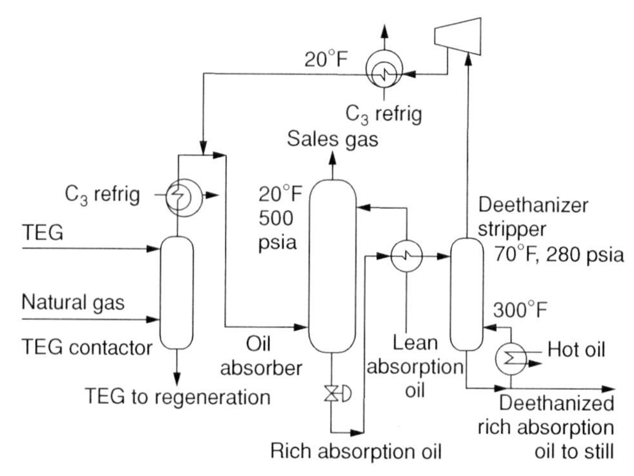

Figure 2.11 Gas plant experiencing freezing problem. (From Ref. 224. Reprinted with permission from *Hydrocarbon Processing* by Gulf Publishing Co. All rights reserved.)

in a triethylene glycol (TEG) dehydrator, then chilled and fed to the absorber. In the absorber, absorption oil absorbed heavier components (LPG and gasoline) from the gas. The rich absorption oil was stripped in the deethanizer to remove any absorbed methane and ethane.

History Initially, the deethanizer overhead was sent to the fuel gas. At a later time, it became economical to recover the heavier components contained in the stream. A compressor and chiller were added. The deethanizer overhead was compressed, chilled, and recycled to the absorber feed (Fig. 2.11).

Problem Following the modifications, the tubes of the chiller downstream of the deethanizer overhead compressor plugged due to freezing after every 8 hours or so. The compressor would trip each time on high discharge pressure.

Cause Initially, the TEG system dehydrated the gas below the dew point required for proper plant operation. Small quantities of water vapor not removed by the TEG contactor entered the absorber. This water vapor was partially absorbed by the absorption oil and entered the deethanizer. In the deethanizer, the water was stripped due to the high bottom temperature and exited in the overhead vapor stream. Here this water became concentrated because the deethanizer overhead was only a small fraction of the gas stream. When the deethanizer overhead vapor was chilled, this water froze. Any water going past the chiller eventually returned to this overhead system because it had no other way out. Ice accumulated in the chiller until it plugged the tubes.

Solution For many years, the solution was to thaw the chiller once per shift for about 2 hours. During this period, the deethanizer overhead was flared. Eventually, a small package TEG dehydrator was added at the discharge of the deethanizer overhead compressor. This completely eliminated the freezing problem. Payout was 1 month by salvaging the flared gases.

Moral When modifying a plant, the designer must look not only at the modification itself but also at the interaction of the modification with the existing system.

CASE STUDY 2.19 IMPURITY BUILDUP IN ETHANOL TOWER

Installation A tower producing beverage-grade ethanol (Fig. 2.12). The tower feed was 15-proof (8% vol.) ethanol from the bottom of an extractive distillation tower. The main product was 190-proof (95% vol.) ethanol–water azeotrope drawn from tray 57. Tower bottom was water. Most of the water was recycled as solvent to the extractive distillation tower; the balance was sewered. A heads purge, containing mainly ethanol but also lights, was removed from the tower overhead and recycled to the fusel oil decanter. Also sent to the fusel oil decanter was a fusel oil stream drawn from tray 49 at a constant flow rate. This stream was mainly ethanol (150 proof, i.e., 75% vol.) but also consisted of amyl alcohol, propanol, butanol, as well as some heavy ketones, aldehydes, and esters. In the fusel oil decanter, the heads and fusel oil products were mixed with cold water, then decanted. The aqueous layer was returned to the extractive distillation tower while the organic layer was the fusel oil product.

Problem Every 2–3 days, the product alcohol developed a smell that was not right. There were no signs of flooding when this happened.

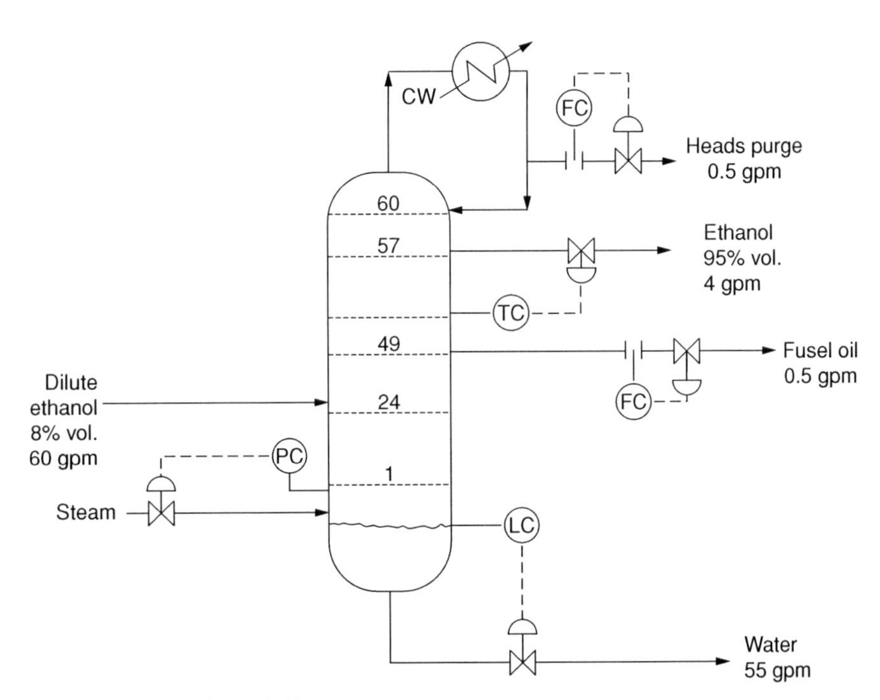

Figure 2.12 Ethanol tower experiencing impurity buildup.

Theory The smell is an indication of a buildup of impurity in the tower. The experience that it takes 2–3 days to build up suggests a slow accumulation rate. It is unknown whether the impurity is from the heads stream or from the fusel oil stream. One possibility is a water-soluble component that goes with the heads or fusel oil stream, then the aqueous phase of the decanter, from which it gets recycled and ends in the feed to the tower.

Cure When the smell was encountered, fresh feed to the system was reduced while the product, fusel oil, and heads were maintained at their normal flow rates. After 3–4 hours, the smell cleared and did not come back until 2–3 days later.

Related Experience Feedstock to ethanol–water distillation normally came from a process that hydrated ethylene. This feedstock was lean in fusel oil and heads and was readily handled by the tower. Occasionally, the plant would pick a low-cost feedstock from fermentation, which was rich in impurities. The tower then experienced hiccups and instabilities due to component accumulation. The solution was *dP* measurement between draw-offs. A *dP* rise indicated accumulation and was countered by increasing heads or fusel oil draw rates.

CASE STUDY 2.20 INTERREBOILER INDUCES STUBBORN HYDRATES IN A C_2 SPLITTER

Henry Z. Kister, Tom C. Hower, Paulo R. de Melo Freitas, and Joaõ Nery, reference 276. Reproduced with permission.

An interreboiler can interact with and aggravate a hydrate problem. This case study describes how an interreboiler converted a routine hydrate problem into a stubborn, nasty one.

Background Light HCs at high pressures and low temperatures (below about 50–70°F) can combine with water to form solid icelike crystalline molecules known as hydrates. Inside cold towers, hydrates precipitate out of the liquid and plug tray holes and valves. When enough holes or valves plug, the tray becomes restricted. Liquid accumulates above the plugged tray and the tower floods. The flooding can be recognized by a rise in tower differential pressure.

Hydrates frequently occur in C_2 splitters. In most (but not all) high-pressure C_2 splitters, plugging of trays due to hydrates initiates below the feed and can be recognized as a rise in the bottom-section differential pressure. Hydrates are prevented by drying the tower feed to less than 1 ppm of moisture. If some moisture still finds its way in, hydrates can be eliminated by injecting methanol into the tower. The methanol dissolves the hydrates just like antifreeze dissolves ice. Since methanol is far less volatile than light HCs, it normally exits the tower bottom after having dissolved the hydrates.

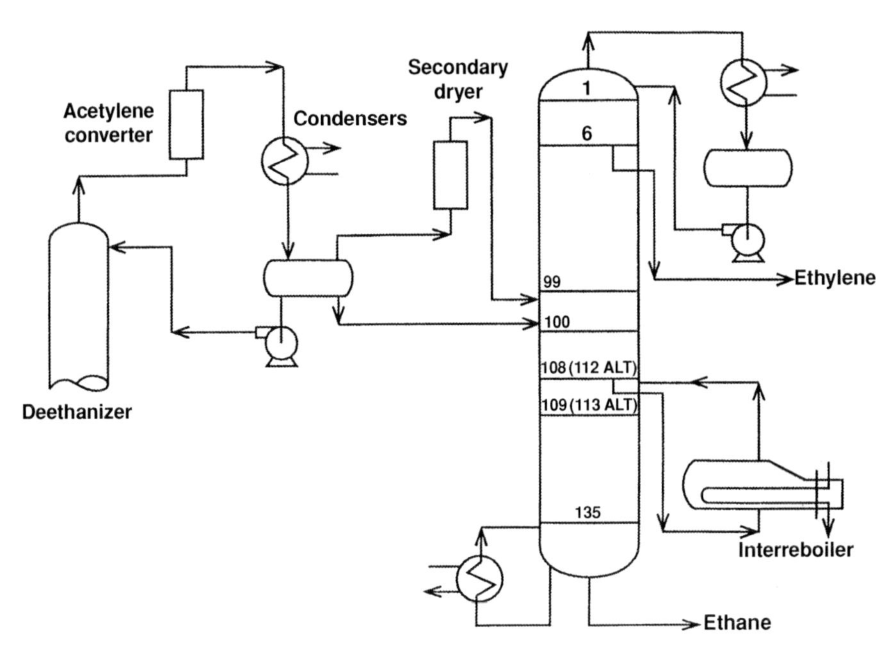

Figure 2.13 A C_2 splitter with interreboiler that experienced stubborn hydrates. (From Ref. 276. Reproduced with permission. Copyright © (1996) AIChE. All rights reserved.)

Installation A C_2 splitter (Fig. 2.13) received a vapor feed and a liquid feed, both from the deethanizer reflux drum. All the tower feed was dried by a primary dryer upstream (not shown) which dried the feed to the plant cold section to less than 1 ppm moisture. The vapor feed to the C_2 splitter passed through a secondary dryer before entering the tower. Both vapor and liquid feeds entered the C_2 splitter via different nozzles. The trays contained rectangular valves.

Eight trays below the feed, on tray 108, there was a liquid draw-off to a kettle interreboiler at ground level. Vapor from the interreboiler was returned to the vapor space between trays 108 and 109. There was an alternate liquid offtake to the interreboiler on tray 112 with a vapor return between trays 112 and 113.

History Following initial start-up, the C_2 splitter operated well without hydrates or any other problem for about 18 months. The secondary dryer was not regenerated even once. At that time, with practically no change in operating conditions of the deethanizer or C_2 splitter, there was a rise in the differential pressure of the bottom section of the C_2 splitter. The rise occurred for no apparent reason. The differential pressure came down again after methanol was injected into the splitter feed.

From then on, hydrates occurred two to three times per week. A hydrate was recognized by a rise in differential pressure in the stripping section accompanied by a reduction in bottom flow rate. The rectifying section differential pressure did not rise.

Living with Hydrates Steps taken to help the C_2 splitter live with the hydrates included the following:

- Methanol injection into the column feed was stepped up. In addition to injection when a hydrate was observed, about 200 liters of methanol was injected into the tower once a week to remove residual hydrates. Excessive injection of methanol had to be avoided because the methanol ended up in the cracking furnaces. Some of it passed through the furnaces uncracked and ended up in the C_3 product.
- The secondary dryer was regenerated. An on-line analyzer continuously measured less than 1 ppm moisture both at the inlet to and outlet from the secondary dryer. Following the first hydrate, this dryer (which was never previously regenerated) was regenerated once per month. The regeneration temperature profile showed no signs of moisture, with temperatures at the inlet and outlet remaining unchanged throughout the regeneration.
- The feed was 80% vapor, 20% liquid. Only the vapor feed passed through the secondary dryer. Following the first hydrate, the liquid feed route was blocked in, so that the feed became all vapor and all of it passed through the secondary dryer. This operation was sustained for 20 days, during which hydrates occurred two to three times a week. After 20 days the feed was returned to normal (80% vapor, 20% liquid) without any effects on hydrate frequency.

Although some progress was made, the frequency of hydrates still remained far too high.

Focus on the Interreboilrer The above experience suggested that the hydrates were quite insensitive to moisture in the feed. The feed did not appear to bring in much new moisture. The little moisture contained in the feed appeared to be trapped and accumulating inside the C_2 splitter. Once inside, it almost appeared as if the moisture was moving from section to section, without going out.

A key observation was that temperatures near the interreboiler went down every time a hydrate occurred. Tray 107, normally at -21°C, cooled to -23 to -25°C when a hydrate occurred. The interreboiler inlet and outlet temperatures also dropped by 2°C. Other temperatures around the C_2 splitter did not change much.

Column diameter, tray spacing, and tray design were the same for the trays above and below the interreboiler. With the interreboiler supplying about 30% of the tower heat duty, the vapor and liquid traffic between the feed and the interreboiler far exceed those in the section below. A flood due to hydrates was therefore expected to initiate above the interreboiler. This was confirmed by the relatively small rise in bottom-section pressure drop during the hydrates, suggesting that many of the stripping section trays—probably those below the interreboiler—were not flooded.

Theory The above supports hydrates between the interreboiler and the feed. When methanol was injected to dissolve the hydrate, much of the methanol and dissolved hydrate would be trapped on the interreboiler trap-out tray. From there it would flow into the kettle reboiler. Since water and methanol are less volatile than the C_2's, they would stay in the kettle. Over a period of time, the water would batch distill back into the C_2 splitter. The moisture would again form hydrates, and so on. This chain of events can only be discontinued by draining the methanol–water mixture from the interreboiler.

Solution Draining of methanol–water from the interreboiler was initiated each time methanol was injected. The distance between the feed point and the interreboiler was increased by lowering the interreboiler draw and return points from tray 108 to tray 112. The quantity of routine (once-per-week) dose of methanol was stepped up to about 500 liters.

The month after instituting this new procedure was hydrate free. At the end of this month, routine methanol dosing was discontinued. No hydrates were observed for another 3–4 months. After that, a high differential pressure was observed in the bottom of the C_2 splitter roughly once every 6 months, mostly due to unrelated events such as level control problems causing base-level rise above the reboiler return nozzle into the trays.

CASE STUDY 2.21 SIPHONING IN DECANTER OUTLET PIPES

Henry Z. Kister and James F. Litchfield, reference 260. Reprinted courtesy of *Chemical Engineering*

Installation The decanter in Figure 2.14*a* separated a light liquid phase that was recycled back to a reactor from a heavy liquid phase that went to product distillation and purification. The decanter was a 4 × 8-ft horizontal drum that provided well in excess of an hour residence time for phase separation. The maximum liquid level in the decanter was set by the light-phase 3-in. draw-off nozzle, which was located in the decanter head, 6 in. below the top of the decanter. The heavy phase was drawn at the bottom of the decanter. After leaving the decanter, the heavy phase flowed through a block valve, through an isolation control valve, and then up through a seal loop. The elevation at the top of the seal loop was an inch or two lower than the elevation of the light-phase draw nozzle (Fig. 2.14*a*). A 1-inch pressure balance line connected the top of the seal loop to the decanter vapor space.

After leaving the decanter, both phases flowed to their respective surge tanks at grade level, which was about 50 ft below the decanter elevation.

Problems With the block and isolation valves wide open, the decanter was susceptible to siphoning through the seal loop, creating erratic flow in this system. The seal loop had siphoned as much as 70% of the decanter liquid. Because of the erratic flow, the decanter was unable to operate at its design temperature. This decanter also did a poor job of handling liquid surges from upstream.

Initial Modifications The pressure balance line size was increased from 1 to 2 in. The seal loop pipe size was increased from 2 to 4 in. These modifications helped, but the erratic flow persisted.

An operating action that mitigated the siphoning was closing the seal loop block valve halfway. This operation mode, however, was not desirable because upon a feed surge some heavy phase was carried over into the light phase. Also, closing the valve had little effect on the decanter operating temperature.

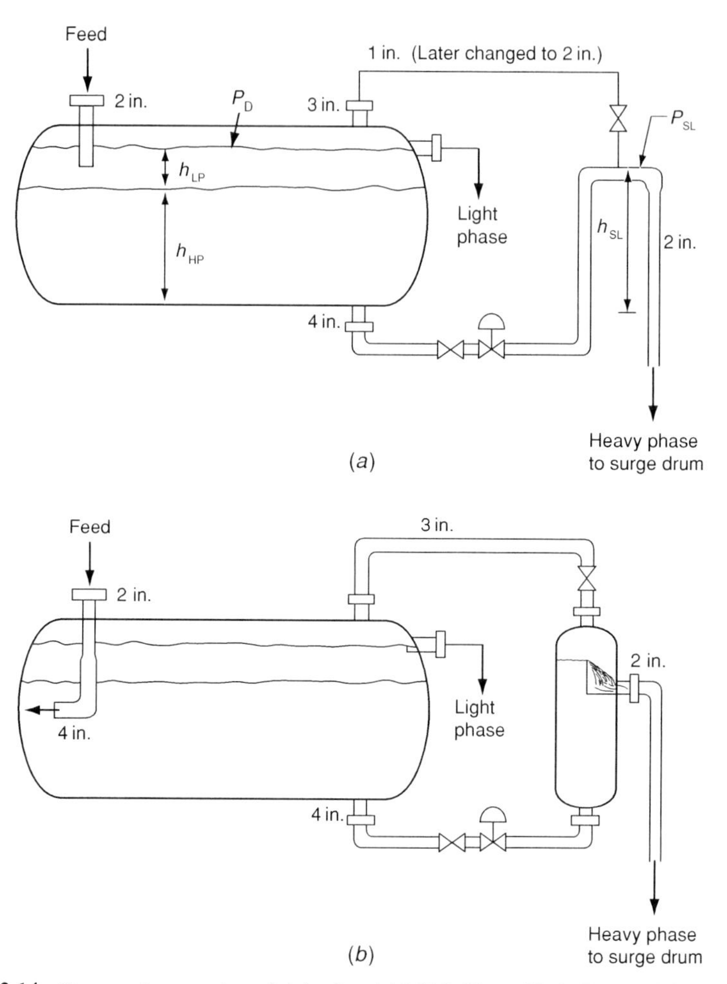

Figure 2.14 Decanter that experienced siphoning: (*a*) initial; (*b*) modified. (From Ref. 260. Reprinted courtesy of *Chemical Engineering*.)

Hydraulic Balance With the pressure balance line doing its job, the static pressure at the top of the seal loop equals the static pressure in the decanter vapor space. The small (1–2 in.) elevation difference between the liquid level in the decanter and the top of the seal loop gave enough driving force to overcome the friction head losses in the seal loop at normal flows. Calculations showed that the flow resistance through the seal loop (including the open valves) was extremely small.

The 50 ft elevation drop from the top of the seal loop to grade exerted strong suction at the top of the seal loop. With good pressure balancing, enough vapor from the top of the decanter would have entrained in the rundown line liquid, raising friction in the rundown line and making the pressure at the top of the seal loop the same as that in the vapor space of the decanter. Conversely, if there were no pressure balancing, the suction at the top of the seal loop would have caused the liquid flow to rapidly increase and siphon out the decanter.

The observation that siphoning was taking place means that the pressure balance line was not fully achieving its intended function. Increasing the line size from 1 to 2 in. helped but did not go the full length. Throttling of the liquid valve between the decanter and seal loop increased the pressure difference between the two, which dampened surging by further increasing vapor flow to the seal loop. However, it became apparent that better vapor balancing and siphon breaking were required.

Cure The seal loop was replaced by a 1 × 4-ft vertical drum (Fig. 2.14*b*) that gave good siphon breaking and pressure equalization with the decanter. The siphon breaking was achieved by drawing the heavy phase from a side sump into which liquid could only enter by overflowing a chordal weir. A 3-in. line was used to balance the pressures between the top of the drum and the decanter vapor space.

Two other modifications were implemented to improve the decanter operation. To mitigate turbulence and short circuiting in the decanter, the internal feed line in the decanter was increased from 2 to 4 in. and the feed was discharged against the head (Fig. 2.14*b*). In addition, for better control of the light-phase thickness, a weir was installed in the light phase just upstream of the outlet nozzle.

Result Following these modifications, the erratic flow problem was mitigated. Most important, stabilization of the flow enabled the decanter to operate at its design temperature, which was 10–20°F lower than the premodification temperature, leading to major improvement in phase separation.

CASE STUDY 2.22 HICCUPS IN AZEOTROPIC DISTILLATION TOWER

Installation Feed to an azeotropic distillation tower (Fig. 2.15) was a homogeneous organics–water azeotrope. The tower used benzene entrainer to enhance the volatility of water. Tower bottoms was the dehydrated organics. Overhead from the tower contained the water, benzene, and a small concentration of organics. The overhead was condensed into two liquid phases which were then decanted. The water phase, which contained a small concentration of organics and benzene, was sent to the stripper. The HC phase was the benzene entrainer, which was recycled to the tower.

Problem Hiccups were experienced in the tower. After 2 days of smooth running, the tower emptied itself out from the bottom, giving an off-specification bottom product for a while. Once it emptied itself out, the column returned to normal operation and was good for another couple of days before emptying itself out again.

Analysis Hiccups are normally a symptom of the accumulation of an intermediate component in the tower. The component cannot find a way out of the tower at a rate fast enough to match the rate at which it enters the tower. It builds up in the tower

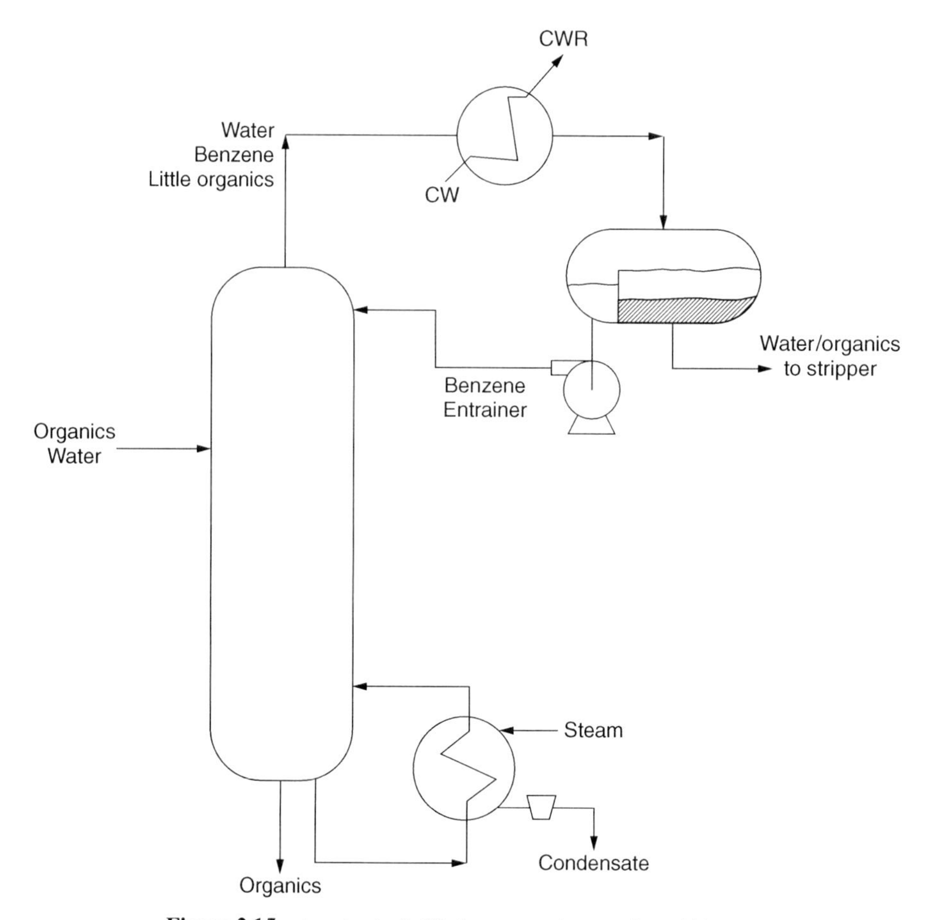

Figure 2.15 Azeotropic distillation system that experienced hiccups.

until the tower floods. Emptying the column out, either by carryover from the top or by dumping out of the bottom, clears the flood and the offending component.

Testing Extensive testing identified the offending component to be an aromatic alcohol which was present in small quantities in the feed. The component was repelled by the organics in the tower, and this enhanced its volatility, sending it up. The alcohol had affinity both to water and to benzene, probably more to the benzene than to the water. In the decanter, most of it was extracted by the benzene and returned to the tower.

Solution Water was added to the decanter. This extracted the component out of the benzene and into the water phase. This fully eliminated the hiccups.

Another Plant In a related experience, a tower dehydrated ethanol (bottom product) using cyclohexane entrainer. Tower overhead containing the cyclohexane entrainer, water, and a small amount of ethanol, was condensed, then decanted, using a process scheme similar to that in Figure 2.15. Periodically, the tower experienced poor decanting, which was overcome by adding water to the decanter. It is believed that some compound that was soluble in both the water and the cyclohexane was accumulating near the top of the tower. Adding water extracted it out of the system and restored good decanter action.

Related Experience (Contributed by E. J. (Jim) Morris, Consultant, Houston, Texas) The final tower in a formaldehyde purification train separated anhydrous formaldehyde gas overhead from aqueous formaldehyde bottoms. Acetone entrainer was used to enhance relative volatility between formaldehyde and water. Overhead from the tower, containing anhydrous formaldehyde and acetone entrainer, was partially condensed, then separated into the anhydrous formaldehyde gas product and an acetone reflux stream that was returned to the tower. An intermediate-boiling, bright yellow diacetyl (2,3 diketobutane) impurity built up between the aqueous formaldehyde bottoms and the acetone-formaldehyde overhead, causing the column to periodically "puke." The diacetyl was trapped across the upper section of the column, and colored the acetone reflux bright yellow. The problem was solved by adding a liquid side-draw from the tray where the impurity was at its highest concentration, determined by visual colorimetric analysis of liquid samples from the trays.

CASE STUDY 2.23 HICCUPS IN AN EXTRACTIVE DISTILLATION TOWER

Contributed by Rian Reyneke, Sasol, Secunda, Gauteng, South Africa

Installation As part of a piloting program, a 900-mm-ID (Internal Diameter) extractive distillation column received a wide range of C_2–C_6 aldehydes and ketones and C_1–C_6 alcohols in the feed. It used water solvent to separate the aldehydes and ketones as the overhead product from alcohols which formed the bottom product.

Problem Under some conditions, hiccups were experienced in the tower. The cause appears to be a heavy ketone. The problem tended to occur when this heavy ketone was present in the feed at a concentration of 1%, and at the same time the water-to-feed ratio was low. A typical occurrence was as follows: For the first 12 hours after this ketone showed up in the feed at 1% concentration, none of it was detected either in the top or in the tower bottom. At the end of 12 hours, the column appeared to empty itself out from either the top or the bottom. The temperatures in the tower shot up just before the hiccup. From then on, the cycle repeated every 2 hours. Analysis of the reflux drum liquid immediately after the hiccup found a relatively high concentration of heavy ketones.

Bench-Scale Tests At an earlier step in the piloting program, the separation was tested in a 50-mm-ID Oldershaw column and a similar behavior was observed.

After some initial period of stable operation the column would start foaming severely, hiccup, return to semistability for a while, and then repeat the same pattern. A second liquid phase would be observed on a few trays close to the top under certain operating conditions.

Modeling Physical properties were checked and the separation was simulated using a VLLE model. The simulation showed that two liquid phases would exist on most of the rectification trays. According to the model, the second liquid phase was rich in heavy ketones and aldehydes. The model showed that the water-to-feed ratio needed to be about double the original, and the reflux ratio needed to be significantly reduced, to avoid the formation of the second liquid phase.

Solution The changed operating conditions (double the previous water-to-feed ratio and the lower reflux ratio) were tested in another, 100-mm-ID pilot column. This time the column operated stably for an extended period of time, with all heavy ketones coming out in the distillate product.

Postmortem One theory is that of heavy ketone accumulation without foaming. The ketones had high boiling points and tended to come out in the bottom. Water addition made them volatile, but if the water-to-feed ratio was low, not volatile enough to leave in the overheads. They therefore built throughout the tower, eventually hiccuping. When the water rate was increased, these ketones became more volatile. Similarly, increasing the reflux and subcooling it tended to push heavy ketones down the tower. So the heavy ketones got stuck somewhere in the middle, where they built up until they led to a hiccup. Higher water, lower reflux rate, and less subcooling increased water concentration, which increased volatility of the heavy ketones, which induced enough of the heavy ketones to exit in the overheads without excessive buildup in the tower.

A second theory postulates a similar buildup mechanism to that described above but adds foaming to the explanation. Foaming is often experienced in extractive distillation (250, 510), and the Oldershaw tests provide evidence supporting foaming in the present system. Work by Ross and Nishioka (414) shows that foam stability is maximum at the plait point, that is, just before a solution breaks into two liquid phases. According to this theory, the buildup of the heavy ketone initiated foaming just below the point at which the solution broke into two liquid phases [once two liquid phases are present, the foaming tendency drops, as one phase serves as antifoam to the others (414)]. The foaming then caused the hiccups.

Related Experience On an entirely different continent, DMF was the solvent in a tall extractive distillation tower in which butadiene (bottom product) was separated from a C_4 stream. While the saturated C_4's exited in the tower overhead, some of the 2-butenes accumulated, causing instability. The solution was to raise boilup, allowing all the 2-butenes and some butadiene escape in the overhead. The escaping butadiene was recovered downstream.

Chapter 3

Energy Savings and Thermal Effects

Fractionation issues featured very low on the distillation malfunctions list for the last half century (255). Only two issues rated a mention, subcooling issues and heat integration issues. Neither of these made it to the top 20 distillation malfunctions. This matches the author's experience. Neither issue is generally too troublesome.

One major issue with subcooling has been causing enhanced internal condensation and reflux, which has hydraulically overloaded trays, packing, or liquid distributors. In some cases, this has caused insufficient reflux or loading in the section above. Another major issue with subcooling has been excessive quenching at the inlet zone, diverting light components into the section below with consequent product losses, excessive reboil requirement, or component accumulation.

Heat integration generates complexity and operability issues, which generated imbalances and "spins." Control problems, especially with preheaters, were also reported, but these are grouped under a different heading. Most of the cases came from refineries and olefin/gas towers, where a high degree of heat integration is practiced. Many cases involved the simpler forms of heat integration such as preheaters, interreboilers, and recycle loops. In some of these cases, the fix was as simple as bypassing a stream around the preheater or bypassing a smaller feed stream around the tower.

CASE STUDY 3.1 EXCESS PREHEAT BOTTLENECK CAPACITY

Henry Z. Kister and Tom C. Hower, reference 263. Reproduced with permission.

Installation An olefins plant C_2 splitter (Fig. 3.1). Feed to the column was a vapor ethylene–ethane mixture with minor quantities of other components. Top product was polymer-grade ethylene, while bottom product was ethane, which was recycled to the plant's cracking furnaces as a cracking feedstock. The main requirement of the

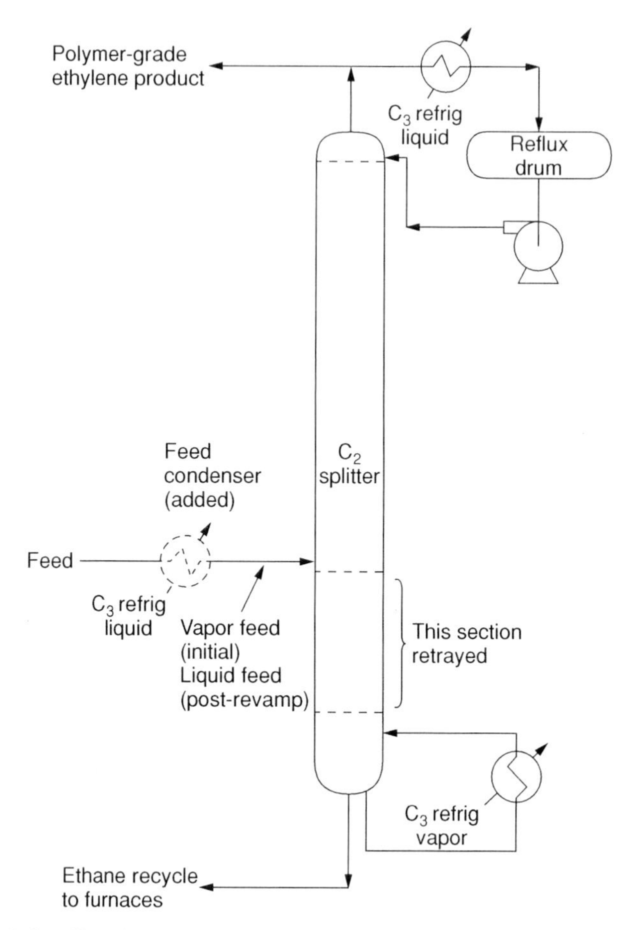

Figure 3.1 A C_2 splitter that was bottlenecked by excess preheat, showing debottlenecking modifications. (From Ref. 263. Reproduced with permission.)

column was to produce on-specification ethylene. There was an economic incentive to minimize the amount of ethylene in the column bottom stream, but in this case the bottom flow rate was small and minimizing the loss of ethylene to that stream was not critical. The rectifying section of the column contained about three to four times as many trays as the stripping section.

Problem Following previous revamps, the plant was operated at 135% of its initial design capacity. Field experience indicated that at this rate the C_2 splitter operated right at its hydraulic capacity limit, both in the top and bottom sections. This was confirmed by calculation.

The plant capacity was to be further raised to 150% of its original capacity, the C_2 splitter being one of the major bottlenecks. Since downtime and lost capacity were extremely costly, the proposed solution had to positively assure that the desired capacity increase would be achieved and ethylene purity would be maintained. A preliminary revamp study concluded that replacing the column internals alone could not positively assure that both these objectives would be simultaneously met. The only solution that appeared capable of positively achieving both objectives was to add a 40-tray section in series with the existing column, which would enable reflux and reboil to be reduced and allow for greater throughput. This solution required large capital expenditure and had a negative impact on the payout of the planned revamp.

Solution The idea that solved the problem with relatively little expense is shown in Figure 3.1. A feed condenser was added which lowered the vapor and liquid loads in the rectifying section sufficiently to ensure this section was capable of processing the increased throughput. This, however, considerably loaded up the bottom section of the column. To accommodate the greater loads, the sieve trays in the bottom section were replaced by dual-flow trays (i.e., sieve trays without downcomers). This type of tray is capable of achieving significantly greater capacity than a normal sieve tray, often at the penalty of a slightly lower efficiency and a somewhat lower turndown. The loss in efficiency, however, only occurred in the small stripping section and could be tolerated since it did not affect the purity of the ethylene product.

Postmortem The revamped column (Fig. 3.1) achieved 150% of its initial design capacity while producing on-specification ethylene product. Ethylene losses in the bottom stream increased from about 1 to 1.8%, which represented a minor economic loss, especially when one considers the small bottoms flow rate.

CASE STUDY 3.2 A COLUMN REVAMP THAT TAUGHT SEVERAL LESSONS

Contributed by Ron F. Olsson and Michelle Roberson, Celanese Corp.

Figure 3.2*a* shows a two-column system separating a light organic A from water. The first column is a stripper that removes some of the water as the bottom stream. The second column is a still producing a 90% pure A in the overhead and a bottom water stream.

The still operates at a lower pressure. Overhead vapor from the stripper reboils the still. A steam reboiler supplements the boiling. Both the condensed stripper overhead (liquid) and the uncondensed stripper overhead (vapor) enter the still on tray 30. The revamp objective was to maximize throughput and minimize energy usage.

Revamp One energy inefficiency can be detected immediately. The still reboiler makes use of only a small fraction of the overhead vapor. There is plenty more available, and there is little need for the auxiliary steam. Further, since component A constitutes only a small fraction of the still feed, less vaporization should help separation. This was confirmed by calculation and is further discussed below. Indeed, it was decided to replace the reboiler by a larger one.

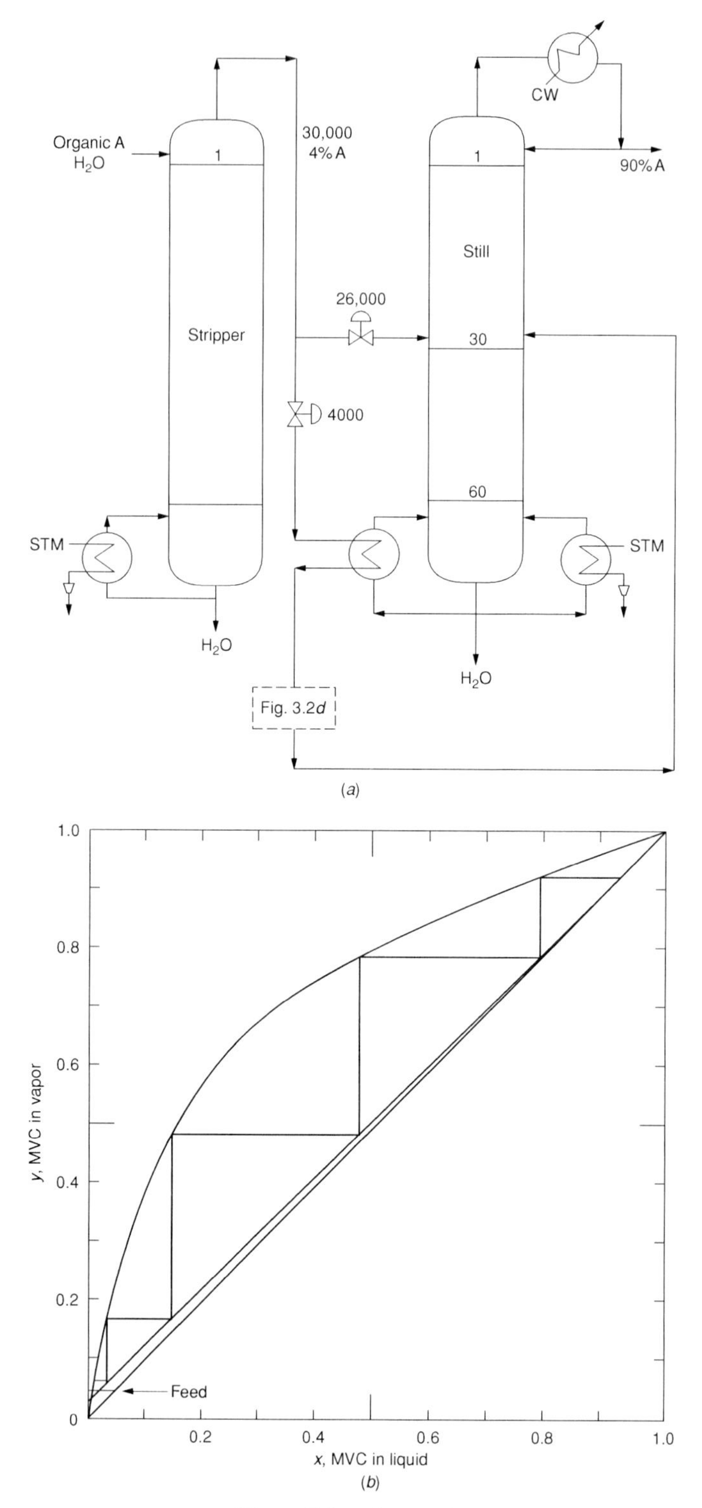

Figure 3.2 *(Continued)*

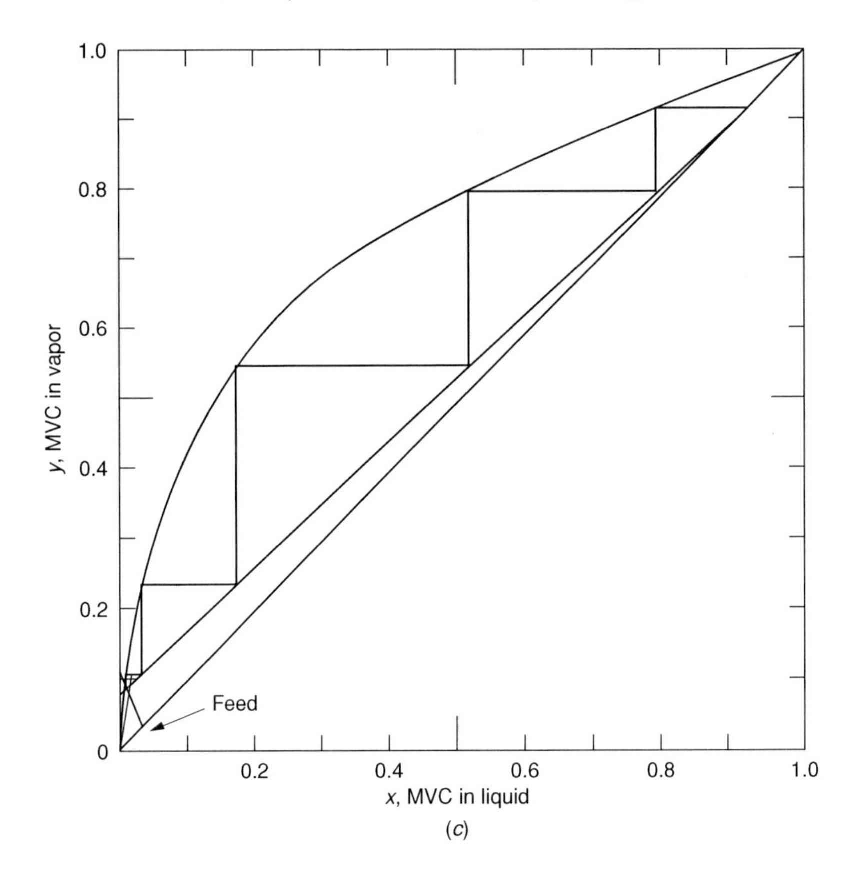

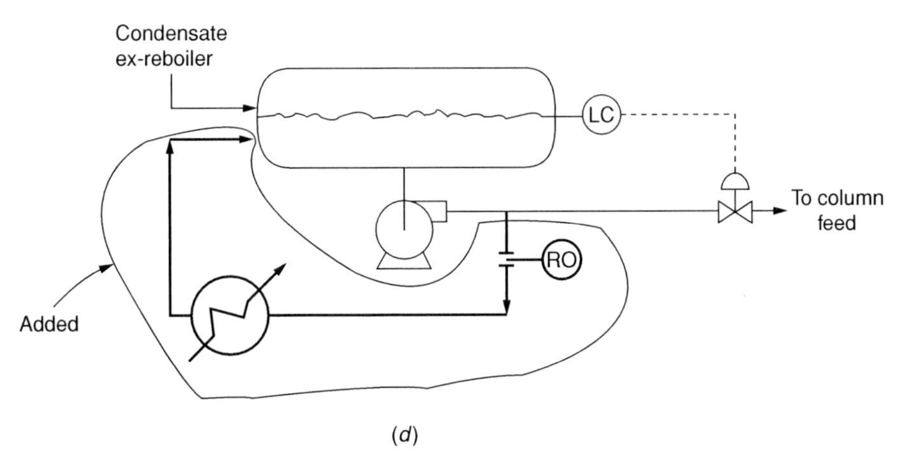

Figure 3.2 Column revamp that taught several lessons: (*a*) two-tower separation scheme; (*b*) McCabe–Thiele diagram, showing excess reflux requirement with dew point feed; (*c*) McCabe–Thiele diagram for a partially vaporized feed, giving much lower reflux requirement; (*d*) reboiler condensate cooling that overcame pump problem.

A McCabe–Thiele diagram (341, described in detail in Ref. 251) (Fig.3.2*b*) readily shows a second inefficiency. The high feed vaporization led to the component balance (operating) line lying very close to the 45° diagonal. The liquid-to-vapor molar ratio (L/V) in the rectifying section of this diagram was 0.96, giving a reflux-to-distillate ratio (R/D) of 25. This in turn led to excess reflux requirement, which led to a capacity bottleneck in the rectifying section of the still. The total energy usage due to the pinch was also higher than it needed to be, but here it was available.

Figure 3.2*c* shows how with a smaller degree of feed vaporization the component balance line steepens. The L/V in this diagram declined to 0.92, giving a much lower R/D of 12.5. Reflux rates could thus be largely reduced and the rectifying section debottlenecked. Here is an additional argument for using a larger still reboiler/feed condenser and eliminating the steam reboiler.

The initial attempt at revamp fell short of meeting design expectations. Even though a larger reboiler was installed, the 3-in. control valve in the vapor line to the reboiler was not upgraded. It reached critical flow and became the bottleneck to increasing boil-up. This valve was later replaced by a 6-in. valve to overcome the bottleneck. The lesson learned here is that it is not sufficient to devise an effective revamp/energy-saving scheme. Such a scheme needs to be fully engineered.

When the larger reboiler was installed, it was appreciated that liquid flow through the condensate pump will increase severalfold and net positive suction head (NPSH) problems are likely. Reboiler condensate gravity-flowed into a tank, from where it was pumped to the still on level control (Fig. 3.2*d*). Elevating the drum would have helped somewhat but was limited by the possibility of flooding some of the reboiler tubes. A very clever idea solved this problem. A slip stream was taken from the pump discharge via a restriction orifice (Fig. 3.2*d*), cooled, and returned to the drum. The extent of cooling required was quite small—about 10°F. The cooler was an unused heat exchanger that was available. The lower condensate temperature was sufficient to overcome the NPSH problem. The lower temperature was also beneficial to reducing reflux in the still!

CASE STUDY 3.3 BYPASSING A FEED AROUND THE TOWER

Henry Z. Kister and Tom C. Hower, reference 263. Reproduced with permission.

Installation An olefins plant debutanizer which separated butadiene and butenes as the top product from pyrolysis gasoline that left in the column bottoms. The bottom product flowed to a hydrogenation reactor after being preheated by reactor effluent in the reactor feed–effluent exchanger (Fig. 3.3). The column received two feeds. The smaller feed stream, which entered at a higher point up the column, contained most of the C_4. The lower feed contained less C_4 but was larger in quantity than the top feed.

Problem Toward the end of a run and about 6 months prior to a scheduled shutdown, fouling at the bottom of the column caused it to reach a capacity limitation.

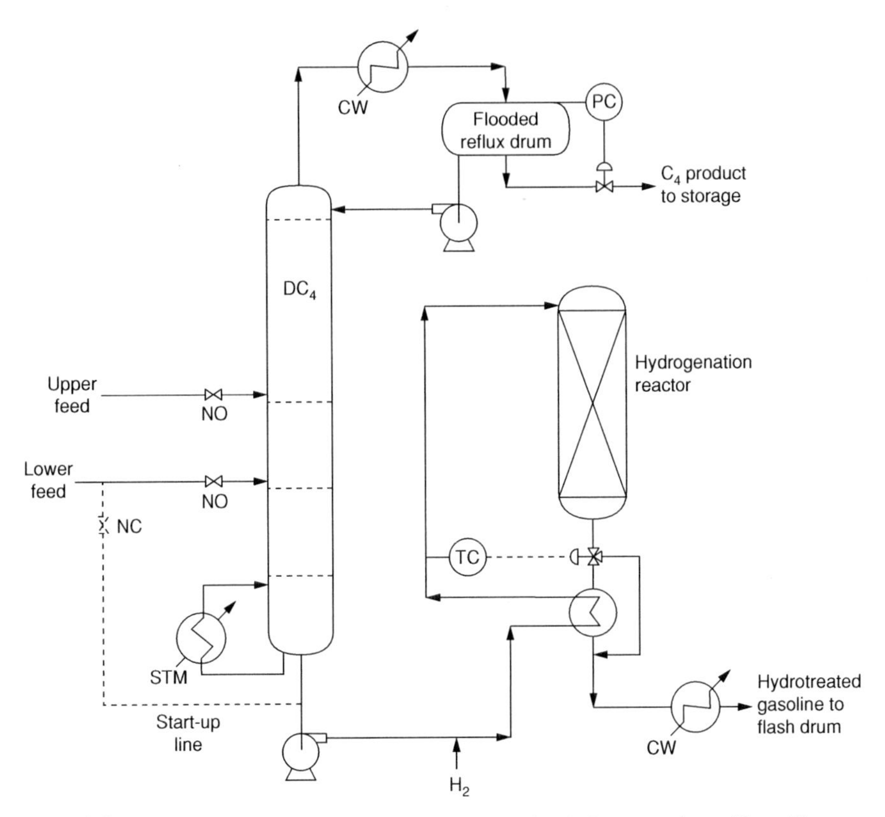

Figure 3.3 Bypassing lower feed around debutanizer solved end-of-run capacity problem. (From Ref. 263. Reproduced with permission. Copyright © (1987) AIChE. All rights reserved.)

This resulted in excessive heavies in the C_4 product, although the column bottom stream remained on specification for C_4. The off-specification C_4 make could not be sold and had to be flared. The loss of C_4 make was costly but was a preferred short-term solution to shutting the plant down or cutting plant rates, which would have been even more costly because of lost production. Flaring the C_4 make, however, was only acceptable as a short-term solution because of environmental considerations. The plant was facing a shutdown unless the problem could be quickly resolved.

Options Analysis of the column feed streams revealed that the lower feed contained only about 8% C_4, half of which was butadiene. A physical inspection of the piping revealed that some of the start-up lines could be utilized to provide a route through which the lower feed could be diverted to the bottom of the column. The possibility of bypassing this lower feed around the column was then considered. The main fear was that the butadiene contained in that stream would disable the hydrogenation reactor to an extent that it would become inoperable.

Nevertheless, it was decided to go ahead with the operation, realizing there was little to lose. The cost of a new charge of reactor catalyst was negligible compared to the cost of either the flared C_4 make or a plant shutdown.

A shutdown would be required whether the operation failed or was not carried out at all, while success would have avoided the need to shut down or flare.

Cure Prior to bypassing the lower feed, the upstream plant was trimmed to minimize the butadiene in that stream. This reduced the concentration of butadiene to about 2–3%, giving about 1–2% butadiene in the hydrogenation reactor feed.

Bypassing the lower feed unloaded the column, and the C_4 make achieved its purity specifications. The greater quantity of butadiene in the feed did not disable the catalyst, although it shortened its life to an extent that one additional catalyst charge was required. This was considered a relatively minor expense.

One surprising side effect of this operation was a great reduction in overall steam consumption, which resulted from unloading the debutanizer. The steam savings achieved in the following 6-month period (to the next scheduled shutdown) was alone more than sufficient to pay for the new catalyst charge.

CASE STUDY 3.4 HEAT INTEGRATION SPIN

Installation A multifeed olefins plant demethanizer equipped with an internal condenser (Fig. 3.4). The tower capacity was bottlenecked by downcomer limitation between feeds 2 and 3. Overhead vapor from the internal condenser was superheated by cooling feed 1, then further heated by cooling feeds to drums 1 and 2 in the demethanizer cold box exchangers.

Problem Flooding occurred at higher plant rates, recognized by high *dP* above feed 2. Once flooding initiated, the *dP* kept getting worse, the tower became less stable, but the loss of ethylene in the overhead remained low. To get out of flood, the operators reduced the top reflux by cutting back on condenser refrigerant. Reducing the reflux, however, doubled the ethylene losses in the overhead stream.

Analysis At flooding, temperature T_1 dropped sharply, probably due to massive carryover. This carryover cooled feed 1 further and enhanced liquid condensation into drums 1 and 2. The higher, colder feeds raised the liquid loads entering the bottleneck region, which aggravated the flooding, which in turn generated more carryover from the overhead, and so on. This is termed a *heat integration spin*. To get out of the spin, refrigeration to the condenser was cut, which reduced the cooling in the overhead exchangers and thus the condensation in drums 1 and 2 and the cooling of feed 1. Final cure was by debottlenecking the tower.

CASE STUDY 3.5 CHANGE IN CUT POINT FLOODS TOWER

Installation A refinery crude tower (Fig. 3.5) producing an overhead naphtha product, kerosene and diesel side products, and a straight-run resid bottom product. The tower contained two cooling pumparounds (PAs). In each PA, a liquid stream was drawn from the tower, cooled, then returned to the tower. Most of the cooling

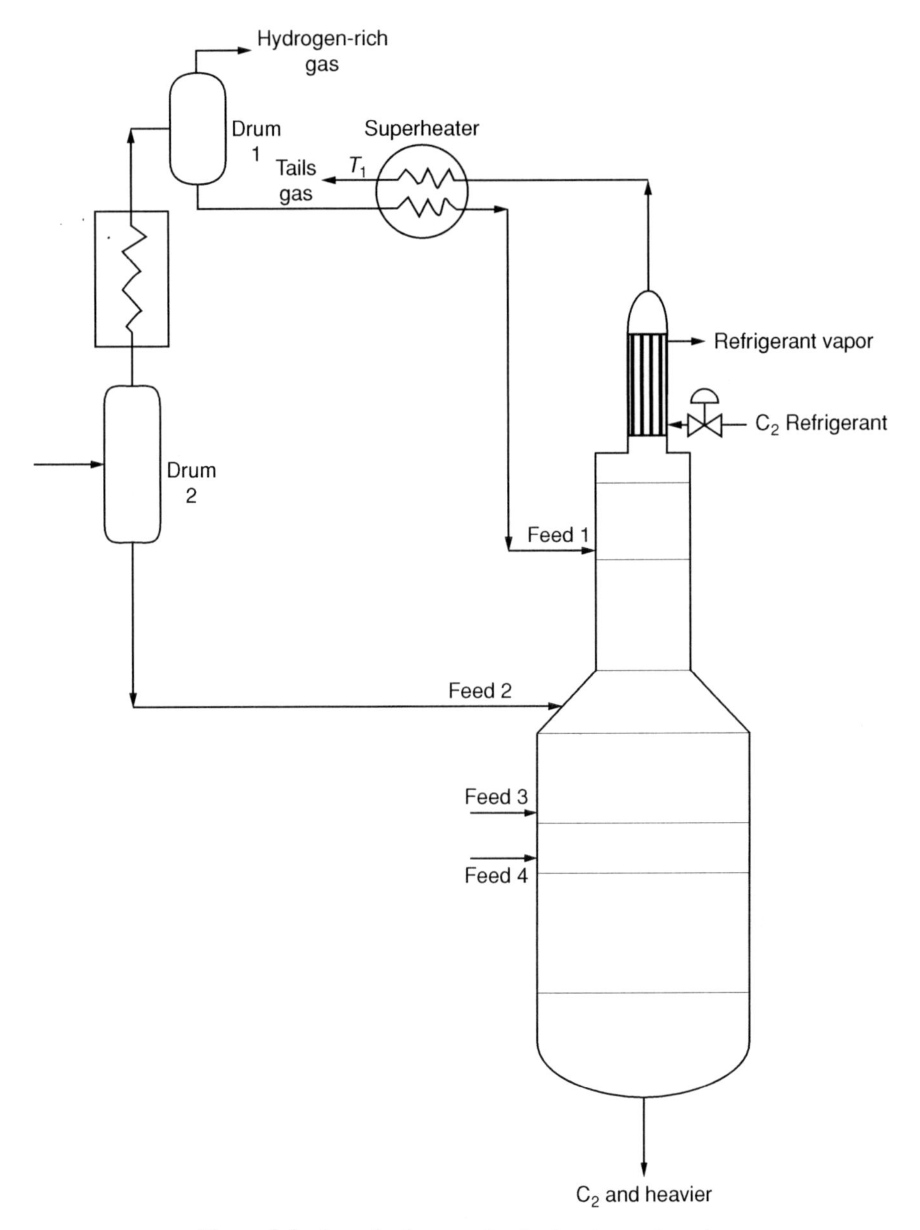

Figure 3.4 Demethanizer experiencing heat integration spin.

was done by preheating tower feed in the crude preheat train (Fig. 3.5). Reference 236 has detailed description of crude tower process schemes.

Problem At times it was economical to lower the kerosene–naphtha cut point (i.e., to send the heavier components in the naphtha into the kerosene product). However, lowering the cut point initiated flooding in the top section, limiting tower feed rates.

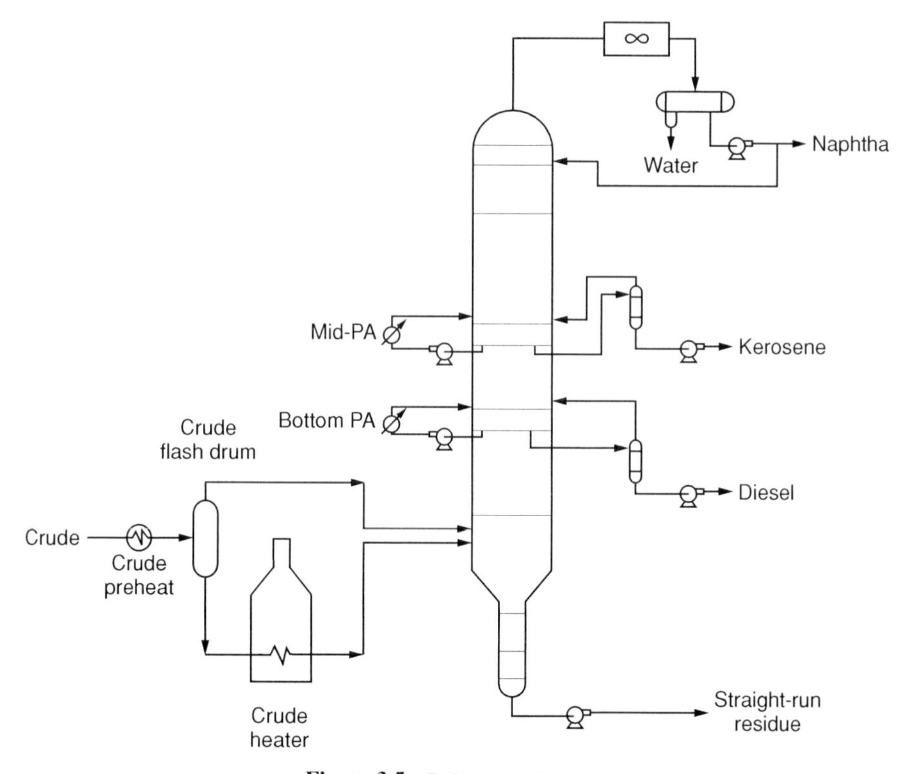

Figure 3.5 Refinery crude tower.

Cause Lowering the cut point reduced the naphtha make and increased the kerosene make. The inclusion of the heavier components of the naphtha in the kerosene lowered the boiling points of both and therefore also the mid-PA draw temperature. At the lower mid-PA temperature, the PA coolers could not preheat the crude as much. Both PA cooler duty and crude preheat were reduced.

Since tower feed was on temperature control, the reduced preheat was compensated by harder firing of the heater. Overall, the tower heat input was not largely changed. With heat removal in the mid-PA curtailed, the only place where the heat duty could be removed was the overhead condenser. To get there, the vapor traffic above the mid-PA had to increase. The trays above the mid-PA were heavily loaded even in normal operation, and the additional vapor traffic sent them into jet flood.

CASE STUDY 3.6 SIMULATION DIAGNOSES HEAT REMOVAL BOTTLENECK

Contributed by Gerald L. Kaes, Kaes Enterprises, Inc., Colbert, Georgia

Installation An FCC main fractionator, taking hot, superheated vapor feed from the reactor. The tower fractionated the feed into a top naphtha product, a number of side cuts, and a bottom decant oil (DO) product. Heat was removed from the tower in the overhead condenser as well as in a number of PAs. In each PA an externally cooled liquid stream was circulated through a few trays, condensing rising vapor by direct contact.

History The FCC unit was being expanded for higher throughput. Initially, it was proposed to add a new C_3/C_4 splitter to the unit, which would be reboiled using the heavy cycle oil (HCO) PA. With this in mind, the FCC main fractionator was evaluated and found to need no modifications to tower internals. This plan to add a C_3/C_4 splitter was later scrapped to cut costs when the expansion was reduced in scope.

Problem Upon increasing feed to the unit beyond the pre-revamp rates, flooding was experienced in the sections above the HCO PA. The column overhead temperature increased, raising reflux, which aggravated (or possibly induced) the flooding. Tower operation became erratic.

Analysis The pre-revamp tower simulation was adapted to the post-revamp conditions. This led to the discovery that the tower had a heat balance problem. The elimination of the C_3/C_4 splitter from the expansion took away a major heat sink from the HCO PA. The additional FCC charge raised the heat removal requirement of the HCO PA, but it was not possible to remove sufficient heat in the quench and HCO PAs. The only place where the additional heat could be removed was in the column overhead condenser. To get there, more uncondensed vapor ascended past the HCO PA, raising vapor traffic and tower overhead temperature. To compensate, the tower top-temperature controller raised the reflux. The higher vapor and liquid traffic initiated flooding near the top of the fractionator.

Cure Additional heat exchanger duties were added to the HCO PA, and flooding above was eliminated.

CASE STUDY 3.7 REMEMBER THE HEAT BALANCE

Contributed by Mark E. Harrison, Eastman Chemical, Kingsport, Tennessee

Installation A packed rectifier was designed to remove low-boiling components from much higher boiling products for recycling to an upstream operation. Feed entered below the packed section. Since the feed contained a varying mixture of high-boiling components (which caused the base temperature to fluctuate), the reboiler steam was flow controlled instead of base-temperature controlled. Some high or intermediate-boiling components were allowed to pass overhead. Occasionally, the operator trimmed the steam flow to control the amount of high-boiling components taken overhead.

Problem At design feed rates, the column operated as expected, but it flooded when the feed was shut off.

Troubleshooting A review of the design heat and material balances indicated that the feed was cold, requiring about 40% of the reboiler heat duty just to heat the high-boiling components in the feed to the bottom temperature. The packed column was sized for a vapor boil-up corresponding to the remaining 60% of the reboiler duty. It was then recognized that flooding occurred only when the feed was reduced or shut off. Because the reboiler steam was flow controlled, shutting off the feed removed the heat sink for 40% of the reboiler duty. The entire reboiler duty would then translate into boil-up equivalent to 166% of the column design capacity.

Cure An upper limit to the column pressure drop was established based on operation short of flooding. The control system was modified to manipulate reboiler steam flow to maintain a column pressure drop below the limit. This provided steam flow control during normal operation (i.e., with feed) but reduced the reboiler steam flow to curtail the higher boil-up when the feed was reduced or shut off.

Outcome Reducing the feed no longer started column flooding.

Chapter 4

Tower Sizing and Material Selection Affect Performance

No single tower sizing issue featured high on the distillation malfunctions list for the last half century (255), but five issues found spots between the 20th and 31st places on the list, emphasizing the significance of a variety of issues related to sizing. This matches the author's experience.

Low liquid loads handling difficulties in tray towers were the top issue. Practically all of these described one out of two problems: either leakage of liquid from the tray deck, causing the trays to dry out, or vapor breaking into downcomers, causing difficulties (even making it impossible) to establish a downcomer seal. In many cases, inability to seal the downcomer made it impossible for liquid to descend and led to flooded trays above the unsealed downcomer.

Low vapor loads have been just as troublesome, leading to excessive tray weep. Surprisingly, the majority of cases took place in valve trays, which are inherently more weep resistant than sieve trays. Cures included blanking or replacing with leak-resistant valve units and increasing the vapor loads. The number of weeping case studies in the last decade is well below the number in the previous four decades.

Undersizing trays and troublesome tray layouts have also been troublesome. Undersizing downcomer inlet areas has been a major issue. A restriction at an inlet weir, insufficient hole area, incorrect number of passes, and nonstandard design features also contributed troublesome case histories.

Turning to packing, poor efficiency for reasons other than liquid or vapor maldistribution was reported in a variety of cases, mostly recent. One issue has been packed beds that are too long, accentuating maldistribution and lowering efficiency. In wash sections of refinery vacuum towers, excessive bed length has led to drying up and coking. In other cases, a unique system characteristic, such as high pressure in structured packings, high hydrogen concentration, high viscosity or surface tension, or oil layers on packing in aqueous service, caused loss of efficiency.

Supports, holddowns, or tower manholes were troublesome in several packed towers. The major issue has been insufficient open area on the support or holddown causing a capacity restriction. Packing migration through the supports,

---

interference of I-beams, and maldistribution due to a manhole in the bed also caused problems.

CASE STUDY 4.1 EXTREMELY SMALL DOWNCOMERS INDUCE PREMATURE FLOOD

Contributed by Chris Wallsgrove

Installation A 10-ft-ID caustic wash tower (Fig. 4.1*a*) removing a small quantity of CO_2 and traces of H_2S from light HC gases at about 200–300 psig. The tower contained 40 single-pass valve trays at 24 in. tray spacing. Each tray was equipped with a 6-in.-wide downcomer. Overheads from the column flowed via a knockout drum to a compressor. The plant was at its initial start-up.

Problem At low plant rates the column operated well. As rates were increased to 60–70% of design, massive entrainment into the knockout drum was observed. This could not be tolerated because of the risk of damaging compressor blades.

Investigation The symptoms suggested premature flooding. Column differential pressure was not being measured. Calculations indicated the column was adequately sized. However, it was suspected that some degradation of the caustic solution could have occurred near the bottom of the column, giving rise to the formation of a viscous emulsion. This, together with the extremely small downcomers (2% of tower area), was believed to have caused premature flooding.

Solution A larger downcomer area was sought to overcome the premature flooding problem. Retraying with larger downcomer trays appeared the logical solution but would have caused a 6-week delay and a week-long shutdown. Downtime was extremely expensive and undesirable. As an alternative solution, the plant and start-up team decided to construct home-made downpipes (Fig. 4.1*b*) in the bottom few trays, where liquid load was greatest and potential for emulsion formation was highest. The pipes were constructed by "burning" a few holes in the tray near the outlet weir and fitting 6-in. vertical pipes in them. The pipes protruded a small distance above the tray, simulating a weir, and extended to a small distance above the tray below, simulating a downcomer apron area. The decision was made to go ahead because there was little to lose; the worst that could happen was mechanical damage to a few trays, but new trays would have been needed anyway to enlarge the downcomers.

Implementation The bottom few trays were modified (Fig. 4.1*b*). This was done in a 3-day shutdown. When the column came back on-line, the problem completely disappeared, and the column attained full rates without any problem.

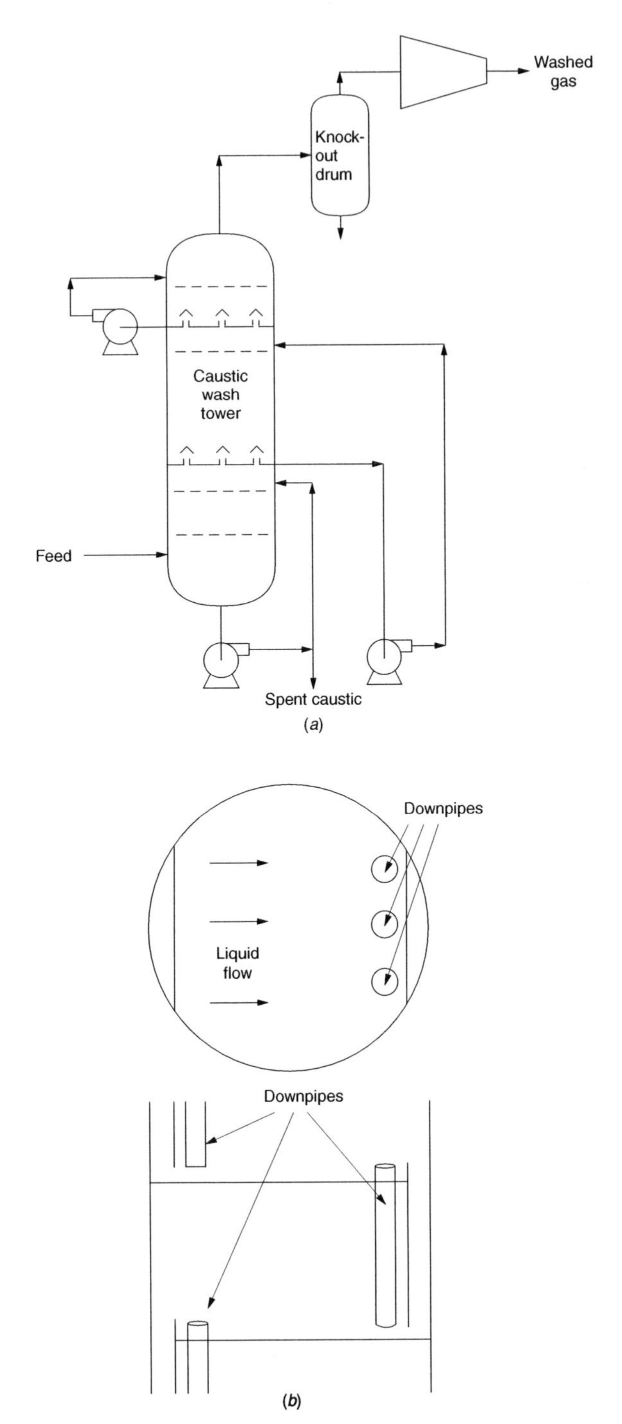

Figure 4.1 Caustic wash tower that experienced premature flood: (*a*) tower schematic; (*b*) addition of downpipes increased downflow area and mitigated flooding.

CASE STUDY 4.2 EXTREMELY SMALL DOWNCOMERS FLOOD PREMATURELY

Installation A diethanol amine (DEA) absorber, removing small quantities of acid gases from a HC gas at 500 psig. The tower was 8 ft ID and contained 30 single-pass valve trays at 24 in. tray spacing. Weir heights were 4 in., downcomer clearances 1.5 in. Downcomer width was 6 in.

Problem Liquid carryover in the tower overheads was experienced with vapor loads above 60% of the design and with liquid loads ranging from 30 to 100% of design. Stable operation of the column without liquid carryover could only be achieved at a liquid circulation rate of 25% of the design.

Troubleshooting Initially, the problem was diagnosed as foaming. The foaming was confirmed by laboratory tests. Antifoam was added. Laboratory tests verified that the antifoam effectively suppressed the foam. However, even though the foaming appeared to be mitigated, the carryover continued without improvement. The absorber was shut down and inspected. The trays, downcomers, and tower were clean and free of polymer or blockages. A few valves were missing, but nothing that could explain the carryover. The tray and downcomer dimensions were all per design, and so was the fabrication and assembly.

The tray design was checked both by the engineering contractor and by the tray vendor. At full rates, the trays were designed to operate at 80% of jet flood and 68% of downcomer choke flood, both after allowing for a system factor of 0.7. Calculated downcomer backup was 9.5 in. of clear liquid at full loads. So the trays were not close to any calculated limit even at maximum loads, while the carryover was observed at much lower loads. Downcomer apparent residence time at design rates was 9 seconds, which is well within good design criteria (250).

Cause A close review of tray dimensions revealed that the downcomer widths were minimal. A width of 6 in. in an 8-ft-ID tower produces a downcomer of the shape of a long and narrow slot. This geometry increases the friction resistance to liquid downflow and to upflow of disengaging vapor, an effect seldom accounted for in normal sizing procedures. Normal sizing procedures therefore give optimistic area predictions for "narrow-slot" downcomers. Further, such long and narrow slots become extremely sensitive to foaming and to conditions where vapor–liquid disengagement is difficult, such as high-pressure towers.

In this tower, the design downcomer velocity was 0.27 ft/s. Maximum downcomer velocity sizing criteria listed in Ref. 250 suggest that this velocity would have been somewhat aggressive but not far outside the recommended range of 0.2–0.25 ft/s for a high-foaming application such as an amine absorber, assuming a normal downcomer. The field tests showed that the maximum velocity at which satisfactory operation was experienced was 0.07 ft/s. The difference was due to the narrow-slot geometry.

To prevent the narrow-slot geometry, it was recommended (250) to avoid downcomers smaller than 5–8% of the column cross-sectional area. The reference emphasizes that adhering to this rule is most important in superatmospheric services and where there is a tendency to foam. With a downcomer area of 2.6% of the tower cross section area, the DEA absorber violated this rule by a large margin, which caused the premature flood.

Solution The solution would have been to expand the downcomers. This was proposed but not implemented. The reason was that the gas entering the absorber contained much less acid gas than originally expected and the tower was not really needed. So taking it out of service was more economical than making the modification.

Another Tower One of the obstacles during the troubleshooting investigation was that a short time later this DEA absorber design was slightly scaled down, but otherwise directly duplicated, to another tower in identical service and process conditions (slightly lower throughput). This other tower was started up and reached full production loads without experiencing a bottleneck. The other tower was 7 ft ID, and design downcomer width was 6 in. It took digging through some correspondence to find that a late design change by the engineering subcontractor increased the downcomer width from 6 to 11 in., which tripled the downcomer area to 7.5% of the tower cross-sectional area—a seemingly minor change that made all the difference between a tower that worked and one that did not.

CASE STUDY 4.3 DUMPING LEADS TO FLUCTUATIONS IN A DEPROPANIZER

Henry Z. Kister and Tom C. Hower, reference 263. Reproduced with permission. Copyright © (1987) AIChE. All rights reserved

Installation An olefins plant heat-pumped front-end depropanizer (Fig. 4.2*a*). Top-section diameter was about twice as large as bottom section. Feed was mostly vapor, but some liquid was condensed in the feed chiller.

Problem The column was unstable, and both pressure drop and bottom level fluctuated periodically. The period of fluctuation was about 30 seconds. Amplitude of fluctuation significantly increased as plant rates were raised.

Investigation Flooding checks were carried out on the column, by both the operator and the designer. These showed that the upper section was at least 20% below flood, and the bottom section was 40% below flood. The pressure drop, although fluctuating, was not excessively high and did not appear to rise rapidly with an increase in plant rates. When reflux rate was increased, the column became more stable. During the winter, when colder refrigerant was available, stability also improved.

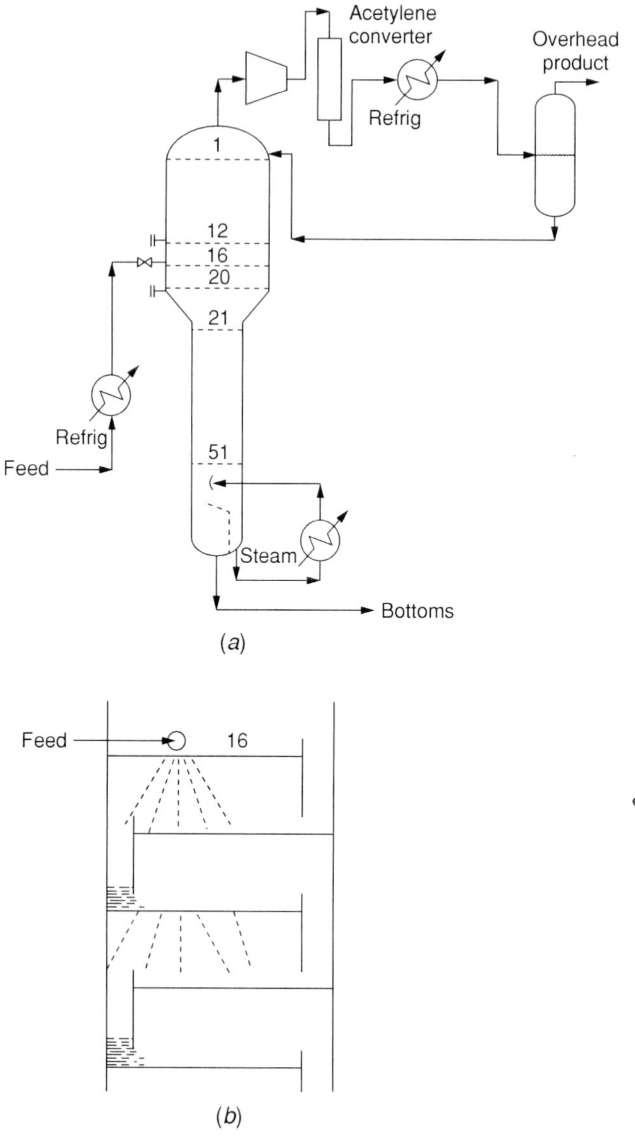

Figure 4.2 Tray dumping in olefins heat-pumped depropanizer: (*a*) depropanizer schematic; (*b*) dumping theory. (From Ref. 263. Reproduced with permission. Copyright © (1987) AIChE. All rights reserved.)

Gamma scans showed that the trays were performing normally. Gamma scan time studies were implemented to study the fluctuations. On a number of trays, the source and detector were placed just above the tray liquid level, and the amount of radiation was recorded over a period of time. It was established that on all plates the liquid level increased by 1 in. during the pressure kicks and then quickly dropped back to normal.

Theories The flooding theory, although having some evidence against it, was not completely discounted. However, an alternative theory was also formulated.

A check on trays 16–20 clearly indicated that these operate under dumping conditions. Column loading in this section was the same as in the bottom section of the column, but the trays were much larger. One possibility was that, while most of the liquid "rained" during dumping, some could have found its way to the downcomer and accumulated until a seal was established. When this took place, the resistance to vapor flow increased, and so did the vapor velocity. The higher velocity would then blow the seal, and the process of sealing and unsealing would then repeat (Fig. 4.2*b*). This theory explained the reduction in fluctuation during higher liquid loads because under these conditions the downcomer seal tended to stabilize.

Cures As this column operation could not be interrupted during the normal running of the plant, it was economical to cater to both theories. The cures implemented during the next scheduled plant shutdown were as follows:

1. Feed point was changed to tray 21.
2. The hole diameter of trays in the upper section was increased from $\frac{1}{2}$ to $\frac{5}{8}$ in., thus effecting a fractional hole area increase.

No more fluctuations occurred after this, even though column vapor rates were at times far greater than the increase achievable from increasing the fractional hole area.

CASE STUDY 4.4 LOW DEPROPANIZER FEED CAPACITY

Contributed by W. Randall Hollowell, CITGO, Lake Charles, Louisiana

Installation A butylenes H_2SO_4 alkylation unit with relatively low propane content in the feed. The alkylation unit refrigeration compressor discharge was condensed and split between depropanizer feed and bypass. Bypassed condensate and depropanizer bottoms were combined and recycled to the reactors. The depropanizer contained trays with movable valves.

Problem Only 10% of the compressor condensate was fed to the depropanizer, with the other 90% bypassed. Propane accumulated in the system, raising pressure in the deisobutanizer overhead accumulator and requiring flaring from there. Alkylate quality, operating costs, and unit capacity were all degraded by very high propane levels. This had been the standard operation for the unit for many years.

Investigation The depropanizer reflux was set at a fixed rate. Feed rate was fixed to ensure that the propane product met the specification for butane content. It took much searching to dig up the design rates for the tower. When located, reflux and overhead product flows were each found to be 25% of design and feed was 10% of design. Total compressor condensate was about 110% of design rate.

Theory A rough calculation suggested that the top and bottom trays were operating at 25% and at 15–20% of the design loads, respectively. Being an old design, it was believed that the design loads were a generous margin below the maximum hydraulic loads of the trays. It therefore appeared that the trays were operating at such a high turndown that tray efficiencies were very poor.

Testing The tower needed to be operated at design reflux and feed rates to test this theory. There was strong resistance to make such a large change in operation because it could potentially throw the propane product sphere out of specification.

A plan was devised to maintain product specifications for the test. First the reflux rate would be raised until either the reflux plus propane product flow reached design, or the condenser capacity was reached. Then the feed rate would be gradually raised until the design feed rate, the reboiler capacity, or the maximum allowable propane impurity would be reached.

Solution The evening shift agreed to try the plan, starting in midevening after cooling-water temperatures had dropped. Design condenser flows were reached, then feed rate was increased up to design. By midnight, unit propane levels were dropping rapidly. By morning, propane levels were about one-fourth of prior levels. The propane product stayed on specification throughout. Propane accumulation in the unit was never a problem again.

Moral Fractionator capacity may appear to be limited when tray turndown impairs efficiency. This pseudocapacity should be challenged by a careful test at conditions where tray efficiency should be good.

CASE STUDY 4.5 MINOR TRAY DESIGN CHANGES ELIMINATE CAPACITY BOTTLENECK

Contributed by Eric Cole, Koch-Glitsch LP, Wichita, Kansas

This case describes a process modification that threatened to lower the capacity of an existing high-pressure deethanizer and minor tray modifications that averted the threat.

Installation A refinery deethanizer that had been in service for several years.

History The tower was normally operated near the hydraulic limit of the stripping section trays. The simulation indicated that the maximum vapor and liquid loadings were two or three trays above the bottom. Under normal operating conditions, the highest loaded trays operated at 84% of downcomer choke flood, and their downcomer backup was at almost 12 in., which is in excess of typical design limits (155). The existing trays were four-pass trays on 24 in. tray spacing with movable conventional valves.

Problem An intended change to the plant heat integration was to discontinue preheating the deethanizer feed. This would have raised the internal loads in the stripping section and limit plant throughput. With no preheat, the highest loaded trays would be required to operate at 92% of downcomer choke flood and with downcomer backup of about 13 in. A retrofit with high-capacity trays was considered, but there was no budget for such a revamp for at least 2 years. A turnaround was coming up, so minor tray modifications could be implemented.

Analysis A hydraulic analysis showed that the column was limited by a few trays in the bottom of the stripping section and the side downcomers were much closer to the downcomer choke and downcomer backup limits than the center or off-center downcomers. The percent downcomer choke flood was almost 20% higher, and the calculated backup was 2 in. higher, for the side downcomers.

Solution Minor modifications were implemented to the bottom eight trays.

The existing 3-in. outlet weirs were replaced with ¾-in. weirs on the bottom eight trays to reduce the liquid height on the trays, and therefore, the downcomer backup. In addition, blocks that closed off 30% of the opening were installed on the outside bottom of the off-center downcomers. The purpose of the blocks was to divert some of the liquid away from the heavily loaded side downcomers and direct it toward the center downcomers.

The hydraulic calculations for the modified trays predicted that for the no-preheat case the percent downcomer choke would be reduced from 92 to 80% and the downcomer backup would be reduced from 13 to about 9.5 in.

Results After the modifications, an immediate increase in column capacity was observed. A test in the preheated operating mode indicated that the modified trays had about 7.5% higher capacity than before.

Lessons A limited budget may reduce the number of options available, but it does not necessarily eliminate all options. Often even a minor change to the tray design can have a significant impact on the column operation.

CASE STUDY 4.6 ESTABLISHING DOWNCOMER SEAL CAN BE DIFFICULT

Henry Z. Kister and Tom C. Hower, references 256, 263. Reproduced with permission.

Installation A low-temperature distillation column equipped with sieve trays. The column was piped so that the entire vapor feed stream always passed through the column. The liquid could be used as reflux to the column or be bypassed around the trayed section of the column and join the vapor feed (the column then simply acted as a flash drum). The column was designed so that all the liquid could be fed to the top (Fig. 4.3*a*). Overhead product was superheated in the product heater.

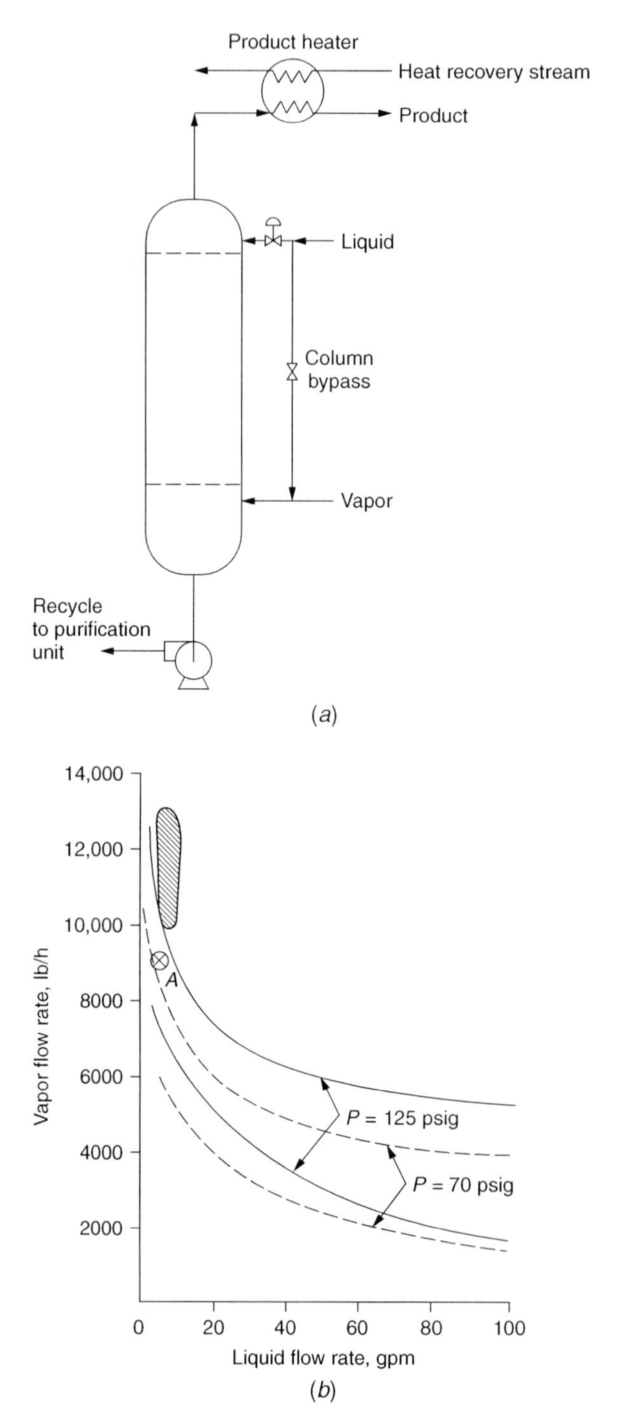

Figure 4.3 Sealing problem in low-temperature olefins tower: (*a*) tower schematic; (*b*) start-up stability diagram, based on Ref. 256. (From Ref. 263. Reproduced with permission. Copyright © (1987) AIChE. All rights reserved.)

Problem Prior to start-up, vapor entered the column while reflux flowed into the vapor feed through the bypass. To establish column action, the bypass was closed and then the reflux control valve was slowly opened. When the reflux control valve was opened, both product and heat recovery stream outlet temperatures would significantly drop. This indicated liquid entrainment because the heat recovery stream was unable to provide sufficient heat to vaporize a significant quantity of liquid. The presence of liquid in the top product and the low temperature of the product leaving the superheater could not be tolerated because of metallurgical limitations downstream. The column was operated as a flash drum, with subsequent loss of the heavy component to the overhead stream.

Analysis The problem was diagnosed to be a downcomer sealing problem. A start-up stability diagram was constructed (Fig. 4.3*b*), showing the range of liquid and vapor rates at which the column can be satisfactorily started (256).

The analysis was based on mathematically modeling the downcomer as a pipe. On this basis, the vapor rate required for satisfactory start-up at a given liquid rate was calculated. It was found that at a pressure of 70 psig, vapor rates which fall below the lower dashed curve in Figure 4.3*b* were too low to stop all the liquid from dumping through the tray perforations. Liquid would not reach the downcomer, and a seal could not be established. Vapor rates above the upper dashed curve in Figure 4.3*b* were too high to permit liquid to descend the downcomer; above this curve the vapor would "blast" the liquid out of the pipe. Satisfactory start-up at 70 psig could only be achieved in the area between the two dashed curves.

Solution Previous start-up attempts took place in the shaded area in Figure 4.3*b* at the normal operating pressure of 70 psig. It is clear that the start-up flow rates were well outside the satisfactory start-up range.

Increasing the column pressure to 125 psig brought the range of start-up flow rates closer to the upper stability limit (the upper solid curve in Fig. 4.3*b*). With some plant trimming, a reduction in vapor flow rates was achieved, and this brought the operating point to within the stability limits. The column was started up at point *A*.

Once the column was started up, the downcomers became sealed and the upper curve ceased to be a limit. The pressure and vapor rate were returned to their normal design values.

CASE STUDY 4.7 A TROUBLESOME PROCESS WATER STRIPPER

Henry Z. Kister and Tom C. Hower, reference 263. Reproduced with permission.

Installation A process water stripper, which stripped heavy HCs from process water.

History A second-hand 28-in.-ID column was packed with 2-in. CS Pall rings. At start-up, the column achieved its design separation and just achieved its design capacity.

A number of months later, the capacity started falling off, although good separation was still achieved. When the column was opened up, several of the rings had disappeared while others were reduced to fractions. A log of the system pH indicated that it was normally about 4 and sometimes fell to 2.

It was decided that ceramic packing was required. The column was repacked with 2-in. ceramic saddles and was returned to service. Upon restart, it became apparent that the column fell short of its initial design capacity, probably because of the lower capacity of ceramic saddles compared to metal Pall rings. However, the drop in capacity was not great and could be tolerated. A few months later, a further drop in capacity was observed. When the column was reopened, there were saddle fractions of all shapes and sizes produced by packing breakage due to turbulence. Figure 4.4 shows similarly-damaged ceramic packing that came out of another tower.

The decision was to replace the packing with SS packing. Two-inch SS Pall rings were specified, but these were in short supply and the column had to be quickly returned to service. One-inch SS Lessing rings were available from a second-hand column, but these could only fill half the column. It was decided to go ahead and use them. When the column was restarted, design capacity again was not achieved, although separation was good.

By this time, the column itself (which was constructed of CS) was suffering from a multitude of problems, including corrosion, erosion, and leakage. At the same time, an additional capacity increase was required. Another second-hand column, 36 in. ID, trayed, and fabricated from SS was available. After modifying its internals, it was used to replace the existing packed column.

Figure 4.4 Breakage of ceramic packings, similar to that described here. (Copyright Eastman Chemical Company. Used with permission.)

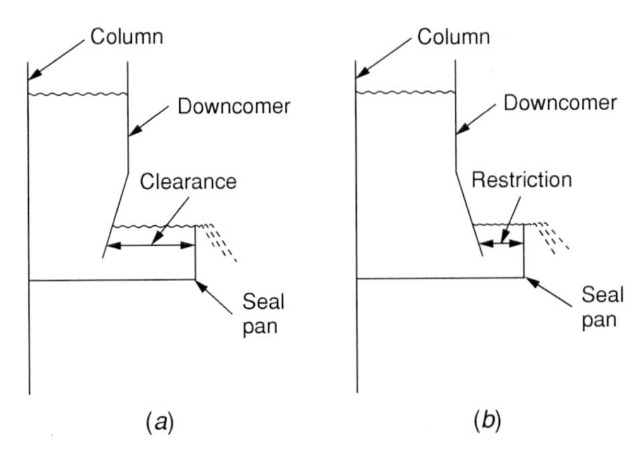

Figure 4.5 Incorrectly installed downcomer causes premature flooding. (From Ref. 263. Reproduced with permission. Copyright © (1987) AIChE. All rights reserved.)

An inspection showed that the old column was fabricated from two sections welded to each other across a 4-in. ring. Two inches of this ring projected outside; the other 2 in. projected inside. It is unknown whether the inside projection was there to boost mechanical strength or to serve as a primitive wall wiper. The packing was not discontinued near the ring, giving an effective internal diameter 4 in. smaller than the apparent column diameter, with a corresponding cross-sectional area reduction of 26% at that location. This ring surely did not help the column in achieving its design capacity.

Installing the trayed column did not spell the end of the problems. Although capacity was about tripled, it experienced operational difficulties. Pressure drop was high, and the column bottom level cycled, increasing suddenly, then dropping, over a period of 50 seconds. When the column was opened up, it was discovered that the bottom downcomer was installed backward, causing a restriction between the bottom of the downcomer and the seal-pan wall (Fig. 4.5). This caused liquid buildup in the downcomer and onto the tray above. Fortunately, the liquid buildup did not propagate too far up the column. When the buildup on the bottom tray was significant, the tray would dump momentarily, clearing the liquid buildup and causing a high level in the bottom sump. The cycle would then repeat.

After modifying the bottom downcomer, the column finally achieved trouble-free operation.

CASE STUDY 4.8 DOES YOUR DISTILLATION SIMULATION REFLECT THE REAL WORLD?

Henry Z. Kister, S. G. (Chell) Chellappan and Charles E. Spivey, references 254, 275. Reproduced with permission. Copyright © (1995) AIChE. All rights reserved

In this case study, the question "Does your simulation reflect the real world?" turned out to be the key for averting failure and succeeding in a column debottlenecking. It also taught that in hydrogen-rich systems, random packing efficiencies can be much lower than those in HC or organic systems. Reference 254 contains more details, while reference 275 is a comprehensive analysis including operating data.

Installation An olefins plant demethanizer was being debottlenecked for a capacity increase of 35%. Due to the large increase in capacity, it was essential to collect operating data and prepare a simulation that correctly reflects the data. This simulation was the basis for the debottleneck.

The demethanizer (Fig. 4.6*a*) consisted of three packed towers in series: T5, T15, and T16. Feed gas at about 500 psia was chilled by progressively colder levels of refrigerant. Condensed liquids were collected in knockout drums and fed to the demethanizer. The demethanizer bottoms contained ethylene, ethane, less than 1% propylene, and less than 100 ppm methane. The demethanizer overhead was mostly methane and hydrogen, with a small fraction of ethylene. This overhead was compressed and then chilled in the partial condenser E2. Some of the liquid condensed in E2 was returned to the demethanizer as reflux. The rest, as well as the uncondensed vapor, was the overhead product. Additional reflux was condensed in the demethanizer intercondenser E1.

Testing Field data were collected over a 1-week period. Data for the most-steady demethanizer operation were analyzed in detail. Key as-measured data are shown in Figure 4.6*a*.

A thorough analysis of the data (275) showed good closure of mass and component balances. Based on the measured vapor compositions of T5, T15, and T16 overheads, dew points were calculated and found to agree well with measured temperatures. Liquid composition analyses at T16 and T15 bottoms were discarded because they gave bubble points 30–100°F higher than measured temperatures, probably due to flashing of lights prior to analysis. Temperature and pressure readings were consistent throughout.

Flows from many of the flash drums to the demethanizer as well as liquid leaving T15 and T16 were metered. These were compared to flows determined from flash calculations for the demethanizer feed chillers based on the overall mass and component balances. Some of these measured flows agreed with our calculations while others did not. Making reasonable changes in feed composition to the demethanizer chillers and in the flash temperatures did not mitigate the discrepancies. To mitigate them, the process-side outlet temperatures of some chillers would have needed to fall below the refrigerant temperature, which is impossible. We have seen similar discrepancies in flow measurements of other multifeed demethanizers. We therefore concluded that the discrepancies reflect meter inaccuracies and the flash calculations were more reliable.

Simulation Model 1 Reliable VLE prediction is essential for a dependable simulation. Through the use of the proprietary information, we were able to closely predict the experimental VLE data available for the hydrogen–methane–ethylene system.

The overall tower HETP is determined from field measurements by adjusting the number of stages in the simulation until the simulation matches measured composition

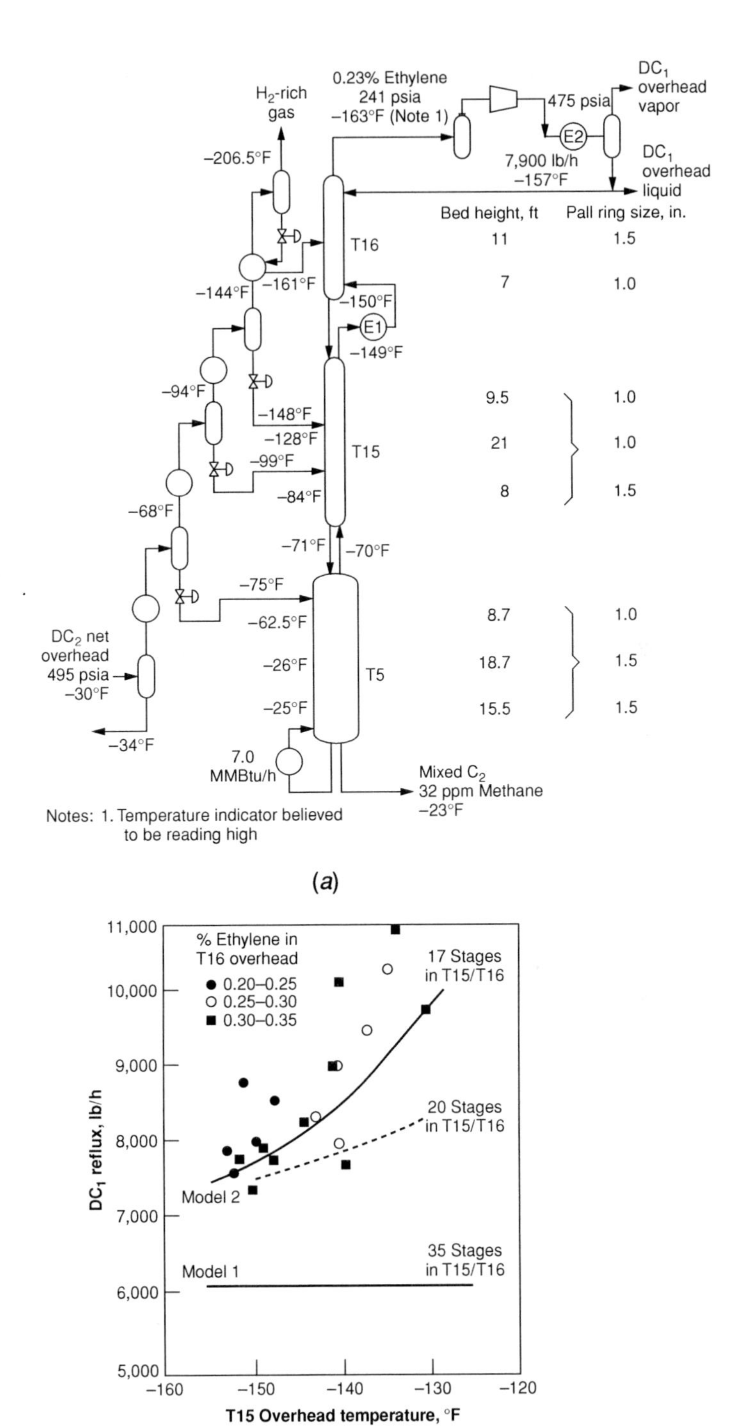

Figure 4.6 Simulation of olefins plant demethanizer: (*a*) flow sheet showing as-measured data; (*b*) demethanizer reflux versus T15 overhead temperature: data versus simulation predictions. (From Ref. 254, 275. Reproduced with permission. Copyright © (1995) AIChE. All rights reserved.)

and temperature profiles and flows. The overall tower HETP is the total tower packed height divided by the number of stages simulated for the tower. It is desirable to obtain section HETPs, but field measurements often are too crude to permit a good breakdown (317).

Bed-to-bed changes in packing diameters and packed heights in this demethanizer made data analysis difficult. Generally, for a given packing shape, packing HETP increases with larger packing diameter (54, 251). In the presence of liquid maldistribution, packing HETP also rises with greater bed height (353, 550). The rise is small when a bed develops less than three to four theoretical stages (353) but accelerates as the number of stages increases. Because the demethanizer distributors were far from perfect, HETP was expected to depend on bed height.

The "basic HETP" concept originally proposed by Zuiderweg et al. (549, 550) was used to account for bed-to-bed variations in packing diameter and bed height. A 1-in. Pall ring was postulated to have a basic HETP that it would achieve under perfect distribution. The actual HETP for each bed was the basic HETP times a multiplier accounting for the larger packing (if applicable) and for expected deviations from perfect distribution. The multiplier for each bed was derived from reliable published correlations that predict the effects of packing diameter and deviations from perfect distribution on HETP. The basic HETP was varied in the simulation until it matched the plant data.

The simulation (model 1) closely matched the measured temperature profile and the measured T15 and T5 overhead vapor compositions. The measured T16 overhead composition and T5 bottom composition were specified in the simulation, making the boil-up and the reflux dependent variables. The simulation closely matched the measured reboiler duty, but the simulated reflux flow rate, 6100 lb/h, was 23% less than measured. The lower simulated reflux was difficult to explain. One hypothesis was a low-temperature metering problem, similar to that experienced with some of the feed flows.

The best match to plant data was with 56 theoretical stages in the demethanizer towers, giving a basic HETP of 14 in. for 1-in. Pall rings and 20 in. for 1.5-in. Pall rings. These values are well inline with HETPs obtained in test columns and under excellent distribution conditions. Maldistribution would have reduced stages in the demethanizer by about 30%, giving HETPs of the order of 18 in. for 1-in. Pall rings and 26 in. for 1.5-in. Pall rings. These higher values are typical of HETPs observed in commercial practice (251).

Simulation Model 2 Although model 1 matched the plant data well, we were concerned about the discrepancy in reflux flow. Raising the basic HETP provided only a marginal increase in reflux while making T5 overhead vapor heavier than measured and also mismatching the measured and simulated temperature profiles. The T5 overhead composition was extremely sensitive to the number of T5 stages. To match the measured T5 overhead composition, 19–22 stages were required in T5. We settled on 21.

We now postulated a higher basic HETP in T15 and T16 than in T5. Model 1 had a total of 35 theoretical stages in T15 and T16. To raise the reflux to the measured

7900 lb/h, this number of stages needed to be roughly halved. With 20 stages, the simulated reflux was 7700 lb/h; with 17 stages, 8100 lb/h; and with 14 stages, well above 9000 lb/h. It became apparent that a simulation that matched the measured reflux needed to use very low efficiencies in T15 and T16. A new simulation (model 2) emerged. For T5, model 2 had an identical number of stages (that is, 21) as model 1. For T15 and T16, model 2 postulated a basic HETP high enough to give 17 theoretical stages.

Model 2 retained the close match to measured composition and temperature profiles previously produced by model 1. In addition, the reflux rate simulated by model 2 agreed with the measured reflux. For T5, HETPs predicted by models 1 and 2 were identical and well inline with typical HETPs (251) and correlation predictions. For T15 and T16, the basic HETPs predicted by model 2 are 35 in. for 1-in. Pall rings and 52 in. for 1.5-in. Pall rings, about double the expected values.

Model 1 or Model 2? While model 1 assumes a uniform basic HETP for 1-in. Pall rings throughout the demethanizer, model 2 postulates one basic HETP for 1-in. Pall rings in T5 and another throughout T15 and T16.

Model 1 made more sense, giving T15 and T16 HETPs that line up much better with typical values, as well as with correlation and supplier predictions. There appeared to be no reason why the HETP of 1-in. Pall rings in T5 should be half that in T15 and T16. On the other hand, model 2 provided a better match to the measured demethanizer reflux. Because correct model selection was crucial for revamp design, we intensified our examination of field data in search of clues validating or invalidating either model.

A study of operating logs provided a major clue. The T15 overhead temperature showed large day-to-day variations, of the order of up to 20°F. Warming up reflected a rise in the ethylene content of T15 overheads. During the same period, the ethylene in T16 overhead stayed reasonably constant at 0.2–0.35 mol %. It appeared that warming in T15 was well countered by the plant operators, who would raise reflux to maintain T16 overhead purity.

The curves in Figure 4.6*b* show reflux changes to counter a rise in T15 overhead temperature as simulated by models 1 and 2. With a high number of stages (model 1) the reflux changes are far smaller than with a low number of stages (model 2). The points plotted are a daily log of the average reflux rate versus average T15 overhead temperature. Model 2 gives a far better simulation, confirming its superiority to model 1.

Figure 4.6*b* conclusively validated model 2 and invalidated model 1. The existence of a low-efficiency region in the top bed of T16, possibly extending to the remaining bed of T16 and to T15, was established.

Revamp Existing packings and distributors were replaced as necessary to accommodate the higher throughput. Distributors, vapor inlets, and two-phase inlets were surveyed, and any sources of severe maldistribution were eliminated. General upgrading of distributors to improve distribution quality was found expensive and uneconomical. Details of the modifications are found elsewhere (275).

Post-revamp measured data (275) showed that the demethanizer operated close to revamp design (model 2) conditions. The comparison also shows that discarded model 1 would have predicted far lower ethylene in T16 overhead than measured post-revamp. Had model 1 been used, the demethanizer would have fallen far short of achieving its design expectations. A correct choice between the two models required the type of analysis of field data and scrutiny of operating logs undertaken here.

Postmortem The cause of the poor packing efficiency in T15 and T16 was not understood at the time. We presented five alternative explanations (275), none of which turned out to be correct. At a later time, Weiland (529) provided what we consider to be the correct explanation. Weiland noted that little hydrogen entered T5 while concentrations of hydrogen in T15 and T16 were high. Hydrogen is a fast-diffusing molecule. Its fast movement can drag heavy molecules from the liquid film on the packing into the vapor, a mechanism sometimes referred to as "reverse diffusion" (251). This counters the mass transfer process and lowers efficiency. This mechanism is detrimental to film mass transfer, such as that on packing, but has much less influence on turbulent mass transfer, such as that on trays.

In the years following this experience, the author encountered several other cases of unexpectedly low efficiency in packed hydrogen-rich towers. Weiland's theory also explains several experiences of unsuccessful replacements of trays by packings in hydrogen-rich services. The key to success in this case study was the reliance on measurements, not predictions. Models, expert opinions, and supplier predictions and theories often lead to incorrect simulations. Good measured data are the only reliable reflection of the real world.

CASE STUDY 4.9 FLOOD TESTING OF A PACKED VACUUM TOWER

Henry Z. Kister, Rusty Rhoad, and Kimberley A. Hoyt, reference 273.

A recent paper (258) lists 19 different definitions of flooding in packed towers that were used by different authors. These definitions are based on a variety of symptoms such as excessive entrainment, a sharp rise in pressure drop, a high pressure drop, liquid accumulation in the packing, loss of separation, and loss of stability. With so many symptoms, one would expect flooding to be readily identifiable in a field test.

This was not so in the flooding test below. In fact, before we processed the test data, we were asked if we had reached the flood point. All we could answer was "we do not know."

Problem A specialty chemical vacuum tower (Fig. 24.1*a*) containing three beds of structured packing separated a HK component from an IK in its lower section. While the bottom product was on specification, the vapor side draw contained 10% HK, compared to 1% design.

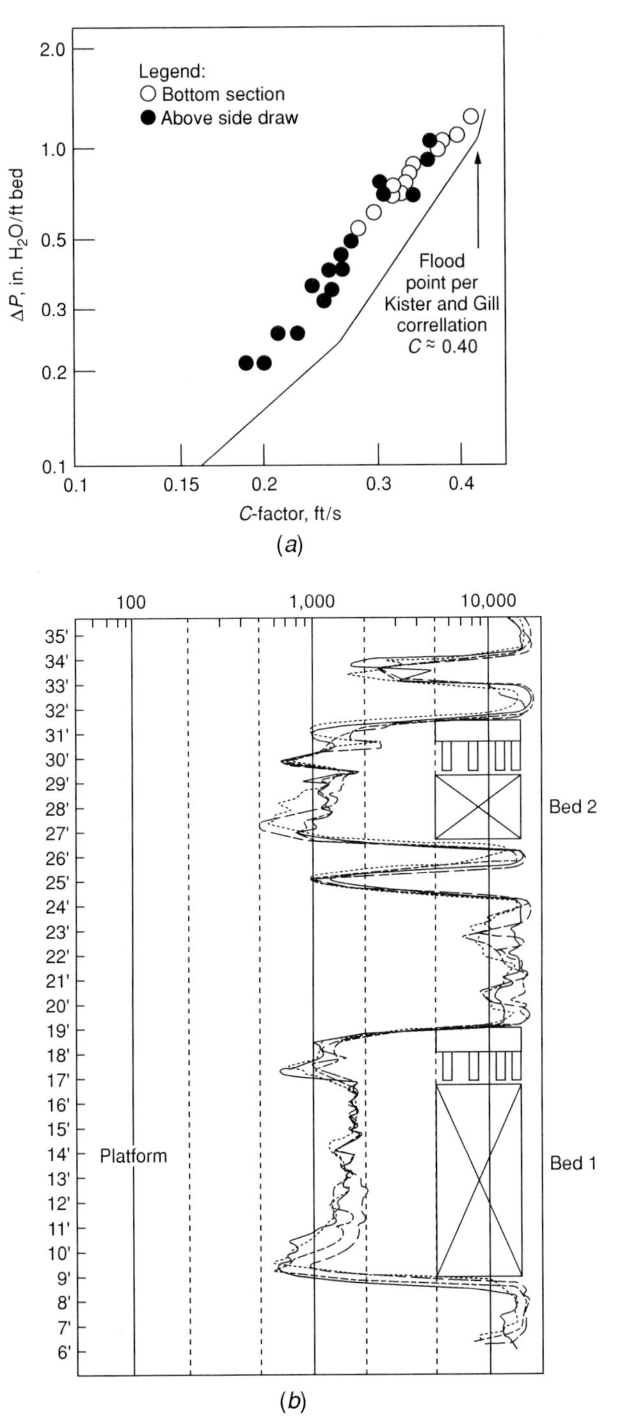

Figure 4.7 Flood testing of packed vacuum tower: (*a*) pressure drop measured in test versus *C*-factor; (*b*) gamma scans of beds below feed taken above suspected flood point; (*c*) gamma scans of beds below feed taken below suspected flood point; (*d*) temperature profile at flood test versus *C*-factor. (From Ref. 273. Reproduced with permission. Copyright © (1996) AIChE. All rights reserved.)

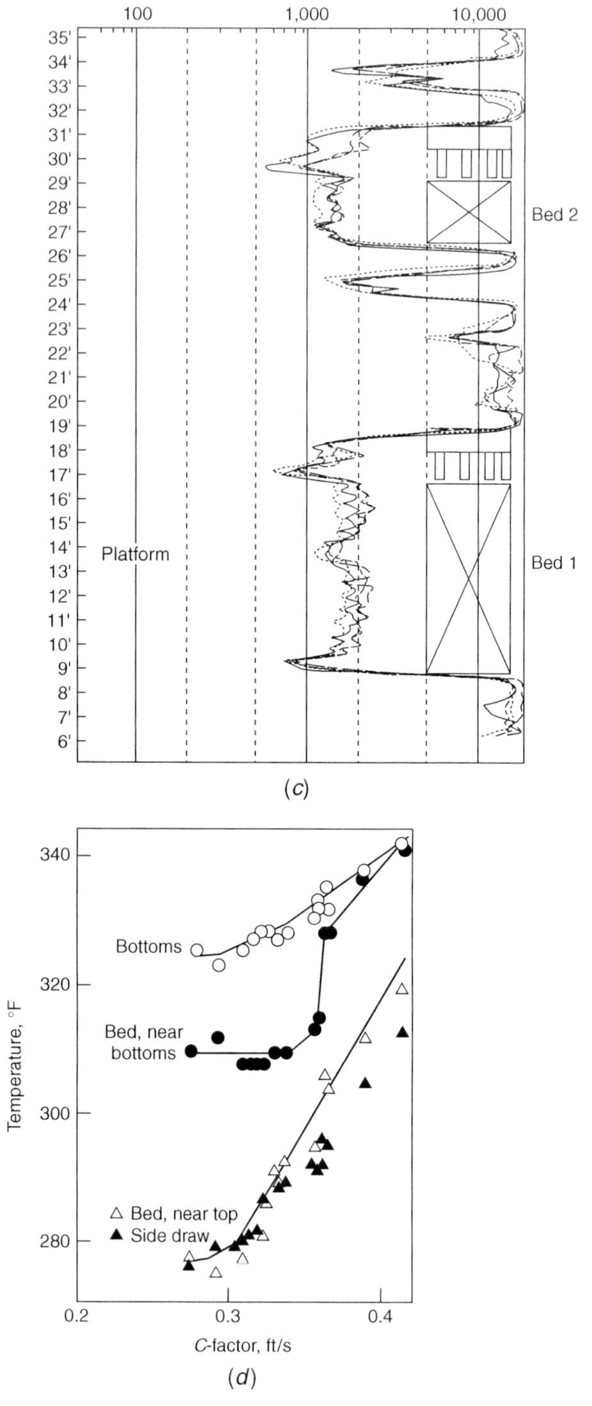

Figure 4.7 *(Continued)*

Testing Initial suspicion fell on the structured packing or distributors. Calculations showed that, at normal operating rates, the highest hydraulic loadings were in the bottom bed, and the bed operated at about 60% of flood, so it was not near a capacity limit. There were grounds to suspect plugging or maldistribution in the packing. To gain insight, a test raised boil-up and reflux until flooding was reached. Flooding and maldistribution were monitored by pressure drop measurements and gamma scans.

A classic key flood symptom in both tray and packed towers (250) is a point of inflection in the curve of pressure drop versus vapor rate, that is, a sharp rise of pressure drop with a rise in vapor rate. The pressure drop rise is due to liquid accumulation. Figure 4.7*a* plots pressure drops measured in the test versus the vapor rate. The vapor rate is expressed as the *C*-factor, C_s (ft/s), which is a density-corrected superficial vapor velocity, given by

$$C_s = U_s \left(\frac{\rho_G}{\rho_L - \rho_G} \right)^{1/2} \tag{1}$$

where U_s is superficial vapor velocity (ft/s) (based on tower cross-sectional area), ρ is density (lb/ft^3), and the subscripts G and L denote gas and liquid, respectively. The hollow circles in Figure 4.7*a* are pressure drops in the bottom bed plotted against the *C*-factor for the bottom bed. The filled circles are the total pressure drops for the upper two beds plotted against the *C*-factor for these beds. The *C*-factor was highest in the bottom bed. Above the vapor side draw, the *C*-factor became lower and stayed reasonably uniform throughout the upper two beds.

The pressure drop–*C*-factor relationship in the bottom bed was much the same as in the upper two beds. This is not surprising because all three beds contained the same type and size of structured packing. However, this suggests the absence of large-scale plugging. Such plugging is likely to affect one of the beds more than the others, and the difference would show on the pressure drop–*C*-factor plot.

It is apparent from Figure 4.7*a* that pressure drop rose monotonously with vapor rate. Neither a point of inflection nor a sharp rise in pressure drop was visible. This would argue against flooding during the tests.

Another key flood symptom is "excessive" pressure drop. Kister and Gill (257) show that for both random and structured packings flooding begins ("incipient flooding") when

$$\Delta P = 0.115 F_p^{0.7} \tag{2}$$

where ΔP is pressure drop (in. H_2O/ft packing) and F_p is the packing factor (ft^{-1}).

For the packing in this tower, F_p was 21 (251), so Equation 2 predicted a flood pressure drop of 1.0 in. H_2O/ft packing. From the test pressure drop data in Figure 4.7*a*, this coincided with a *C*-factor of 0.36 ft/s, suggesting that at the higher *C*-factors flooding initiated in the bottom bed and in the bed above.

The lower line in Figure 4.7*a* is flood and pressure drop predictions from the interpolation method of Kister and Gill (251, 257). It shows a slightly lower pressure drop and higher capacity than those measured. The flood *C*-factor predicted from this procedure is 0.40 ft/s, about 10% higher than that inferred from the pressure drop measurement. This is well within the accuracy of the prediction method. Also, packings may suffer from a small amount of solids accumulation, which would raise the pressure drop and lower the flood point.

In summary, the rule of thumb given by Equation 2 suggests a flood point at a *C*-factor of 0.36. This agreed well with calculation but was contradicted by the lack of a steep rise in pressure drop.

Gamma Scans A grid gamma scan (four equal chords equidistant from the tower center) of the two packed beds below the feed (Fig. 4.7*b*) was shot at a *C*-factor of 0.37 ft/s. The four chords overlay well near the top of the two beds, indicating good liquid distribution. The chords separate near the bottom of each of the two beds, indicating maldistribution in the lower portions of the beds. The scan does not show any conclusive signs of flooding. The low gamma-ray transmission at the bottom of each bed may mean accumulation of liquid, but it also may mean a high liquid flow due to maldistribution.

To assist with gamma scan interpretation, a second grid scan (Fig. 4.7*c*) was shot well below the suspected flood point (at a *C*-factor of 0.3 ft/s). In this scan, the region of low transmission at the bottom of each bed disappeared. Also, the liquid distribution appears quite uniform (albeit not perfect).

The maldistribution at the higher loadings (Fig. 4.7*b*) occurred at the bottom of the bed, not near the top, suggesting that the maldistribution did not initiate at the liquid distributor. In both Figures 4.7*b* and 4.7*c*, liquid entering the bed (near the top) appears quite uniform, implying that the low transmission near the bottom of the beds in Figure 4.7*b* is due to liquid accumulation. This accumulation is a symptom of incipient flood.

Figure 4.7*b* suggests that the liquid accumulation generates channeling; the liquid accumulates only along two chords. Presumably, the vapor keeps rising through the areas in which liquid does not accumulate. This means that the vapor does not need to penetrate through a static head of liquid, which may explain why no steep rise in pressure drop was apparent at incipient flooding.

In summary, a lone grid scan was unable to diagnose flooding. However, comparing grid scans at unflooded and suspected flooded conditions did signal symptoms of flooding. These symptoms were seen at the same point where the pressure drop became excessive according to Equation 2. While the combined evidence is still not conclusive, it does point to flooding at a *C*-factor of 0.36 ft/s.

Temperature Profile Figure 4.7*d* is a plot of bottom bed temperatures versus the *C*-factor. There were good temperature spreads at the low *C*-factors. As vapor (and liquid) loads were raised, the temperature spreads diminished. This indicates a smaller efficiency as loads are increased, a behavior that is experienced with some structured packings (251) such as the one in this tower.

At a *C*-factor of 0.36 ft/s, the near-bottom bed temperature rose sharply and became the same as the bottom temperature, indicating no separation between these two points. The efficiency sharply dropped. This drop coincided with the point where liquid accumulation was seen on the gamma scans and where tower pressure drop became excessive per Equation 2. The efficiency drop was due to the channeling or entrainment induced by the liquid accumulation.

Flood Point The sharp rise in middle temperature at a *C*-factor of 0.36 was the clincher. Together with the pressure drop and gamma scan results, the evidence for flooding at a *C*-factor of 0.36 ft/s became conclusive.

Epilogue The tests confirmed that the flood point was close to predictions and the pressure drop was only slightly higher than expected. The pressure drop–*C*-factor relationship in the bottom bed was much the same as in the upper two beds, which argued against large-scale plugging. No liquid maldistribution was apparent below the flood point. The poor separation experienced in the packings at normal rates therefore appeared to be neither due to maldistribution nor due to plugging. If there was some plugging, it would be slight at most.

This leaves the question of what caused the poor separation between the IK and HK component. The answer is in Case Study 2.8.

Morals These capacity tests teach several invaluable lessons about flood of packed vacuum towers—lessons that we have verified in other installations. First, for vacuum packed towers, the pressure drop may not rise sharply upon flooding. Here, the loss of separation (e.g., as seen by our temperature spreads) is probably the best flood indicator. Second, gamma scans can be inconclusive if shot only at high rates. Taking a low-rate scan as well can render the results far more valuable and conclusive. Finally, Equation 2 gave good prediction of the actual flood point. Cases 1.14 and 1.15 discuss several additional cases of successful applications of this equation.

CASE STUDY 4.10 IN SPECIAL APPLICATIONS, SPRAY TOWERS DO BETTER THAN PACKINGS

Installation A grass-roots 16-in.-ID tails tower containing a 6-ft bed of 1-in. polypropylene saddles used town water to absorb near-pure HCl gas ("by-gas") that remained unabsorbed after an upstream falling-film absorber. The tails tower produced dilute hydrochloric acid of a desired concentration that was recycled to the falling-film absorber. Off gas from the tails tower was designed to contain less than 2% of the total HCl fed to the falling-film absorption system. This off gas flowed to a 6-in.-ID vent scrubber containing an 8-ft bed of $\frac{5}{8}$-in. polypropylene Pall rings that used city water to remove the residual HCl before the off gas went to vent.

Problem The seal loops around the primary falling-film absorber and tails tower experienced frequent siphoning, resulting in pressure surges, loss of seal, and instability. The pressure surges frequently collapsed the packing support in the tails tower. Each time, pieces of packing were scattered, finding their way to all sections of piping, causing blockages, requiring cleaning, and frustrating attempts to troubleshoot for the root causes of the siphoning. Attempts to boost the mechanical strength of the supports helped but did not solve the problem.

Solution The packing and support were removed from the tails tower. This had the additional benefit of testing and denying a theory that interference of the tails tower packing with "breathing" contributed to the siphoning. The liquid distributor was replaced by spray nozzles. Nozzle position and orientation were adjusted in situ by running city water at the design flow rate with the tower open and inspecting and visually optimizing the spray pattern. This eliminated the packing collapse and permitted uninterrupted search for the root cause of siphoning at the expense of what was perceived to be some loss of HCl. A detailed account of troubleshooting of the siphoning in the falling-film absorber is found elsewhere (269).

Packing or Spray Only? After the siphons were eliminated and the falling-film absorber stabilized, it was intended to reinstate the packing in the tails tower. However, plant tests showed negligible HCl in the vent scrubber overhead gas. The HCl in the water leaving the vent scrubber was 0.05% of the HCl entering the falling-film absorber (design 2% maximum). The tails tower with the spray-only was achieving much better than the design absorption (which was based on packing). There was no need to reinstate the packing, and the robustness of the spray-only system during upset saved much downtime.

At one time, the vent scrubber was taken off-line and inspected. Its packing support was found dislodged and lying on the scrubber bottom. All the packing except for a handful of pieces had been washed away. It is believed that these occurred while the system experienced pressure surges. The vent scrubber successfully operated as a spray tower all along.

Chapter 5

Feed Entry Pitfalls in Tray Towers

Troublesome feed arrangements to tray towers barely made it into the top 20 of the distillation malfunctions for the last half century (255). In the author's experience, issues with these arrangements deserve a higher spot on the list, probably just outside of the top 10.

Maldistribution of feed into multipass trays, mostly in large towers, has been a major source of capacity and separation bottlenecks. Feed entries that induced vapor or flashing feed into downcomers or led to vaporization inside downcomers have been sources of severe capacity bottlenecks. Obstruction of downcomer entrance by the feed pipes have generated premature flood in many instances. Finally, inadequate liquid or vapor distribution to low-pressure-drop trays (such as dual-flow trays) has led to poor separation and capacity bottlenecks.

CASE STUDY 5.1 FLASHING FEED GENERATES A 12-YEAR BOTTLENECK

Henry Z. Kister, Tom C. Hower, Paulo R. de Melo Freitas, and Joaõ Nery, reference 276. Reproduced with permission. Copyright © (1996) AIChE. All rights reserved

In this case study, a flashing feed entering a downcomer bottlenecked an entire olefins plant for 12 years. This bottleneck survived three unsuccessful fix attempts by a major engineering contractor who failed to study plant data and look at gamma scans.

Installation A demethanizer (Fig. 5.1*a*) separated methane and a small amount of hydrogen as the overhead product from C_2 and heavier HCs as the bottom product. The tower had a small-diameter upper section containing 23 valve trays and a larger diameter bottom section containing 44 valve trays and also an interreboiler. The demethanizer received four feeds. Three of these entered the upper section above trays 7, 11, and 17. The fourth feed entered the tower swage just below tray 23.

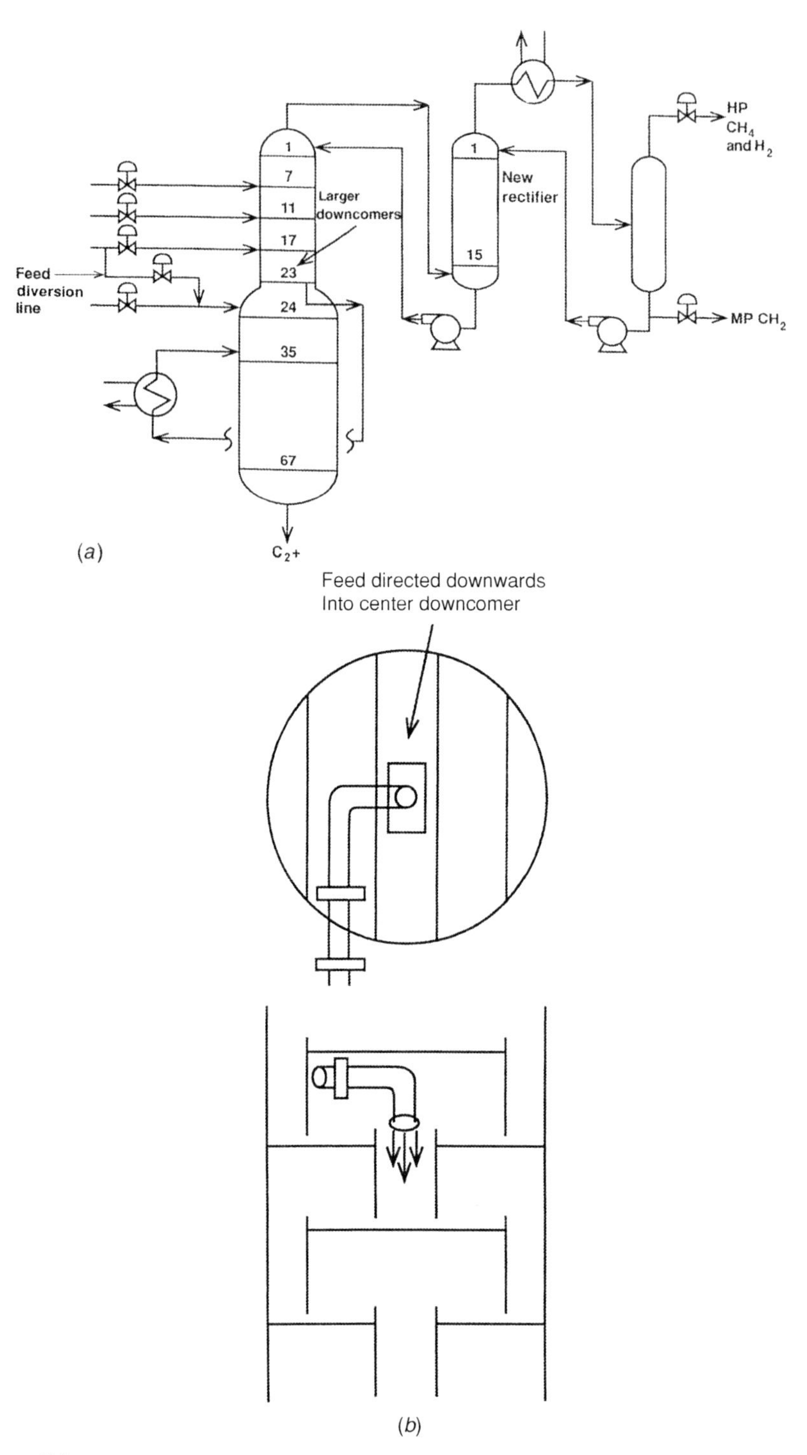

Figure 5.1 Demethanizer bottleneck that persisted 12 years: (*a*) schematic showing tower and modificatiions performed over the years in attempt to debottleneck it; (*b*) feed entry that caused bottleneck. (From Ref. 276. Reproduced with permission.)

History The demethanizer started up in 1978, with the plant producing 46 tons/h ethylene. It operated well at full reflux and an overhead pressure of 30 bars. As plant throughout was increased to 50 tons/h ethylene, the pressure drop in the upper section of the demethanizer (tray 23 upward) rose and the column became unstable. The demethanizer became a plant bottleneck.

In an attempt to debottleneck the tower in 1983 and 1986, the following modifications were performed (Fig. 5.1*a*) at the recommendation of a major engineering contractor:

- Downcomers from trays 17–23 (and later those from trays 10–16) were expanded by 14%.
- A new 15-tray rectifier was added in series with the existing demethanizer to reduce ethylene losses.

The modifications gained nothing. The demethanizer remained bottlenecked at 50 tons/h ethylene. The new rectifier achieved very little. The ethylene losses remained the same. The vapor temperature into the rectifier was much the same as the vapor temperature at the rectifier overhead.

Operational Debottleneck At the capacity limit, there was a rise in differential pressure, massive entrainment (observed as the liquid accumulation in the bottom of the rectifier), and a loss of separation (seen both by an increase in ethylene losses and by warming of the top section). These were symptoms of flooding initiating somewhere above tray 23 and propagating upward.

As the feeds were mostly liquid (by weight), the peak hydraulic loads in the upper section of the demethanizer were between trays 17 and 23. To debottleneck this section at the 1986 turnaround, a line was connected from the tray 17 feed into the feed to tray 23 (Fig. 5.1*a*). A portion of the tray 17 feed was diverted into the lower feed point so it bypassed the small-diameter section. With the diversion valve 50% open, a plant throughput of 57 tons/h ethylene was achieved.

1995 Debottleneck Although the feed diversion line was effective, it was a control nightmare and an inefficiency. In the next debottleneck, an attempt was made to raise throughput by about 20% with the feed diversion line closed.

A debottlenecking study by a major engineering contractor recommended replacing all the demethanizer trays by random packing. A critical evaluation by the plant revealed that in the large-diameter (bottom) part, the existing trays had plenty of capacity and there was no need to replace them. On the other hand, the top section was a bottleneck, so there replacing trays by packing appeared justified.

Just prior to the modifications, plant personnel became aware of field data (275) of low random-packing efficiencies at the top section of a demethanizer under conditions of good vapor and liquid distribution. The low efficiency was attributed to the high hydrogen concentration (see Case Study 4.8). Naturally this became a concern. A task force was formed to critically evaluate the need for packing in all upper sections of the demethanizer.

A hydraulic analysis showed that at the debottleneck throughput, trays 1–10 and 11–16 would operate at 40 and 55%, respectively, of the closest flood limit.

They were capable of handling double the current throughput. Replacing trays 1–16 by packings, therefore, would achieve no capacity gain, would lose separation, and would cost money. Plans to replace trays 1–16 by packings were therefore scrapped (even though the packings and distributors have already been purchased).

At the debottleneck throughput, trays 17–23 approached downcomer choke. This section could be debottlenecked either by a retray or by replacing trays by packings. Replacing trays by packings was preferred because the packings have already been purchased. Also, the loss of separation in this section would be minimal, both because this section contained only a few stages and because this section had the least hydrogen concentration. It was therefore decided to proceed with replacing trays 17–23 by 2-in. random packings.

Bottleneck Identification The hydraulic analysis reopened the search for the root cause of the bottleneck. While downcomer choke on trays 17–23 was the closest flood limit, it was not expected to be reached until the throughput was raised by 20% and the feed diversion line was closed. In practice, the bottleneck was observed without raising throughput and with the feed diversion line 50% open. Calculations showed that at these loads, trays 17–23 should have been operating at 60% of downcomer choke and nowhere near any other flood limit.

Another strong argument against downcomer choke was that in 1983 enlarging the downcomer top area by 14% did not affect the column bottleneck. Had the bottleneck been downcomer choke, the larger downcomer top area should have led to an improvement.

Tower gamma scans were closely examined. The scans showed flood initiation around tray 17, not tray 23. This was unexpected since the hydraulic loadings on tray 23 were higher than on tray 17. Had the tower encountered a tray or downcomer limitation, it should have initiated in the higher loading region, that is, tray 23.

Tray 17 was a feed tray. Points of transition, like feed trays, are spots where major tower bottlenecks often initiate (255). A drawing review (Fig. 5.1*b*) showed flashing feed entering the downcomer. The literature (248, 250, 304) recommends against this feed entry for flashing feeds. For the tray 17 feed, the vapor content may appear small, about 0.6 weight percent, but 0.6% by weight is 10% by volume, a vapor fraction far too large to be entered into a downcomer. This vapor disengaged from the liquid and impeded the descent of liquid, bringing about a premature downcomer choke limitation.

The tower bottleneck therefore turned out to be neither the tray nor downcomer capacity. It was poor introduction of feed. The problem was not unique to tray 17. The feed entries on trays 11 and 7 were just as poor. However, trays 7 and 11 had not encountered a bottleneck yet, possibly due to their large margin away from downcomer choke.

Solution The feed arrangements on trays 7, 11, and 17 were replaced by well-designed feed arrangements. In addition, trays 17–23 were replaced by 2-in. modern packings.

The column restarted in May 1995. The bottleneck completely disappeared. The feed diversion line never needed to be opened again. The 12-year-old feed enrty bottleneck ceased to exist.

CASE STUDY 5.2 FLASHING FEED ENTRY CAN MAKE OR BREAK A TOWER

Contributed by Ashraf Lakha, Koch-Glitsch, Stoke-on-Trent, United Kingdom

Installation A crude stabilizer, 2.6 m ID above the feed with 25 one-pass valve trays, and 4.3 m ID below the feed with 25 four-pass valve trays. The tower contained three water draw chimney trays. Of these, tray 23A (2 actual trays below the feed) was not being used to draw water due to a change in operating philosophy. The tower also had interreboilers at trays 11 and 19.

Problem The column was designed to process 1150 m^3/h of unstabilized crude but was having problems achieving more than 950 m^3/h.

Troubleshooting Initial suspicion fell on the trays, but a hydraulic evaluation found them adequate for 1150 m^3/h. A gamma scan showed flooding predominantly in the feed region (tray 25) and just above it in the conical section, and also at the unused water draw tray 23A. These areas were singled out for detailed study.

Focus on Feed Arrangement Flashing feed containing a substantial vapor fraction entered the tower via two pipes discharging directly into two off-center false downcomers (Fig. 5.2*a*). No drawings were available for the feed pipes, but photographs showed only 12 holes at the underside of each feed pipe. Hole diameters appeared small, but actual hole dimensions were not available. A rough calculation showed that hole area was totally inadequate for the amount and vapor fraction of the feed. The very high pressure drop (about 0.75 kg/cm^2) measured across the feed pipe confirmed the excessive feed velocity. The entry of this high-velocity flashing feed directly into the false downcomers was likely to have turned the liquid pool in the false downcomers into high-turbulence froth or spray and could have led to vapor–liquid slugs. There is no way that adequate vapor–liquid separation could have occurred within the false downcomers. Consequently, froth stacked up within and above the false downcomers, possibly initiating froth carryover onto the tray above the feed. The gamma scans supported this theory.

The openings in the transition seal pan above the feed tray were directly above the false downcomers (Fig. 5.2*a*). The froth stack-up inside the false downcomers may have interfered with the liquid coming out of the transition seal pan, causing it to back up and thereby contributing to premature flooding of the trays above the feed.

The above analysis identified the feed entry arrangement as the root cause of the flooding of the trays just above the feed. It does not explain the bottleneck seen on the trays below.

Focus on Water Draw Chimney Tray 23A The tray (Fig. 5.2*b*) used tall overflow weirs to maximize residence time for HC–water separation. The overflow weirs discharged into two side downcomers, which terminated on tray 23 below. Tray 23 was a four-pass valve tray rotated at 90° to tray 24. The side downcomers from the chimney tray to tray 23 extended all the way to the floor of tray 23 and were closed at the floor

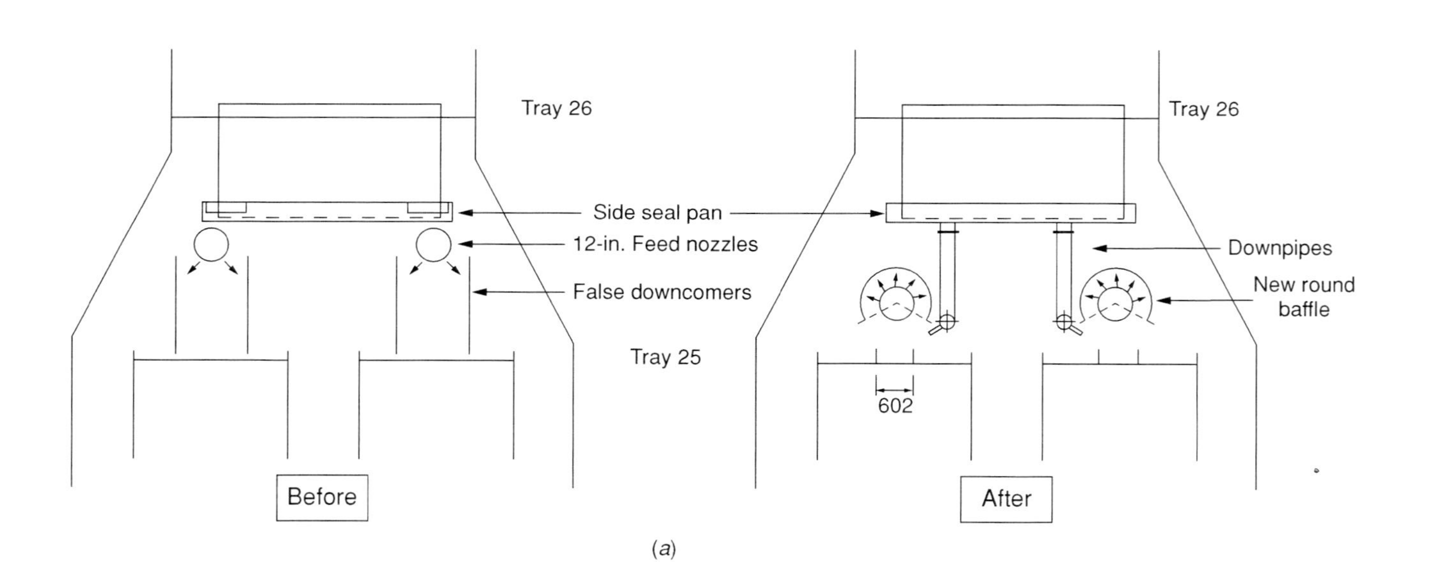

Figure 5.2 Causes and cures of premature flood in crude stablizer: (*a*) flashing feed arrangment; (*b*) water draw chimney tray.

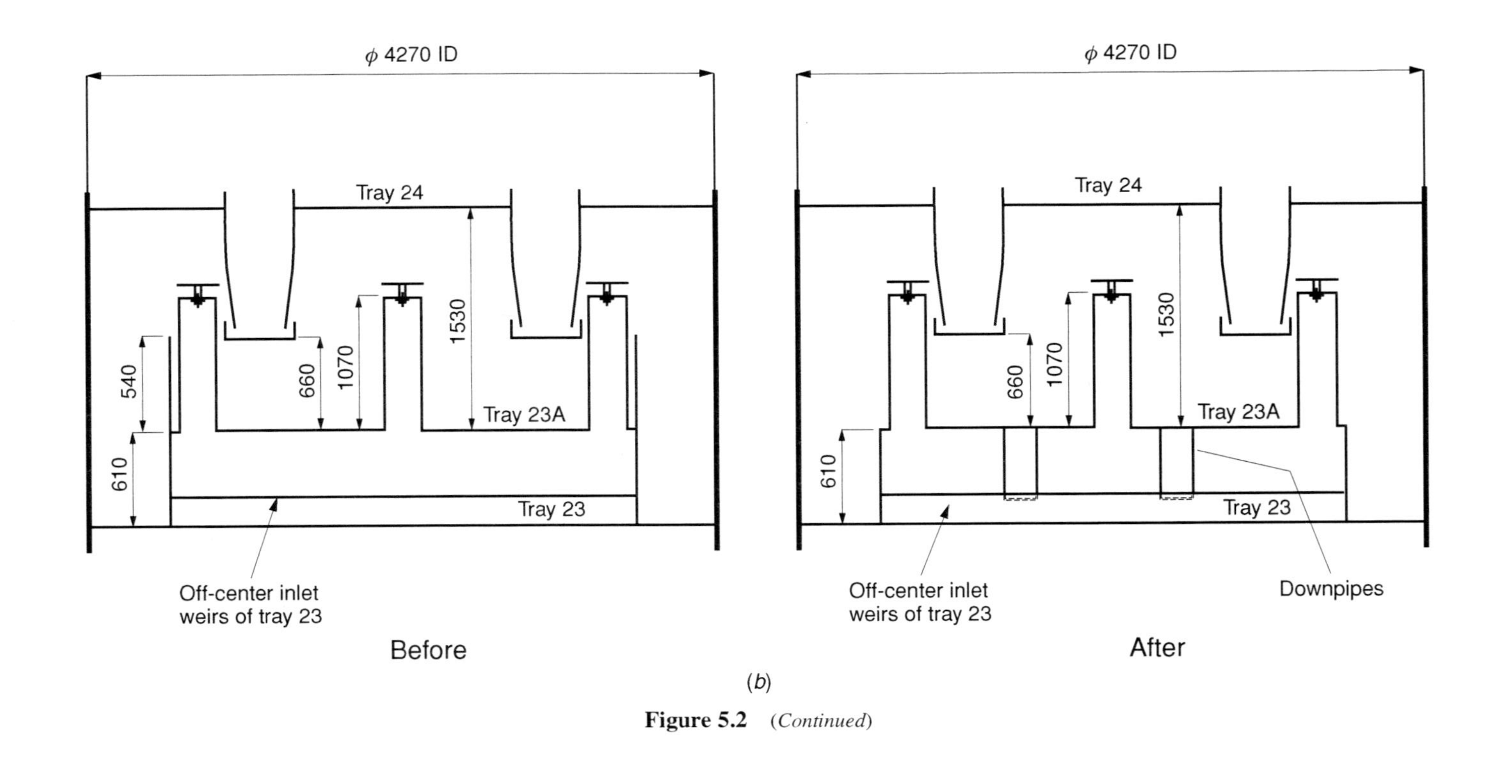

(*b*)

Figure 5.2 (*Continued*)

of tray 23. The only openings through which liquid could exit these downcomers were two narrow slots 100 × 610 mm at the bottom of each side downcomer, that opened to the blanked off-center panels on tray 23. The total exit area was 0.25 m^2.

At the operating conditions, the clear liquid velocity through the slots was a huge 1.3 m/s. Just the friction losses through the slots contributed 270 mm of clear liquid backup in the downcomers. Additional backup due to chimney pressure drop and clear liquid height on tray 23 below (which was probably enhanced due to turbulence generated by the high liquid inlet velocity) could have impeded liquid descent into the downcomers. This could have induced liquid rise above the top of the chimneys. The gamma scans showed flooding on the chimney tray and no clear vapor space above it, supporting the likely rise of the liquid level to the top of the chimneys.

The top of the chimney tray overflow weirs was only 180 mm below the seal pan overflow from the tray above. At the operating liquid rate, the liquid head above the overflow weirs was calculated to be approximately 90 mm. This left a narrow (90-mm) gap between the liquid level on the chimney tray and the seal pan overflow. It is most likely that before reaching the top of the chimneys the rising liquid on the chimney tray would have induced additional backup of liquid in the downcomers from tray 24, bringing them closer to the flood point. Once liquid level on the chimney tray exceeded the chimney height, the ascending vapor entrained some of the liquid, further loading up the downcomer from tray 24, until it flooded. This mechanism is supported by the observation that tray 24 was flooded with froth density much higher than that above the chimneys of tray 23A. Above the chimney tray, the scans showed about 1 m of dense froth with less dense froth above it.

The above analysis identified the narrow liquid exit slots and the excessively tall overflow weirs on the draw tray (in relation to the seal pan overflow and chimney heights) as the root cause of flooding below the feed.

Solution The following modifications were performed to overcome the problem.

The existing feed pipes were replaced with well-designed feed pipes, each with a round baffle above (Fig. 5.2*a*). The round baffle covered the feed pipe. The pipe openings pointed toward the baffle. The baffle separated the vapor from liquid, with the liquid dropping down. The false downcomers on the feed tray were replaced with inlet weirs.

The existing water draw tray 23A was replaced with a new one with some key additional features (Fig. 5.2*b*):

- Four 8-in. liquid downpipes were added to boost liquid downflow. The downpipes extended from the center of the chimney tray to the blanked off-center panels of tray 23 below.
- The overflow weir height on the draw tray was eliminated to provide more distance between the liquid on the chimney tray and the top of the chimneys and seal pan overflows above.
- Riser open area was increased compared to that on the existing draw tray.

Some other minor modifications were also carried out on the return pipes from the interreboilers at trays 11 and 19.

Result After the above modifications, the tower achieved the design capacity of 1150 m^3/h of unstabilized crude with no further problems.

Moral It is essential to correctly design flashing feed inlets and chimney trays.

CASE STUDY 5.3 FLASHING FEED PIPING BOTTLENECKS DEMETHANIZER

Contributed by David P. Kurtz, Koch-Glitsch LP, Wichita, Kansas

Installation A cryogenic gas plant demethanizer operating at about 260 psi. Column cross-sectional area above the feed was about twice that below the feed. Feed to the tower was the exhaust of a turboexpander. The flashing feed was about 90% vapor by weight.

Problem The column flooded below design rates, sending its products off specification.

Investigation A simulation based on tower operation was prepared. Based on the simulation, tray hydraulics were checked. The calculations showed that above the feed the trays operated at 80% of jet flood and at 65% of downcomer flood. Beneath the feed, the trays operated at 60% of the closest flood limit that was jet flood. So all trays operated at comfortable margins from any flood limits.

Gamma scans showed that the trays were flooded above the feed. This shifted the focus to the feed zone.

Feed Entry The flashing feed entered the cone section via a large-diameter H-pipe with slots open toward the downcomers from the tray above (Fig. 5.3). The

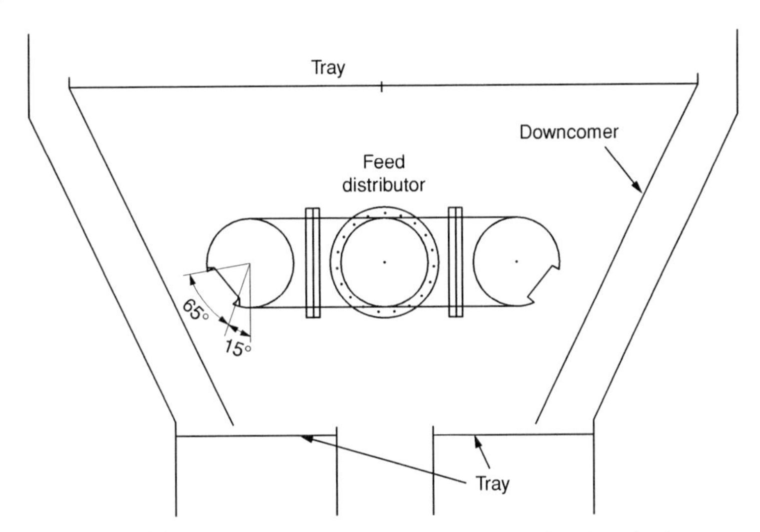

Figure 5.3 Flashing feed entry that bottlenecked gas plant demethanizer.

downcomer followed the cone walls. The velocity of the two-phase mixture in the pipe and slots was within good practice guidelines, but with the discharge angle and the downcomer wall angle, a good portion of the feed was deflected directly upward by the downcomer wall, generating high local entrainment (Fig. 5.3). The large-diameter H-pipe blocked about 50% of the tower area. At the narrowest point, the free area for vapor rise was about 40% of that available at the trays above. The high vapor velocity created by the H-pipe kept the entrainment created by the flashing feed from disengaging before reaching the tray above, causing the tray above the H-pipe to prematurely flood.

Cure The large H-pipe was removed and replaced with a vane distributor. For additional assurance the valve tray below the feed was replaced with a chimney tray. The tower achieved design capacity after the modifications.

CASE STUDY 5.4 FLASHING FEED ENTRY CAN BOTTLENECK A TOWER

Installation A new 3-ft-ID light HC tower equipped with proprietary dual-flow trays. The tower contained 30 trays, 11 above the feed at 18 in. spacing and 19 below the feed at 24 in. spacing. The feed was 23% vapor by weight, and the tower operated at 70 psig at the top. Tray open area was 25% of the tower area below the feed, 32% above. The holes were $^1/_2$ in. diameter.

Problem The tower experienced poor separation and premature flooding. The trays were designed for 55% efficiency and did not achieve this. The flood was identified by high *dP* and showed on gamma scans.

Hydraulic Analysis The scans as well as operating parameters showed that the flooding initiated about 10 trays below the feed. Tray hydraulics is expressed in terms of the C-factor, given by

$$C_S = U_S\sqrt{\frac{\rho_G}{\rho_L - \rho_G}}$$

where U_S is the superficial velocity (ft/s), ρ is density (lb/ft^3), and the subscripts G and L denote gas and liquid densities.

In the bottom section of the tower the C-factor was estimated at 0.18 ft/s at a liquid load of 20 gpm/ft^2, which should be well below flood for these high-capacity trays. Above the feed the C-factor was higher, but the liquid rate was much smaller, and the trays should also have operated a comfortable margin away from flood. There were no signs of fouling.

The gamma scans showed some strange behavior. The tower was scanned along the 90°–270° centerline, if the feed inlet is at 0°. The unflooded scan shows no liquid in the vapor spaces of the 10 trays right below the feed. This is strange because the trays were dual-flow trays and operated at a good liquid load (20 gpm/ft^2), so their weep should clearly show but did not. The froth heights were only about 6 in. which again should not be close to flood.

Fix Attempt The feed distributor was a perforated pipe with perforations pointing straight down. In a fix attempt, the licensor replaced the distributor by one that had holes pointing down and slots pointing up. A round baffle was placed above the slots to redirect the vapor down so that it does not impinge on the tray above. This feed distributor was modified probably to induce segregation of vapor from liquid. There was no improvement when the tower returned to service after distributor modification.

Root Cause Based on the entire hole area and the mixture density, the two-phase flow velocity from the holes and slots was 18 ft/s. Directed downward, it is quite conceivable that both the vapor and liquid descended a good distance due to the high velocity. The vapor eventually turned around and went back up. This recycle overloaded the lower trays and reduced their efficiency.

The fix attempt achieved little improvement because at high fluid velocities, cutting slots on top of a pipe achieves little vapor–liquid segregation. The high-velocity two-phase feed issuing from the slots was then directed downward by the round baffle, ending in the region below the feed, just like it did prior to the modification.

Cure The trays were replaced by another vendor's high-capacity trays, both top and bottom. The feed distributor was also replaced by a new, well-designed distributor. The new trays achieved the separation and the required capacity with no further problems.

CASE STUDY 5.5 A GOOD TURN ELIMINATES HYDRAULIC HAMMER

Contributed by Mark E. Harrison, Eastman Chemical, Kingsport, Tennessee

Problem A water-hammer-type pounding at the column feed point was violently shaking the column and the connecting piping. The column was operating at about only 30% design rate.

Troubleshooting The location of the noise suggested a problem with the feed pipe. A check of the design drawings indicated that the feed pipe and feed sparger were somewhat oversized, especially at the 30% feed rate. The sparger discharge-orifice velocity was calculated to be less than 1 ft/s. The feed was subcooled and far from its bubble point, so flashing in the sparger could be ruled out. One postulation was that feed liquid was running out of the upstream orifices, allowing vapor to enter the feed sparger through open downstream orifices, and that the condensation of this vapor in the feed sparger was causing a hydraulic hammering.

Corrective Action One solution might have been to plug some of the orifices to raise the discharge velocity to several feet per second. However, to keep velocities below 6 ft/s at design feed rates, the following remedy was implemented: the feed pipe was turned so that the discharge orifices were on top of the pipe; this ensured that the sparger remained full of liquid at low feed rates; additionally, a deflector bar was installed above the orifices to keep feed from impinging on the tray above.

Outcome The hydraulic hammer was eliminated.

CASE STUDY 5.6 DISTRIBUTION KEY TO GOOD SHED DECK HEAT TRANSFER

Henry Z. Kister and Samuel Schwartz, reference 262

Installation Three intercooler quench towers in the cracked gas compressor train of an olefins plant (Fig. 5.4*a*). In each tower, compressor discharge gas was cooled by direct contact with cold circulating water. Each quench tower contained 12 rows of shed decks, 9 sheds per row, 700 mm apart, with each row rotated 90° to the one below.

Problem The gas was cooled to only 6–12°C above the entering water. This temperature approach is poor considering the large number of shed decks. This gave hot gas outlet temperatures in summer.

History For many years, the problem increased plant energy consumption but did not restrict production. Following plant debottleneck to high throughputs, the hot gas temperatures began to bottleneck the compressor capacity in summer, which in turn restricted ethylene production.

Inspection during a brief shutdown showed plugging of orifices in the pipe distributors spreading liquid to the shed decks. The holes were enlarged from the original 10 mm to about 12–13 mm. This eliminated the plugging, as observed in the next turnaround inspection, but did not change the temperature approaches and gas outlet temperatures.

Consideration was given to replacing the shed decks by packings. This was rejected due to concerns about the much greater sensitivity of packings to maldistribution and fouling than that of shed decks. In this system, separation of two liquid phases and the possibility of flashing upon pressure letdown were special challenges to gravity distributors, while plugging and polymerization were already experienced. There were more concerns, and these were detailed elsewhere (262). Many of these concerns could have been alleviated, but at a high cost, and the reliability issues would not have fully gone away. The approach elected was to improve the shed decks.

Liquid Distributor Review Figure 5.4*b* shows the double pipe distributor supplying liquid to the shed decks in each tower. About half the liquid was dumped along the tower east–west centerline. The other half was dumped along two narrow east–west chords, halfway between the east–west centerline and the north or south

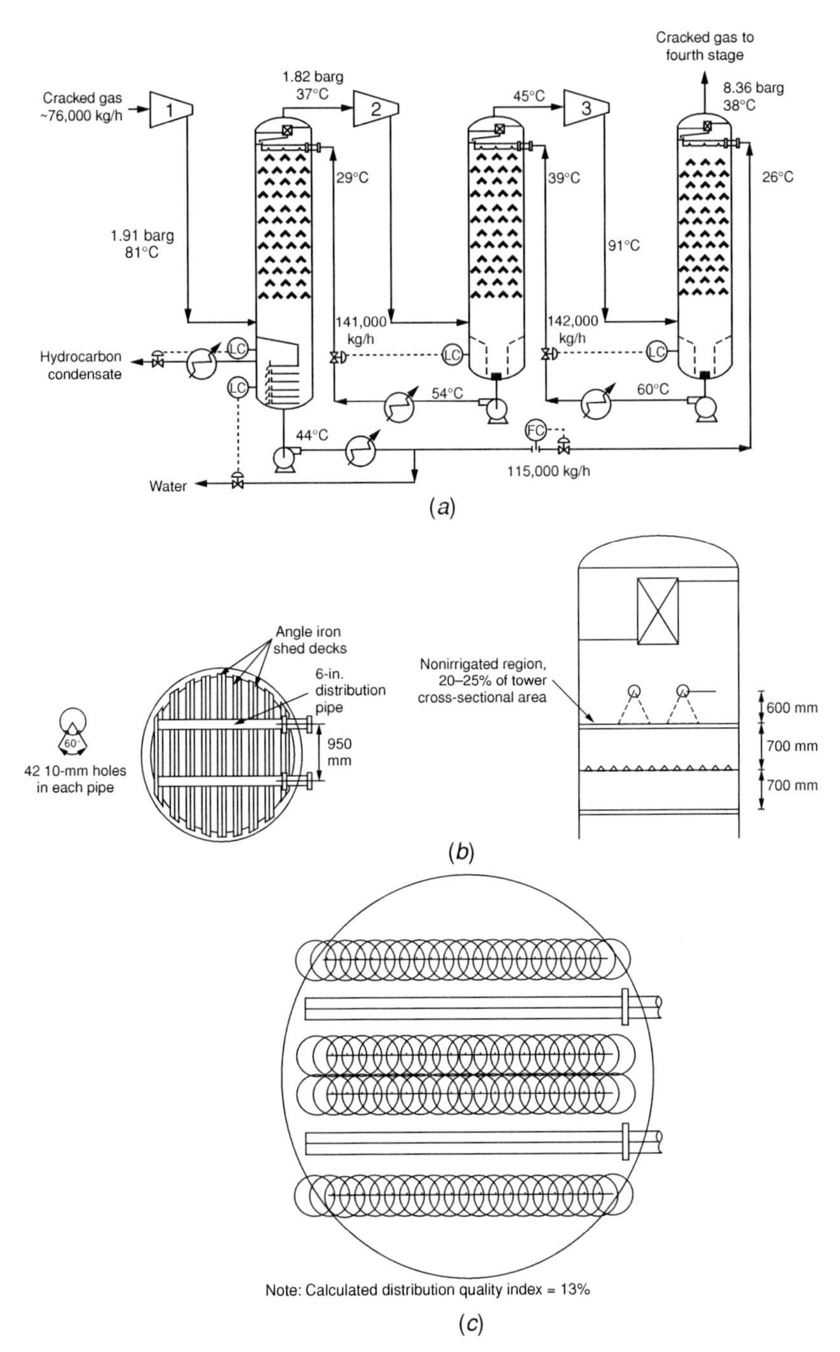

Figure 5.4 Liquid distribution problem in three intercooler quench towers: (*a*) process schematic, showing plant operating data (winter); (*b*) double-pipe distributor introducing liquid to shed decks; (*c*) circle analysis of irrigation quality using method of Moore and Rukovena (353). (From Ref. 262. Reprinted courtesy of Penn Well Corp.)

end. At the same time, the north and south ends as well as the regions under the pipes remained unirrigated.

Figure 5.4*c* is a distributor irrigation quality analysis using the method of Moore and Rukovena (353), which is described in Kister's book (250) and has been strongly recommended (250, 382) for distributor evaluation. It clearly shows the dry areas at the north and south ends and under the feed pipes. Based on this method, we calculated a distributor quality index of 13%, which is extremely poor. An index of 90% is typical for high-quality distributors, 75–90% for intermediate-quality distributors.

Other Deficiencies Inlet gas velocities were about 30–40% higher than those recommended (353, 473) for packed towers. Shed decks are more tolerant to gas maldistribution, but there was room for improvement.

The shed decks used were 100 mm wide with 140-mm gaps between the sheds. These were far too large to give good liquid curtains. Standard angle irons with 50-mm-wide sides and 50-mm gaps give much better liquid curtains.

Modifications In each tower, the liquid distribution pipes were replaced by a well-designed spray distributor. To prevent spray nozzles plugging, two full-size filters with no bypass were installed in each quench circuit, in accordance with recommended practice (250). At each tower feed, a simple gas distribution deflector was added, shown in detail elsewhere (262). The large shed decks were replaced by the smaller decks described above. The cost of all these modifications was minor.

Commissioning Prior to start-up, the system was tested by filling the bottom sump of the first-stage quench tower with water and using the normal pumps to circulate water from tower to tower in the normal circulation route (Fig. 5.4*a*). The manholes were kept open, and spray action was watched. The test revealed several leaking flanges, and these were repaired. Two nozzles produced nonhomogenous sprays, and those were replaced by two new nozzles. The tests ascertained that the new spray distributors were operating as intended.

When the quench towers were started up, the filters repeatedly plugged by coke and polymer that spalled off from the piping. The filters needed cleaning every 2–3 at the beginning. After a few days, the problem went away. Presently, under normal operation, the filters need cleaning once every 3 months. In the event of a compressor surge, the filters require more frequent cleaning.

Results The approaches between the gas temperatures leaving the quench towers and the quench water entering the quench towers were halved in the first and second towers and declined by over 10°C in the third. The new approaches ranged from 1.5 to 4°C, compared to 6–12°C previously. While the other modifications helped, it is our evaluation that eliminating the poor liquid distribution was by far the major factor behind the improvement.

Chapter 6

Packed-Tower Liquid Distributors: Number 6 On The Top 10 Malfunctions

After the tower base, the liquid distributor is the second most troublesome internal in a distillation tower (255). The number of liquid distributor malfunctions reported in the last decade is almost double that in the preceding four decades, probably because of the wide use of packed towers in the industry over the last few decades.

More distributor malfunctions have been reported in chemical towers than in refinery and olefins/gas towers, probably due to the comparatively wider application of packings in chemicals. For chemical towers alone (excluding refinery and olefins/gas plant towers), liquid distributor malfunctions outnumber any of the other malfunctions, including plugging, tower base, internal damage, and abnormal operation. Liquid distributors are the top malfunction in chemical towers.

The two major liquid distributor issues (255) have been plugging and overflow. Good filtration and use of fouling-resistant distributors were successful cures. While plugging is a common cause of overflow, only a few of the reported cases of overflow were due to plugging. Distributor overloading by excessive liquid loads, insufficient orifice area, and hydraulic problems with the feed entry into a distributor caused the rest of the overflow cases.

The next major issues (255) have been poor irrigation quality, fabrication/installation mishaps, and feed entry malfunctions. It is surprising that poor irrigation quality accounts for only 20% of liquid distributor malfunctions. The literature on liquid distributors has focused on optimizing irrigation quality, yet other more troublesome items, such as plugging and overflow prevention, received little attention. Further, the number of irrigation quality malfunctions reported is on the decline, suggesting the industry has learned to produce good irrigation, at least in most cases. On the other hand, cases of distributor overflow, fabrication and installation mishaps, and feed entry problems are sharply on the rise, so the industry should focus on improving these.

Feed entry malfunctions have been particularly troublesome. Excessive velocities, splashing, and poor orientation of the feed pipes caused poor performance of entire towers. The problems escalate when handling a flashing feed. It is amazing how many times a flashing feed enters the tower via a liquid distributor. This type of entry has been a fertile source of case studies.

Poor hole pattern and distributor damage come next (255). Damage is the only distributor issue for which the number of refinery malfunctions outnumbered the chemicals. Some issues such as irrigation quality and hole pattern appear infrequently in refinery towers.

Redistribution, out-of-levelness, insufficient mixing, and interference with hold-downs constitute the remaining issues (255). Distributor out-of-levelness, which is frequently suspected when a tower malperforms, is one of the minor issues. Finally, insufficient mixing is a size-related issue that is seldom troublesome in smaller towers (<5 m ID) but rises in significance with larger diameter.

Lessons learned from the cases outlined in this book strongly support recommendations by Olsson (382) for minimizing distributor malfunctions. Olsson advocates critically examining the fouling potential and absence of vaporization in streams entering the distributor, testing distributors by running water through them at the design rates, either in the shop or in situ, and, finally, ensuring adequate process inspection. A review of the cases presented in the literature and in this book suggests that Olsson's measures would have prevented more than 80–90% of the reported liquid distributor malfunctions.

CASE STUDY 6.1 MALDISTRIBUTION CAN ORIGINATE FROM A MULTITUDE OF SOURCES

Contributed by Ron F. Olsson, Celanese Corp.

Installation A 6.5-ft-ID stripper with 35 sieve trays on 18 in. tray spacing in a foaming service. The tower produced a bottom stream with 600 ppm impurities. It was desired to minimize impurities in the bottom and to raise capacity.

Operating History Tower capacity was limited by downcomer backup flooding induced by foaming. The tower could not operate without costly antifoam addition. To increase tower capacity and improve staging, the trays were replaced by two packed beds, 22 and 28 ft tall, of 1.25-in. rings. A vapor distributor was installed beneath the bottom bed. Once the tower was returned to service, capacity was higher and the tower operated stably for the first 10 days without antifoam injection. However, the impurity in the bottom increased to 1000 ppm and varied widely. Calculations showed packing HETP of 6 ft versus 2–2.5 ft design.

Troubleshooting I Using contact pyrometers, temperatures were measured around the tower periphery at several elevations. On the north and west quadrants, the temperatures were steady and consistent with the predicted temperature profile.

In the south, temperatures were steady but 10–20°F lower than expected. In the east, temperature varied plus or minus 50°F in cycles.

Hydraulic checks of tower internals showed that the pressure drop across the vapor distributor was far too low (about 0.25 in. of liquid) to give effective vapor distribution.

The tower was shut down and inspected. The redistributor and liquid distributor were found mechanically sound and level within 0.1 in. along the entire tower diameter. No evidence was found for plugging in either the distributor or the packing, and there were no crushed packings.

A design flaw was detected at the reboiler return entry. There were two equal reboiler return inlets, each equipped with a V-baffle (Fig. 6.1*a*), closed at the top and bottom and open on the sides. The V-baffles were separated by 8 in. Assuming each V-baffle splits the reboiler return equally, about half the reboiler return vapor would issue into the restricted space between the baffles on the east side (Fig. 6.1*a*) of the tower, causing a vapor jet to rise up the tower in the east. This was the side of the tower where the temperatures cycled.

To improve vapor distribution, the following modifications were performed:

- The V-baffles were modified by installing a vertical partition plate between them and a plate over the top (Fig. 6.1*b*) to keep the two reboiler return streams from impinging on each other and jetting up the east side of the tower.
- Smaller orifices were installed over the vapor chimneys to raise pressure drop from 0.25 to 4 in. of liquid.

Following these modifications, the irregularity in the temperature profile was eliminated. A temperature survey after start-up showed each quadrant to be operating at the predicted temperatures. However, no improvement in separation was observed other than reduced variability of the bottom composition.

Troubleshooting II Digging into the history of packing the tower, it was found that the packing was loaded into the tower by directly dropping from the tower manholes, apparently at the recommendation of the vendor's representative. This technique constitutes poor practice, goes against recommendations in the literature (50, 250), and leads to hill formation (Fig. 6.2*a*). It was decided to remove the packings and repack the tower using the recommended chute-and-sock method (Fig. 6.2*b*). To ensure better evenness, after every 2ft of packing loading a person layered the packing with a rake. The person stood on a sheet of plywood to avoid crushing the packing. When the tower was repacked, 7% (or 110 ft^3) more packings were required, indicating that there were significant void spaces within the original bed which contributed to maldistribution.

An in situ water test was conducted in which water was fed into the top distributor and sampled at the bottom of each bed. The test showed gross maldistribution in both beds prior to repacking. Leaks were detected between the distributor pan and the support plate and were eliminated by improving gasketing. A major improvement in liquid distribution was observed in a second in situ water test following the repack.

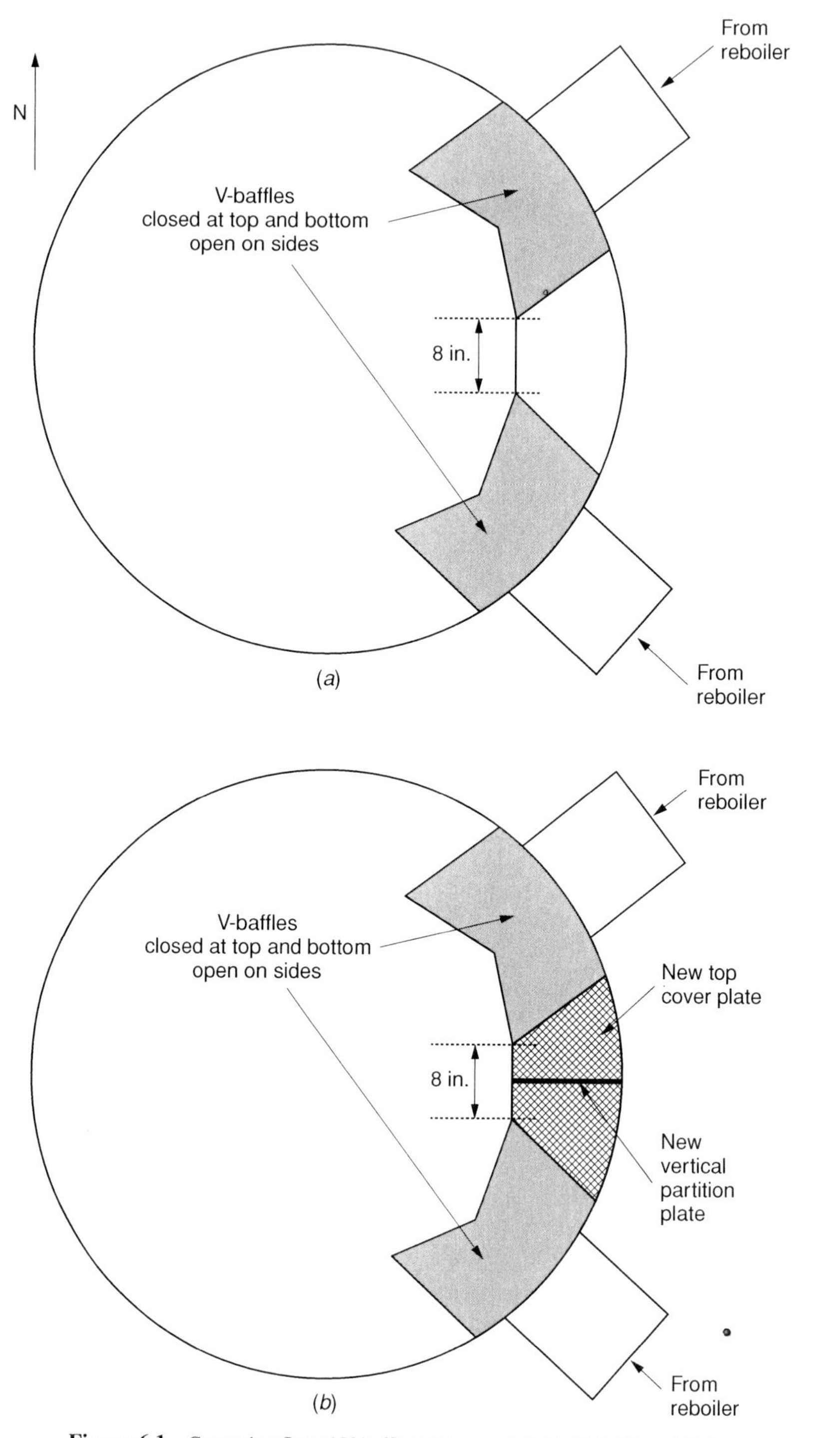

Figure 6.1 Correcting flawed V-baffle arrangement: (*a*) initial; (*b*) modified.

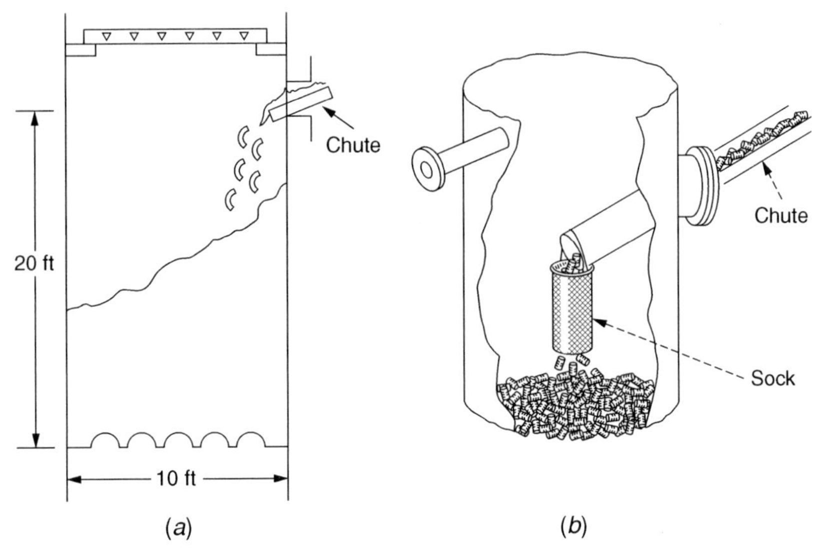

Figure 6.2 Random packing installation techniques: (*a*) poor, promoting hill formation; (*b*) good, chute-and-sock method. (*a*) Reprinted with permission from Ref. 250. Copyright © 1990 by McGraw Hill. (*b*) From Ref. 82. Reprinted courtesy of *Chemical Engineering*.

Once the tower was returned to service, the HETP decreased from 6 to 4 ft. The repacking gave a major improvement to the separation in the tower. The bottom impurity dropped from 1000 to 600 ppm.

Troubleshooting III The improved separation was very short-lived. Over the first week in operation, the good separation gradually deteriorated, then leveled off at a high bottom impurity. Gamma scans revealed massive liquid maldistribution. The scans also showed liquid level on the top orifice pan distributor that was 2 ft high. The vapor risers were only 1 ft tall, and the calculated liquid height was 6 in.

The possibility of solids in the feed was explored. During the design, Operations personnel was consulted and reported no solids in the feed. Samples were drawn and showed coloration (Fig. 6.3) but no solid deposits. On that basis, the design did not provide filtration on the feed. The turnaround inspection also showed no evidence for solid buildup.

Figure 6.3 shows that the coloration completely disappeared upon feed filtration, conclusively proving that the coloration was due to solids. Allowing the samples to settle over a period of a week or two (this test was performed during the troubleshooting but not during the design) also showed a layer of solids in the bottom of the sample jar. When Operations personnel simply report "no solids" in the feed, their concise statement should be interpreted as "we have not seen solids in the feed, nor have we experienced fouling issues with our existing (at that time, trayed) internals." It is believed that at the turnaround the solids disappeared due to the shutdown wash of the tower prior to personnel entry.

Figure 6.3 Feed samples before and after filtration.

The solids problem was eliminated by the addition of a feed filter. Figure 6.3 shows the difference in feed appearance before and after filter addition. Following the filter addition, the improved separation (an HETP of 4 ft and a bottom impurity level of 600 ppm) was consistently achieved.

Troubleshooting IV Even though efficiency and separation were better, they still fell short of design. Gamma scans showed that liquid maldistribution was much better but still not good.

The orifice pan distributor and redistributor were evaluated using the Moore and Rukovena method (353, described also in Ref. 250). The analysis gave a distribution quality index (DQI) of 45% (Fig. 6.4*a*), compared to over 90% for high-quality

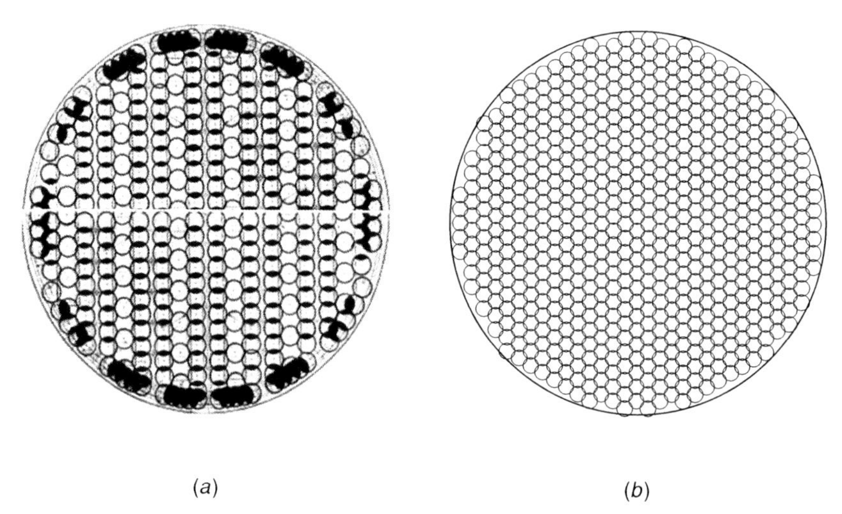

Figure 6.4 Distributor evaluation using Moore and Rukovena method (353): (*a*) old distributor, 45% DOI, poor; (*b*) new distributor, 90% DQI, good.

distributors and 75–90% for intermediate-quality distributors. About 25% of the tower cross-sectional area was found unirrigated, while another 17% of the tower cross-sectional area was overirrigated. The distributors' hole patterns were very irregular. On this basis, it was decided to replace the distributors by high-performance distributors designed for a DQI exceeding 90%.

Two other modifications were implemented. The bottom bed was 28 ft tall, which is well in excess of the criterion recommended in the literature of limiting bed height to 20 ft maximum (82, 250, 304), particularly with small packings. It was decided to split it into two beds with a redistributor between. Chimney trays were used to collect and mix the liquid before directing it into the redistributors.

Following these modifications, bottom impurity was lowered to 100–200 ppm.

Lessons Learned

1. Always double check the design of distributors and internals. Perform Moore and Rukovena distributor quality analysis on the vendor drawings as part of this check.
2. Specify high-quality (90% or better) distributors. They are higher priced but pay for themselves in reduced bed heights or better than expected staging which reduces utilities.
3. Flow test the distributors with water at the vendor shop and, if possible, also in situ.
4. For pan distributors, gasket the area between the pan and the support plate.
5. Check for vapor distribution requirements.
6. Limit bed depths to 20 ft maximum.
7. Load random packing using the chute-and-sock method, layering it every 2 ft with a rake.

CASE STUDY 6.2 IMPROVED DISTRIBUTION AND PUMPAROUNDS CUT EMISSIONS

Contributed by Ron F. Olsson, Celanese Corp.

Installation Process vents containing about 0.5% organics were water scrubbed. About 80–90% of the organics was in the form of dust. The tower was 13.5 ft ID and contained three 10-ft-tall beds of 2- or 3-in. plastic saddles (Fig. 6.5). Liquid from the bottom of each bed was recirculated to the top of the bed, so each bed was a separate PA system. Liquid was distributed to the beds via V-notch distributors. All distributors were handling similar liquid flow rates and were of the same design.

Problem The goal was to reduce organic emissions from the tower. Typically, there was 200–300 ppm in the vent gas. Calculations showed that the tower was achieving only 1.5 theoretical stages, where at least three stages should have been achieved, one

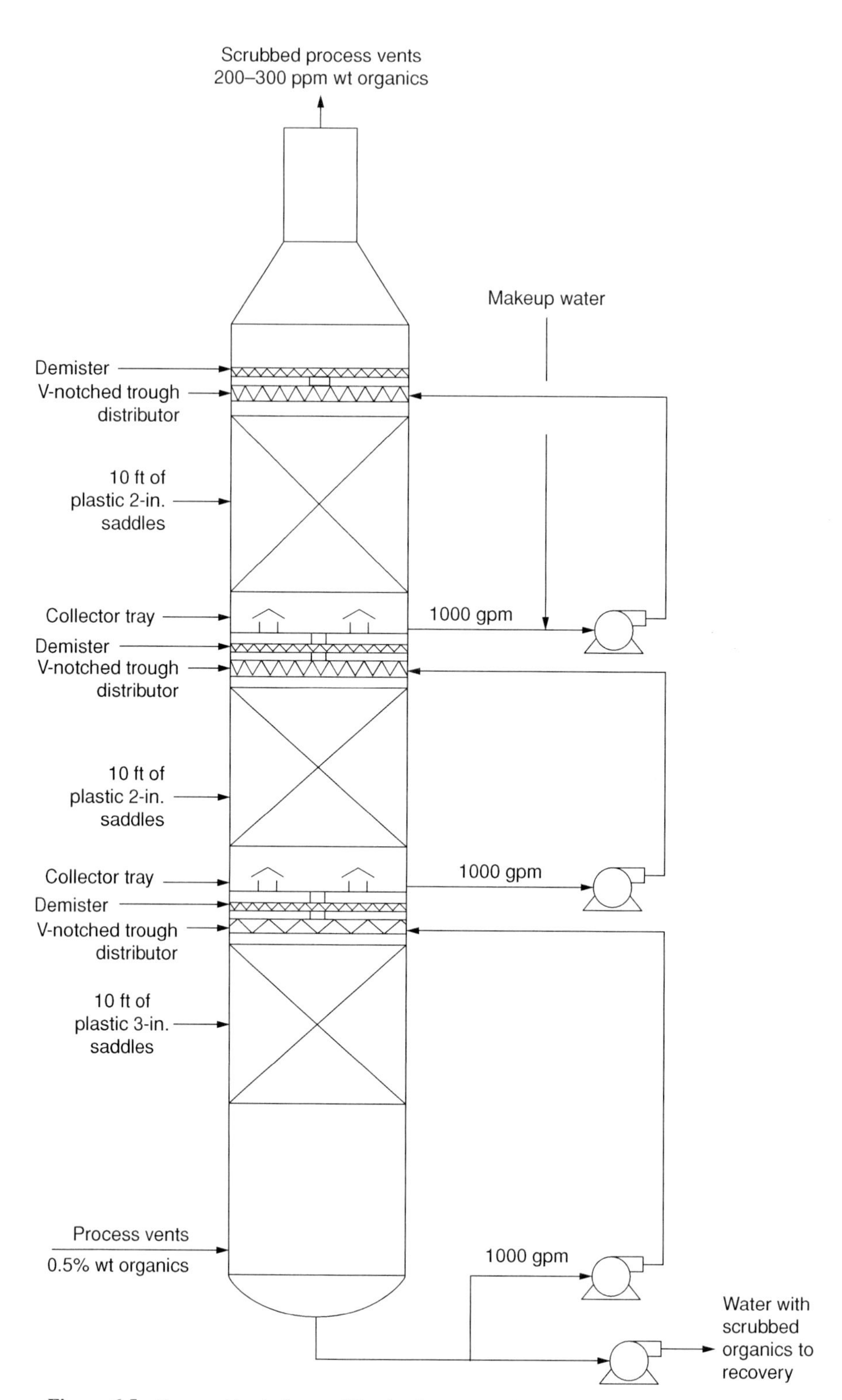

Figure 6.5 Vent scrubber before modification (supports and bed limiters for each bed not shown).

for each pumparound. Test data showed that scrubbing efficiency was particularly poor in the middle bed.

Liquid Distribution Evaluation Distribution quality was analyzed using the method of Moore and Rukovena (353, described in detail in Ref. 250). The distribution quality index (DQI) calculated for all three distributors was less than 50%, compared to over 90% for high-quality distribution and 75–90% for intermediate-quality distribution. The distributors provided only two to three distribution points per square foot of bed cross section, which is low compared to the recommended minimum of four to six distribution points per square foot (82, 250, 473). In addition, the distributor V-notches were at the low end of their operating range, which made them extremely sensitive to levelness.

This analysis confirmed that there was much room for improving liquid distribution. Modifications were planned (below) for improving liquid distribution.

Internals Inspection When the tower was shut down to implement the modifications below, some additional defects were observed. The demister above the bottom bed had a large hole in it, about 2 ft in diameter. The rest of the demister was full of dust. It is likely that much of the vapor channeled through the hole, bypassing the plugged section. This vapor channeling is the likely cause of the very poor scrubbing efficiency observed in the middle bed, which would have been most severely affected. This channeling would have also propagated, at least to some extent, to the other two beds, contributing to their low efficiencies as well.

The inspection found that the packing depth of the bottom bed was 3 ft below the bed limiter. This contributes to maldistribution and reduced efficiencies. Finally, water marks showed that distributor troughs were dry at one end of the tower, indicating some out-of-levelness.

Modifications Modifications were implemented to improve distribution and internals reliability and maximize scrubbing capability. The following changes were implemented:

- The 2- and 3-in. plastic saddles were replaced by 1-in. modern metal packing. The smaller size doubles the surface area, the metal wets better than plastic in this aqueous service, and the smaller size also gives higher pressure drop that improves vapor distribution.
- The low-quality liquid distributors were replaced by fouling-resistant high-quality distributors. The new distributors provided 10 drip points per square foot of tower cross-sectional area and were evaluated to give a DQI exceeding 90%. The distributors had a compartmentalized design, so they were insensitive to levelness.
- Intermediate demisters were eliminated. Plugging and damage of these were sources of vapor maldistribution that affected the entire tower. A new demister was installed at the top of the scrubber.

- The tower was reconfigured to give five PA loops instead of the previous three. This means shorter beds (only slightly, see below). This added two equilibrium stages for improved scrubbing.
- Tower internals were rearranged to maximize packed bed depth. This took advantage of tower height previously underutilized as well as that occupied by the demisters. As a result, each of the lower four beds were 8 ft tall, and the top bed was 9.5 ft tall, increasing the total packed height from 30 to 41.5 ft.

Results The organics contents of the scrubbed gas went down by factor of 100 and is now well below 5 ppm wt.

Prior to the changes, the water in all PAs looked milky, due to the presence of suspended organic solids. Following the changes, water was milky only in the bottom PA. The water in the upper PA was clear.

The improved separation made it possible to cut back on water purge rates, which in turn reduced water makeup and feed rates to the recovery unit.

CASE STUDY 6.3 KEEPING SOLIDS OUT OF PACKING DISTRIBUTORS

Contributed by Pamela Tokerud, Koch-Glitsch LP, Wichita, Kansas

Problem A packed distillation tower was demonstrating a 25% efficiency shortfall. Post start-up test runs confirmed that the column was not meeting the process design specifications. Gamma scans highlighted that distribution was nonideal.

Troubleshooting Theories focused on both liquid and vapor maldistribution. Detailed review of the liquid and vapor distributors and the feed arrangement showed their design was adequate. The liquid distributors consisted of a parting box that collected the entering liquid (in the case of the feed, together with liquid collected from the bed above) and metered it to laterals (troughs) below. Liquid issuing from orifices in the laterals irrigated the packing below. Distributor flow tests performed prior to shipment revealed no significant maldistribution.

Observations through the upper sight glass showed that liquid from the main header of the reflux distributor issued from the header orifices with a horizontal momentum in the direction of the reflux flow in the header. It was decided to equip each orifice on the reflux header with a short, vertical pipe to eliminate the horizontal momentum.

In addition, the laterals in both the feed and reflux distributors were operating at a higher liquid head (as observed through the sight glasses) than designed and demonstrated in the flow test. Plans for the next shutdown included an on-site flow test of each distributor to measure flows and heads, and to sample individual drip point flows.

Tower feed, originating in a storage tank, was filtered by two full-size filters in parallel, one on-line, the other off-line. Each filter had two filtration stages in series.

Each stage contained a basket with an internal filtering screen. The second-stage filter was designed to remove finer particles. In addition, spare filter baskets for each stage were available and utilized during the trade-outs and cleaning process. The filters were reported to be in place. Field measurements indicated minimal pressure drop across the filters.

Inspection Upon entrance into the column, distributor fouling was observed. Field flow testing found both the feed distributors and the stripping section redistributors to have a minimum of 10% pluggage of the orifices.

Solution The distributors were cleaned. Modifications were made to the reflux distributor pipe. In addition, modifications were made to the vapor distributor and collectors; however, from later experiences elsewhere, this was shown to have had minor contribution to the overall performance improvement.

After all modifications were completed and during review of the start-up procedures, it was learned that the screens had been cut out of the filters baskets, although the baskets were in place, during the original start-up. The screens were cut out to avoid replacing the filter baskets several times each hour. The importance of filtration was then recognized, and the procedures were followed in this and subsequent start-ups. Frequent changes of the filtration baskets was required and tolerated until the system was clean (about 1 day). Continual proper use of the filters has resulted in the column achieving the process specifications.

Morals

- In a troubleshooting investigation, it is essential to have all the facts to ascertain a correct diagnosis.
- Parallel full-size external filters are essential on all solid-containing lines entering packed towers (reflux and feed) to prevent fouling of the distributors and packing.
- It is important to thoroughly flush all lines and auxiliary equipment from high to low points before connection to the column to avoid introduction of corrosion products, construction debris, dirt, and other materials upon start-up.
- Differences between the use of parallel external and internal filters can mean the difference between operating or not.

CASE STUDY 6.4 PLUGGED DISTRIBUTORS

Tower A Liquid to a 30-in. ID scrubbing tower with random packings contained some metal catalyst carried over from an upstream reactor. The liquid was filtered, but the filter openings were not much smaller than the $^1/_4$-in. orifices of the liquid distributor. The distributor plugged up, giving an extremely poor tower performance.

The pipe distributor was replaced by a single spray nozzle. The nozzle was placed a good distance (around 2 ft) above the packings. Before start-up, the nozzle was water tested in the tower to ensure that it gave a homogenous spray and good liquid irrigation. The tower worked well since.

Tower B A trough distributor at the top of a refinery tower plugged with corrosion products and salts that formed on the upper head of the tower and, and when agglomerated, spalled off and fell into the troughs. To prevent recurrence, trough covers were added. The tower worked well since.

Tower C The HETPs were twice the expected due to plugging of $\frac{5}{32}$-in. holes in a distributor irrigating structured packings in aromatic isomer separation. The plugging was caused by a small quantity of rust, which plugged more than half the distributor holes. System cleanup and distributor modifications prevented recurrence.

Towers D and E In two different cases, construction debris left in the line was carried by the process fluid into a packing spray distributor and plugged the header and sprays. The debris included gloves, rags, and cardboard. In one of these, the spray plugging led to coking of the wash section in a refinery vacuum tower.

CASE STUDY 6.5 DISTRIBUTOR OVERFLOWS

Liquid overflow is one of the prime causes of distributor failures (255). Figures 6.6*a* and *b* are photographs of distributor overflow taken in an in situ water test. During operation, vapor passing through the risers entrained the overflowing liquid onto the bed above, initiating premature flood. Despite experiences like this, many engineers do not recognize the detrimental effect of distributor overflows. This case study is dedicated to those engineers.

Service Direct-contact cooling and condensation of HC gases. Cooled water was pumped to the top of the tower, where it irrigated a bed of random packings using a trough distributor. The gas was cooled from 140 to 100°F.

Experience Two modern random packings have much the same surface area per unit volume and are often considered equivalent in mass transfer service. Yet heat transfer data that we collected from more than half a dozen installations of each in this heat transfer service showed that packing A gave about 20–30% higher volumetric heat transfer coefficient than packing B.

Debottleneck In debottlenecking one large tower, cost considerations overrode the heat transfer benefits. The quote from the B vendor was $30,000 cheaper, and the bed was tall enough to achieve the design approach of 5°F with the lower heat transfer coefficient. The B packing was therefore selected.

Distributor Test Following fabrication, the liquid distributor was water tested at the vendor's shop. In this service, liquid distributors are quite standard, and most

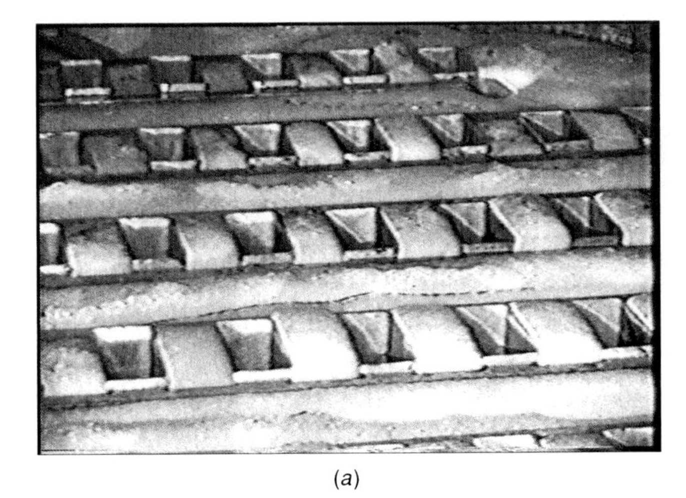

(a)

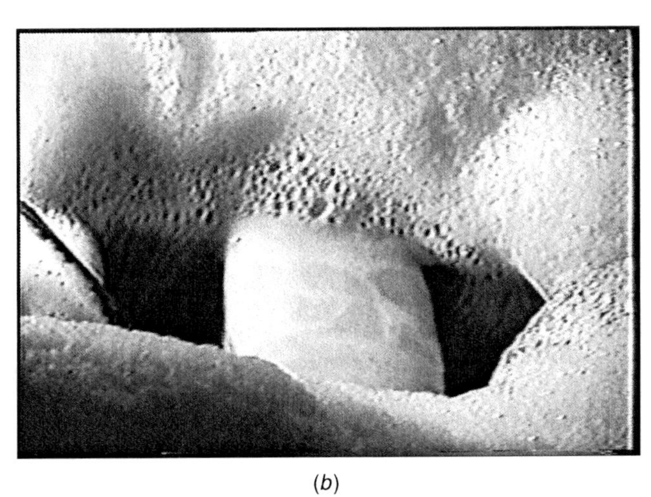

(b)

Figure 6.6 Packing distributor overflow as photographed during in situ water test: (*a*) froth/liquid reaching top of distributor vapor risers; (*b*) liquid pouring into vapor risers. During operation, vapor flowed through risers and entrained overflowing liquid.

users (we were told) elect to skip the test to save money and advance schedules. We decided to proceed with the water test.

When water was circulated at the design flow rate, the water level was flush with the top of the troughs. This means that in service, any aeration, or pressure drop, or liquid HC entry, or waves, or turbulence, will cause liquid overflow into the vapor passages. This overflow will cause maldistribution at best. At worst, the rising gas will carry it over into the overhead line.

The easiest solution was to extend the troughs by about 4 in. There was unanimous agreement between the client, the vendor's technical staff, and us that this modification was necessary. The next day, the vendor's project manager became involved. He objected, "The distributor will work as is. It is designed the way we always design our distributors, and we are guaranteeing the performance of the distributor and packing. We see no need to change. If the client wishes to proceed with the taller troughs, the cost is $50,000."

The debottlenecking team decided to pay the price and proceed with the change. At about the same time, a similar distributor from the A vendor was water tested for another project and passed with only very minor changes. The cheaper quote cost more in this case.

Performance The B packing with the modified distributor was put into service. The measured temperature approach between the overhead gas and coolant was 2°F, which was half the design and the lowest we have seen for any tower containing the B packing. The heat transfer coefficient was the highest we have seen for any tower containing the B packing and of the same order as that produced by the A packing.

Epilogue If the project manager's statement is precise, it provides the explanation to why the A packing outperformed the B packing in this service. Distributor overflow is death to distribution.

CASE STUDY 6.6 A HATLESS VAPOR RISER PREVENTS PROPER SCRUBBING

Contributed by Mark E. Harrison, Eastman Chemical, Kingsport, Tennessee

Installation A 2-ft-diameter scrubber was designed to remove acetic acid from a process off-gas stream. Water, the scrubbing fluid, was fed to the top of a single packed bed.

Problem Excessive acetic acid emissions were causing unacceptable losses and odors.

Troubleshooting Varying the water flow to the scrubber between 20 and 200% of design did little to improve scrubbing efficiency. There were no indications of flooding. A higher scrubbing efficiency was expected, based on packing heights and water rates of similar columns. Consequently, poor liquid distribution was suspected.

Cause Separate review of column specifications and internal drawings did not suggest any problems. Figure 6.7*a* is a simplified sketch of the feed pipe entry showing liquid discharged onto the center of the distributor. Figure 6.7*b* is a simplified sketch of the distributor plan featuring a center, hatless chimney. By itself, each sketch shows a sound design. Combining the two (Fig. 6.7*c*) readily shows the design oversight.

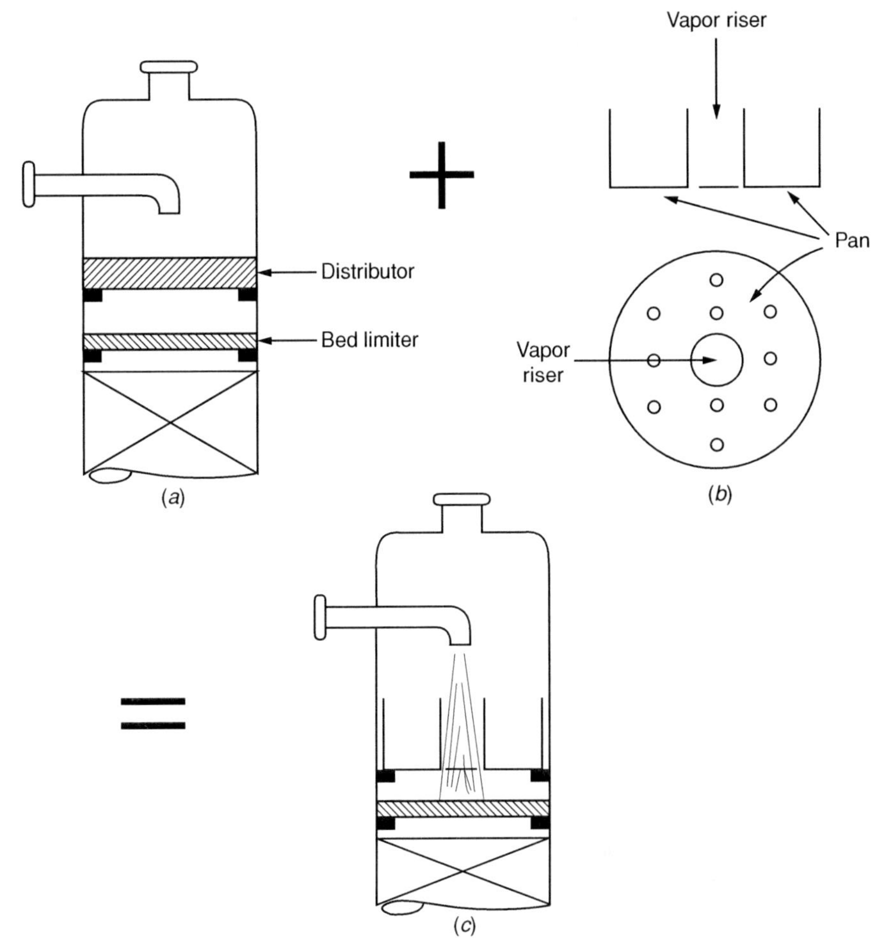

Figure 6.7 Acetic acid scrubber that experienced problems: (*a*) feed pipe entry, OK; (*b*) distributor plan, OK; (*c*) combining (*a*) and (*b*), not OK.

The scrubbing liquid was discharged down the chimney and bypassed the distributor. The inspection confirmed this conclusion.

Solution Calculations showed that the annular vapor space around the distributor was capable of handling the vapor flow. Therefore, the riser was blanked off. Of course, alternative solutions included relocating the feed pipe or the riser, or putting a hat on the riser.

Outcome Started up again, the scrubber removed essentially all the acetic acid with the design water rate.

CASE STUDY 6.7 FEED PIPES NEED PROPER CHANGES WHEN REPLACING TRAYS BY PACKINGS

Installation An amine absorber removing hydrogen sulfide from a HC gas stream.

Problem Absorption was poor following replacement of single-pass trays by packing.

Cause Gamma scans showed very poor distribution, initiating at the liquid distributor and persisting throughout the bed. Figure 6.8, a photograph of the lean-amine feed pipe entry, shows the major cause. When the tower had trays, the feed pipe stretched across the tower, discharging liquid to the tray inlet seal area. When the trays were replaced by packing, the top tray was replaced by the orifice pan distributor shown in Figure 6.8. The feed pipe was not modified. So instead of discharging the lean amine onto the tray seal area, it discharged it above the hatless vapor risers, directing a significant portion into a vapor riser.

Installation The photograph shows that during installation, the baffle on the feed pipe apparently interfered with a distributor riser. The installers addressed this problem by simply bending the riser, which caused some deformation to the tray floor.

Solution Calculations showed that well-designed trays would achieve the same capacity and separation that would have been achieved had the packing operated properly. Besides the maldistribution, the packing experienced fouling and compression. The added reliability and robustness of trays in this application made it attractive to replace the packing by well-designed trays. The tower then achieved its separation and capacity objectives with no further problems.

Moral A proper process inspection prior to start-up could have flagged this flaw.

Figure 6.8 Lean-amine feed pipe discharging liquid into distributor riser.

CASE STUDY 6.8 SLUG FLOW IN A DEBUTANIZER FEED PIPE

Contributed by Dave Ferguson, Quest Tru Tec LP, La Porte, Texas

Installation A debutanizer at a gas plant. The column had two packed beds. The feed was mostly vapor, entering onto a gallery distributor between the beds. The column had just been restarted after a major turnaround where trays were removed and packing installed.

Problem The column was unstable and had a high pressure drop. Flooding was suspected.

Investigation The rates at which the instability initiated were well below design. Several gamma scans were performed at various rates. Each scan of the top bed showed that the top 2 ft of the bed was denser than the rest of the top bed. The scans of the feed distributor (located above the bottom bed) showed it to be heavily loaded on some scans and normally loaded on other scans.

The gamma scans of the bottom bed showed it to have a higher than normal density compared to the top bed even though the packing was the same. The height of the bed was where it was expected, ruling out crushed packing. The strange thing about the density of the bottom bed was that it was not consistent. In each scan the density of the bed varied 3–4 lb/ft^3 at various elevations. For example, one scan showed the apparent density 2 ft below the top of the bed to be 6 lb/ft^3 while the density 4 ft below the top of the bed was 10 lb/ft^3. The change in density between the two elevations was smooth. Moreover, a subsequent scan showed that the density 2 ft below the top of the bed was now 9 lb/ft^3 and the density 4 ft below the top of the bed was 6 lb/ft^3. This phenomenon was observed throughout the bottom bed.

Theories The theory about the top bed was that some fouling material had been flushed into the column with the reflux during start-up. This was causing liquid flooding in the top 2 ft of the bed.

The theory about the bottom bed was that for some reason the distributor was overfilling on a cyclical basis. This overflowing of the distributor was resulting in waves of liquid flooding down through the bed. The cause of the overfilling of the distributor was not apparent from the gamma scans.

The gamma scans did suggest something to a consultant plant management had hired. He theorized that the feed pipe was too large and that the feed was not sufficient to keep the liquid phase moving up the pipe. Liquid was therefore collecting in the bottom of the vertical section of the feed pipe until a large slug developed, restricting the vapor flow enough to lift the slug of liquid into the column. When the slug of liquid hit the gallery distributor, it overwhelmed the distributior, poured down the chimneys, and flooded the distributor. The slug of liquid then hit the top of the bed, creating the "wave" action seen in the gamma scans.

To test this theory, a gamma source and detector were positioned across the elbow of the pipe. The intensity of the radiation passing through the pipe was monitored over time. Rapid cycling of the intensity of the radiation indicated that liquid was collecting and then disappearing, confirming the consultant's theory.

Cures The consultant recommended that the diameter of the feed pipe be reduced. The plant shut down the unit, installed a smaller diameter feed pipe, and opened the top manway to look at the top packed bed. The top bed was fouled with rust and other trash. This was cleared and the vessel restarted. The column started normally and operated properly at design rates.

CASE STUDY 6.9 SLUG FLOW IN FEED PIPE

Installation A 3-ft-ID random-packed stripper using a flashing feed.

Problem Tower base level oscillated and stripping was poor.

Troubleshooting A level gage was installed on the feed distributor. It showed cycling of the liquid level in the distributor, with peaks at about 2 ft of liquid, well above the top of the vapor risers, and valleys showing zero level. This confirmed slugging of the tower feed.

Cure The feed pipe was replaced by one with smaller diameter. No more cycling occurred.

CASE STUDY 6.10 COLLECTOR DRIP BYPASSES DISTRIBUTOR

Installation Chemical vacuum tower operating at low liquid rates.

Problem Separation was poor. Plant data gave an HETP of 1.5 m with 1 in. random packings that normally give HETPs lower than 0.5 m.

Investigation Prior to the start-up, the distributors were thoroughly inspected, checked, and leveled. All dimensions were consistent with the drawings, and nothing strange was observed.

In the next turnaround, the internals were reinspected. A collector between two beds was observed to contain drain holes. Instead of running into the distributor, the liquid dripped out of the collector via the drain holes, much of it bypassing the distributor.

Cure Seal welding the drain holes gave a major efficiency improvement.

CASE STUDY 6.11 HOW NOT TO MODIFY A LIQUID DISTRIBUTOR

Contributed by Mark E. Harrison, Eastman Chemical, Kingsport, Tennessee

Problem A single-bed packed column had operated virtually unheeded for several years. An engineer working with the unit for the first time noticed that the column was not achieving a relatively simple separation.

Troubleshooting There was no indication of flooding. The unit was to be shut down soon, so the decision was made to inspect the reflux distributor. This revealed the handiwork of an earlier troubleshooter: Several 4-in.-diameter holes cut into the floor of the distributor. It appears that the previous troubleshooter had decided that the distributor had been limiting vapor flow and had forgotten to install vapor risers in the new vapor flow paths. These holes were allowing all the liquid to bypass the original distribution orifices.

Corrective Action Vapor risers were added to the 4-in. holes. Hydraulic calculations indicated that the original orifice count and size were slightly excessive for the current liquid flow rates (as well as for the lower flow rates that would result from improved column efficiency). Rather than have some of the orifices plugged, it was decided to have short overflow tubes installed above one-third of the orifices; thus, liquid would pass through these tubes at high flow rates but not at low flow rates.

Outcome Yields and purity were increased considerably.

CASE STUDY 6.12 TRACER ANALYSIS LEADS TO A HOLE IN A DISTRIBUTOR

Contributed by Matt Darwood, Tracerco, Billingham, Cleveland, United Kingdom

Problem The pressure drop through the packed section of a large petrochemical column was higher than design. Samples indicated poor mass transfer in this zone of the column.

Investigation Due to the large diameter and high packing density, a standard gamma scan was unable to determine the condition of the packing. This initial scan confirmed that both the bed and the distributor were in their correct positions and there had not been a collapse of the packing. The scan also confirmed that the liquid in the base of the vessel was at the correct level and did not impinge on the packing.

A radioisotope tracer study using a short-half-life tracer in a suitable chemical form was carried out. The tracer was injected into the liquid feed before entering the packed bed. Two rings of detectors were placed on the outside of the vessel about

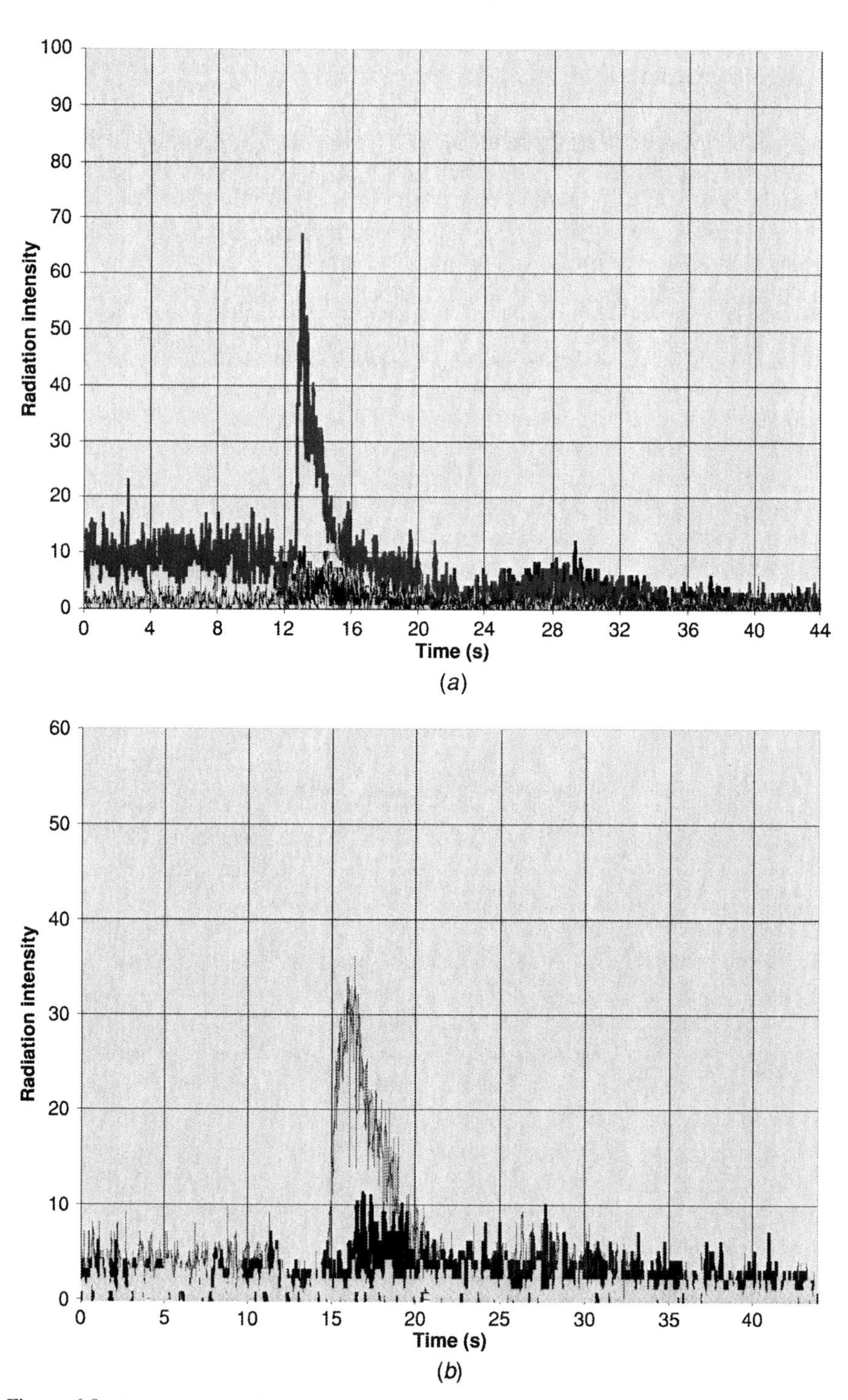

Figure 6.9 Tracer pulse study showing output for four detectors at each elevation. Sharp peaks on north detector only point to preferential flow there: (*a*) top ring of detectors; (*b*) bottom ring of detectors.

2 m apart. Each ring consisted of four detectors, each at 90° from the next. The study (Fig. 6.9) clearly showed a preferential liquid flow down the north side of the column, from the top to the bottom.

Analysis The fact that the preferential flow occurred throughout the entire bed suggested liquid maldistribution before the liquid reached the packing. Since the initial gamma scans had indicated that the distributor was in place, it was suspected that distributor damage could have occurred, allowing preferential liquid flow down the north side of the bed.

Cure A replacement distributor was purchased and the column was opened. The distributor was found to have a large hole in it on the north side, allowing liquid to preferentially flow down that area. The distributor was replaced, the column was restarted, and both the pressure drop and separation in the system achieved expectations.

Moral Using radioisotope tracer studies and gamma scans is an excellent way to diagnose packed-bed problems.

CASE STUDY 6.13 TILTED DISTRIBUTORS GIVE POOR IRRIGATION

Contributed by Mark E. Harrison, Eastman Chemical, Kingsport, Tennessee

Installation A packed column was installed to separate methanol and water from ethylene glycol. The separation is easy, but a control problem was expected because of the sharp temperature profile and the difficulty of controlling such a profile in a packed column having a low liquid holdup. In comparison with trayed columns, the temperature and composition profiles move much faster through packings and therefore become sensitive to small changes in the heat balance.

Problem Soon after the column was commissioned, it was noticed that in stripping all the methanol and water from the ethylene glycol bottoms an excessive amount of ethylene glycol was being taken overhead.

Troubleshooting Increasing the reflux did not seem to improve the separation. Because column control remained steady, it appeared that the column efficiency was much lower than had been expected. A check revealed that the drawings did not indicate how the top orifice pan liquid distributor was to be attached to its support ring. One explanation for the lower efficiency was that an upset had dislocated the distributor. Another was that the distributor's orifices were plugged, causing poor distribution.

With a contact pyrometer having a probe sharp enough to penetrate through the insulation to the vessel wall, the column's radial and vertical temperature profiles were measured. One side of the column was colder than the other side for several feet below the reflux distributor, confirming the hypothesis of liquid maldistribution. A shutdown inspection of the column showed that the reflux distributor was tilted from its support. This caused it to dump all the reflux down one side (the cold side) of the column.

Cure The distributor was securely and evenly clamped to its support ring. Started up, the column achieved the desired separation. A check showed the radial temperatures just below the distributor to be uniform in profile.

Chapter 7

Vapor Maldistribution in Trays and Packings

It is easier to distribute vapor than liquid because vapor spreads much easier. However, vapor-distributing devices used in the industry are far more primitive than liquid distributors. The result is that vapor distribution is not trouble free.

Vapor maldistribution problems are most frequently encountered in packed towers because packing pressure drop is too low to adequately straighten a maldistributed vapor. Tray towers are not immune either, especially when the tray pressure drop is low (often in an attempt to maximize capacity).

The major source of vapor maldistribution has been undersized gas inlet and reboiler return nozzles, leading to the entry of high-velocity jets into the tower. These jets persist through low-pressure-drop devices such as packings. Installing vapor distributors and improving vapor distributor designs, even inlet baffles, have alleviated many of these problems. One situation where inlet baffles and vapor distributors have often failed is in some refinery main fractionators (typically in FCC main fractionators), where the entering vapor is highly reactive and superheated, rapidly coking dry surfaces in its path.

Other vapor maldistribution problems in packings include interference of I-beams, local quenching at the entrance of a cold feed, and overflow in poorly drained vapor-distributing chimney trays. In trays, one vapor maldistribution issue has been the onset of vapor cross-flow channeling, accompanied by a loss of efficiency and capacity, due to excessive hydraulic gradients in low-pressure-drop trays. Another issue has been obstructions causing uneven vapor split into multipass trays.

CASE STUDY 7.1 OVERFLOWING VAPOR DISTRIBUTOR CAUSES PACKING FLOOD

Contributed by Henry Z. Kister, Fluor, Aliso Viejo, California, and Norman P. Lieberman, Process Improvement Engineering, Metairie, Louisiana

Distillation Troubleshooting. By Henry Z. Kister

Installation A C_3/C_4 splitter in a refinery FCC unit operating at 250 psig top pressure. Splitter overhead provided feed to a cumene unit and had a tight (0.2 % maximum) isobutane specification. Tower bottom went to an alkylation unit and could tolerate C_3 impurities, but any C_3 impurity in the bottom was an economic loss. It was desired to maximize C_3 recovery in the top product.

History In an attempt to maximize C_3 recovery, the 20 trays in the top section were replaced by two beds of 1-in. modern random packing (Fig. 7.1*a*). Instead of improving, recovery worsened. With the trays, the tower produced 4000 barrels per day (BPD) of specification C_3 at a reflux of 18,000–20,000 BPD, with 3% propylene in the bottoms. With the packing, reflux needed to be kept below 14,000–15,000 BPD to make specification on the top product. Recovery declined, with the bottom C_3 content rising to 8–9%. Attempts to operate at higher reflux made the top product off specification.

A study by the vendor found insufficient open area in the packing supports. The packing holddown had a mesh backing that could also have restricted capacity. The supports were replaced, the holddown mesh removed, and the packings replaced by higher capacity packings. The upper bed was replaced by No. 2 (approximately 65–70 ft^2/ft^3 specific surface area) structured packings, while the bottom bed was replaced by $1\frac{1}{2}$-in. random packings. The tower was thoroughly inspected, and all appeared in good order. Upon return to service, the results were disappointing: There was little improvement.

With the packing modifications giving little improvement, the vendor then directed attention to the trayed section below the feed. This section contained three-pass valve trays. These trays were modified. Weir lengths were modified to ascertain even liquid distribution. Antijump baffles were added. Despite these changes, there was little improvement.

Troubleshooting The vendor spared no effort in attempting to resolve the problem. After the failure of the initial efforts, a task force was assembled which included engineers from the refinery, the vendor, and consultants. We served on the task force.

Three differential pressure transmitters were installed, one across the top bed, one across the bottom bed, and one across the trays. The pressure drop behavior was studied with the top-product flow rate held constant while the reflux rate varied (Fig. 7.2*a*). The pressure drop data were corrected for the vapor static head and represent the dynamic pressure losses alone. Due to the subtraction of the large static head, the plotted data represent the difference between two large numbers. While their absolute values can be in error due to the subtraction, the trends are adequately reflected in Figure 7.2*a*.

During the test, the tray pressure drop remained constant at 0.085 psi per tray regardless of the reflux flow rate, which confirmed that the flood was not initiating in the bottom section. Figure 7.2*a* shows that the top-bed pressure drop closely followed a square-law behavior. This behavior is expected. On the other hand, the bottom-bed pressure drop escalates faster than a square law, suggesting the possibility of liquid accumulation in that bed.

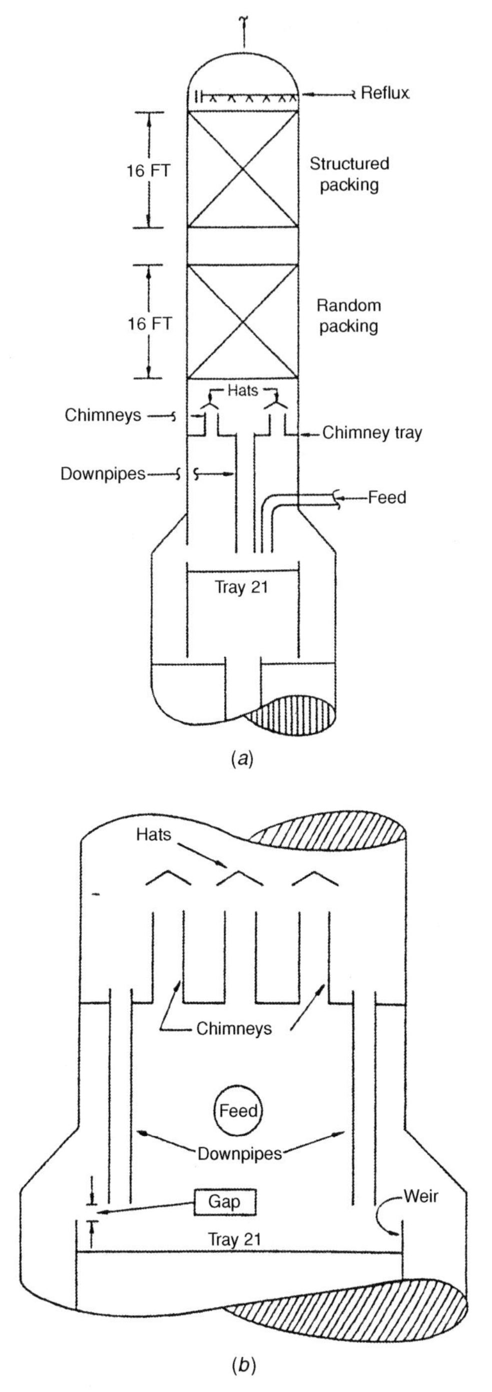

Figure 7.1 A C_3/C_4 splitter that experienced problems following packing retrofit: (*a*) arrangement of retrofitted sections; (*b*) close-up focusing on vapor distributor chimney tray showing unsealed gap and undersized downpipes from tray. (Reprinted with permission from Ref. 304. Copyright © 1988 by Gulf Publishing Co. All rights reserved.)

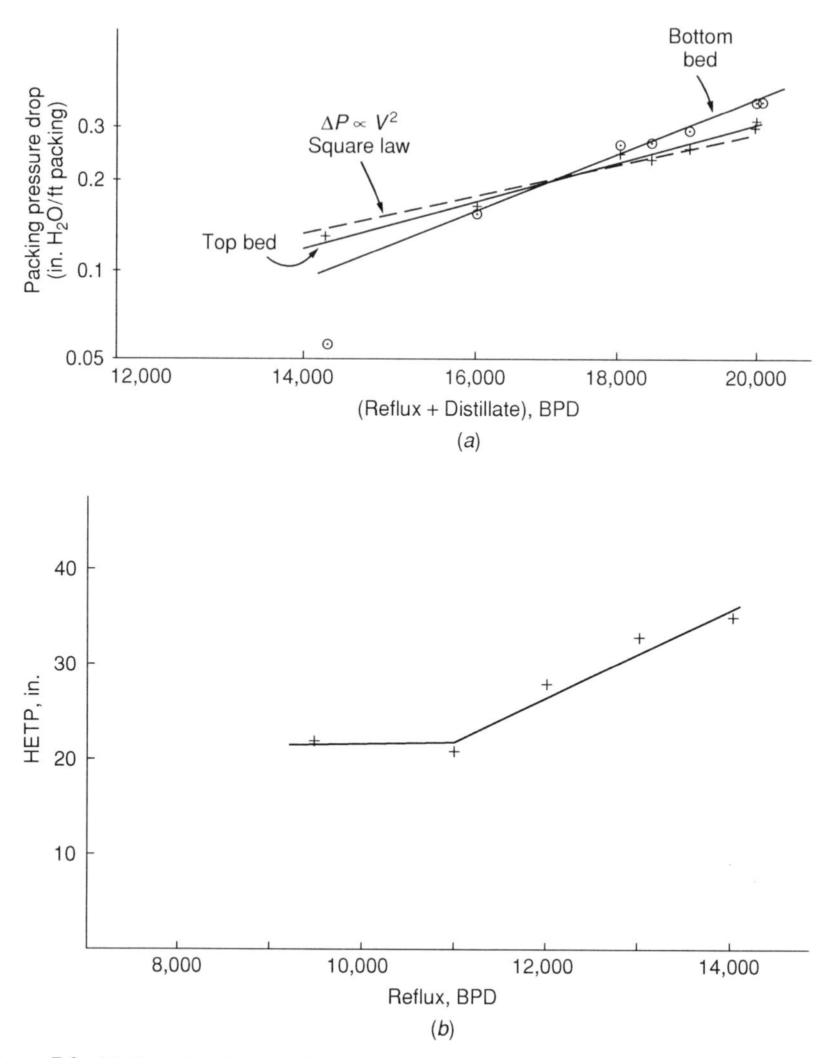

Figure 7.2 Field test data for packed sections of C_3/C_4 splitter: (*a*) measured packing pressure drop versus load (pressure drop on log scale, load on square scale); (*b*) HETP from test data versus reflux flow rate.

Figure 7.2*b* shows HETPs calculated from plant data as a function of reflux. The HETP is calculated over the entire top section. Fig. 7.2*b* shows a rapid increase in HETP as reflux is raised beyond 11,000 BPD. This increase of HETP with reflux is abnormal for either random or structured packings. The only plausible explanation for this behavior is flooding.

A combination of the above results is strong evidence for flooding initiating at the bottom bed. However, this bed had already undergone a change in packing from a 1-in. to a 1.5-in. random packing. If the flood initiated in the packings, the larger packings should have improved capacity, at least somewhat.

Focus on the Vapor Distributor The only unmodified internal in the lower part of the tower was the vapor distributor. The vapor distributor was a chimney tray equipped with 6-in.-tall chimneys (Fig. 7.1*b*). Liquid from the chimney tray drained via six 4.73-in.-ID downpipes that extended from 1 in. above the chimney tray floor to 1 in. above the outlet weir elevation on tray 21.

A close examination of the chimney tray revealed the following flaws:

- The downpipes were badly undersized. At a reflux of 11,000 BPD (the last good efficiency) the liquid flow rate was 364 gpm, giving a velocity of about 500 gpm/ft^2, or 1.1 ft/s, at the inlet to the pipes. This is more than twice the maximum downcomer velocity recommended for this type of system (250). Further, the highly recommended (250) self-venting flow correlation of Figure 24.1*c* (438, 447) predicts that six pipes 4.73 in. ID will handle a maximum self-venting flow of 362 gpm. This number coincides exactly with the liquid flow through the pipes at a reflux of 11,000 BPD, the highest reflux at which no efficiency loss occurred. At higher reflux rates, the downpipes are likely to choke, backing liquid up on the chimney tray.
- The downpipes were unsealed. These unsealed downpipes could permit vapor to force its way up the downpipes and destroy the hydrostatic head of liquid that would ordinarily build up in the downpipes and promote liquid drainage from the chimney tray. Anyone who has ever siphoned gasoline out of a car can visualize the problem.
- At a reflux plus distillate rate of 15,000 BPD, the entrance head loss of liquid flowing from the chimney tray into the downpipes was 3½ in. of liquid. (Entrance head loss is a consequence of the acceleration of liquid as it enters a nozzle.)
- At the above rate, the pressure drop of the upflowing vapors as they passed through the chimneys and out under the hats was equivalent to 1½ in. of liquid.

In summary, with all the above flaws there is no way that the vapor distributor would be able to operate at more than 11,000 BPD reflux without its liquid level exceeding the top of its short chimneys. Once reaching the top of the chimneys, the vapor will blow this liquid up the bed and flood it. This was the root cause of flooding in the bottom bed.

Other Problems Even with the lowest HETP, achieved at 11,000 BPD reflux, the separation was not much better than the separation previously achieved by the trays. The top distributor design was reviewed, and it was shown to deliver more liquid to the center of the tower than to the peripheral regions. Prior to going into the tower, it was not water tested in the shop. No redistributor existed between the two beds.

Overall, it was felt that some improvements in separation were achievable and worth going for.

The Fix The following modifications were made:

- *To eliminate the flooding*: Two chimney tray panels were replaced by panels with much larger downcomers that were properly sealed in the liquid of the tray below.
- *To improve distribution*: The trough-orifice reflux distributor was replaced by an orifice pan distributor. A redistributor was added at the top of the lower bed.
- *To improve separation*: The 1.5-in. random packings in the lower bed were replaced by No. 2 structured packings. The feed point was lowered 10 trays, and the trays in this region were replaced by new two-pass valve trays. This change was made to convert some stripping stages (which the tower had an excess of) into some desperately needed rectifying stages.

Results The chimney tray modifications successfully eliminated the flooding. Following the fix, HETP was in the 22–26-in. range and no longer rose with reflux. At the same time, the extensive distribution improvements did not enhance the separation a great deal.

Overall, the revamped tower achieved four to five more stages than it did with the trays prior to the first modification. The additional stages are attributed almost entirely to the lower feed point. Changing the trays to structured packings, even after all else in the system was debugged, did not significantly improve separation.

Morals

- Overflowing vapor distributors cause premature flood and poor separation.
- Undersized and unsealed downpipes are a prescription for disaster.
- Structured packings have little to offer in high-pressure, high-liquid-rate distillation.
- It is important to have liquid distributors water tested prior to installation in the tower.

CASE STUDY 7.2 VAPOR CROSS-FLOW CHANNELING

Installation An ethylene dichloride (EDC) rectifier 14 ft ID containing two-pass valve trays at 18 in. tray spacing. Feed to the rectifier came from a reactor and entered below the bottom tray. The feed nozzle was close to the bottom tray. The tower had 54 trays, with a product draw 6 trays below the top. The tower operated just above atmospheric pressure.

Problem The tower achieved 70% of the design throughput before beginning to flood.

Troubleshooting The tower was gamma scanned under unflooded conditions along a chord parallel and close to the center downcomer. As expected, the scan showed some entrainment, tall spray heights, and trays on the verge of jet flooding.

Near the onset of flood, the tower experienced oscillations. The oscillation amplitude was about 10% of the bottom sump level. The oscillation period was about 4 min. Since a typical tray time constant is 0.1 min, 4 min is a typical travel time of an oscillation from the highest loading area (above the product draw) to the tower bottom.

Vapor Cross-Flow Channeling (VCFC) Check Previous work (246) showed that VCFC always occurs when the following conditions occur simultaneously:

- A large open area on the tray. With moving valve trays that have sharp-edge orifices, standard open-slot area is 14–15% of the tray active area. Open-slot areas exceeding 15 % are large. A recent report (195) showed that with a very high ratio of flow path length to tray spacing, VCFC can occur even when slot area is 15% of the active area.
- A high ratio (>2) of flow path length to tray spacing.
- A high liquid flow rate (>5–6 gpm/in. of outlet weir).
- A pressure of 70 psia or less.

A check of the EDC tower versus these criteria showed the following:

- The valves used were long-legged, uncaged valves. The open-slot area of these valves was 20% of the tray active area. The process licensor had a very tight pressure drop specification on that tower. To meet the pressure drop specification, the vendor increased the valve density and slot height, giving a slot area well beyond the standard 14–15%. The prime condition for VCFC is therefore easily satisfied. A 20% open-slot area is a huge open area.
- The flow path length was 5.5 ft, the tray spacing 1.5 ft, giving a huge ratio of 3.7 of flow path length to tray spacing. This condition, again, is easily satisfied.
- The liquid rate was around 5 gpm/in. of outlet weir, just enough to meet this condition.
- The pressure was slightly above atmospheric, well below 70 psia.

With all four conditions met and two of them exceeded by far, there is little doubt that the trays experienced VCFC. The oscillations experienced on the verge of flood are another symptom of VCFC that we have seen in many towers. Vapor cross-flow channeling has been the cause of premature flood as well as loss of tray efficiency and fully explains the observations here.

The proximity of the vapor inlet to the bottom tray could possibly have aggravated the channeling. However, the inlet velocity was relatively low for trays, with a $\rho_G V_N^2$ of 1000 [where ρ_G is gas density (lb/ft^3) and V_N is nozzle gas velocity (ft/s)], suggesting that, if it did play a role, it was not major.

Cure The tower was retrayed by high-capacity trays. Modifications were also made to the vapor entry. No more premature flooding occurred.

CASE STUDY 7.3 CENTER DOWNCOMER OBSTRUCTS BOTTOM FEED

Installation A refinery amine absorber initially containing one-pass trays. To improve capacity, the tower was retrayed with two-pass trays.

Problem Following restart, tower efficiency was poor.

Cause The new bottom tray had a center downcomer which was oriented at 90° to the vapor inlet (Fig. 7.3*a*). The bottom of the seal pan from the downcomer was only a short distance above the liquid level. Most of the incoming vapor was therefore channeled into the panel on the nozzle side.

Solution A tunnel was cut through the downcomer to allow vapor equalization (Fig. 7.3*b*). No more problems occurred.

CASE STUDY 7.4 CHANNELING INITIATING AT A CHIMNEY TRAY

Henry Z. Kister, Kirk F. Larson, John M. Burke, Rick J. Callejas, and Fred Dunbar, references 264, 272. Reprinted with permission from *Hydrocarbon Processing*, by Gulf Publishing Co., all rights reserved

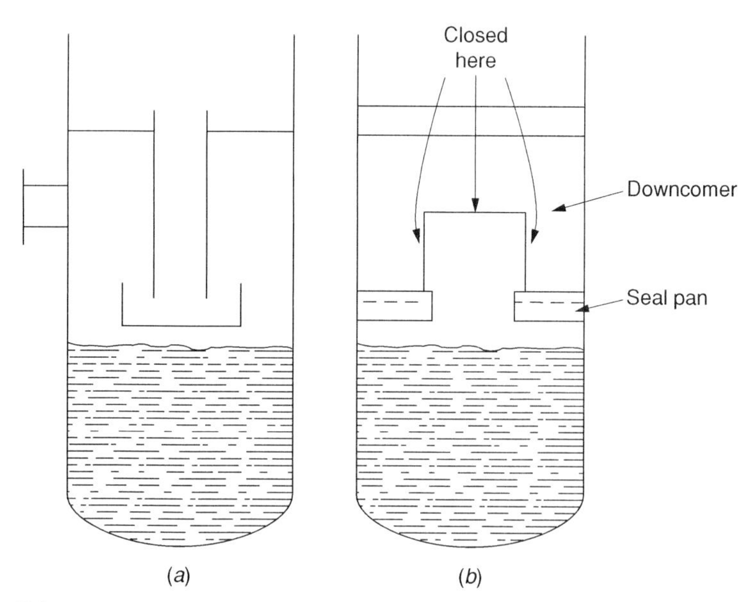

Figure 7.3 Center downcomer that obstructed vapor flow: (*a*) initial; (*b*) tunnel through downcomer that eliminated obstruction.

Installation A water quench tower in an olefins plant (Fig. 7.4*a*). It contained three PA sections. In each, the ascending gas was cooled by contacting progressively colder water. The top section contained 10 two-pass, fixed-valve trays and a chimney tray. The chimney tray is shown in detail in Figure 7.4*b*. It contained perforations for passing the liquid collected from the trays to the shed decks below. The quality of irrigation is not critical, both because over half the liquid to the decks comes from the sprays and because shed decks are not very sensitive to liquid distribution at high liquid rates.

Problem The tower always adequately removed the heat from the gas. The bottom water temperature was close to design. The temperature approach between the tower overhead gas and the entering quench water was better than design, at times as low as 1–2°C. The problem experienced was that at ethylene product rates exceeding 105% of the design, significant entrainment occurred from the top of the tower. With certain feedstocks, and at lower rates, no entrainment was observed. The entrainment was highly undesirable because entrained liquid collected in the cracked gas compressor suction drum. This drum was designed to run dry and had a high-level trip set at a relatively low level. Activation of this high-level trip would trip the compressor and the entire olefins plant.

Gamma Scans Tower testing included quantitative analysis of extensive gamma scans. The scans showed jet flooding initiating on trays 10 and 8, the center-to-side flow trays just above the chimney tray. The flooding was concentrated near the center of the trays, with outlet sections of the same trays appearing lightly loaded. This, plus other observations of spray height gradients and entrainment gradients, suggested that vapor preferentially channeled through the center of the trays.

The gamma scans showed that the chimney tray contained froth that exceeded the chimney height. These scans also showed hydraulic gradients on the chimney tray, with froth heights at the tray inlets (near the side downcomers) exceeding the froth heights near the center. Therefore, liquid preferentially overflowed the risers near the chimney tray liquid inlets, with gas rising preferentially near the center of the tray.

It may appear surprising to talk about chimney tray froth heights rather than chimney tray liquid heights. The scans showed that at the inlets to the chimney tray (just outside the inlet weirs) the chimney-tray liquid was just as aerated (possibly more) than the tray above. Scans of other sections of the chimney tray also clearly showed aeration. This aeration is believed to be frothing due to the waterfall over the 200-mm inlet weirs. At that point the inlet liquid head was calculated to be about 120 mm. Since chimney height is 250 mm, it would not take much aeration for the froth to rise above the chimney. Furthermore, once some head builds up above the chimneys, vapor may be induced to rise up the perforations at the bottom of the chimney tray. This intensifies the overflow and aeration.

Channeling Due to the hydraulic gradients on the chimney tray, most of the liquid overflow down the chimneys occurred near the side downcomers. Gas channeled preferentially through the chimneys near the center of the tray. Had the valve trays

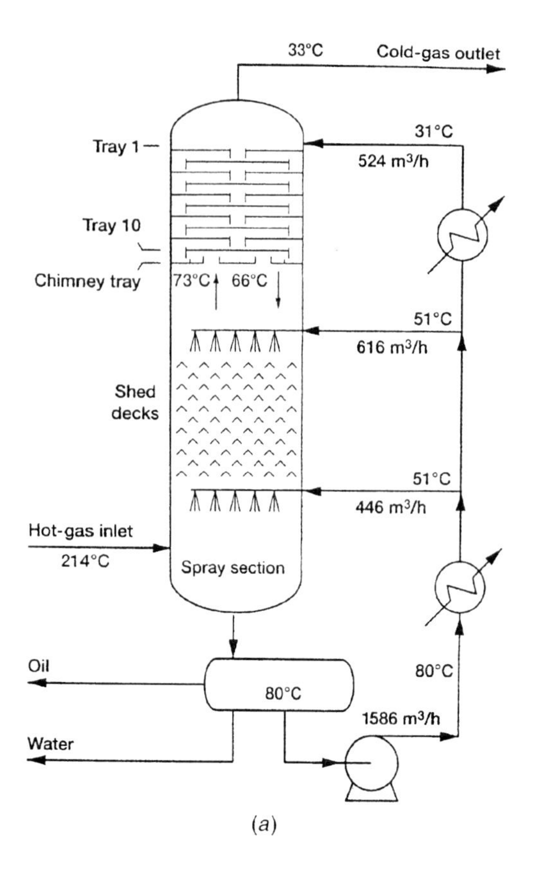

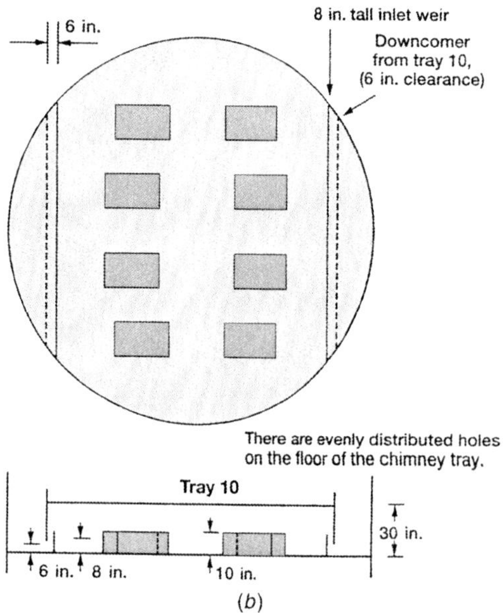

Figure 7.4 Vapor channeling in olefins water quench tower: (*a*) schematic of tower; (*b*) simplified plan and elevation of water quench tower chimney tray. (From Ref. 264. Reprinted with permission from *Hydrocarbon Processing* by Gulf Publishing Co. All rights reserved.)

above contained a small slot area, the gas channeling would have been mitigated. However, the open-slot area of the fixed-valve trays was particularly large (18.5% of the active area), and this large area did not mitigate channeling. Most of the gas continued to rise near the center of the trays, creating regions of tall sprays and high entrainment. Some entrainment ended in the tower overhead.

Cure The chimney tray was modified to eliminate the waterfall and to prevent aeration and overflow. A quarter of the fixed valves on the trays were blanked off, reducing the open-slot area from 18.5 to 13–14% to mitigate tray channeling. After these modifications, entrainment was no longer observed.

Chapter 8

Tower Base Level and Reboiler Return: Number 2 on the Top 10 Malfunctions

Layton Kitterman, one of the all-time greats in distillation troubleshooting, estimated that 50% of the problems in the tower originate in this region (277). Over 100 reported case histories (255) verify that, indeed, more problems initiate at the tower base than in any other tower region, although the actual percentage is lower than 50. Plugging/coking has been the only malfunction for which a higher number of cases had been reported. The trend in tower base incidents suggests little improvement over the years (255), so it will continue to be a major troublespot.

Half the case studies reported were liquid levels rising above the reboiler return inlet or the bottom gas feed (255). Faulty level measurement or control tops the causes of these high levels. Restriction in the outlet line (this includes loss of bottom pump, obstruction by debris, and undersized outlets) is another cause. A third major cause is excessive pressure drop in a kettle reboiler circuit, with liquid level in the tower base backing up beyond the reboiler return to overcome the pressure drop (Section 23.4).

In the majority of cases, high tower base levels caused tower flooding, instability, and poor separation. Less frequently, vapor slugging through the liquid also caused tray or packing uplift and damage.

The corrective measures to prevent excessive tower base level are those recommended by Ellingsen (123): reliable level monitoring, often with redundant instrumentation, and good sump design. One added measure (255) is avoiding excessive pressure drop in a kettle reboiler circuit.

Impingement by the reboiler return or incoming gas is next, albeit far less troublesome than the high base levels. This issue seems to be on the increase in the last decade. Several of the case studies reported severe local corrosion due to gas flinging liquid at the tower shell in alkaline absorbers fed with CO_2-rich gas, mostly in ammonia plants. This calls for special caution with the design of gas inlets into these towers. Troublesome experiences were also reported with inlet gas impingement on

Distillation Troubleshooting. By Henry Z. Kister

liquid level, instruments, the bottom tray, the seal pan overflow, and the inlet from a second reboiler.

Other tower base issues include gas entrainment in the bottom liquid, low base levels, insufficient surge, and problems with preferential baffles at the tower base. Gas entrainment and insufficient surge led to instability, or pump cavitation, or contributed to base-level rise above the vapor inlet. Low base levels appear to be particularly troublesome in chemical towers. In services distilling unstable compounds like peroxides, low base levels induced excessive temperatures or peroxide concentration, either of which led to explosions. A total loss of liquid level induced vapor flow out of the bottom, which overpressured storages. Preferential baffle issues led to starving the reboiler of liquid with an accompanying capacity restriction, or to level control and instability problems.

Two tower base troublespots are discussed under different headings: water-induced pressure surges initiating at the tower base due to undrained stripping steam lines (Section 13.5), and leaking draws to once-through thermosiphon reboilers (Section 23.2). Both of these mainly affect refinery towers.

CASE STUDY 8.1 BASE LIQUID LEVEL CAN MAKE OR BREAK A FRACTIONATOR

Contributed by Mark Pilling, Sulzer Chemtech USA, Tulsa, Oklahoma

Installation Main fractionator for a refinery distillate hydrotreater.

Problem The column was unstable. Pressure drop fluctuated periodically, and bottoms level read empty. Periodic banging noise was heard from the column. Kerosene product was off specification with too heavy of an end point.

Investigation Efforts were made to check the bottoms system and to establish a level in the bottoms. The level controller was checked and validated. The level glass showed no level. Draws and reflux rates were adjusted to try and get kerosene product on specification with no success.

Cause After repeated inspections of the piping and level control system, a very small black neoprene construction blind was found in the top side of the level controller piping. It was not detected earlier as it looked very similar to a gasket.

Cure To avoid shutting down the column, a temporary tubing line was run from the vent on the level bridle to a higher point on the column. This fixed the level reading, which then showed the level to be full. The level was brought down to normal and the banging sound went away. The banging was hammering within the column due to the high level above the reboiler return. Later inspection of the column showed the bottom several trays to be damaged from the previous incidents of high level.

Not Out of the Woods Yet Lowering the liquid level did not lower the kerosene product end point, which remained too high and could not be brought back on specification. Finally, a sample was taken immediately off the column rather than from the product rundown line. This sample was not taken earlier as it was very hot and difficult to obtain. The sample directly off the column showed excellent distillation properties which were well within specifications. Further investigation showed a leaking valve between the diesel and kerosene products upstream of the normal product rundown sample point. This leaking valve was repaired and the column finally achieved design rates and product specifications.

Morals

- All blinds should have long handles and should be tagged and recorded upon installation.
- Product lines between the column and sample point must be inspected prior to start-up and during troubleshooting. Any opportunity for product contamination (e.g., valves, exchangers) should be reviewed and investigated.

CASE STUDY 8.2 HIGH-LIQUID-LEVEL DAMAGE

Installation A refinery coker fractionator. Feed to the tower was a vapor–liquid mixture (by volume almost all vapor) that entered above the bottom sump (no stripping section). The section immediately above the feed was a wash section equipped with fouling-resistant grid packing. Above this there was a PA section equipped with fouling-resistant structured packings. Above the PA there were four trays, then a diesel draw. Ten trays above the diesel draw was a jet fuel draw. There were a total of 34 trays in the tower.

Experience In one incident, liquid built up above the vapor inlet. From column *dP* readings, it is estimated that the liquid head above the inlet was about 30 ft. The vapor slugged through the liquid, uplifting many panels off their supports (Fig. 8.3*b* in Case Study 8.5). One side of the two-pass trays was damaged much more than the other. On this side most of the panels were missing. It looks like the vapor slug selected a preferential path up the tower on that side. There were big open spaces on the jet fuel draw tray, which explained why following the incident the refinery could not draw much jet fuel.

The top five trays were not damaged at all. On the side that received less damage, a total of 12 trays were not damaged. The grid and packed sections were not damaged.

CASE STUDY 8.3 EVENT TIMING ANALYSIS DIAGNOSES HIGH-LIQUID-LEVEL DAMAGE

Contributed by Mark E. Harrison, Eastman Chemical, Kingsport, Tennessee

Apparent Problem The troubleshooter was called in to the control room because "the tower did not make on-spec products." A look at the operating charts (Fig. 8.1*a*) showed that the tower was flooded. This is evidenced both by the *dP* and the reduction in bottom flow.

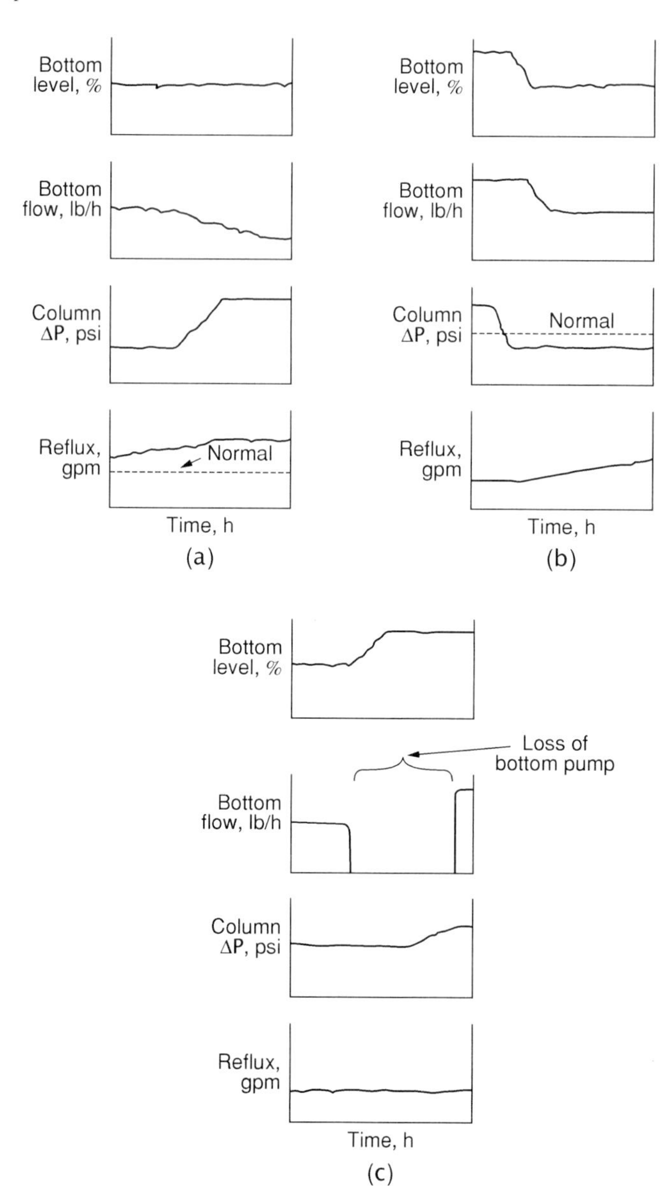

Figure 8.1 Operating charts for high-liquid-level damage incident: (*a*) final charts, showing flood in tower; (*b*) intermediate charts, showing rise of reflux; (*c*) initial charts, showing high liquid level that caused tray collapse.

Troubleshooting The operating team was aware that the tower was flooded. They stated that the tower products were off specification even before the column flooded. Figure 8.1*a* shows the reflux coming up even before the tower flooded, obviously with the intent to improve product purity.

Rolling back the charts to the beginning of the reflux rise (Fig. 8.1*b*) indicates that at that point the tower again was flooded. This is evidenced by the high *dP*, but this time the base level and bottom flow rate were also high. With an ordinary flood, the bottom flow rate and base level tend to decline as liquid accumulates in the tower.

Root Cause Figure 8.1*c* shows the initial event. There was a temporary loss of the bottom pump. As a result, the base liquid level went up. The bottom-level indicator initially showed a level increase but then leveled off as it reached the maximum of its range. The liquid level kept on rising, now into the trays, as evidenced by the rise in the tower *dP*.

Some time later the pump came back on-line and intensely pumped out the base liquid. The liquid accumulation in the tower ceased, and the *dP* first leveled off, then started to decrease (Fig. 8.1*b*). Soon the bottom level came back to normal. But the *dP* did not fall back to normal; it fell to a value below normal, suggesting some trays collapsed or were damaged during the high-liquid-level event. At the same time, the reflux started increasing to meet the reduced purity with the reduced number of trays. Further increases in reflux and boil-up brought the tower to flood (Fig. 8.1*a*), still without getting the product back on specification.

Morals

- In a troubleshooting investigation, always study the history and sequence of events.
- Beware of tray and packing damage due to high base levels.

CASE STUDY 8.4 CAN IMPROVED LEVEL MONITORING AVOID HIGH-LEVEL DAMAGE?

Installation A crude fractionator (Fig. 8.2*a*). Feed to the tower was a high-velocity vapor–liquid mixture (by volume almost all vapor) that entered via a downward-sloping vapor horn. The horn was closed on the top and sides but open at the bottom and swirled the feed along the shell. The stripping section was enclosed in a smaller diameter internal "can" which contained a bed of random packing. Stripping steam was supplied via a sparger with holes directed at 45° downward along the entire length. The sparger entered about midway between the lower and upper level taps.

History The eight trays above the feed experienced repeated episodes of damage due to high base levels. At different times panels were dislodged and deformed, and manways were uplifted. Strengthening the trays helped but did not mitigate the incidents.

Incorrect measurement and false indication of the base level were a constant headache, especially during start-up and while returning from an outage, when liquid loads and levels varied widely. Gamma scans showed liquid levels above the steam inlet while the level transmitter was reading in the middle of its range. The scans also

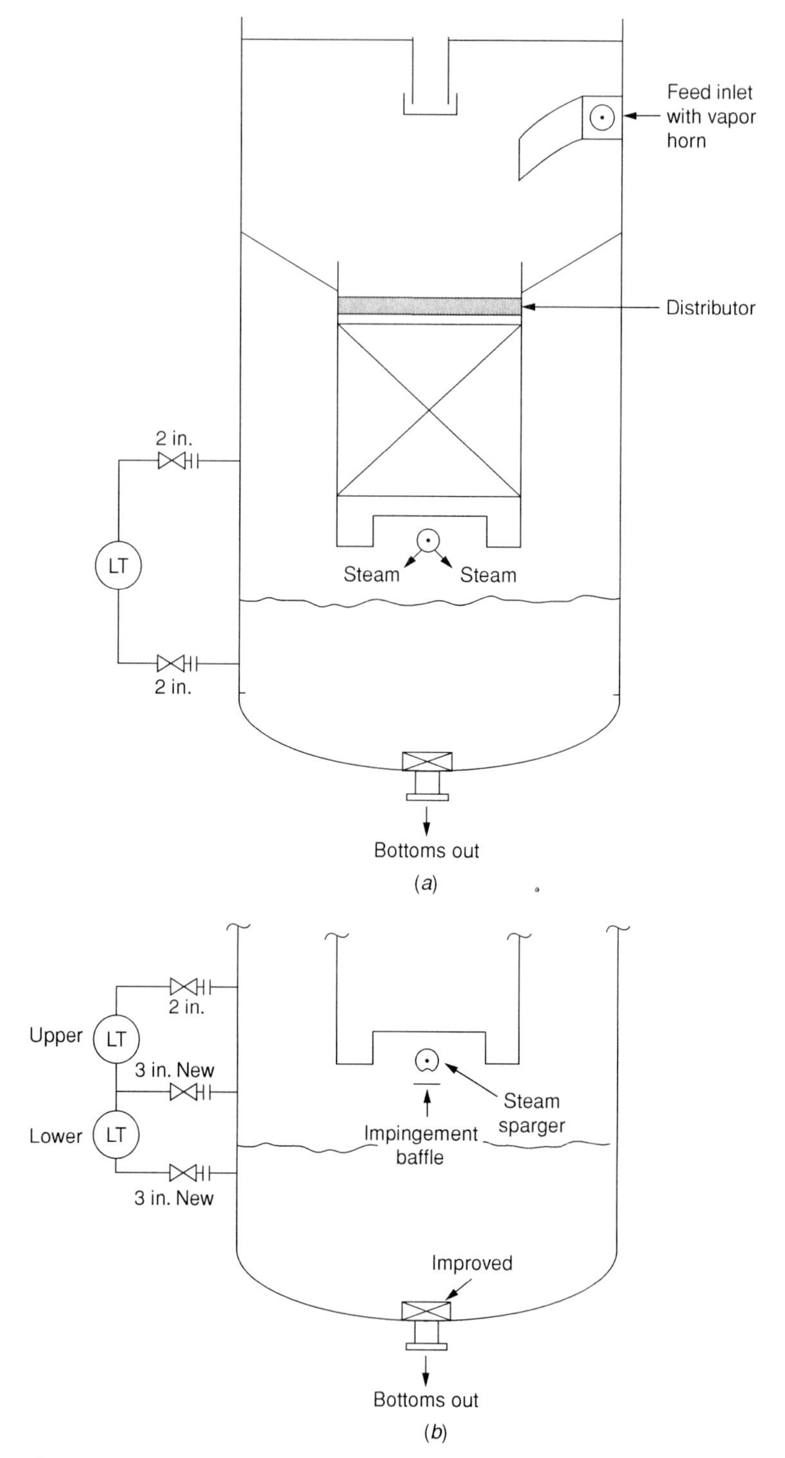

Figure 8.2 Improving base-level monitoring to reduce high-liquid-level damage incidents: (*a*) initial system; (*b*) modifications to level monitoring.

showed many incidents in which the liquid level rose above the top of the can. At one test, the level transmitter reading was allowed to change in steps from near empty to near full while gamma scans that tracked the changes showed a level exceeding the steam inlet throughout.

Cause Once the liquid level reached the bottom of the steam sparger, it became aerated due to steam bubbling into the liquid. Aeration lowers the liquid specific gravity (SG), which reduces the hydrostatic head on the level transmitter. So when the liquid level exceeded about 50% of the nozzle span, it began reading misleadingly low. Here the situation would be even worse because the steam left the sparger at 45° toward the liquid surface, hitting the liquid surface at high velocities. This impact aerated the liquid even before the level reached the bottom of the sparger. Roughly, significant aeration began when the base level exceeded about 30% of the level nozzle span. Somewhere, possibly just above 30% of the span, the level measurement started varying with aeration, becoming unreliable and misleading.

Once the base level exceeded the top of the cutout at the bottom of the can, its rate of rise sped up. The rising liquid trapped a big bubble of vapor in the annulus between the tower shell and the can. From the cutout up, the liquid only accumulated in the can. The bubble was slightly compressed in a "diving bell" effect. The volume of the can was quite small and the vapor fraction relatively high, allowing little room for liquid accumulation. It was estimated that, if the bottom flow was interrupted, it would have taken half a minute to fill the can with frothy liquid. During this time the annulus accumulated very little liquid because of the trapped vapor bubble. Gamma scans showed that the annulus always remained empty while the liquid level rose above the can. Once the liquid level rose to or approached the feed inlet, the vapor–liquid mix, entering at more than 100 ft/s, entrained and slugged it up the tower, causing the tray damage incidents.

Since the upper level nozzle was in the annulus, the froth buildup in the can raised the static pressure at that upper nozzle as much as it did at the lower level nozzle. So when the sump level rose above the top of the cutout, the level indication became even more misleading.

Solution The key was to properly monitor the base liquid level and always keep it below the steam inlet. A good practice is to have the upper tap of a liquid level below the vapor inlet so that it always measures nonaerated liquid. To achieve this, the following modifications were implemented (Fig. 8.2*b*).

- The steam sparger was modified to circumvent impingement on the liquid level. The new sparger had downward-pointing perforations, only beneath the can, with a flat impingement baffle underneath to deflect the incoming steam sideways.
- A new nozzle was installed about 6 in. below the impingement baffle under the modified steam sparger. This nozzle became the upper level tap of the main (lower) level transmitter.
- A new (upper) level transmitter was added between the new nozzle and what used to be the upper tap of the main transmitter. This upper transmitter normally

read zero. Any positive reading on this transmitter would indicate a liquid-level rise heading into the can.

- To prevent plugging with some fouling crudes, the main transmitter nozzles were enlarged to 3 in.
- The vortex breaker was improved.

Results Following the modifications, the level could be comfortably operated at around 50%. The transmitter new reading coincided with the level measured by gamma scans. Incidents of damage due to high liquid level were alleviated.

CASE STUDY 8.5 HIGH-BASE-LEVEL DAMAGE INCIDENTS

Base-liquid-level rise above the reboiler return or vapor feed nozzle has been one of the most common causes of tray and packing damage in chemical plant and refinery towers (123, 255). Figure 8.3 shows the end results in a number of incidents. Several other cases are given below.

Tower A A 3-ft-ID random-packed stripper used gas as the stripping medium. At one time, bottom pump problems caused liquid level rise above the gas inlet and up the column. The high-level alarm was activated, but no action was taken. The end result was crushed packings, with many pieces extruding through the bed limiter and ending in the distributor.

Tower B A cryogenic rectifier with tower feed gas superheated to about 150°F. During one outage, gas feed was interrupted. Liquid was dumped to the tower base, with the liquid level covering a few trays. When gas was reintroduced, several of the lower trays were lifted off their supports. Here the slugging action induced by vapor passage through liquid was intensified by the rapid vaporization taking place as the superheated vapor boiled off the liquid.

Tower C A cryogenic rectifier separating methane, carbon monoxide, and hydrogen. One start-up, the liquid level covered several bottom trays; then superheated gas at about –70°F entered the tower via the bottom vapor inlet. As in tower B, the slugging action was augmented by the rapid vaporization, lifting all the lower trays in the tower off their supports.

Tower D A large-diameter ethylbenzene–styrene tower was filled with liquid well above the reboiler return nozzle. The reboiler was then started up. A violent vapor slug resulted that lifted trays off their supports, damaging all the trays in the tower.

Tower E Water level in the base of a chemical tower rose above the reboiler return nozzle when the reboiler was started up. The vapor slugged through the liquid, lifting a few trays off their supports. It is believed that a layer of insoluble organics of

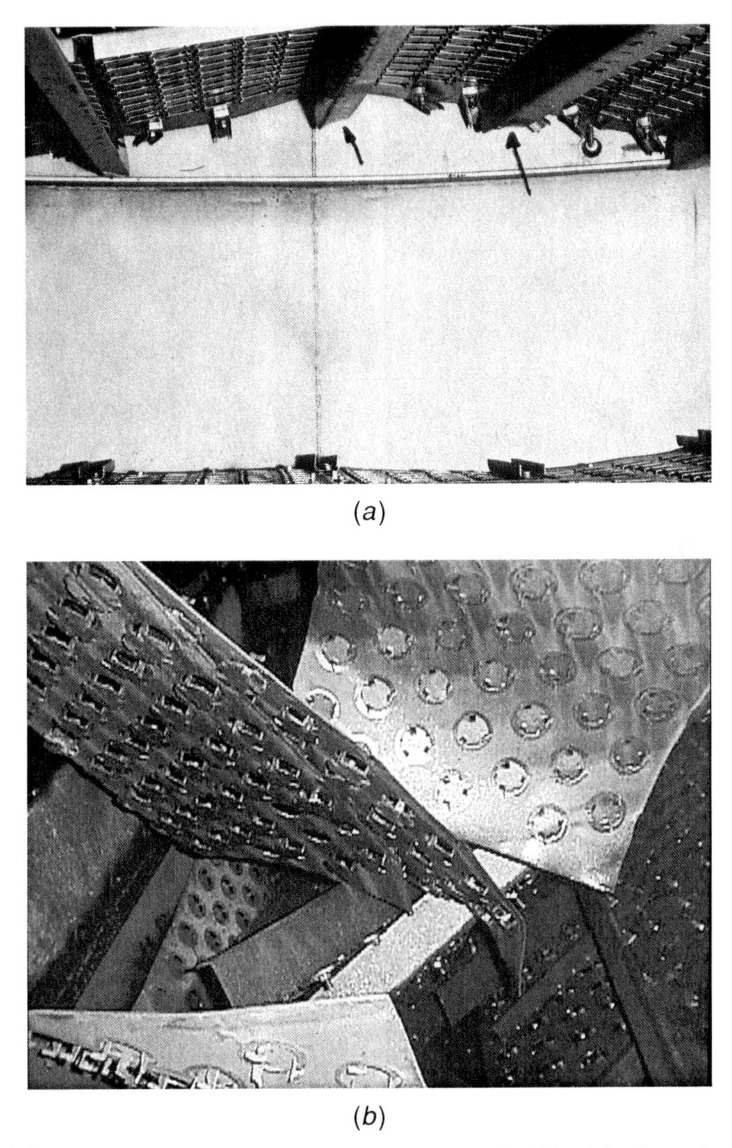

(*a*)

(*b*)

Figure 8.3 Tray and packing uplift, most likely induced by base-liquid-level rise above reboiler return or vapor inlet nozzle: (*a*) showing uplifted tray panels; (*b*) showing tray panel displacement; (*c*) showing collapse of packing support, dumping bed of packing into tower base; (*d*) showing structured packing uplift that also damaged spray distributor. ((*a*, *c*) Copyright Eastman Chemical Company. Used with permission. (*d*) From Ref. 114. Reprinted with permission from *Hydrocarbon Processing* by Gulf Publishing Co. All rights reserved.)

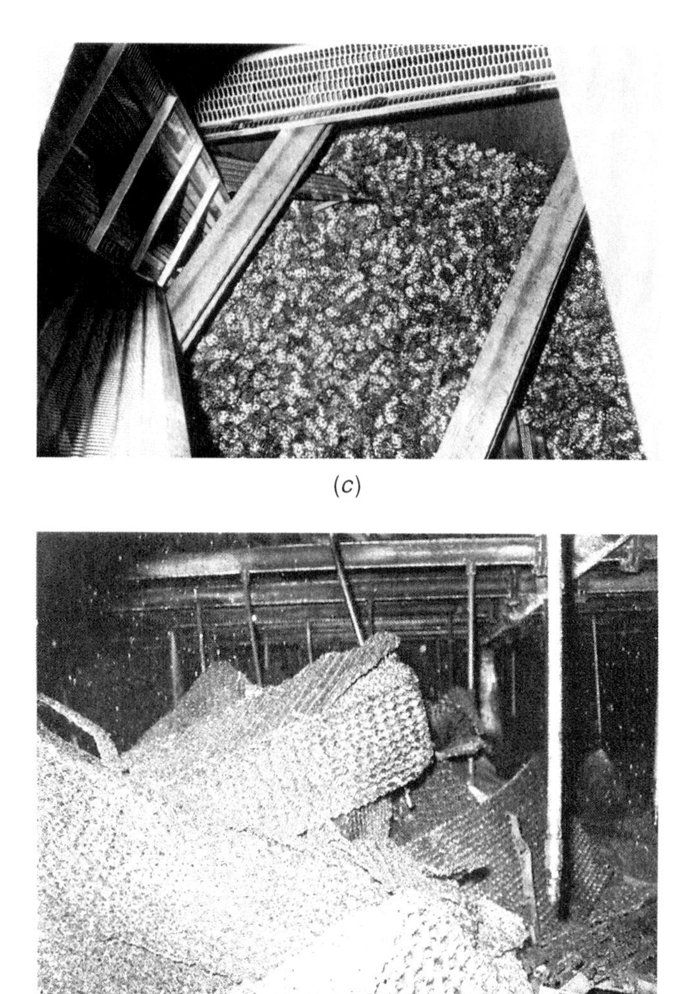

(c)

(d)

Figure 8.3 (*Continued*)

low specific gravity was above the water, giving a low base level reading that misled operators.

Tower F A petrochemical tower was reboiled by high-pressure steam. During a steam emergency, the steam supply was interrupted. Liquid from the trays dumped. When the steam supply was reinstated, the reboiler return was submerged under more than 5 ft of liquid. The steam came back at quite a good rate, uplifting the bottom 20 trays off their supports. Recurrence was effectively prevented by improved level monitoring and by installing a reliability switch (two-thirds voting) that would trip reboiler steam on high base levels.

Tower G A chemical tower packed with random plastic packing experienced bottom liquid level well above the reboiler return. The base-level indicator read 100%. The high-level alarm came on and stated "bad reading," meaning that the level was too high. The operators understood that statement to mean that the meter was bad and took no action. The slugging of reboiler return vapor through the liquid bent upward part of the vapor distributor beneath the bottom bed. Pieces of packing from the bed were blown into the liquid distributor above.

Tower H A sour-water stripper experienced recurrent disintegration of the random packing support plate due to base-level rise above the reboiler return nozzle. Increasing the strength of the packing support did not eliminate the problem.

Tower I A stripper in a chemical plant using live steam had problems with the bottom level transmitter. The base level climbed above the steam inlet nozzle, causing trays to be lifted off their supports and valve floats to blow off their holes. The level transmitter was replaced, and the tower was retrayed with heavy-duty fixed-valve trays. The tower operated for more than 8 years without recurrence.

Tower J In a 7-ft-ID chemical tower, base liquid level rising above the reboiler return led to uplifting of manways and panels in the one-pass trays above the feed but no damage to the few two-pass trays below the feed.

CASE STUDY 8.6 REBOILER RETURN IMPINGEMENT ON LIQUID LEVEL DESTABILIZES TOWER

Installation Isomer separation tower containing more than 80 trays. Feed to the tower was 22 m^3/h, of which 21 m^3/h ended in the top product and 1 m^3/h in the bottom product. Boil-up was about 240 m^3/h. This boil-up was supplied by two parallel horizontal thermosiphon reboilers. Including circulation, liquid flow rate to the reboilers was 800 m^3/h.

The tower base contained a preferential baffle that divided it into a reboiler draw compartment and a bottoms draw compartment. Liquid from the bottom center downcomer descended via downpipes to the lower part of the reboiler draw compartment. The idea was to send all the tray liquid to the reboiler and none into the bottom draw compartment in order to improve the reboiler temperature difference and to gain a fraction of a theoretical stage. The reboiler returns entered the tower against shielding baffles, which diverted the fluid downward. These shielding baffles were closed at the top and sides and open only at the bottom.

Problem The system could not be operated unless the liquid level exceeded the top of the preferential baffle. This precluded the use of the tower level control and destabilized the tower. In addition, the reboiler outlet temperature would occasionally shoot up to 170°C, resulting in a major upset.

Investigation A gamma scan showed that, while the bottom draw compartment was full of liquid, the liquid level in the reboiler draw compartment was quite low. This was the opposite from the way the tower base should have been working, that is, a full reboiler draw compartment overflowing into the bottom draw compartment (Fig. 8.4).

If the level in the reboiler draw compartment dropped too low, the bottom of the downpipes would be exposed and their liquid seal lost. Vapor would then impede liquid descent in the downpipes. This, together with the low liquid head, would reduce reboiler circulation, causing the reboiler return temperatures to shoot up.

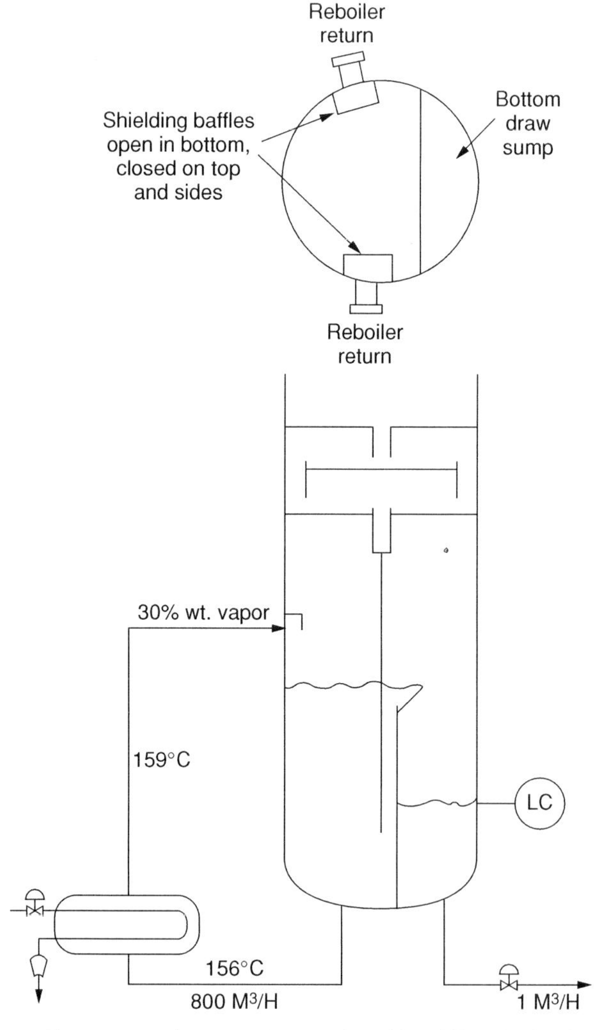

Figure 8.4 Base of isomer separation tower that experienced instability, showing base preferential baffle and reboiler return shielding baffles.

Cause There are two major flaws in the tower base design in Figure 8.4. First, reboiler returns should not impinge on the liquid level. In this system, the shielding baffles directed reboiler return vapor onto the liquid level, which pumped (or vapor lifted) liquid from the reboiler draw compartment into the bottom draw compartment. Second, Figure 8.4 shows that the reboiler draw flow rate was 800 times the bottom draw rate. Thus the preferential baffle overflow must not exceed a tiny fraction of the total liquid flow rate. Any minor leakage, splashing, waves, or vapor pumping across the baffle is likely to lead to higher overflows. The result would be buildup of liquid to the top of the bottom draw compartment. The two flaws combined to give the tower operating problem.

Cure and Prevention Placing the tower on dynamic matrix control (DMC) successfully avoided the spiking of the reboiler outlet temperatures. The new control cut reboiler steam as soon as the reboiler outlet temperature started to increase.

The rest of the problem was not cured because a short time after the problem was diagnosed the tower was taken out of service. In other towers experiencing similar problems, however, a cure was achieved by eliminating the downward impingement of the reboiler return on the liquid level. When reboiler draw rate is hundreds or thousands times larger that the bottom flow rate, it is best to eliminate the preferential baffle altogether or at least seal weld it and notch it to prevent excessive liquid flow across it.

CASE STUDY 8.7 INSUFFICIENT SURGE CAUSES INSTABILITY

Contributed by Neil Yeoman, Koch-Glitsch, LP, Merrick, New York

Installation An 18-in.-ID specialty chemicals tower, removing undesirable heavy ends from a process, was revamped for a capacity increase. Modifications to the tower included replacing its random packing with structured packing and increasing the reboiler return nozzle from 6 to 12 in. Based on some prior experience, there had been significant concern about the size of the reboiler return nozzle and use of a 12-in. nozzle was a very conservative response to that concern. The structured packing was expected to enhance capacity while achieving the previous separation.

Problem The tower was not achieving the desired (and expected) separation.

Preliminary Diagnosis The initial plant consensus was that the problem was related to a failure to properly utilize the reboiler return nozzle and that some modification of the piping between the reboiler and the tower was what was required. This conclusion was driven at least in part by the same experience that caused the reboiler return nozzle to be increased from 6 to 12 in. Initial focus was on selecting the arrangement of the piping to be changed.

Consulting I arrived at the site late in the afternoon. I was shown the tower, which was not operating at the time, in anticipation of quickly implementing the changes upon which it was expected agreement would be reached the following day. I entered the control room but, due to a cloud of secrecy, was not permitted to view the control panel. The only documents the plant was prepared to supply were some construction drawings showing the recent modifications. I could not get process flow diagrams or P&IDs. I did not know what species were being processed by the tower.

Drawing Review Boil-up to the tower was supplied by a steam-heated forced-circulation reboiler. Liquid exited the tower via a 3-in. line and was pumped via another 3-in. line to the tube side of a two-pass horizontal shell-and-tube heat exchanger. The two-phase effluent from the reboiler exited the second pass via a 6-in. line which was swaged up to 12 in. about halfway along its travel to the tower. The steam to the shell side of the reboiler was controlled by the base level in the tower. The bottoms product was removed from the pump discharge upstream of the reboiler. The bottoms product was small compared to the reboiler feed and was removed on flow control.

Review of the drawings disclosed that the tower base level was normally operated below the bottom tangent line of the tower. It was either in the very shallow bottom head or in the outlet piping, most likely the latter. This had occurred because of the change in size of the reboiler return nozzle from 6 to 12 in. The tower originally only had a few inches of height provided for liquid surge, and this had been taken away by the reboiler return nozzle size increase, which was made eccentrically. It was my conclusion that the problem was instability caused by the lack of liquid surge in the tower.

Solution To solve this problem, I proposed that a small surge tank be installed alongside the tower; that the liquid from the tower drain into the new surge tank; that the pump take suction from the new surge tank; that the reboiler return flow be diverted to the new surge tank; that the tank be so designed as to separate the vapor and liquid entering from the reboiler; and that the vapor so separated pass from the new surge tank to the tower via a 12-in. line and the 12-in. reboiler return nozzle on the tower. The surge tank I proposed would be vertical, 30 in. in diameter, and 60 in. tangent to tangent. Plant personnel believed that they had sections of 30-in.-diameter pipe from which the tank could be easily fabricated. In case they did not, I offered a 24-in.-diameter alternative. In a pinch they could also have used an 18-in.-diameter alternative.

My proposal was accepted. Not only did it deal with the problem of lack of surge, but also it provided a superconservative solution to the concern about proper flow to the reboiler return nozzle. One of the concerns we had was that the 12-in. piping upstream of the tower nozzle was too large and there was uncontrolled separation of phases in the piping. With the new tank in place, the line from the reboiler to the new tank could be made smaller and uniform flow assured between the reboiler and the tank. I recommended that the two-phase inlet to the new surge tank be 8 in. in diameter. Since only vapor would flow to the tower, the biggest concern had disappeared.

Epilogue I was told that the recommendations I had made would be accepted in total and that the changes would be made in probably no more than a week. When I followed up about 2 weeks later, I was told that the modifications had been made on time, the tower had been restarted, and plant personnel were now very happy with its performance.

CASE STUDY 8.8 BAFFLING BAFFLES

Contributed by Chris Wallsgrove

Installation A caustic scrubber absorbing small quantities of carbon dioxide and hydrogen sulfide from light HC gas in a grass-roots ethylene plant. The tower contained three recirculating loops of caustic solution, each at a different caustic strength. Dilute caustic circulated through the lower section and the tower base. The tower was fabricated of stress-relieved CS and contained SS single-pass valve trays, 12 trays in each loop.

The tower base contained an angled or sloped baffle (Fig. 8.5, not to scale) which directed the descending caustic solution to one side of the base division baffle. The division baffle divided the tower base into two compartments: a larger compartment that served as the reservoir for the bottom caustic recirculation pumps and a smaller compartment from which a level controller regulated the flow of excess "spent" caustic out of the system. Excess solution overflowed the division baffle into the smaller compartment. In principle, the function of the baffles was similar to that frequently used to maintain constant head for thermosiphon reboilers (250).

The incoming cracked gas entered the tower opposite the angled baffle, which turned the gas upward toward the trays in order to contact the descending caustic solution.

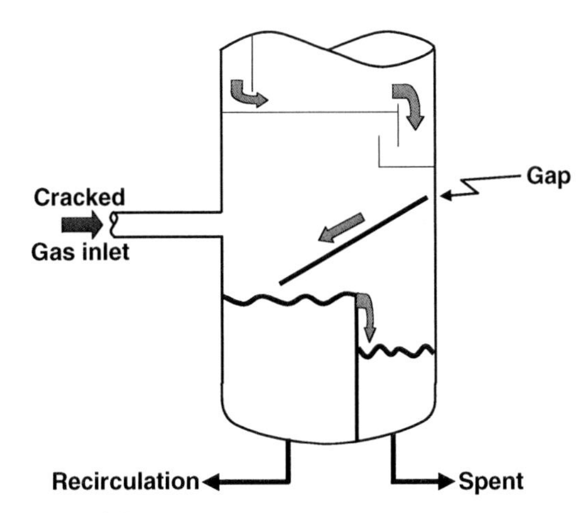

Figure 8.5 Base of caustic scrubber, showing angled baffle.

Problem From initial start-up, the consumption of fresh caustic and the make of spent caustic grossly exceeded design. At high gas rates the bottom-caustic recirculation pump frequently lost suction, which precluded achieving design rates. Despite this the tower achieved satisfactory acid gas removal.

Troubleshooting The only conceivable explanation was that much of the descending caustic solution found a path into the "spent" side. The division baffle had been rigorously inspected during precommissioning. It had no open hatchway or other hole which could have allowed excess caustic flow into the spent side. A variety of theories were proposed, including damage to the division or angled baffle and incorrect orientation or assembly. None of these were possible to test or correct on-line. The plant was shut down and the tower opened for inspection.

Findings The large angled baffle opposite the gas entry nozzle was installed after the tower shell was erected. It was specified as removable in order to gain access for inspection to the spent side. Because the tower was stress relieved in the fabricator shop, the angled baffle panels were designed to be attached to clips. The clips were welded to the shell in the shop, prior to shell stress relieving. This circumvented the need to weld the baffle to the shell in the field, which could have negated the shop stress relief. As such, the angled baffle was not sealed to the shell. In fact, the tower out-of-roundness and the baffle tolerances were such that an irregular gap of up to 25 mm existed between the shell and the baffle!

It was concluded that the incoming gas was blowing the descending liquid UP the angled baffle such that an accumulation of liquid existed at the wall–baffle junction. This liquid drained through the inadvertent gap into the spent side of the tower base. The higher the gas inlet velocity, the higher the tendency to blow liquid uphill and allow it to thus drain away. This was so prevalent at high gas rates that the recirculation side was starved of liquid and the pumps therefore lost suction.

Solution A strip of CS was welded around the shell–baffle junction to totally seal the gap. The CS tower shell did not need to be totally stress relieved because the caustic solution was very dilute under all conceivable circumstances and the tower operated at close to ambient temperature. Thus the mechanical design had been overly conservative.

Upon restart the tower performed fully in accordance with the design.

CASE STUDY 8.9 A 7-FT VORTEX

Contributed by Ron F. Olsson, Celanese Corp.

Installation A 6.5-ft-ID tower containing random packings. Boil-up was supplied by a vertical thermosiphon reboiler.

Problems When pushed to high rates, the tower appeared to experience a premature reboiler limitation.

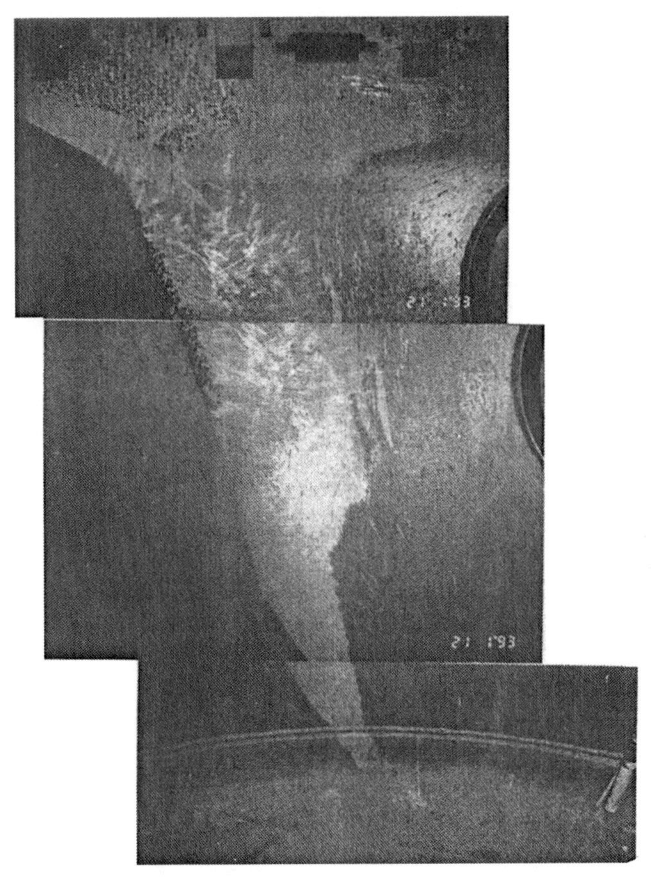

Figure 8.6 Base of tower at turnaround. The shiny path in the dark coloration was caused by a giant vortex.

Investigation A gamma scan showed a high degree of aeration in the tower base liquid. Since the tower was handling a foaming system, the aeration was attributed to foaming.

Inspection When the tower was opened in the next turnaround, the inside of the tower shell near the base of the column was observed to have a dark coloration, except for a shiny path carved by a tornado initiating at the bottom outlet nozzle and rising 7 ft up the walls (Fig. 8.6). It was realized that the observed aeration was not foaming but a giant vortex.

Cure A vortex breaker was added. The reboiler limitation was no longer observed after this.

Chapter 9

Chimney Tray Malfunctions: Part of Number 7 on the Top 10 Malfunctions

Intermediate draws are the third most troublesome internal in the tower (255), following the tower base/reboiler return (Chapter 8) and packing liquid distributors (Chapter 6). In the last decade, the number of reported intermediate-draw malfunctions has experienced a rapid rise. It appears that either the design of intermediate draws is becoming a forgotten art or pushing towers to maximum capacities is unveiling flaws and bottlenecks previously hidden in oversized towers. In any case, there is much room for improvement. Good and bad practices for intermediate-draw design are described elsewhere (250). Intermediate-draw malfunctions are by far most troublesome in refinery towers because of the large number of intermediate draws in each refinery main fractionator.

About half of the reported cases occurred in chimney trays (this chapter), the other half in downcomer trap-outs (including draw boxes, Chapter 10). Leakage and overflow, that is, undesirable liquid descent from a chimney tray, top the chimney trays malfunctions. These leaks led to losses in product recovery, yielded off-specification products, overloaded towers and vacuum systems, and caused pump cavitation. Mass and energy balances have been invaluable in diagnosing these leaks. Restriction to the liquid removal from chimney trays is also a major issue, often initiating premature flooding in the tower. Undersized outlet lines, downpipes, and inadequate degassing are the most common causes of the restriction.

Level measurement problems, predominantly on chimney trays in refinery vacuum towers, suggest caution when interpreting level readings in these towers. Thermal expansion has also been troublesome, particularly in this service. Other malfunctions include chimneys impeding liquid flow, interference of vapor from chimneys with incoming liquid, coking, fouling, freezing, and damage.

Distillation Troubleshooting. By Henry Z. Kister

CASE STUDY 9.1 HEAT BALANCES CAN IDENTIFY TOTAL DRAW LEAKS

Installation A refinery vacuum tower (Fig. 9.1) contained two PAs. Each PA is a direct-contact condenser. This condenser consisted of a packed bed and a total-draw chimney tray underneath. Liquid from each chimney tray was pumped and cooled, and

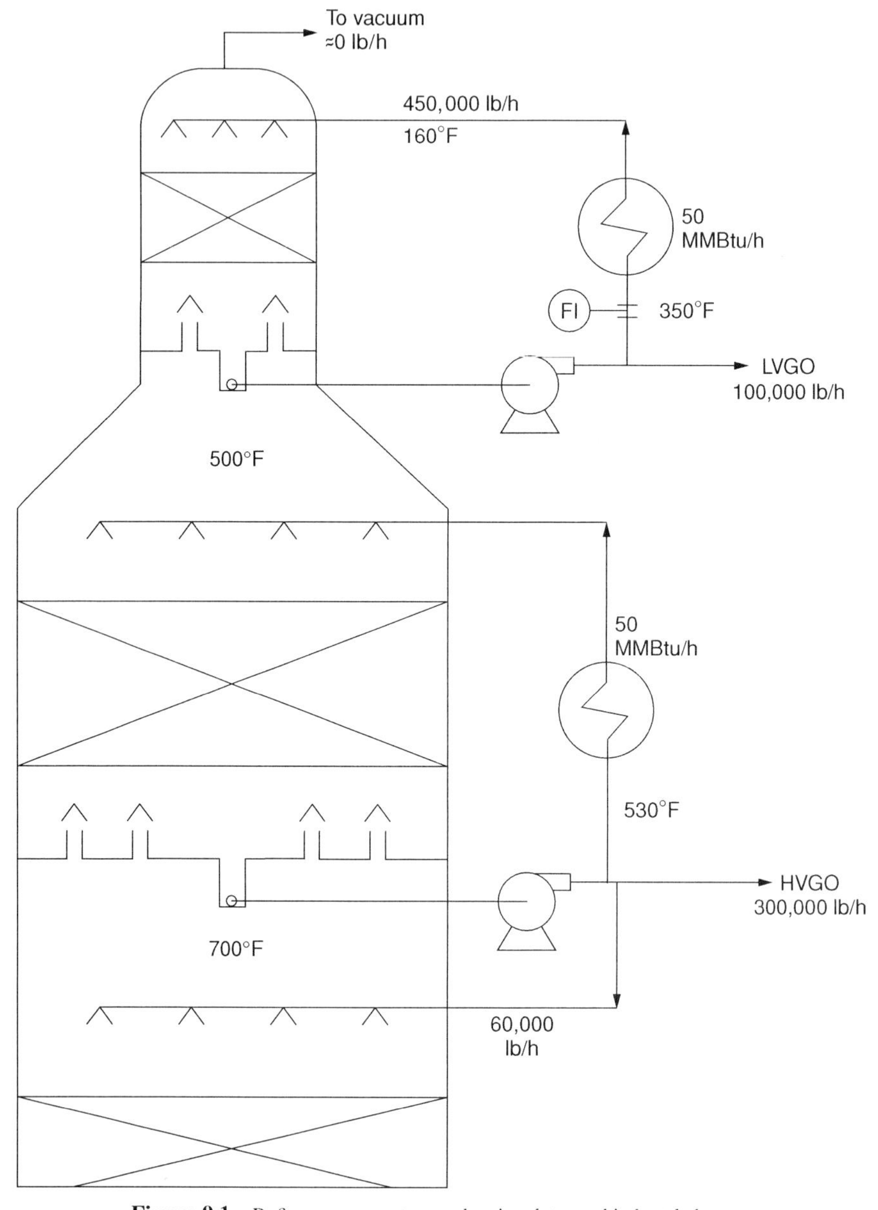

Figure 9.1 Refinery vacuum tower, showing data used in heat balances.

the cold liquid was distributed to the packing, where it cooled and condensed the ascending vapor. The upper, light vacuum gas oil (LVGO) PA was a total condenser (this is common in dry vacuum towers that use no steam). All the vapor entering the top bed was condensed and drawn from the upper chimney tray to become the LVGO product.

The lower heavy vacuum gas oil (HVGO) PA was a partial condenser. It operated much hotter. Uncondensed vapor from this section ascended to the LVGO section. All the condensed vapor was withdrawn from the lower chimney tray. Most of the condensate was the HVGO product, but some was returned to the tower as reflux to the wash section below.

Problem The HVGO temperatures were lower than expected, which bottlenecked heat transfer in the HVGO coolers. As a result, an additional heat load was shifted to the smaller LVGO coolers, which were bottlenecked as well. With cooling maximized on both LVGO and HVGO circuits, the vacuum system was struggling to keep vacuum.

Troubleshooting A heat balance was compiled on the LVGO section. Since it was a total condenser (the chimney tray was a total-draw tray), the flow rate of vapor in equalled the flow rate of LVGO product out. The LVGO section heat duty is therefore determined as

$$Q_{\text{LVGO}} = (H_{\text{V},500^\circ\text{F}} - h_{\text{L},350^\circ\text{F}})\, M_{\text{LVGO}} = 200\ \text{Btu/lb} \times 100{,}000\ \text{lb/h}$$
$$= 20\ \text{MMBtu/h}$$

where Q is the heat duty (Btu/h), H_V and h_L are vapor and liquid enthalpies (Btu/lb), and M_{LVGO} is the LVGO product flow rate (lb/h). Altogether, 20 MMBtu/h was removed from the vapor in this section. The LVGO section heat duty can also be calculated from the heat removed in the coolers:

$$Q_{\text{LVGO}} = M_{\text{PA}} C_{\text{P}}(350^\circ\text{F} - 160^\circ\text{F}) = 450{,}000 \times 0.6 \times 190 \approx 50\ \text{MMBtu/h}$$

where M_{PA} is the pumparound flow rate (as measured on the flowmeter; Fig. 9.1) (lb/h), and C_{P} is the specific heat (Btu/h lb°F).

The heat removed by the coolers was much larger than that required to condense the LVGO product flow rate. Instruments were checked. Flow rates were compared to values calculated from control valve openings, pump curves, and spray nozzle pressure drops. The checks verified that none of the measured numbers were in gross error. There was nothing in the measurements that would even get close to explaining the large difference between the heat duties.

Analysis The heat balance is based on the assumption that all the vapor entering the LVGO section exits as LVGO product. This assumption is valid only if condensed liquid is not escaping some alternate route. There are two plausible alternate routes: entrainment from the top of the tower or leakage/overflow from the chimney tray.

Entrainment from the top of the tower can readily be detected by slop in the ejector steam condensate. In this tower, little slop was produced. This leaves leakage/overflow from the chimney tray as the only plausible explanation. The leak/overflow would be quite large. At 200 Btu/lb enthalpy difference between entering vapor and condensate (above) and a heat duty of 50 MMBtu/h, the total condensate make in the LVGO section is

$$M_{\text{Condensate}} = 50 \times 200/10^6 = 250{,}000\ \text{lb/h}$$

So a total 250,000 lb/h LVGO was condensed. Of this, 100,000 lb/h became the LVGO product. The balance leaked or overflowed down the chimney tray.

The leaking or overflowing LVGO ended up in the HVGO section, where it lowered the bubble point. This lowered the HVGO draw temperature, which bottlenecked the HVGO coolers.

Cure In the next turnaround, the chimney tray was seal welded. No more problems occurred.

CASE STUDY 9.2 ANOTHER LEAKING TOTAL-DRAW CHIMNEY TRAY

Installation A refinery vacuum tower (Fig. 9.2). The tower was similar to that in Case 9.1, except that the PA coolers also cooled the LVGO and HVGO products.

Problem The tower could not be operated without a bleed of HVGO into the HVGO (Fig. 9.2). Closure of the manual valve on the bleed line caused a loss of liquid level on the upper chimney tray and the level valve on the LVGO product closed. The tower was forced to operate with a bleed.

A second problem was the vacuum in the tower. The tower was operating close to the ejector capacity limit and had problems maintaining vacuum.

Testing In the next turnaround the LVGO draw tray was water tested. There was no sign of leakage in the test.

Diagnosis The need to use an HVGO bleed is proof that the chimney tray is leaking. The observation of level loss when the bleed is closed is strong evidence that the problem is leakage, not overflow, in this case.

A water test would conclusively identify a leak but is inconclusive for affirming no leaks during operation. Thermal expansion at elevated temperature may initiate joint leakage not apparent at ambient conditions. Good seal welding of the tray joints is needed to positively eliminate leakage.

A Beneficial Leak? Initially, it was attempted to minimize the HVGO bleed, which was run at 1000 BPD. At one time, the bleed was raised to 2000 BPD. Raising the bleed to 2000 BPD unexpectedly solved the vacuum problem, permitting full vacuum to be achieved. Apparently, the HVGO bleed served as an absorption oil that helped absorb lighter components from the overhead vapor and unloaded the ejectors. Product purity was not an issue because the LVGO product was small and was blended with the HVGO. So the leak turned out beneficial.

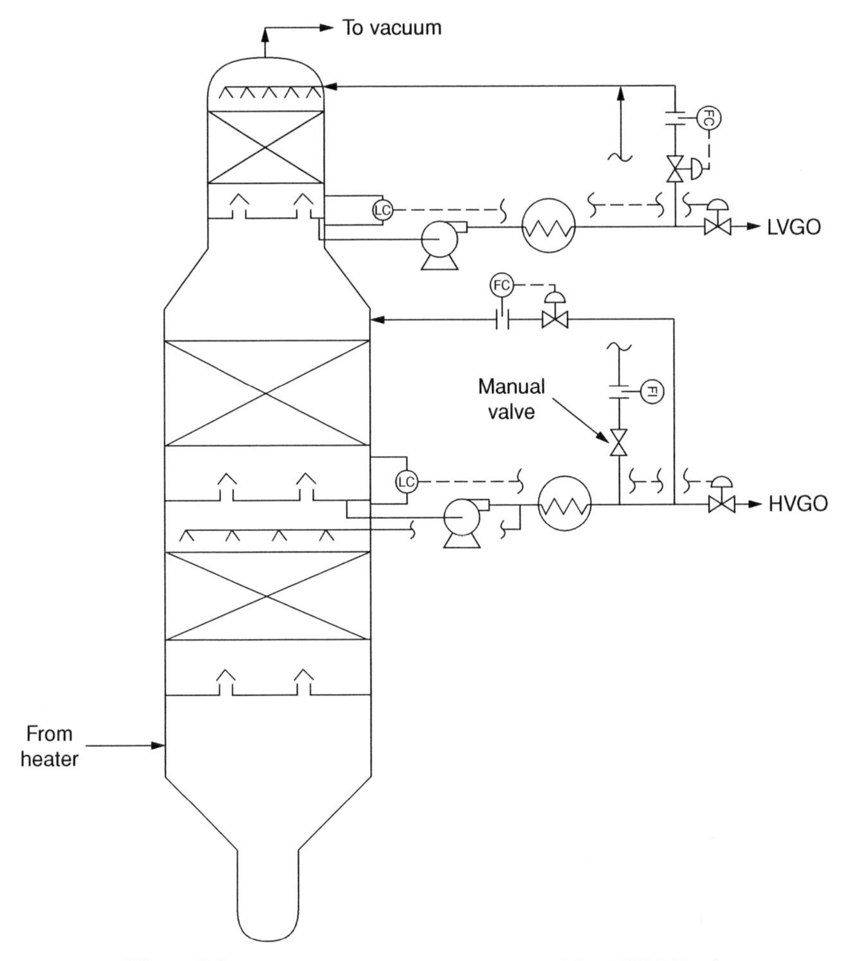

Figure 9.2 Refinery vacuum tower, showing HVGO to LVGO bleed.

CASE STUDY 9.3 CHIMNEY TRAY OVERFLOW TARNISHES SUCCESSFUL REVAMP

Installation A refinery vacuum tower similar to that in Case Study 9.1. Focusing on midtower, the liquid collected on the upper chimney tray was split three ways: product HVGO, reflux to the wash section below (through the lower spray nozzles in Fig. 9.3), and circulating PA, which was cooled and sprayed to the top of the HVGO bed.

In the wash bed, HVGO reflux washed heavy ends ("asphaltenes") and organometallic compounds from the rising vapor. Spent reflux, "overflash," from the lower chimney tray was routed to the flash zone, where it combined with the liquid portion of the 50%/50% weight vapor–liquid feed, forming the resid bottom product. There was no stripping section (this was a "dry" tower). The resid product was worth a lot less than any of the tower distillates (LVGO, HVGO).

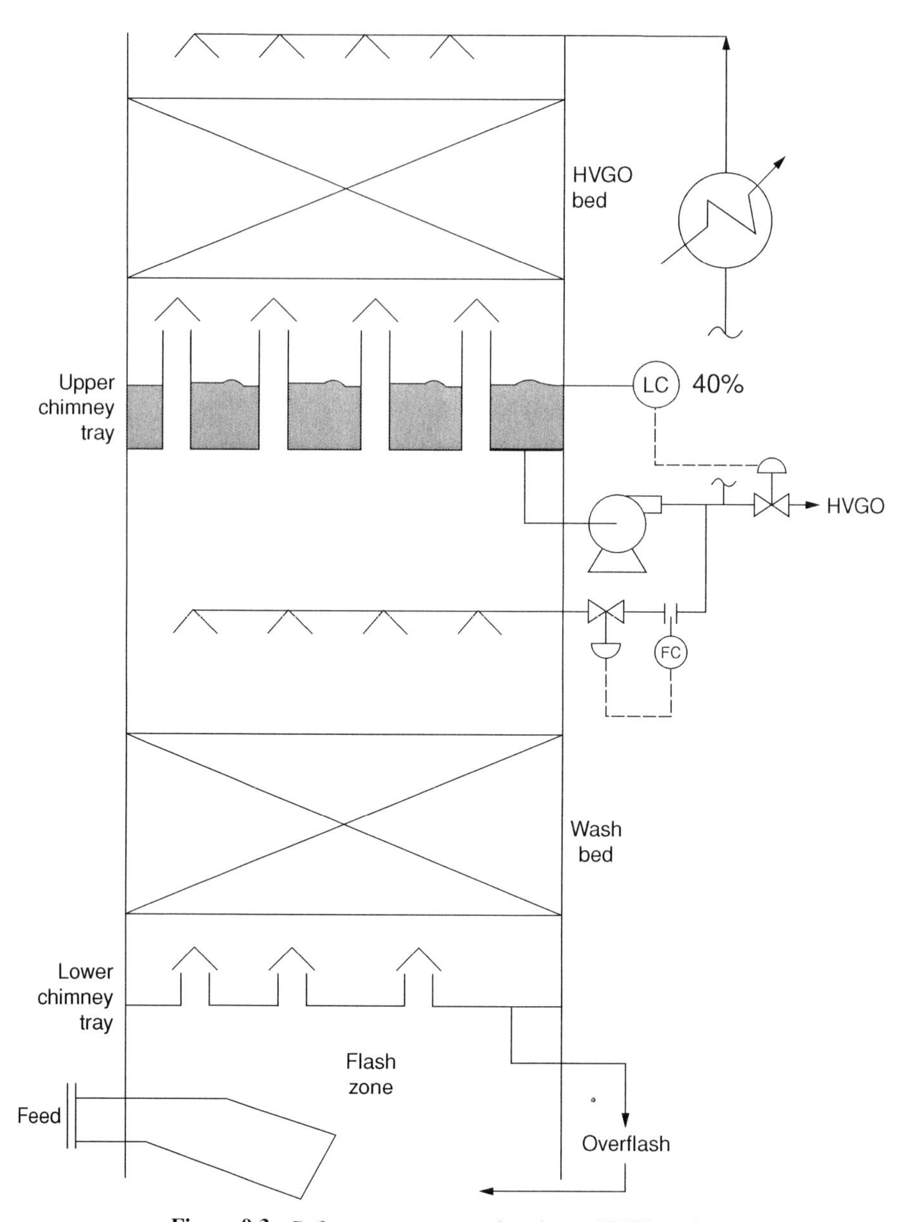

Figure 9.3 Refinery vacuum tower, focusing on HVGO section.

Debottleneck The tower was recovering 15,000 BPD of HVGO. It was desired to increase this quantity. A debottleneck study showed that increasing the size of the inlet line ("transfer line") from 12 to 24 in. and improving the feed inlet device ("vapor horn") will enhance HVGO recovery to 18,000 BPD. Payout for these modifications was short and they were implemented in the next turnaround. At the same time, the wash spray nozzles and the wash grid packing, both of which were coked, were replaced. The upper chimney tray was seal welded and water tested. No leakage was apparent at the test.

Problem Upon restart, the tower could not recover even the previous 15,000 BPD of HVGO product. It recovered about 13,000 BPD of HVGO, and the recovery was gradually declining, reaching 12,000 BPD when the tests below started. The lost HVGO ended in the resid, and the resid make was up 2000–3000 BPD. Instead of recovering more HVGO, the tower was making more resid.

Troubleshooting In troubleshooting multi-side-draw towers (such as refinery fractionators) that experience product recovery problems, the first step is to search for the coexistence of the following three symptoms:

- The tower recovers less of an upper product.
- The tower recovers more of a lower product.
- The temperatures in the section between these two products are low.

There is only one physical phenomenon that can simultaneously produce these three symptoms: Some of the upper product descends down the tower, ending up in the lower product. The analogous behavior in a simple two-product tower is having some of the top product end in the bottom. In this case, the wash bed temperatures after the turnaround were about 80°F colder than before. So all three symptoms existed. This meant HVGO product was going through the wash bed into the tower bottom.

To verify, a heat balance was compiled on the HVGO section, similar to that described in Case Study 9.1. Based on the cooler duty, the total HVGO product condensed (not including the wash) was 22,000 BPD, not 13,000 BPD. The meters were checked and validated. The heat balance finding was real. A total of 9000 BPD of HVGO was disappearing. Nine thousand barrels are a lot of barrels, and they do not just vanish. There are not many hiding places. The only conceivable route for them to get out and to the wash zone was by leak or overflow from the upper chimney tray.

The chimney tray level was checked. The transmitter read 40%, which was halfway up the 36-in.-tall chimneys. The level glass confirmed the transmitter reading. Forty percent was the level at which the chimney tray was always operated. There were some changes to the crude feedstock, but these were minor. Gamma scans taken 8 months earlier showed the liquid level at 50–60% of the chimney height. There was no apparent reason to suspect overflow. Likewise, there was no reason to suspect leakage. The chimney tray was seal welded and water tested. The start-up records were reviewed. There were no reports of upsets or pressure surges that could have damaged the tray, and it was too close to the start-up to suspect corrosion damage.

With all the logic arguing against, the fact remains that every day 9000 barrels of HVGO found a path to the resid via the wash section. Heat balances do not lie if based on correct instrument readings.

Cure There is one positive way of testing for overflow: Draw more product and watch the HVGO pump. Once the level falls and the pump cavitates, there is no overflow.

The level control on the HVGO chimney tray was placed on "manual" and the HVGO product valve opened, increasing the HVGO draw rate. As the valve was opened, the resid make dropped and the wash bed warmed up.

The HVGO draw rate was increased first to 15,000 BPD, then to 17,000, 19,000, 20,000, and eventually 22,000 BPD. When the HVGO draw rate reached 22,000 BPD, the draw tray level was lost and the pump cavitated. The HVGO draw was then reduced to 20,000 BPD. It ran stably at that draw rate. A short time later, the operators returned the level control to automatic with a set point at 30% (whatever it meant). At that level, it has operated stably since, drawing 20,000 BPD of HVGO.

CASE STUDY 9.4 LEAKING CHIMNEY TRAY UPSETS FCC FRACTIONATOR HEAT BALANCE

Contributed by W. Randall Hollowell, CITGO, Lake Charles, Louisiana

Installation A grass-roots FCC main fractionator, over 20 ft ID. Hot reactor effluent feed was desuperheated by direct contact with cooled recirculating liquid, "slurry pumparound" (SPA). The heat absorbed was used for steam generation, no preheat. Above the SPA was a short wash section in which a small reflux removed entrained slurry and heavy ends from the ascending tower vapor. Above the wash section was a seal-welded chimney tray with a single 5-ft-tall chimney. This chimney tray collected the first distillate side product, heavy cycle oil (HCO). There was a fractionation section above the HCO chimney tray. This zone was refluxed by the next higher side product, intermediate cycle oil (ICO), initiating at a higher up chimney tray. The ICO PA condensed the ICO product (recycled back to the reactor) and provided dropback reflux.

Problem There was lower steam generation from the SPA than expected. Trays near the top of the tower had much higher loadings than expected and top-tray reflux rate was much higher than expected. This caused flooding of the top trays at high conversion and high feed rates.

Investigation The SPA steam production observed was 35 MMBtu/h lower than it should have been by the tower heat and mass balance. This corresponded to about one-third to one-half of the SPA heat duty. This could only be explained by a leak of 950 barrels per hour (BPH) through the HCO chimney tray, down into the tower wash section. This leaking HCO was revaporized in the SPA section, taking away 35 MMBtu/h from the duty available for steam generation. The 35 MMBtu/h was eventually removed by the overhead reflux, causing high loads in the top of the tower.

A confirmation of the leak was obtained by stopping flow to and from the HCO chimney tray (upsetting the tower for a short time) and determining the leak rate from the slope of the dropping level. To perform this test, first the level on the ICO chimney tray above was lowered as much as possible. Then the ICO reflux to the ICO/HCO fractionation section was stopped (allowing the level to build up on the ICO chimney tray), and simultaneously, the HCO dropback was stopped. The slope estimated a 970-BPH leak rate, in good agreement with the heat balance.

Some HCO material went up into the ICO during the test. The leak maintained a very high wash rate below the HCO tray. Thus the HCO was still relatively light. There was little danger of coking the trays and no dirty or very high boiling oil went into the ICO and thus the reactor.

A check of operator logs indicated that, while the tower was inventoried during initial start-up, it was filled to the top with oil. As liquid ascended inside of the 5-ft-tall chimney, it exerted upward pressure against the floor of the HCO chimney tray. The upward pressure would have peaked at 5 ft of liquid (equivalent to 1.5–2 psi) when the vapor passages in the chimney completely filled with ascending oil. This was well above the uplift resistance of the chimney tray. This oil pressure is believed to have ripped out part of the bottom seal weld and bowed up the floor plate.

Solution Until the first shutdown, the HCO level set point on the chimnet tray was reduced from 50 to 30%, reducing the leak rate by 25%. This eliminated flooding of top trays and increased steam production. When the tower was opened, the floor of the HCO chimney tray was found to be buckled up. It was repaired and the tower heat balance was restored.

Morals

- Start-up inventory should be monitored so that gross overfilling will not occur even when level indicators are not working.
- Heat and material balances are fundamental to understanding and troubleshooting tower operations.
- Plant tests are invaluable for validating theories.

CASE STUDY 9.5 FLAT HATS CAN INDUCE LEAKS

Installation A caustic wash tower in an olefins plant absorbs small quantities of acid gas (CO_2, H_2S) from a HC gas stream. The tower was equipped with sieve trays. The tower contained a number of caustic circulation circuits. Each circuit consisted of several trays. At the bottom of each circuit caustic was collected by a total-draw chimney tray. From the chimney tray, the caustic was pumped back to the tray at the top of the circuit.

Problem During commissioning, each circuit was tested by water circulation. During the test, it was impossible to hold level on the draw trays, and the pumps were losing prime.

Investigation The tower manholes were opened. A massive downpour was observed through the chimneys. The hats above the chimneys were entirely flat. Liquid collecting on these hats flowed under the hat and dropped into the chimneys (Fig. 9.4). The liquid descended onto the hats by weep from the holes of the sieve tray above.

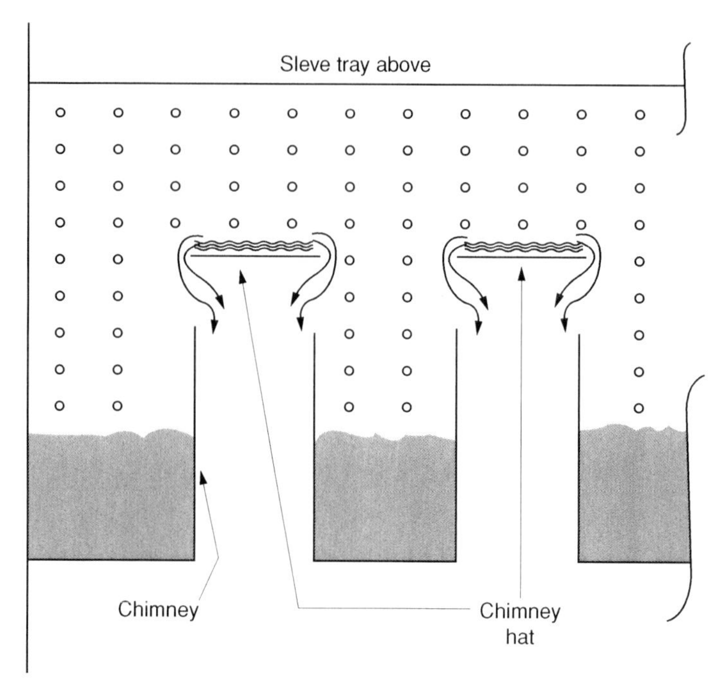

Figure 9.4 Liquid downpour through chimneys with flat hats as seen during commissioning water test.

Solution The problem was eliminated when vapor flow was established once the tower was put in operation.

Morals

- Flat hats are not a good idea for chimney trays.
- Water tests are great, but good engineering judgment is needed for their interpretation.

CASE STUDY 9.6 HYDRAULIC GRADIENT ON A CHIMNEY TRAY

Henry Z. Kister, Betzalel Blum, and Tibor Rosenzweig, reference 264. Reprinted with permission from *Hydrocarbon Processing*, by Gulf Publishing Co., all rights reserved

A refinery vacuum tower experienced HVGO loss to the residue leaving from the bottom of the tower. The lighter gas oil reduced the viscosity of the residue, making it unfit for asphalt. Penetration was high (~500 mm, compared to ~50 mm for good asphalt).

The cause was the HVGO total-draw tray (Fig. 9.5*a*). With the long edges of the chimneys perpendicular to the liquid flow, the flow area around each chimney was small. This generated a large hydraulic gradient. The liquid built up to the chimney height on the chimneys opposite from the outlet draw and overflowed into the chimneys and then onto the tower bottom.

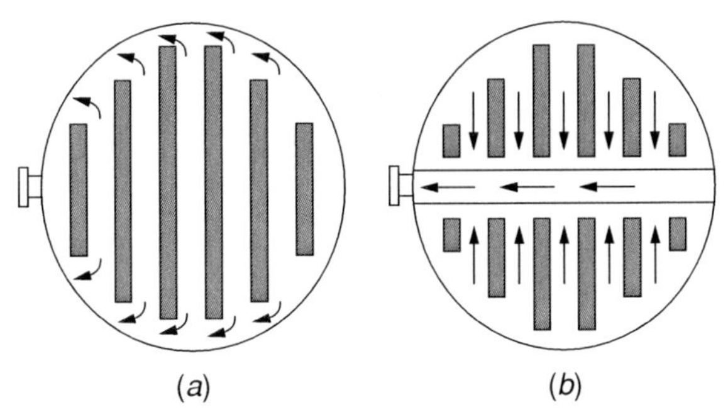

Figure 9.5 Chimney arrangement on HGVO draw tray and liquid movement: (*a*) in actual tower, leading to high hydraulic gradients, (*b*) good design practice, minimizes hydraulic gradients. (From Ref. 264. Reprinted with permission from *Hydrocarbon Processing* by Gulf Publishing Co. All rights reserved.)

A good chimney tray design is shown in Figure 9.5*b*. This design minimizes hydraulic gradient. In this existing tower, a much easier and cheaper solution was to rotate the tray in Figure 9.5*a* by 90°. This solved the problem. No more HVGO is lost to the residue, and penetration of the residue has been good since the modification.

CASE STUDY 9.7 "LEAK-PROOF" CHIMNEY TRAYS IN AN FCC MAIN FRACTIONATOR

Henry Z. Kister, Betzalel Blum, and Tibor Rosenzweig, reference 264. Reprinted with permission from *Hydrocarbon Processing*, by Gulf Publishing Co., all rights reserved

As part of the revamp to maximize capacity of the FCC main fractionator in Case Study 9.9, the LCO and HCO draw/PA offtake trays were replaced by total-draw chimney trays. The purpose was to minimize reflux to the section below. Excess reflux represents additional liquid and vapor recycle that consumes capacity. The reflux was minimized by careful monitoring and control while avoiding any fluctuating leakage or overflow from the chimney tray above. Each chimney tray was to be seal welded.

A schematic of either chimney tray is shown in Figure 9.6*a*. Liquid from the two-pass tray above descended via side downcomers, which terminated in seal pans. All liquid from the chimney tray was drawn from a sump (not shown) located beneath the chimney tray. The downcomers from the chimney tray to the section below were converted to overflows by raising outlet weir heights from about 350 to 610 mm. Normal liquid level on the chimney trays was about 300 mm and the overflow downcomers were inactive. However, should an upset occur and the chimney tray liquid level exceed 610 mm, the liquid would overflow into the downcomers.

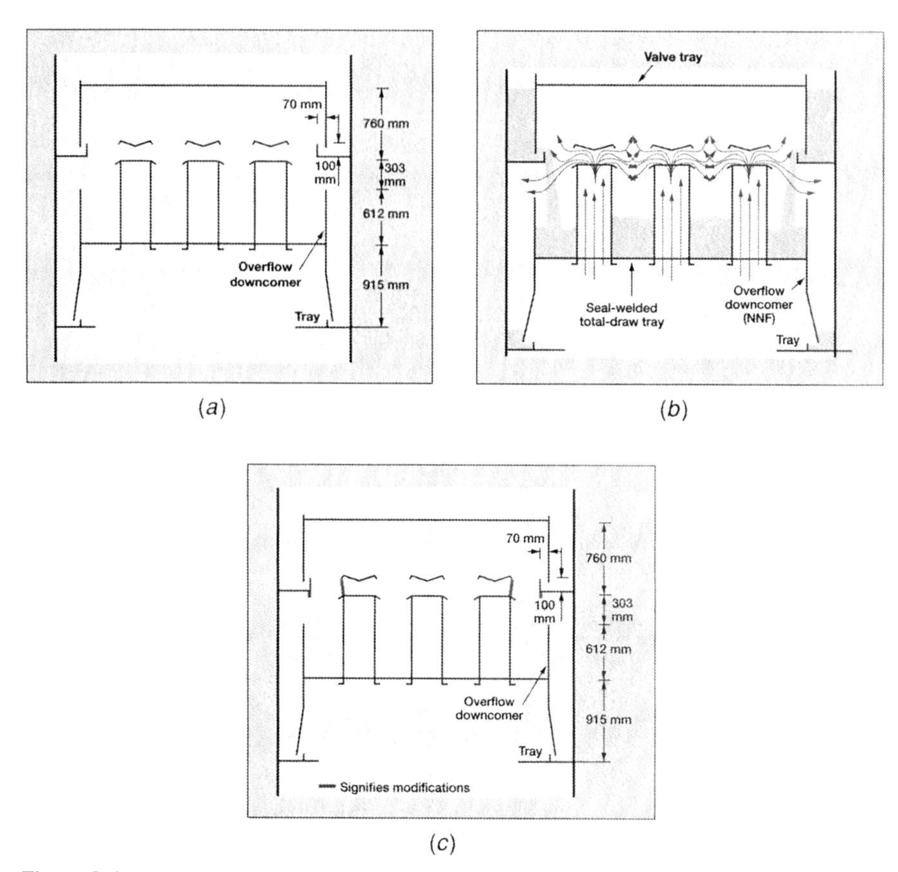

Figure 9.6 Total-draw chimney tray: (*a*) initial design; (*b*) expected flow patterns; (*c*) modifications to circumvent liquid bypass around chimney tray. (From Ref. 264. Reprinted with permission from *Hydrocarbon Processing* by Gulf Publishing Co. All rights reserved.)

At the design stage, the seal pans and the chimney tray were on different drawings, which had been approved for fabrication. A last-minute drawing review put together the sketch in Figure 9.6*b*, which revealed a major flaw. The gas issuing from the outside chimneys and blowing toward the tower wall would blow liquid descending from the seal pan directly into the overflow downcomers. Thus, despite the seal welding of the chimney tray, liquid would bypass it.

Figure 9.6*c* shows how the problem was circumvented. The openings of the outside chimneys that would blow gas toward the wall were closed. A 25-mm vertical drain lip was installed at the bottom of each seal pan to prevent the issuing liquid from crawling underneath and ending in the overflow downcomers.

Moral When it comes to troubleshooting points of transition (feeds, draw-offs, bottom sumps, chimney trays), you do not need an expert. You need a sketch.

CASE STUDY 9.8 LIQUID-LEVEL MEASUREMENT ON A CHIMNEY TRAY

Henry Z. Kister, Betzalel Blum, and Tibor Rosenzweig, reference 264. Reprinted with permission from *Hydrocarbon Processing*, by Gulf Publishing Co., all rights reserved

In applications such as HVGO draw trays in refinery vacuum towers, it is imperative to avoid any leakage to the section below. In this service, any leakage represents loss of high-priced distillate into low-priced residue. The section below is packed. Any leakage is maldistributed and therefore contributes nothing to the packing separation. To avoid leakage, chimney trays in this service are almost totally seal welded. The problem is that in large-diameter towers, if all the joints are seal welded, including those to the support ring, the tray may buckle due to thermal expansion.

Lieberman presented a couple of techniques that can overcome this problem (304, 313). One of these is shown in Figure 9.7*a*. The joint of the tray to the tray support ring is not seal welded, and the bolt holes allow tray movement, thus circumventing the buckling. To prevent leakage, the support ring is covered with an angle iron approximately 80 mm by 80 mm. The angle iron is rolled to the tower diameter

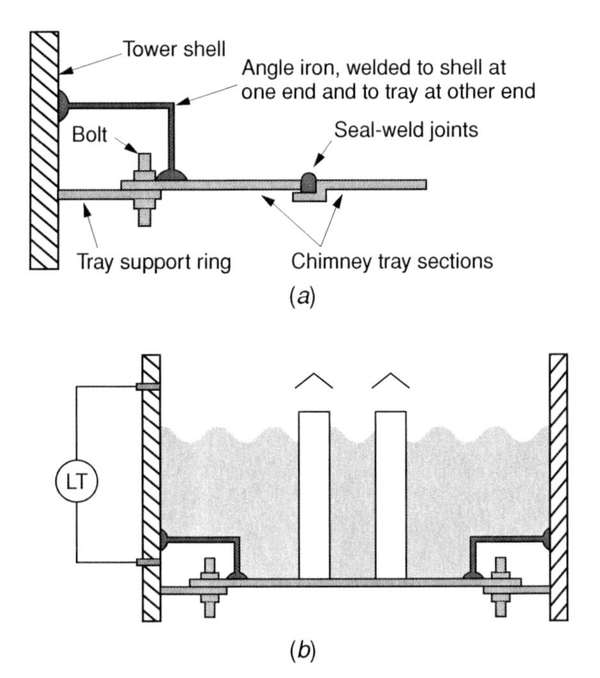

Figure 9.7 Technique to prevent chimney tray leakage while permitting thermal expansion: (*a*) principles of technique (contributed courtesy of Norman P. Lieberman, private communication); (*b*) incorrect application of technique that disabled level transmitter. (From Ref. 264. Reprinted with permission from *Hydrocarbon Processing* by Gulf Publishing Co. All rights reserved.)

providing a snug fit at the tower wall. One edge of the angle iron is seal welded to the shell; the other is seal welded to the tray. At the right thickness, this angle iron is flexible enough to handle the thermal expansion and shields the support-ring joint from the liquid.

In one tower, a chimney tray was revamped with the angle iron technique (Figure 9.7*a*). Upon restart, the tray appeared to be dry despite the seal welding. A review showed that the angle iron was installed over the lower level transmitter nozzle (Figure 9.7*b*), disabling it from seeing the liquid level. The problem was overcome by measuring the static head in the outlet pipe, a technique described by Martin (330).

CASE STUDY 9.9 A CHIMNEY TRAY BOTTLENECKING FCC MAIN FRACTIONATOR

Henry Z. Kister, Betzalel Blum, and Tibor Rosenzweig, reference 264. Reprinted with permission from *Hydrocarbon Processing*, by Gulf Publishing Co., all rights reserved

Installation A FCC fractionator originally designed to handle 15,000 BPD. Over the years, it had been debottlenecked to 24,000 BPD.

Problem As the feed rates were raised to about 24,000 BPD, a tray limitation was encountered in the gasoline–LCO fractionation section (Fig. 9.8*a*). At a feed rate of 20,000 BPD, there was good separation of gasoline from LCO. As the feed rate was raised, the separation worsened. Since the refinery had to keep the gasoline end point on specification, this separation difficulty lost gasoline to the LCO. At 24,000 BPD, nearly 4% of the gasoline was lost to the LCO. The overlap between the 5% ASTM D86 point of the LCO and the 95% ASTM D86 point of the gasoline was 15°C. At the higher rates, there were some signs of flooding in the upper part of the tower, with pressure drop apparently high. The evidence for flooding was not very conclusive since a detailed pressure survey, as normally recommended, was not conducted.

Troubleshooting The gasoline–LCO fractionation section contained one-pass conventional valve trays at 610 mm spacing. Underneath there was an LCO PA section that contained two-pass conventional valve trays at 760 mm tray spacing. Hydraulic calculations were done using both a proprietary method and a published method that we consider reliable for valve trays. The results predicted that at 24,000 BPD of feed the trays in these sections should operate at 80% of flood and should not experience a bottleneck. There was nothing to suggest any flooding further down the tower.

Figure 9.8*b* is a gamma scan of the upper 13 trays with the tower running at 24,000 BPD. The scan shows normal operation below tray 7. From tray 7 up, considerable entrainment is apparent in the vapor spaces. There is some uncertainty as to whether the entrainment actually started at tray 7 or at the chimney tray, due to the possibility of interference in the vapor space above tray 7. Froth heights on trays 6 and up were also higher than those on the trays below.

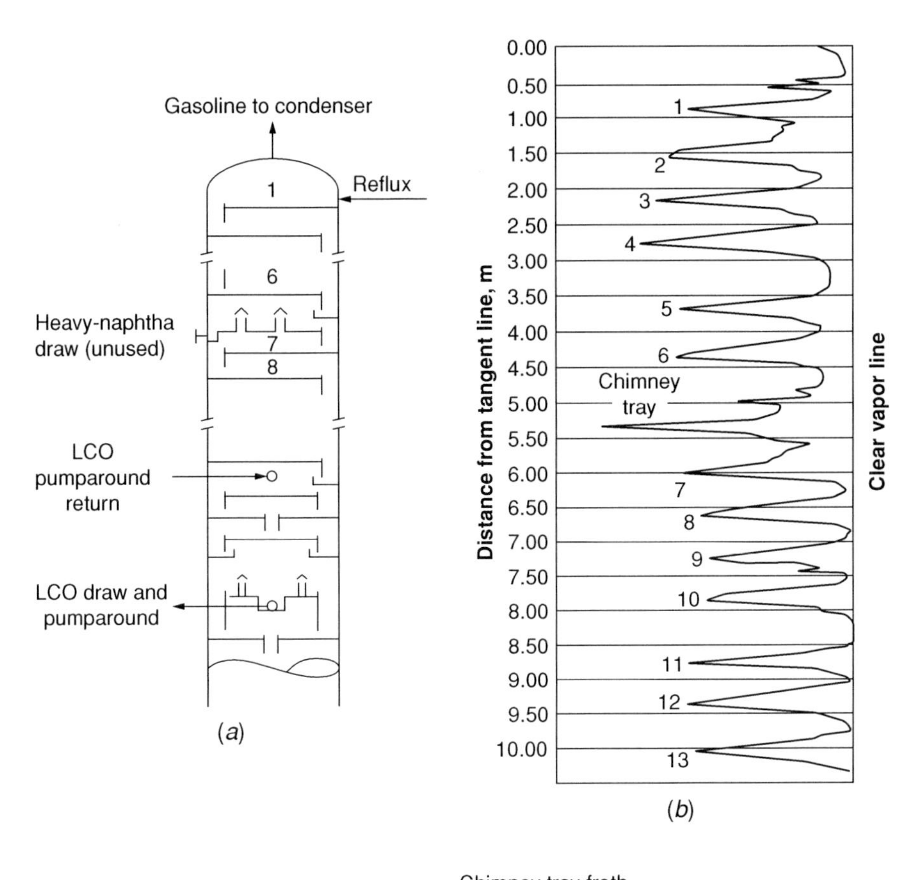

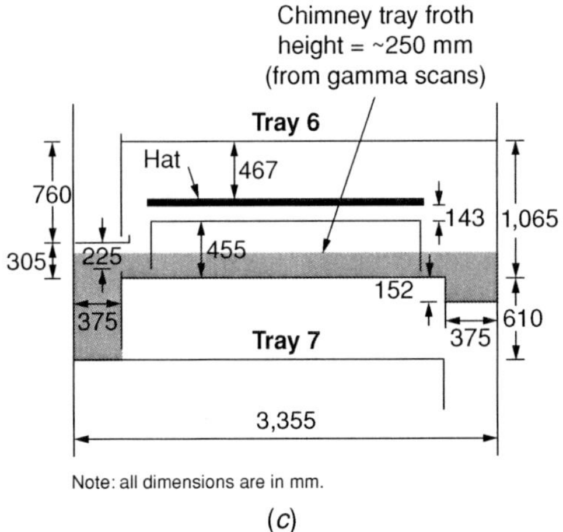

Figure 9.8 Chimney tray bottlenecking FCC main fractionator: (*a*) top section of FCC main fractionator; (*b*) gamma scan of top 13 trays of FCC main fractionator; (*c*) elevation sketch of chimney tray beneath tray 6, showing liquid level. (From Ref. 264. Reprinted with permission from *Hydrocarbon Processing* by Gulf Publishing Co. All rights reserved.)

Figure 9.8*c* is a sketch drawn to scale of the region where the entrainment initiated. Between trays 6 and 7, there was a chimney tray that was originally intended for drawing heavy naphtha. In the mode of operation during the test, no heavy naphtha was drawn from the tray.

Focus on Chimney Tray From the gamma scans, we estimated that the chimney tray was holding 250 mm of aerated liquid. This is surprisingly high considering that the outlet weir was only 80 mm tall and that on the chimney tray there should be very little mixing of vapor and liquid. For some reason, aerated liquid was backing up on this chimney tray about 170 mm above the weir. Such a backup is unlikely to be caused by a hydraulic gradient because the liquid from tray 6 entered the chimney tray just upstream of the outlet downcomer (Fig. 9.8*c*).

A close observation of Fig. 9.8*c* shows that the vertical height of the opening at the downcomer inlet is narrow. With the seal pan descending to 305 mm above the chimney tray floor and the weir rising 80 mm above the floor, the open window at the downcomer entrance is 225 mm wide. Had the chimney tray contained pure liquid, the window would have been large enough to allow the liquid to descend. However, the impact of liquid cascading from the seal pan onto the chimney tray aerated the liquid entering the downcomer, generating froth. The volume of this froth is much larger than the liquid volume. Consequently, the narrow window at the downcomer entrance is too small for all of the froth to descend. Furthermore, vapor bubbles disengaging from that froth in the downcomer travel backward through the same narrow window. This backflow of vapor further resists the descent of the chimney tray froth. The froth builds up on the chimney tray, and when reaching the chimneys, it is entrained onto the next tray. This condition is a variation of the normal downcomer choke mechanism, but in this instance, it is caused by the obstruction of the downcomer entrance by the seal pan.

A similar mechanism occurs at the entrance to the downcomer from tray 7 to tray 8. The heavy-naphtha draw sump is the obstruction. This vertical window opening is somewhat larger (408 mm), but the tray dispersion is much frothier. The vapor disengaging from the downcomer from tray 7 is turned directly backward by the obstruction, blowing against the tray liquid movement. This vapor backflow is known to have caused entrainment and premature flooding in trays of long flow paths, such as the gasoline–LCO fractionation trays.

It is difficult to state which of the described two phenomena was the most likely cause of the tower bottleneck. Choking at the chimney tray downcomer appears more likely, but choking at the tray 7 inlet is supported by the gamma scan. In either case, the downcomer entrance obstruction generated at the chimney tray was the culprit.

Cure In the next turnaround the chimney tray was eliminated. To maximize fractionator capacity, the trays in the gasoline–LCO fractionation section were replaced by high-capacity trays at larger tray spacing. Following the turnaround, the gasoline–LCO separation bottleneck disappeared. Gasoline is no longer lost to the LCO, and the refinery sees a gap of 10°C between the 5% ASTM D86 point of the LCO and the 95% ASTM D86 point of the gasoline.

Chapter 10

Draw-Off Malfunctions (Non-Chimney Tray) Part of Number 7 on the Top 10 Malfunctions

As stated at the beginning of Chapter 9, intermediate draws are the third most troublesome internal in the tower (255), following the tower base/reboiler return (Chapter 8) and packing liquid distributors (Chapter 6). About half of the reported cases occurred in chimney trays (Chapter 9), the other half in downcomer trap-outs (including draw boxes), this chapter.

Vapor chokes in liquid draw lines from downcomer trap-outs are the prime issue, inducing premature flooding in the tower, reduced product yield, instability, and pump cavitation. Trapped gas bubbles choke outlet lines or aggravate a restriction problem. Undersized outlet lines, control valves, and inadequate degassing are the most common causes of the restriction. Resizing outlet lines to obey Simpson's rule for self-venting flow (438, 447) and properly degassing the liquid have been common cures.

Leakage at the trap-out tray, that is, undesirable liquid descent from the tray containing the trap-out, is another major issue. These leaks lowered product recovery, induced off-specification products, overloaded towers and vacuum systems, and caused pump cavitation. Blanking tray openings and cracks has at times been successful, but the more effective and more common cure was to replace the downcomer trap-out by a seal-welded chimney tray.

Other downcomer trap-out issues include damage, plugging, obstruction of downcomer entrance, poor draw box hydraulics, and incorrect installation.

With vapor side draws, the main issue reported was liquid entrainment, either from the tray below or by weep from a tray or collector weep hole above. Reflux drums have been relatively trouble free, with the main issues being liquid levels, undersized or plugged product lines, and poor separation of a second liquid phase.

CASE STUDY 10.1 CHOKING OF DOWNCOMER TRAP-OUT LINE

Installation Side draw of a refinery fractionator. The side draw removed 400 gpm of liquid from a downcomer trap-out (Fig. 10.1) via an 8-in. nozzle and 8-in. rundown line.

Problem Premature flooding and instability are experienced in the section immediately above the side draw.

Attempted Fix The problem was diagnosed to be presence of vapor in the side draw. In an attempt to solve the problem, the vent line was enlarged 1–6 in. (Fig. 10.1). This did not help.

Analysis Downcomer trap-outs seldom incorporate enough degassing time for complete vapor bubble disengagement from the draw liquid. As a result, the trap-out liquid remains aerated; that is, it contains entrapped gas bubbles. In gravity systems, nozzles and rundown lines handling aerated liquid need to be sized for self-venting flow. Figure 24.1*c* in Case Study 24.1(438, 447) is a highly recommended (250) sizing correlation. It predicts that an 8-in. nozzle and rundown line can handle a self-venting flow of up to 226 gpm. Attempting to run higher liquid flows will cause vapor to choke the outlet nozzle and rundown line. The rest of the liquid will back up in the downcomer. When the backup is high enough, there may be sufficient residence time to degas the liquid, and the degassed liquid will siphon out. Once the liquid siphons

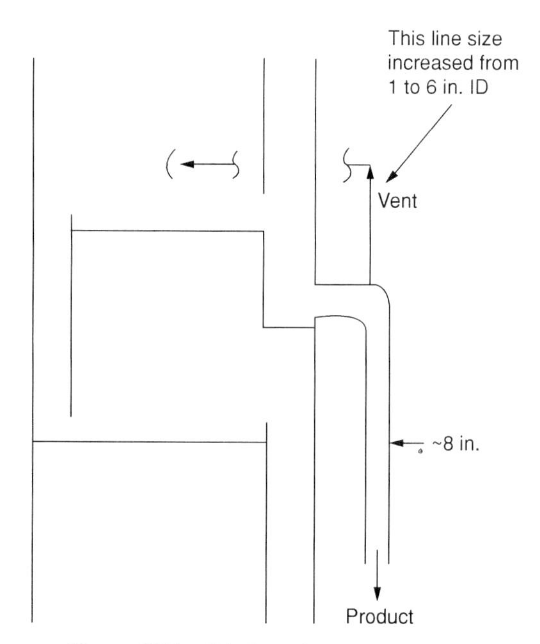

Figure 10.1 Side draw of refinery fractionator.

out, the liquid buildup will restart. This in turn leads to instability and possibly to liquid accumulation and flooding on the trays above. Line venting has been used with different degrees of effectiveness to relieve choking in rundown lines. A little venting is often beneficial, but as seen in this case, enlarging the vent size does not help. There have even been incidents where vents cause vapor to be sucked into the draw line, aggravating the instability.

Cure The best practice is to size both the nozzle and rundown line for self-venting flow per the method in Figure 24.1*c*. When adhered to, there is no need for a line vent, since all the disengaging vapor is capable of ascending back to the downcomer and eventually is released on the next tray up. Figure 24.1*c* gives that for 400 gpm draw rate a 10-in. line and nozzle are required.

CASE STUDY 10.2 FRACTIONATOR DRAW INSTABILITY

Installation Figure 10.2 shows a section of an FCC main fractionator. Trays 12 up fractionated naphtha from LCO. The LCO product was drawn from a sump under the tray 12 downcomer. Trays 13 and 14 were used for direct-contact partial condensing

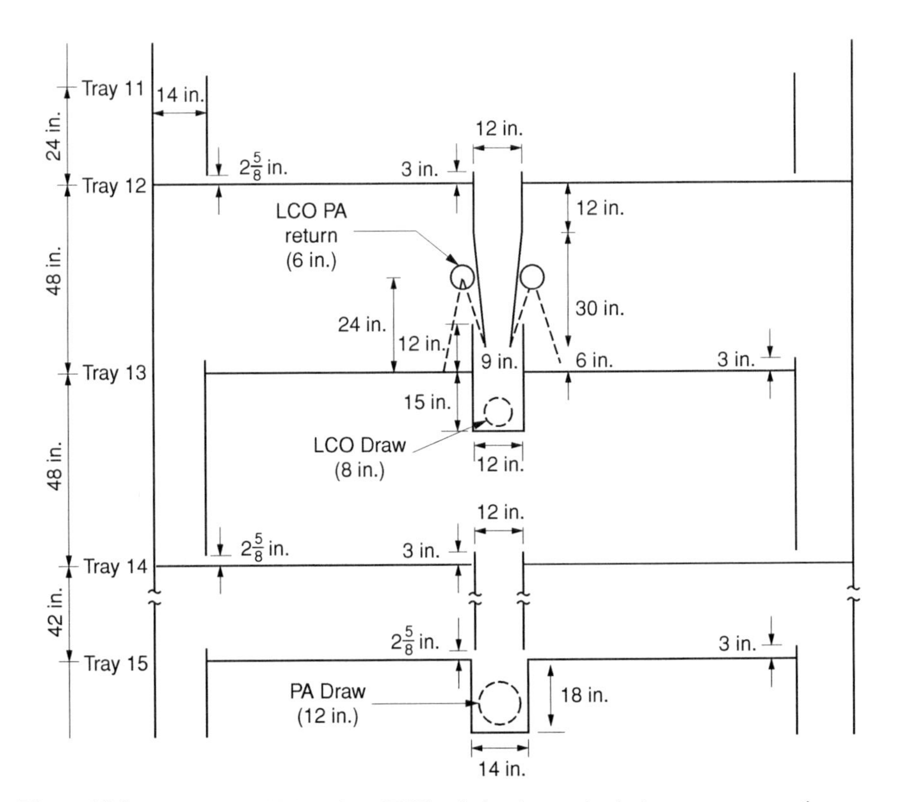

Figure 10.2 LCO draw and PA section of FCC main fractionator that had very narrow operating range.

of ascending vapor. The condensing was achieved by drawing a circulating PA liquid stream from a sump right under tray 15 and pumping it through coolers. The cooled PA returned to the tower near the inlet of tray 13.

Problem The tower had a very narrow stable operating range. When the LCO draw rate was about 7500 BPD, the tower was stable. Raising the LCO draw rate beyond 8000 BPD caused the PA pump to cavitate and the temperature on tray 12 to spike. Reducing the LCO draw rate below 7200 BPD dropped the temperature difference between the PA draw and tray 12 from a regular 120 to 40–80°F, and there was a high *dP*, suggesting flooding.

Changing the reflux to the top of the tower had a similar impact. Reducing the reflux (e.g., to raise the end point of the gasoline) heated tray 12 and caused the PA pump to lose suction. Increasing the reflux reduced the temperature difference between the PA draw and tray 12 and initiated flood. Manipulating the LCO draw rate had a much greater impact than changing the reflux.

Losing the PA pump had greater adverse effects than flooding. This is because the PA was used for reboiling the deethanizer stripper.

Investigation The *dP* was measured over the top 20 trays. It normally read 2 psi and when flooded went up to 3.5 psi. The high *dP* was seen when cutting back on LCO draw but not when the pump lost suction.

The PA return partly entered the LCO draw sump and partly flowed directly to tray 13 (Fig. 10.2). At the low LCO draw rate, liquid overflowed the LCO draw sump onto tray 13. Raising LCO draw rates reduced the overflow onto tray 13. Upon further increase, some PA liquid was drawn as LCO. When the LCO draw rate was excessive, too much of the PA liquid was drawn as LCO, starving the PA draw and cavitating the PA pump. This was observed when LCO draw rates exceeded 8000 BPD.

At intermediate LCO rates (~7500 BPD), the LCO draw temperature was 425°F, compared to a tray 12 temperature of 438°F. This means that some of the subcooled PA return was mixed with the liquid descending from tray 12. The subcooling of the LCO draw would quench any vapor bubbles present in the draw liquid.

As the LCO draw rate was reduced, the flow rate of liquid descending from tray 12 approached and finally exceeded the LCO draw rate. When this happened, the liquid from the tray 12 downcomer overflowed the sump and prevented entry of the subcooled PA return liquid into the LCO draw sump. This coincided with observations by operating personnel that as they cut back the LCO draw rate the LCO draw temperature approached that of tray 12. When flooding initiated, the LCO draw temperature was very close to that of tray 12, indicating no subcooling.

The downcomer liquid contains vapor bubbles. To allow bubble disengagement from the liquid, rundown lines from downcomer draws need to be sized for self-venting flow. The LCO draw line was grossly undersized for self-venting flow per the correlation in Figure 24.1*c* (438, 447), which had been strongly recommended (250). So the rundown line choked and backed liquid up the downcomer and the tower, which caused the flooding. As the LCO draw rate was increased, subcooled PA return was drawn in, quenched the vapor bubbles in the LCO draw sump, and converted the aerated liquid into nonaerated liquid. For nonaerated liquid, the draw line size was adequate, so the draw started working.

Solution A low-cost fix was to replace the draw line by a 12-in. line (including a new nozzle). However, there were economic incentives to replace trays by packing in this tower, and this was implemented. During the retrofit, the draw sump was replaced by a chimney tray, which provided enough degassing time to eliminate the draw aeration.

CASE STUDY 10.3 A NONLEAKING DRAW TRAY

Henry Z. Kister, Betzalel Blum, and Tibor Rosenzweig, reference 264. Reprinted with permission from *Hydrocarbon Processing*, by Gulf Publishing Co., all rights reserved

A feed preparation tower for lube oil received a heavy gas oil feed and produced four products (Fig. 10.3): bottom cut, heavy-intermediate cut, light-intermediate cut, and overhead cut.

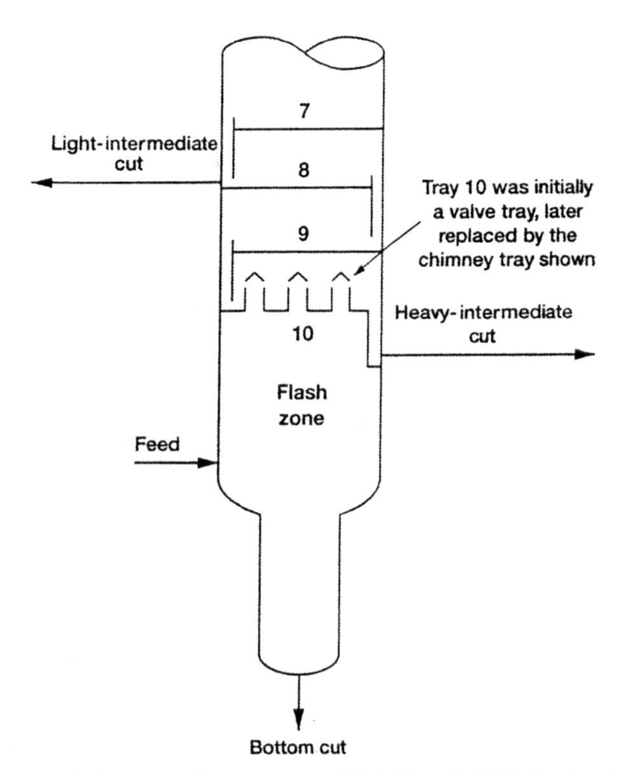

Figure 10.3 Lube oil feed preparation tower, as modified. (From Ref. 264. Reprinted with permission from *Hydrocarbon Processing* by Gulf Publishing Co. All rights reserved.)

The heavy-intermediate cut was drawn from the bottom tray immediately above the flash zone. There used to be trays between the tower bottom and flash zone, with steam injected to strip the bottom cut. However, the steam was wet, and water entry caused pressure surges that repeatedly dislodged the stripping trays. So the steam injection and stripping trays were eliminated, leaving only the flash zone between the heavy-intermediate cut draw and the tower bottom.

The heavy-intermediate cut was withdrawn from tray 10, the bottom tray in the section above the flash zone. This tray used to be a conventional valve tray. When this tray was a valve tray, it was impossible to withdraw a heavy-intermediate cut from the tower. This heavy-intermediate cut leaked through tray 10 into the tower bottom, lowering the bottom cut viscosity well below specification. It is believed that the high gas velocities kept all the valves open, with the tray liquid leaking through the large open area, a mechanism postulated by Bolles (58).

To catch the heavy-intermediate cut, the bottom valve tray was replaced by a seal-welded chimney tray. This tray eliminated the valve leakage. After changing the trays, an intermediate heavy product could be drawn, and the viscosity of the bottom cut met all specifications.

CASE STUDY 10.4 LEAK TESTS ARE KEY TO PRODUCT RECOVERY

Installation An atmospheric crude distillation tower. The lowest side product drawn from the tower was a heavy diesel stream. The heavy diesel was drawn on level control from a draw sump located at the end of a center pan (Fig. 10.4*a*) just below a two-pass sieve tray. The diesel draw was a total draw. Leakage from the draw tray, pan, or sump ended in the resid leaving the tower bottoms.

Leak Tests The draw pan and sump were leak tested during the turnaround. Initially, the water could not even fill the sump due to heavy leakage from the sides of the sump, where the sump walls were connected to the column straps. The sides of the sump had come apart from the straps, with gaps of up to 5 mm. These gaps had previously been packed with gasket tape, but much of the tape disappeared.

The gaps were temporarily repacked, and a second leak test was performed. This time the level rose up to the first seam in the channel sections which form the pan walls. The channel sections were added during a debottleneck, 11 years earlier, in order to deepen the draw pan. The channel sections (Fig. 10.4*b*) were probably designed for a different purpose; they were slotted and fabricated from CS (the pan and sump were SS). These dissimilar sections in both size and material were seal welded together, incorporating the slots as part of the pan wall, thereby allowing significant leakage (>10 m^3/h). The flow path of water through these slots is marked in Figure 10.4*b*. The different rates of thermal expansion of the two metals at high temperature led to bulging, forming gap in one section. There was also severe leakage from a large gap between the column straps and the draw pan walls.

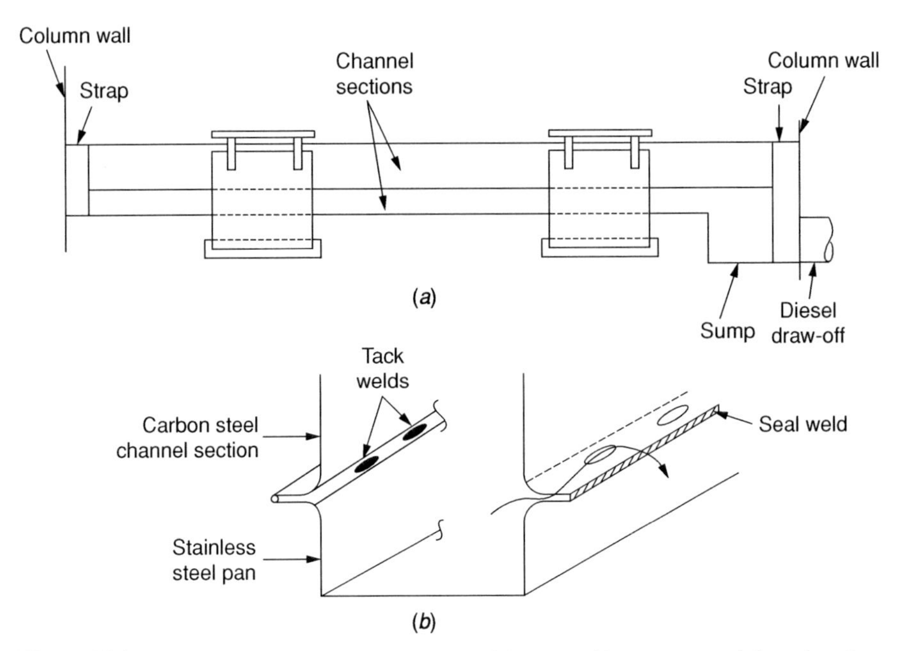

Figure 10.4 Draw pan that leaked: (*a*) arrangement of draw pan; (*b*) arrangement of channel sections in draw pan (arrow indicates water leakage path through slots).

Cure Several gaps were welded. A high-temperature sealant was used to fill some of the openings that could not be welded. Small squares of metal were tack welded over the slots. A final leak test gave a leakage rate of 1 in. in 2 min, which is 10 times higher than the number recommended in *Distillation Operation* [1 in. in 20 min. (250)], but an order of magnitude better than the high leakage previously observed.

Results For at least 11 years, high leakage rates were experienced. Here the hidden flaw became the norm. Improvement came about only after the leak tests that gave insight to the true performance of the draw.

The repairs to this draw pan have solved major operating problems which the tower experienced for 11 years. It was impossible to hold a steady level at greater than 30% in the draw pan. The leakage caused poor diesel recovery, much of it degrading to resid, which showed up as a low diesel cloud point. Prior to the repairs, the diesel cloud point was low, seldom reaching 10°C. Following the repairs, the tower produced 15°C cloud point diesel.

Lessons Leak tests should be performed on critical draw trays every turnaround. Careful design of critical draw trays is essential for high recovery of valuable products.

CASE STUDY 10.5 DOWNCOMER UNSEALING AT DRAW PAN

Installation A refinery crude tower which had four side-draw product streams. Each side draw was removed from a downcomer trap-out (Fig.10.5*a*). Under each draw-off was a PA, which is a direct-contact internal partial condenser. A cold circulating liquid stream entered at the top of the PA section, condensed a portion of the rising vapor, and was withdrawn at a higher temperature some trays down. The hot PA was then cooled in external coolers and returned to the top of the PA section. Excess liquid reflux (from above the side draw) together with condensate formed the reflux to the section beneath the PA.

Problem To optimize heat recovery, it was desirable to maximize PA duty. When the plant maximized duty, the downcomer carrying liquid to the side draw became unsealed, causing major disturbances to the tower. To circumvent the problem, the operators ran the tower with much higher than required internal reflux at the expense of reduced heat recovery.

Analysis When the PA heat duty is increased, the tower heat balance is maintained by a reduction in reflux from the section above. This reduces the amount of liquid overflowing the seal pan. When the seal pan overflow is small, a small increase in product flow or a small reduction in reflux from the section above can dry it up completely. The product outlet drains the seal pan. Once the seal pan dries up, there is nothing stopping tower vapor from ascending the downcomer. This interrupts liquid downflow, floods the trays above, and generates a major disturbance. To circumvent, the operators increase internal reflux at the expense of lower heat recovery at the PA.

Solution The seal pan was modified to prevent the downcomer from becoming unsealed (Fig.10.5*b*). No more disturbances occurred after this, and PA heat removal could be optimized without causing tower instability.

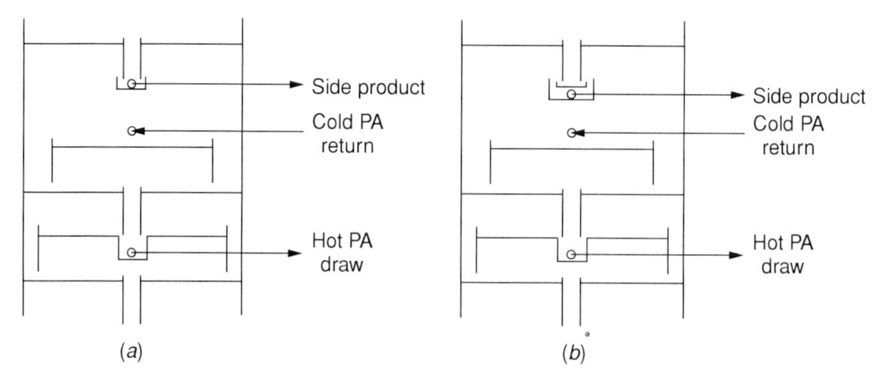

Figure 10.5 Side-draw and PA arrangement in crude tower: (*a*) initial, led to downcomer unsealing when PA duty maximized; (*b*) modified, no unsealing.

CASE STUDY 10.6 LIQUID ENTRAINMENT IN VAPOR DRAW

Installation Isostripper in a refinery HF alkylation unit. The tower separated C_4 HCs and lighter components from a bottom alkylate stream. Isobutane was drawn 12 trays below the top as a vapor side draw. The side-draw geometry is shown in Figure 10.6. The tower contained four-pass sieve trays.

Problem The tower worked well at rates less than 50% of flood. At higher rates, heavy components appeared in the side-draw samples. As tower rates came up, so did the heavies concentration.

Cause Figure 10.6 shows that the bottom of the draw nozzles was 18 in. above the tray floor. Fractionation Research movies (140) show that when a tray operates at about 50% of jet flood spray height typically reaches 12–15 in. It is conceivable that at higher rates some of the spray was aspirated by the vapor draw.

Improvement A box was built around the vapor draw nozzles which allowed vapor to reach the nozzle from above but not from below. This improved operation, allowing the trays to operate at up to 70% of flood without a rise in heavies content of the side draw. As tower hydraulic loads were raised above this, the heavies problem reoccurred.

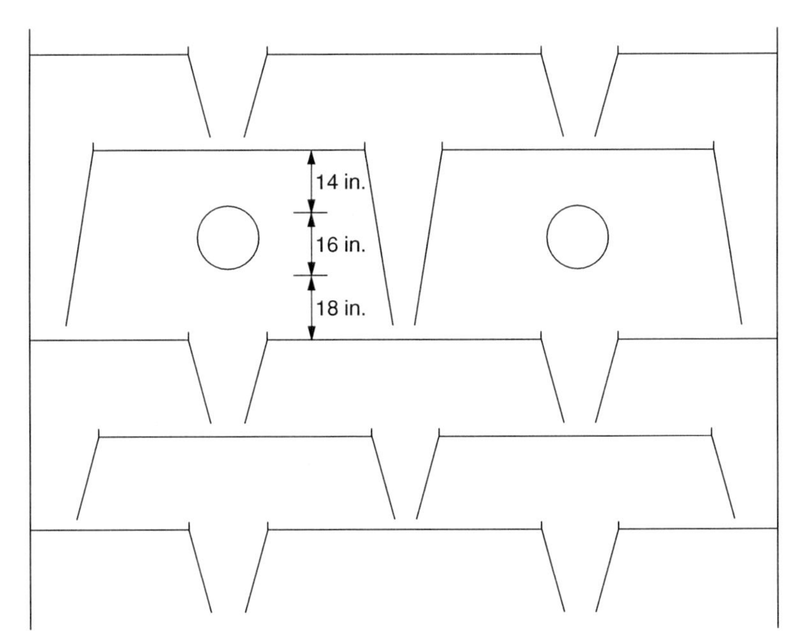

Figure 10.6 iC_4 vapor draw nozzles in alkylation isostripper.

Related Experience A similar problem was experienced in an HF alkylation main fractionator. The problem was successfully solved by replacing the tray below the draw by a chimney tray and installing baffles diverting potential weep from the tray above away from the draw.

CASE STUDY 10.7 WEEP INTO A VAPOR SIDE DRAW

Installation A tower containing three beds of structured packings. Liquid leaving each bed was collected by a chevron liquid collector and from there flowed into the redistributor below (e.g., Fig. 10.7).

Problem Side-product quality had been questionable for several months. Small amounts of high-boiling components were always present in the side draw regardless of various manipulations by operating personnel.

Troubleshooting The tower was equipped with several view ports. One pair of view ports was located between the redistributor and collector in the side-draw

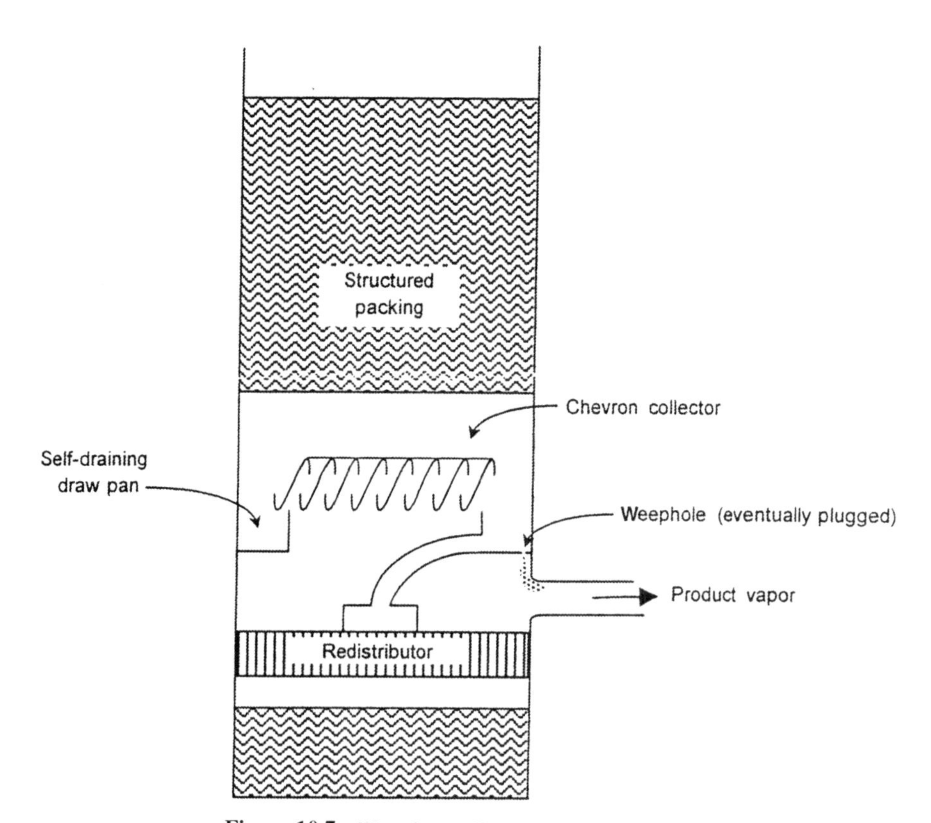

Figure 10.7 Weep from collector into vapor side draw.

region. The two ports were at 90° to each other, so they permitted full view of the side draw.

Although difficult to see, a close observation revealed a stain on the column wall above the vapor outlet but not below. Even more difficult to see, yet visible when one looked for it, was liquid (actually a localized mist) being aspirated into the side-draw nozzle. The liquid originated in a weep hole in the chevron collector above. Each channel of the chevron collector (Fig. 10.7) had a weep hole, and one of the weep holes was directly above the side-draw nozzle. This heavies-rich liquid weep was aspirated directly into the side product and contaminated it.

Solution The weep holes were plugged. Since the collector channels were self-draining, there was no need for weep holes. Following this modification the desired product quality was achieved.

One Lesson Viewing ports are invaluable for troubleshooting, and two (preferably at 90° to each other) are better than one. These should be incorporated whenever safety and environmental considerations permit. For those who look in: You will probably see your reflection until you cover your head (and port) with a jacket or dark cloth.

CASE STUDY 10.8 AERATION DESTABILIZES REFLUX FLOW

Henry Z. Kister, Fluor, and James F. Litchfield, reference 260. Reprinted courtesy of *Chemical Engineering*.

Installation Reflux to a chemical tower flowed by gravity from a 30-in.-ID vertical accumulator through a vortex flowmeter and a flow control valve. Accumulator level was not automatically controlled.

Operation Initially, the column reflux was very erratic. This erratic behavior was dampened by the operators by keeping the control valve wide open and running the reflux accumulator at approximately 10% liquid level. However, over a period of time, the reflux flow rate dropped off and the liquid level in the accumulator rose. The normal reflux flow rate was reestablished by stroking the control valve two or three times. This mode of operation destabilized the column and was an operating nuisance. The control valve was examined several times during brief outages. Each time the valve was found clean and in good condition.

Cause Accumulator drawings were reviewed. At the 10% liquid level, the entering liquid feed dropped about 6 ft into a shallow pool of liquid at the bottom (Fig. 10.8). The liquid level was only about 18–20-inches above the liquid outlet nozzle. The water-falling liquid entrained vapor as it penetrated the shallow pool of liquid, creating some very fine vapor bubbles. Most of these bubbles got entrained in the

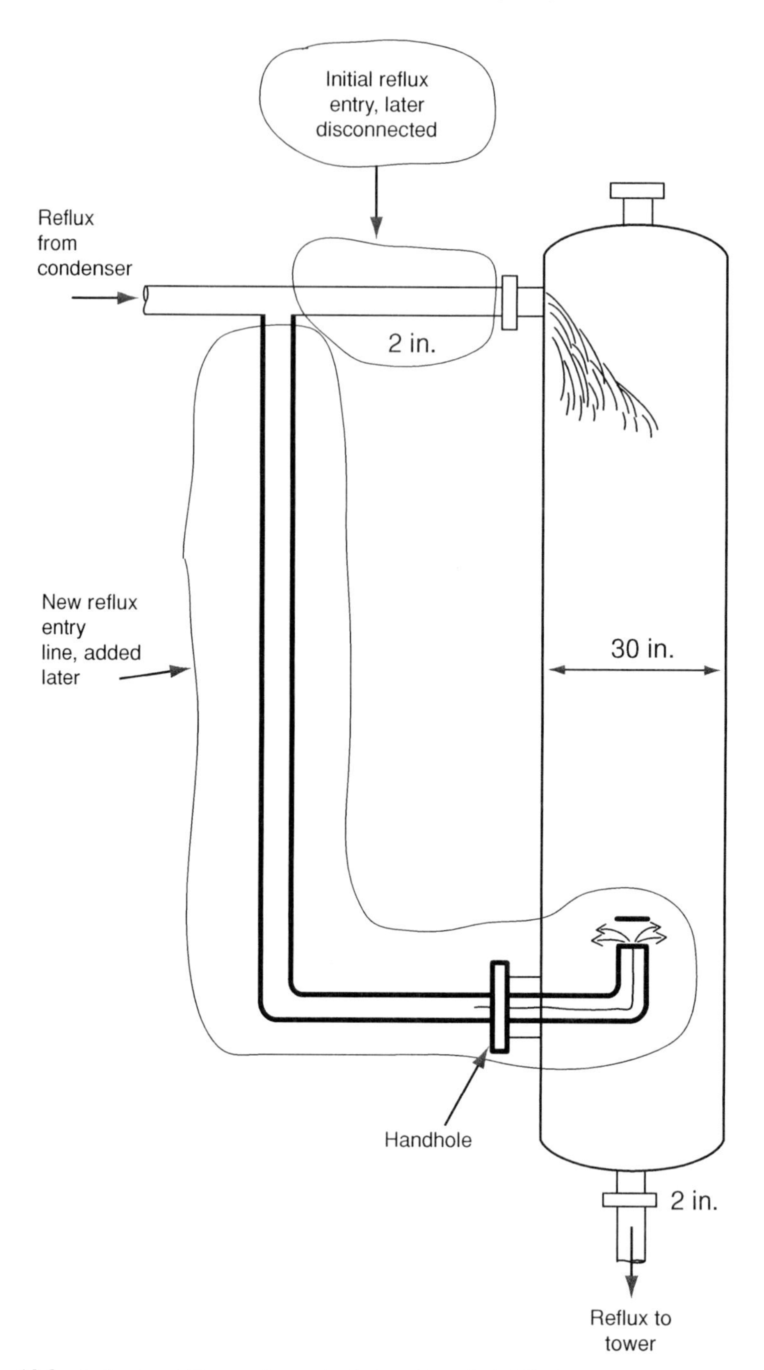

Figure 10.8 Reflux instability problem and solution. (From Ref. 260. Reprinted courtesy of *Chemical Engineering*.)

discharging liquid and some later got trapped at the control valve inlet. These trapped bubbles limited the flow rate through the valve.

Reflux Pipe Design The reflux pipe was 2 in. in diameter. This line diameter is generously sized for draining nonaerated liquid but is too small for draining the aerated liquid. Aerated liquid draining requires rundown lines that are sized for self-venting flow, that is, a flow in which liquid descends while any entrapped vapor bubbles disengage upward. Figure 24.1*c* is a correlation for self-venting flow (438, 447) that was highly recommended (250). Based on the correlation, a 2-in. line can drain up to 7 gpm aerated liquid. Since the reflux flow rate was 12 gpm, the balance 5 gpm would accumulate in the reflux accumulator, raising its level. At the same time, trapped gas would reduce the flow area through the valve and line, reducing the reflux flow rate. Stroking the valve vented the trapped bubbles and siphoned the accumulating liquid out of the accumulator.

Initially, the accumulator was operated at much higher levels than 10% and experienced a far more erratic operation. At higher liquid levels, the waterfall height would diminish. This and the greater pool depth would prevent vapor bubbles from reaching the accumulator outlet. The accumulator bottom liquid would degas, reverting to nonaerated liquid. This nonaerated liquid would easily siphon out, and the accumulator level would rapidly drop. Once siphoned out, the waterfall would again aerate the bottom liquid, and the aerated liquid flow would resume. The back-and-forth switches between aerated liquid flow and siphoning caused the initial erratic behavior.

Cure There was an 8-in. hand-hole 15 in. above the bottom tangent line of the drum. The 2-in. feed line was rerouted to enter the drum by passing through the hand-hole cover (Fig. 10.8). The feed pipe was extended to the drum centerline and then bent upward, discharging upward against a flat horizontal deflector baffle. This baffle redirected the incoming liquid, spreading it sideways. This eliminated the waterfall and aeration and fully restored reflux stability.

Chapter 11

Tower Assembly Mishaps: Number 5 on the Top 10 Malfunctions

Assembly mishaps are in the fifth spot among distillation malfunctions (255). In 1997, a distillation malfunction survey (245) singled out assembly mishaps as the fastest growing malfunction, with the number of malfunctions reported between 1990 and 1997 more than double the number of malfunctions between 1950 and 1990. The good news is that this growth has leveled off. It appears that the industry took corrective action after noticing the alarming rise in assembly mishaps. Many major organizations have initiated systematic and thorough tower process inspection programs, and these are paying good dividends.

The largest number of reported assembly mishaps is for packing liquid distributors. Most of these cases are recent. This is one area where inspections can be improved. Incorrect packing assembly is another major issue, more troublesome in some less common packing assemblies (e.g., breakage of ceramic random packing, collapse of poorly assembled grid beds, unsupervised installation of structured packings). So these should not reflect negatively on the majority of packing assemblies. The lesson is that good practice is to have the supplier supervise structured packing installation (including the distributors) and to exercise special caution in specific situations like dumping ceramic packings, fastening grid, and deciding whether to leave the tray support rings in the towers retrofitted by packing.

Improper tightening of nuts, bolts, and clamps and incorrect assembly of tray panels are, as can be expected, near the top on the list of assembly mishaps and deserve to be on the checklist of every tower inspector. Debris left in the column and incorrect materials of construction also belong on the checklist. Other malfunctions that have been frequently encountered and constitute items that process inspectors should focus on include flow passage obstruction and internal misorientation in feed and draw areas; leakage from "leak-proof" and "leak-resistant" collector trays (these

Distillation Troubleshooting. By Henry Z. Kister

should be water tested at turnarounds); downcomer clearances improperly set; and tray manways left unbolted.

CASE STUDY 11.1 SHOULD VALVE FLOATS BE REMOVED BEFORE BLANKING?

Contributed by Ron F. Olsson, Celanese Corp.

Installation A 10.5-ft-ID chemical column equipped with one-pass trays containing moving uncaged valves. The trays had no major support beams. Tray panels were laid parallel to the liquid flow (Fig. 11.1) and were supported by integral trusses also parallel to the liquid flow.

History The tower was to be operated at considerable turndown, so many rows of valves needed blanking. To minimize downtime, it was decided to leave the valve floats in the holes and to install the blanking strips over them, so the strips keep the floats shut. Selected rows of valves were to be blanked, all rows perpendicular to the liquid flow, in accordance with good blanking practice (250). The blanking strips specified were 12 gage, 3 in. wide. The length of each blanking strip was specified to equal the panel width, so the number of strips per row equaled the number of panels.

Problem Each fabricated blanking strip was about 5 ft long, which covered half the tower width and stretched over a number of panels. Each strip was bolted down to the tray floor by two bolts, one near each end. When the bolts were tightened, the bulging valve floats caused the blanking strips to bend in the shape of a W. The bent strips did not hold down any valve floats, except for those right next to the bolts. The problem was discovered upon inspection during installation and, fortunately enough, before too many trays were blanked.

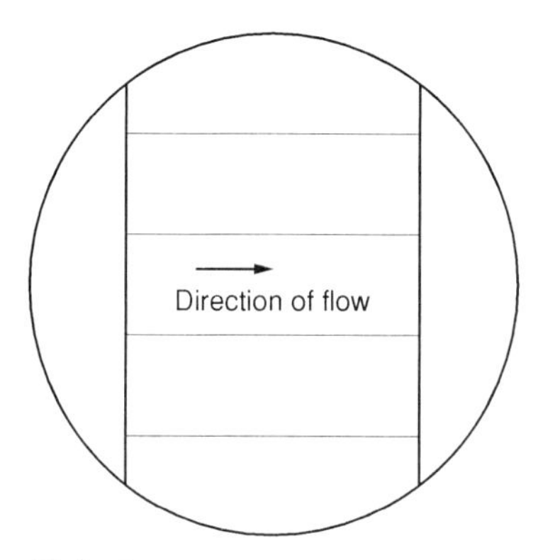

Figure 11.1 Tray panels laid out parallel to liquid flow.

Cure More bolts were added to hold the blanking strips down. At the end, the strips looked like WWW, but they properly held down the valve floats.

Moral It is best to remove floats before blanking valves.

CASE STUDY 11.2 DIRECTIONAL VALVE INSTALLATION

Background Figure 11.2 shows directional valves typical of those used on some high-capacity trays. Their purpose is to issue some of the vapor with a horizontal velocity component in the direction of liquid flow on the tray, providing a forward "push" for the liquid. This enhances tray capacity.

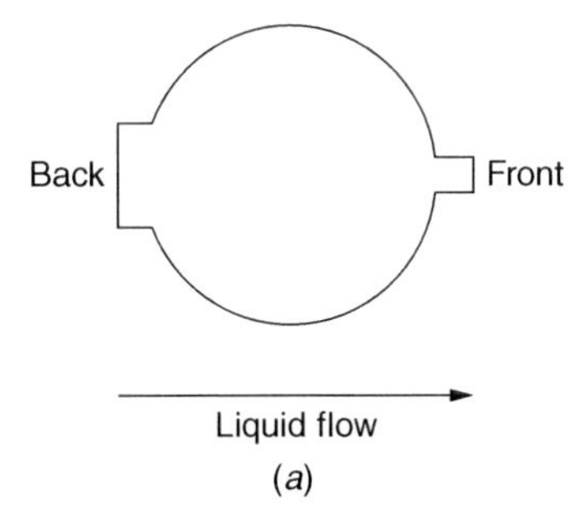

(*a*)

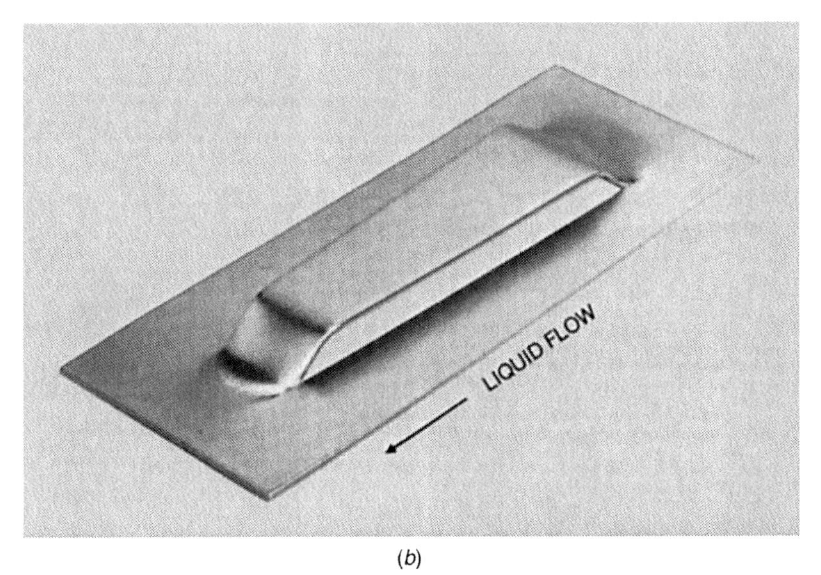

(*b*)

Figure 11.2 Directional valves: (*a*) round; (*b*) rectangular. [(*b*) Reprinted courtesy of Sulzer Chemtech.]

During installation, all directional valves need to be pointing in the correct direction. Backward installation, that is, with the larger opening facing liquid flow, lowers capacity and also increases weeping.

Tower A Several panels with round fixed directional valves were installed backward. Caught by inspection and corrected before commissioning.

Tower B At a turnaround, rectangular fixed valves were found to be pointing in the right direction, except those on the manway panels, which pointed backward. The manways were installed some days after the rest of the trays were assembled. There was no major effect on performance, but the tower was not operated close to its limit.

CASE STUDY 11.3 CAN PICKET FENCE WEIRS CAUSE EARLY FLOODING?

Contributed by W. Randall Hollowell, CITGO, Lake Charles, Louisiana

Installation An atmospheric crude fractionator had PA, reflux, and diesel side-stripper feed drawn from the same tray. The tower had three four-pass PA trays. The trays above the PA had very low liquid rates. One turnaround, picket fence weirs were added to the trays above the PA to avoid drying and improve liquid distribution.

Problem After restart, the tower flooded at a lower crude rate than it had historically. The flooding appeared to initiate at the PA.

Cause When the tower was opened for turnaround, it was found that picket fence weirs had been installed on the top PA tray. At the heavy liquid loading of a PA tray, the picket fence weirs led to excessive weir loads, causing premature flooding.

Solution The picket fence weirs were removed from the top PA tray, and the historically higher crude rate was reachieved.

Moral A good process inspection can save a lot of grief.

CASE STUDY 11.4 INSPECTING SEAL PANS IS A MUST

Contributed by Goran Z. Tošić, HIP-Petrohemija, Pančevo, Serbia

Installation An ethylene plant 5-ft-ID trayed depropanizer stripper separating C_3 and lighter from C_4 and heavier HCs. The tower operated well for 20 years.

Problem Upon restart after a turnaround, it was impossible to establish normal reboil. This resulted in low tower pressure (5.8 barg, normal 6.7 barg), low differential pressure (approximately 0.05 bar, normal 0.18 bar), low level in the reflux drum, low bottom temperature, and poor temperature profile in the column. The low temperatures led to excessive C_3 in the plant C_4 product.

Depropanizer Base (Fig. 11.3) Bottom downcomer liquid entered a seal pan. Liquid to the tower vertical circulating thermosiphon reboiler was drawn from the seal pan via a 12-in. nozzle N6. The vapor/liquid reboiler return entered the tower via a 16-in. nozzle N5. Seal pan overflow went into the tower bottom sump. This arrangement preferentially diverted the tray 1 liquid into the reboiler and the reboiler return liquid into the bottom sump. Sump level was controlled by a level transmitter mounted between nozzles A and B

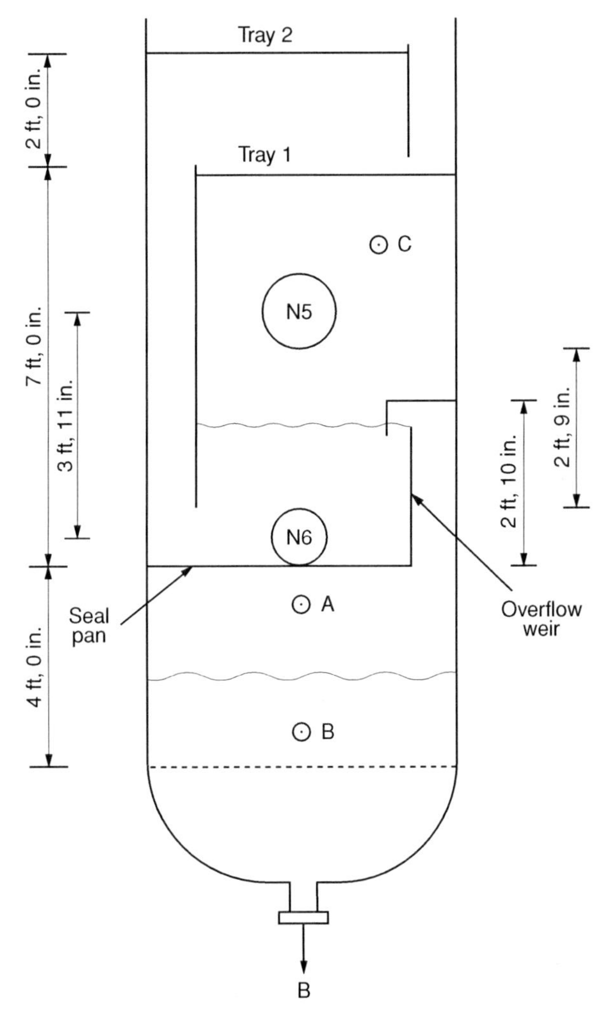

Figure 11.3 Depropanizer base.

Troubleshooting A hydraulic analysis established that, if the overflow weir fell out, only 55% of the liquid entering the seal pan would reach the reboiler; the balance would endup in the bottom sump. This would starve the reboiler of liquid supply, impede boil-up, and lead to the operating temperatures and pressures observed in the tower.

Living with the Problem The tower had a spare reboiler. Twice the operating reboiler was exchanged for a clean spare with little improvement. An attempt was made to raise liquid level above 100%. To monitor the higher levels, the upper tap of the level transmitter was changed from nozzle A to nozzle C, above the reboiler return. This was unsuccessful, probably because of the very small elevation difference (Fig. 11.3) between N5 and N6. Levels below the top of the reboiler draw starved the reboiler of liquid, while levels exceeding, even approaching, the reboiler return would initiate liquid carryover (by the vapor) up the tower and tray flooding. The range in between was far too narrow for satisfactory operation.

Inspection The tower was shut down. The seal pan overflow weir was found fallen out, as diagnosed by the hydraulic analysis. The inspection showed that the overflow weir was not installed properly 10 years earlier. It was either worn out or damaged during hydrotesting or start-up and fell out.

Solution The overflow weir was repaired and reinstalled. Good tower performance was reinstated.

Morals

- Bottom seal pans must be properly inspected every turnaround.
- Whenever practical, hydrotests should be done prior to the tower final inspection.
- A hydraulic analysis is an invaluable troubleshooting tool.
- A short elevation difference between draw and return nozzles to circulating thermosiphon reboilers can reduce the tower ability to handle malfunctions. The baffle arrangement in Figure 12.2 is far more robust and could have circumvented the need to shut down upon malfunction.

CASE STUDY 11.5 A GOOD SIMULATION LEADS TO OPEN MANWAYS

Constributed by Gerald L. Kaes, KAES Enterprises, Inc., Colbert, Georgia

Installation A new glycol dehydration tower in a natural gas plant. The tower contacted natural gas with dry glycol to absorb water vapor out of the gas. The overhead gas needed to meet a water dew point specification prior to entering a natural gas pipeline.

Problem The overhead gas dew point was much higher than design.

Investigation Several changes in operation were tried unsuccessfully, including increasing the glycol circulation, decreasing the water content of the regenerated glycol, and changing the temperature of the inlet glycol to the dehydration tower. The trays were gamma scanned and appeared to be in place with no apparent damage. The tower was checked and was not out of plumb. The tower was modeled with a simulator program. The performance of the tower corresponded to about one theoretical tray, much lower than the design of four or more theoretical stages out of the 10 actual trays in the tower.

Cure When the unit was taken off stream and the tower opened, it was discovered that all the tray manways in the tower were left open. Thus the column was performing as a flash drum, agreeing with the simulation.

CASE STUDY 11.6 LUBE OIL VACUUM TOWER PROBLEM

Contributed by Yuri Ratovski and Oleg Karpilovskiy, Koch-Glitsch, Moscow, Russia

Installation A new lube oil vacuum column with structured packing was installed to produce vacuum diesel oil, four lube fractions, and heavy vacuum residue. Heater coils and transfer line were revamped for the new operating conditions.

Problem The column successfully started up. All lube oils except the heaviest one satisfied all specifications. The temperature difference between heater outlet and tower inlet was more than 40°C compared to the design 12°C. The flash point of the vacuum residue was below the specification value. The yield of heavy distillate was low. The heavy distillate was black, indicating that something was wrong with the wash zone.

Investigation First the problem with transfer line temperature drop was considered. The gate valves at heater outlet, installed due to local safety requirements, were checked and found to be partly closed. This produced high pressure drop and consequently temperature drop. Opening the valves solved the problem—temperature drop decreased and heavy distillate yield increased. The quality of heavy distillate and residue was still bad.

A complete set of process data was collected and a tower simulation was prepared. The tower operated at significantly lower pressure than design. A wash zone flooding hypothesis was proposed. Tower top pressure was raised to eliminate the suspected flooding. There was no effect on heavy distillate and residue quality.

One possible reason for the low vacuum residue flash point was cracking in the tower bottoms. Refinery personnel considered cracking highly unlikely because they used quench to decrease the bottoms temperature, and the temperature was quite low, about 330°C. On the other hand, the residence time was enormous, more than 1 hour, and this was theorized to induce cracking. The tower did not have a smaller diameter in the stripping zone, and reducing residence time required operating the liquid level below the lower level nozzle, which was impractical. Based on this theory, quench flow rate was further increased. This successfully raised the residue flash point, at least to specification value, although still much below the usual values for such towers. Heavy distillate was still black.

Inspection The vacuum tower was shut down after 6 months of operation. The tower was opened and wash zone internals were inspected. With one exception, all internals in the wash zone were OK without deformation or plugging. The exception was an open internal manway. The wash zone was enclosed inside a sleeve, the diameter of which was smaller than the tower diameter. There were horizontal panels on top of the sleeve preventing vapor from the flash zone from bypassing the wash section packings. These panels had two manways, one on each side of the tower. One manway was closed, the other was open. The manway cover, with bolts ready for mounting, was found to be leaning against the tower wall.

Reason of Bad Lube Oil Quality Entrained drops of heavy liquid together with vapor from the feed inlet distributor were coming up directly through the open internal manway, bypassing the wash bed packing. This entrainment ended in the heavy lube fraction, giving it the black color. Once the manway cover was reinstalled, the quality of the heavy lube oil improved and its color specification was achieved.

Morals

- The closing of internal manways must be fully checked.
- Valves at heater outlet should be avoided where possible. If required for safety, use low-pressure-drop valves which can be completely opened. Check if they are fully open.
- Residence time of vacuum residue in tower bottoms should be minimized.

CASE STUDY 11.7 DEBRIS IN LIQUID DISTRIBUTOR CAUSES ENTRAINMENT

Henry Z. Kister and Tom C. Hower, reference 263. Reproduced with permission.

Installation A water-wash column equipped with two-pass sieve trays. The reflux entered the column through a perforated-pipe reflux distributor.

Problem Following a plant shutdown in which the plant was modified to increase capacity, excessive entrainment was observed from the column.

Investigation The entrainment rate was measured by closing drains and timing a rise in level in a knockout pot downstream of the tower. Calculations were carried out using a number of entrainment correlations. The most conservative of those predicted less than about a third of the measured entrainment rate.

During low-rate operation periods, the entrainment stopped. Increasing the liquid flow rate to the top of the column also reduced entrainment. Gamma scans showed nothing abnormal about the column, except that the amount of liquid flowing over one of the top outlet weirs appeared greater than the amount of liquid flowing over the other.

Cause When the column was opened in a subsequent shutdown, a welding rag was found inside and halfway along the reflux distributor. The rag restricted liquid flow to some sections of the distributor. Either liquid maldistribution on the top tray or high-velocity jets issuing from the distributor perforations upstream of the restriction could have caused the entrainment.

The rag was believed to have been left in the reflux line, which was modified during the shutdown. Upon restart, the water flow carried the rag into the reflux distributor, where it remained until it was discovered. Removing the rag solved the problem.

CASE STUDY 11.8 POOR RANDOM PACKING INSTALLATION LOSES CAPACITY, FRACTIONATION

Contributed by Dave Simpson, Koch-Glitsch UK, Stoke-on-Trent, England

Installation This was a new process of which the design contractor had no direct experience. The distillation column consisted of several beds of random packing with alternative feed points between beds. Assumptions on both the number of theoretical stages required and the expected HETP of the packing had been made and generous safety margins included.

Problem At start-up it quickly became apparent that the column could achieve neither the fractionation required nor the desired throughput capacity. The fractionation was very poor and nowhere close to being acceptable regardless of feed rate or reflux ratio. Changing the feed point had little influence. Feed rate was limited to less than 70% of design.

Troubleshooting The process design was reexamined, instrumentation and operating parameters were checked, and the hydraulic design of the packing and the associated internals were critically reviewed. Nothing was found that could explain the shortfall of performance. Plant data were studied, but again nothing obvious presented itself. The tower was shut down for inspection.

Inspection The inspection revealed several gross errors with the installation of the packing and internals. First and most seriously the packing had been installed so that the beds were packed right up to the underside of the distributors. Second the distributors were installed at 90° to the correct orientations, and third the gaskets for the distributors were not properly installed.

Analysis Installing the packing without leaving a space, typically 150 mm, below the distributors was the root cause of the premature flood. This space is essential for the vapor to travel from the top of the packing to the risers in the distributors without entraining the liquid falling from the distributor to the top of the packing. The flooding resulted in maldistribution and loss of fractionation. The distributors' incorrect orientation meant that at the reflux and feed points liquid was introduced to the wrong areas of the distributors, causing maldistribution and entrainment. This caused further loss of efficiency. The incorrect gasket installation meant that the distributors, which were deck type, leaked around the support ring and along the deck joints, again causing maldistribution and loss of efficiency.

Troubleshooting Made Easy In this case there were no differential pressure instruments and the absolute pressure indicators were not capable of showing the pressure drop accurately. Also, no gamma scans were performed. Lack of these two invaluable diagnostic tools impeded the troubleshooting.

Solution The packing and distributors were reinstalled correctly and the column design performance was achieved, meeting all design requirements.

Lessons Tower internals installation should be carried out or at least supervised by qualified personnel. Tower internals installation should be closely inspected by process and/or operation personnel. Inspection checklists are useful.

Random packing is supplied with an excess amount to allow for settling, site-handling losses, and damage. It is typical to have some left over and the beds should not be topped up but installed only to the correct heights.

CASE STUDY 11.9 COMING TO GRIPS WITH RANDOM PACKING HANDLING

Two-inch metal Pall rings were to be loaded into a tower in Southern Louisiana during the summer. The rings arrived at the site a few weeks prior to installation, packaged in 5-ft^3 cardboard boxes. A vendor specification requiring protection of the packings was not adhered to, and the cardboard boxes were mounted on pellets in the laydown yard.

Over the next few weeks, rain and the oil layer on the packings weakened the cardboard. Several boxes tore apart, pouring packings out. The soil under the pellets turned into mud in heavy rain. Mud penetrated some boxes and stuck to the packings. Packings that fell out of disintegrated boxes were stuck in the mud.

When this was discovered, boxes that were still in good shape were picked up and placed on pellets under heavy tarps in nonmuddy regions. Packings from damaged boxes that were still clean or reasonably clean were emptied into large (about 16 × 8-ft) wooden crates, padded, and covered by transparent heavy plastic sheets. These measures were effective. The covered cardboard boxes held and the packings in the crates suffered no further crudding.

One problem was experienced in the crates. The transparent cover generated "hot-house" conditions when the sun shone. The top packing layer was so hot that one could not touch it. The oil layer on the hot packing flowed down, generating a shallow pool of oil at the bottom of each crate.

An attempt was made to clean the mudded packings by hydroblasting. This was very effective; all the mud was removed. However, the hydroblasting deformed the rings to the point of becoming useless (Fig. 11.4). All packings hydroblasted needed to be thrown out.

The next attempt at cleaning placed the mudded packing in crates on the back of a flat-top truck. The truck was taken to a car wash, which was instructed to use water sprays with no detergent. Detergent was avoided due to concerns that it may induce foaming in service. It needed several washes per load to get most of the mud off, but some dirt stuck. The washed rings were spread out on plastic sheets to dry in the sun. The pieces that were still dirty were hand picked and returned to the wash.

Everyone that handled the packings received hand cuts, some quite deep. Many packing pieces were quite sharp and went right through the safety gloves worn by the handlers. The problem was aggravated by the rings supplied not fully closing (Fig. 11.4).

From the large wooden crates, the packings were dug out in 1-ft^3 buckets. This process was the source of many hand cuts. During the process, some debris was found lodged amidst the packings, mainly pieces of cardboard and dirt from the original boxes. These needed removing as the packings were placed on the hopper. The hopper

Figure 11.4 Pall rings deformed by hydroblast cleaning.

was lifted by crane and emptied onto a chute, and the packings were loaded into the tower using the chute-and-sock method (82, 250; Fig. 6.2*b*), with a person inside the tower raking the pieces as they came in.

The loading hopper was another source of dirt. When lowered to the ground, the hopper mouth often struck the mud. Some of this mud found its way into the next load of packings that entered the tower. The hopper mouth was wiped as well as possible, but the process was not perfect.

When the packings were loaded from the hopper to the chute, pieces fell to the ground. These pieces were very sharp, and there were people working below. The area was cleared as soon as the hazard was recognized. Also, while loading the packings, it was discovered that the handhole at the bottom of one bed, which is used to empty the packings, was open and had nothing but a plastic sheet over it. Luckily, the plastic sheet was strong enough to hold the weight of the packings above it. The handholes were quickly bolted once this was noticed.

Epilogue Despite the eventful packing-handling process, there was a happy end. Close inspection and supervision eliminated the problems early and at minimum cost. The construction proceeded without delay. The mud and crud removal was effective, and the tower fully achieved its design separation and capacity.

Moral This case demonstrates the multitude of pitfalls often encountered while loading tower packings. Close inspection and supervision are the best tools for circumventing them.

CASE STUDY 11.10 STRUCTURED PACKING INSTALLATION

Column A A 3-ft-ID chemical tower was packed with wire-mesh structured packing without the vendor supervision. Upon start-up, the tower achieved very poor separation. Investigation showed that a 3-in. gap was left between the packing and the tower shell, apparently with the intention of providing space for vapor disengagement. After the tower was repacked to give the packing a snug fit to the shell, separation was achieved.

Column B This installation was supervised by a vendor representative as well as the author. Both went together for a quick dinner. Upon return, they observed the installers using a large hammer to bang in a "brick" of structured packings so that it would squeeze in between two existing layers. Fortunately this was caught before damage was done. From then on the vendor representative and the author went to dinner at staggered times, making sure supervision remained continuous.

Column C This installation was not supervised by the vendor. During installation, the packed bed grew in length by 3 in. This reduced the space between the liquid distributor and the packings from 4 in to 1 in. The installer wanted to cut the top 3 in., but the packing vendor objected. Eventually, a new layer of structured packing was supplied by the vendor at extra cost and an installation delay.

Column D Installed packing sloped from a high point. The high point was caused by a thermocouple pushing some bricks up. An in situ water test showed liquid maldistribution. The bed needed repacking.

Column E The packing vendor supervisor was late to arrive, so it was decided to proceed with the installation without him. The installers could not put the structured packing bricks together without leaving several gaps at each level. Eventually the vendor supervisor arrived and upon inspection ordered removal of all the installed packing and a repack. The supervised repack had no gaps in the packings.

Moral Have all structured packing installations performed, or at least thoroughly supervised, by the vendor.

CASE STUDY 11.11 CORRECT FEED INTO PARTING BOXES

Installation Hot-pot absorber using hot potassium carbonate to absorb CO_2 from process gases in an ammonia plant. The tower contained three beds of random packing. The lean solvent was distributed to the packing via a V-notched trough distributor with a single parting box (Fig. 11.5*a*). The parting box received the liquid feed from a perforated pipe parallel to the parting box and mounted a few inches above it. The pipe contained bottom perforations.

Problem Carbon dioxide concentration in the overhead gas was six times higher than the design.

Investigation The engineering contractor, process licensor, and packing vendor developed their own theories, each blaming the others for the failure. Theories included incorrect packing specification, insufficient redistribution, inherent poor distribution of V-notch distributors, fouling, and foaming.

Eventually, the tower was shut down and inspected. The distributor was found rotated 90° from its design orientation. Therefore, the feed pipe that was supposed to be parallel to the parting box was at a right angle to it (Fig. 11.5*b*). Thus most of the liquid issuing out of the feed pipe bypassed both the parting box and the distributor and rained directly onto the packing.

Cure The distributor was rotated by 90°. The tower easily achieved the design separation afterward.

Moral Finger-pointing does not solve plant problems. Inspection and field tests do.

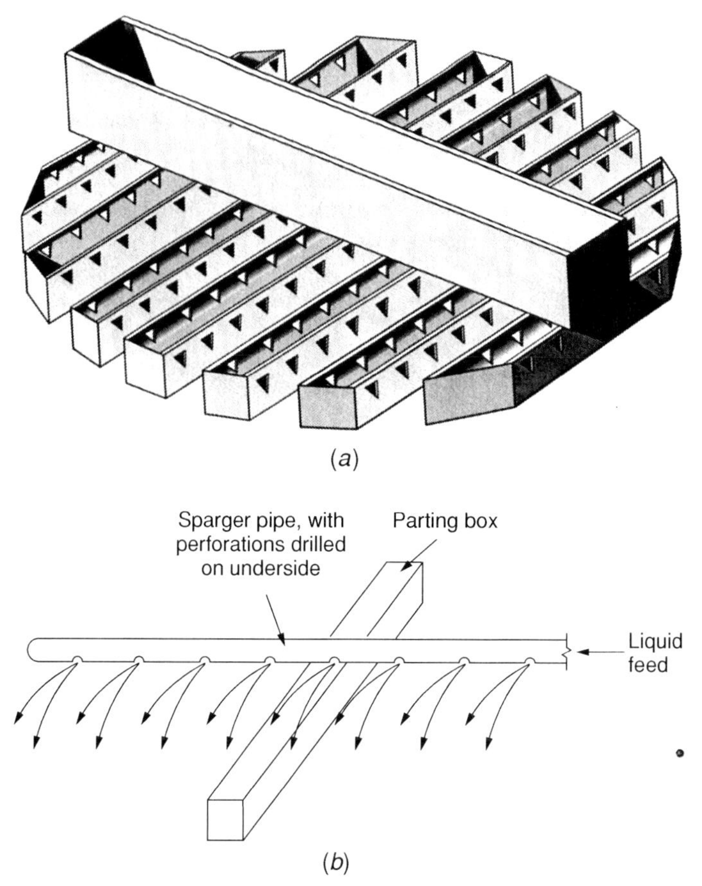

Figure 11.5 Liquid distributor feed problem in absorber: (*a*) V-notched trough distributor with single parting box; (*b*) feed pipe at right angles to parting box, incorrect. [(*a*) Reprinted courtesy of Koch-Glitsch LP.]

CASE STUDY 11.12 INVERTED CHIMNEY HATS

Contributed by Pamela Tokerud, Koch-Glitsch LP, Wichita, Kansas

Installation This was to be a simple replacement of an orifice distributor with a high-performance orifice distributor for increased throughput. The column height was limited and the existing vessel support ring was to be reused. The existing bed of structured packing was evaluated and determined adequate for the higher throughput.

Problem After installation of the new high-performance orifice distributor, the column could not achieve pre-revamp capacity.

Investigation The drawings were reviewed to ascertain there were no errors in the design. The distributor was designed correctly, with the right orifice size and approximately 20% open area for vapor flow. Discussions then focused on the installation of the distributor. The distributor hats were removable due to limitations imposed by the manhole access. The installer attached the hats upside down—resulting in approximately a 2% open area for vapor flow.

Solution Upon entrance into the column it was confirmed that the hats were incorrectly installed. This was corrected and the column exceeded design capacity.

Morals

- Installers should be knowledgeable of the equipment.
- Final inspection of installed equipment is critical.

CASE STUDY 11.13 PROBLEMS WITH FABRICATION AND INSTALLATION OF PACKING LIQUID DISTRIBUTORS

Distributor A About half of the distributor troughs contained holes punched in the direction of liquid flow; the other half had holes punched in the reverse direction. Showed up as different liquid heads in distributor troughs during a distributor flow test. Some of the troughs needed to be rebuilt.

Distributor B This distributor had troughs connected by liquid equalizing channels. The channels were to ensure uniform liquid head throughout the distributor. A water test at the vendor shop showed unequal heads in the troughs due to insufficient equalization. The channels needed rebuilding.

Distributor C Liquid entered a parting box from a feed pipe with bottom perforations mounted parallel to the parting box and just above it (Fig. 11.6*a*). The liquid left the pipe perforations with a horizontal momentum in the direction of pipe flow. This momentum induced horizontal flow in the parting box. Reflection by the end wall of the parting box generated a backward wave. The back-and-forth movement led to sloshing of the liquid in the box and overflow. The problem was identified in a distributor flow test and was corrected by adding short tubes at the pipe perforations to eliminate the horizontal momentum (Fig. 11.6*b*).

Distributor D Similar to the experience with distributor C, the feed entered the parting box from several perforations at the underside of the feed pipe. An in situ water test found that the liquid jets issuing from the pipe hit the liquid surface in the parting box and splashed over the sides, missing the distributor altogether. The fix was adding short pipes that issued the feed liquid under the liquid levels in the parting box (Fig. 11.6*c*).

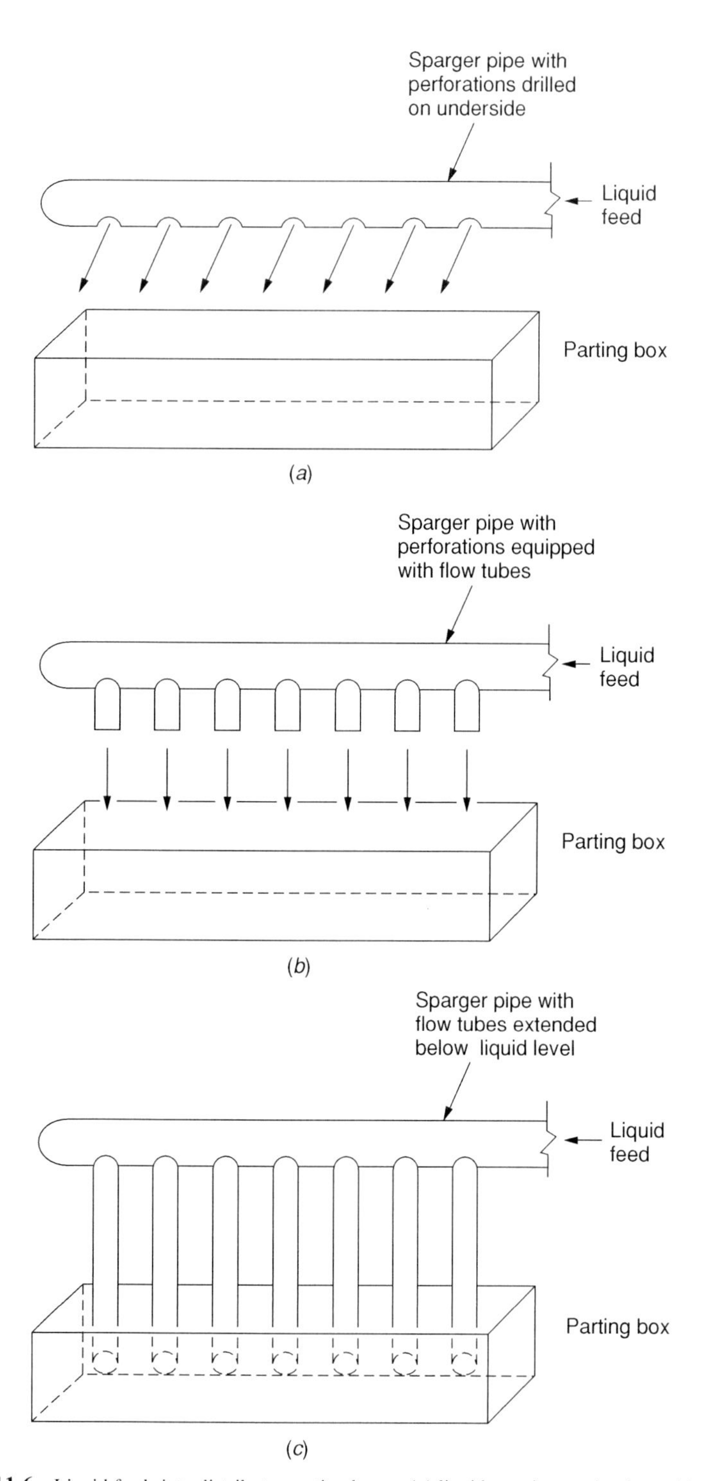

Figure 11.6 Liquid feeds into distributor parting boxes: (*a*) liquid entering parting box with horizontal momentum; (*b*) adding short tubes eliminates horizontal momentum; (*c*) adding longer tubes issues liquid feed under liquid level and eliminates both horizontal momentum and liquid splash.

Distributor E Feed liquid entered the distributor parting box via a single pipe that entered the tower horizontally, then elled down, discharging the liquid feed downward a few inches above the parting box. In situ water tests discovered that the liquid jet from the pipe hit the liquid surface in the parting box very hard, causing a large quantity of liquid to jump over the side of the parting box and to miss the distributor altogether. The jump was eliminated by extending the pipe below the liquid level.

Distributor F A tower operating at very low liquid rates achieved poor separation. A temperature survey showed very poor liquid distribution. The tower was shut down. An in situ water test showed that most of the liquid was passing through the opening of a few missing bolts in the distributor floor. Installing the bolts reinstated good operation.

Distributor G The distributor had drip tubes with openings 1 in. above the distributor floor. Elevated openings give the distributor a better plugging resistance and are common in practice. After one distributor flow test, the last inch of liquid was observed to stagnate. The distributor floor contained only one small ($\frac{1}{8}$-in.) drain hole, and this took hours to drain the last inch of liquid. Fixed by adding drain holes.

Moral Distributor flow tests are invaluable.

CASE STUDY 11.14 ONE HEAT EXCHANGER CAUSING PROBLEMS IN TWO TOWERS

Tom C. Hower and Henry Z. Kister, reference 224. Reprinted with permission from *Hydrocarbon Processing*, by Gulf Publishing Co., all rights reserved

This case describes how a design error that converted a countercurrent heat exchanger into a cocurrent exchanger caused poor separation in both an absorber and a deethanizer.

Installation This case occurred in a gas plant using an absorption–regeneration process for recovering HCs heavier than methane from the gas (Fig. 11.7). Inlet gas was contacted with lean absorption oil in the absorber. The overhead product was the sales gas (methane with some ethane). The bottom product was the rich oil, which contained ethane, LPG, and gasoline absorbed in the absorption oil. After the rich oil left the column, it was mixed with the absorber feed, chilled in the bottom step chiller E2, and flashed. This added one theoretical stage to the absorber. Rich oil leaving the flash drum chilled the lean oil in exchanger E3, then entered the deethanizer. Overhead product from the deethanizer was the methane and some of the ethane absorbed in the oil. Some of the deethanizer overhead was sent to the fuel. The balance was compressed and recycled to the absorber feed. The deethanizer bottoms contained the absorption oil, part of the ethane, the LPG, and the gasoline. The LPG and gasoline were separated from the absorption oil in the still (not shown). Hot absorption oil from the still was cooled by reboiling the deethanizer, then by an air cooler. It was

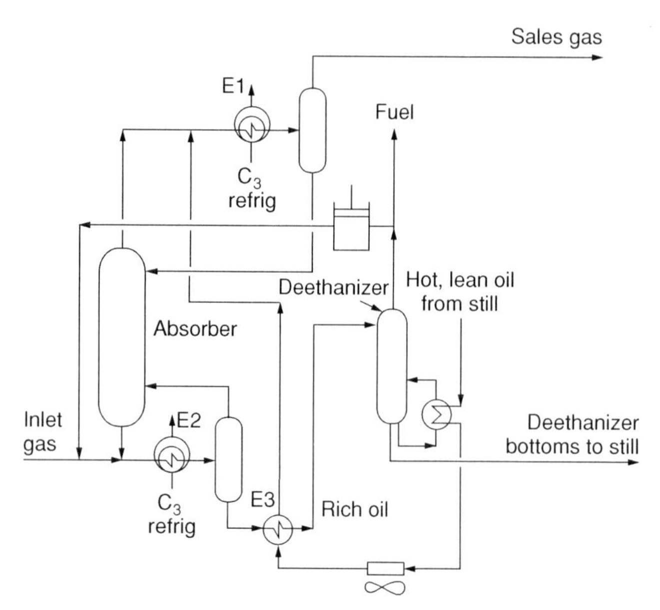

Figure 11.7 Absorption–regeneration process for recovering heavy HC from natural gas. (Reprinted with permission from *Hydrocarbon Processing* by Gulf Publishing Co. All rights reserved.)

then chilled by heat exchange with the rich oil in the interchanger E3. The lean oil was then presaturated by mixing with the absorber overhead. This permitted the heat of absorption to be removed in chiller E1. The chilled mixture was separated, and the presaturated rich oil flowed to the top of the absorber.

Problem Both the absorber and the deethanizer appeared to be operating at poor efficiencies. The sales gas contained an excessive quantity of C_3 and heavier HCs. The deethanizer bottoms contained an excessive amount of ethane and methane.

Investigation Operating conditions were compared to the design conditions. It was discovered that the rich oil leaving the rich oil–lean oil interchanger E3 was about 80°F colder than in the design while the lean oil leaving the interchanger E3 was about 120°F hotter than in the design. The interchanger E3 was closely examined. It was then found that the interchanger was initially specified as a countercurrent exchanger but due to a design error was built as a cocurrent exchanger. This caused the poor heat transfer.

Solution One side of the interchanger E3 was repiped to convert it into a countercurrent exchanger. This solved the problem.

Postmortem Poor heat removal from the lean oil caused excessive temperatures in the absorber and the absorber overhead system, resulting in an escape of heavies in the absorber overhead product. The excessive chilling of the deethanizer feed could not be matched by the reboiler heat input, causing lights to escape in the deethanizer bottoms.

Moral Checking and double-checking of design details can save a lot of headaches during start-up.

CASE STUDY 11.15 LIQUID LEG IN VENT LINE LEADS TO TOWER UPSET

Installation Tower bottoms were pumped into storage which was vented into the tower vapor space near the feed point. The vent line had a low leg, which looks like a seal loop (solid line in Fig. 11.8). The presence of the low leg was an installation error (the line was designed without a low point) incurred as the line was routed into the pipe rack.

Problem The storage tank started to build pressure. The buildup would be due to liquid accumulation in the seal loop, backpressuring the tank. The tank pressure would build up as high as 50 psig. When the pressure built high enough, it would blow all the liquid into the tower. The tower pressure went up and reboil was lost. It was impossible to operate the tower.

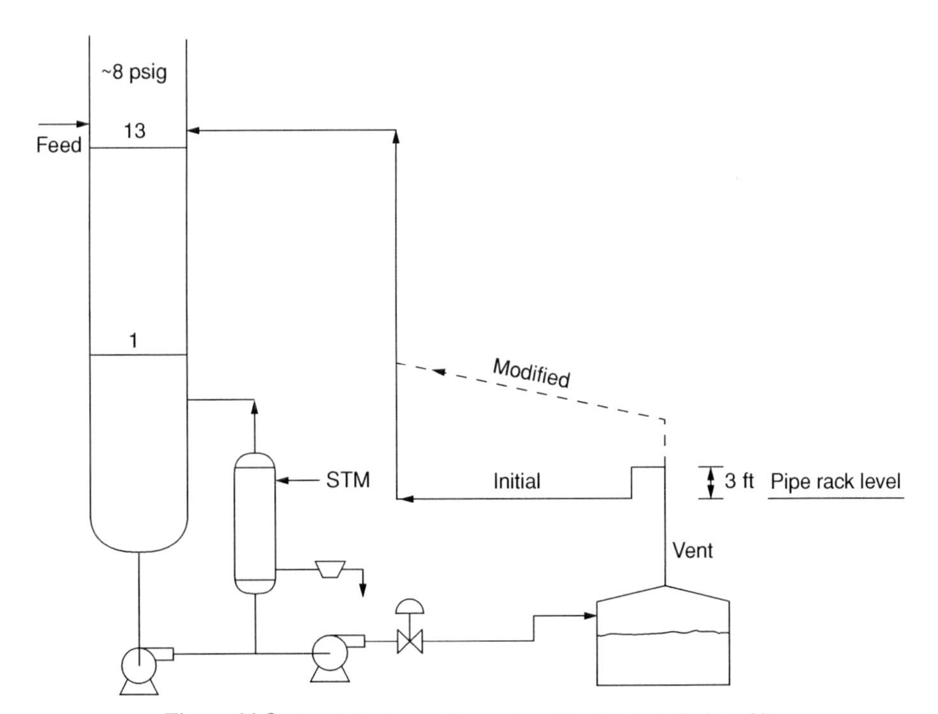

Figure 11.8 Liquid leg in vent line and modification that eliminated it.

Solution The seal leg was eliminated. The horizontal section of the vent line was inclined so that it drained back to the storage tank (dashed line in Fig. 11.8). Following the modification the problem disappeared.

CASE STUDY 11.16 IS YOUR COOLING WATER FLOWING BACKWARD?

Installation The main unit in a new C_3 purification plant was a C_3 splitter. The condenser of this tower was the prime consumer of cooling water in the plant. The cooling-water circuit is shown in Fig. 11.9.

Commissioning Prior to start-up, the cooling-water circuit was commissioned. All seemed in order. The pump was pumping, the pressure gage was reading close to design, and water was flowing through the condenser and cooling tower return sprays. The only problem was that the flowmeter read nothing; in fact, the needle was pegged below the zero mark. Instrumentation was a major headache on this plant, and it surprised no one to see another incorrect instrument.

Troubleshooting Checks by an expert instrument engineer found that the flow transmitter was functioning properly. He then raised the question whether the cooling water may be flowing backward. Initially this was considered a joke, but as unlikely as it appeared, it was decided to test it. Valve B was closed and vent valve V was opened. A jet of water shot up about 30 ft in the air. Almost immediately the jet started to lose height, and within moments it stopped. There was no pressure at the vent. At

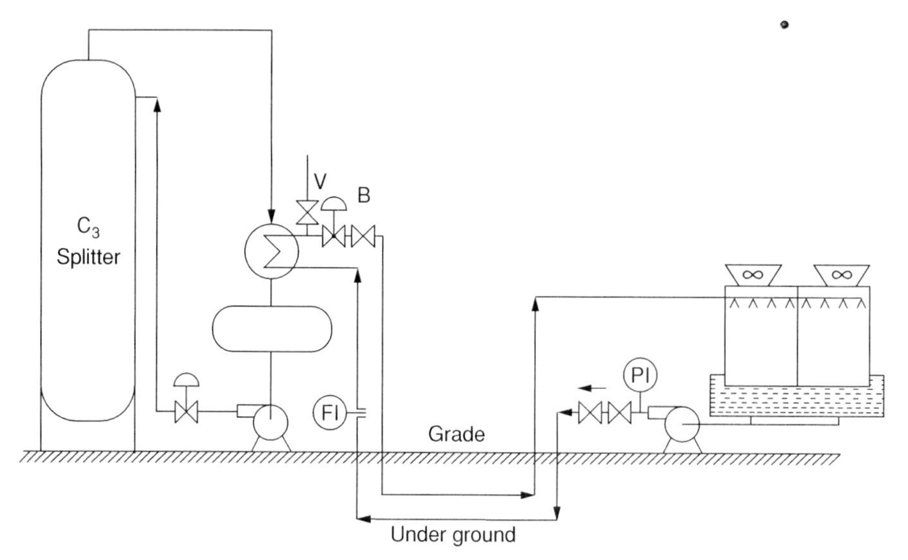

Figure 11.9 A C_3 splitter unit cooling-water circuit.

the same time, the pump continued pumping and its discharge pressure rose slightly. The test verified that the pump discharge was not reaching the condenser.

Mystery Explained It was then decided to reopen the trench that housed the underground water pipes. The pump discharge was found connected to the condenser water return line, while the cooling tower return line was connected to the condenser water supply line. The pumping route was therefore from the pump, through valve B, through the condenser, backward through the flowmeter, and through the sprays.

Cure The piping connections were interchanged.

Moral In a troubleshooting investigation, theory testing should begin with theories that are easiest to prove or disprove, almost irrespective of how likely or unlikely they appear.

Chapter 12

Difficulties During Start-Up, Shutdown, Commissioning, and Abnormal Operation: Number 4 on the Top 10 Malfunctions

Close to 100 reported incidents place commissioning, start-up, shutdown, and abnormal operation fourth on the list of distillation tower malfunctions (255). These incidents were spread evenly throughout chemical, refinery, and olefins/gas towers. Some of these incidents were accidents, involving fatalities, injury, and major damage. More commonly, these incidents led to plugging/coking, internals damage, product loss, and prolonged downtimes. The good news is that over the last decade abnormal operation incidents appear to be on the decline (245, 252, 255). The industry has made good progress in reducing these incidents. This progress appears to be ongoing and can be attributed to greater emphasis on safety by most major corporations. Hazops and "what-if" analyses, safety audits, improved procedures, and extensive safety training have all contributed to this very welcome progress.

The top issues are blinding/unblinding and backflow (255). There is some overlap between blinding/unblinding and backflow; in several cases, poor blinding led to a backflow incident. Both blinding and backflow incidents led to chemical releases, explosions, fires, and personnel injuries. High-pressure absorbers account for quite a few of the backflow incidents. Here loss or shutdown of the lean solvent pump resulted in backflow of high-pressure gas into the lean solvent line, from where it found a path to atmosphere or storage. Several other incidents reported flow from storage or flare into the tower while maintenance was in progress. Some blinding incidents involved valves that were plugged or frozen.

Water removal has been a top issue in refinery fractionators. Water removal incidents are closely linked to water-induced pressure surges, which tops the causes

Distillation Troubleshooting. By Henry Z. Kister

of tower internals damage (Chapter 13). About half of the reported cases of water-induced pressure surges were induced by start-up, shutdown, and abnormal operation; vice versa, most of the water removal failures resulted in pressure surges. Refineries implement special procedures to remove water prior to start-up of their hot-oil fractionators, but if something goes wrong, a pressure surge often results.

Washing and steam/water operations are common commissioning operations that are quite troublesome and have contributed to several case histories. Most malfunctions in washing led to fouling and corrosion, but in some cases, washing liberated toxic gas or transported chemicals into undesirable spots. Most steam/water operation incidents either caused overheating, or formed a condensation zone in which rapid depressuring took place, leading to either vacuum and implosion or excessive flows and internals damage as vapor from above and below rushed toward the depressured zone.

For all four operations (blinding/unblinding, backflow, washing, steam/water operations) the number of incidents reported over the last decade appears to be on the decline (255). Also, the reported incidents for all four are evenly split between refinery and chemical towers. These four operations plus water removal account for about 70% of the reported abnormal operation incidents.

Several cases of overheating were reported—none in the last decade. Some cases resulted from steaming, but there are also other causes, such as failure of the cooling medium in a heat-integrated system during an outage. Other abnormal operation incidents involved pressuring or depressuring, overchilling, purging, and cooling. Pressuring and depressuring caused internals damage if too rapid or if performed backward via valve trays (Chapter 22; in particular, Sections 22.4 and 22.6 in Distillation Troubleshooting Data Base). The main cooling incidents involved condensation, which induced air into the tower or formed a zone of rapid depressurization. The malfunctions of purging are varied.

Overchilling deserves special discussion. While all the abnormal operation malfunctions above show a marked decline in the last decade, overchilling shows a rise (255). The majority of the reported overchilling cases occurred in olefins or gas plant towers. Conversely, for towers in this industry, overchilling is the major abnormal operation malfunction. Moreover, overchilling had led to brittle failure, releasing vapor clouds, which had been responsible for major explosions accompanied by loss of life, injuries, and major destruction. The rise in overchilling case histories is the only setback, yet a major concern, to the progress achieved in reducing abnormal operation malfunctions.

CASE STUDY 12.1 COMMISSIONING OF LEAN-OIL STILL REBOILER

Tom C. Hower and Henry Z. Kister, reference 225. Reprinted with permission from *Hydrocarbon Processing*, by Gulf Publishing Co., all rights reserved

This case describes a troublesome experience with the commissioning of a fired reboiler circuit of a lean-oil still, where a poor blinding practice, start-up without proper instrumentation, and incorrect interpretation of observations combined to hinder the diagnosis of the problem.

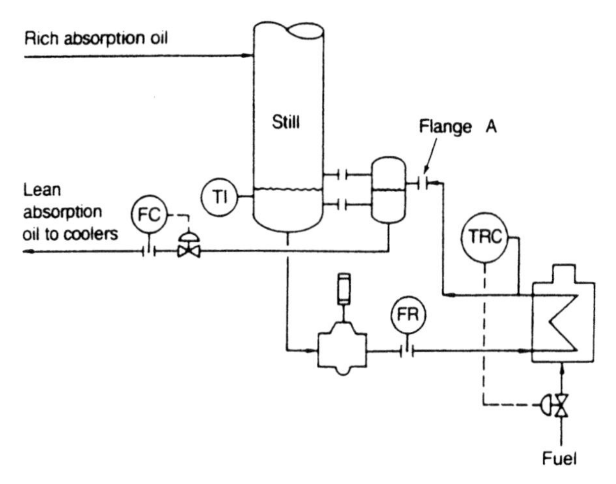

Figure 12.1 Still reboiler commissioning circuit. (From Ref. 225. Reprinted with permission from *Hydrocarbon Processing* by Gulf Publishing Co. All rights reserved.)

Installation This case occurred in a still separating LPG and gasoline from absorption oil in a natural gas plant (Fig. 12.1). The still was reboiled by a fired heater. Fuel to the heater was temperature controlled at the reboiler outlet. The temperature sensor was located close to the heater. A flash drum separated the reboiler outlet mixture into vapor and liquid phases. The vapor entered the column, while the liquid was partially withdrawn as the bottom product and partially returned to the tower. Liquid to the reboiler was pumped by an in-line centrifugal pump located close to the still so that the line between the still and the pump was very short. The liquid flow to the heater was metered.

Commissioning At the time that the problem occurred, the column was being commissioned. Absorption oil (void of LPG and gasoline) was being circulated through the still and heater. The heater was commissioned. The heater outlet temperature recorder/controller (TRC) was operating on automatic control and showing a temperature of 140°F. The heater outlet temperature was being gradually raised at a slow rate toward the final target temperature of 540°F. The flow recorder (FR) in the heater circuit (Fig. 12.1) was not commissioned yet. There was a temperature indicator in the bottom of the column, but the thermometer was broken, and its replacement had not yet arrived.

Problem Once the reboiler outlet temperature reached 350°F, it would not rise any further, no matter how hard the heater was fired.

Troubleshooting The heater was inspected and was found to be firing quite hard. The pump was checked and appeared to be operating quite well. The short pump suction line was hot. These appeared to suggest that the pump, heater, and circulation were working properly. The pump discharge pipe was insulated and could not be checked. Accessible valves and flanges were physically inspected from close up. Inaccessible ones were inspected from the ground. No problems were detected.

In an attempt to diagnose the problem, the flowmeter in the pump circuit was commissioned. It read zero flow. This reading was disbelieved and was attributed to a likely instrument problem.

It was then decided to provide ladders to access inaccessible flanges and look at them close up. Upon inspection of the flange at the flash drum inlet (flange A), it was observed that its gasket looked rugged, as if it was cut with an acetylene torch. It was then realized that this was not a gasket but an isolation blind without a handle. The blind remained in the flange since hydrotesting.

Analysis The troubleshooting efforts were impeded by several observations that suggested proper circulation. First, the heater outlet temperature was responsive to increases in heater fuel rates until it reached 350°F. The temperature sensor was located close to the heater; as the fuel rate was raised, the sensor became hotter due to heat conduction. When it reached about 350°F, losses from the heater offset the conduction and the temperature could no longer be raised. Second, the pump suction pipe was hot. However, this heat came not come from the heater as was presumed but from a dead-headed pump. Finally, upon inspection from the ground, nothing unusual was noticed about the drum inlet flange, which was 20 ft above grade; the blind appeared as if it were a gasket.

Solution The system was depressured and the blind removed. The pump was examined and found to be damaged after being dead-headed for 10 hours. The pump wear rings and seals were replaced. The plant was very lucky not to have damaged heater tubes, as the heater was firing very hard with the fuel control valve widely open.

Morals

- Blinds should have long handles and tags. Good blinding practices for distillation systems are described elsewhere (250).
- Instrumentation must be operational before the system is commissioned.
- In a troubleshooting investigation, the obvious interpretation of an observation may not be the correct one.

CASE STUDY 12.2 REVERSE FLOW LEADS TO CORROSION AND FLOODING

Contributed by Mark Pilling, Sulzer Chemtech, Tulsa, Oklahoma

Installation Stabilizer for a refinery C_5/C_6 isomerization unit with standard sieve trays.

Problem Column had worked well at its initial running period. Pressure drop was normal and operation was adequate. Following a shutdown due to a power failure, the column became unstable at moderate to high rates. Pressure drop fluctuated periodically. Column would cycle, periodically emptying tower into the overhead receiver. It was strongly suspected that the column was damaged during the power failure shutdown.

Cause Upon opening of the tower, the answer was obvious. The three trays immediately above the feed were severely damaged and lay upon the feed tray. The damage was caused by hydrochloric acid entry. Hydrogen chloride was present in the stabilizer, but there were no apparent sources of water to form hydrochloric acid. The tower was designed to run bone dry and had CS internals.

There was a caustic scrubber downstream for neutralizing the hydrogen chloride in the stabilizer off gas. The scrubber gas was then sent to the fuel gas. During the shutdown, the stabilizer pressure fell below the fuel gas pressure. This forced the caustic backward into the stabilizer through a faulty check valve. The introduction of the aqueous caustic into the stabilizer created a highly acidic system once the caustic was consumed. Aggressive corrosion resulted. The damaged trays lying upon the feed tray caused premature flooding.

Cure Trays and check valve were replaced. Stabilizer operation returned to normal.

Morals

- Beware of reverse flow, especially during outages.
- Check valves cannot be relied on to prevent reverse flow.

CASE STUDY 12.3 CAUSTIC WASH CAN DISSOLVE DEPOSITS

Caustic wash has been found effective in dissolving deposited solids. It works best with polar deposits such as acidic deposits and metal salts of corrosion products.

There are a few pitfalls that may render the caustic wash ineffective:

- Experience had been good with caustic concentration 1% to 4% weight, typically about 2%.

  If too strong, the solution may initially dissolve the deposits well but then form a coat on the surface that may inhibit further dissolving. This is analogous to rust formation.

  Similarly, if too weak, the caustic may become watery and ineffective. It is important to recognize that as the caustic dissolves the deposits it becomes consumed. It needs to be made up to maintain concentration.
- It is imperative to perform bench tests to determine the potential effectiveness of a caustic wash and the best concentration. This is easily accomplished using samples of the deposits and a beaker and experimenting with a variety of caustic concentrations.

  Caustic wash is not a cure-all. There are systems where it does not help. Similarly, there may be only a very limited concentration range where such a wash is effective. Experimentation in the laboratory often makes the difference between a successful and an ineffective caustic wash.

- Caustic wash appears to work fastest around 200°F. It will still work at lower temperatures, but at a slower rate.

Following are case studies where the caustic wash has been effective.

Case 1 During an operation upset, a scrubber scrubbing formaldehyde dust dried out. Following the upset, high differential pressure was experienced across the scrubber. Plugging with formaldehyde dust was suspected.

In an attempt to dissolve the dust, water flow was increased to the tower. Formaldehyde dust is acidic (pH 3–4). Theoretically, the dust should have dissolved in the water wash, but it did not. The differential pressure remained high following the water wash.

Next, caustic was injected to the top of the tower. A pH of about 12 was maintained in the tower. Within one shift the tower went back to normal, with differential pressure becoming low again.

Case 2 A new process had a tower equipped with an internal reboiler. The reboiler initially worked well but slowly crudded up. Salts deposited on the outside of the tubes. The tubes were U-tubes in a "bathtub" arrangement. (This bathtub arrangement is described on p. 460 of Ref. 250.) The salts were metal salts of corrosion products. The salts were not acidic; they were neutral and very hard.

Mechanical cleaning of the tubes using 10,000-psi water jets was attempted. It was found ineffective in penetrating the tube bundle and cleaning it. Caustic washing was attempted next. It turned out that the caustic wash was effective only at around the 2% concentration. Once the concentration fell below 1%, the caustic would become inactive. It took five cycles of wash before no more caustic appeared to be consumed and the solution was at 2%. This signified that the salts were fully consumed. Once the tubes were pulled out, all the deposits were gone and tubes were shiny metal.

Case 3 Random packings were used for more than 12 years in a vacuum tower in acrylate service. The packing run length was 2 years, restricted by polymerization. At the shutdown, the packings were removed and replaced by a new load. Afterward, the fouled packings were soaked in a 4% caustic solution. This softened the polymer. The softened polymer was then removed by a water wash.

CASE STUDY 12.4 ON-LINE WASH OVERCOMES SALT PLUGGING

Contributed by Betzalel Blum, Oil Refineries Ltd., Haifa, Israel

This case describes experiences with on-line water wash of the top trays in both an FCC main fractionator (MF) and a reformer stabilizer without allowing the water to go down. In both cases, the procedure is much the same.

Hot condensate at 90–95°C is injected into the reflux via a ¾-in. line. A batch of about 2 minutes is injected followed by an hour wait before the next batch. Upon injection, the tower top temperature decreases. Normally, the top temperature of the FCC MF runs at about 132°C. When salting out is experienced on the top tray, the top temperature drops to 120°C. The water injection causes a further drop of the top temperature. The water injection is continued until the temperature shows a further drop of 10°C, down to 110°C, but no colder than this. As soon as the temperature drops to 110°C, the water is closed.

Going below 110°C generates the risk of an upset and needs to be avoided. During the wash, the tower pressure goes up due to the vaporization of water. There is also a sudden overload of the overhead system and a jump up in pressure with possible lifting of the relief valve. The receiver appears to get a chunk of liquid.

It is important to watch the level in the reflux boot and to open or be ready to open the bypass around the boot-level control valve. It is also important to watch the gas leaving the reflux drum. If it goes to a compressor, a problem may result, especially if the off gas causes liquid carryover from the drum. No problems had been experienced with reflux pump seals during the wash.

Prior to the operation, throughput to the unit is reduced to the minimum turndown. The water wash is done with fairly low tower loads in order to be able to handle upsets should they happen. All the salt may not clear up on the first injection, and repeating the procedure may be required. With the FCC MF, some of the water comes out via the heavy-naphtha route.

Following the initial wash, the refinery uses a preventive water injection, following the same procedure, once per 2 months.

CASE STUDY 12.5 SIMULATION IDENTIFIES DRAW PAN DAMAGE

Contributed by Gerald L. Kaes, KAES Enterprises, Inc., Colbert, Georgia

Installation An FCC MF, receiving hot, superheated vapor feed from the reactor. Before ascending to the fractionation zones, this vapor was quenched in the lower part of the fractionator by direct contact with a cooled circulating liquid slurry. Fractionator products were a naphtha top product, a light cycle oil (LCO) side cut, and a heavy DO bottoms. The distillates are far more valuable than the DO.

Problem The refinery lost all electrical power. Within a few hours after power was restored, it became apparent that the FCC MF was not performing properly. The operators could not hold a liquid level on the LCO draw tray. This caused the DO yield to sharply rise and the LCO yield to drop.

Troubleshooting Tray gamma scans revealed that the trays were in place with no apparent tray damage. The simulation model developed for the column during normal operation was used to duplicate the new temperatures and product flows. It confirmed that there was a sizable flow of LCO from the LCO draw tray down to the column quench zone below, suggesting leakage from the LCO draw pan. Upon opening the column, it was discovered that the rapid heating of the tower (when the refinery lost power and the quench circulation suddenly stopped flowing) had broken the attachment between the LCO draw pan to the column shell. Light cycle oil poured down the tower between the column shell and the draw pan.

Cure Reattaching the draw pan to the tower shell reinstated good operation.

CASE STUDY 12.6 UNIQUE CONTROL PROBLEM IN TOTAL-REFLUX START-UPS

Installation Chemical tower equipped with a circulating thermosiphon reboiler. A preferential baffle in the tower bottom separated the reboiler compartment from the bottom compartment (Fig. 12.2). This baffle preferentially diverted tray liquid to the reboiler draw compartment and reboiler return liquid to the bottom draw compartment. This improves reboiler log mean temperature difference (LMTD) and tower mass transfer.

Problem The tower was started on total reflux. There was no bottom flow so the liquid level in the bottom compartment was meaningless. There was a liquid level somewhere in the reboiler compartment, but this level was not monitored and

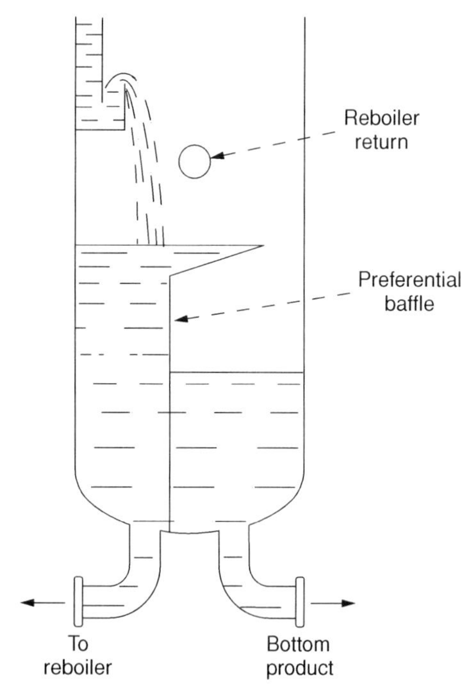

Figure 12.2 Preferential baffle in tower bottom, separating reboiler draw compartment from bottom draw compartment. (Reprinted with permission from Ref. 250. Copyright © 1990 by McGraw-Hill.)

often went dry. When it did, boiling ceased and liquid from the trays dumped. This replenished liquid supply to the reboiler, generating a vapor surge. The result was violent swings during total-reflux operation.

Solution An upper level tap was installed above the top of the baffle, just below the reboiler return elevation. During total-reflux operation, the bottom level was run above the top of the baffle and was manipulated by varying boil-up.

An alternate solution that effectively prevented similar problems is installing a valved line connecting the reboiler draw line with the bottom draw line. The valve is normally shut. During total-reflux operation, the valve is opened, equalizing the levels in the reboiler and bottom draw compartments.

Chapter 13

Water-Induced Pressure Surges: Part of Number 3 on the Top 10 Malfunctions

Close to 100 reported incidents place tray, packing, and tower damage third on the list of distillation tower malfunctions (255). Most of the causes are discussed in Chapter 22, with damage due to high liquid levels discussed in Section 8.3.

In refineries, one cause towers above the others: water-induced pressure surges (255). These are described in this chapter. Water-induced pressure surges are not unique to refineries, and some cases were reported in other petrochemical applications where a pocket of water enters a hot tower containing HCs or other water-insoluble organics. However, there is no denying that the large refinery fractionators are the major troublespots. The good news is that these pressure surges are very much on the decline (253, 255). Much of the progress here can be attributed to AMOCO, which experienced its share of pressure surges in the 1960s. AMOCO investigated these cases very thoroughly and shared its experiences and lessons learned with the industry by publishing three superb booklets: *Hazard of Water*, *Hazard of Steam*, and *Safe Ups and Downs* (2–4). Following the recent merger between BP and AMOCO, BP has now incorporated these booklets into BP's series of safety booklets, which is publically available from the Institution of Chemical Engineers (IChemE), Rugby, England.

The key to prevention is keeping the water out and using "heavy-duty design" (442) in the affected regions. The leading route of water entry is undrained stripping steam lines, but other causes are not far behind. These include (253, 255) water in feed/slop, accumulated water in transfer lines to the tower and in heater passes, water accumulation in dead pockets, water pockets in pump or spare pump lines, condensed steam or refluxed water reaching hot sections, and hot oil entering a water-filled region. The vast majority of the reported case histories came from the refinery vacuum, crude, FCC, and coker main fractionators. In these services, water-induced pressure surges accounted for between a third and half of the reported damage incidents (253).

Distillation Troubleshooting. By Henry Z. Kister

CASE STUDY 13.1 SIDE-STRIPPER PRESSURE SURGE CAN DAMAGE MAIN FRACTIONATOR

Installation A refinery preflash vacuum fractionator removing heavy diesel and gas oil from the bottoms of an atmospheric crude tower. The tower operated under slight vacuum. The gas oil was removed as a side draw and was stripped by 150 psig steam in a side stripper. Due to upsets in the steam system, the stripping steam was often wet. The steam pressure letdown was quite close to the stripper.

Experience On one occasion, an explosion was heard; then the fractionator lost separation. When the tower was opened, trays above the stripper were found blown upward, and those below the stripper were blown downward. The damage was extensive, both to the trays and supports, with damaged tray panels badly warped.

Cure Stripping steam use was discontinued. This was at the expense of losing diesel to the gas oil.

CASE STUDY 13.2 DAMAGE DUE TO WATER ENTRY INTO HOT TOWERS

Pressure surges due to water entry into a hot tower containing heavy oils or other high-boiling, water-immiscible compounds have been the most common cause of tray and packing damage in refinery fractionators (250, 255). Figure 13.1 shows the end results in a number of incidents. Several other cases are given below.

Tower A An olefins oil quench tower that received feed from several hot reactors (furnaces). One of the furnaces was being brought on-line after decoking. There was a pocket of undrained water in its outlet piping, and this pocket was blown into the tower. The pocket rapidly vaporized upon entry into the tower, generating an explosion that ripped trays in great force.

Tower B Another olefins oil quench tower that received feed from several hot reactors (furnaces). With this tower, it was a repeated experience to find the upper trays uplifted and collapsed. The most likely cause was the pumping of HC liquids collected in the flare knockout drum into the middle of this tower. Occasionally these HCs contained water, which rapidly vaporized upon tower entry, causing uplift of the trays above. Following the rerouting of the pumpback to another location in the plant that could tolerate water, no more tray collapses were experienced.

Tower C This tower was in hot aromatic HC service. A spare pump was being hooked to the tower. There was a pocket of water in the pump piping. The pocket reached the tower that was operating under deep vacuum. The water pocket rapidly vaporized, causing extensive internals damage.

(*a*)

(*b*)

Figure 13.1 Tray damage by water-induced pressure surge incidents in refinery towers: (*a*) major damage in catalytic cracker fractionator; (*b*) uplifted panels of vacuum tower stripping trays; (*c*) uplifted manway, bottom tray of atmospheric crude fractionator. [(*a*) Reprinted with permission from *Hazards of Water*. Copyright 1984 by Amoco Oil Company. (*b*) From Ref. 201. Reprinted with permission from *Hydrocarbon Processing* by Gulf Publishing Co. All rights reserved.]

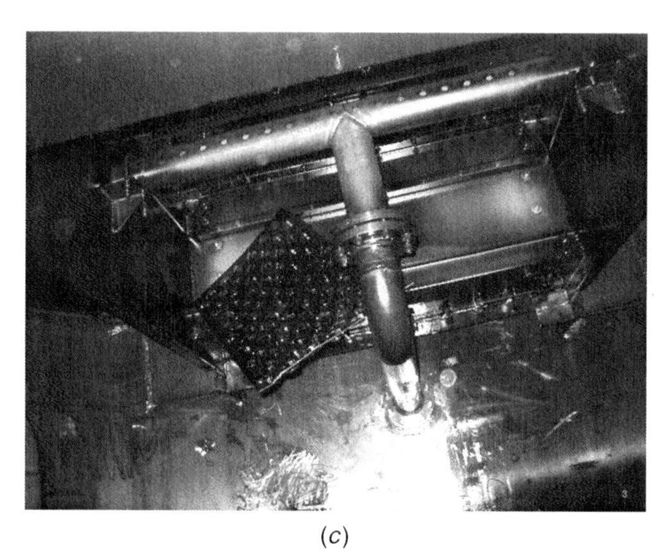

(c)

Figure 13.1 *(Continued)*

Tower D A tower in hot organic service operated under vacuum with a bottom temperature of about 400°F. Water entered the tower via a leaking valve, causing a pressure surge that damaged trays.

Tower E A spare pump was started in a refinery crude tower PA. A pocket of water lying in the pump piping reached the tower, causing valve trays below to bend downward.

Tower F A refinery vacuum tower experienced a recurrent loss of trays in the stripping section due to wet stripping steam. One turnaround, the wet steam was replaced by superheated steam and the trays by grid. The stripping section held well after this.

Tower G A refinery crude tower experienced frequent pressure spikes. The spikes often caused their PA pumps to lose their prime. To correct, steam trapping and condensate draining from the stripping steam header were improved. This eliminated the spikes.

Tower H A refinery vacuum tower used a HVGO PA to cool hot ascending vapors from about 670 to about 500°F by direct contact over a bed of packings. The HVGO PA was externally cooled, then sprayed onto the packing. One of the PA coolers was a steam generator. At one time, this exchanger leaked. This caused the tower vacuum to deteriorate but no internals damage. The exchanger was blocked in shortly after and the vacuum reinstated. Gradual leak development, atomization of water into small drops in the sprays, and small-quantity leaking were some of the explanations offered for the lack of damage.

CASE STUDY 13.3 INTERFACE CONTROL LEADS TO PRESSURE SURGE IN QUENCH TOWER

Contributed by Chris Wallsgrove

Installation A 28-ft-diameter ethylene plant oil quench tower cooled reactor (cracking furnace) effluent gas (Fig. 13.2). The tower had a lower PA circuit that cooled the incoming gas from about 200–250°C to 150–160°C over 12 disk and donut trays. The quench oil PA was regulated to maintain a temperature of about 200°C at the tower bottom. The gas was further cooled in the upper tower section by direct contact with a heavy gasoline reflux over 15 valve trays. The tower top temperature was maintained, by manipulating the gasoline reflux, at around 105°C, high enough to prevent water condensation as well as to prevent heavy ends from going overhead, which would discolor and/or raise the end point of the heavy gasoline.

Overhead gas from the oil quench tower flowed to the water quench tower (Fig. 13.2), where it was further cooled by direct contact with two externally cooled circulating water PAs. Water vapor (about a third of the gas weight), as well as heavy gasoline vapor, contained in the reactor effluent gas, was condensed in this water quench tower. The mixture of circulating and condensed water and heavy gasoline was separated into a water phase and a heavy gasoline phase in a large settler vessel at the base of the water quench tower. Much of this heavy gasoline became the reflux

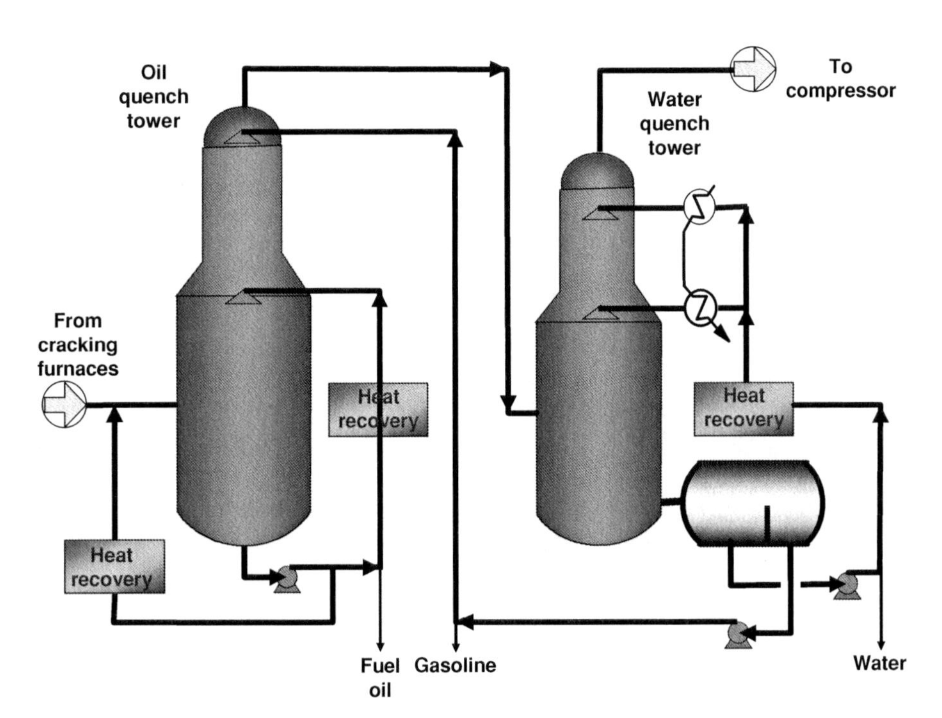

Figure 13.2 Olefins plant quench oil and quench water towers.

to the top section of the oil quench tower. Good water–gasoline separation requires large settling time, internal baffles that minimize turbulence and promote phase separation, and ultimately good interface level control. Minor water entrainment into the gasoline and gasoline into the water is not harmful and is normal. The minor quantity of water entrained in the gasoline reflux is evaporated on the top few valve trays of the oil quench tower, as its overhead temperature is controlled to be above the water dew point.

All instruments in the entire quench oil system are flushed with a light clean oil, as the quench oil (and to a certain extent the raw heavy gasoline) has a high concentration of reactive multi–olefinic/asphaltic components which polymerize and gum up.

History The plant had been in routine operation when a process control upset caused carryover of heavy components from the oil quench tower, contaminating the water quench system. This is not particularly abnormal for ethylene plants, and the upset was recovered from routinely.

What was not known was that the water–gasoline separator interface level transmitter has been contaminated with gum/polymer and started to give a false low interface level. The system started to remove less net water to try to restore the interface level set point. The transmitter continued to show a lower level. This false interface level was corroborated by field gauge glass readings which were vague (or indeterminate) due to black quench oil contamination on the inside of the glasses. Eventually the water level rose to the point where it flowed over the main weir into the gasoline compartment, at which point the oil quench tower reflux was 100% water rather than gasoline.

Nothing dramatic was noticed in the control room at this point, except that the pressure drop across the oil quench tower suddenly increased to its top range. In addition the bottoms temperature started to rise. Then the reactor (cracking furnace) outlets started to give high-pressure alarms. It was obvious that there was a major restriction in the oil quench tower, as field pressure measurements showed that the gas path through the tower was experiencing a very high pressure drop. Attempts to clear the "blockage," which at first was thought to be regular flooding, failed and later that night the plant was shut down.

Upon tower entry, all the disk-and-donut trays were found displaced upward, forming a solid plug of distorted/twisted metal at the level of the penultimate donut tray. The disks were primarily supported, in the design, by 8-in. pipe "spacers" which were welded to the inner periphery of each donut and to the edge of each disk. These pipes were crumpled, or compacted similar to a child's "bendy" drinking straw, or broken completely clear of their respective trays. The tower was virtually plugged with a compacted mass of displaced trays. One senior engineer remarked, "We could not have achieved this effect with dynamite!"

Postmortem Water reflux, instead of the design gasoline, had entered the tower and some had continued down as liquid water until a slug of it dropped into the hot oil in the tower base. The heating effect of the rising hot cracked gas, which is sufficient to evaporate most of the normal reflux gasoline and minor water contaminant, was overwhelmed by the much higher latent heat of evaporation of this gross water reflux. In the tower base the water acted as water always does when dropped into hot oil, but on a massive scale.

Moral The plant was down for almost 3 weeks, teaching a very hard-learned lesson about interface level transmitters.

Chapter 14

Explosions, Fires, and Chemical Releases: Number 10 on the Top 10 Malfunctions

Chemical explosions were in the 10th spot among distillation malfunctions based on the number of case histories (255). In terms of real losses, including loss of human life and suffering of victims and families, as well as widespread damage, equipment, and production losses, losses from these accidents far surpassed losses from any other distillation malfunction. Mitigating these is therefore the highest priority on any checklist aimed at preventing distillation malfunctions.

The term *chemical explosion* is used here to distinguish from explosions due to rapid vaporization, for example, when a pocket of water enters a hot-oil tower (Chapter 13). Just over half of these chemical explosions in our survey (255) were initiated by exothermic decomposition reactions. Of these, about two-thirds occurred in ethylene oxide, peroxide, and nitro compound towers. The rest came from a variety of towers.

Decomposition-initiated explosions are associated with specific services. In these services, excessive temperatures (either a hot spot or a high tower base temperature) or excessive concentration of an unstable component initiated the decomposition. In some cases, the excessive temperature resulted from a rise in pressure due to rapid generation of noncondensables by a decomposition reaction. In others, precipitation or low base levels led to the concentration of an unstable component at the hot temperature. Catalysis by metal or catalyst fines and air leaks has also contributed to some decomposition explosions. The good news about decomposition explosions is that the number of case histories reported appears to be on the way down (255). Credit for this very welcome trend is due to all those who have worked hard over the years to improve safety in these industries.

Line fractures is the next leading cause of chemical explosions, with about one-quarter of the case histories in our survey (255). The vast majority of cases were of

Distillation Troubleshooting. By Henry Z. Kister

lines carrying light HCs ranging from C_1 to C_4, and their fracture led to the formation of vapor clouds that ignited and exploded. Unlike decomposition reaction explosions, which appear to be on the decline, the number of line fracture explosions appears to be holding steady (255). Half the reported cases came from either gas or olefins towers, indicating that line fracture is a major issue there. The other half came from refinery towers.

Less common, yet important causes of explosions in towers are commissioning operations, HC releases, and violent chemical reactions. Most of these case histories appear on the decline (255). In all the reported commissioning cases, an operation such as purging, flushing, or deinventorying led to the formation of an explosive mixture. Some of the explosions resulting from HC/chemical release were due to release of C_4 HCs trapped in a plugged or frozen valve.

Fires that did not lead to explosions were in the 20th spot among tower malfunctions (255). Counting some of the case histories published recently (e.g., 124, 125, 136, 336, 337, 392), over half of these were structured packing fires while a tower was open for maintenance during turnaround. The reported number of packing fires has been on a rapid rise. In almost all, pyrophoric or combustible deposits in the packings and/or hot work played a role. These fires destroyed packings, sometimes also damaging tower shells. Fortunately, to date the author is not aware of anyone hurt. Most packing fires reported in the literature took place in refinery towers, although many occurred in chemical and petrochemical towers. An excellent paper by Bouck (62) reviews the chemistry behind the refinery fires and presents many of the cases and solutions practiced by the industry. Preventive measures practiced with various degrees of effectiveness were surveyed by the fractionation Research Inc. Design Practices Committee (139). These include good shutdown washes, keeping packing wet at turnarounds, avoiding hot work near structured packings, and using fire-retardant metallurgy.

Other causes of nonexplosion column fires included line fracture, unexpected backflow, opening the tower before complete cooling or removal of combustibles, and atmospheric relief that was ignited.

Chemical release to the atmosphere from distillation and absorption towers is just below the 20th spot among tower malfunctions (255). Causes included inadvertent venting or draining to the atmosphere, unexpected backflow, runaway reactions, cooling-water loss or vessel boilover, and sudden clearing of trapped chemicals. Atmospheric release incidents appear to be on the decline, probably due to the tighter requirements on safety and the environment in recent years.

Lessons learnt (255) emphasize the requirement for extremely cautious design, operation, and maintenance in towers handling compounds prone to exothermic runaway decomposition or violent reactions and in light HC (especially C_1–C_4) towers. Lessons drawn from previous accidents and near misses must be incorporated into existing and new facilities. Although other services reported fewer explosions, the possibility of their occurrence should always be considered and the appropriate preventive measures incorporated.

CASE STUDY 14.1 PREVENTING STRUCTURED PACKING FIRES

Installation Top fractionation section of a refinery crude fractionator separating lighter from heavier naphtha. The section contained commercial structured packings, around 75 ft^2/ft^3 surface area, fabricated out of 410 SS.

Experience When the tower was opened up at the turnaround, white smoke appeared to be coming out of the overhead line. At first it was thought that it was steam, but soon it was identified as a mixture of CO_2 and SO_2. Once the tower was entered, it was realized that the top two layers of structured packings in this bed burnt away. The cause of the fire was pyrophoric sulfur deposits on the packings. Detailed description of the mechanisms is found elsewhere (62).

Prevention To prevent recurrence, the 410 SS packings were replaced by an identical bed fabricated by the more fire resistant 316 SS. Attention is paid to cool the tower as much as practical prior to manhole opening. Water is sprayed on the top of the packing for the entire period in which the tower is open during the turnaround. If there is need to do work underneath, the sprays are temporarily turned off, but for no longer than an hour or at most two at one time. The sprays cool the packings and keep them wet. Additional measures may also have been implemented.

Result No recurrence occurred for close to 10 years.

CASE STUDY 14.2 PREVENTING STRUCTURED PACKING FIRES

Installation Wash section of a refinery crude tower containing SS structured packing about 100 ft^2/ft^3 surface area. Liquid to the packing was distributed by a spray distributor.

Experience At a turnaround the packing caught fire, generating temperatures hot enough to melt the packings. Fortunately, the fire remained local, was quickly put out, and did not damage tower wall. Investigation found that the fire was caused by pyrophoric deposits. One of the spray nozzles was plugged and the bed was not properly washed before opening the manholes.

Prevention Shutdown procedure was modified to include a water wash at high flow rates. The water wash continues while the manholes are open to ensure the packing is kept wet.

Result No recurrence occurred in the next turnaround.

CASE STUDY 14.3 OTHER PACKING FIRE EXPERIENCES

Tower A A chemical deep-vacuum tower in hot organics service experienced repeated turnaround packing fires. The tower contained stainless steel wire-mesh structured packings. At one time a fire was even experienced during operation due to a major flange leak. Recurrences were eliminated for near a decade by installing a battery of thermocouples that monitor for hot spots inside the packings. Upon detecting a hot spot, the fire is immediately snuffed out with inert gas.

Tower B A pharmaceutical deep-vacuum tower experienced repeated packing fires. The tower contained stainless steel wire-mesh structured packings. The tower operated in campaigns. Switchover from one campaign to another took place while the tower remained under vacuum, and some of the fires occurred during these switchovers. Hot organics on the packing, air leaks, and presence of oxidants, are believed to have played a role. Temperature rise to above 800°C was observed. Preventive measures included installing many additional temperature indicators and improving fire-snuffing capability.

Tower C A turnaround fire occurred in a water stripper containing random packing. The packing was gunked up with pyrophoric deposits. Both packing and shell were damaged. Fighting the fire was difficult, with success finally achieved by snuffing with steam.

Tower D A structured packing fire occurred in a large refinery fractionator. The fire occurred several days after the tower was first opened. Welding was performed beneath the packing, and although sparks are unlikely to have hit the packing, the rising heat and fumes could have led to the ignition of pyrophoric deposits in the packing. An alert operator sounded the alarm when he noticed a rise in bed temperature, and got the people out in time. Immediately afterwards the tower was drenched with water. The fire was extinguished prior to shell damage.

Tower E A large aromatics tower containing carbon steel structured packing was shut down and water-washed. Following the wash, an air purge was initiated to prepare for personnel entry. It was then noticed that the air leaving the tower was oxygen-deficient, indicating rapid oxidation of the packing.

Chapter 15

Undesired Reactions in Towers

Undesired chemical reactions were unplaced among the distillation malfunctions (255), mainly because many of them led to chemical explosions, which hold a prime spot on the list and are discussed at length in Chapter 14. The more benign reactions are discussed in this chapter.

The leading causes of the undesirable reactions are identical to those that produce explosions (Chapter 14). These include excessive bottom temperatures and hot spots, frequently conducive to decomposition and dehydration reactions; concentration of reactive components; reactive chemicals from extraneous sources or left over from commissioning; catalyst fines, rust, and tower materials that catalyze reactions; long residence times; excessive additives; and air leaks. Common consequences include product loss, product contamination, fouling, foaming, and accumulation.

CASE STUDY 15.1 LOWERING BOTTOM TEMPERATURE CAN STOP REACTION

Contributed By Frank Wetherill (Retired), C. F. Braun, Inc, Alhambra, California

Installation A chemical plant producing a heavy, water-soluble glycol. Effluent from a front-end hydrogenation reactor was purified into the desired product in a four-column separation plant, shown as the solid lines in Figure 15.1.

The first two columns separated water from the feed. Final traces of water were removed in the flash drum overheads. The intermediate column removed the low-boiling organic impurities. Finally, the product was separated from the high-boiling residues in the product column. The high boilers were mostly other alcohols and aldehydes as well as some inorganic salts.

Problem Although the water content of the feed was removed in the first two columns and flash drum, the final product consistently showed a small water content. It lowered the product freeze point and was confirmed by chemical analysis.

Distillation Troubleshooting. By Henry Z. Kister

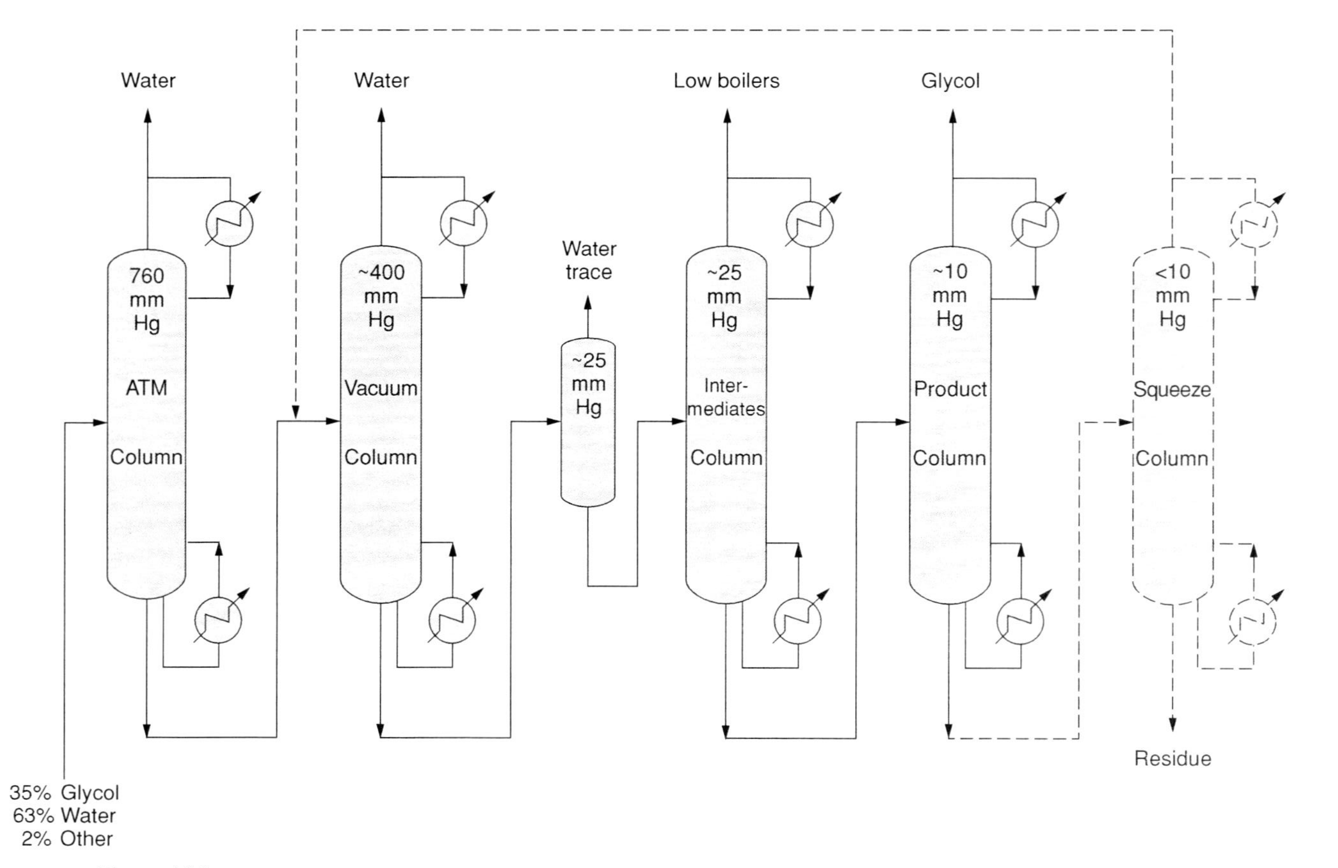

Figure 15.1 Glycol purification train (solid lines) and modification (dashed lines) that eliminated problem of water in product.

Investigation The feed to the product column was analyzed and was found to contain no water. The column was reboiled by high-pressure steam and condensed by cooling water, and it was speculated that one of these exchangers could have been leaking. To check this, the product column was placed on total reflux, and the overhead analyzed periodically for water. These tests showed an increasing quantity of water in the product with time, even though no feed entered the column. It was concluded that the problem occurred in the product column.

A leak in either the condenser or the reboiler was suspected. The product column was shut down, and the exchangers were pressure tested. Neither exchanger showed any signs of leakage.

Solution Following the investigation, a new theory was formulated. The bottom temperature was run at about 360°F. It was economical to run this temperature as hot as possible to maximize glycol product recovery from the residue. As most of the heavy impurities were alcohols and aldehydes, it was believed that these might undergo condensation reactions to form acetals and water. Such a reaction would tend to be promoted by higher temperatures.

To suppress these possible reactions, column base temperature was lowered to 320°F. At this lower temperature, the problem disappeared. However, product recovery also dropped because the lower bottom temperature allowed more product loss with the residue.

To counter the reduced recovery, the process was modified by adding a new residue-vaporizing ("squeezing") column. The modification is shown as the dashed lines in Figure 15.1. Bottoms from the product column entered this new column, which stripped additional product from the residue under more severe conditions. Water and other volatile impurities could be tolerated in the overhead of this column, because the overhead stream was recycled back to the vacuum column.

The residue squeezing column was a small, simple column and required little investment. In addition to solving the problem, this modification also led to a higher overall plant product recovery than ever before, because it enabled additional stripping of product previously lost in the product column bottoms.

Related Experience In one ethylene dichloride plant, HCl was separated in the front of the separation train. Despite this, HCl appeared in the product tower. The feed was sampled and found free of HCl. Further investigation revealed that excessive temperature at the tower bottom could trigger a temperature-sensitive reaction that formed HCl, and the HCl distilled upward. The excessive temperatures were caused by hot spots in the reboilers. Elimination of the hot spots eliminated the HCl.

CASE STUDY 15.2 REACTION, AZEOTROPING, ACCUMULATION, AND FOAMING

Installation Formaldehyde column.

History Tower flooding took place at 70% of the design feed rates. Gamma scans verified downcomer flooding initiating around the feed point. Enlarging the downcomers increased tower capacity to 90% of the design rates. Antifoam injection raised tower capacity to design rates but was undesirable because the antifoam ended in the product.

Investigation Extensive plant sampling supplemented by bench-scale tests suggested dehydration reactions between methanol and formaldehyde in the feed region formed oxygenated compounds. These oxygenated compounds were relatively high boilers, but they azeotroped with water, and the azeotropes boiled in the 85–95°C range, compared to a top tray temperature of 65°C, a feed tray temperature of 95°C, and a bottom temperature of 105°C. At this intermediate boiling point, these azeotropes accumulated just above the feed. Both field and bench-scale tower tests showed that the oxygenated compounds in the feed region concentrated to 10–30% weight over several hours.

Theory The accumulation of these azeotropes initiated foaming. It is possible that low solubility of the oxygenated compounds in the tray liquid brought the solution close to its plait point, a point where the foaming potential is maximized.

Plant tests showed that the frequency of foam incidents directly increased with the concentration of formic acid in the tower feed. Acid is known to catalyze the oxygenate-forming dehydration reactions, supporting the theory that oxygenate concentration induces foaming. This theory is further supported by bench-scale tower tests that showed that the flooding commenced in the region where the oxygenate concentration peaked. Finally, bench-scale tests successfully eliminated the flooding either by temporarily raising the tower top temperature sufficiently to allow the azeotropes to escape in the overhead or by dumping the tower, thus allowing the azeotropes to escape in the bottom.

Cure The most effective route to alleviating foaming was minimizing the acidity of the tower feed.

CASE STUDY 15.3 DO NOT PREJUDGE THE DESIRABILITY OF A REACTION

Installation Chemical tower equipped with ceramic random packings.

Experience The tower packings were repeatedly chewed up by traces of hydrogen fluoride. To avoid the packing damage, the packings were replaced by alternative materials that resist hydrogen fluoride attacks.

When the tower returned to service, the fluoride was found in the bottom product, making it off specification. It appeared that the reaction of the hydrogen fluoride with the ceramics produced some volatile compound that escaped in the tower overhead, where it could be tolerated.

Epilogue The new packings were replaced by ceramic packings.

Moral Not all packing-destroying reactions are undesirable.

Chapter 16

Foaming

Foaming is 11th among distillation malfunctions (255). Foaming is a service-specific phenomenon. About one-third of the cases reported were in ethanolamine absorbers and regenerators that absorb acid gases such as H_2S and/or CO_2 from predominantly HC gases. Another 10% were also in acid gas absorption service, but using alternative solvents such as hot potassium carbonate (hot pot), caustic, and sulfinol. Another 10% were in absorbers that use a HC solvent to absorb gasoline and LPG from HC gases. The number of foaming incidents appears to be holding steady, with no sign of growth or decline (255).

Case histories of foaming were also reported in each of the following services:

Chemical: aldehydes, soapy water/polyalcohol oligomer, solutions close to their plait points, extractive distillations, solvent residue batch still, ammonia stripper, dimethylformamide (DMF) absorber (mono-olefins separation from diolefins), cold water H_2S contactor (heavy-water GS process).

Refinery: crude preflash, crude stripping, visbreaker fractionator, coker fractionator, solvent deasphalting, hydrocracker depropanizer.

Gas: glycol contactor.

Olefins: high-pressure condensate stripper.

In about a third of the reported cases, solids catalyzed foaming. Although not specifically reported, solids could have catalyzed foaming in many other cases. In many cases, the foaming was caused or catalyzed by an additive such as a corrosion inhibitor. Hydrocarbon condensation into aqueous solutions, certain feedstocks, small downcomers, and low temperatures were reported to promote foaming.

Three cures have been successful for foaming problems: eliminating the foam-causing chemical (e.g., by eliminating the additive or by filtering out the solids that catalyzed the foams), injecting antifoam, and debottlencking downcomers (using larger downcomers, reducing downcomer backup, or replacing trays by random packings). Injecting antifoam has not always been effective, and some experimentation with the inhibitor type and concentration is often required.

Distillation Troubleshooting. By Henry Z. Kister

CASE STUDY 16.1 CONCLUSIVE TEST FOR FOAMING

Henry Z. Kister and Tom C. Hower, reference 263. Reproduced with permission.

Installation A Benfield process hot-pot absorber. The absorber uses hot potassium carbonate solution to absorb CO_2 and H_2S from concentrated sour gas (Fig. 16.1). Because of the corrosive environment, the column was packed with 2-in. polypropylene Raschig rings.

History The absorber plugged on start-up. It was losing capacity during the first few days until it was virtually plugged. When it was opened, plant personnel observed that the plastic rings had melted, thus plugging the column. The heat of reaction caused

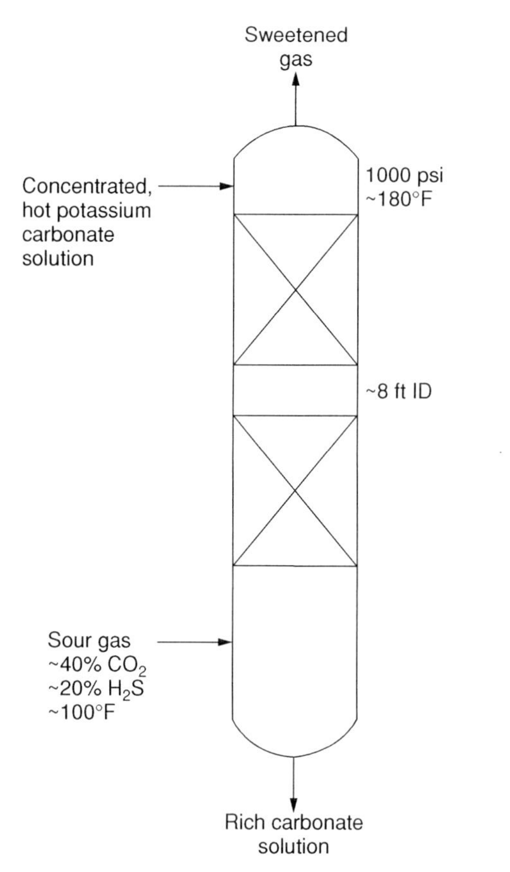

Figure 16.1 Benfield process hot-pot absorber. (From Ref. 263. Reproduced with permission. Copyright © (1987) AIChE. All rights reserved.)

the packings to melt. It is possible that the melting started in a few hot spots; once started, the hot spots moved to other points, thus melting the whole bed.

The plastic rings were replaced by SS Raschig rings. When the column was restarted, it operated well but flooded below design rates. Flooding was recognized by liquid carryover into the overhead stream.

The carbonate solution was checked for foaming tendency. This appeared reasonably small. Shortly later, it was realized that, although the solution itself did not have a large foaming tendency, the gas contained a surface-active agent, which was injected downhole into the gas well as a corrosion inhibitor. When this material came in contact with the solution, it foamed, causing premature flooding.

An antifoam injection at a concentration of 10 ppm was started. Field tests at solution temperature and atmospheric pressure showed that this concentration was sufficient to eliminate foaming. Once the injection began, column capacity was increased by 50%, but it was still short of design capacity. It was then decided to repeat the foaming tests under actual operating conditions. These were carried out in a level glass set up in the field so it could be operated at system conditions. When gas was bubbled through the glass at 1000 psi and 180°F (process conditions), the solution broke into foam.

Field tests at system conditions showed that an antifoam injection of 1000 ppm was needed to suppress the foaming. When antifoam injection into the column was increased to this concentration, design capacity was reached.

CASE STUDY 16.2 POOR OPERATION OF AMINE ABSORBER

Contributed by Mark Pilling, Sulzer Chemtech USA, Tulsa, Oklahoma

Installation Refinery sulfur plant amine absorber packed with random packing.

Problem The column was not meeting product specifications. The pressure drop was also higher than predicted.

Investigation A turnaround inspection showed that a previous troubleshooting attempt had removed every other trough from the liquid distributor. The reasoning for this is unknown. A lean-amine sample showed the amine to be black and opaque. Additional testing indicated that the feed to the absorber had some liquid HCs in it as well.

Theories The poor performance could have been caused by a variety of reasons. The most likely cause was foaming in the column from either the poor quality amine or the condensation of liquid HCs in the tower. The poor liquid distribution from the modified distributor would not have helped the situation.

Cures The distributor was installed per design with all the troughs. The packing was replaced with a similar type and size. The amine was replaced. The feed was kept free of HC liquids. Tower operation was then back on specification upon start-up.

Morals

- Poor amine quality and HC condensation are common causes of foaming.
- Altering equipment in a haphazard fashion can be troublesome.

CASE STUDY 16.3 TOO MUCH ANTIFOAM IS WORSE THAN TOO LITTLE

Tom C. Hower and Henry Z. Kister, reference 224. Reprinted with permission from *Hydrocarbon Processing*, by Gulf Publishing Co., all rights reserved

Installation An amine absorber in a gas plant. The tower was operating below maximum capacity.

History At relatively high rates, some carryover of liquid from the top of the absorber was apparent. Foaming was suspected. In a small-scale test, bubbling gas into a sample of solution in a laboratory beaker confirmed the suspicion and indicated strong foaming. Upon addition of a silicone water- based antifoam to the beaker, the foam subsided. It was decided to add the same antifoam to the tower. Antifoam and an antifoam injection pump were ordered.

To try to save the cost of an antifoam injection pump, it was attempted to add the antifoam batchwise at the lean-amine pump suction. Calculation showed 1 gal of antifoam was required. As soon as this antifoam reached the tower, there was massive carryover of liquid from the top of the tower, overhead gas went sour, and the tower bottom level was lost. It was then realized that adding excessive quantities of antifoam can promote foaming instead of suppressing it. An antifoam is a surface-active component and therefore can be conducive to foaming when injected in excessive quantities.

Cure The gas flow to the absorber was interrupted while the amine was kept circulating. The absorber–regenerator system contained a lean-amine surge tank, and its content was mixed with the system inventory, thus diluting the antifoam. After 12 hours, it was attempted to resume gas to the absorber. Foaming reoccurred. Apparently, the antifoam was still not sufficiently dilute. The gas flow was interrupted again. Amine circulation and mixing with the surge tank contents were continued for another 12 hours. A second attempt was then made to resume gas to the absorber. This time it was successful, and no further foaming was experienced.

Moral From then on, the antifoam injection pump was used for all further antifoam additions.

CASE STUDY 16.4 STATIC MIXER HELPS ANTIFOAM INJECTION

Contributed By Ron F. Olsson, Celanese Corp.

It was recommended (250) that when a foaming inhibitor is applied in a foaming system it should be injected upstream of a point of high turbulence. Good locations are pump suction or upstream of a pump discharge letdown valve.

In one column, a foaming inhibitor was injected just upstream of the reflux control valve. The inhibitor was effective in inhibiting the foaming but not in completely mitigating it.

In an attempt to improve mixing, it was decided to install a static mixer just downstream of the inhibitor injection point, upstream of the reflux control valve. It was postulated that the inhibitor is a high-viscosity (about 3000-cP, almost solid) fluid, and a static mixer will therefore be able to improve the inhibitor dispersion.

The modification made a day and night difference in column performance. The foaming problem was completely mitigated and the inhibitor consumption could be reduced by an order of magnitude. Payout for the static mixer was less than a month.

Counterexperience In a completely different service and plant, a foaming inhibitor was injected into the feed pump suction with a static mixer in-line. The static mixer plugged. There was a filter upstream of the injection point, but it did not save the mixer. The solution in this case was to take the mixer out.

CASE STUDY 16.5 GAMMA SCANS DIAGNOSE FOAMING

Installation An aldehyde tower flooding at rates well below design.

Analysis Gamma scans showed that the flood started just below the feed tray and progressed up the tower. The tower loads were lowered to just below incipient flooding, and a scan was shot through the downcomers on one side of the tower, parallel to the outlet weirs. The scan (Fig. 16.2*a*) shows an inflection point halfway up the downcomer, with an extremely uniform vapor–liquid mixture above. This uniformity provided strong evidence for foaming in the downcomer. The foam persisted all the way to the next tray, indicating that flood was about to initiate.

Cure Initially, antifoam was injected, which substantially increased tower capacity. A scan at the higher rates with antifoam injection (the dashed curve in Fig. 16.2*b*) shows that the foam layer disappeared. Continuous injection of antifoam was not acceptable for process reasons, so the downcomers were enlarged. This further increased capacity. A scan at the highest rates after the downcomer was enlarged (the solid curve in Fig. 16.2*b*) showed that the enlarged downcomers (as well as the tower) were no longer close to a capacity limit.

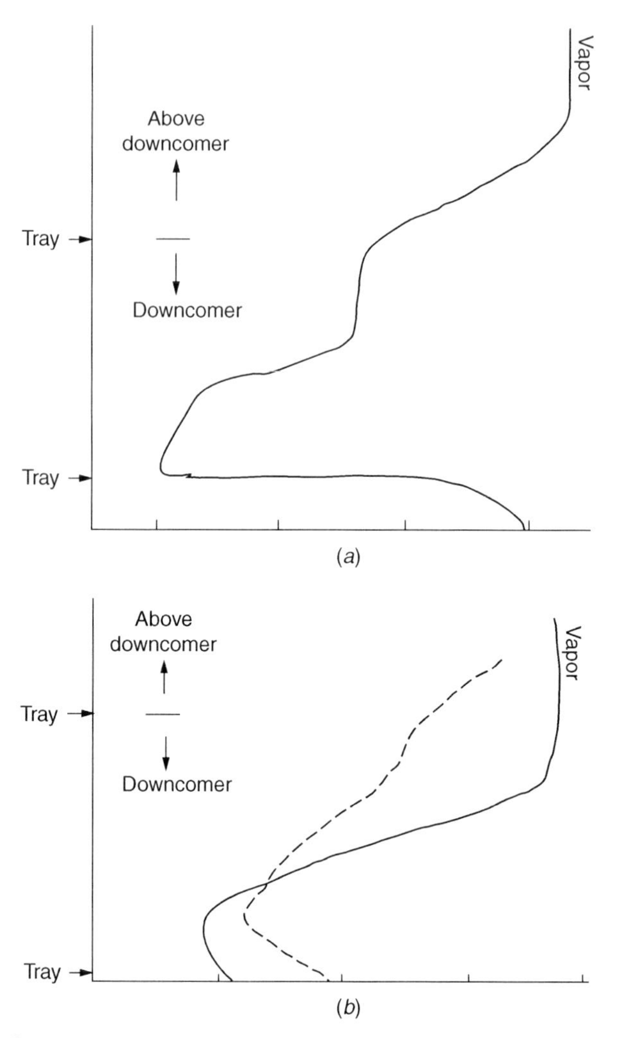

Figure 16.2 Gamma scans shot through side downcomers of aldehydre column: (*a*) at incipient flooding, showing foam layer extending to next tray; (*b*) after corrective actions. Dashed line is following antifoam injection and higher rates. Solid line is after downcomers were enlarged, no antifoam injection, higher rates.

CASE STUDY 16.6 LOW DOWNCOMER VELOCITIES ARE CRITICAL FOR FOAMING SYSTEMS

Installation A grass-roots refinery preflash tower operating at 80 psia. Preheated crude, which contained about 15% vapor by weight at about 450°F, entered the tower six trays above the bottom. Condensed overhead vapor was the naphtha product. Some of the condensate was refluxed to the tower to condense out heavies and maintain the naphtha on specification. The tower bottom was the less volatile portion of the crude, which flowed to the crude heater, thence to the crude tower. The lower trays of the preflash tower used a small quantity of stripping steam to strip lights out of the crude to enhance naphtha recovery. The tower was equipped with valve trays containing round, uncaged, moving valves at 24 in. tray spacing.

Problem The tower flooded, producing black naphtha, at rates well below design.

Analysis Preflash drums and bottom sections of preflash towers are highly foaming systems. For highly foaming systems, recommended criteria listed in Ref. 250 limit maximum liquid superficial velocities at downcomer inlets to 0.2–0.25 ft/s. Experience acquired by the author since the time of this writing is that, although generally reasonable, this criterion can be somewhat optimistic for some highly foaming systems. For the trays below the feed of this preflash tower, the actual liquid velocity at downcomer inlet was 0.4 ft/s, well in excess of the maximum-velocity criterion.

Barber and Wijn (39) measured foam heights in a vertical pilot-scale preflash drum mounted parallel to the main preflash drum in an actual refinery process train. Their measurements show that the foam heights at the base of the drum are a linear function of the downward liquid velocity in the drum. At a downward liquid velocity as low as 0.05 ft/s, they measured a foam height of 24 in. Assuming the above extends to the downcomers below the feed, for the tray spacing of 24 in. a downcomer inlet liquid superficial velocity of 0.05 ft/s would produce enough foam to carry over into the next tray and flood the tower. At the actual downcomer inlet liquid velocity of 0.4 ft/s the tower would be totally flooded.

Cure The current downcomers already occupied a large portion (25%) of the total tower cross-sectional area, so there was little scope for enlarging them to sufficiently reduce the downcomer velocities. A much more attractive solution was to simply remove the trays below the feed and eliminate the stripping steam. This permitted utilizing the entire tower cross section for liquid descent, giving a liquid downward velocity of 0.1 ft/s. Once the tower returned to service with the trays below the feed removed, the flooding was eliminated and the naphtha came on specification. The reduced recovery penalty due to elimination of the stripping trays was barely noticeable.

CASE STUDY 16.7 ENLARGED DOWNCOMER CLEARANCES MITIGATE FOAMING

Contributed by Geent Hangx, DSM, Geleen, The Netherlands, E. Frank Wijn, Consultant, Purmerend, The Netherlands, and Henry Z. Kister, Fluor, Aliso Viejo, California

Installation An olefins plant condensate stripper stripping C_2 and lighter components from gasoline condensate collected from the knockout drums of the cracked gas compressor. Pressure was 11 bars and bottom temperature 62°C. The tower was

1.9 m ID and contained two-pass valve trays with round, uncaged moving valves. There was a water draw-off chimney tray 11 trays below the top. Figure 16.3 shows key dimensions of the trays and chimney tray. The overflow weir of that tray was 100 mm below the top of the chimneys. The water draw sump was mounted directly above the downcomer from tray 12.

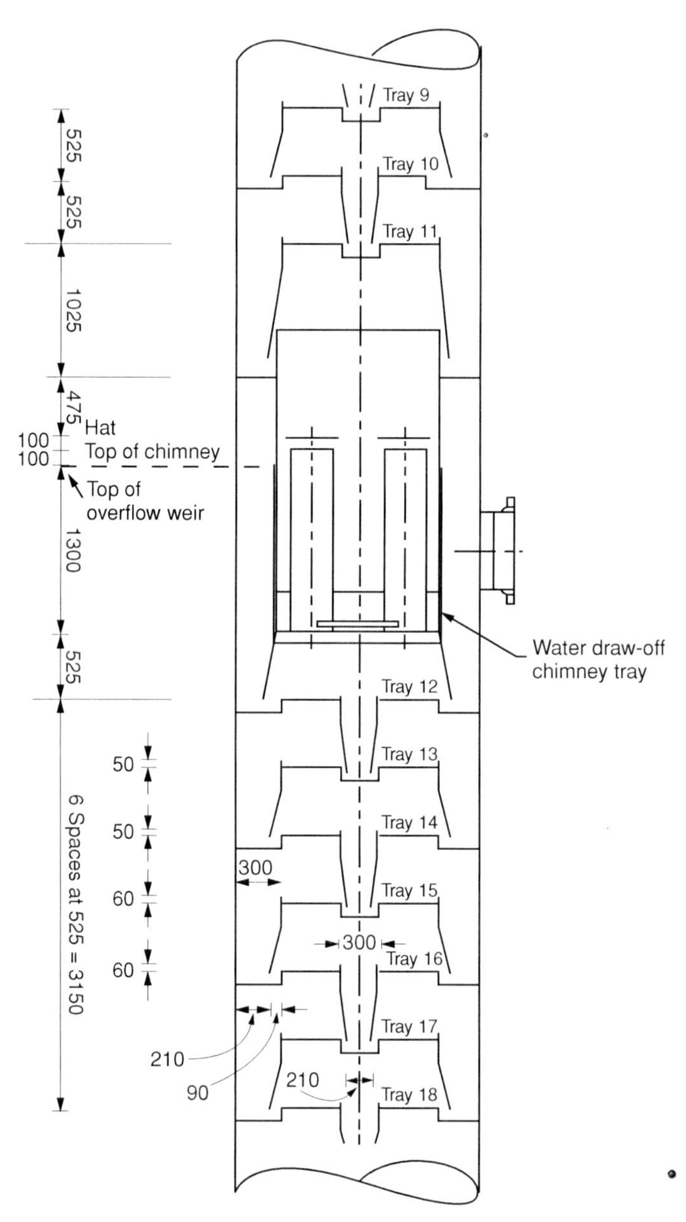

Figure 16.3 Key dimensions of condensate stripper trays in region of flood initiation.

Problem The tower flooded prematurely. Upon flooding, pressure drop would shoot up, and at certain rates liquid carryover was observed. Small changes in vapor rates had a big influence on pressure drop near the flood point.

Hydraulic calculations showed that at the maximum operable conditions the tower was at about 40% of jet flood and the clear liquid backup in the downcomers was about 30% of the total downcomer height. Downcomer inlet velocity was low, about 0.062 m/s. The premature flood rendered the tower a plant capacity bottleneck.

Gamma Scans Gamma scans of tray active areas at the maximum operable conditions (tower not flooded) showed low froth heights (about 200 mm) and clear vapor spaces above all the trays.

A second active area scan was performed when the tower was flooded. It showed flooding on the five trays below the chimney tray, on the chimney tray, and one tray above. The vapor space of tray 12, immediately below the chimney tray, was filled with foamy dispersion. The dispersions on trays 13 and 14 below appeared foamy, but to a lesser extent. Trays 13–16 showed progressively clearer vapor spaces as one descended. Trays 17 and 18 were not flooded, showed clear vapor spaces, but their froth heights approached the tray spacing. Below these, the trays had froth heights typically 100 mm taller than in the unflooded scans (i.e., about 300 mm) and had clear vapor spaces above. There was little distinction in froth appearance between trays with side and center downcomers.

In the second (flooded) scan, the chimney tray was flooded with foamy material building up about 1 m above the top of the chimneys. Tray 11 was flooded, tray 10 approached flood, but the nine trays above appeared identical to the unflooded scan.

The above pattern resembled that observed in an older tower used earlier in the same service. This tower also flooded prematurely, and gamma scans showed flood initiating near the water draw chimney tray. In this tower, the flooding was spread over many more trays above and below the water draw.

Analysis A simulation showed that the highest vapor and liquid loadings were right at the bottom trays. Near the chimney tray, the vapor load was 20% lower and the liquid load 10% lower than at the bottom tray. This argued against a regular flood, which should have initiated at the tower bottom. This also argued against tray flood. With the much higher vapor loading, a tray flood should initiate in the bottom, not near the chimney tray.

Foaming Theory Kler, Brierley, and Del Cerro (278) reported foaming in condensate stripping service, so foaming is conceivable. Ross and Nishioka (414) found that the highest potential for foaming is near the plait point, that is, where a solution is still homogenous, but is near the point of breaking into two liquid phases. In this tower, the liquid leaving the chimney tray was saturated with dissolved water but should contain no free water (any free water should have been removed by the chimney tray). So, the peak foaming tendency in the tower was right below the chimney tray. As one descended the tower, the dissolved water was stripped, progressively making the liquid less foamy. Above the chimney tray, the foaminess would greatly diminish due to the presence of two liquid phases, as one phase served as an antifoam to the other (414). The observation that the flood initiated near the chimney tray is well in line with the possibility of foaming. The foamy appearance of the froths on tray 12 and the neighboring trays also supports foaming.

In an attempt to address foaming, antifoam was injected into the stripper feed, but the flood did not respond to it. It is likely, though, that most of the antifoam did not reach the foaming zone. This is because antifoams are usually solid compounds, and would tend to settle near the bottom of the chimney tray, from where they would be removed in the water.

Foaming bottlenecks towers via a downcomer flood mechanism. There are three possibilities: downcomer backup, downcomer choke, or both.

The calculated downcomer backup was somewhat on the low side, requiring the aeration factor to be low, around 30% for the downcomer froth to reach the tray above. Such low aeration factor, however, is conceivable (251). Also, the foam-flood reported by Kler, Brierley, and Del Cerro (278) in a similar system proceeded via downcomer backup. Finally, the high sensitivity of the flood to vapor changes supported a downcomer backup rather than a downcomer choke mechanism. Downcomer choke was also conceivable, with the downcomer inlet velocity of 0.057 m/s close to maximum allowable for a foaming system (0.061–0.076 m/s). Further, liquid entry into the downcomers and vapor disengagement from them were somewhat obstructed in this tower by the seal pans from the trays above, which would aggravate downcomer choke.

Alternative Theories One alternative theory postulated that a water leak from the chimney tray flashed in the tray 12 downcomer and choked it. Another theory argued that the flood was caused by some nonstandard features in the chimney tray design. Neither of these theories could explain the success of the solution (below).

Solution A hydraulic study predicted that increasing downcomer clearances as well as a number of other downcomer-debottlenecking modifications could alleviate the flood and raise tower capacity. The economics favored implementing simple, inexpensive modifications at the next turnaround. The only modification that was implemented was shortening the downcomers by 50 mm, thus increasing the clearances under the downcomers from 60 to 110 mm.

Results Following restart, the bottleneck disappeared and the tower fully handled plant loads. It is unknown whether foaming will reoccur at higher rates. The fact that increasing downcomer clearances alone raised tower capacity is strong evidence supporting a foaming bottleneck proceeding via a downcomer backup flood mechanism.

CASE STUDY 16.8 HARDWARE CHANGES DEBOTTLENECK FOAMING

Tower A A chemical tower experienced foaming and responded to antifoam injection. The tower contained valve trays at 12 in. spacing. Many of the valves were blanked due to expected oversizing under nonfoaming conditions. Removing the blanking strips alleviated the foaming and raised capacity.

Tower B This 3-ft-ID refinery tower contained blanked valve trays at 18 in. spacing. The tower experienced foaming which was eliminated by removing the blanking strips just as for tower A.

Chapter 17

The Tower as a Filter: Part A. Causes of Plugging—Number 1 on the Top 10 Malfunctions

With well above 100 case histories, plugging/coking is the undisputed leader of tower malfunctions (255). The number of plugging/coking incidents reported over the last decade suggests that these problems are neither easing off nor declining (252, 255). Plugging/coking is here to stay and most likely will continue to top the tower malfunction list. More plugging/coking cases have been reported in refineries than in chemical plants, probably due to the incidence of coking, which is a major problem in refineries but uncommon in chemical towers.

Among the causes of plugging, coking was identified (255) as the leader. These will be discussed separately in Chapter 19. This chapter focuses on the other causes of plugging.

Scale and corrosion products closely follow coking as the leading cause of plugging (255). The case histories show no rapid growth or decline in the number of these. Scale and corrosion products appear to be more of a problem in refinery and olefins/gas towers than in chemical towers. Precipitation or salting out follows closely behind. Precipitation appears to have become more of a problem recently, possibly due to a trend to use lower quality feedstocks and to minimize plant effluent, and affects both chemical and refinery towers. The next two common causes of plugging are solids in the tower feed and polymerization. Solids in the feed and polymer formation are more of a problem in chemical than in refinery towers. In fact, no polymer formation case histories have been reported in refinery towers.

CASE STUDY 17.1 PACKED-BED DAMAGE

Contributed by Matt Darwood, Tracerco, Billingham, Cleveland, United Kingdom

Distillation Troubleshooting. By Henry Z. Kister

Installation A vacuum distillation column in a large oil refinery contained six packed beds above the feed. In the section affected, each bed contained structured packing and had a chimney tray below it. A downcomer from each chimney tray directed reflux into a gravity liquid distributor that distributed it to the bed below.

Problem The separation efficiency deteriorated over a period of time, causing downstream processing issues and greater separation costs.

Investigation A gamma scan confirmed that the packed beds as well as the chimney trays were in their correct locations and the distributors were positioned correctly and holding liquid. The scan also showed that the third bed from the top (bed 3) had a large density gradient from top to bottom. With no significant change in liquid load, the scan showed a gradual density increase by approximately 25% over the 2 meter length.

Further investigation of bed 3 using grid gamma scans confirmed the existence of a "hole" at the top of the bed that proceeded down in an inverted-cone shape. The results indicated that some of the debris from the packing had collected in the bottom section of the bed. The debris from the damaged packing explained the density change. The hole explains the poor mass transfer in the bed.

Initially it was unknown what had caused the packing to fail so dramatically and quickly (it was only installed 19 months before). Further analysis revealed it was a materials-of-construction issue. In the preceding turnaround, bed 3 was retrofitted with new structured packings from a different vendor. Following the revamp, the bed was operated at significantly higher temperatures. The new packings corroded rapidly at the new process conditions.

Cure During the next shutdown, the whole of bed 3 was replaced with fresh packing of upgraded metallurgy. The beds below bed 3 were inspected for debris buildup, but only very small traces of debris were found.

Summary Gamma scanning is an excellent tool to investigate the mechanical integrity of packed beds while remaining on-line.

CASE STUDY 17.2 FOULING OF WIRE-MESH STRUCTURED PACKINGS

Installation Specialty chemical nonaqueous distillation. Tower was 4 ft ID and contained high-efficiency wire-mesh packings similar to the Sulzer BX. Due to the fouling potential of the feed, it was filtered.

Problem Measured HETPs were 25 in. for the rectifying reaction and 40–60 in. for the stripping section. Design HETP was 12 in., and even lower HETPs are often achievable with this packing.

History Shortly after initial start-up, the plant experienced problems of the filters plugging every few hours. This was an operation-and-maintenance nuisance, so the filters were removed.

Following filter removal, the plant experienced problems with its feed distributor. Every time the tower was taken off-line and inspected, half the holes were found blocked. To overcome this problem, plant personnel designed and installed new, fouling-resistant distributors. Upon restart, the rectifying section HETP improved to 16 in., but the stripping section HETP remained high.

At that point the packings had been in service for 4 years. Throughout the 4 years, the plant did not look at the condition of the packings. To alleviate possible fouling, every now and then they ran steam through the bed.

Analysis Without the filters, first the distributors, then the bed itself, act like filters, with the end result of maldistribution and poor performance.

Moral If a fouling potential exists, filters must be used to avoid plugging of distributors and packings. Wire-mesh packings are sensitive to fouling and should be avoided in fouling services.

Chapter 18

The Tower as a Filter: Part B. Location of Plugging—Number 1 on the Top 10 Malfunctions

With well above 100 case histories, plugging/coking is the undisputed leader of tower malfunctions (255). The number of plugging/coking incidents reported over the last decade suggests that these problems are neither easing off nor declining (252, 255). Plugging/coking is here to stay and most likely will continue to top the tower malfunction list.

Chapter 17 focuses on the causes of plugging. Chapter 19 addresses coking, which was identified (255) as the leading cause. This chapter focuses on the location of plugging. The case histories are evenly split between packed and tray towers.

Both packings and distributors plug. Generally, the case histories are evenly split between plugged packings and plugged distributors. Using fouling-resistant distributors (sometimes at the price of somewhat lower distribution quality under clean conditions), retrofitting with large-opening, fouling-resistant packings, and replacing packings by fouling-resistant trays have been successful cures.

In trays, cases of plugged active areas outnumber those of plugged downcomers by more than 2 to 1, suggesting that tray design for fouling service should focus on enhancing fouling resistance in the active areas. Numerous cases reported improved fouling resistance by retrofitting active areas with larger holes or larger valves. Sticking of moving valves, especially when operated at low loads, is a major issue, often overcome by retrofit with large fixed valves or sieve decks with large perforations.

Although most plugs take place in the tower, draw lines, instrument lines, and feed lines also plug. Line plugging appears to be less of a problem in chemical towers than in refinery and olefins/gas towers.

In many cases, tower plugging is confined to a limited zone, opening a path for living with the problem. A fouling-resistant internal, which may trade off efficiency

Distillation Troubleshooting. By Henry Z. Kister

by high fouling resistance, may only be required in the limited zone, permitting high-efficiency devices (which may be far less fouling resistant) to be used throughout the rest of the tower. In some cases, on-line washing and hot-tapping bypasses have gotten around a limited plugged zone. In the cases of limited-zone plugging, correct diagnosis has been key to successful solution. Gamma scans and pressure surveys have been most useful here.

CASE STUDY 18.1 VALVE TRAYS IN STICKY CHEMICALS SERVICE AT HIGH RATES

Service Sticky, highly fouling specialty chemicals.

History The plant had bubble-cap trays in several towers in this service. These experienced severe fouling near the bottom of the caps, leading to short run lengths. The bubble-cap trays were replaced by valve trays with uncaged valves. This eliminated the tray-fouling problem. Sticking of the valves had not been a problem, but this can partly be because high-throughput operation was constantly sustained through the towers. The biggest remaining problem is blockage of the downcomers. The plant gets about a year run length out of its towers, at the end of which the downcomers plug, lead to tower flooding, and need cleaning.

The plant often finds valves that pop out of their seats, but this does not have a significant effect on performance, again probably due to the high-load operation. Popped-out valves sometimes caused problems at the bottom pumps. Otherwise, at the turnaround the plant collects the popped-out valves and puts them back using a pair of pliers.

CASE STUDY 18.2 FOULING BEHIND INTERRUPTER BARS AND INLET WEIRS

Installation A tower in an aromatic derivative plant equipped with round valve trays. The trays contained "interrupter bars," that is, inlet weirs $\frac{1}{2}$–$\frac{3}{4}$ in. tall installed just upstream of the first row of valves, whose function is to impart downward movement to the liquid entering the trays. This downward movement was shown effective (37) in minimizing inlet weep from valve trays.

Problem The trays experienced polymerization fouling.

Troubleshooting Shutdown washes wiped out all the evidence that could point to the cause. Little polymer was visible once the tower was opened.

One turnaround, a boroscope was inserted into the tower after it was depressurized but prior to decontamination. The boroscope was inserted via a thermocouple nozzle. People close to the nozzle needed to wear masks and appropriate protective clothing, and many other safety precautions were implemented. The boroscope saw that the trays were clean, with the main polymer accumulation taking place in the stagnant zones behind the interrupter bars.

Solution The interrupter bars were removed, and the polymerization problem disappeared.

Another Plant A refinery water stripper experienced severe corrosion. Corrosion products accumulated behind the inlet weirs and plugged the downcomers. Removing the inlet weirs eliminated the plugging.

CASE STUDY 18.3 EFFECT OF TRAY HOLE SIZE ON FOULING

Smaller holes or moving valves have greater tendency to plug than larger holes or fixed valves (250). Nonetheless, even trays with larger openings that have good track records in fouling environments may plug over a period of time (e.g., Fig. 18.1). Following are some experiences.

Towers A and B Sieve trays with $\frac{1}{2}$-in. holes experienced severe fouling problems in (i) an olefins oil quench tower and (ii) a refinery HF alkylation unit isostripper. Replacement by sieve trays with 1-in. holes completely eliminated the problem.

Tower C In a retray of a refinery HF alkylation unit isostripper, sieve trays with 1-in. holes were replaced by sieve trays with $\frac{3}{4}$-in. holes. Within 2 weeks of start-up the trays plugged with fluoride deposits, initiating tower flooding and instability. The trays were replaced by sieve trays with 1-in. holes with no further problem.

Tower D Sieve trays with $\frac{1}{4}$-in. holes experienced scaling problems in aqueous service. The hole diameter became drastically smaller over a period of time. Problem was mitigated by retraying with $\frac{1}{2}$-in. holes.

Tower E Sieve trays with $\frac{3}{8}$-in. holes experienced plugging in a mildly fouling chemical tower. Replacement by sieve trays with $\frac{1}{2}$-in. holes was a major improvement. It was amazing how such a relatively small change could effect such a significant improvement.

Tower F Sieve trays with $\frac{1}{2}$-in. holes were achieving 9–10 months run length in a highly polymerizing service. At the end of this period the downcomers would plug. Polymer built right above the downcomer entrance on the tower walls would spall off and plug the downcomer when enough was built.

In an attempt to increase run length, the trays were replaced by proprietary trays that push liquid toward the dead zones and the downcomer inlets. The new trays had "mini" type of fixed valves. These plugged after 6 months, with polymer forming

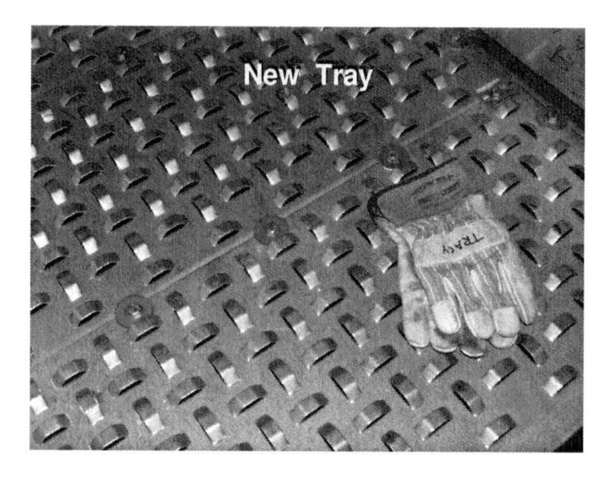

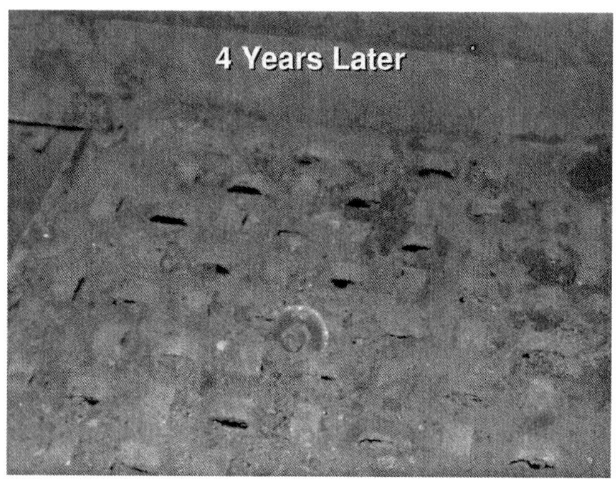

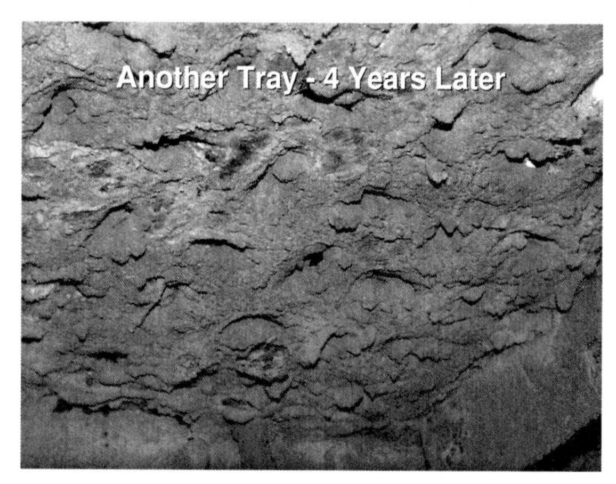

Figure 18.1 Even fouling-resistant trays are not immune to plugging under severe fouling conditions. These photos show active areas of one of the more fouling-resistant tray types in the industry plug up over a 4-year run. The plugging is believed to have been caused by severe salting out, a common cause of plugging.

under the fixed valves. Following this experience, the proprietary trays were replaced by conventional fouling-resistant trays.

CASE STUDY 18.4 VALVE STICKING: NUMEROUS EXPERIENCES

Tower A Two strippers in a refinery FCC unit were operating in parallel. One was continuously operated at high loads and contained rectangular valves. This tower seldom experienced blockage due to salting out. It required a water wash every few years. The other tower was used as a flywheel, with fluctuating loads through its round valves. This tower experienced severe salting out, requiring a water wash every 3–4 months.

Tower B Following a switch in crudes to a heavy, high-metal crude, a platformer unit depentanizer started experiencing salting-out problems. The salting out occurred around the valve caps. The switch of crudes was accompanied by a reduction of loads through the depentanizer.

Tower C Sticking of valve floats to sludge and polymer at the base of the valves was a problem in an olefins caustic wash tower. It led to high pressure drop and reduced caustic utilization. Problem was alleviated by retraying with sieve trays.

Tower D One refinery fractionator experienced severe valve sticking in the lowest five trays. The bottom three trays were in PA service, the next two were the bottom trays of a fractionation section. The problem was solved by removing the round valve floats from the bottom five trays.

Tower E A tower in fouling acrylate service was lucky to achieve 6 months run length, constrained by sticking of round valves. The sticking led to flooding. Replacement by long (5-in.) rectangular fixed valves increased the run length to at least 18 months (no feedback since).

Tower F Sticking of round valves was experienced in a tower in fouling chemical service, leading to flooding and short run lengths. The problem went away after the valve trays were replaced by dual-flow trays.

Towers G and H Valve trays operated at high loads in sticky chemical service experienced no sticking (Case Study 18.1). In another experience, severe and sticky chemicals in a butadiene plant were handled by trays with round valves without sticking. Here too the towers were continuously operated at high loads and the run length was limited by plugging in the downcomers.

Tower I A superfractionator in petrochemical service was hydrotested using treated river water. Following the test, the tower was drained but not dried. It remained boxed in for a few months. Upon start-up, the tower capacity was severely restricted, so it was shut down and opened up. Sludge and corrosion products were present on the tray floors, presumably from the hydrotesting. Valve floats were stuck in the sludge at the base of the valves. Both valves and decks were fabricated from 410 SS. The stuck floats were mechanically reopened.

Tower J A tower equipped with CS valve trays was left idle and open to the atmosphere with its top head removed for several months. Upon restart, valve sticking limited throughput through the tower to 70% of the design. The tower was shut down again and chemically washed until all the rust was removed. It achieved full throughput afterward.

Tower K Floats of caged round valves stuck to the tops of the cages in a refinery fractionator. These valves were stuck open, not shut.

CASE STUDY 18.5 PLUGGING INCIDENT: TRAYS Versus STRUCTURED PACKINGS

Installation An olefins oil quench tower separates light HCs and gasoline as the overhead product from heavy, hot fuel oil leaving from the bottom. Most of the fuel oil is injected as quench to cool reactor ("furnace") effluent upstream and from there returns to the tower. The tower contained a splash-deck PA lower section and a valve tray upper section.

Experience For several years, the tower operated without any fouling incidents. Later, the valve trays were replaced by high-capacity structured packings. The purpose was to reduce pressure drop.

The modified tower worked well until an incident of a viscosity runaway in which the fuel oil in the tower bottom was overheated ("cooked"). When normal operation resumed, the tower pressure drop was much higher than before. Later, the pressure drop rose again, but this rise could not be correlated with any incident. The final pressure drop reached was 5 psi over 10 ft of packing. Liquid carryover from the top of the tower was also observed.

The packing was gamma scanned. The grid gamma scan showed much the same density along all four chords and no apparent signs of plugging. However, at low liquid loads (the bed operated at about 1 gpm/ft^2) and large tower diameters, gamma ray absorption by the liquid is small, so the equal density reflected the bed integrity much more than the liquid distribution.

When the tower was opened, the packings were filled with coke fines. No chunky coke deposits were found. It is suspected that the fines were carried over from the bottom during the cooking incident.

Cure The packing was removed and valve trays reinstated. The new trays were designed with more open area to minimize pressure drop. No more plugging occurred after this.

CASE STUDY 18.6 PLUGGING INCIDENT: PACKING Versus PACKING

Larger packings have a greater fouling resistance than smaller ones (251). Nonetheless, even packings with larger openings that have good track records in fouling environments may plug over a period of time (e.g., Fig. 18.2). Following are some experiences.

Tower A A 3-ft-ID acid gas regenerator containing 1-in. Pall rings experienced run lengths of 1–3 months limited by plugging. Replacing by 2-in. modern

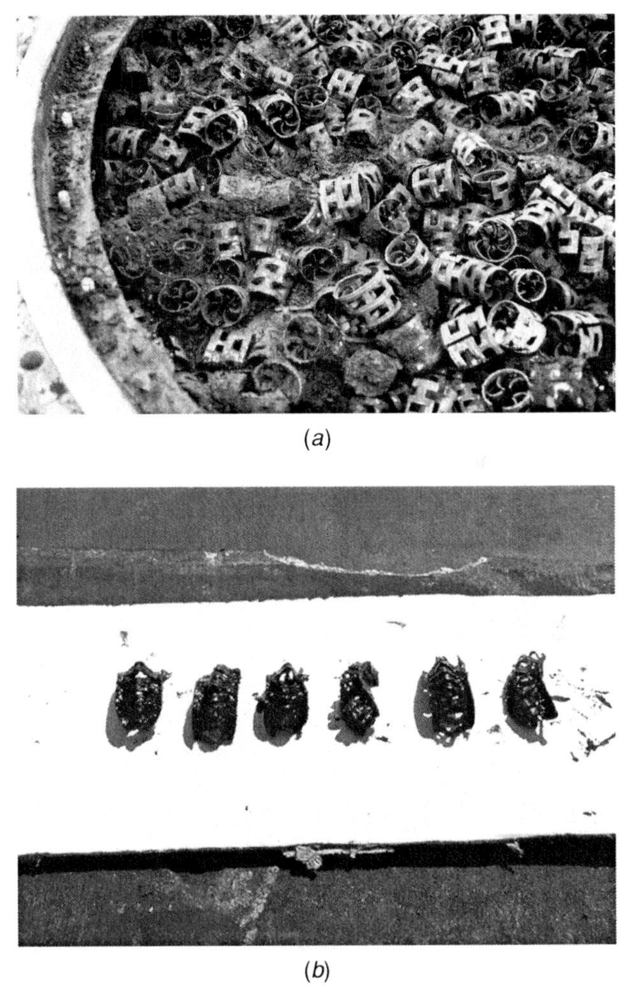

Figure 18.2 Fouling can reduce the open area of a packing and induce premature flood: (*a*) deposits inside tower; (*b*) fouled and deformed pieces removed from tower. [(*a*) Copyright Eastman Chemical Company. Used with permission.]

("third-generation") random packing of good fouling resistance and improved metallurgy raised run length to 1–2 years.

Tower B A large-opening flat plastic random packing experienced severe plugging in aqueous service. Problem was mitigated by replacing by a large, open Pall ring derivative packing.

CASE STUDY 18.7 PLUGGING IN A PACKED-TOWER GAS INLET

Tom C. Hower and Henry Z. Kister, reference 224. Reprinted with permission from *Hydrocarbon Processing*, by Gulf Publishing Co., all rights reserved

This case describes how plugging near the bottom of a packed tower was mitigated by stacking the bottom foot of packings.

Installation An acetylene plant aftercooler tower cooled compressor discharge gases (Fig. 18.3*a*). The gases originated from the acetylene reactor and contained some heavy unsaturated components. The aftercooler contained one 40-ft bed packed with 2-in. ceramic Raschig rings.

Problem The heavy components formed polymeric compounds that would plug the rings. Plugging was most severe at the bottom foot or so of the bed, where the hot gas first contacted the packing. Column run length between cleanings was approximately 4 months.

Initial Attempt at Solution Following one cleaning it was decided to go through the pain of stacking all the rings, one on top of the other in a stacked staggered arrangement (Fig. 18.3*b*). Upon restart, the column failed to cool the gas. Stacking the rings led to channeling of vapor and liquid through separate passages, with little heat transfer between the phases.

Solution The packing was removed and reinstalled. This time, only the bottom foot (about six layers) of packing was stacked. The rest was randomly packed using a wet-packing technique. Upon restart, good cooling was achieved, and the plugging problem was practically eliminated. Column run length between cleanings increased from 4 months to 2 years.

Postmortem It is believed that the polymeric compounds stuck to the sides of the rings in the bottom layers. When the rings were randomly packed, deposits would accumulate in the packing interstices and low-velocity areas until they plugged the bottom layers. Stacking the bottom layers eliminated the interstices near the bottom of the bed and permitted continuous washing of the polymer in that region.

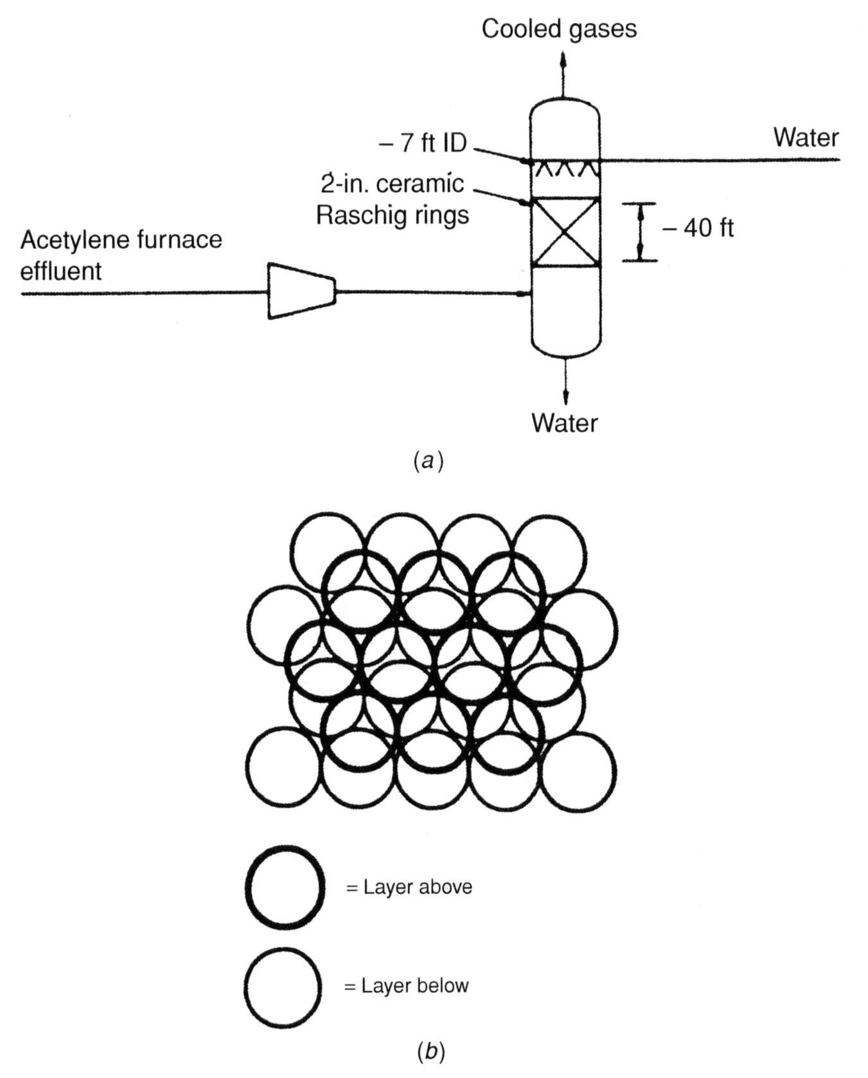

Figure 18.3 Acetylene plant compressor aftercooler column that experienced plugging near bottom: (*a*) tower schematic; (*b*) staggered stacked arrangement in aftercooler packed bed. (From Ref. 224. Reprinted with permission from *Hydrocarbon Processing* by Gulf Publishing Co. All rights reserved.)

CASE STUDY 18.8 OVERCOMING TOP-TRAY PLUGGING IN A CRUDE FRACTIONATOR

Installation Refinery crude oil fractionator, 70,000 BPD capacity.

Problem The top tray in the tower plugged. This restricted tower capacity to less than 50,000 BPD.

Troubleshooting Gamma scans showed liquid stacking on the top tray. There was little liquid on the tray below. The downcomer from the top tray to the tray below was virtually empty. The next tray down had a few of inches of liquid, probably due to vapor condensation. On the trays below, the liquid inventory was progressively higher, eventually reaching normal values.

Theory The plugging on the top tray increased the resistance to vapor, making vapor ascent through the downcomer easier. At some time, possibly following a reduction in liquid flow, vapor broke through the downcomer liquid seal. Once the vapor broke through, liquid could no longer descend into the downcomer and was entrained over the top of the tower.

Cure A liquid inlet nozzle was hot tapped into the second tray, bypassing the top tray. This restored capacity to 70,000 BPD. Over a period of time, the second tray fouled too. Another liquid inlet was hot tapped on the third tray.

CASE STUDY 18.9 PARTIALLY PLUGGED KETTLE DRAW DOES NOT IMPAIR TOWER OPERATION

Installation An amine regenerator stripping small amounts of H_2S and CO_2 out of a rich-amine solution. Bottom product was the lean amine. The tower was equipped with two-pass trays containing uncaged, moving valves.

The bottom downcomer was a "trousers" type (Fig. 18.4). Each leg of the trousers terminated in a seal pan. The seal pan on each leg overflowed into a sump, from which liquid gravity-flowed into a kettle reboiler. Vapor from the reboiler returned via a pipe distributor, which passed through the central gap in the downcomer. The pipe distributor contained one row of 4-in. holes on each side of the pipe, with the holes directed downwards at 45°. There was an equal number of holes on each side of the downcomer and none in the gap through the downcomer.

The kettle reboiler had an overflow baffle from the boiling region to the kettle draw sump. Liquid from the kettle draw sump gravity flowed into the base of the regenerator tower (Fig. 18.5).

Problem Following a steam emergency, liquid flow from the tower to the reboiler was lost, causing loss of boil-up. Boil-up was quickly reestablished by raising liquid level above 100%. While this was an acceptable short-term solution, it was uncertain if it could be sustained, and a crash shutdown was contemplated.

Troubleshooting Gamma scans showed normal operation on all trays except for the bottom two. The second tray from the bottom appeared damaged and holding little liquid. Froth height and density on the bottom tray were higher than on upper trays possibly due to debris from the second tray. The base liquid level changed with

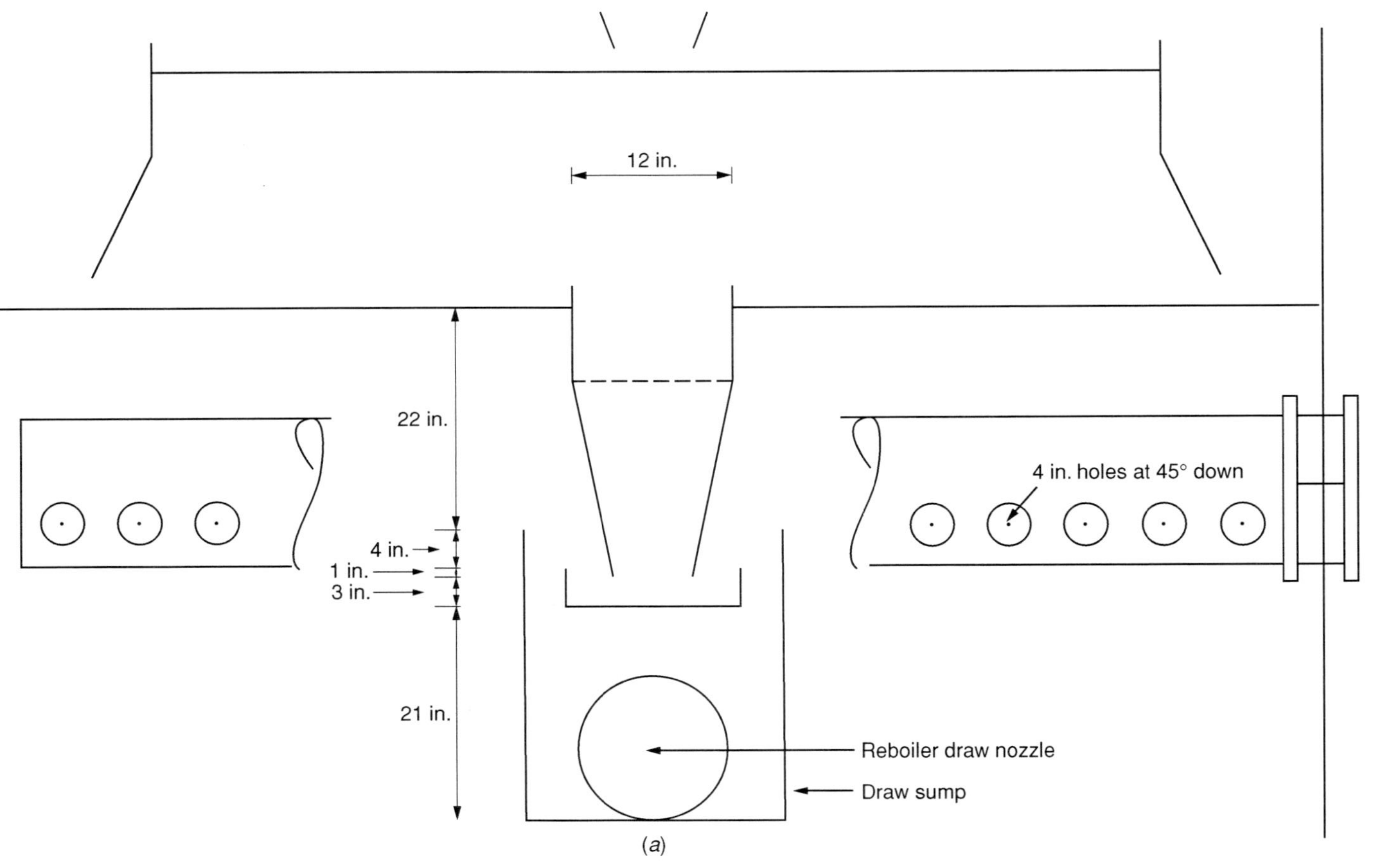

Figure 18.4 Amine regenerator bottom downcomer, reboiler draw, and reboiler return: (*a*) 0–180° view; (*b*) 90–270° view.

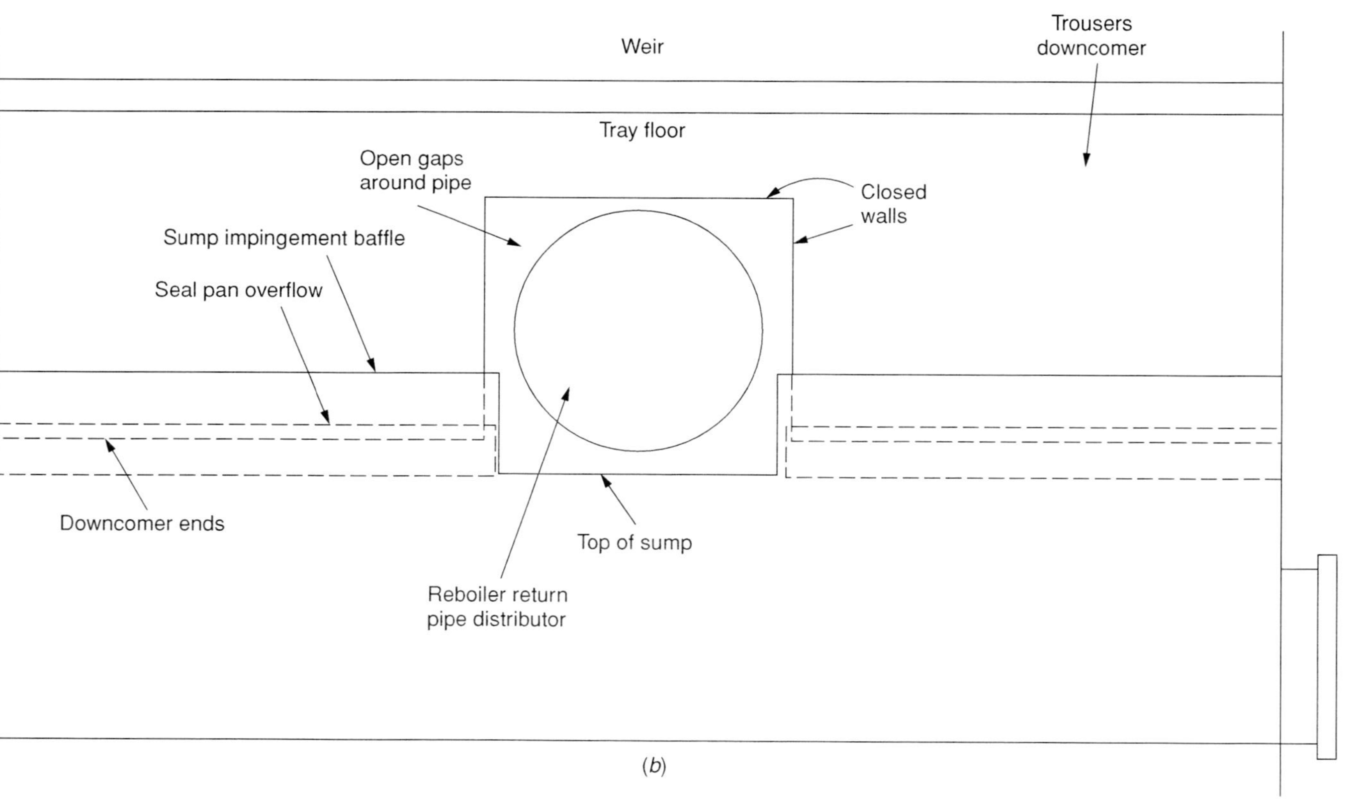

Figure 18.4 *(Continued)*

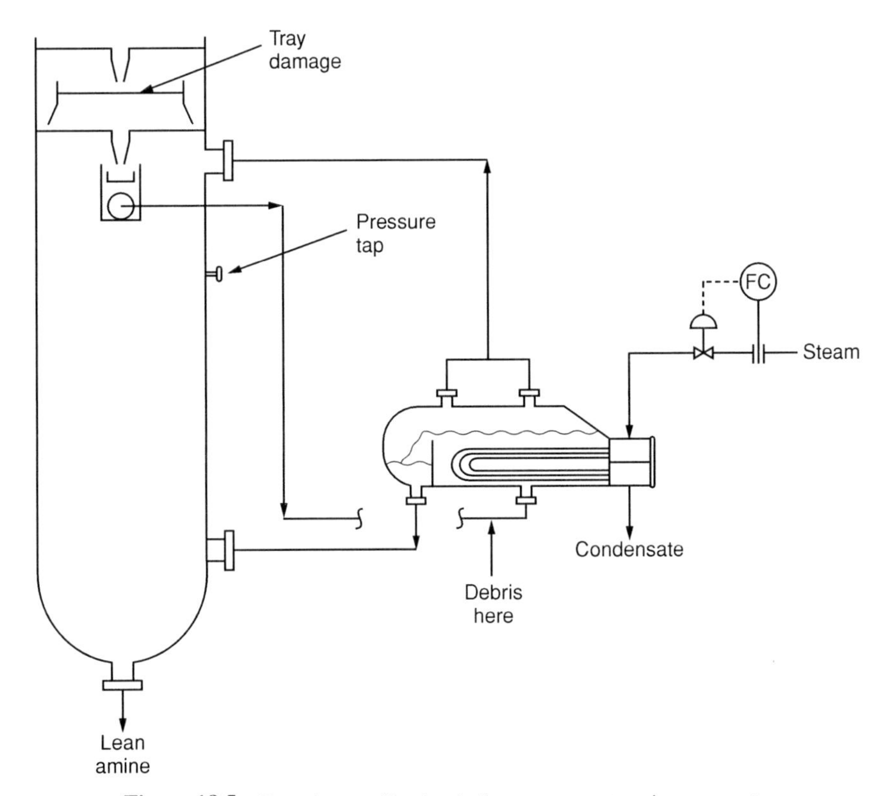

Figure 18.5 Tower base and kettle reboiler arrangement, amine regenerator.

time, hovering around the elevation of the reboiler draw nozzle. Neutron backscatter confirmed liquid in the reboiler draw line and vapor in the vapor return line. Neutron backscatter also showed that the liquid draw sump was overflowing. X-rays showed presence of tray components, such as valves and clips in the line from the tower to the reboiler. X-rays also identified a tray component in the draw box.

Theory The above suggests that the second bottom tray became damaged during the steam emergency, and debris from this tray restricted flow from the draw sump to the reboiler. Raising the level in the bottom of the tower raised the level in the kettle draw compartment. When this level exceeded the kettle overflow baffle, backflow from the tower base reestablished full liquid supply to the kettle.

Cure Initial reaction was to shut the tower down, repair the damaged tray, and clear the debris. A closer analysis revealed that there was little need to shut down. Only one tray was damaged, liquid was getting to the reboiler, and the stripping was good. Stable and satisfactory operation could be sustained with the high liquid level at the tower base. The level was maintained reasonably constant by using the tower *dP* transmitter to monitor the liquid height.

There were some concerns that the liquid level rise above the reboiler return entry would cause tray uplift and further damage, as is often experienced in regenerators. However, here the reboiler vapor reentered the tower not via a bare large-diameter nozzle, but via a multitude of small holes in the pipe distributor. If submerged, vapor from the holes was likely to bubble or jet through the liquid but was unlikely to slug through it. It is these slugs that lead to tray uplift, not the bubbling/jetting action likely here.

Result At the time of this writing, the tower has been operating well at the high level for over a year.

Chapter 19

Coking: Number 1 on the Top 10 Malfunctions

With well above 100 case histories, plugging/coking is the undisputed leader of tower malfunctions (255). Among the causes of plugging, coking was identified (255) as the leader. The incidence of coking is uncommon in nonrefinery towers, but has been a major problem in refinery fractionators. It persists despite the highly fouling-resistant hardware like grids used in services prone to coking. Coking incidents grew rapidly in the 1990s, with less than 20% reported earlier. This rapid growth reflects refiners' shift towards "deep cutting" of the crude residue, that is, maximizing distillate recovery out of crudes by raising temperature, lowering pressure, and minimizing wash bed reflux in the refinery vacuum towers.

About two-thirds of the reported coking incidents occurred in refinery vacuum towers. Of these, about two-thirds were due to insufficient reflux to the wash bed. This wash bed removes the heavy ends ("asphaltenes") and organometallic compounds from the hot feed vapor by contacting the vapor with a volatile reflux stream. If insufficient, dry spots form in the wash bed and coke up. The problem of insufficient reflux reflects a learning curve problem associated with deep cutting the residue, a technology that emerged over the last decade.

A major lesson learned is that excess vaporization occurs in wash beds that are too tall and/or contain packings that are too efficient. In either case, the additional stages intensify the vaporization of the wash oil, leaving little liquid to reach and wet the lower sections of the bed. These lower sections of the bed dry and coke. Poor modeling and simulation are other causes of coking. The heavy ends of the crude must be correctly characterized in the simulation. The use of simulated distillation (169, 236) has improved characterization of the heavy ends. Golden et al. (169) and Trompiz (487) taught the industry that correct modeling of the feed entry to the tower is mandatory for a good simulation. Their procedure, now the standard of the industry, is to model the feed by a series of flash steps that correctly represents the physical sequence of steps between the heater outlet and flash zone. When the above lessons are overlooked, the simulation underestimates wash oil vaporization, leading to the drying and coking.

Distillation Troubleshooting. By Henry Z. Kister

Other causes of reported coking incidents in vacuum towers were also concentrated in and around the wash zone. Causes include poorly designed, damaged, or untested wash section spray nozzles and spray headers; excessive liquid residence times at hot temperatures (either in the bottom sump or on overflash chimney tray); level control problems (mostly on the overflash chimney tray); maldistribution of vapor from the flash zone; and excessive heater outlet temperatures.

Other refinery main fractionators in which coking has been troublesome are FCC, coker, visbreaker, and atmospheric crude. In all these, the problems have been most severe near the feed, where the hot vapor or vapor/liquid feed enters. The FCC main fractionators appear to feature in more case studies. Their problems were fueled by a 1990s trend to replace the shed decks and disk and donut trays that had been traditionally used in the slurry PA section above the feed by grid packing. Shed decks and disk and donut trays are far less sensitive to vapor or liquid maldistribution than grid and therefore far less prone to coking during upsets. Many such retrofits introduced coking issues where none had been previously seen. Aggravating these is the hot, superheated, reactive feed that enters the tower at high velocities. This feed is very difficult to evenly distribute, as any redistribution baffles in its path easily grow coke, as reported in some case studies.

CASE STUDY 19.1 COKING IN A TALL, EFFICIENT WASH ZONE

Contributed by Henry Z. Kister and Robert F. Beckman, Fluor, Aliso Viejo, California

Installation A refinery vacuum tower processing 80,000-BPD of atmospheric tower resid derived from a mixture of Middle East crudes. Distillate was hydrocracker feed vacuum gas oil and bottom product was penetration-grade asphalt. To minimize metals in the HVGO, the wash bed was 12 ft long and contained high-efficiency Y-type structured packings, 50 ft^2/ft^3 surface area.

Problem and History The tower experienced chronic coking and high-pressure drop in the wash zone. Wash bed pressure drop rose 10–15 mm Hg over 1 year. Each yearly turnaround the wash section packing was found coked and needed replacing. This repeated three to four times. The tower was unable to achieve its design throughput. Slop wax production was about twice that expected. In an effort to reduce entrainment from the flash zone, the original simple 90° vapor horn at the feed inlet was replaced by a 360°, proprietary, state-of-the-art vapor horn. The converse was achieved, and entrainment actually increased. Throughout all this, the quality of the HVGO remained good, with metal (Ni + V) content of less than 1 ppm.

Root Cause The high-efficiency structured packing provided four to five separation stages. The wash oil was volatile HVGO distillate, which can be evaporated by the hot rising vapor. Over four to five separation stages, more than 90% of the 1 gpm/ft^2 wash oil rate would evaporate, leaving less than 0.1 gpm/ft^2 to reach the bottom of the bed. In wash beds of vacuum towers, with flash zones at 750°F, the minimum liquid flow rate required to keep the packing wet and to prevent coking is 0.2 gpm/ft^2. The large evaporation taking place in the long, efficient packed bed lowered the liquid rate well below the minimum wetting rate, leading to coking.

Wash bed coking raised pressure drop and flash zone pressure, which lowered HVGO recovery. The coking also generated local flooding and maldistribution, reducing wash bed efficiency. Simulations showed that the coked wash bed achieved only one or two separation stages. The efficiency loss drastically reduced evaporation, allowing liquid to descend, which prevented further deterioration. The one or two separation stages were enough to keep the HVGO on specificaton for metals.

Figure 19.1 is an energy balance on the wash bed compiled from operating data when it was coked. At that time coking reduced the bed separation to one theoretical stage. Only about 66% of the wash oil evaporated, allowing about 0.3 gpm/ft^2 to reach the bottom.

The energy balance also confirmed the high entrainment from the flash zone. Of the 113,500 lb/h metered overflash, only 53,800 lb/h was spent wash. The balance 59,700 lb/h was entrainment. This approximate figure was confirmed by a metals balance on the overflash.

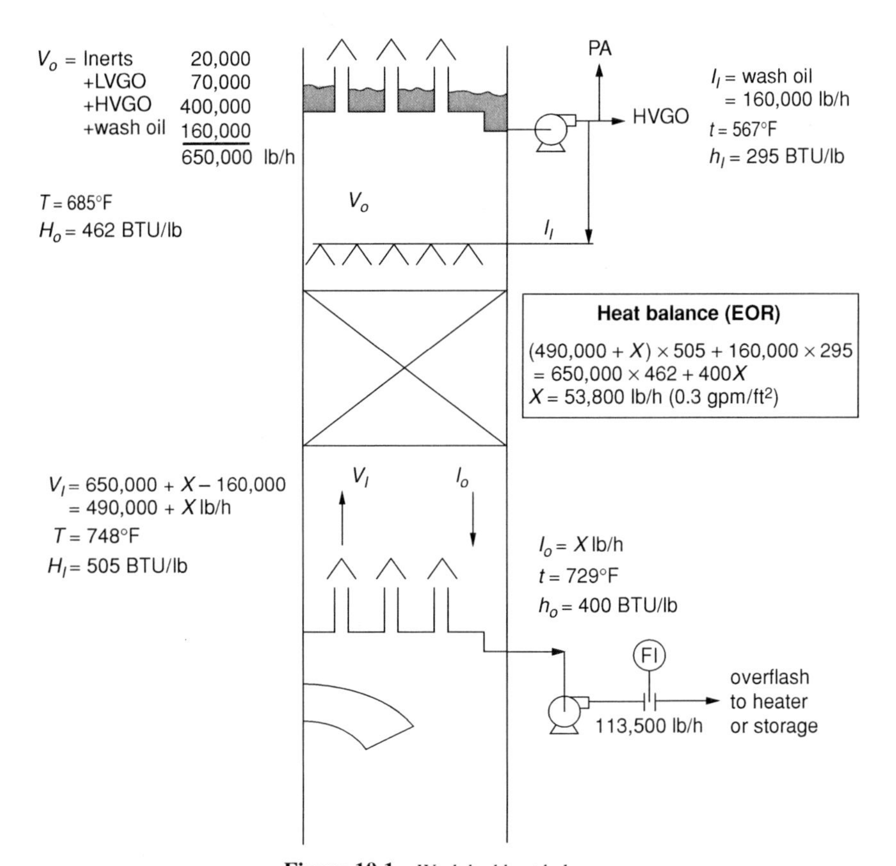

Figure 19.1 Wash bed heat balance.

The high entrainment initiated in the undersized transfer line brought feed to the tower. Calculated velocity at the 42-in. tower inlet nozzle was 130% of sonic. This high velocity led to the formation of fine droplets that entrain easily. Normally, much of the entrainment would have been knocked out in the flash zone, given sufficient space to settle. Instead, the vapor–liquid mixture issued horizontally into the hollow throat inside the 360° vapor horn. The 360° horn occupied 45% of the tower cross-sectional area, leaving only 55% of the tower cross-sectional area for vapor–liquid disengagement. The very high vapor velocity through the throat area exceeded the system limit flood velocity, resulting in massive carryover of liquid droplets. This caused the high slop wax production.

Cure The 12-ft wash bed was shortened to 6 ft. The high-efficiency packings in it were replaced by 4 ft of grid packings followed by 2 ft of low-efficiency (40-ft^2/ft^3 X-type) structured packings. The number of separation stages was thus reduced from 4–5 to between 1 and 1.5. The spray distributor was redesigned for better bed coverage.

With fewer stages, evaporation declined from over 90% of the wash oil rate to about 67% of the wash oil rate, allowing much more liquid to reach the bottom of the bed. This, in turn, allowed a cut in wash oil rate. As shown in Figure 19.1, vaporized wash oil adds up to the vapor load V_O. Reducing wash oil rate and the fraction vaporized reduces V_O, which unloads the tower and permits higher capacity.

To reduce flash zone entrainment, the transfer line and feed inlet nozzle were enlarged. A new 150° tapering vapor horn, with liquid removal, replaced the 360° horn. The new horn occupied only 20% of the tower cross-sectional area and removed most of the liquid before reaching the throat. With the larger throat area the upward velocity through the throat was below the system limit flood velocity, permitting good droplet disengagement and preventing massive carryover.

Results The modified tower had operated continuously for more than 2 years with no signs of coking, nor an apparent increase in wash bed pressure drop (which was regularly monitored). Entrainment from the flash zone was more than halved, and slop-wax make became as expected. Tower capacity increased, and flash zone pressure decreased. Despite the drastic reduction in stages, there was no increase in the metals content in the HVGO.

Moral In refinery vacuum towers processing low-metal crudes, tall and efficient wash zones can lead to coking and poor performance. State-of-the-art vapor horns can be more troublesome than well-designed, simple, less sophisticated horns.

CASE STUDY 19.2 TOO MANY STAGES LEAD TO WASH BED COKING

Installation A refinery vacuum tower, about 30 ft ID, was processing low-metal crudes. Distillates HVGO, used for FCC feed, and LVGO, mixed into the diesel pool, were separated from bottom resid. The tower contained a 4.5-ft wash bed of high-efficiency structured packings.

History The wash bed experienced chronic coking. Coking increased wash bed pressure drop from about 4 mm Hg to more than 10 mm Hg, which raised flash zone pressure and reduced HVGO yield. Coking also caused maldistribution and flooding in the wash bed, which worsened separation and incurred further losses in HVGO yield. Upon coking, metals (Ni + V) in the HVGO rose from the normal 0.4 ppm to about 0.8 ppm, and the HVGO/resid cutpoint was lowered about 50°F. The HVGO yield declined from about 72% to 66%, with the resid picking up the lost yield. These changes favored frequent cleaning turnarounds, roughly every year to year and a half.

Both the packing and distributor coked. Packing and spray nozzles were changed every turnaround. Packings from three different suppliers were tried. In all cases at least the top 3 ft of bed consisted of high-efficiency (Y-type) structured packings, the other 1.5 ft being either the same or high-surface area grid. In one turnaround, the spray distributor was replaced by a high-performance gravity distributor advocated for wash sections. This distributor repeatedly coked too.

Root Cause The wash oil flow rate was 0.7 gpm/ft^2. The proximity of a tower upper capacity limit precluded using higher wash rates. The high-efficiency grid/packing in the wash bed was estimated to provide about 1.7 separation stages when clean. Previous packings were estimated to provide the same, even better staging. With 1.7 stages, about 85% of the wash oil would evaporate, leaving a liquid flow rate of about 0.1 gpm/ft^2 at the bottom of the bed. In wash beds, with flash zones at 750°F (which was the case here), the minimum liquid flow rate required to keep the packing wet and prevent coking is 0.2 gpm/ft^2. The large evaporation taking place in the efficient packing lowered the liquid rate below the minimum wetting rate, leading to coking.

Poor simulation of the tower contributed to the coking. With the HVGO draw only one or two stages above the feed, it is critical to correctly model the feed entry. The good practice (169, 236) is to model the tower feed as a series of flashes which correctly describes the physical sequence of steps at the feed entry, as proposed by Golden et al. (169, 175) and Trompiz (487). Conventional modeling, which simply feeds the tower one or two stages below the draw, tends to underestimate evaporation in the wash bed. The packing supplier estimated 60% evaporation of wash oil, presumably on the basis of a conventional model. This would have left an adequate wash rate of 0.25 gpm/ft^2 at the bottom of the packed bed and circumvented coking. Correct modeling usually gives much higher evaporation rates for 1.7 stages.

Cure The 3 ft of structured packing of 70 ft^2/ft^3 specific surface area was replaced by larger, less efficient structured packing of 40 ft^2/ft^3 specific surface area. An 8-in. layer of higher efficiency grid was replaced by lower efficiency grid. This lowered the number of stages in the bed from 1.7 to about 1. This reduced vaporization from 85% to about 67% of the wash oil, leaving 0.23 gpm/ft^2 near the bottom of the bed.

Results Over more than 2 years in operation, coking was not observed and the wash bed pressure drop did not rise. Extensive plant tests indicated that the percent vaporization of the wash oil with the packing was around 70%, which compared well with the expected 67%, and that the modifications opened the door for further optimization, increasing tower throughput and HVGO yield beyond the previously achievable values. Metal content of the HVGO was practically unchanged compared to the previous uncoked operation.

CASE STUDY 19.3 VACUUM TOWER COKING

Tower A This tower was equipped with structured packing throughout, except for the stripping section that contained trays. Over a period of time, there was an increase in wash bed *dP*, indicating coking, although the HVGO color did not change. The coking was due to high liquid level on the slop wax chimney tray induced by level control problems.

In the next turnaround, an overflow was installed on the slop wax chimney tray to make sure the liquid level did not exceed the top of the riser. This eliminated the *dP* rise over the next run. However, the resid yield went up 1–2%, indicating the overflow was continuously active.

Tower B This tower experienced chronic coking in a short wash bed and spray nozzles. Cause was an operating procedure that would not commence wash oil to the bed until about a week after start-up. No more plugging occurred after procedure was modified.

Tower C A refinery vacuum tower processing low-metal crudes had a 4-ft tall wash bed containing grid. The HVGO product was black. Next turnaround, the grid was inspected from above and below and appeared clean. Upon return to service the HVGO color did not improve. Next turnaround, the grid bed was dismantled. The top and bottom grid layers were clean, but the middle of the bed was coked solid. It appears that the top of the bed was kept wet by the wash while the bottom of the bed was kept wet by entainment of crude from the flash zone. In the middle of the bed, much of the wash was vaporized, while the entrainment was removed by the lower few inches of bed height, causing drying which led to coke.

CASE STUDY 19.4 COKING OF GRID IN FCC MAIN FRACTIONATORS

Typically, the zone above the feed of an FCC main fractionator is a slurry PA, using externally cooled, circulating slurry to desuperheat incoming reactive, hot reactor effluent gas. Fouling-resistant internals such as shed decks, disk and donut trays, and grid packing are common in this application. Of these, grids have been more prone to coking due to their sensitivity to maldistribution, as in the cases below.

Tower A Liquid to the grid was supplied by a notched trough distributor (Fig. 19.2). The notches discharged the liquid into "flow tubes," that is, triangular baffles that directed the liquid down.

During one start-up, there was extensive carryover of reactor catalyst into the main fractionator. Much of the catalyst ended in the slurry distributor, causing

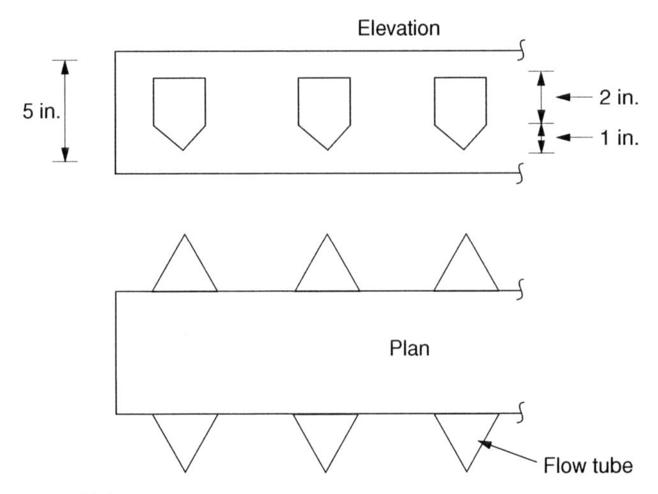

Figure 19.2 Notched trough distributor that supplied slurry to grid packing.

maldistribution of liquid to the grid. The maldistribution in turn led to extensive coking of the grid. Coking of the grid led to capacity loss, carryover of slurry into the trayed section above, and extensive fouling there. The coking in the grid was so extensive that it took about 2 weeks to jack-hammer the grid out at the next turnaround.

Tower B Coking occurred throughout the vertical length of the slurry PA grid bed on the side of the tower directly across the vapor inlet nozzle. It appeared that the incoming high-velocity jet of superheated vapor issuing from the bare nozzle hit the tower wall across and was deflected upward, preferentially rising on that side. This generated a hot region that formed coke. The problem could have been aggravated by a manhole in the grid bed in the coked region. The stagnant flow at the manhole was conducive for coking.

Tower C This tower had a history of nonuniform coking in the slurry PA grid bed. There were episodes of liquid slurry carryover into the trayed section above. The coking was believed to be induced by vapor and liquid maldistribution. The liquid distributor was extremely complex, and so were the pans bringing the wash liquid down to the slurry section.

Tower D Disk and donut trays were replaced by grids in the slurry PA. There were concerns about gas maldistribution to the grid bed due to the high gas inlet velocity and the short vertical height (about 5 ft, which was less than one-third of the tower diameter) between the top of the gas inlet nozzle and the bottom of the grid bed. To alleviate, a gas distributor was added at the gas inlet nozzle. The gas distributor was a rectangular box with a sloped bottom so its area tapered from the gas inlet nozzle to the opposite wall. The bottom had large openings, about 1.5 by 1 foot, with downward vanes at the end of each opening. Upon restart, the distributor pressure drop rose from 2 to 6 psi over the first month in operation. When taken off-line shortly later, the vanes were totally coked, and there was coke inside the distributor. No coke was observed on the top or bottom of the grid bed. Removing the distributor gave a much better operation.

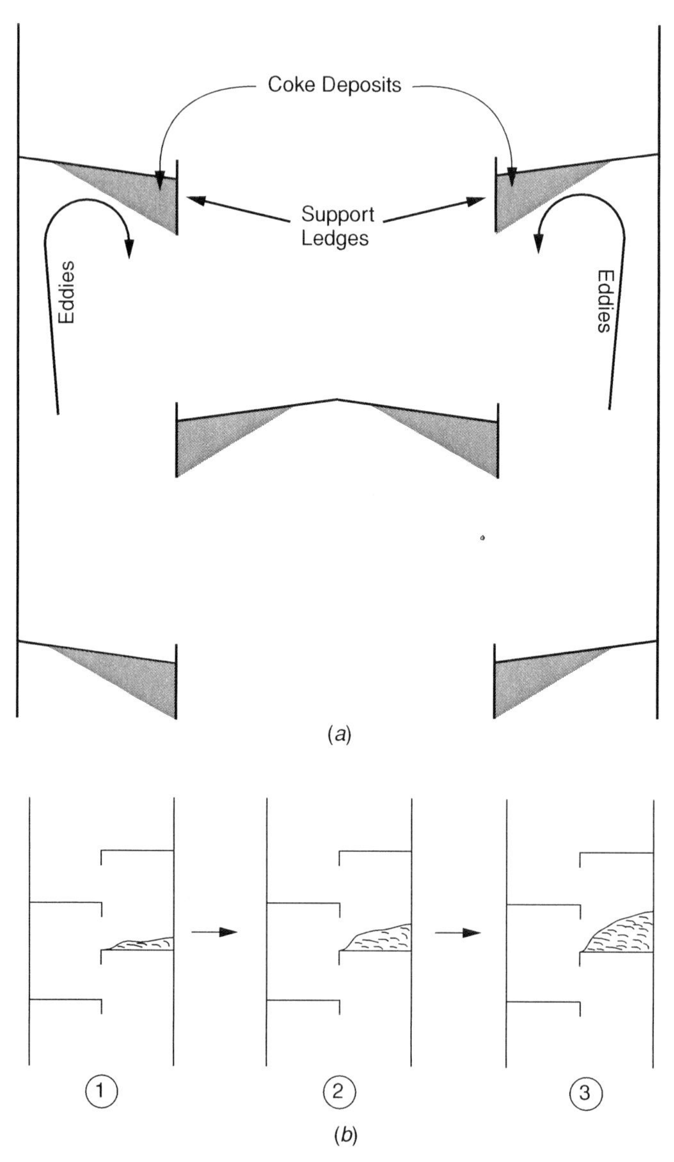

Figure 19.3 Coking of baffle trays: (*a*) eddies deposit coke behind support ledges; (*b*) progressive buildup of coke on baffle trays.

CASE STUDY 19.5 COKING OF BAFFLE TRAYS

Tower A The slurry section of an FCC main fractionator quenched superheated, reactive effluent from the reactor by direct contact with a circulating slurry stream that was cooled outside of the tower. The slurry section contained segmental baffle trays.

Coke deposits built up under the segmental baffle trays. The buildup was most severe behind the support ledges. It is most likely that eddies caused this deposition (Fig. 19.3*a*). The deposits grew quite heavy and had caused several tray sections to collapse. These collapses, however, did not lead to an apparent loss of performance or to a shorter run length. The tower lasted 4 years between turnarounds.

Tower B The stripping section of a refinery vacuum tower contained segmental baffle trays. The tower received a reactive feed from a processing unit, which had a high coking tendency.

Coke deposits built up under the segmental baffle trays. Like tower A, the buildup was most severe behind the support ledges and was most likely caused by eddies (Fig. 19.3*a*). When the deposits grew thick, chunks of coke spalled off and got stuck in the bottom pump strainers. Often a chunk of spalled coke would block the pump suction and cause it to trip.

Tower C A tower in sticky chemical service contained flat segmental baffle trays at 12 in. spacing. The trays plugged over a few months. Plugging began at the back of the tray and progressed upward until it interfered with the flow window (Fig. 19.3*b*). Occasionally some solids spalled off and blocked the bottom liquid offtake. Like towers A and B, it is likely that the deposits built up under the support ledges, then dropped onto the trays and became cemented there. Since the baffle trays were not sloped, the solids did not slide off the trays. At the turnaround the trays were cleaned by hydroblasting.

Chapter 20

Leaks

Based on the reported number of case histories, leaks are in the13th spot in the list of distillation malfunctions (255). In terms of real losses, including loss of human life, suffering of victims and families, damage to environment and equipment, and production losses, consequences of many leaks surpass most of those incurred by other distillation malfunctions. Mitigating leaks should therefore be a top priority for distillation design and operation. Leaks are equally troublesome in chemical, refinery, and gas/olefins towers, and their incidence appears to show neither growth nor decline.

Just less than half the reported case histories were heat exchanger leaks. Of these, about one-half were reboiler tube leaks; the other half were leaks in preheaters and PA exchangers. There was only one reported case of condenser tube leak. Most of the exchanger tube leaks led to product or utility contamination. Radioactive tracer techniques, good product analysis, and heat balances, effectively diagnosed many exchanger leaks. Some exchanger leaks led to instability and capacity loss. Occasionally, there were severe consequences such as explosion due to overchilling, fire due to rupture of a fired reboiler tube, and a pressure surge.

Many cases were reported of leaks of chemicals to the atmosphere or air into the tower. Of the reported atmospheric leaks, about a third led to explosions or fires, while about half of the discharges of flammable materials remained near misses. Fewer case histories were reported of chemicals leaking in/out of the tower from/to other equipment, but these too often led to severe consequences such as an explosion, a fatal accident, and major damage. Finally, there have been a few cases reported of seal/oil leaks from pumps and compressors. The consequences of these were comparatively less severe, although far from benign. One caught fire and another led to a pressure surge.

It appears that all leaks, especially leaks into or out of the tower, whether to/from the atmosphere or to/from other equipment, are some of the prime and most severe safety hazards in towers.

Distillation Troubleshooting. By Henry Z. Kister

CASE STUDY 20.1 TRACERS DIAGNOSE LEAKING REBOILER

Contributed by Matt Darwood, Tracerco, Billingham, Cleveland, United Kingdom

Installation A distillation column in an aniline plant that had been uprated a number of times since its initial commissioning.

Problem A number of quality issues had arisen in the tower which indicated that significant levels of water had started to ingress into the system. It was suspected that either one or both of the reboilers were leaking.

Investigation To identify which, if any, of the two reboilers were leaking steam/water into the process system, a tracer study using short-half-life radioisotopes was instigated. The radioisotope tracers were chemically and thermally stable and had no effect on this and downstream processes. The study required no preparation of the reboilers or the surrounding equipment and was carried out on-line.

The tracer was injected into the steam line feeding both reboilers using a high-pressure injection rig and the radioisotope tracer was detected using detectors strategically placed around the reboilers' various exit pipes (Fig. 20.1). The data from each detector were sent back to and stored on a central data collection system which immediately showed the results. The results from the tracer study confirmed that the newer of the two reboilers was leaking at approximately 1.5% of the total steam flow. The other reboiler was not leaking (or at least not at a rate greater than 0.4%, the detection level). Tracer injection and detection, which took approximately 20 min to

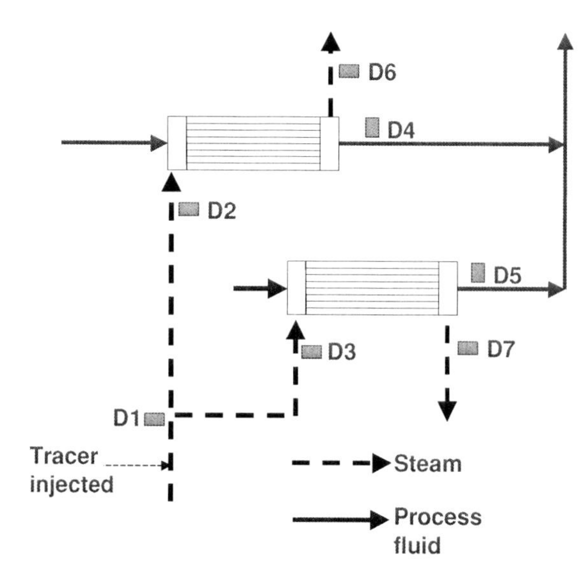

Figure 20.1 Location of detectors for reboiler leak troubleshooting.

carry out, were then repeated and the same results were obtained.

Following the tracer study, a thorough mass and water balance was compiled and confirmed that a leakage of 1.5% from the reboiler was the likely source of the water ingress into the system.

Cure The leaking reboiler was taken off-line at the next planned maintenance and was repaired. Following this the product quality issue was resolved.

Moral Radioisotope tracer technology is an excellent way to rapidly and reliably identify leaking reboilers or condensers nonintrusively while the process remains on-line.

CASE STUDY 20.2 PREHEATER LEAK IDENTIFIED FROM A SIMPLE FIELD TEST

Installation Feed to a hydrotreater fractionator passed through a series of flashes (Fig. 20.2). The flashes split the feed into three feed streams. The lower feed was

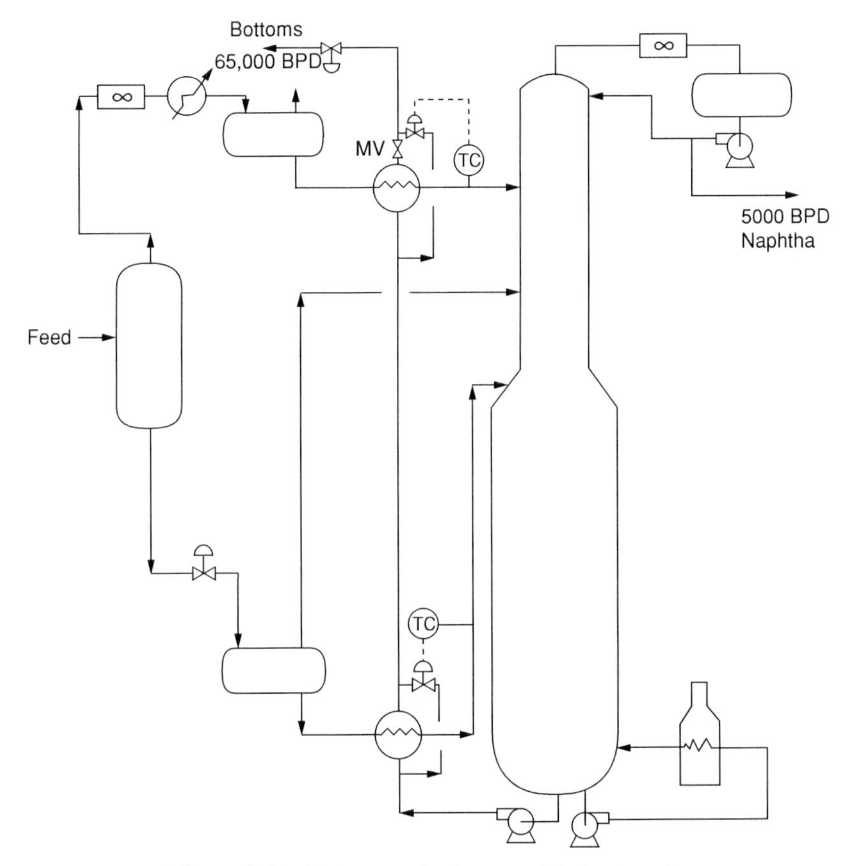

Figure 20.2 Hydrotreater fractionator and feeds schematic.

the main tower feed. The upper and lower feeds were liquids bottoms from knockout drums. On their way to the tower the upper and lower feeds were preheated by heat exchange with the tower bottoms. The temperatures of these feed streams were automatically controlled by manipulating bypass flows on the heating sides of the preheaters, as shown in Figure 20.2.

Problem The tower returned to service following a debottleneck in which the tray section above the top feed was retrofitted by high-capacity trays. Upon restart, the tower flooded at well below the design loads. The flash point of the bottom product was low while the end point of the top product was high, indicating poor separation.

Investigation Tower operation was simulated. There was a good match between all measured and simulated values, except for the reboiler duty. The simulated duty was much less than measured duty. The possibility of feed leaking into the bottom product was considered. The bottom product was analyzed for feed components that usually end up in the overhead, but those were not found. The possibility of exchanger leak was discounted.

Field Observation At one time, the bypass around the upper feed preheater was fairly wide open, with the upper feed leaving the exchanger still running hot. In an attempt to help the bypass, the manual valve MV downstream of the upper feed preheater was throttled. This should have reduced the feed temperature. Instead, the opposite was observed: As the manual valve was throttled, the temperature of the upper feed to the tower actually increased. The only plausible explanation to this behavior is massive leakage from the bottom side into the feed side of the upper feed preheater.

Once this was realized, the tower was resimulated with the leak described above. With a leak of 19,000–20,000 BPD, the simulation fully matched plant data, including reboiler duty.

Cure At the next opportunity, the system was taken off-line and the leak repaired. Following the fix, the flooding disappeared and separation was good.

Some Lessons Simple field observations and comparison of simulation to good field data are invaluable troubleshooting tools. Heat balances around individual heat exchangers are also extremely useful in identifying leaks and could have supplied an invaluable clue in this case.

CASE STUDY 20.3 SEVERAL LEAKS IN ONE HEAT EXCHANGE SYSTEM

Installation An olefins plant oil quench tower received feed from many hot reactors (furnaces). The tower cooled the gas by direct contact with a circulating oil PA. The PA took the oil from the tower bottom, cooled it in a battery of heat exchangers, and returned it to the middle of the tower, where it was sprayed onto shed decks. The gas entered the tower at 340°F and was cooled in the PA section to 260°F. The PA cooled the oil in two naphtha feed preheaters, a boiler feed water preheater, an air cooler, and a water cooler.

Problem At different times during the time period 10–14 years after initial start-up, each of the exchangers except for the air cooler developed major tube leaks. Most were due to flow-induced vibrations.

Causes and Cures Leak of the boiler-feed water exchanger caused extensive fouling of the steam drums and required a plant shutdown. The cooler was modified to prevent further damage and changed to cooling-water/quench oil service. The leak in the water cooler took place at about the same time and caused some fouling in the cooling-water system. This cooler was also modified, repaired, and returned to the same service. The naphtha/quench oil exchangers leaked at another time. The leak led to excessive quantities of "gasoline" produced, which were picked up by the plant daily mass balance. These exchangers were isolated, and their bundles were changed in the next turnaround.

CASE STUDY 20.4 BOTTOM LEAK DISRUPTS FLOW IN UPPER PUMPAROUND

Installation Rosin column in a tall oil refinery. Tall oil is oil derived from trees, which is a by-product of pulp and paper manufacturing. The rosin column fractionates the oil into various resin and fatty acid cuts. The column had a bottom product and three side-draw products. The column had two separate PA cooling loops at the top to remove heat and provide reflux. The column contained packings and was operated at deep vacuum.

The bottom side-draw (BSD) product was drawn off via a 4-in. isolation valve on the column, which was opened and closed depending on whether or not production of the BSD product is needed. Approximately 6–12 in. downstream from the isolation valve was a hard-piped steam blowout line, used to clear the line prior to start-up or following shutdown of the BSD system. The BSD product line was 316 SS while the steam blowout line, including the nipple from the last valve to the BSD product line, was CS.

Problem Over time, exposure to the various acids present in the BSD product corroded the nipple connecting the steam line to the BSD product line.

Leakage Incident At one time, the BSD system was started by blowing the line out with steam, opening the 4-in. valve to the column, and starting the BSD pump. Approximately 2 hours later, an alarm sounded for low flow in the upper cooling loop of the column. It was thought that a strainer was plugged on the loop pump, so the operator switched to the spare pump. This did not correct the low-flow situation.

At this point, it was noted that the top pressure in the column had climbed to 50 mm Hg from 37 mm Hg. The initiation of this increase coincided with starting the BSD system. The operator then shut down the BSD system and closed the 4-in. isolation valve to the column. Pressure in the column began to fall and, after about an hour, the column pressure and the upper loop flow stabilized at their normal values.

Following proper isolation and lock-out, insulation was removed from the lines in question. After tapping on the nipple gently, a $\frac{1}{4}$-in.-diameter hole formed. When the nipple was cut out, it was apparent that the entire nipple was very thin and close to failing even more dramatically. It was appreciated that even though CS is the accepted material for steam lines, this material is not suitable for a nipple exposed to acids. The corroded nipple was replaced by a 316 SS nipple.

Postmortem This was an interesting incident in that a small vacuum leak near the bottom of a distillation column was detected by a disruption to PA flow in the upper cooling loop. The increase in pressure raised the equilibrium temperatures inside the column. Since the heat input to the column was not changed, some of the vapor normally condensing in the upper cooling loop was condensed lower in the column. The upper cooling loop did not have enough vapor entering to sustain the PA flow.

Moral This incident is a good reminder of the delicate balance that is ongoing within distillation columns and how seemingly small problems can cause large upsets.

Chapter 21

Relief and Failure

Based on the reported number of case histories, overpressure relief issues are just above the 20th spot in the list of distillation malfunctions (255). Just like leaks (Chapter 20) and explosions (Chapter 14), in terms of real losses, including loss of human life, suffering of victims and families, damage to environment and equipment, and production losses, consequences of relief issues are often among the most severe incurred by distillation malfunctions. Mitigating relief issues should therefore be a top priority for distillation design and operation. Relief issues have been equally troublesome in chemical, refinery, and gas/olefins towers. Their incidence appears to show a slight, almost insignificant decline (255).

Topping the relief issues is correctly setting the relief requirements. In some cases, small modifications to controls, steam supply, or vacuum breaking gas entry, permitted large reduction in relief requirements. In others, towers blew up because their relief capabilities were short of the relief loads. Finally, in some cases, tray damage resulted from a relief condition inducing excessive tower internal flow rates.

A surprisingly large number of mostly refinery cases reported overpressure in the tower or in downstream equipment due to the unexpected presence of lights or a second liquid phase. Other cases were reported in which hazardous materials were discharged to the atmosphere from a relief valve, including HC liquids and gases that caught fire. Finally, in some reported cases relief valves were incorrectly set.

Incidents were reported in which equipment failure led to either an accident or a near miss. Line rupture failures were considered separately in Chapter 14. Cases in this chapter report trips not activating or incorrectly set as well as incidents induced by pump failure, loss of power, vacuum, or instruments.

CASE STUDY 21.1 ATMOSPHERIC CRUDE TOWER RELIEF TO ATMOSPHERE AND OVERPRESSURE

Contributed by W. Randall Hollowell, CITGO, Lake Charles, Louisiana

Installation An atmospheric crude fractionator (same as described in Case Study 1.13) distilled kerosene and lighter oil overhead, over 40% volume of the

Distillation Troubleshooting. By Henry Z. Kister

light crude charge. There were two parallel banks of three in-series, double split-flow condensers. The top condenser shell had 29 in. ID and 20 ft tube length. A normal pressure drop of up to 6 psig resulted in short tube life from flow-induced vibration damage in the top half of the bundle.

There was a large PA reflux in the middle of the tower that condensed a large diesel stream and a heavy gas oil stream.

A pressure control valve (PCV) on the overhead accumulator opened to the flare header at 30 psig. Two 6 in. × 8 in. pressure safety valves (PSVs) on the tower top were set for 50 and 53 psig. The PSVs had atmospheric discharges. Typical hot-day pressures were 23 psig in the overhead accumulator and 32 psig on the tower overhead.

Problem The PA reflux pump failed and the spare could not be started. This greatly increased the overhead vapor rate. The overhead apparently plugged between the tower and the overhead accumulator drum, causing discharge from the PSVs on the tower overhead. The overhead PSVs discharged to the atmosphere, causing a brief, heavy rain of condensed oil centered about $\frac{1}{8}$ mile downwind from the tower. A much larger area received a lighter rain of oil.

Investigation The bundles in the top condenser were found to be bowed downward. The bundles had been rotated 180° to improve tube life. Circular, inlet impingement baffle plates that were formerly below the top nozzles had been rotated to above the outlet nozzles. Bowing of the bundles apparently caused the impingement baffles to block off the bottom nozzles.

The tower top PSVs were not designed for a blocked overhead; thus the tower overpressure was greater than design. Tests did not find any significant bulging or other vessel damage.

The release discharged into a warm, cloudless, sunny afternoon. These conditions minimized the amount and volatility of the oil rain. Although there were no tests made, the condensed oil rain appeared to be more viscous than kerosene and had no naphtha odor; thus it may have had a high flash point. The wind direction was optimum in that it was away from the other process units, away from the crude unit fired heaters, and avoided the central steam boilers.

Operator responses were excellent. Crude charge was quickly discontinued, with emergency steam added to the unit to protect heater tubes and to help vaporize the vented oil.

Solution Condenser bundles had to have impingement baffles removed before they were rotated and reinstalled. The PSV capacities and requirements were thoroughly reviewed.

Moral The plant was lucky that ignition of this rain did not occur, as it would have caused a major disaster. This occurred before the Occupational Safety and Health Administration (OSHA) regulations existed. Any difference in the arrangement of the same hardware can affect relief requirements and must be covered by management of change (MOC) and by careful HAZOP analysis.

CASE STUDY 21.2 RELIEF ACTION CAUSES TRAY DAMAGE

Contributed by Tak Yanagi, Consultant, Monterey Park, California

Installation A test column was protected from overpressure by a 10-psi rupture disc installed in the overhead line from the column.

Experience When the rupture disc blew, the top tray was uplifted off its supports.

Moral Beware of oversizing relief devices, as they can lead to vapor loads that induce uplift greater than the tray supports are designed to handle.

Related Experience A tower equipped with valve trays processed hot, water-insoluble organics and had a relief valve below the bottom tray. Upon start-up, a pocket of water entered the tower bottom. It quickly vaporized and generated a pressure surge that lifted the relief valve. The relief valve sharply dropped the bottom pressure, making it lower than the pressure further up the column. This bent downward and deformed the lower trays.

Chapter 22

Tray, Packing, and Tower Damage: Part of Number 3 on the Top 10 Malfunctions

Close to 100 reported incidents place tray, packing, and tower damage third in the list of distillation tower malfunctions (255). The number of damage incidents appears to be on the decline (255), suggesting good progress. Tower internals damage is equally troublesome in refinery and chemical towers but appears less of a problem in olefins/gas plant towers.

The two prime causes of tray, packing, and tower damage are excluded from this chapter. Water-induced pressure surges, accounting for about a quarter of the reported damage incidents, almost all from refinery fractionators, is the subject of Chapter 13. The top cause of damage in chemical towers (123), and also troublesome in refinery towers, is excessive liquid levels in the tower base, discussed in Section 8.3. Also, damage due to chemical explosions and fires (Chapter 14), relief and failure (Chapter 21), and damaged packing distributors (Section 6.7) were excluded from this chapter.

The remaining causes of tower internals damage are the subject of this chapter. These include insufficient mechanical strength; uplift due to rapid upward gas surge; downward force on trays (particularly valve trays); flow-induced vibrations; poor assembly or fabrication; and popping of valves out of valve trays (mainly legged valves). The standard uplift resistance of trays is relatively low, suggesting that in services prone to damage, "heavy-duty" internals design, as described in Shiveler's article (442), can offer much improvement. The rapid upward gas surges and downward forces appear particularly troublesome in chemical towers. Lessons drawn from many of these incidents can improve start-up, shutdown, and commissioning procedures. Flow-induced vibrations can often be circumvented by double-locking nuts, improved supports, and avoiding operation near the weep point. Detailed guidelines for dealing with this phenomenon are in excellent papers by Brierley et al. (72), Winter (539), and Summers (474). Good inspection can often detect poor assembly and fabrication, and changing valve type has effectively eliminated popout issues.

Distillation Troubleshooting. By Henry Z. Kister

There are other, somewhat less common causes of tray and packing damage, including breakage of packings (mostly ceramics), melting/softening of packings (mostly plastic), downcomers bowed or compressed, and compressor surges.

Many of the damage incidents took place during abnormal operation, such as start-up, shutdown, commissioning, and outages. During these operations, special caution is required to prevent water entry into a hot-oil tower, excessive base level, rapid pressuring or depressuring that can uplift trays, abrupt step-up of cold water into a steam-filled tower, downward pressuring on valve trays, and overheating of plastic packings.

CASE STUDY 22.1 SHORT TRAY HOLDDOWN CLIPS UNABLE TO RESIST A PRESSURE SURGE

Installation An atmospheric crude distillation tower. Right above the feed to the tower there was a packed wash bed. The sections above this bed contained sieve trays. The bottom tray was equipped with a draw sump from which heavy diesel was drawn on level control. The diesel draw was a total draw, with some of the diesel pumped back as reflux to the wash bed.

Problem During an upset, the tower experienced a pressure surge, possibly resulting from a high liquid level in the tower base. Following the surge, the diesel make dropped, the resid make increased, and the diesel cloud point decreased by 4°C, compared to previous operation.

Interpretation In a tower with side draws, such as a refinery fractionator, the symptom of less of the upper product, more of the lower product, and lighter upper product means that some of the upper product weeps or leaks past the draw point. If this follows a pressure surge, damage is implicated.

Cause When the tower was opened in the next turnaround, about two-thirds of the south side of the bottom tray was found to be lifted and had become lodged across the remainder of the tray, covering part of the weir to the draw-off pan (Fig. 22.1*a*).

Cause of Uplift Surprisingly, the uplifted section and the rest of the tray were undamaged, except for a slight upward bulging on the tray. It appeared that the tray holddown clips did not extend far enough onto the tray to properly secure the tray sections. During the pressure surge, these clips had not been able to resist the pressure surge, allowing the tray to lift. An example of the short holddown clips is in Figure 22.1*b*. Similar incidents have taken place in the past on other trays.

Cure The bulged sections were straightened and reinstalled using larger holddown clips (Fig. 22.1*c*). These larger holddown clips were found effective in resisting uplift due to pressure surges in the tower. The northern section of the same diesel draw tray was fastened by the larger clips during the pressure surge and was not damaged (the damage was confined to the southern side that was fastened by three small clips).

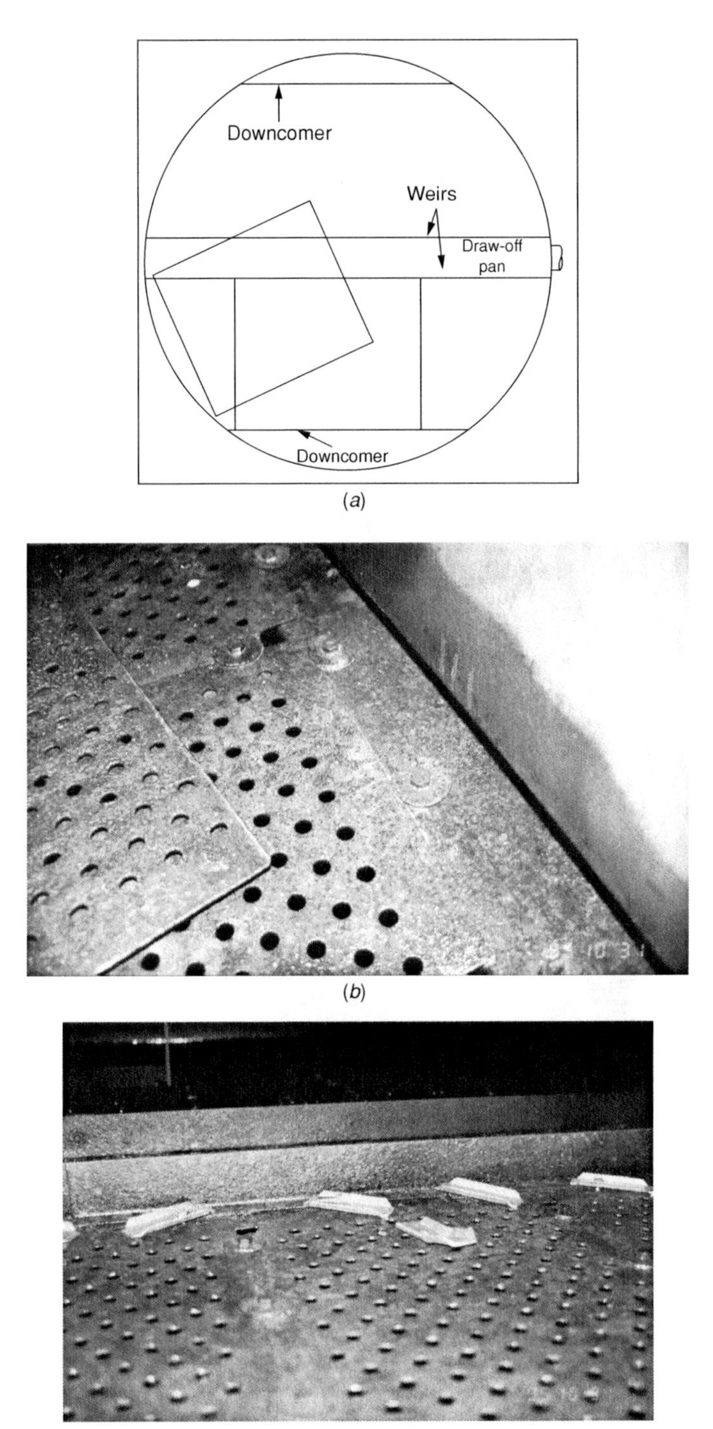

Figure 22.1 Crude tower tray damage incident: (*a*) uplifted and displaced panel; (*b*) examples of short holddown clips that were unable to resist pressure surge; (*c*) examples of larger tray holddown clips that effectively resisted pressure surges.

Moral Heavy-duty design can eliminate many problems in towers prone to pressure surges.

CASE STUDY 22.2 UPLIFTING OF POORLY FASTENED TRAYS

Tom C. Hower and Henry Z. Kister, reference 224. Reprinted with permission from *Hydrocarbon Processing*, by Gulf Publishing Co., all rights reserved

Installation A 6-ft-diameter absorber using acetone to absorb acetylene from a gas that contained mainly hydrogen, carbon monoxide, and carbon dioxide. The tower contained bubble-cap trays for many years. To increase capacity, the bubble-cap trays were replaced by sieve trays. The revamped tower worked well for a few months.

Problem A major plant upset occurred in which gas feed to the column went from full flow to no flow and back to full flow within about a minute. Following the upset the absorber stopped working. Overhead gas was the same as feed gas and bottom solvent was the same as the incoming solvent. The column was shut down, and the trays were found stacked at the bottom of the column.

Investigation This type of upset occurred from time to time in the plant but never caused dislodgment of the bubble-cap trays. The bubble-cap trays were supported as shown in Figures 22.2*a* and 22.2*b*. The tray panels were hooked down through the support ring, which prevented uplift.

Once the column had been in service, welding was no longer permitted inside for fear of heavy-metal acetylides being present. In particular, mercury and copper acetylides are extremely unstable, and it was known that there was at least some mercury (blown out from instruments) in the system. For this reason, no hooks were welded on to the sieve trays (Fig. 22.2*c*). During normal operation, the trays were held down by the weight of the tray and the weight of the liquid on the tray. This, however, was not sufficient for preventing uplift during a major upset.

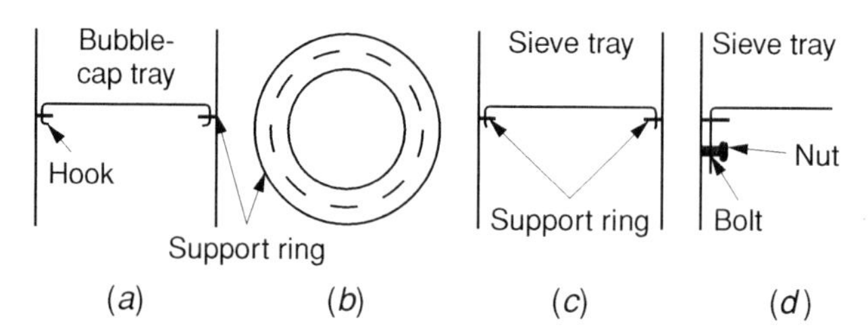

Figure 22.2 Support of bubble-cap and sieve trays in acetylene absorber: (*a*) support of bubble-cap trays, elevation; (*b*) support ring; (*c*) support of sieve trays; (*d*) modifications to sieve tray supports. (From Ref. 224. Reprinted with permission from *Hydrocarbon Processing* by Gulf Publishing Co. All rights reserved.)

Solution Holes were drilled in the vertical lip of the trays, and bolts were inserted in these holes all the way to the tower shell (Fig. 22.2*d*). A nut was fastened to each bolt. This provided two levels of protection against uplift: additional friction between the bolt and the wall and, more importantly, a restriction of upward movement by the nut/bolt assembly. The trays were never dislodged again.

CASE STUDY 22.3 PACKING COLLAPSE DUE TO QUENCHING AND RAPID BOILING

Installation A wastewater stripper that stripped organics and HC's out of the plant wastewater. The stripper used direct steam injection and contained two 20-ft-tall packed beds with 1-in. Pall rings fabricated from polyvinylidene fluoride (PVDF) plastic.

Experience During start-up, water feed to the tower was stepped up. Following this, there was a sudden drop in tower pressure drop to zero and a simultaneous sudden rise in bottom sump level. A gamma scan some time later showed that the bottom bed was gone. A calculation showed that the observed sump level increase was equal to the volume of material in the packing of the bottom bed. Amazingly, the tower continued to achieve good stripping after the bottom bed was gone.

PostMortem It is most likely that the stepping up of the cold feed caused rapid quenching, generating a local vacuum that induced rapid flow of steam toward the quench zone. This rapid flow caused the bottom packing support to collapse.

Related Experience A water quench tower in an olefins gas cracker used a cooled circulating quench water loop to cool hot reactor effluent gases. At one time, quench water circulation was lost. During the outage, the packing surfaces heated up. When the quench water circulation was reinstated, rapid boiling took place, leading to a pressure surge that uplifted about a third of the bed. Surprisingly, the tower continued to operate.

CASE STUDY 22.4 RAPID PRESSURE FALL AT START-UP

Process Description Feed to a 10-ft-ID, 20-tray column (Fig. 22.3*a*) was 35 wt % A, 45 wt % B, and 20 wt % heavies and entered at tray 10. Overhead liquid product from this tower was 60% A, 40% B, and little heavies, while the bottom contained 75% B, all the heavies, and little A. Product A is a much lower boiler, with an atmospheric boiling point of 47°F. The atmospheric boiling point of B is 270°F. The trays were single-pass sieve trays.

Start-Up Problem At start-up, the feed mixture was charged into the tower then the reboiler (vertical thermosiphon) was started up. Initially, only A boiled off. Upon further heating of the feed mixture, B started to boil over. As soon as it did, it condensed

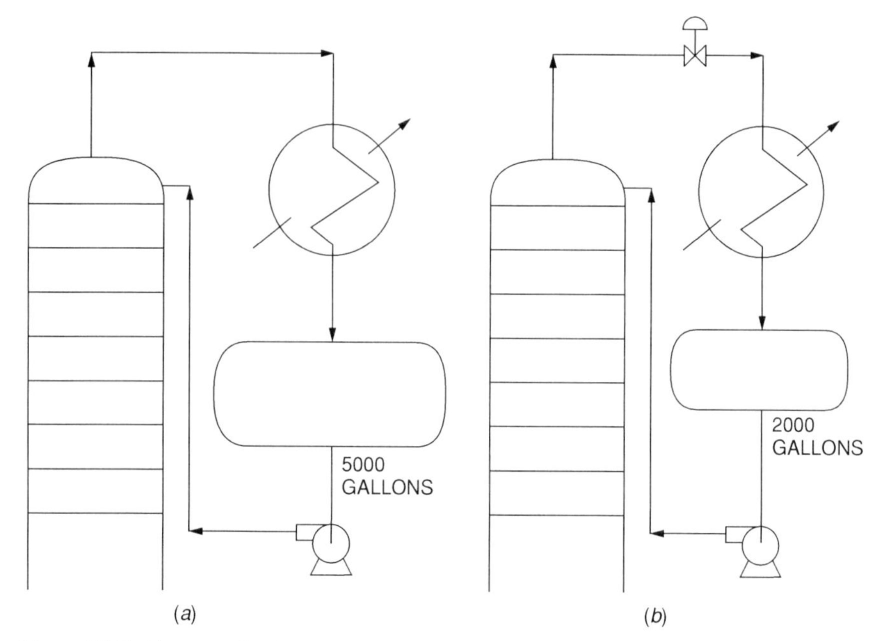

Figure 22.3 Tower overhead system that experienced rapid pressure fall at start-up and its modification: (*a*) initial; (*b*) modified.

easily and, upon condensation, absorbed much of the A in the overhead loop. In the words of operating personnel, "B absorbs A like a sponge." The tower pressure dropped "like a rock" from 22 to 15 psia within seconds. The drop in pressure sucked liquid from the trays into the overhead system. The 5000-gal reflux drum filled in 10 seconds. This caused repeated occurrences of tray damage, most of the time from the feed tray up. There had been instances where trays below the feed also got damaged. The damage direction was upward.

Solution The plant made two modifications to alleviate the problem (Fig. 22.3*b*):

(i) The 5000-gal reflux drum was replaced by a 2000-gal drum. This minimized the start-up absorption.

(ii) A control valve was installed in the overhead line from the tower to the condenser. This helped keep the pressure in the tower. This control valve had a limiter that did not permit it to close beyond 20% open, in order not to draw vacuum in the reflux drum.

CASE STUDY 22.5 TRAY UPLIFT DURING COMPRESSOR START-UP

Installation An olefins caustic wash tower located at an intermediate stage of the cracked gas compressor. The tower was equipped with sieve trays at 24 in. tray spacing, $\frac{1}{2}$ in. hole diameter, and hole area 8% of the active area.

Problem There were recurrent incidents of dislodging the bottom trays in the tower. Up to eight trays were dislodged. The trays were bowed upward before they were dislodged. Replacing the trays and adding stiffeners did not solve the problem.

Possible Causes One likely cause was a pressure surge when the cracked gas compressor was restarted after a trip or an outage. This pressure surge was most likely to occur during an opening of a check valve in the feed line to the tower. The sudden step-up in gas flow could have dislodged the trays. Another likely cause was a pressure surge due to liquid level in the tower base rising above the feed nozzle. In at least some of the incidents there was evidence of liquid level rising above the feed nozzle during the start-up. In one incident, plugging of the spent caustic line was the cause of the high level.

Solution The bottom section was retrayed with "heavy-duty" trays that have an uplift resistance of 1–2 psi. This mitigated the problem.

CASE STUDY 22.6 INTERNAL DAMAGE DURING HOOK-UP OF VACUUM EQUIPMENT

Contributed by Mark E. Harrison, Eastman Chemical, Kingsport, Tennessee

Loss of Vacuum Pump An interruption in the operation of a vacuum pump resulted in the partial loss of vacuum. Initially, the vacuum remained unchanged because the unit setting the system pressure is the condenser, not the vacuum pump. The vacuum pump (and pressure control valve) regulate the inerts in the condenser and thus the condenser effectiveness. A check valve prevented the loss of column vacuum due to vapor backflow from the vent system.

Over a period of time, the column boil-up continued to push the normal inert load into the condenser. Since the vacuum pump was not operational, these inerts could not be removed. They accumulated in the condenser and blanketed the tubes. The column pressure began to rise to the limit capped by liquid boiling temperatures and the temperature of the reboiler heating fluid. Over about 20 minutes, the column pressure rose roughly by 200–400 mm Hg.

Vacuum Pump Return to Service When returned to service, the vacuum pump that may have taken hours to evacuate the inert gases from the column at start-up almost instantly evacuated the few cubic feet of inert gas blanketing the condenser surface area, restoring process condensing, almost immediately regaining the vacuum. This sudden drop in column pressure, compounded by the column having served as a heat sink at the higher, upset pressure, produced an excessive boil-up surge. The surge scream/roar through the trays was heard loud and clear in the immediate area. Fortunately, no trays were damaged on this occasion, but this mechanism could easily damage tower internals.

Prevention This surge can be minimized by manually opening the vacuum control recycle stream before returning the vacuum pump to service and slowly ramping to the desired pressure. A caution note was added to the procedure to prevent recurrence.

CASE STUDY 22.7 VALVE POP-OUT: NUMEROUS EXPERIENCES

Valve trays often experience problems of valve floats "popping out" of their seats and finding their way into the bottom of the tower. Figure 22.4 provides some illustration. Here are some experiences with valve tray pop-out.

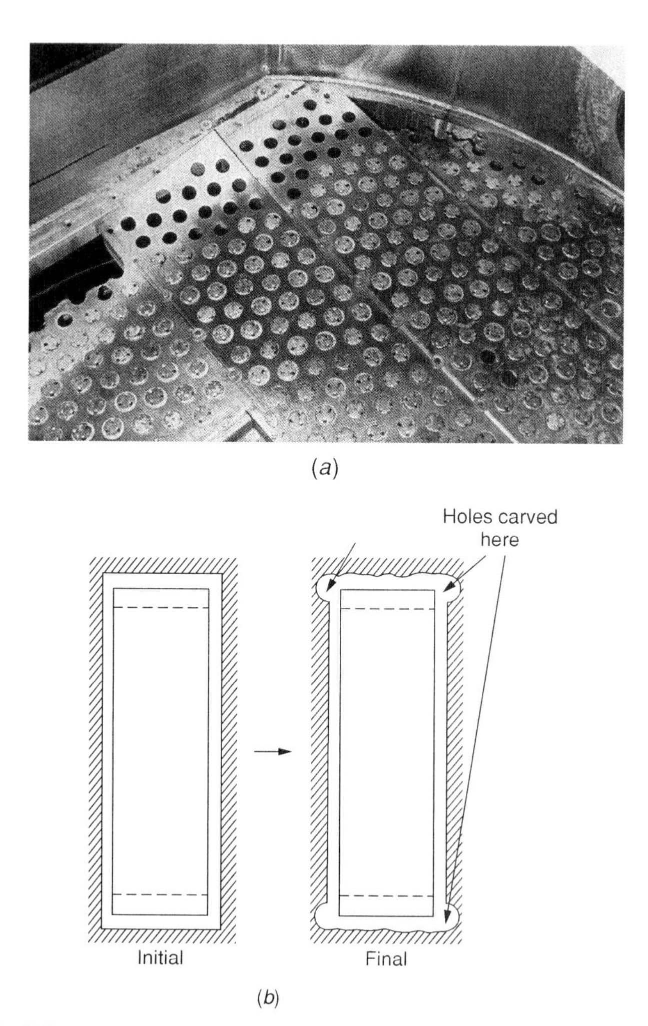

Figure 22.4 Valve pop-out: (*a*) corrosion had destroyed the valves on feed side of this tray; (*b*) pop-out of rectangular valves. [(*a*) Copyright Eastman Chemical Company. Used with permission.]

Tower A A large fraction of the round valve floats were found in the bottom of an ethylene oxide tower at the turnaround. The problem was not realized until the column was inspected. The tower worked well to the last minute. High-throughput operation was always sustained in the tower. Cause of the pop-out was corrosion of the valve legs.

The vendor could not supply new valves in time for the scheduled restart, so it was decided to make new legs in the workshop and attach them to the collected floats. These home-made valves lasted a very short time and were in the bottom of the tower at the next turnaround.

Tower B In one olefins debutanizer valves were lost from the trays. When replaced, the legs of the new valves were bent. Upon restart, the valves could not open, causing a capacity bottleneck. The symptoms were similar to plugged trays. That happened both in a debutanizer and in a depropanizer stripper in one plant. An identical experience happened in a refinery tower by a completely different company on a different continent.

Towers C and D A large fraction of round valves in two different chemical services popped out of their holes due to leg corrosion. There was little effect on performance. High-throughput operation was always sustained. The only adverse effect was the nuisance of having to reinsert valve floats into the holes at the turnaround.

Tower E About a quarter of the tower valve trays in chemical service lost about 80% of the valve floats without any noticeable change in performance. The tower was continuously operated at high rates.

Tower F In a tower making a high-purity chemical separation, a considerable fraction of the round valve floats popped out. The problem was noticed when the tower lost separation at turndown. The same trays with the popped-out valves worked fine at normal rates.

Tower G Monel valve trays near the top of a refinery crude fractionator badly corroded. The round valves corroded near the seats, with a large fraction of floats popping out. These valve floats traveled down and ended in pump suction, interfering with pump action. Replacing the corroded trays with new monel trays containing valves of larger openings eliminated the problem, at least for a few years.

Tower H Round valve floats popped out of the trays, found their way into the bottom outlet, and damaged the impeller of the bottom pump. To prevent recurrence, the bottom offtake was extended a few inches above the sump floor.

Tower I A chemical tower experienced severe corrosion at the bends of the round valve legs. As a result, the legs fell off and the valve floats popped out. The problem was mitigated by using long-radius bends at the legs. These reduced the stress concentration, making the bends less sensitive to corrosion attacks.

Tower J In one refinery fractionator, about 20% of the round, uncaged valves were found popped out every turnaround. This did not affect operation but was a maintenance nightmare every turnaround. The tower was retrayed with round, caged valve trays of heavy-gage construction. Following the retray, the pop-out was eliminated, with only a handful of cages throughout the entire fractionator found popped out in the next turnaround.

Tower K Pop-out of round valve floats occurred in several towers in one plant in corrosive chemical service. None of the towers experienced poor performance, but the floats jammed the reboiler inlet in one tower, forcing a premature shutdown, and damaged the bottom pump in another. In one tower, replacing by rectangular valves improved pop-out resistance.

Tower L Spinning of valves in the trays of a refinery crude fractionator caused wear on the tray holes, enlarging the holes and changing their shapes from round to elliptical. This resulted in valve floats popping out with their legs unbroken. Many popped-out floats ended in the pump suction.

Tower M Corrosion of the upper CS valve trays of a refinery crude fractionator resulted in most floats popping out. This did not affect performance. The trays worked fine.

Tower N Every turnaround, about 15% to 20% of the valves throughout a coker fractionator were found either displaced or stuck open due to corrosion.

Tower O Long rectangular floats (about 5 in.) popped out of valve holes in a tower in corrosive petrochemical service. The bottom bend of the legs corroded, and the floats popped out backward. The problem was solved by retraying with fixed valve trays.

Tower P Pop-out of long (5 in.) rectangular valve floats occurred in a tower in corrosive chemical service. The floats were heavy and chunky, and their legs pounded the tray floor, which led to carving of the holes harboring the floats (Fig. 22.4*b*). Once the holes became elongated, the valves popped out. Many floats popped out with unbroken legs.

Tower Q About 50% of the short rectangular (60-mm-long) valve floats popped out in a tower handling corrosive chemicals. These rectangular floats are only 5.5 mm wider than the harboring holes, so less than 3 mm on either side of the float overlaps with the deck metal. Once the edges of the floats thinned out due to corrosion, there was nothing to keep them in the holes, and they popped down.

CASE STUDY 22.8 VAPOR GAP DAMAGE

Installation A vacuum specialty chemicals column 10 ft ID equipped with valve trays. Melting point of the chemicals was about 60°F lower than operating temperature. Column had 30 trays, with feed entering 10 trays from the top.

History Tower had been in operation for 2 years. During these 2 years, there were many start-ups and shutdowns, and liquid sometimes filled the lower part of the tower.

Problem When the column was opened during one shutdown, the 16 lower valve trays had been distorted downward, with the main support beam bent and twisted into a V-shape about 1 in. deep. Solidification in several downcomers was observed.

Theory This problem is typical of a vapor gap damage (Fig. 22.5). A vapor gap was formed under the bottom tray, subjecting the bottom tray to the full hydrostatic head of the liquid above. The tray bent, allowing the liquid to drain, which moved the vapor gap to the tray above and so on.

The vapor gap was probably initiated by a blocked downcomer in the bottom section, with liquid accumulating on the trays above. This could have occurred by one of two mechanisms:

(i) The bottom downcomer was blocked, with liquid accumulating above it. When reboiler heating started, the vapor formed slugged through the trays, and a vapor gap was formed.

(ii) A blockage occurred anywhere in the bottom section. If the rate of drawing vacuum is of the same order as the rate of hydrostatic head buildup on the trays, liquid accumulates on several trays without causing damage. This continues until there is enough pressure difference to draw a vapor slug through the liquid column or until there is sufficient pressure difference downward to cause damage to the tray with the blocked downcomer.

Cure Draw vacuum and bring column to temperature prior to feed introduction at start-up. Alternatively, heat the feed as much as possible before introducing it to the column.

Related Experience One tower in plugging service accumulated liquid to the height of several trays during start-up. When this was recognized, the operators started the bottom pump, widely opening the bottom valve to quickly drain the liquid. It is believed that the bottom downcomer was plugged, so little liquid drained from the bottom valve tray. A vapor gap formed under the bottom tray, subjecting it to the full hydrostatic head of the liquid above. The tray bent, allowing the liquid to drain, which moved the vapor gap to the next tray and so on. The end result was bending down of six bottom valve trays.

CASE STUDY 22.9 LOSS OF VACUUM DAMAGES TRAYS

Contributed by Mark E. Harrison, Eastman Chemical, Kingsport, Tennessee

Installation A chemical tower about 4 ft ID equipped with bubble-cap trays.

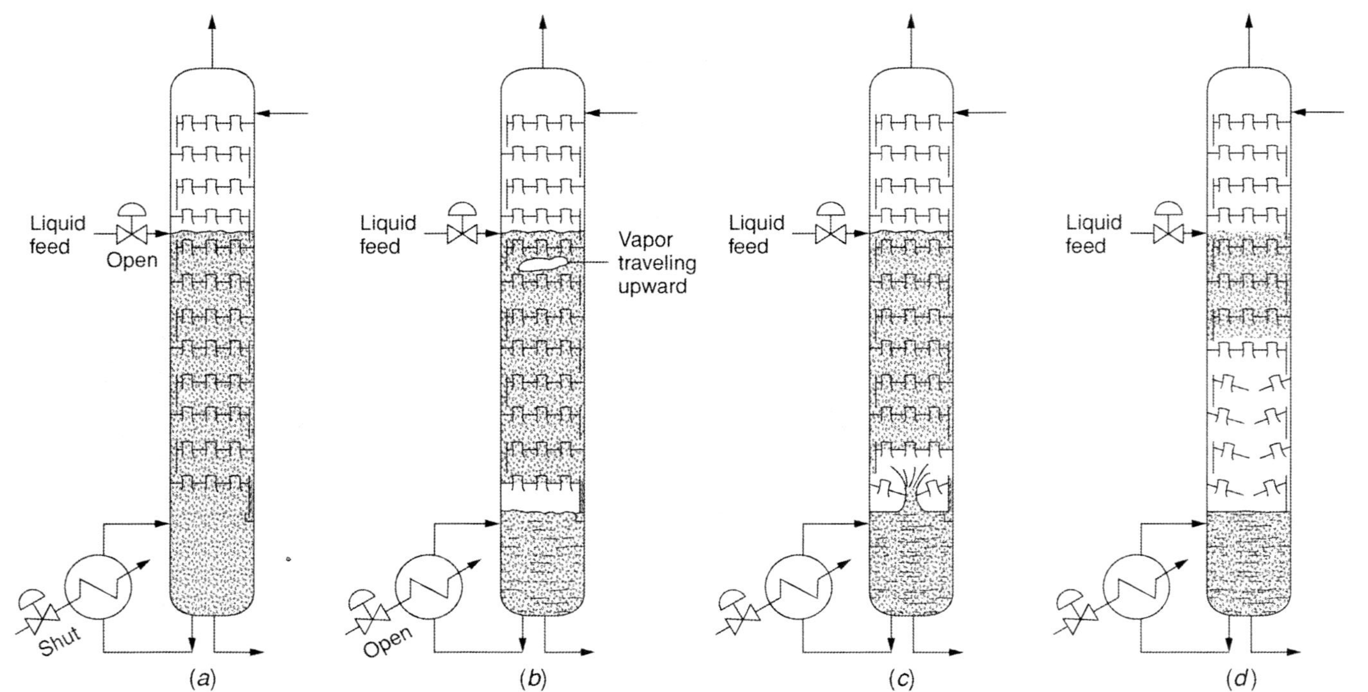

Figure 22.5 Typical vapor gap problem. (*a*) Lower part of column flooded (full of liquid). (*b*) Reboiler started. Liquid "drains" into reboiler and vaporizes. Vapor travels up. Liquid travels too slowly into sump. Vapor gap forms under bottom tray. (*c*) Bottom tray fails when attempting to support liquid column above it. When it fails, vapor gap shifts upward. Second tray now attempts to support liquid column. (*d*) End result. (Reprinted with permission from Ref. 250. Copyright © 1990 by McGraw-Hill.)

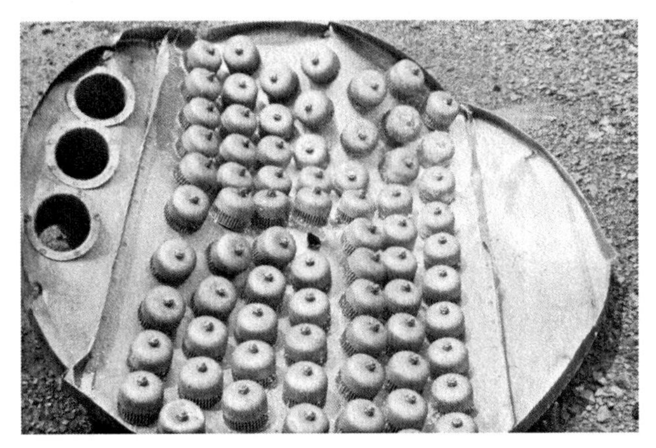

Figure 22.6 Downward vapor surge collapses bubble-cap trays. (Copyright © Eastman Chemical Company. Used with Permission.)

History A sudden loss of vacuum from the top of the tower generated vapor backflow through the upper trays. The bubble caps resisted the downward flow, as they are designed for upward vapor flow and not downward liquid flow. The liquid on the trays and in the downcomers further impeded the vapor downflow. This differential pressure downward across the trays damaged several trays (Fig. 22.6).

CASE STUDY 22.10 FOULING AND DAMAGE IN AN EXTRACTIVE DISTILLATION ALDEHYDE COLUMN

Contributed by Rian Reyneke, Sasol, Secunda, Gauteng, South Africa

Installation An extractive distillation column separating aldehydes as an overhead product from alcohols and ketones as the bottom product using water at the extractive agent (Fig. 22.7*a*). The column contained 60 two-pass sieve trays with a relatively small fractional hole area (about 6.7% of the bubbling area). Water entered at 50°C onto tray 58, and feed entered tray 42. The column operated at 100 kPag at the top.

Problem Tray damage was experienced on trays 58 down to 1. Trays were bent downward. Support ledges (not I-beams, Fig. 22.7*b*), right under the seal areas, were bent downward 70–100 mm deep. The damage did not vary greatly from tray to tray, although the damage was slightly worse around trays 58, 42, and some other locations further down.

Operating History After initial start-up, the tower worked well for 12 months. About that time the plant changed its their antifoam injection procedure. Initially, the plant mixed water with antifoam in a mixing drum. With this system, the plant experienced a problem of antifoam separating out in the mixing drum. To overcome, plant personnel decided to eliminate the water-mixing step and to directly dose the tower with antifoam. They installed a new pump designed to inject a much smaller amount of concentrated antifoam. Although they injected a much smaller amount,

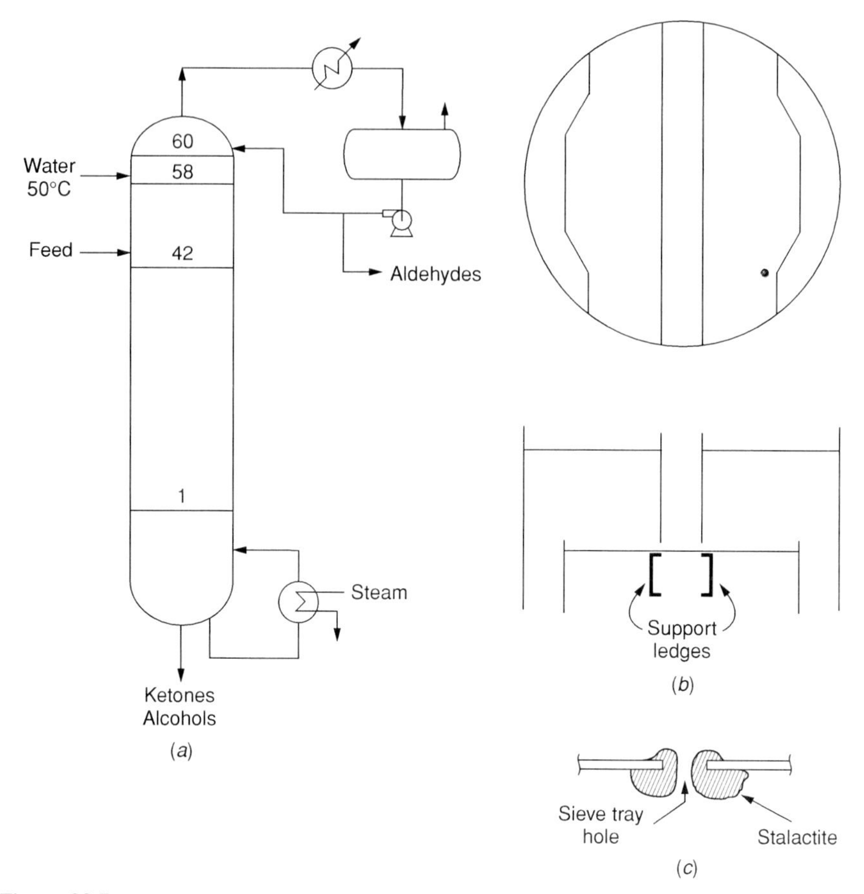

Figure 22.7 Extractive distillation aldehyde column that experienced damage and fouling: (*a*) tower schematic; (*b*) tray and support layout; (*c*) deposit formation on upper trays.

due to the high concentration they tended to overdose, possibly by about 10 times (compared to the needed antifoam dose).

Initially, column pressure drop was about 40 kPa over the 60 trays. Shortly after the changes to the antifoam injection, it went up about 20 kPa. Separation became worse but was still acceptable. One of the aldehydes started showing in the bottom, and the plant had to increase the water to ensure it is removed in the overhead.

A second, unrelated problem often experienced in the tower was steam emergencies. These occurred from time to time.

A third, also unrelated problem was that there were a few incidents when the bottom pump would trip, causing the bottom part of the column to fill with liquid, since the extraction water would continue running. By the time the pump would restart, several trays would be covered with liquid.

Turnaround Inspection Upon turnaround inspection, some of the upper trays appeared to have 10% to 20% of their holes blocked. The blockage was with material that looked like silicon cement. Also, the two trays above the feed appeared to be extensively blocked. The holes on these trays appeared blocked with a precipitate that grew underneath similar to stalactites (Fig. 22.7*c*). There was not much blockage below the feed. No blockages were observed in any downcomers.

Analysis The observation that the extent of damage was reasonably uniform throughout the column is evidence against a link between the plugging and the tray damage. The plugging was confined to the region above the feed and would explain the rise in dP and decline in separation experienced during the run, but not the mode of tray damage.

The observation that the damage was reasonably uniform also argues against vapor gap damage upon pump restart. Rapid pumping of liquid covering several trays is known to damage trays by a vapor gap mechanism (250, Case Study 22.8), but such damage is most severe near the tower bottom and diminishes as one ascends in the tower.

Theory It is most likely that the tray damage occurred during one of the steam emergencies or the column shutdown. The steam generates the boil-up that keeps up the tower pressure. When the steam supply is interrupted, the 50°C water would very rapidly quench the vapor remaining in the tower. To prevent vacuum, nitrogen is introduced as a vacuum relief into the overhead line. This nitrogen vacuum break line is small, $1\frac{1}{2}$–2 in., so it will take some time before it repressures the tower. Meanwhile, the rapid condensation generates a local vacuum, which sucks nitrogen and other non condensables from the overhead system. These gases exert a downward force on the trays as they move to pressure up the quenched regions. The downcomers at that time are still filled with liquid which is slow to drain, so the descending gases can only move via the small hole area. Descending tray and downcomer liquid would compete with the gas for the tray opening, further restricting vapor downward movement and increasing the downward force on the trays. The weeping tray liquid will promote quenching and will tend to render quenching uniform throughout the column, except for regions of higher vapor inventory such as near the feed and near the bottom tray. These, indeed, are the regions where the damage was slightly greater.

Recurrence Prevention Based on this theory, recurrence can be prevented by interrupting the water feed whenever the reboiler steam is interrupted.

CASE STUDY 22.11 TRAY DAMAGE BY GAS LIFTING OF REFLUX DRUM LIQUID

Contributed by Christo M. van den Heever, Mass and Heat Transfer Technology (Pty) Ltd., Johannesburg, South Africa

Installation Tower in petrochemical service (Fig. 22.8) was being precommissioned. The 2.2-m-ID top section contained 53 one-pass valve trays. The 2.9-m-ID bottom section had 7 two-pass valve trays. The overhead vapor line descended 35.7 m from the top of the tower to the overhead condenser. The condensate line entered the reflux drum below the liquid level, as shown in Figure 22.8. The tower pressure control valve was in the overhead line upstream of the condenser.

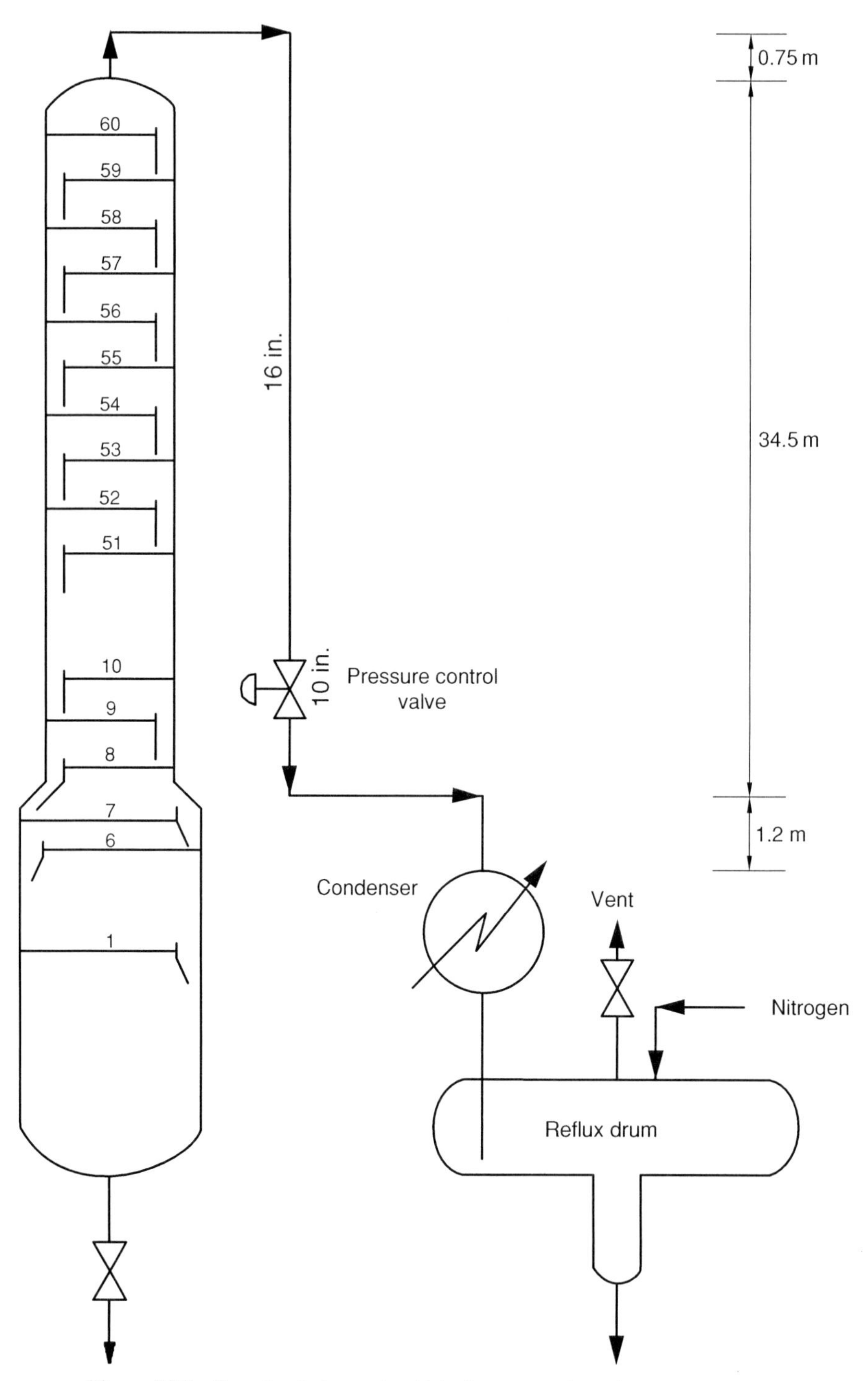

Figure 22.8 Petrochemical tower in which all trays experienced downward damage.

Initial Run The tower was initially operated on hexane and water only. The reflux drum had hexane with about 20 ppm of water. Tower pressure was 282 kPag at the bottom, 255 kPag at the top, and 280–290 kPag at the reflux drum. Column top and bottom temperatures were 117° and 128°C, respectively.

Steam Failure Upon steam failure, the column was shut down. About 12 hours later, the tower was found to be under a slight vacuum. Operating personnel tried to repressure the column via the reflux drum. At this point, the reflux drum was under slight positive pressure. Column top and bottom temperatures were 42° and 52°C, respectively.

The reflux drum was pressured to 160 kPag with nitrogen, then vented to the atmosphere. Only vapor was vented and no liquid was released via the vent pipe.

The reflux drum again was pressurized, this time to 250 kPag. As soon as this pressure was reached, there was a rumbling sound from inside the column from top to bottom. The sound was also described as "a domino-effect" from top to bottom. The reflux drum pressure dropped sharply to 0 kPag. The column bottom level rose from 26 to 89%. The column bottom temperature dropped from 52° to 46°C.

Over the next 30 min, the reflux drum pressure rose linearly from 0 to 50 kPag. The column top pressure followed after a 10-min lag, reaching 40 kPag after 30 min.

Restart The tower was restarted once steam supply resumed, but the operation appeared defective. The tower pressure drop did not exceed 10 kPa. Top and bottom temperatures appeared the same at 119°C. The plant was stopped and the tower opened for inspection.

Tray Damage All trays in the tower were damaged by a downward-acting force. Most of the damage was confined to the manway side of the trays. The top tray experienced the most damage and was virtually removed from the support ring. This tray was also severely tilted with the lowest portion on the manway side and the highest portion on the opposite side. Numerous downcomer floors have been ripped out of position while others were severely bent.

There was evidence supporting sequential progression of the damage from the top downward. On top of each tray's manway side, there was a series of two parallel scratch marks that were likely to have been made by the tray support ring clamps from the tray above. It appears that the manway side of the tray above collapsed first, then hit the tray below it with a downward force which made the scratch marks (the clamp has two vertical parallel legs at its bottom which is the most likely "etching pen").

Cause Tray pressure drop changed from 27 to 10 kPa and tower temperature difference changed from 10 to 0°C from the period before the steam failure to that after. This conclusively shows that the tray damage took place during the steam failure incident. The only conceivable mechanism during that period is associated with the pressuring up of the reflux drum. The following theory was postulated and matches all the facts and observations:

- When the reflux drum was pressured to 250 kPag, the liquid in the drum was forced up via the dip pipe and condenser into the overhead vapor line. Although the dip pipe had a small hole at its top to act as a siphon breaker, it was at least partially blocked off. When inspected afterward, the hole was partially blocked with only a 3-mm welding rod tip being able to pass through the hole. Such a small hole is much too small to equalize pressures.

- The reflux drum can hold up to 7.6 m^3 of liquid. The total overhead line and condensing system volume up to the reflux drum is 6.6 m^3. If the overhead line was full with liquid hexane, the static pressure of this liquid would be 270 kPa. If the column pressure was approximately –20 kPag, or if the liquid in the drum would not be enough to fill the entire overhead line, the pressure difference between the drum (at 250 kPag) and the tower would be sufficient to lift the reflux drum liquid all the way back to the top of the tower via the overhead pipe.
- When the liquid level in the reflux drum dropped to the bottom of the dip pipe, nitrogen entered the dip pipe and gas-lifted the liquid. As liquid was displaced from the overhead line, the static head declined and the quantity of nitrogen entering rose, accelerating the liquid remaining in the overhead line.
- The initial liquid transferred into the tower was at a volumetric rate equal to the nitrogen introduced into the reflux drum. A calculation showed that a rate of 580 m^3/h corresponds to a force equal to 4 tons of hexane liquid hitting the top tray at about 5 m/s.
- It is most likely that either the liquid "bullet" or the downward flow of nitrogen that rapidly followed would find the weakest section of tray and bend it downward. The rumbling sound heard was the sound of the impact as one tray hit the one below. This also explains the domino effect.
- The sudden increase in bottom level from 26 to 89% corresponds to a volumetric increase of 8.5 m^3, most of it being liquid originating from the reflux drum. When that bottom level rose, reflux drum pressure dropped to 0 kPag.

Moral Vacuum relief should be from the bottom up, not from the top down. Controlling tower pressure with a valve in the vapor line to the condenser can be troublesome during abnormal operation. If used, a hot-vapor bypass from the overhead line to the vapor space of the reflux drum should be added and would have acted to equalize pressures in this case. The hot-vapor bypass typically has a control valve that maintains a desired pressure difference between the tower overhead and the reflux drum.

CASE STUDY 22.12 TRAY DAMAGE AS A RESULT OF STEAMOUT FOLLOWED BY A WATER WASH

Contributed by Lars Kjellander, Perstorp Oxo AB, Stenungsund, Sweden

Installation In a chemical plant, two columns were arranged in series for an isomer separation (Fig. 22.9). The first column had 61 valve trays, the second contained six beds of structured packing, creating a total of approximately 100 theoretical trays. This separation required a high reflux rate, so the system was built for high liquid

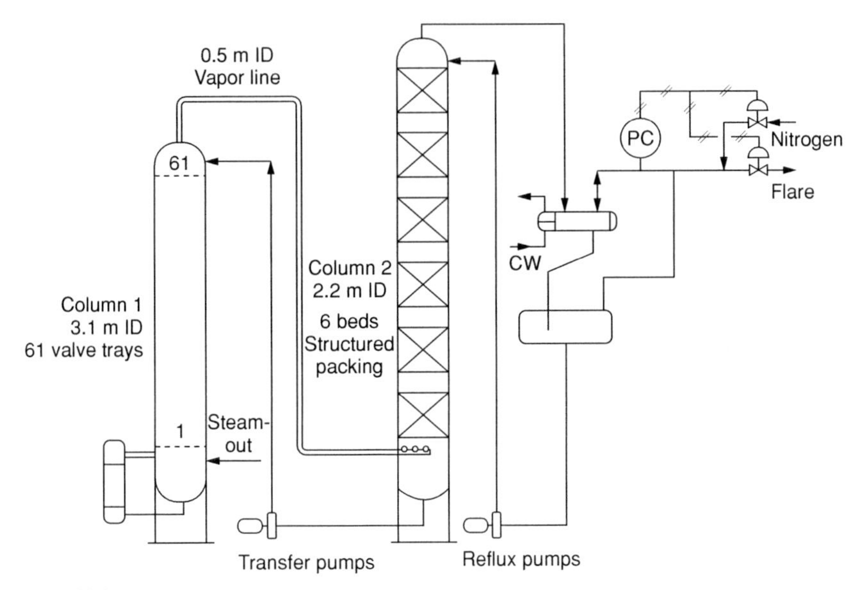

Figure 22.9 Two towers in series for isomer separation. Trays in column 1 experienced downward damage.

flows. In preparation for a maintenance turnaround, the columns were to be steamed and washed with water.

This Happened:

1. Steam was injected to the bottom of column 1 and both columns were steamed.
2. Cold water was pumped from the reflux drum into the packed column. The steam in this column condensed and was replaced with inert gas. The packed column became a large gas reservoir.
3. When the liquid level started to rise in the packed column, the bottom pumps were started and a large flow of water was pumped to the top tray of column 1.
4. The remaining steam in the trayed column now condensed as the water traveled downward, and inert gas rushed in via the vapor line. All 61 trays were more or less damaged, the trays and downcomers being pushed and bent downward by the inert gas trying to get down.

In this case the trays were not bolted the standard way, but were welded to the rings and bars, so they remained in place. All but one could be reused after a major job to straighten and flatten the trays.

CASE STUDY 22.13 RAPID CONDENSING AT FEED ZONE DAMAGES TRAYS

Installation A sour-water stripper equipped with moving-valve trays. Sour water fed to the tower was preheated by heat exchange with the tower bottoms.

Problem Trays were damaged during the run. Trays immediately above the feed were bent downward while those immediately below the feed were bent upward. The damage was greater on the trays above the feed.

Cause The damage was experienced when the feed/bottom interchanger was fouled, so the feed was highly subcooled. At one time, the feed flow rate was rapidly stepped up. This caused rapid condensation of steam and vapor at the feed region, generating a low-pressure region there. Vapor rushed from above and below toward the condensation zone. On trays above the feed, the downflow was impeded by closing of valve floats. Strong downward forces resulted and are believed to have bent down the trays immediately above the feed. Likewise, upward vapor rush toward the condensation zone is believed to have bent up the trays just below the feed. Below the feed, the upward force was somewhat relieved by the open valves, which explains the lesser damage.

CASE STUDY 22.14 PREVENTING WATER STRIPPER DAMAGE

Contributed by Marlo Rose, Irving Oil, Saint John, NB, Canada

Installation A refinery sour water stripper with single-pass moving valve trays, removing small quantities of hydrogen sulfide and ammonia out of waste process water. Reflux was condensed in a direct-contact heat transfer section at the top of the tower using an externally-cooled top pumparound loop. Overhead gas left the tower at 200°F.

Problem During its first three years of operation, the tower was shut down seven times due to severe fouling of the feed/effluent exchangers and the reboiler. During four of the seven startups, trays were dislodged.

Troubleshooting Following one unsuccessful start-up attempt, the causes for dislodging trays were investigated. A sour water stripper in another unit did not experience similar start-up problems due to differences in design, including sieve trays instead of moving valve trays.

Three possible causes were identified. First, the trays had low mechanical strength. They were attached to the support rings by butterfly clips that proved incapable of withstanding even minor start-up instabilities. Secondly, the column was commissioned using only a single component—firewater. Variations in pressure and temperature during start-up caused water to flash to steam or steam to collapse to water. This induced instability, local surges, and vacuum in sections of the column, which in turn exerted large forces on the trays and dislodged them. Thirdly, during the switch from firewater to sour water, the stripper bottom was routed from the sewer to a degasser until ammonia concentration came on spec. The degasser route bypassed the bottom level control valve, so the level needed to be controlled on a manual block valve. Both the switchover and the manual control destabilized the tower base, drew down the bottom level, pulling a vacuum, and dislodging trays.

Solution With the causes of start-up damage identified, a solution was formulated. First, the stripper bottom tray, which was most susceptible to damage, was strengthened by adding two stabilizer bars. These were bars perpendicular to the liquid flow, bolted to the support ring, to which the tray was bolted to in several spots. Schedule constraints precluded adding stabilizer bars to more trays. Secondly, nitrogen was introduced into the stripper along with firewater to create a multi-component mixture in the tower. This provided a means of breaking the vacuum in the column, keeping the column constantly pressurized, as well as dampening the surges and collapses caused by water/steam flashing and condensing. Thirdly, it was recognized that it was unnecessary to divert the sour water to the degasser during the introduction of sour water to the system. Continuous reboiling rendered the potential risk of high ammonia in the stripped water minimal, making it possible to divert the stripped water from sewer to an equalization pond before it came on-spec. The pond route allowed the stripped water to go through the level control valve, eliminating the stability issues associated with transitioning to manual block valve control.

Two other important changes were also made to the start-up procedure. A continuous supply of firewater was introduced to the reboiler inlet to ensure uninterrupted reboiling throughout the start-up. In past start-ups, the reboiler was commissioned after the top pumparound circulation was established. This caused difficulty in establishing boilup, and at time loss of the reboiler. Finally the start-up procedure was split into two parts: startup/circulation using firewater, and swinging unit from firewater to stripped water. Splitting the procedure reduced the risk of introducing sour water before the system was stabilized and steady, and made it clearer and easier to understand and follow.

Results The revised procedure was flawlessly executed at the next startup. There were no further damage incidents for the following year, which was the time of writing this case.

CASE STUDY 22.15 PREVENTING ANOTHER WATER STRIPPER DAMAGE

Contributed by Prakash Karpe, San Francisco, CA

Installation A 6 ft ID refinery sour water stripper with 43 single-pass fixed valve trays, removing small quantities of hydrogen sulfide and ammonia out of waste process water. Feed entered tray 7, tray 1 being the top tray. Reflux was condensed in a direct-contact heat transfer section at the top of the tower using an externally-cooled top pumparound loop. The heat transfer section consisted of the top 5 trays in the tower and a chimney tray (CT) which collected the liquid from the 5th tray and sent it to the pump. Overhead gas left the tower at 200°F.

History Initially it was impossible to hold a liquid level on the CT. The tower was shut down and the chimney tray was partly seal welded, partly gasketed. Upon restart, the tower worked well for two days, but then the CT started losing level again.

Figure 22.10 Downward-buckled top tray in sour water stripper.

A 10–12 gpm water makeup to the CT was initiated. Over the next week the water makeup needed to be gradually increased to avoid loss of level on the CT, reaching about 50 gpm after 7–10 days. The tower was again shut down. Many of the gaskets were found displaced, blown out and torn. An in-situ water test showed that the tray was leaking badly, easily accounting for the observed leak rates. The remainder of the gaskets were removed, and the CT was fully seal-welded. Re-testing showed no leakage.

When the tower was opened, tray damage was observed. The top tray was buckled downward (Figure 22.10) , with the buckled tray pulling the downcomer with it. The second tray was buckled upward. The feed tray, tray 7, was buckled downwards. The tray below the feed, tray 8, was only slightly damaged downwards.

Troubleshooting The causes for the trays buckling were investigated. A sour water stripper in another unit, which did not have a top pumparound, did not experience similar start-up problems.

It was noticed that the trays in the stripper did not have integral trusses, nor were they designed for "heavy duty". In this service, steam-water damage is not uncommon, and heavy-duty design can improve robustness. Even more significant, the tower was started up on industrial water (similar in quality to firewater), not on sour water, both at the pumparound and the feed. Correct design temperatures were maintained during the startup. However, as the pumparound draw tray started losing level, initially firewater was added to the suction of the pumparound pump to maintain level. The firewater was at 60°F, much colder than tower temperature. Any abrupt stepping up of the firewater is likely to rapidly condense the steam at the steam-water contact zone, creating a local vacuum, which can exert large forces on the trays and buckle them. From the damage sustained, it appears that the water/steam contact zone where local vacuum was generated upon water step-up was just below the top tray. The cold water probably got below the top tray by weeping, which is consistent with the relatively low reboiler steam flows during startup.

The downward damage at the feed tray suggests a different mechanism. Here both trays 7 and 8 were damaged downwards, suggesting possible rapid local flashing at the feed. This is likely to have taken place after the switchover to sour water feed, and could have been caused by a pocket of hydrocarbons in the feed that suddenly flashed upon contact with the tray liquid, which is essentially boiling water. The switchover is most fertile time for such an incident, because the water inside the tower is hydrocarbon-free when suddenly hit with gas-containing and possibly hydrocarbon-containing sour water. Also, the sour water feed was at about 200°F, compared to the leaking firewater at 60°F, which would be conducive to flashing upon switchover.

Prevention The startup procedure was modified to start the tower up on 200°F sour water rather than industrial water. In the pumparound loop, 220°F boiler feedwater was used instead of firewater. Higher steam rates were prescribed for startup to prevent tray weep. The tower overhead pressure control was kept open on manual during startup, giving the tower a chance to "breathe". Pumparound and feeds were to be ramped up and sudden step-ups avoided.

Results The revised procedure was flawlessly executed at the next startup. There were no loss of CT level nor damage incidents at that startup.

CASE STUDY 22.16 RETRAYING MITIGATES FLOW-INDUCED VIBRATIONS

Installation A 5-ft-ID solvent recovery tower. Feed at 180–185°F contained 80–90% water, the balance being organics, including a heavy water-soluble alcohol, MEK, and C_6–C_8 hydrocarbons. The tower recovered the organics as overhead and side products. The tower bottoms was water. The tower operated at a slight positive pressure. The tower contained 44 sieve trays at 12 in. spacing. The feed entered 18 trays above the bottom. Tray panels were parallel to liquid flow, as shown in Figure 11.1.

History The tower had its first turnaround after 6 months in service. During the 6 months it experienced some shutdowns. It had operated at about 40–50% of the design feed and reboiler steam rates.

During the run, the only operating problem experienced was periodic foaming that took place in the last half of the run. The foaming appears to have been caused by accumulation of a heavy alcohol just above the feed tray. The foam was seen in the sight glasses and raised tower dP from 1.5 to 5 psi. Opening the side draw above the feed tray and temporarily cutting back on steam allowed the heavy alcohol to escape and returned operation to normal.

Damage At the turnaround, cracking was observed at the corners of many trays. The cracking took place under the bolts that held pieces of trays together. More than 50% of the holddown clips were cracked. Pieces of tray fell out. Broken trays were found all through the tower, with the highest concentration just above the feed. The first 5–6 trays just above the feed were broken quite badly. The top 10 trays were in good shape—not much breakage there. Just below the feed and near the bottom, breakage was observed.

Evaluation A metallurgical evaluation determined that the breakage was caused by fatigue over many vibration cycles. The plant installed vibration-monitoring equipment that determined the vibration amplitude (in./s). The worst vibrations were measured just above the feed, with the region below the feed being next worst. The vibrations were strong enough to be felt simply by putting a hand on the tower wall. The region from the middle to the top of the tower showed the least vibrations.

Hydraulic Checks The trays were operating at very low vapor velocities. The C-factor C_B (ft/s) based on the bubbling area was very low at about 0.05 ft/s (design 0.1 ft/s), where

$$C_B = U_B \sqrt{\frac{\rho_G}{\rho_L - \rho_G}}$$

where U_B is the vapor velocity based on the bubbling area (ft/s), ρ is the density (lb/ft^3), and the subscripts G and L denote gas and liquid.

The hole F-factor F_H was also very low, 4–5 ft/s $(\text{lb/ft}^3)^{0.5}$, where

$$F_H = U_H \sqrt{\rho_G}$$

where U_H denotes hole velocity (ft/s).

The liquid loads on the trays were also low, of the order of 1–2 gpm/in. of outlet weir. The tray hole areas were 7–9% of the bubbling areas.

The low C-factors and hole F-factors are conducive to weeping and, indeed, much weeping was observed through the sight glasses. Brierley et al. (72) and Summers (474) determined that flow-induced vibrations are most severe under weeping conditions, well in line with the observations in this case.

Cure Brierley et al. (72) and Summers (474) advocate moving process conditions away from weeping as cure to a flow-induced vibration problem. This approach was implemented in this tower by replacing the sieve trays by valve trays.

Following the retray, the measured vibrations were far less. The sight glasses showed good tray action and much less weeping. There were no more incidents of tray damage due to flow-induced vibrations in this tower.

Related Experience Every turnaround, trays were found in the bottom of a 5-ft-ID vacuum stripper. Neither the trays nor the nuts and bolts showed any signs of damage. It appeared as if a gremlin inside the column undid all the nuts and bolts. The problem was most likely due to flow-induced vibrations. It was fully cured by using double-locking nuts.

Chapter 23

Reboilers That Did Not Work: Number 9 on the Top 10 Malfunctions

Reboilers are the most troublesome auxiliary in a distillation system and are rated ninth among distillation tower malfunctions (255). The number of case histories shows neither growth nor decline. Fewer reboiler malfunctions were reported for chemical towers compared to refineries and gas plants. This is because two of the more troublesome reboiler types, the once-through thermosiphon and the kettle reboiler, are less common in chemical plants.

Surprisingly, circulating thermosiphons, by far the most common type of reboiler, account for only about one-fifth of the troublesome case histories. This indicates quite a trouble-free performance, which has characterized this reboiler type. The malfunctions reported are varied. They include excessive circulation causing loss of heat transfer or tower flooding; insufficient ΔT and resulting pinches; surging due to presence of a small quantity of low boilers in the tower base; and others.

Even though kettle reboilers are far less common than thermosiphons, the number of kettle reboiler malfunctions reported exceeds that of circulating thermosiphons. Excess pressure drop in kettle reboiler circuits is the dominant malfunction (over 80% of the case histories), causing liquid to back up in the tower base beyond the reboiler return elevation. This high liquid level leads to premature flood and capacity loss. Kettle reboilers whose pressure drops are OK are seldom troublesome. The cure is to correctly compile force balances for these reboilers.

A very similar situation applies to once-through thermosiphons. The relatively high number of case histories for this very uncommon reboiler type signifies a very troublesome reboiler. With these reboilers, bottom-tray liquid is collected by a sump or draw pan, then flows through the reboiler into the tower base (Figure 23.3). The bottom product is reboiler effluent liquid collected at the tower base. Any liquid

Distillation Troubleshooting. By Henry Z. Kister

leaking or weeping from the sump or bottom tray shortcuts the reboiler into the tower base, which starves the reboiler of liquid. This leakage is the most common problem with the once-through thermosiphon reboilers, as evidenced by over 70% of the case histories. Common cure here is to improve the leak resistance of the draw tray, sometimes replacing it by a seal-welded chimney tray.

Internal reboilers and side reboilers have been troublesome, with a high number of reported case histories relative to their limited application in the industry. The main issues with side reboilers have been liquid draw and vapor return problems and inability to start. With internal reboilers, frothing due to boiling at the reboiler initiated flood or interfered with level controls.

With forced-circulation reboilers, only a few problems were reported, mainly issues with NPSH and the return lines. Finally, many cases concern the condensing side of reboilers heated by latent heat, where accumulation of noncondensables or problems with condensate draining are occasionally troublesome. No heating-side malfunctions were reported for the heating side of sensible-heated reboilers.

CASE STUDY 23.1 REBOILER SURGING

Installation A high-pressure tower was removing light ends from process chemicals. Tower bottom at about 180°C consisted of water-insoluble organic high boilers with a small fraction of water. The tower was equipped with a vertical thermosiphon reboiler.

Problem The tower experienced cycles of boil-up oscillations.

Mechanism Figure 23.1*a* shows operating charts of a typical reboiler surging incident (250). A similar sequence occurred here. The boiling point of the organics approached the heating medium temperature. Over a period of time or due to operation changes, the concentration of water in the tower base was depleted. The depletion of the low-boiling water raised the base temperature, causing the reboiler temperature difference to decline. The boiling was largely ceased. With little boil-up, lights-rich liquid from the trays dumped into the base. The lights-rich liquid restored the reboiler temperature difference, the reboiler began to boil again, and tower pressure surged. After stabilizing, the lights and water were slowly boiled off and depleted from the base and the cycle repeated.

Cure The reboiler draw-off was kept at the bottom of the tower base. The bottom draw-off was elevated, so the new bottom draw-off was always taken about a foot above the bottom (Fig. 23.1*b*), so that water accumulated below the bottom draw nozzle and went back to the reboiler. This ensured the reboiler always received some water. This eliminated the surging.

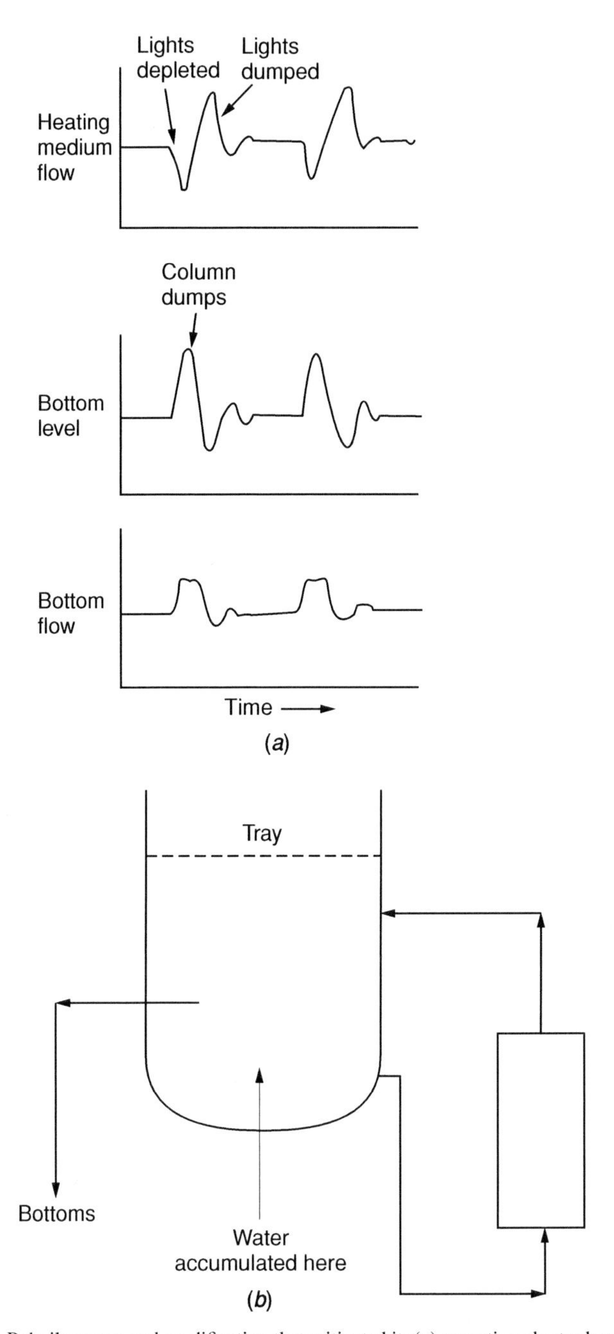

Figure 23.1 Reboiler surge and modification that mitigated it: (*a*) operating charts showing reboiler surge in another case study; (*b*) raising bottom draw-off to avoid depletion of water mitigated surge here. [(*a*) Reprinted with permission from Ref. 250. Copyright © 1990 by McGraw-Hill.]

CASE STUDY 23.2 SEPARATION OF TWO LIQUID PHASES IN A REBOILER

Tom C. Hower and Henry Z. Kister, reference 225. Reprinted with permission from *Hydrocarbon Processing*, by Gulf Publishing Co., all rights reserved

This case describes a troublesome experience of water settling in the base of the reboiler of a gasoline stripper.

Installation A small 17-in.-ID tower stripping light HCs from gasoline. Feed to the column contained a small amount of water.

History The original installation had a forced-circulation reboiler (Fig. 23.2*a*). To increase reboiler stripping capacity without modifying the bottom pump and piping, the reboiler was replaced by a slightly larger thermosiphon reboiler (Fig. 23.2*b*). At the same time, reboiler controls were upgraded.

Problem Every week or so, the new thermosiphon stopped working altogether, and vapor flow to the column was interrupted. The operator would then drain the bottom of the reboiler. The liquid drained was always water. The draining would continue until all the water was drained and gasoline started coming out. The operator would then shut the drain and return the column to normal operation. This problem never occurred with the old forced-circulation reboiler.

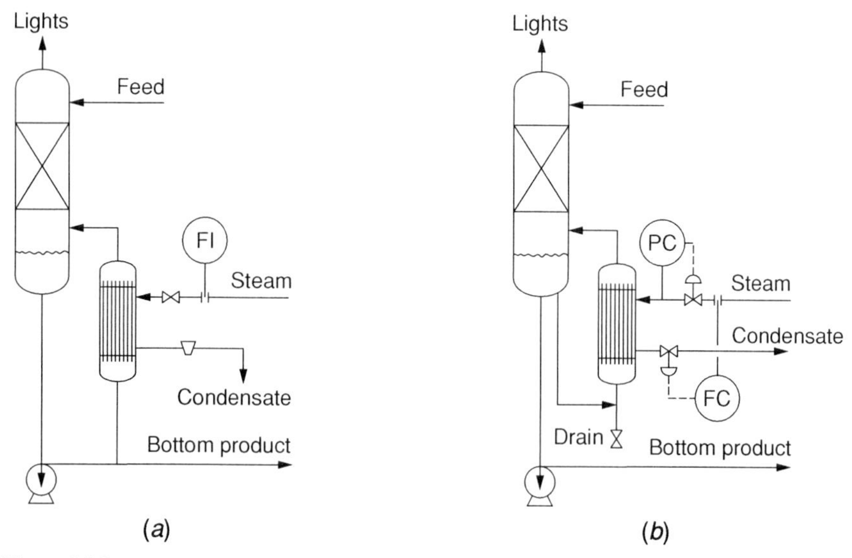

Figure 23.2 Water settling in gasoline stripper reboiler: (*a*) pre-revamp forced-circulation reboiler, no water settling; (*b*) post-revamp thermosiphon reboiler, water-settling experienced. (From Ref. 225. Reprinted with permission from *Hydrocarbon Processing* by Gulf Publishing Co. All rights reserved.)

Reboiler Due to a head restriction, the new reboiler had short (6-ft-long) tubes. To accommodate the heat transfer area requirement, the reboiler diameter was larger than the column diameter (19 in. ID). The reboiler base cross-sectional area was therefore 25% larger than the column cross section area, and the reboiler tube area was more than half the column cross section area. At maximum design conditions, the liquid velocity was about 200 feet per hour (fph) at the reboiler base and about 400 fph inside the reboiler tubes before vaporization began.

Theory The top of the tower was too cold to permit substantial quantities of water to escape in the overhead vapor. Therefore, water entering the stripper reached the bottom. Some water left the column with the bottom stream, the rest sought the lowest points, that is, the base of the reboiler. According to Lieberman's rules of thumb (304), a velocity of the order of 100 fph can be used for the design of gasoline–water gravity settlers when incomplete settling is satisfactory. The velocity at the reboiler base was only twice that, so it is conceivable that some settling occurred. The setting was assisted by the new control valve in the condensate outlet. This control valve flooded the lower part of the reboiler shell with condensate. No vaporization took place in the tubes submerged in the flooded region, and tube liquid velocity remained low in that region. This gave more residence time for settling.

As water accumulated over a period of time, more water would reach the reboiler tubes. Since water has a latent heat much greater than gasoline, fractional vaporization in the reboiler decreased, leading to a lower density difference for driving the thermosiphon. Also, accumulation of liquid water on the reboiler side acted to lessen the thermosiphon driving head. The lower density difference and lower driving head led to a lower circulation rate through the reboiler, which promoted more water settling. This further slowed circulation; eventually, thermosiphon action ceased altogether, and any remaining water settled. This problem never occurred with the forced-circulation reboiler because of the greater velocities experienced through its base and tubes.

Solution The solution would have been to revert to a forced-circulation reboiler. This was never implemented because the plant could live with the problem and the economic incentive for correcting it was relatively low.

Another much simpler solution would have been to install a bucket trap at the reboiler drain valve with a float set to work for oil and water separation. This solution was pointed out later by a person who had a related experience.

CASE STUDY 23.3 LEAKING DRAW TRAY MAKES ONCE-THROUGH REBOILER START-UP DIFFICULT

Installation A refinery stripper equipped with a once-through horizontal thermosiphon reboiler at ground level. The tower was equipped with venturi valve trays. Liquid to the reboiler was supplied from a draw pan at the bottom tray (Fig. 23.3).

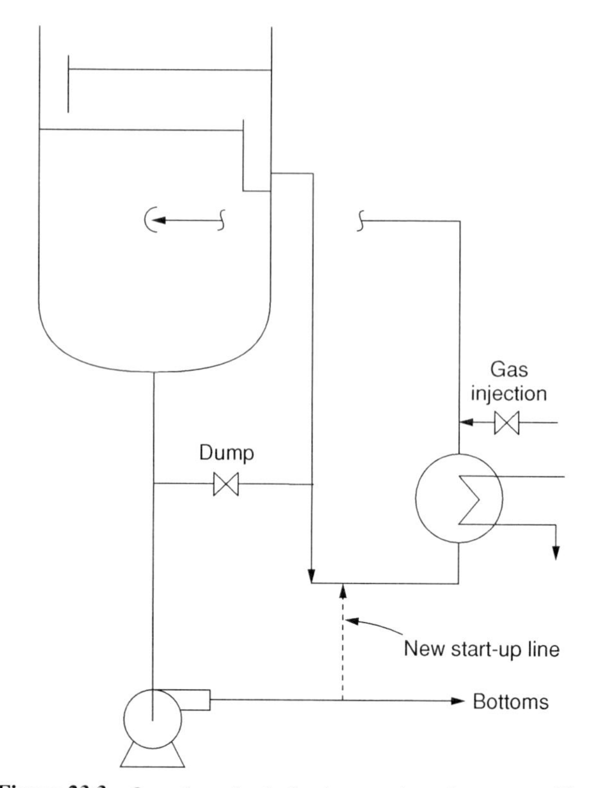

Figure 23.3 Once-through reboiler that experienced start-up problem.

Problem This reboiler was extremely difficult to start.

Cause Before reboiler start-up, there was no vapor to hold liquid on the trays. Bottom-tray liquid wept through the valves and most of it bypassed the draw sump. Supply of liquid to the reboiler was therefore minimal and insufficient to start thermosiphon action. With little vapor generated at the reboiler, there was little to resist the weep and divert liquid to the draw sump. Venturi valves weep much more than the standard, sharp-orifice valves, making it more difficult for the bottom tray to resist weeping and hold liquid during the start-up.

Solution In an attempt to initiate thermosiphon circulation, gas was injected downstream of the reboiler (Fig. 23.3). This was not successful, possibly because the liquid supply to the reboiler was too small. The gas injection was then discontinued and the dump line connecting the bottom sump with the reboiler liquid line (Fig. 23.3) was opened in an attempt to increase liquid supply to the reboiler. This was not successful either, this time possibly because of the low liquid head. No attempt was made to inject gas while the dump line was open. Instead, a new line was connected from the bottoms pump discharge into the reboiler inlet (Fig. 23.3). This gave the reboiler the needed liquid supply at an adequate head. The reboiler started thermosiphoning and generating vapor. The vapor resisted bottom-tray weep, forcing liquid to flow across the tray into the draw pan. The normal liquid supply route to the reboiler was thus established. The start-up line was no longer needed (until the next start-up) and was blocked in, and normal operation was established.

CASE STUDY 23.4 LIQUID-STARVED ONCE-THROUGH REBOILER

Installation Liquid to a once-through thermosiphon reboiler came from a draw pan located beneath the seal pan from the bottom downcomer. Draw pan overflow was 4 in. away from and at the same elevation as the seal pan overflow (Fig. 23.4*a*).

Problem Reboiler capacity was prematurely limited.

Cause Some liquid from the seal pan bypassed the opening of the draw pan and overflowed into the bottom sump.

Cure The overflow pan was extended to provide a wider (12-in. instead of 4-in.) opening for the entering liquid. The draw pan overflow weir was raised to an elevation 2 in. above the seal pan overflow (Fig. 23.4*b*). This eliminated the problem.

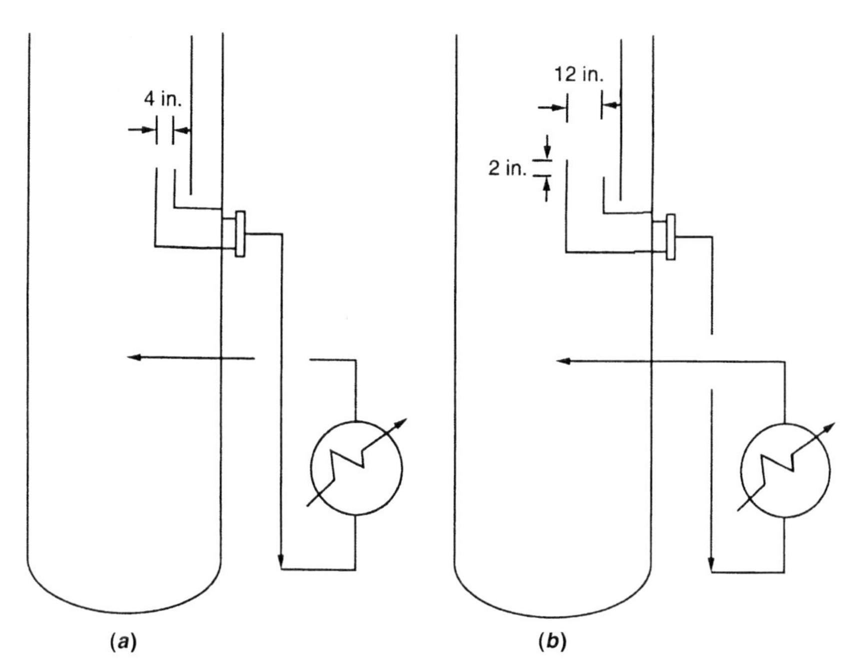

Figure 23.4 Liquid draw arrangements to a once-through thermosiphon reboiler: (*a*) reboiler starved of liquid; (*b*) reboiler not starved of liquid.

Related Experience In one refinery tower, tray panels were uplifted from the bottom tray, bypassing a large portion of the liquid around the once-through thermosiphon reboiler. To overcome, liquid level in the tower was raised to the bottom tray.

CASE STUDY 23.5 SURGING IN A EXTRACTIVE DISTILLATION REBOILER SYSTEM

Contributed by Petr Lenfeld, Koch-Glitsch s.r.o., Brno, Czech Republic

Installation An extractive distillation unit separated a monomer from a mixture of olefins. A secondary extractive distillation column C3 washed out acetylenes from the raw monomer (Fig. 23.5*a*). The column was fed with gaseous HCs from the discharge of a compressor. Extractive solvent entered several trays below the reflux return. Overhead product, the raw monomer, was reprocessed in the standard column train. Bottom product, solvent rich with acetylenes, fed downstream desorption and solvent regeneration columns. The column C3 was reboiled with two thermosiphon reboilers—main steam-heated reboiler R1, and economizer R2 heated with column bottoms. Both reboilers were once-through to prevent long thermal exposure and polymerization.

Problem The column C3 suffered pressure surges that always started about a month after the turnaround. The surges amplitude increased with time. All the surrounding equipment (feed compressor, downstream columns) were infected with the surges as well. The surges could be eliminated only by gradual reduction of column (and unit) feed rate, and this was necessary to keep product quality on specification. At a rate reduced to approximately 50% capacity, the unit was usually shut down. After the shutdown, steamout, and washing as well as cleaning of the R2 economizer, the column got back its original capacity for a month, and then the feed had to be gradually reduced again. The pressure surges caused plant operation in 3-month cycles for many years.

Investigation Close to the shutdown, process data were collected at "normal" operation with reduced feed and at "surge mode" induced by reflux rate increase.

The normal operation test showed nothing strange except a periodic pulsation of the liquid flow to the shell of the R2 reboiler. At the surge mode, both the column top and bottom pressures showed simultaneous peaks of the same amplitude at a period of 8 min. Reboiler R1 steam flow rate and steam pressure also showed the peaks, even though the steam control valve was stagnant. Although in auto, this valve was slow to react. A short time later, the bottom temperature fell steeply, opening the valve. Eventually, opening the valve restored the original bottom temperature, but this happened about half an hour later.

Then the unit was shut down and the column was inspected. The downcomer above the solvent feed was found plugged with polymer. Some trays were found with valves stuck closed.

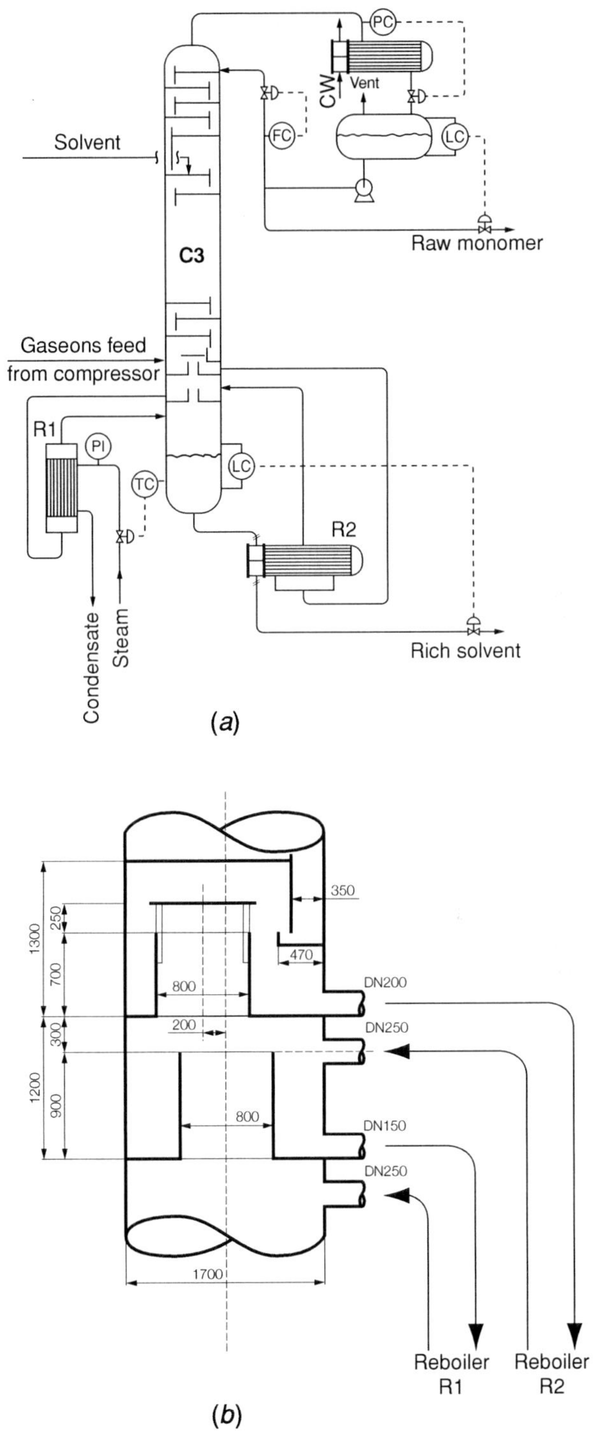

Figure 23.5 Extractive distillation tower showing two-reboiler arrangement: (*a*) tower and reboiler arrangement; (*b*) close-up sketch showing key dimensions in millimeters at two chimney trays supplying liquid to reboilers, roughly to scale.

Flood Theory The first suspicion was that the surges were caused by column flooding due to plugged downcomers and valves stuck closed. Accumulated operating experience, however, produced much evidence to deny the flood theory, including the following:

- Flooding means liquid accumulation on trays, usually accompanied by pressure drop increase. In this case, after the reflux rate increase, the pressure drop remained constant (although the top and bottom pressures rose) and the liquid reached the column bottom, as evidenced by the steep fall in the bottom temperature.
- Every shutdown, the column was inspected at the manholes, and if nothing strange was found, it was closed down again with no cleaning. Only economizer R2 was mechanically cleaned, simply because it was a horizontal thermosiphon and therefore easy to clean. Since the polymer in the downcomers was insoluble in water, simple column steamout and washing could not explain the performance improvement upon restart.
- Pressure surged simultaneously at both top and bottom. If there was any restriction in the column (flooding, stuck valves), the peaks would show amplitude reduction and time delay as restriction dampens surges.

The above arguments shifted the focus to the reboiler.

Reboiler Theory A new theory was formulated postulating that reboiler fouling was the root cause of the fluctuations. Both reboilers suffered from fouling and tube plugging, especially the economizer R2, where holes had to be cut in the head baffles to allow the column bottom product to pass through. The shell side of R2 was also dirty but did not plug. Gradual fouling slowly decreased the heat transfer rate in R2.

Chimney Tray Dumping The lower heat duty in R2 reduced the thermosiphon pumping action. The observed inlet flow fluctuations in the normal test could have resulted from slug flow in the reboiler outlet pipe. Because both reboilers are once-through, R1 was fed by R2 outlet liquid only. If R2 was plugged, the thermosiphon lift was insufficient, or R2 suffered from fluctuations, not all the liquid would pass through R2. The upper chimney tray then would start to overflow.

Each chimney tray had only one big chimney, 800 mm diameter, near the tray center (Fig. 23.5*b*). The vapor velocities through the chimneys were low, so any overflowing liquid would dump down the chimneys. Because there was no hat over the lower chimney, almost all the liquid overflowing the upper chimney would fall directly into the column bottom. The R1 reboiler would be starved of liquid and periodically dry out. Bundles cleaning at shutdowns restored the R2 heat duty sufficient to lift enough liquid to the R1 inlet, stopping the chimney overflow. This theory explained all the above observations, although its explanation of intensity of the oscillations remains somewhat unconvincing.

Lights Depletion This is likely to be a supplementary mechanism capable of explaining the intense fluctuations. It may also be used as a stand-alone reboiler theory, but this is less likely.

As is typical in extractive distillation, liquid leaving the trays consisted of high boilers with a small amount of low boilers. Upon fouling, R2 did not boil off enough lights, so they reached R1. This would cause an upsurge in the R1 duty, steam pressure, steam flow rate, and vapor flow up the tower. The vigorous boiling would increase the liquid offtake from the lower chimney tray, overstrip the lights off the lower trays, while the vapor surge could impede liquid descent down the chimney. This would retard boiling in R1 and would accelerate its dry-out. With the lights boiled off, the R1 vapor generation would dip. The column would dump, and lights-rich liquid would return and again accumulate at the lower chimney tray. The cycle would then repeat. This reboiler surging mechanism is described in detail in Case Study 23.1 and Ref. 250.

R2 Return The centerline of the R2 return nozzle was at the top elevation of the lower chimney (Fig. 23.5*b*), so some of the liquid returning in the two-phase mixture from R2 dumped down the lower chimney even under normal operation. This by itself could not account for the surging, since surging occurred only during the later part of the 3-month cycle, while the R2 return issue existed throughout the cycle. However, it was a contributor that accelerated the R1 dry-out and/or lights depletion above.

Solution A proposal to add a hat to the lower chimney tray was not implemented because of time pressure. Next shutdown both reboilers were retubed with new bundles because of heavy fouling. The surging disappeared. No shutdown was necessary for the following year. The problem is likely to reappear when the reboilers foul. The observation that reboiler retubing suppotred the reboiler theory and denied the flooding theory.

Morals

- Once-through reboilers are sensitive to hydraulics
- Hats over the chimneys of a chimney tray is a small investment that can promote trouble-free operation.

CASE STUDY 23.6 REBOILER FEED BLOCKAGE

Contributed by Matt Darwood, Tracerco, Billingham, Cleveland, United Kingdom

Installation A petrochemical splitter column equipped with a kettle reboiler.

Problem The pressure drop across the bottom section of the column was surging periodically. Temperatures around the reboilers were changing erratically.

Investigation A gamma scan confirmed that the trays were in their correct location. Froth heights and entrainment levels that the gamma scans showed on the bottom two trays were much higher than those on the trays above, even though the vapor and liquid traffic should have been the same. The liquid level in the base of the column was also higher than expected. There was a level indicator on the tower base, but it was not working properly and gave erratic results, so it was little help in the investigation.

Next, gamma scan time studies were used to pinpoint the onset and origin of flooding. A fixed gamma ray source and detector ("densitometer") was positioned at various locations: one below tray 1 (above the liquid in the base but below the vapor return), one between trays 1 and 2, and one between trays 2 and 3. The density of the material at these points was measured over a period of time. The results indicated that the liquid in the base rose and then decreased, causing a rise in the traffic on both trays 1 and 2.

The liquid in the base of the column was rising to a level approaching the vapor return line. The reboiler return vapor blew liquid from the base up the column.

Cause Pressure drops in the kettle inlet and outlet lines were checked and found to be adequate. This left line blockage as the only conceivable way the liquid in the base could be rising above the reboiler return.

A thorough investigation of the pipework leading into the reboiler was carried out using pipeline gamma scanning. High gamma-ray absorption indicates presence of high-density materials like solids inside the pipe. The scan confirmed the presence of deposits at the bottom of the pipe, approximately 10 cm before a 135° bend which fed the reboiler. Although the deposits did not totally block the pipe, they generated enough pressure drop to cause the liquid level to back up to the reboiler return elevation.

Cure The column was shut down and a new section of pipe was installed, clearing the blockage. A nucleonic level indicator was added at the column base. No more high-pressure drops and liquid surges occurred after this.

Lessons As well as investigating flood initiation, gamma scans can be useful in determining the location and extent of blockages within pipework.

CASE STUDY 23.7 THERMOSIPHON THAT WOULD NOT THERMOSIPHON

Henry Z. Kister, Tom C. Hower, Paulo R. de Melo Freitas, and Joaõ Nery, reference 276. Reproduced with permission.

With integrated heat recovery systems, start-up of a thermosiphon interreboiler can be challenging. This is particularly true if the temperature difference between the heating medium and the process fluid is small. This case shows why.

Installation A cryogenic gas plant demethanizer (Fig. 23.6) separated methane (top product) from ethane and heavier components (bottom product). The column contained three random-packed beds and had two interreboilers. The two interreboilers were contained in a single aluminum plate heat exchanger located at ground level. Both interreboilers were thermosiphon reboilers heated by the tower feed. The lower interreboiler boiled liquid collected from the middle packed bed, while the upper interreboiler boiled liquid collected from the top packed bed.

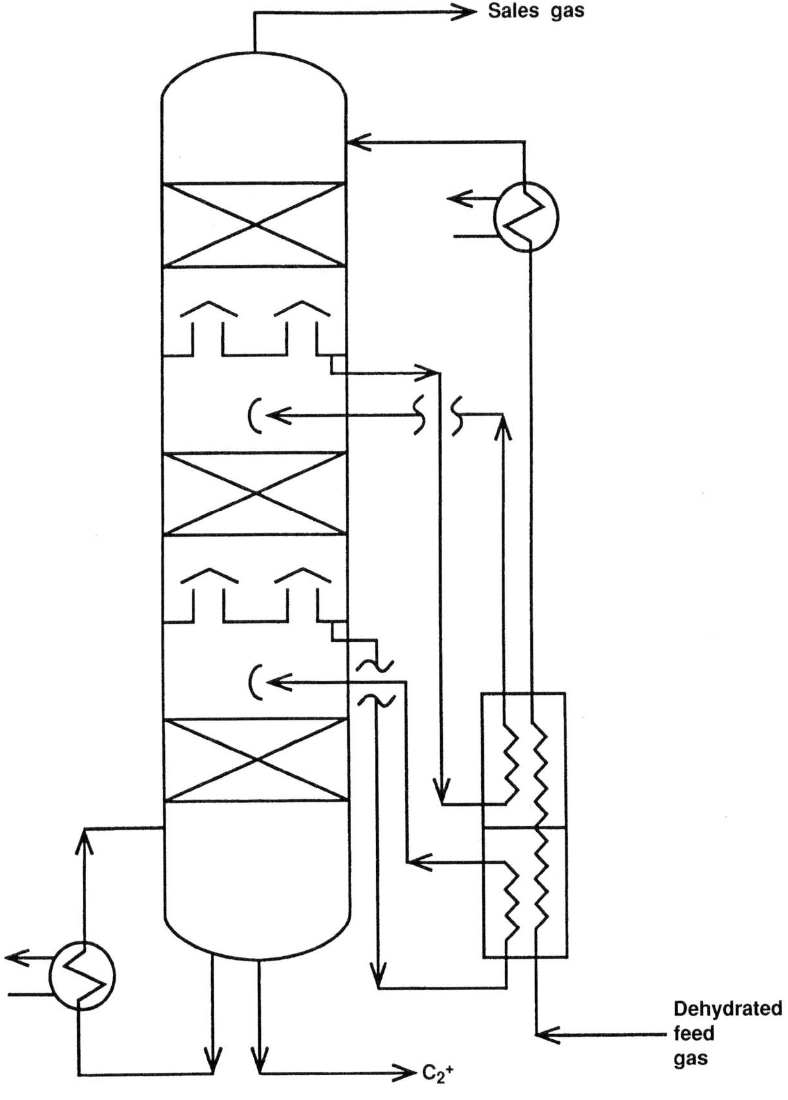

Figure 23.6 Cryogenic gas plant demethanizer with interreboilers. (From Ref. 276. Reproduced with permission. Copyright © AIChE. All rights reserved.)

Problem At start-up, the upper interreboiler would not thermosiphon. On the process side, inlet and outlet temperatures would remain the same. Similarly, on the heating side, inlet and outlet temperatures would remain the same. The symptom was similar to a plugged heat exchanger.

The problem was experienced only with the upper interreboiler. The lower interreboiler started without any problem.

Cause Upon shutdown, the liquid lines to the interreboilers as well as the vapor return lines fill up with stagnant liquid. When the column depressures, lights batch distill out of this liquid, leaving the heavy components behind. Atmospheric heat leakage augments this batch distillation process. The heavies are left in the interreboiler and lines.

Upon restart, the large liquid head in the lines from and to the interreboiler greatly suppresses the boiling point. The boiling point rises further due to the depletion of lights during the shutdown. The net result is a boiling point too high for the heating side to boil.

It is uncertain why only the upper interreboiler experienced this start-up difficulty. The smaller liquid head acting on the lower interreboiler, and therefore the lesser suppression of the boiling point, provides at least a partial explanation. Also, due to the superheat on the heating side of the lower interreboiler, there could be enough temperature difference to initiate boiling action. Finally, the stacking of some liquid head on the bottom chimney tray plus the shorter lines could have permitted enough fresh liquid to get to the lower interreboiler and initiate boiling.

Solution A $\frac{1}{4}$-in. tubing was added to the process line out of the upper interreboiler. The tubing introduced methane (sales gas) into the interreboiler outlet. The methane gas lifted the interreboiler liquid and initiated thermosiphon action.

We have seen similar case studies in a number of plants. In some, the gas was injected into the process line out of the interreboiler, in others into the process line entering the interreboiler. In some cases, bone-dry fuel gas, or even dehydrated feed gas, has been successfully used as alternatives to sales gas for initiating the gas lift.

An alternative method of solving the problem would have been to drain the reboiler liquid to the cold flare. This, however, was undesirable due to product loss. Further, the liquid consisted largely of components in the gasoline range which would have been difficult to vaporize in the cold blowdown drum.

CASE STUDY 23.8 ESTABLISHING THERMOSIPHON ACTION IN A DEMETHANIZER REBOILER

Henry Z. Kister and Tom C. Hower, reference 263. Reproduced with permission.

Installation Vertical thermosiphon reboiler on a gas plant demethanizer using column feed as the heating medium (Fig. 23.7). The column feed leaving the reboiler

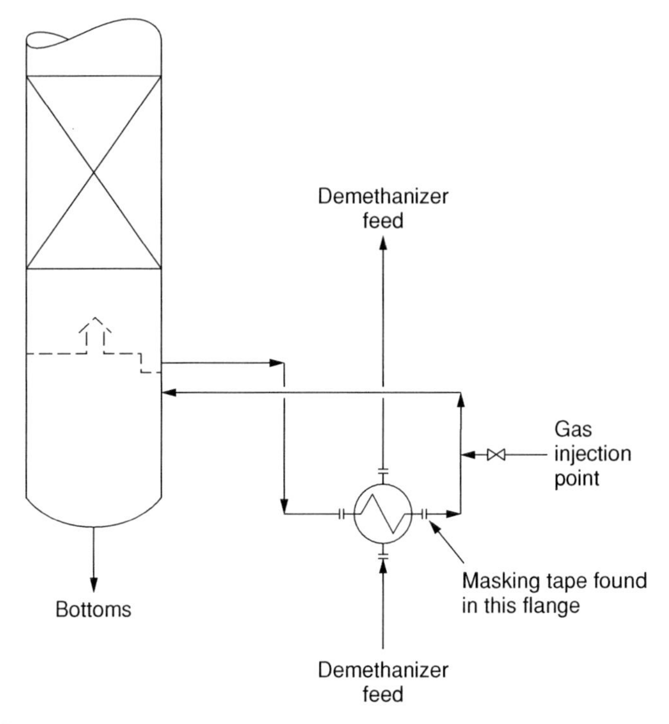

Figure 23.7 Demethanizer reboiler experiencing start-up problems. (From Ref. 263. Reproduced with permission. Copyright © (1987) AIChE. All rights reserved.)

flowed on to be used as the heating medium in a side reboiler. The column was at its initial operating period following plant commissioning.

Problem The start-up of the bottom reboiler tended to be erratic. At times, the reboiler would start thermosiphoning immediately without any problems. At other times, it was necessary to inject gas downstream of the reboiler, which would create a gas-lifting effect, and this would get the thermosiphon started. Yet at other times, even gas lifting would not work, and the thermosiphon could not be established. In almost all cases, once the thermosiphon was established, the column operation was normal, until the next time the column would shut down. The erratic behavior would then be repeated in the next start-up.

Cause In one of the subsequent shutdowns, the reboiler piping was pulled apart. A piece of masking tape used during construction as a gasket cover was found in the reboiler outlet flange. The gasket cover was split in the middle. The split was such that the gasket cover could either stick together or separate, causing the erratic behavior. It was believed that separation sometimes was effected by the rise in pressure following some vaporization at the reboiler and at other times by the relative vacuum pulled by the gas injection downstream of the flange.

Cure The masking tape was removed from the flange. Further use of masking tape as flange covers was banned. Only plastic flange covers were allowed from then on—these have to be removed during construction or no bolts can be installed at the flange.

CASE STUDY 23.9 FILM BOILING

In memory of J. A. (Polecat) Moore (Retired), Union Carbide

Installation Vertical thermosiphon using Dowtherm which entered at 725°F as the heating medium. The process liquid entered at 440°F and the two-phase mixture left the reboiler at 550°F (Fig. 23.8).

Problem Reboiler would only achieve a fraction of the design duty. The problem was caused by film boiling. Dowtherm entering from the bottom side of the reboiler with a temperature difference of about 300°F caused film boiling.

Solution The heating medium flow direction was reversed to avoid film boiling.

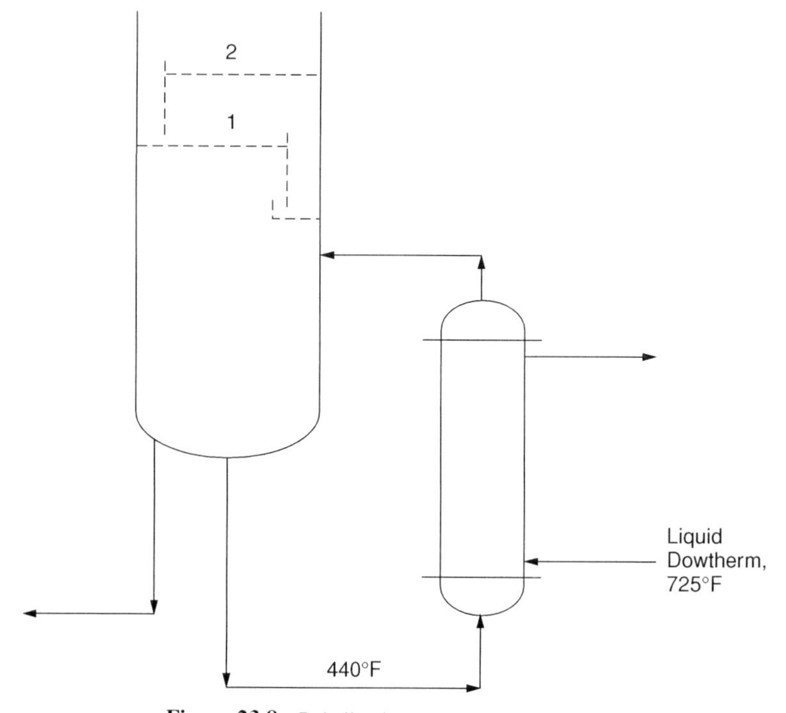

Figure 23.8 Reboiler that experienced film boiling.

CASE STUDY 23.10 LOSS OF CONDENSATE SEAL IN A DEMETHANIZER REBOILER

Henry Z. Kister and Tom C. Hower, reference 263. Reproduced with permission.

Installation Vertical thermosiphon reboiler on a demethanizer using refrigerant vapor as a heating medium. Condensate flow out of the reboiler was controlled by the control tray temperature (Fig. 23.9*a*).

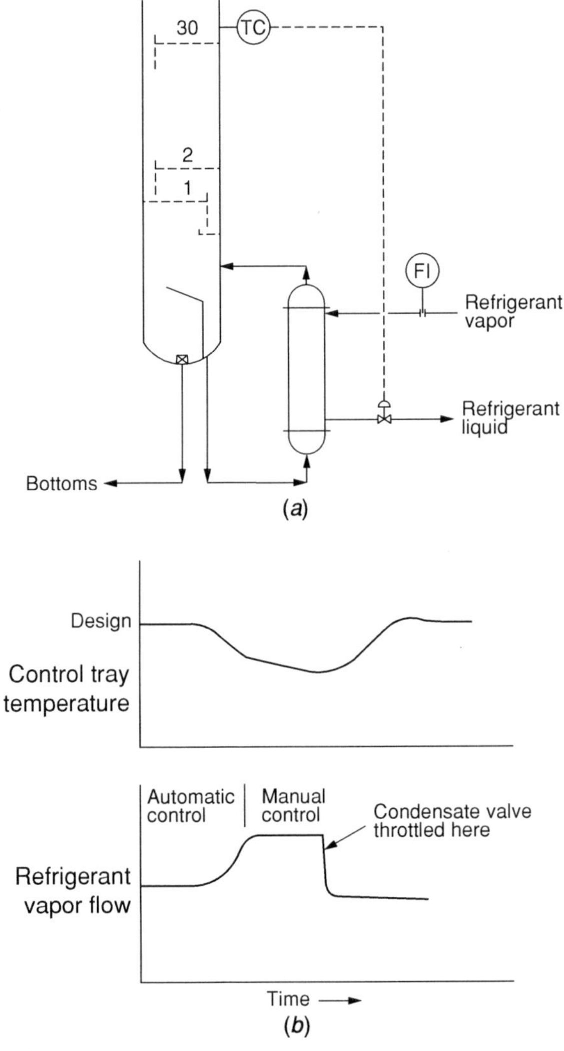

Figure 23.9 Loss of condensate seal in demethanizer reboiler: (*a*) demethanizer and reboiler arrangement; (*b*) operating charts during loss of seal incident. (From Ref. 263. Reproduced with permission. Copyright © (1987) AIChE. All rights reserved.)

Problem At the time, the column operated at low rates. The bottom purity was critical, while the overhead purity was less important, so the column was being over-reboiled. Despite this, bottom product went off specification.

Figure 23.9*b* shows the operating charts. Column control tray temperature slightly dropped, and the controller called for more reboil. As reboil rate was raised, the control temperature continued dropping. This continued until the operator placed the controller on manual. The temperature continued to drop, but at a slower rate. The low control temperature was accompanied by a large increase in methane in the bottom stream.

Analysis The problem was caused by losing the condensate seal. When a control valve is located at the outlet of the reboiler (Fig. 23.9*a*), a liquid level is held in the reboiler shell, which covers a portion of the tubes. This level also ensures the reboiler is liquid sealed, so that no vapor escapes with the condensate. When the rate of flow through the outlet valve exceeds the rate at which vapor condenses in the reboiler, the liquid level, and therefore the liquid seal, may be lost. When this occurs, a further increase in vapor flow rate increases pressure drop in the reboiler inlet lines, reducing the reboiler condensation pressure. This in turn reduces reboiler ΔT and in low-ΔT services (such as the demethanizer reboiler) significantly lowers heat transfer.

Solution The condensate valve was heavily throttled to reestablish the condensate seal (Fig. 23.9*b*). When the seal was reestablished, column operation returned to normal.

Avoiding Recurrence Another plant experiencing a similar problem monitored the condensate temperature. As long as a liquid level is held in the shell, the condensate is subcooled. A rise in the condensate temperature forecasts an imminent loss of the reboiler seal, and the operator can take timely action.

CASE STUDY 23.11 PREVENTING LOSS OF CONDENSATE SEAL

Installation A vertical, steam-heated thermosiphon reboiler. Condensate flow out of the reboiler was manipulated by the tower temperature control. The control system is identical to that in Figure 23.9*a*, except that the heating medium was steam.

Problem When the control valve opened too fast or too far, the liquid condensate seal in the steam chest would often totally drain. Steam would then pass into the condensate system accompanied by loss of heat transfer and hammering. The problem is identical to that in Case Study 23.10.

Solution A level transmitter was installed to monitor the condensate level in the steam chest. A low-level override was added to keep a minimum condensate level in the steam chest (Fig. 23.10). This prevented loss of the condensate seal and breakthrough of steam into the condensate system.

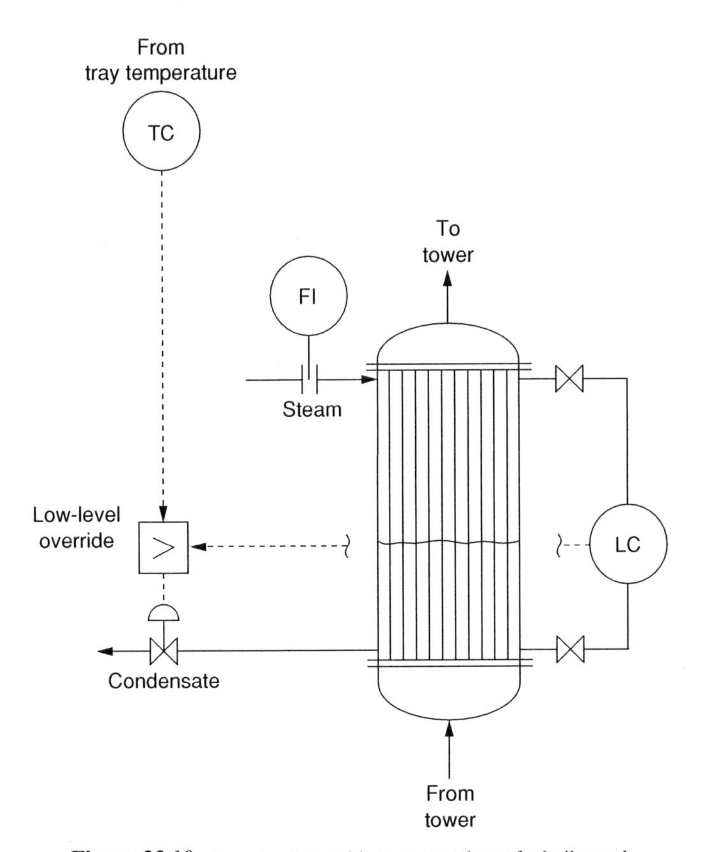

Figure 23.10 Low-level override to prevent loss of reboiler seal.

This solution introduced a new, albeit lesser, problem. The system experienced some back-and-forth switches between the override and the tower temperature controller. When the temperature controller opened the valve too far, the condensate level fell to the minimum, and the override control took over. Once the override took over, the condensate level would rebuild. Once the level increased, the override switched off. Once the override went out, the temperature controller tried to open the valve again, and so on. An operator action was often needed to break the chain, but the upsets produced were far less than the upsets previously produced by losing the condensate seal.

CASE STUDY 23.12 INABILITY TO REMOVE CONDENSATE FROM REBOILER

Installation Bottom product of a light HC tower at 180°F was reboiled by 40-psig steam in a vertical thermosiphon reboiler. Condensate pressure was at 25 psig. Reboiler control valve was in the steam line to the reboiler.

Problem Reboiler limitation and instability due to inability to drain condensate.

Cause The pressure at the reboiler was the condensate header pressure (25 psig) plus about 10 psi of static head to get the reboiler condensate into the condensate header. With steam supply at 40 psig, the margin for control was small, and the valve often went off control. The condensate tended to accumulate in the reboiler and get highly subcooled. The condensate header mixed variable amounts of subcooled condensate with flashing condensate, which in turn caused pressure fluctuations there that were directly transmitted to the condensing side of the reboiler.

Solution During unstable periods, the condensate was routed to the deck.

Footnote Condensate draining to the deck needs to be avoided when there is a risk of the hot condensate causing vaporization of hazardous materials on the sewer system. See Case Study 1.3.

Chapter 24

Condensers That Did Not Work

Condenser malfunctions are in the 15th place among distillation malfunctions, evenly split between chemical, refinery, and olefins/gas towers (255). There appears to be a slight decline in condenser malfunctions. Two major headaches with condensers, namely condenser fouling and corrosion, have been excluded, being primarily functions of the system, impurities, and metallurgy. Fouling and corrosion cases were included only if induced or enhanced by a process, equipment, or operational issue.

A famous statement by Smith (471) is that to troubleshoot a condenser one needs to ask three questions: "Is it clean? Is it vented? Is it drained?" The survey (255) verifies that, indeed, once fouling is excluded, inadequate venting and inadequate condensate removal constitute over half of the reported condenser case histories. Other issues do not get close. Cures are adequately sized, adequately located, and adequately piped vents and drains.

Other issues may be important in specific situations. These include flooding in or entrainment from partial condensers, especially knockback condensers; an unexpected heat curve resulting from Rayleigh condensation in wide-boiling mixtures or in the presence of a second liquid phase; maldistribution between parallel condensers; some condenser hardware issues; and interaction with vacuum or recompression systems.

CASE STUDY 24.1 PRESSURE AND LEVEL SURGING

Henry Z. Kister, Rusty Rhoad, and Kimberley Hoyt, reference 273.

Installation A chemical vacuum distillation column (Fig. 24.1) separated a liquid feed into three product streams. The lights liquid product contains the light key (LK) and lighter components. The heavy-ends bottom-liquid product contained the heavy

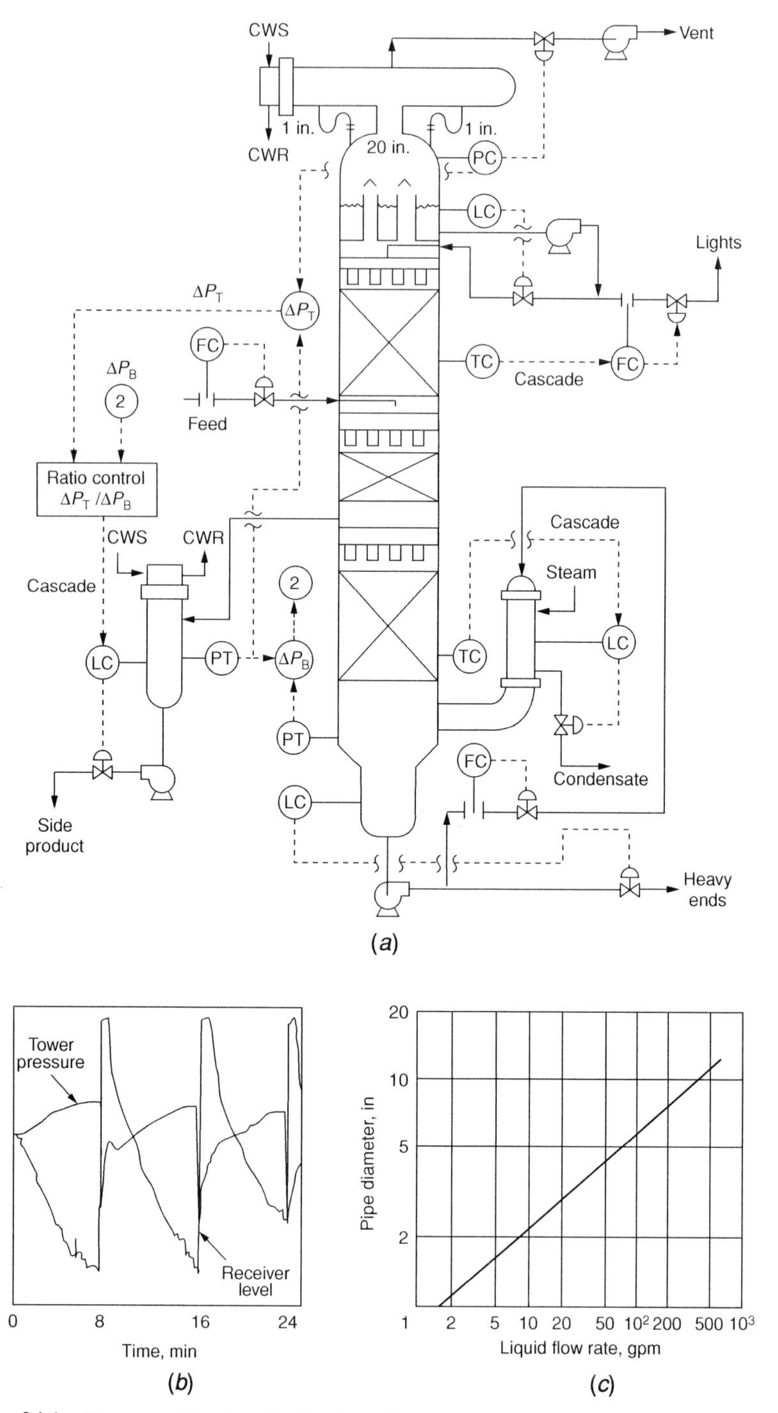

Figure 24.1 Pressure and level surging in chemical vacuum tower: (*a*) schematic of tower and controls; (*b*) operating chart showing surging of tower pressure and reflux receiver level; (*c*) highly recommended correlation for self-venting flow. [(*a*, *b*) From Ref. 273. Reproduced with permission. Copyright © (1996) AIChE. All rights reserved. (*c*) From Ref. 438. Reprinted courtesy of the Institution of Chemical Engineers, United Kingdom.]

key (HK) and heavier components. Intermediate boilers, such as the intermediate key (IK), were drawn as a vapor side product from the stripping section. There was a specification of 0.3% maximum IK in the bottoms and 1.0% maximum HK in the side product.

Pressure at the top of the column was 60 mm Hg absolute. To minimize pressure drop, the 20-in.-ID overhead line was short, with the overhead total condenser mounted directly above the tower (Fig. 24.1*a*). Inerts were vented to the vacuum pump. Reflux and distillate generated in the condenser drained back into the reflux receiver. This receiver was a chimney tray mounted internally at the top of the column. Reflux was pumped from the reflux receiver into the packing liquid distributor. Column pressure was regulated by a control valve in the vent line to the vacuum pump. Distillate flow was adjusted by a cascade control from the temperature in the top packed bed. Reflux was regulated by the receiver level.

The vapor side draw was condensed in the side-draw condenser. The flow of side product was regulated by a level controller on the condensing side of the side condenser. The set point on this level controller was adjusted by the column differential-pressure-ratio controller. This novel control system is discussed in Case Study 26.8.

Boil-up was supplied by a steam-heated falling-film evaporator. The reboiler heat duty was regulated by a level controller on the condensate. The set point on this level controller was adjusted by a temperature controller at the bottom of the packing.

The column contained three beds of high-efficiency structured packing. Liquid from the upper bed was collected and then mixed with the feed in the distributor that supplied liquid to the middle bed. Liquid from the middle bed was collected and then flowed to a distributor that supplied liquid to the bottom bed.

The design shown in Figure 24.1*a* and described above appears specialized—much too extreme for a column that always runs at 60 mm Hg. However, the column operated in campaigns, and during some runs the pressure was 5–10 mm Hg. Thus, this specialized design was mandated to provide the necessary flexibility to handle the various campaigns.

Problem Initially, violent surging was experienced in the tower. The surging was far too violent to permit adequate fractionation.

At feed rates exceeding 50% of design, both the column pressure and the reflux-receiver level periodically surged. Figure 24.1*b* shows a typical operating chart. Column pressure gradually rose while receiver level gradually declined over an 8-min interval. Then, suddenly, the pressure dived while the level surged almost instantaneously. The cycle then repeated.

Analysis Violent surging in a column system often points to the presence of liquid where it is unexpected. With this in mind, the condenser drainage was examined. The condenser generated a total of 20 gpm of reflux plus distillate. There were two potential routes for the condensate to drain: via the 20-in. overhead line or the two 1-in. drain lines.

The column was designed to drain primarily through the 20-in. overhead line, while the 1-in. drain lines were installed for supplemental drainage of liquid held up behind the baffles in the condenser.

As Figure 24.1*a* shows, each 1-in. line had a high and a low point. The low point was needed as a seal loop preventing vapor rise up the line. The purpose of the high point presumably was to elevate the seal loop above the tower entry nozzle. The nozzle was flush with an access platform, so lowering the seal loop would have required cutting a hole in the platform.

Liquid will drain via the 20-in. overhead line if the vapor velocity is low enough to permit liquid to descend. This line can be modeled as a short wetted-wall column. The point at which the vapor velocity becomes too high to allow the liquid to descend therefore coincides with the flood point of the wetted-wall column. The flood point of wetted-wall columns can be calculated reliably from the correlation of Diehl and Koppany (104), as has been recommended by several authors (250, 393, 471). The calculation showed that, at design rates, the vapor velocity in the 20-in. line was 160% of the flood velocity of a wetted-wall column. This means that the vapor velocity was far too high to permit any liquid drainage via that vapor line. At 50% of design, the velocity was low enough to let the liquid drain via the 20-in. line, and the surging subsided.

The second condensate drainage route was the two 1-in. drain lines. A calculation showed that a 6-in. liquid head would have been sufficient to overcome all the frictional pressure drop incurred by 20 gpm of liquid flowing down these lines. Because the condenser was mounted 6 ft above the column, there was plenty of head to drain the liquid via these lines. This assumes, however, that we are dealing with *bona fide* liquid.

Liquid generated in the condenser is not bona fide liquid—it is *aerated* liquid, namely, a liquid containing gas bubbles. For all the vapor bubbles to degas from a nonfoaming liquid, a residence time of 30–60 seconds is needed (250). Given this residence time, an aerated liquid will revert to bona fide liquid. If the residence time available is less than 30 seconds, some gas bubbles will remain entrained in the liquid.

Aerated-liquid rundown lines must permit self-venting flow; that is, the liquid velocity must be low enough to permit gas bubbles to disengage upward. Excessive liquid velocity will drag gas bubbles downward. This will increase resistance to flow or "choke" the line, causing liquid to back up at the line entry. This phenomenon is analogous to tray tower flooding by downcomer choke. In many cases, the extra backup eventually will provide enough residence time to cause most of the bubbles to disengage before they enter the rundown line.

Figure 24.1*c* shows a correlation for self-venting flow (438, 447) that has been highly recommended (250) for nonfoaming systems. It indicates that the two 1-in. lines are capable of draining 3 gpm of aerated liquid. Because the reflux receiver got only 3 gpm and dispensed 20 gpm, its liquid level fell. At the same time, the condenser generated 20 gpm. The undrained 17 gpm built up in the condenser. This buildup submerged some of the heat transfer area. Condensation diminished and column pressure rose. These are the gradual changes seen in Figure 24.1*b*.

Had the high points in the 1-in. lines been absent, an equilibrium would have been established. With the liquid accumulation in the condenser, residence time increases and, therefore, many of the bubbles disengage. Further buildup would cease when the liquid head became high enough and the bubble volume low enough to push the remaining bubbles through. The end result would have been a loss in condenser capacity (as described in Case Study 24.2), but not surging.

With the current system, some bubbles would become trapped at the high point, where they would resist the liquid flow. The accumulation of liquid in the condenser continued until the liquid residence time became high enough to eliminate all vapor bubbles. The liquid had then become real, not aerated, liquid. A further rise in head, probably minor, would remove the vapor bubbles from the high point.

Bona fide liquid would then be present all the way from the condenser to the column inlet. As pointed out earlier, the "real liquid" head that is needed for draining 20 gpm of condensate is 6 in. The head available was well over 6 ft. A siphon formed, which very rapidly drained all the liquid from the condenser to the reflux receiver. The reflux-receiver level shot right up. This draining exposed the previously submerged heat transfer area in the condenser, and the pressure dove. Once all the liquid drained, the whole cycle started again. Similar cycles have been described (214, 269) in entirely different gravity systems.

Solution The two 1-in. lines were replaced by two 3-in. lines. These were large enough to accommodate a total of more than 20 gpm of self-venting flow. The new lines contained a seal loop but no high point. No more surging occurred after this change.

CASE STUDY 24.2 INADEQUATE CONDENSATE REMOVAL

In memory of J. A. (Polecat) Moore (Retired), Union Carbide

Installation A kettle reboiler boiling propane refrigerant in the shell to condense ethylene vapor in the tubes. The condensate flowed by gravity into an accumulator located below the exchanger. From the accumulator, liquid was pumped out (Fig. 24.2).

Problem The exchanger was designed to condense 26,000 lb/h ethylene; in practice, only 15,000–16,000 lb/h was condensed.

Cause The condenser draw compartment provided little residence time for degassing. The ΔT was small, which precluded significant subcooling, so there was no quenching of uncondensed gas. The liquid leaving the condenser was therefore not *bona fide* liquid but *aerated* with gas bubbles. For this aerated liquid, the condenser outlet line needed to be sized for self-venting flow. For 26,000 lb/h of self-venting flow, at least a 6-in. line, preferably an 8-in. one, is the required line size calculated from the self-venting flow correlation (438, 447, Fig. 24.1*c*) that has been highly recommended (250). The 3-in. condenser drain line and nozzles were badly undersized. The undersized drain line caused liquid backup that flooded condenser tubes, lowered its condensing area, and reduced the heat transfer rate.

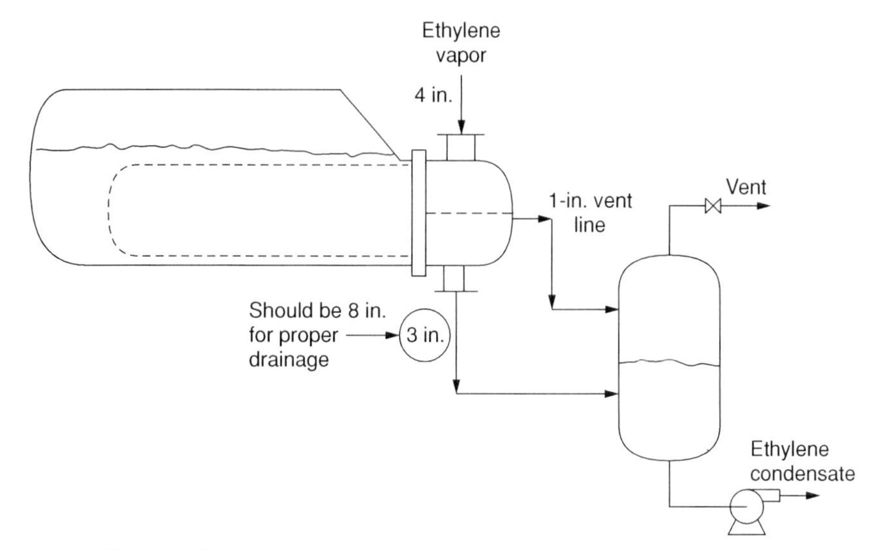

Figure 24.2 Ethylene condenser that fell short of achieving its condensing duty.

Solution The 3-in. condensate nozzle and line were replaced by a 10-in. nozzle and line. The kettle easily achieved its full condensing load following the modification.

CASE STUDY 24.3 NONCONDENSABLES CAN BOTTLENECK CONDENSERS AND TOWERS

Contributed by Ron F. Olsson, Celanese Corp., and Henry Z. Kister, Fluor, Aliso Viejo, California

Installation Overheads from a chemical tower (Fig. 24.3) were condensed in a spray condenser by direct contact with cooled circulating reflux. The condenser drained freely to the reflux drum. Noncondensables from the reflux drum went to the vacuum system.

History Tray changes were made that greatly enhanced tray efficiency at the penalty of slightly lower capacity. Prerevamp calculations showed that there was more than ample capacity in the tower.

Upon restart, the tower was sensitive and erratic. Its performance quickly deteriorated to the point that it flooded at well below the design rates. The plant had problems maintaining vacuum. On suspicion of polymerization and plugging, the tower was shut down. Neither polymer nor plugging was found. However, it was found that a 150-psi nitrogen line was open to the tower. It appears that the nitrogen overloaded the trays and condensing system. The line was shut and the tower returned to service. The tower ran at well above design rates for the next half year.

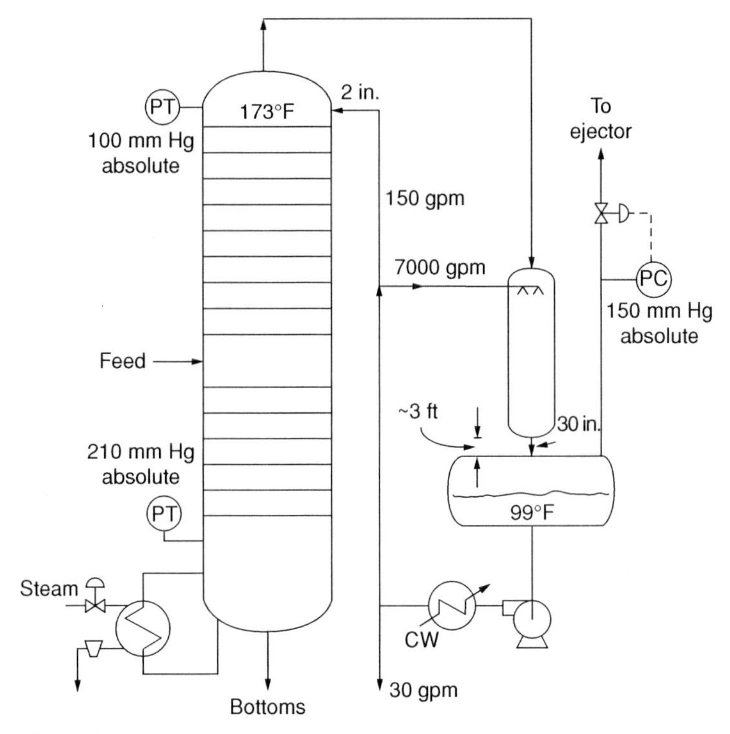

Figure 24.3 Chemical tower and overhead system that experienced fluctuations.

Following this, the unit was shut down for unrelated reasons. Upon restart, severe flooding was experienced. Initially the flooding took place at design loads, but over a week or so, loads had to be lowered to 60–70% of design, and severe flooding was still experienced. The flood was so severe that the reflux drum would go from half full to full within 3 min, suggesting that liquid inventories from the upper trays were being lifted into the reflux drum. Bottom pressure rose from 210 to 330 mm Hg absolute. There were huge pressure fluctuations in the tower. The tower had a nozzle in the intertray vapor spaces about every 10 trays, so a pressure survey with a hand-held pressure gage was conducted. It showed that the pressure drops on all the lower trays in the tower was not high. The pressure drop across the top five trays could not be reliably measured due to the problem with the upper pressure measurement point (see below) but could have been as high as 100 mm Hg.

The tower was again shut down. The top trays were clean, with no signs of plugging or polymerization. During recommissioning, after the tower was brought to vacuum but before liquid was introduced, it was observed that while the bottom and reflux accumulator pressure transmitters read 100 mm Hg absolute, the overhead pressure transmitter read 140 mm Hg absolute. At this time, all three should have read the same. It was realized that the overhead pressure transmitter was reading high. The high reading was caused by excessive nitrogen purge to the transmitter.

Upon adjustment and recalibration, all three pressure transmitters read the same (still with no liquid). It appeared that prior to the shutdown the top pressure transmitter read 120 mm Hg but the top of the tower operated at 80 mm Hg absolute. The lower pressure increased vapor velocities and contributed to the flood. Calculations showed that while at the design top pressure of 140 mm Hg absolute the trays would operate a sufficient margin away from flood, at 80 mm Hg the same trays would operate right at the flood point.

Full restart brought good operation again. The tower still experienced pressure fluctuations, but these were less severe than previously and appeared dependent on noncondensables in the system. One day, there was a slow buildup of pressure in the tower over about 12 hours. Then there was a 20 mm Hg sudden drop in tower top pressure. The drop was accompanied by a 10°F drop in the top-tray temperature and a 5% rise in reflux drum level. Another day, there was a slow rise in tower pressure over 4 hours followed by a 7 mm Hg sudden drop in tower top pressure. The sudden drop was accompanied by smaller top-temperature and drum-level changes. The plant optimized the operation by adjusting the flow of noncondensables to the tower.

Analysis Calculations using the correlation in Figure 24.1*c* shows that the 30-in. pipe at the bottom of the condenser is capable of handling up to 5400 gpm of self-venting flow. The actual liquid flow rate leaving the condenser was much higher, 7200 gpm. This means that some liquid backed up in the condenser and the 30-in. line was running practically liquid full.

The observation that the reflux accumulator was at 150 mm Hg while the tower top was at 100 mm Hg suggests that the liquid in the condenser and the short pipe between the condenser and reflux drum pull part of the vacuum in the tower. Operating records from tower commissioning (after the upper pressure transmitter was repaired) confirmed that the 50 mm Hg pressure difference was established as soon as liquid circulation was established, that is, before vapor was flowing through the tower. This pressure difference was therefore produced by the liquid. The short pipe was 3 ft long. At the normal circulation rate, there was an estimated pressure drop of 1 ft of liquid in the 30-in. pipe (mostly entrance and exit). This leaves 2 ft of net suction, which roughly coincides with the 50 mm Hg pressure difference between the accumulator and the tower.

Due to the liquid backup in the condenser, noncondensables were unable to freely leave from the condenser bottom. They accumulated in the condenser, raising its pressure. With the pressure in the accumulator fixed by the pressure control, the pressure in the tower and condenser would slowly rise until there is enough pressure difference to push enough liquid into the reflux drum and release some of the inerts. Once vented, the tower pressure fell. With large inerts accumulations, the fluctuations were large.

The pressure cycling could have interfered with flooding. The flood could have been induced by a large pressure fluctuation that dropped the tower top pressure. Once flooded, hot liquid was carried over by the overhead vapor. This heavy-rich liquid absorbed some of the lighter components, dropping pressure in the condenser and aggravating the flooding.

Final Cure The pressure controller was relocated from the top of the reflux drum to the tower overhead. This kept a steady pressure in the tower and completely eliminated the pressure fluctuations.

CASE STUDY 24.4 ENTRAINMENT FROM C_3 SPLITTER KNOCKBACK CONDENSER

Tom C. Hower and Henry Z. Kister, reference 225. Reprinted with permission from *Hydrocarbon Processing*, by Gulf Publishing Co., all rights reserved

Installation This case occurred in the overhead system of an olefins plant C_3 splitter. The C_3 splitter overheads were condensed in the overhead condenser and entered the reflux drum. Liquid from the drum was pumped as the column product and reflux. A small amount of uncondensed overhead entered a vent (knockback) condenser (Fig. 24.4). The uncondensed vapor was cooled in the tubes of this condenser to a close approach to the cooling-water temperature. This permitted recovery of as much product as possible from the vent gas. The vent condenser was mounted on top of the reflux drum, so that any condensed liquid would drip back into the drum. The vent stream leaving the vent condenser was sent via a flow controller to a low-pressure system.

Problem At times, when a significant amount of venting was needed, the vent line to the low-pressure system downstream of the flow control valve would ice up. This signified low temperatures in this line, which could not be tolerated due to a metallurgical limitation.

Analysis Knockback condensers are designed for a maximum vapor velocity. Once vapor velocity rises substantially above this maximum, the condenser will no longer

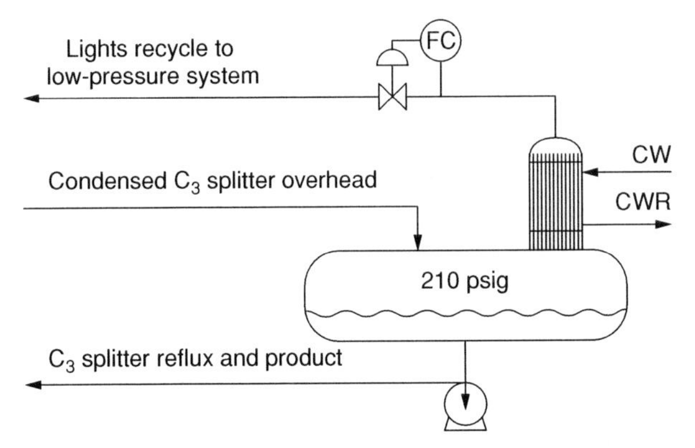

Figure 24.4 A C_3 splitter vent condenser. (Reprinted with permission from *Hydrocarbon Processing* by Gulf Publishing Co. All rights reserved.)

act as a knockback condenser. Instead, the condensed liquid will be carried over by the rising vapor. A method for predicting the maximum velocity is presented elsewhere (104).

In this case, opening the flow control valve too widely induced excessive flow of vapor through the vent condenser. The maximum velocity through the condenser tubes was exceeded, and condensed liquid was carried over. The carried-over liquid chilled upon flashing in the control valve, causing icing of the line downstream of the valve.

Cure A valve limiter, which prevents an excessive opening of the control valve, eliminated this problem.

Moral Excessive vapor flows in knockback condensers lead to entrainment.

CASE STUDY 24.5 EXPERIENCE WITH A KNOCKBACK CONDENSER WITH COOLING-WATER THROTTLING

Installation A chemical tower equipped with a water-cooled knockback internal condenser with condensation in the shell. Condensate at 120°C was collected on a chimney tray (Fig. 24.5). Liquid product was withdrawn from the chimney tray on tray temperature control. Reflux was the liquid overflow from the chimney tray. Tower pressure was controlled by a split-range controller that either manipulated the vent from the tower or throttled the cooling-water return.

Experience At low-rate operation, throttling of the cooling-water raised the cooling-water return temperature above 100°C. The water valve sometimes shut completely trying to maintain pressure, which caused instability and less frequently also boiling of cooling water. The hot temperatures caused severe corrosion. With a measured heat transfer coefficient of more than 100 Btu/h ft^2 °F, fouling was not a problem unless the water boiled off. The condenser was cleaned every few weeks.

At high noncondensable rates, entrainment of liquid in the vent caused problems downstream.

Analysis At low rates, during winter, and when the noncondensables were low, the system had excessive condensation capacity. Tower pressure fell, and the pressure controller throttled cooling-water flow. This raised the cooling-water outlet temperature.

The high noncondensable operation raises a different issue. Knockback condensers are designed for a maximum vapor velocity. Once vapor velocity exceeds this maximum, the condenser no longer acts as a knockback condenser. Instead, the condensed liquid is carried over by the rising vapor. A method for predicting the maximum velocity is presented elsewhere (104).

Opening the pressure control valve too widely induced excessive flow of vapor through the vent condenser. When the maximum vapor velocity in the shell was

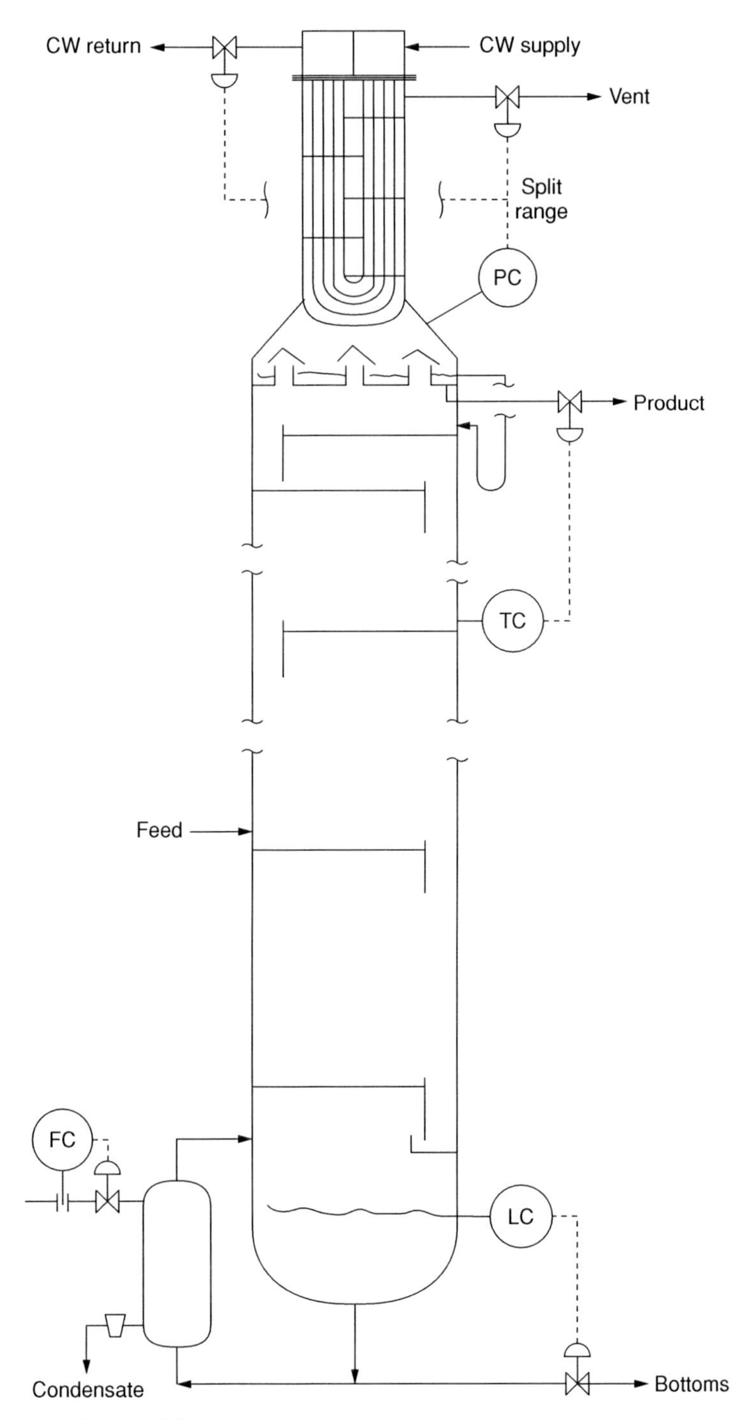

Figure 24.5 Arrangement of tower and internal knock-back condenser.

exceeded, condensed liquid was carried over. Aggravating this problem was the inherent fast tuning of pressure controllers. To keep the tower pressure from fluctuations, pressure control valves are tuned to open and close quickly, causing frequent excursions above the maximum vapor velocity.

Cure To keep the cooling-water control valve from closing, a nitrogen purge was added to the condenser. At low heat duties, this nitrogen inert blanketed the condenser, temporarily lowering its heat transfer coefficient, and permitted operating the cooling-water valve at a large opening. To alleviate the entrainment problem, a valve limiter was installed on the vent control valve, which prevented excessive opening.

Chapter 25

Misleading Measurements: Number 8 on the Top 10 Malfunctions

Misleading measurements are eighth in the list of distillation tower malfunctions (255). Misleading measurements range from those leading to minor headaches when validating a simulation to major contributors to explosions and accidents. The problem is ongoing, with the number of case histories showing neither decline nor growth.

Incorrect readings, plugged instrument taps or lines, incorrect location of instruments, problems with meter and meter tubing installation, and incorrect calibration, are the major issues. In several cases, incorrect levels and control valve position indications led to explosions, fires, and discharge of flammable liquid to the flare or fuel gas. Some of these accidents led to injuries and loss of life; others remained near misses. Other incorrect level indications caused tray damage, pump cavitation, or nonoptimum operation. This emphasizes the importance of independent validation of level measurements, especially when a risk or potential of hazards exists. A "what if" or HAZOP analysis should address the consequences of level measurement failure.

There have been some cases where incorrect temperature measurements, especially in the hot region near the tower base, also led to explosions, overheating, and flammable liquid discharge. Thermowell fouling and thermowell not contacting tower fluid have been the common issues. In most cases, the consequences have been less severe, typically fouling or nonoptimum operation. There have been several reports of pressure, flow, and *dP* instruments reading incorrectly. In most cases, these led to capacity bottlenecks, off-specification products, nonoptimum operation, fouling, and grey hairs on the heads of process engineers attempting to validate simulations.

It is surprising how many variables can fool a level instrument. Presence of froth or foam lowers the liquid specific gravity well below the design density, fooling the instrument into reading low. Similar fooling occurs in aqueous services such as amine absorbers, where either foam or HC condensation lowers the density of the tower base fluid below design. In some services, lights that have lower specific gravity

Distillation Troubleshooting. By Henry Z. Kister

can reach the tower base, especially during start-up or a different campaign, and lead to similar fooling. There were several cases in which an interface level measurement failed, probably due to emulsification, solids, or poor phase settling. Some resulted in explosions and damage. Finally, there were attempts doomed to failure to measure liquid level on partial-draw trays. Fooling of the level instrument is a major issue, with some services (above) more prone to it than others.

The encouraging news is that only relatively few case histories reported absence of a meter when one was needed. In most cases, the meters are there. However, to minimize misleading measurements, they need to be continuously validated and properly installed, checked, calibrated, and inspected.

CASE STUDY 25.1 POOR STEAM EJECTOR PERFORMANCE OR COLUMN VACUUM MEASUREMENT ISSUE?

Contributed by G. X. Chen, Fractionation Research, Inc., Stillwater, Oklahoma

Installation A distillation test column was relocated from the West Coast to Oklahoma during the early 1990s. The process vessels and much of the support equipment were relocated, including 3 in. × 3 in. × 3 in. three-stage, noncondensing, steam jet ejectors manufactured in 1958.

In preparation for the first deep vacuum operation since 1972, the steam jet ejectors were disassembled and inspected in 1994. Routine maintenance, additional brazing of the cracked bronze flange castings, and some cleanup work on the nozzles were performed.

To record atmospheric pressure, a high-quality mercury column barometric pressure gauge was installed in the control room, but corrosion on the glass face of the mercury reservoir impeded adjustment of the zero. A decision was made to use the local airport barometric pressure readings due to the close proximity of the local airport (~1 mile) and the confidence in the quality of the instrumentation required for aviation use. From then on, the barometric pressure gauge in the control room was not relied upon.

Experience During operation, it was very difficult to reach 16 mm Hg absolute at the column overhead. It was uncertain whether or not this pressure was ever reached on the column pressure transmitter. Dew point calculations based on temperature readings at the top of the column indicated a lower pressure than measured by the pressure transmitter. The difference was attributed to the presence of nitrogen near the top of the column, but calculations showed that the quantity of nitrogen introduced into the column by instrument purges was far too small to account for the large observed pressure difference.

Over the next several years, no deep vacuum tests were performed, and the deep vacuum operational difficulties remained unresolved. Prior to the next deep vacuum test, it was decided to upgrade the steam jet ejectors. Per manufacturer's recommendation, the size of the steam jet ejectors was reduced from 3 to 2 in. and latest-technology design improvements were incorporated. The final design included 2 in. × 2 in. × 2 in. steam jet ejectors with 316 SS tails, nozzles, and steam chests and ductile iron bodies using viton gaskets. The modified system was specified to handle a suction load of 10 lb/h of xylene (the vacuum test system) at 5 mm Hg absolute.

Upon start-up, the modified jet ejectors performed very similarly to the previous ones.

Troubleshooting A new, high-precision, certified, vacuum test gauge was purchased, which according to the instructions could measure absolute pressure as long as the gauge was never rezeroed to account for the local elevation. Some time after its arrival to the facility, it was rezeroed to local atmospheric pressure. A new wall-mounted barometric pressure dial gauge was purchased as well but was never trusted due to a discrepancy with the airport barometric pressure reading.

The manufacturer was consulted as to why the steam jet ejectors were not meeting performance criteria. After the manufacturer was convinced that the instrumentation and test methods were acceptable, it was concluded that the most likely problem was wet steam. This explained why both the old and the new steam jet ejectors were performing poorly. The manufacturer offered to test the new steam jet ejectors at its facility at no cost to the company. The ejectors were removed from service and sent to the manufacturer.

In the meantime, the manufacturer instructed how to test for wet steam. The method was to open a valve to let steam escape to the atmosphere and then estimate the distance the steam blew clear from the valve until it turned white. Upon testing, there was virtually no distance observed with clear steam. The steam was white straight from the valve opening. This supported the wet-steam theory.

The high-pressure steam piping supply to the experimental unit traveled approximately 100 ft underground from the boiler before entering the above-ground steam header. Over the previous several years there were some problems with underground piping, and a segment of the insulation of the underground steam piping was known to be damaged. Groundwater was frequently present around the piping, presumably causing condensation in the line, leading to wet steam. The manufacturer's experience had been that the small orifice sizes in the nozzles could clog easily with condensate, causing the ejectors to perform poorly. A steam separator intended to eliminate entrained droplets greater than 10 μm was added on the steam supply line close to the ejectors while the ejectors were being tested at the manufacturer's facility.

The manufacturer's test showed good ejector performance. The shut-off suction supply pressure was lower than 2 mm Hg absolute. This was much lower than all the pressures measured in the column overhead system, which never reached less than 25 mm Hg absolute. This test seemed to confirm the wet-steam theory.

The ejectors were reinstalled and retested. The in-line droplet separators did not improve performance. The piping supplying steam to the ejectors was reconfigured and insulated to minimize wet steam while providing taps for test gauges. No improvement resulted. The local power plant tested the ejectors at its facility using superheated steam. The test showed poor ejector performance despite the superheated steam supply. Two different experts toured the facility and visually observed the steam during a plant steamout. They concurred that wet steam was a very likely culprit.

A literature review suggested that the nozzle in one of the steam jet ejectors could be undersized based on standard design criteria. A modified replacement nozzle was designed to more closely conform to the design criteria in the literature. The nozzle was installed and tested. No improvement resulted.

The steam jet ejector piping was redesigned once again, this time to include individual (<1-μm) in-line droplet separators and pressure regulators for each ejector. This redesign provided precise pressure adjustment along with the best possible quality of steam but still did not improve performance. Additional solutions were considered, including an entirely new system from a competing manufacturer.

Solution During the next low-pressure test, the column pressure measurement was checked with a digital absolute pressure manometer brought on-site from a supporting company. The digital manometer read around 28 mm Hg less than the column overhead pressure transmitter. Investigating the discrepancy led to the discovery that all the airport barometric pressure readings are normalized to a sea-level elevation. Calculations determined a difference of nearly 28 mm Hg between the airport atmospheric pressure reading and the actual local atmospheric pressure at the elevation of the experimental facility.

Once adjusted for this pressure difference, the data taken from the steam jet ejector tests showed performance comparable to that specified by the manufacturer. This discovery was validated by dew point calculations based on the temperature at the head of the column.

Epilogue For 13 years, the column overhead pressure transmitter had been calibrated using the airport barometric pressure reading as a baseline. Good-quality calibration gauges and manometers were used, but there was always a shift on the zero in the calibration of approximately 28 mm Hg. While this reading had been questioned periodically, absolute confidence in this reading was always asserted.

The closest the unit got to correctly diagnosing the problem was in the purchase and use of the precision, certified, vacuum calibration gauge and the barometric pressure gauge. The vacuum gauge was intended as a check for absolute pressure, but some time after its arrival to the facility, it had been rezeroed to local atmospheric pressure, thus ending its effectiveness as an absolute vacuum test gauge. The on-site barometric pressure gauge was never trusted due to the discrepancy with the airport barometric pressure.

Lessons

1. Never trust readings from outside without validation. If instrumentation readings disagree, find out why.
2. Beware of variations in atmospheric pressure when reviewing calibration of vacuum-measuring gauges

CASE STUDY 25.2 INCORRECT READINGS CAN INDUCE UNNECESSARY SHUTDOWNS

Contributed by Chris Wallsgrove

Installation A large propylene fractionator (Fig. 25.1) in a grass-roots olefins complex outside the United States.

Start Up The tower started without incident. During start-up, no attempt was (or ever is) made to optimize or fine tune the tower. The objective is to get on-line, and on specification, with minimum time or material loss.

During start-up, only two controllers worked on automatic (column pressure and reflux drum level). Three instruments gave no reading whatsoever (base level, bottom-product flow, and reboiler heating medium flow). However, by manual operation the column was started, and produced on-specification product 13 days after plant "oil-in." This timing and the instrument problems are typical for this type of system. As plant throughput was raised, some of the experienced operators complained that this column seemed "touchy" in that constant effort was required to maintain the overhead product on specification. This is abnormal for large towers, which in general react slowly and predictably.

Initial Run Once conditions stabilized, it was attempted to put all controllers onto automatic. This involved correcting instrument faults and tuning the controllers. No amount of instrument work could produce a reliable value from the bottom-product

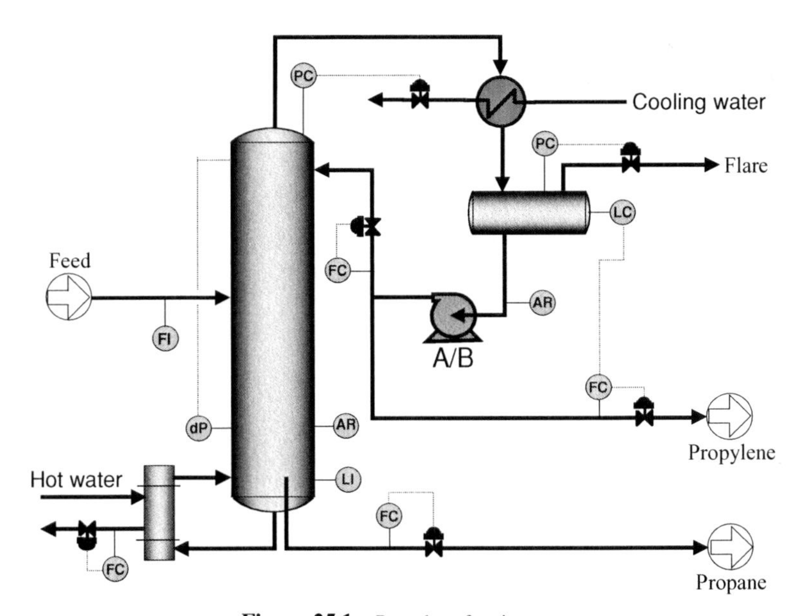

Figure 25.1 Propylene fractionator.

flowmeter, and the reboiler heating medium flow was so unstable that "dead" tuning constants were required. This controller stayed on manual.

Measured values were compared with expected values. This involved considerable extrapolation, as the column operated at only about 50% of design rate with a grossly nondesign feed composition.

This initial comparison found:

- high reflux ratio (>200% design),
- poor separation in that the overhead was only marginally "on specification" and tray 35 (near the base) was far too rich in propylene (2.5 times expected value),
- low column pressure drop, and
- very rapid response to minor changes or upsets (six to eight times faster than predicted).

The problem with the bottom-product flowmeter was finally diagnosed as being partial condensation of the vapor product in the line (it was a bare line at approximately 40°C), giving a mixed phase to the orifice plate. The line was steam traced and insulated, which solved that problem.

Theories The following theories were proposed to explain the poor performance:

- reflux flow reading high, in that the actual reflux ratio was well below design despite contrary indications by the flowmeter;
- tray damage, displacement, or incorrect assembly; or
- remotely possible, the complex four-pass trays losing efficiency due to gas bypassing and liquid maldistribution at turndown.

The reflux flow transmitter, controller, and valve were checked and rechecked. With the major propylene consumer not on-line yet and storage being nearly full, there was an opportunity to shut down the tower and pull the reflux orifice plate. This took 3 days and involved flaring a lot of propylene. Upon inspection, the orifice plate was found to be 100% as it should be. All loop arithmetic was done over and over (e. g., orifice calculation, differential pressure transmitter calculation) and again checked out 100% okay.

The conclusion was therefore jumped at that "it must be tray damage." A plant performance test was run at this time, which occupied all technical efforts. Following this test, the plant was shut down for a mini-turnaround to rectify major problems in other plant areas. It was decided to grab this opportunity to rigorously inspect all the trays in the propylene fractionator. It took 3 weeks to purge and vent this huge tower and to remove over 700 internal tray manways. The inspection found all the internals to be very clean and no tray damage. The top 5 trays (out of more than 180) exceeded the specified "out-of-level" tolerance by several millimeters. The worst were rectified by disassembly and shimming. Upon restart, tower operation did not improve.

Troubleshooting Heat and material balances are invaluable for troubleshooting, so they were compiled both by the computer and manually. Despite intensive work, neither the tower heat nor material balances closed. Compiling component balances proved that the feed laboratory analysis, which was measured by a different analyzer, was consistently and grossly in error for C_4's. This was resolved by using one chromatograph for all analyses around the propylene fractionator. The material balance now closed.

For reliable heat balancing, an accurate kilowatt-hour meter was installed in the power supply to the reflux pump motor. Certified and verified pump curves were obtained from the manufacturer on a "'power-to-motor" basis. Data collected during stable operation showed that the reflux flowmeter consistently read higher than the flow rate determined from the pump curve. The difference varied from 24.4% at low throughput to 32.5% at design rate. The reflux flow orifice had, at this point, been checked twice (which required shutting down the column on both occasions) and was found to be correct. The flow transmitter and all instrumentation had been rigorously checked innumerable times and were proven correct.

A subsequent review of causes for this flow measurement discrepancy revealed that the reflux pump discharge piping class was spiral-wound (seamed) pipe. A sample of similar pipe was located (i.e., same suppliers, same class, same size) and examined. The spiral weld bead on the interior of this pipe protruded 3–4 mm. The reflux flow orifice was located in an orifice run that consisted of approximately 30 m of this pipe upstream and 5 m downstream. The instrumentation installation specification specifically forbade spiral flow in orifice runs. To quote, "spiral flow through a flow orifice plate will result in errors of up to 50% in the measured flow." It was concluded that the spiral weld bead was inducing spiral flow in the liquid going toward the orifice plate.

Cure Proposed solutions included replacing the meter run or installing a flow straightener device. These were expensive and therefore rejected. The cure adopted was to recalibrate the reflux flowmeter to match the flow rate calculated from the pump curves. The operators then raised the actual reflux flow rate to the design value. The heat balance now closed (±5%), and tower operation dramatically improved. A test run verified steady on-specification propylene production at rates exceeding design with no problems at close to the design reflux ratio (based on the "modified" flow calibration).

Morals

- Troubleshooting should always proceed stepwise, starting with the simple and obvious.
- Complete, accurate and reliable data are essential for correct diagnosis.
- Always mistrust, or suspect, new instrumentation.
- Heat and material balances are invaluable troubleshooting tools.
- Experienced people can often spot problems, even if they cannot fully explain or define them. For example, "it doesn't feel right."

- Good theory testing proceeds by first testing those theories which are easiest to prove (or disprove), almost irrespective of how likely (or unlikely) such theories are.
- Avoid making permanent changes until all practical tests are done, reviewed, and digested. Many changes are done on plants, particularly during initial operations, which are not necessary and which do not solve the problem they address. Or worse — create additional problems, including safety risks.

CASE STUDY 25.3 CAN LYING PRESSURE TRANSMITTERS BOTTLENECK TOWER CAPACITY?

Contributed by Dave Simpson, Koch-Glitsch UK, Stoke-on-Trent, England

Installation Two reused tower shells were linked in series to provide one fractionator with nearly 90 trays. The main product was taken as a side draw from the rectifying section. Constrained by the existing column diameters, high-capacity trays were required to handle the hydraulic loads.

Problem Commissioning of the plant was prolonged due to problems with instrumentation, control, and transfer pumps. Once these had been resolved, flooding was observed at well below the design rates, resulting in off-specification products. The flood was confirmed by test runs and gamma scans.

Troubleshooting First suspicion fell upon the high-capacity trays. Many theories were formulated and discarded, including manufacturing error, installation error, and fouling. Soon it was appreciated that there were still unresolved instrumentation issues. The column top-pressure transmitter was used for control. The bottom-pressure transmitter read lower than the top, indicating a pressure gain instead of a pressure drop across the column. This bottom-pressure instrument was ignored because two others near the top of the column supported the control instrument reading. Eventually the three transmitters at the top of the column were inspected and were all found to be installed below their taps. The pipes from the shell to the transmitters filled with condensate. This added a static head to the pressure reading, which was significant due to the relatively low column pressure. So all three read high.

Due to the misleading high-pressure reading, the column operated well below its design operating pressure. This induced excessive volumetric vapor flows in the column, which caused entrainment and eventually flooding at a feed rate well below design.

Solution The pressure instruments were reinstalled correctly (above the taps so that the pipes drained back to the column) and the column was operated at design conditions. The performance was then satisfactory.

Lessons

1. Check all instrumentation thoroughly at an early stage.

2. Do not disbelieve an instrument reading because it does not fit the theory of what the problem is. Fix the instrument. If it can not be fixed, find out why.
3. Climb the column and look to see that everything is as it should be.

CASE STUDY 25.4 MISSING BAFFLE AFFECTS LEVEL TRANSMITTER

Problem A tower experienced a discrepancy between the reading of the level glass and the level transmitter in the tower base.

Cause The level transmitter read incorrectly due to impingement by the reboiler return on the upper nozzle. The drawings showed a V-shaped baffle in front of the reboiler return nozzle. This baffle should have diverted the reboiler return sideways, preventing impingement on the upper level transmitter nozzle right across. However, this baffle was not installed.

CASE STUDY 25.5 BOTTOM-LEVEL TRANSMITTER FOOLED BY FROTH

Installation In two different chemical towers, the upper tap of the base-level transmitter was just below the bottom tray.

Experience The level indicator never showed 100%. The operators did not suspect a high liquid level in the base of the tower. Yet, high liquid levels were experienced. In one of the towers, the high levels led to uplifting of a few trays. In the other, they led to flooding of the lower trays which took a long time to drain out.

Cause When the liquid level rises above the vapor or reboiler return entry, froth is generated. This froth has a lower specific gravity (say about half) than the bottom liquid. The level transmitter is calibrated with the bottom liquid specific gravity and will interpret the froth layer as a lower liquid level. The result is a low-level indication, typically 70–80%, when the froth/liquid level reaches or exceeds the upper tap. When the operators see 70–80%, they are often misled into thinking that all is okay.

Cure Recurrence was prevented by relocating the upper tap of the level transmitter below the bottom of the reboiler return nozzle.

Postmortem Figure 25.2 shows a more sophisticated technique that can more positively identify a high bottom level. The normal level transmitter has its upper tap below the reboiler return nozzle as specified above. A second level transmitter is then installed between the upper tap of the normal level transmitter and a nozzle just below the bottom tray. Normally, this transmitter should read zero. Any positive (nonzero) level reading in the second transmitter is interpreted as bottom liquid level rising above the reboiler return.

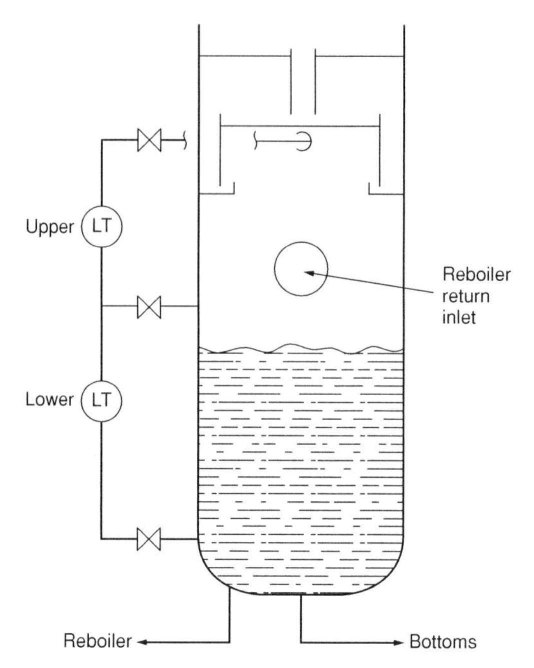

Figure 25.2 Using two level transmitters to detect rise of base liquid level above reboiler return (or vapor feed) inlet.

CASE STUDY 25.6 BOTTOM-LEVEL TRANSMITTER FOOLED BY LIGHT LIQUID

This problem is frequent in chemical towers. Here are some cases:

Tower A The base of a chemical tower usually contained liquid of a specific gravity (SG) close to 1.0. During start-up the base contained lighter organics, with an SG of 0.7–0.8. The level transmitter, calibrated for the heavier liquid, was fooled by the lights and read low. This caused recurrent rises of base level above the reboiler return inlet and flooding that propagated up the tower.

Towers B Similar to tower A, although the SG values were not identical. During start-ups, the bases of several chemical towers in one plant contained liquids of specific gravities much lower than during normal operation. The level transmitters, calibrated for the heavier liquids, were fooled by the lights and read low. This caused recurrent rises of base level above the reboiler return inlet, flooding that propagated up the tower, and recurrent episodes of trays lifting off their supports.

To prevent recurrence, procedures were altered to keep base levels down during start-ups. Many trays were replaced by "heavy-duty" designs. Where implemented, these measures have effectively prevented recurrent tray damage.

Chapter 26

Control System Assembly Difficulties

Three control malfunctions, each with around 30 case histories, hold the 14th, 16th, and 17th spots among distillation malfunctions (255). The three are control system assembly difficulties, temperature and composition control issues, and condenser and pressure control problems. Had the survey lumped them up into one item, "control malfunctions," they would have featured prominently in the 3rd spot on the malfunction list. The survey, as well as this book, preferred to split and itemize them due to the vast differences between the issues. However, it is important to recognize that, despite their low places, control issues feature very prominently on the malfunctions list.

Turning to control system assembly difficulties, most of the case histories come from chemicals and olefins/gas towers, where splits are usually much tighter than between petroleum products in refinery towers. There appears to be neither growth nor decline in these malfunctions.

Over half of the reported control system assembly difficulties stem from violation of three basic synthesis principles. The first is violation of the material balance control principle, discussed extensively by McCune and Gallier (342) and in Shinskey's book (441). Special difficulties have been encountered when adopting the material balance control to towers with side draws. The classic work and good practices for this situation were first described by Luyben (322, 323) and have been presented in many texts (78, 250). The second is violation of what has become known in some circles as "Richardson's rule," which states (410), "Never control a level on a small stream." The third is attempting to simultaneously control two compositions in a two-product column without decoupling the interference between them.

Some of the case histories address the drawbacks of some of the common material balance control schemes. Examples include the slow dynamic response of a scheme that controls tray temperature by manipulating reflux in large tray towers, or the inverse response experienced with the scheme that controls the bottom level by manipulating the reboiler steam.

Distillation Troubleshooting. By Henry Z. Kister

Two approaches have been successful in curing control system assembly problems: The traditional approach diagnoses deficiencies and eliminates them by judicious changes to the control system. The alternative approach, representing the more modern way of addressing the problems, is to replace the conventional control scheme by advanced controls using models and statistical process controls.

CASE STUDY 26.1 C_2 SPLITTER COMPOSITION CONTROLS

Tom C. Hower and Henry Z. Kister, reference 225. Reprinted with permission from *Hydrocarbon Processing*, by Gulf Publishing Co., all rights reserved

This case describes control improvements in a C_2 splitter column by switching the temperature control from reflux to reboil.

Installation An olefins plant C_2 splitter separating ethylene as the top product from ethane as the bottom product.

Control Ethylene was by far the more important product, so the composition was controlled in the top part of the column (Fig. 26.1). The composition controller

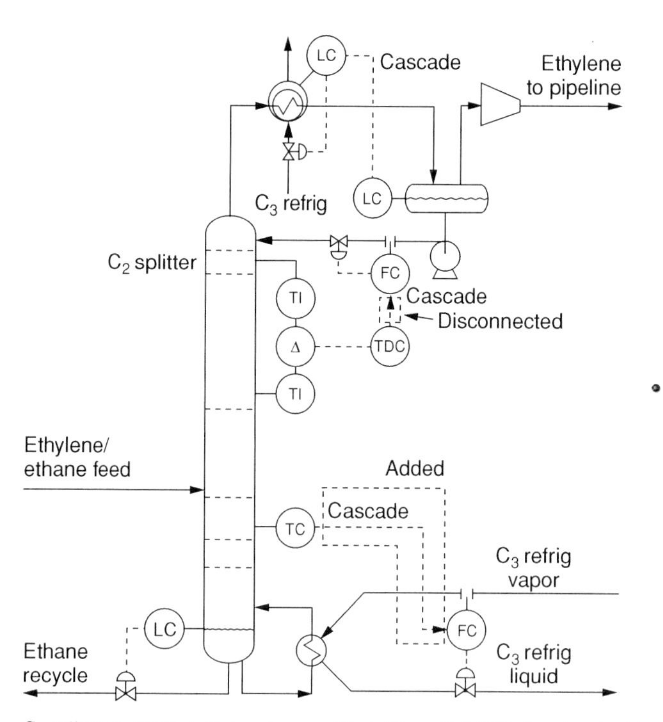

Figure 26.1 C_2 splitter controls, initial and as modified. (From Ref. 225. Reprinted with permission from *Hydrocarbon Processing* by Gulf Publishing Co. All rights reserved.)

was a temperature difference controller. This controller subtracted the top temperature from a tray temperature 50 trays below and used this difference as the control signal. The actual control tray was therefore 50 trays below the top; the top-tray temperature was insensitive to composition. Both the top- and control tray temperatures were equally affected by changes in tower pressure, so that their difference was independent of changes in pressure. The temperature difference was used to prevent changes in pressure from being interpreted by the temperature controller as composition changes. In summary, the top-composition controller was a pressure-compensated temperature controller located 50 trays below the top of the tower.

The top-section temperature controller was cascaded to the reflux flow. Reboil was flow controlled. Bottoms were controlled by the sump level, and overhead product was pressure controlled. Accumulator level was adjusted by varying condensation rate.

Problem Control was slow and sluggish. Since it was important to ensure that ethylene was always on specification, the bottom section was used to accommodate for the control deficiencies. Bottoms ethylene content, which should have been about 1 mol %, varied widely between 0.5 and 8%. The bottom section often ran cold, indicating the escape of ethylene in the bottom. Controller tuning did not solve the problem.

Analysis Consider what happens when the control tray temperature drops. The controller will lower the setting on the reflux flow controller, which in turn will reduce reflux to the top tray. This lowers the level of liquid on the tray, which reduces the flow of liquid into the downcomer. This in turn lowers the liquid entering the second tray and so on. The process needs to repeat through 50 trays before it reaches the control tray. The change will reach the control tray after a significant hydraulic lag. Further, the flow changes set off composition transients, which are far slower. The result is a slow and sluggish response.

Modification There was a temperature indicator 10 trays above the bottom. It was connected to a temperature controller that cascaded to the reboil flow controller. The cascade between the top-section temperature difference controller and the reflux was disconnected. The new control system, with reflux on flow control and reboil on the lower tray temperature control gave good and fast control. Both overhead and bottom products were kept on specification, and both compositions were far more stable than previously.

Reanalysis The modified system controlled composition by changing vapor supply to the column. As distinct from the slow and sluggish propagation of liquid rate changes, vapor rate changes propagate rapidly and simultaneously through the column, giving a good, fast response. In addition, tray 10 is believed to have been a better temperature control tray than the upper tray. The modified system gave better control of both the top and bottom compositions.

Two-Composition Control At a later date, the differential temperature controller was cascaded onto the reflux controller in an attempt to control the top composition using the reflux and the bottom composition using the reboil. This worked extremely well for a couple of days or so, until a slight upset was introduced into the column feed. As soon as the upset occurred, the two temperature controllers started chasing each other, leading to erratic reboil, reflux, and temperature control. This was stopped by the operator by disconnecting the cascade from the top temperature controller to the reflux.

The problem experienced in this case was interaction between the two temperature controllers. This interaction was initiated by the slight upset and from then on could not be stopped while the two controllers operated simultaneously.

Final Solution From then on, the column was always operated with the bottom-section temperature controller cascaded to the reboil flow controller. Reflux entered on flow control. The differential temperature controller was no longer used.

Morals

- In large superfractionators, the fast response of boil-up manipulation is advantageous for composition control
- Interaction of two composition controllers in one column leads to poor control.

CASE STUDY 26.2 CONTROLLING TEMPERATURE AT BOTH ENDS OF A LEAN-OIL STILL

Tom C. Hower and Henry Z. Kister, reference 225. Reprinted with permission from *Hydrocarbon Processing*, by Gulf Publishing Co., all rights reserved

This case describes an unsuccessful attempt to control the temperature of both ends of a lean-oil still.

Installation A natural gas lean-oil still. The still separated gasoline and lighter components as the top product from absorption oil.

Controls The column (Fig. 26.2) had a fired reboiler, with reboiler outlet temperature controlled by manipulating the fuel flow to the reboiler. The column had an air condenser and a flooded reflux drum. Column pressure was controlled by flooding condenser tubes. When the controller called for more pressure, the valve in the product line closed and more tubes were flooded. This reduced condensation rate and raised pressure. The column top-temperature controller manipulated the air condenser's louvers. To reduce temperature, it opened the louvers; this increased condensation and cooled the column.

Problem The column experienced unstable control and erratic operation. Steady pressure could not be maintained. Sometimes the fluctuations were quite violent. At times the column would empty itself out either from the top or from the bottom.

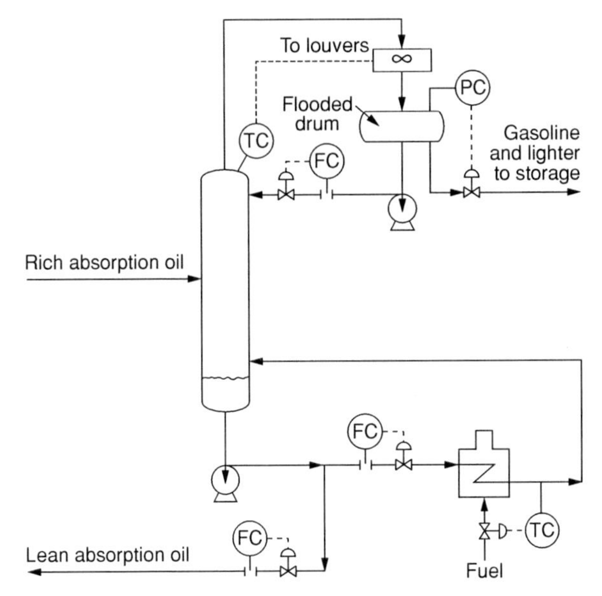

Figure 26.2 Natural gas still experiencing control problems. (From Ref. 225. Reprinted with permission from *Hydrocarbon Processing* by Gulf Publishing Co. All rights reserved.)

Analysis A distillation column is governed by a mass balance and a component balance:

$$F = D + B + \text{accumulation} \quad (1)$$

$$Fz = Dy + Bx + \text{component accumulation} \quad (2)$$

where B = bottom flow rate, lb mol/h
D = distillate flow rate, lb mol/h
F = feed flow rate, lb mol/h
x = concentration of component (e.g., light key) in bottom stream, mole fraction
y = concentration of component (e.g., light key) in the distillate stream, mole fraction
z = concentration of component (e.g., light key) in the feed stream, mole fraction

In a steady-state system, the accumulation terms in Equations 1 and 2 are equal to zero. In the still system shown in Figure 26.2, F and z are fixed by conditions upstream of the column. The flow rate B is fixed by having the still bottom on flow control. Since accumulation is zero, D becomes the difference between F and B (by Equation 1) and is therefore also fixed. Considering Equation 2, it now has only two variables, x and y; all other terms are fixed. Since at steady state the component accumulation is zero, Equation 2 makes x a function of y or vice versa. Therefore, only one of the two—either x or y—can be fixed; the second variable will be a function of the one that is fixed.

In the Figure 26.2 system, each temperature control fixes a composition. Both x and y are fixed independently. To equate both sides of Equation 2 under these conditions, the component accumulation term must become nonzero. This means that the column is no longer at steady state. Failure to achieve steady state was responsible for the control problem experienced.

In the system shown in Figure 26.2, only one temperature can be satisfactorily controlled, while the other must be allowed to drift as needed to avoid component accumulation and maintain steady state. Either the top or bottom temperature can be controlled. Usually, the bottom temperature is more important and is controlled while the top temperature is allowed to drift. Also, experience with top-temperature control in similar installations has been that it tends to be sluggish. This is partly because the action of the top-temperature control needs to be slow to avoid interference with the pressure controller. In addition, louvers become mechanically unreliable as they wear, and are usually best avoided for column condenser control.

Solution Temperature control to the louvers of the air condenser was disconnected. The control problem disappeared.

CASE STUDY 26.3 INVERSE RESPONSE

In this case, the unusual control scheme in Figure 26.3 turned out the only suitable scheme for the control of a tower.

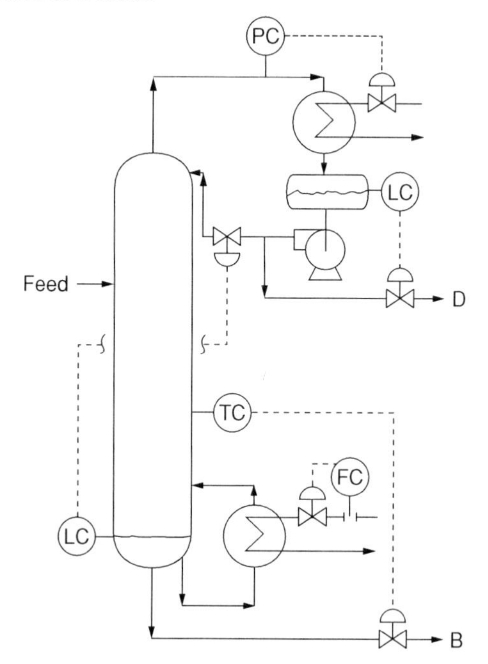

Figure 26.3 Unconventional control scheme that controls bottom level by regulating reflux and turned out to be the only satisfactory scheme for this case. (Reprinted with permission from Ref. 250.) Copyright © 1990 by McGraw-Hill.

Installation A xylene splitter containing 50 valve trays. The bottoms flow rate was much smaller than the feed flow rate and than the boil-up. The column was controlled by the basic scheme shown in Figure 26.4*e*. The reboiler was a forced-circulation fired heater, there was no baffle in the tower base, and the level control valve was in the fuel gas line to the heater. Also, the pressure control was manipulated by a valve in the condensate line leaving the condenser, not by throttling coolant.

Problem When the tower base level went down, the level controller called for more heat to the reboiler. The valve opened and the heating intensified. This, however, did not bring the level down. The level stayed pretty constant for 3–4 minutes. During this time, the column was heating up excessively, causing the top product to go off specification. About 3–4 minutes later, the level finally went down; when it came down, it did so very sharply. This destabilized the tower.

Analysis The symptom described is that of "inverse response," previously reported by Buckley et al. (76, 77) in a large tower containing valve trays using the same control scheme. The phenomenon is also discussed in Refs. 78 and 250. In the froth regime, an increase in vapor flow reduces tray froth density. Froth height above the weir rises, and some of the tray liquid inventory spills over the weir into the downcomers. The expelled liquid ends in the tower base, and bottom level initially rises (76, 77) or stays constant, as it did here. This is opposed to the expected response, and was termed *inverse response* by Buckley et al. The control scheme in Figure 26.4*e* is destabilized by inverse response because the base-level controller keeps increasing reboiler heat until the level comes down again.

Cure The base-level controller was switched from the boil-up heat duty to the reflux flow rate (Fig. 26.3). This solved the problem. The solution is identical to that implemented by Buckley et al. (76, 77) in their tower.

CASE STUDY 26.4 INVERSE RESPONSE WITH NO REFLUX DRUM

Contributed by Lars Kjellander, Perstorp Oxo AB, Stenungsund, Sweden

Installation Chemical tower (Fig. 26.5*a*) with an internal condenser, an internal reboiler, and 29 valve trays. The control scheme shown was similar to that of Figure 26.4*e*, except that there was no direct composition control on the small bottom stream. The flow rate of that bottom stream was manually adjusted by the operators, who would ensure that the top product was on specification while avoiding excessive loss of distillate to the bottom. So the bottom stream was composition controlled via the operator's hands.

Problem The tower experienced inverse response. Following a step up in heat input, the level rose for 3–5 minutes and only then began to fall. This response was almost identical to that first described by Buckley et al. (76, 77) and shown in Figure 16.5, page 505 of *Distillation Operation* (250).

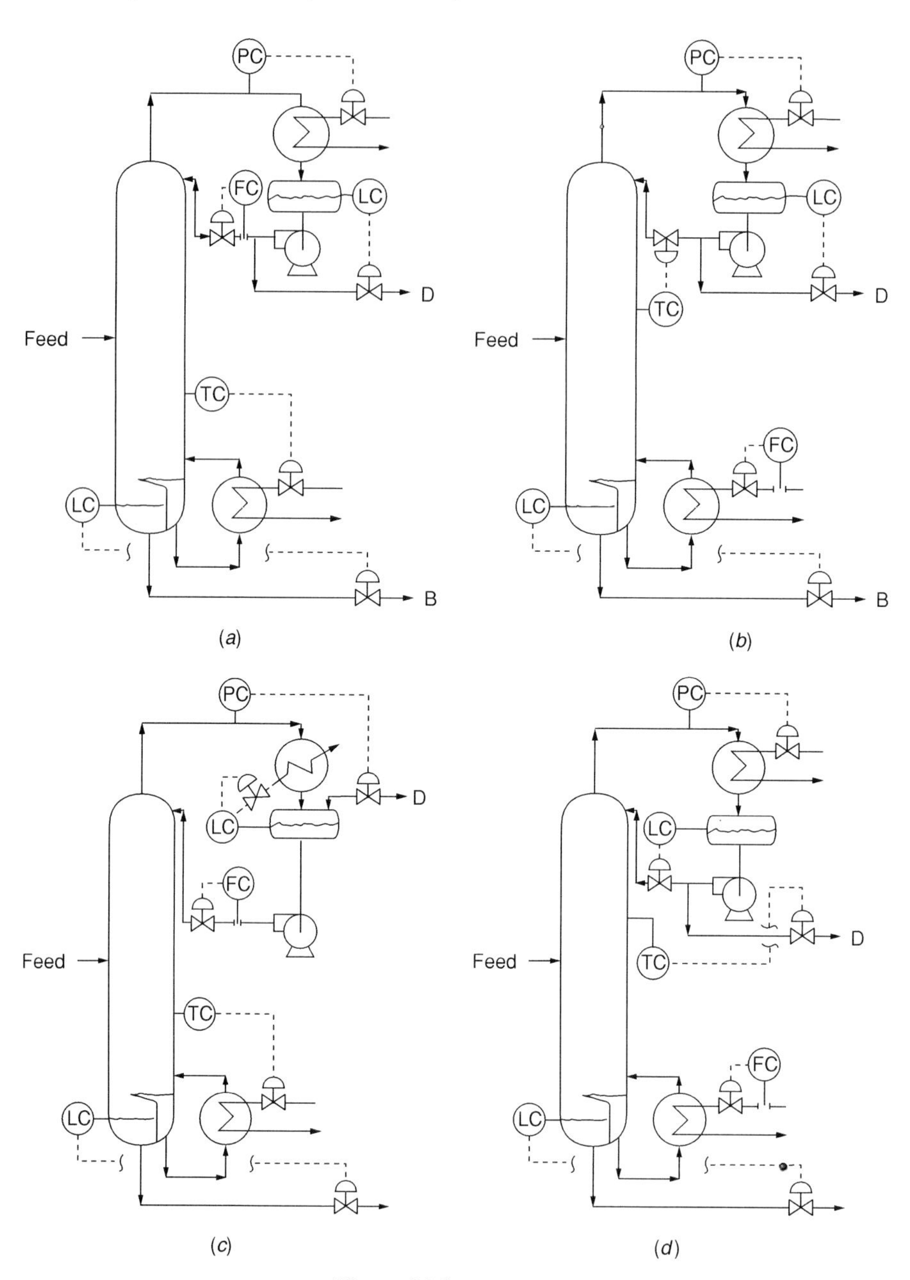

Figure 26.4 (*Coninued*)

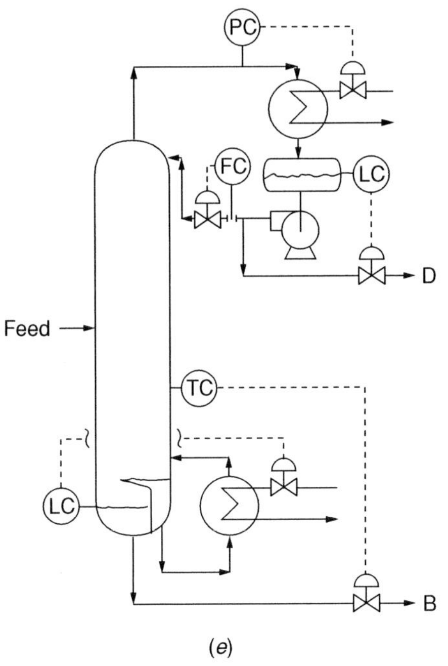

Figure 26.4 Common overall assemblies of material balance control schemes: (*a*) indirect control, composition regulates boil-up; (*b*) indirect control, composition regulates reflux; (*c*) as (*a*) but with vapor product; (*d*) direct control, composition regulates distillate flow; (*e*) direct control, composition regulates bottom flow. The sketches are schematics depicting overall layouts, not individual loops. For instance, the temperature control in (*a*)–(*e*) means a composition controller, which may be a temperature controller, an analyzer controller, a virtual analyzer controller, or a vapor–pressure controller, manipulating the boil-up rate. Similarly, the pressure control in (*a*), (*b*), (*d*), and (*e*) means a pressure control manipulating the condensation rate (either by adjusting the coolant rate, flooding the condenser tubes, or manipulating the inert purge or intake). Also, boil-up control in (*a*)–(*e*) can be by manipulation of reboiler condensate flow instead of reboiler steam flow. (Reprinted with permission from Ref. 250. Copyright © 1990 by McGraw-Hill.)

The usual cause of inverse response is liquid displacement from the trays upon vapor rate heat step-up (see Case Study 26.3). Here there may be an alternative explanation. The liquid in the base of the tower at and above the internal reboiler was present as froth, not clear liquid. The level transmitter measured the liquid head equivalent to the froth height. Upon increase in reboiler heat input, initially the froth height rose, but the froth density did not change greatly. This caused the measured liquid level to rise. Once enough was boiled, the froth density declined and so did the measured liquid head in the base.

First Modification (Fig. 26.5*b*) Was identical to that successfully used by Buckley et al. (76, 77) to eliminate inverse response. The bottom-level control was cascaded to the reflux flow control.

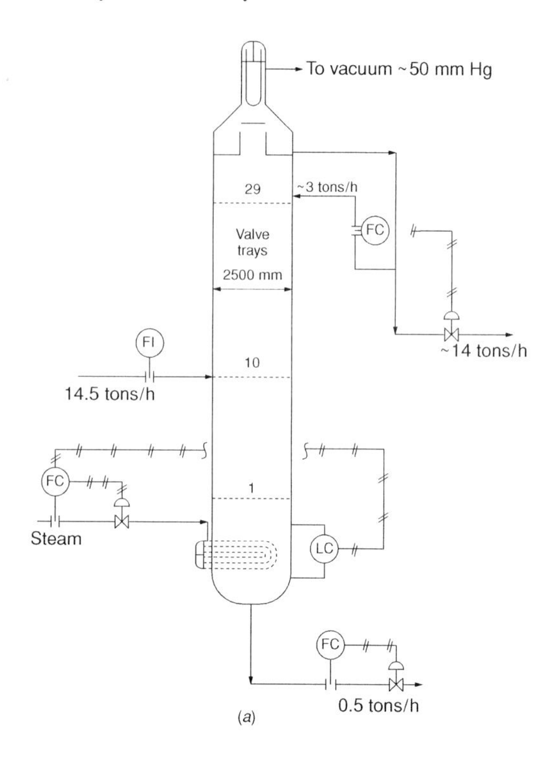

(a)

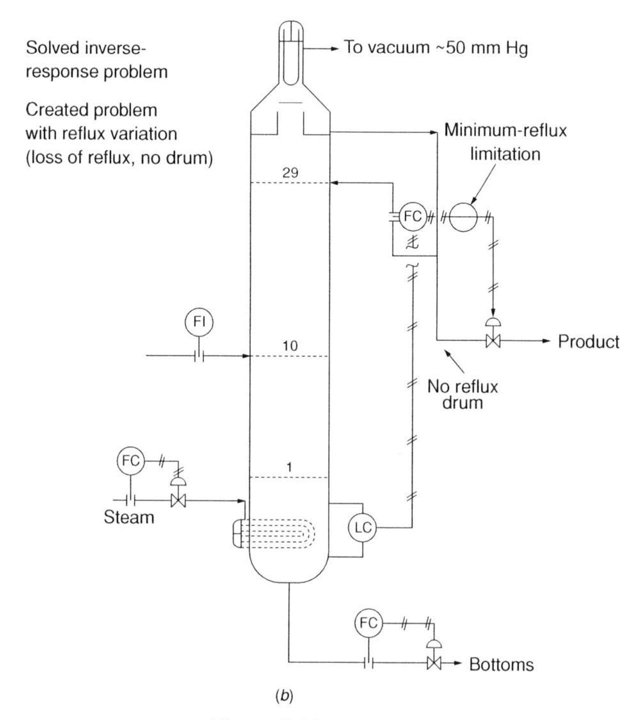

(b)

Figure 26.5 *Continued*

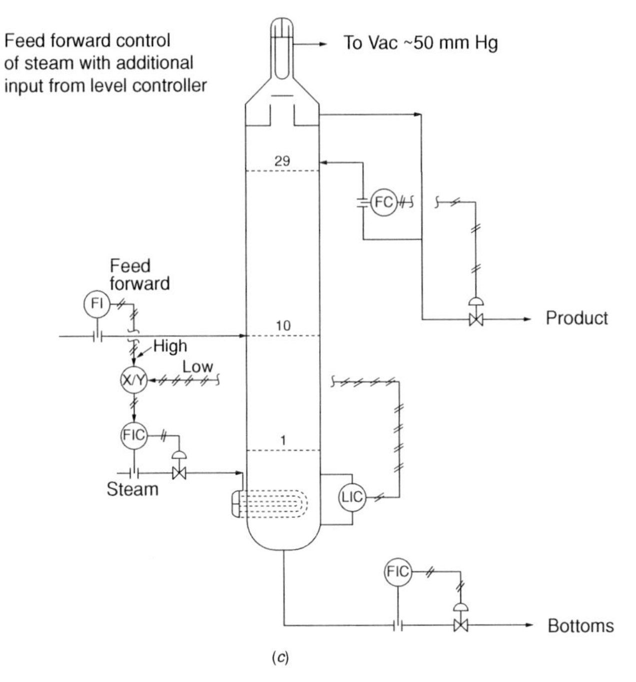

Figure 26.5 Chemical tower experiencing inverse response: (*a*) initial control scheme; (*b*) first modification; (*c*) final solution.

As in Buckley's case, this modification successfully eliminated the inverse-response problem. However, here the solution created a problem of reflux variations, even loss of reflux, due to the absence of a reflux drum. Buckley mentions having to live with an interaction between the bottom and reflux drum-level controls. Here this interaction was far more severe due to the absence of a drum surge to cushion it.

Final Solution (Fig. 26.5*c*) With most of the disturbances being changes in feed flow rate, the steam control was placed on feed forward from the feed flow rate, with additional input from the level control. The feed-forward control had a strong influence upon the steam flow controller, while the level controller had a weaker influence.

The column had been in operation for over 12 years with the final modification and had operated very well, even when the feed rate varied.

CASE STUDY 26.5 REBOILER SWELL

One column was reboiled using a vertical thermosiphon reboiler with an undersized outlet nozzle. The system worked for several years. During these years, tower level was controlled by manipulating the bottom valve. One shutdown, the control system was changed so that the bottom level was hooked to manipulate the steam flow (Similar to the scheme in Fig. 26.4*e*). This change caused the reboiler to stop working "would not thermosiphon." Occasionally, steam would get to the reboiler, but it was erratic and unstable.

The problem was that of a "reboiler swell." Consider an increase in bottom level. The controller would raise the reboiler steam. Due to the undersized outlet nozzle, the pressure inside the reboiler would rise. This would cause liquid to back up from the reboiler into the column. The liquid level in the sump would rise, and the controller would further raise the reboiler steam. This in turn would again raise the level and again raise the steam, generating an unstable response.

The solution to this problem was to revert to the original control scheme in which the sump level control manipulated the bottom flow.

Related Experience Reboiler swell problems were experienced in a column using the control scheme in Figure 26.4*a*. Here the swell did not produce unstable response but caused large level swings. The solution was to use a tight proportional band on the sump level control at the expense of large fluctuations to the bottom flow rate.

CASE STUDY 26.6 BASE BAFFLE INTERACTS WITH HEAT INPUT CONTROL

Installation A heavy-ends chemical tower. Most of the tower feed became the overhead product. The base of the tower (Fig. 26.4*e*) contained a preferential baffle. Heat input to the vertical thermosiphon reboiler was controlled from the level of the bottom-draw compartment, as shown in Figure 26.4*e*.

Problem Tower control was erratic.

Cause Liquid flow to the reboiler, including both vaporization and recirculation, was hundreds of times greater than the small bottoms flow. Therefore, the overflow across the baffle was minute. An increase of reboiler heat completely dried up the overflow. On the other hand, cutback in reboiler heat would induce massive dumping of liquid over the baffle. Both of these gave a jerky level in the bottom-draw compartment, which in turn jerked tower heat input.

Cure The preferential baffle was removed. The tower control was smooth afterward.

Moral Preferential baffles in the tower base are not a good idea when the bottoms flow is hundreds of times smaller than the reboiler flow (see also Case Study 8.6). The problem is magnified when using the control system in Figure 26.4*e*.

CASE STUDY 26.7 GOOD REFLUX CONTROL MINIMIZES CRUDE TOWER OVERFLASH

Installation The lowest side-draw product from a crude fractionator was an atmospheric gas oil (AGO) stream. The side draw was removed from a tray sump (Fig. 26.6*a*), then flowed into a steam side stripper. A small quantity of stripped lights was returned to the tower. Stripper bottom product was the AGO.

Problem To ensure stable operation and satisfactory AGO color, the crude tower was operated with a reflux to the wash trays that was about 5% of the total distillate. This reflux rate is considered excessive and represents high-value AGO product degraded into low-value resid bottom product. Any AGO escaping in the resid also loaded up the vacuum tower downstream.

Analysis The AGO product was drawn on flow control (Fig. 26.6*a*), with the set point adjusted by the operators and later by the advanced control system to maintain a satisfactory AGO color and to prevent process upsets. As the AGO product flow rate exceeded the intended reflux flow rate to the wash trays, the reflux to the wash section became the small difference between two large numbers. Excessive opening of the flow valve, or even small fluctuations in tray liquid flow rate, would drastically diminish the reflux to the wash trays. This in turn caused the AGO product to go black, and even led to downcomer seal loss, which generated a major upset. To give themselves a comfortable operating margin from upsets and off-color AGO, the operators (and later the advanced control) would compensate by cutting back on AGO draw rate. This increased reflux to the wash trays and stabilized operation, but at the expense of the economic loss associated with AGO degradation to resid.

Solution To solve, reflux to the wash section needed to be supplied at a steady flow rate, not as the fluctuating small difference between two large numbers. To achieve, the draw tray was converted into a total-draw chimney tray. Reflux was supplied to the wash section on flow control (Fig. 26.6*b*), slowly adjusted by the advanced control to maintain the required AGO quality.

This modification reduced the reflux to the wash trays from 5% of the total distillate to 2% of the total distillate while producing a more consistent AGO color and quality.

Moral When reflux flow rate to the section below is significantly less than the draw rate above, it should be on flow control, not on level or difference control.

CASE STUDY 26.8 VAPOR SIDEDRAW CONTROL

Henry Z. Kister, Rusty Rhoad, and Kimberley Hoyt, reference 273.
Reproduced with permission. Copyright © (1996) AIChE. All rights reserved

Automatic control of towers with side draws, especially vapor side draws, presents unique challenges. This case took place in the tower described in detail in Case

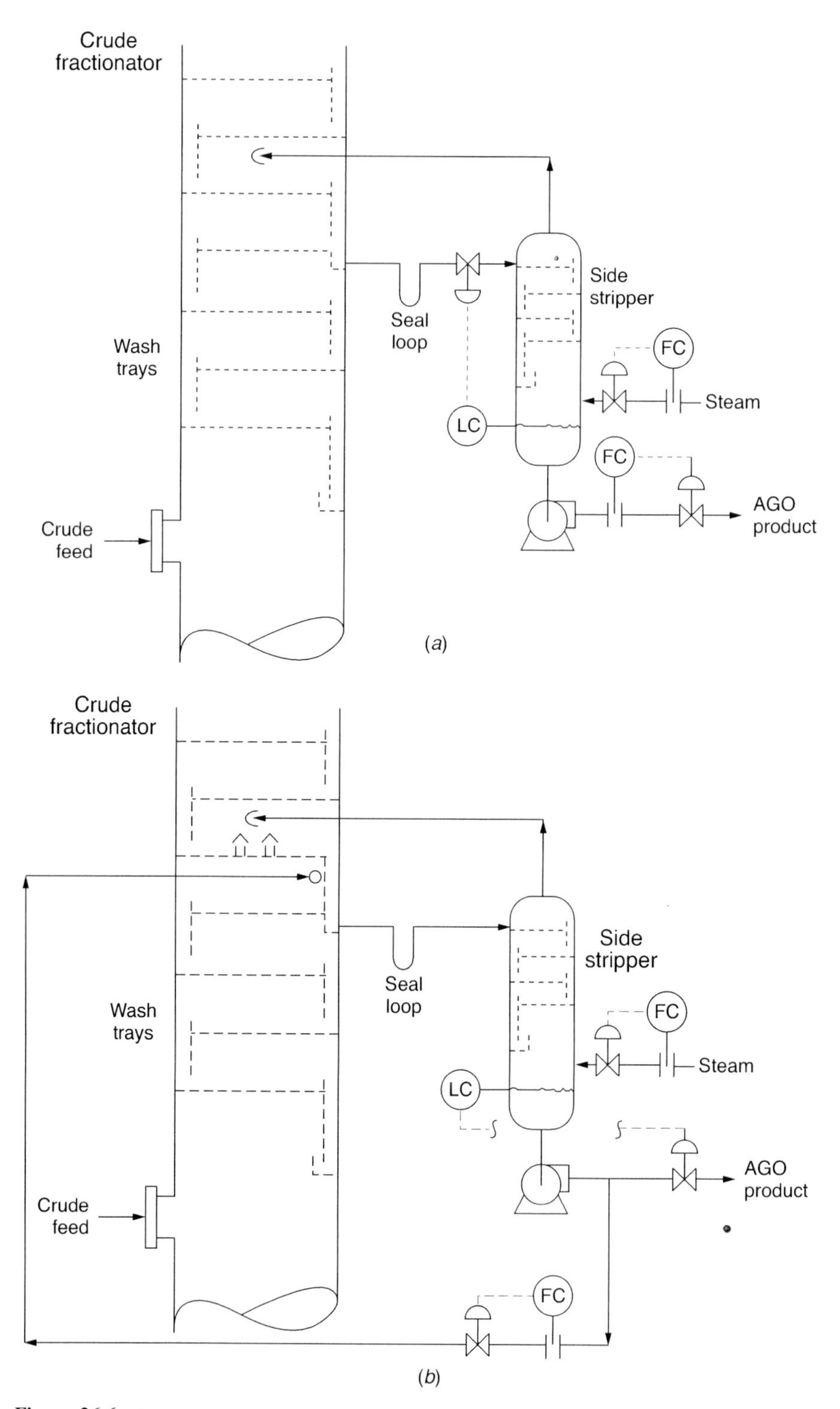

Figure 26.6 Schemes for drawing AGO from crude tower: (*a*) initial, led to excess overflash and loss of AGO to resid; (*b*) modified, minimal loss of AGO to resid.

Study 24.1 after the surging problem was eliminated. Throughout our troubleshooting investigation, the unique control system was looked at very critically but was not faulted.

Material Balance Control Figure 26.4 shows the four common material balance control schemes for simple columns (top liquid product, no side draw). Detailed description is given in Ref. 250. According to Ref. 250, good material balance control in simple columns requires use of one of these four schemes. Three other material balance schemes may be justified in special circumstances but have serious drawbacks. Other alternatives seldom are successful.

Presence of a side draw can be addressed by splitting the column into two half columns. Then, the four material balance control schemes in Figure 26.4 can be considered for each half.

Figure 26.7*a* shows that the controls on the lower half column follow Figure 26.4*a*. This half column essentially is a stripper, and its controls are those used in most strippers.

Figure 26.7*b* shows that the controls on the upper half column follow Figure 26.4*d*. Note that there is no bona fide bottom-level control, but immediate drainage of liquid from the middle bed prevents liquid accumulation, thus acting like a level control.

The Figure 26.4*d* control scheme requires that vapor is introduced into the upper half column on flow control. The total vapor flow consists of vapor generated by the reboiler less vapor withdrawn as side product. One way of keeping this vapor flow constant (250, 322, 323) is by an internal vapor controller (IVC), as shown in Figure 26.7*c*. The steam flow is measured and converted into the reboiler vapor flow by multiplying by the latent heat ratio. The side-product flow is measured and subtracted from the reboiler vapor flow. The difference is the internal vapor rate in the upper column, which then is used to regulate the vapor side-product flow rate.

Alternatively, the internal vapor flow in the upper half column can be controlled by a differential pressure controller across the upper section of the tower. This differential pressure is a strong function of the internal vapor flow rate, as shown by the filled circles in Figure 4.7*a* in Case Study 4.9 (this is the same tower). Keeping the pressure drop constant, therefore, holds the internal vapor rate constant. Like the IVC, the differential pressure controller regulates the vapor side-product flow rate.

A variation of this technique used in this vacuum column is to ratio the pressure drop above the side draw to the pressure drop below the side draw (Fig. 24.1). Like the IVC and differential pressure controllers, the ratio controller regulates the vapor side-product flow rate. Like the differential pressure controller, the ratio controller acts to keep a constant internal vapor flow in the upper half column. In addition, the ratio control improves the response to a feed rate change, but at the expense of potential interaction between the lower and upper temperature controllers. In this column, any such interaction was insignificant.

In both the side condenser and the reboiler, the main controller (the ratio controller in the side condenser and the temperature controller in the reboiler) was cascaded to the condensate liquid level in the exchanger. Raising the liquid level floods some tube

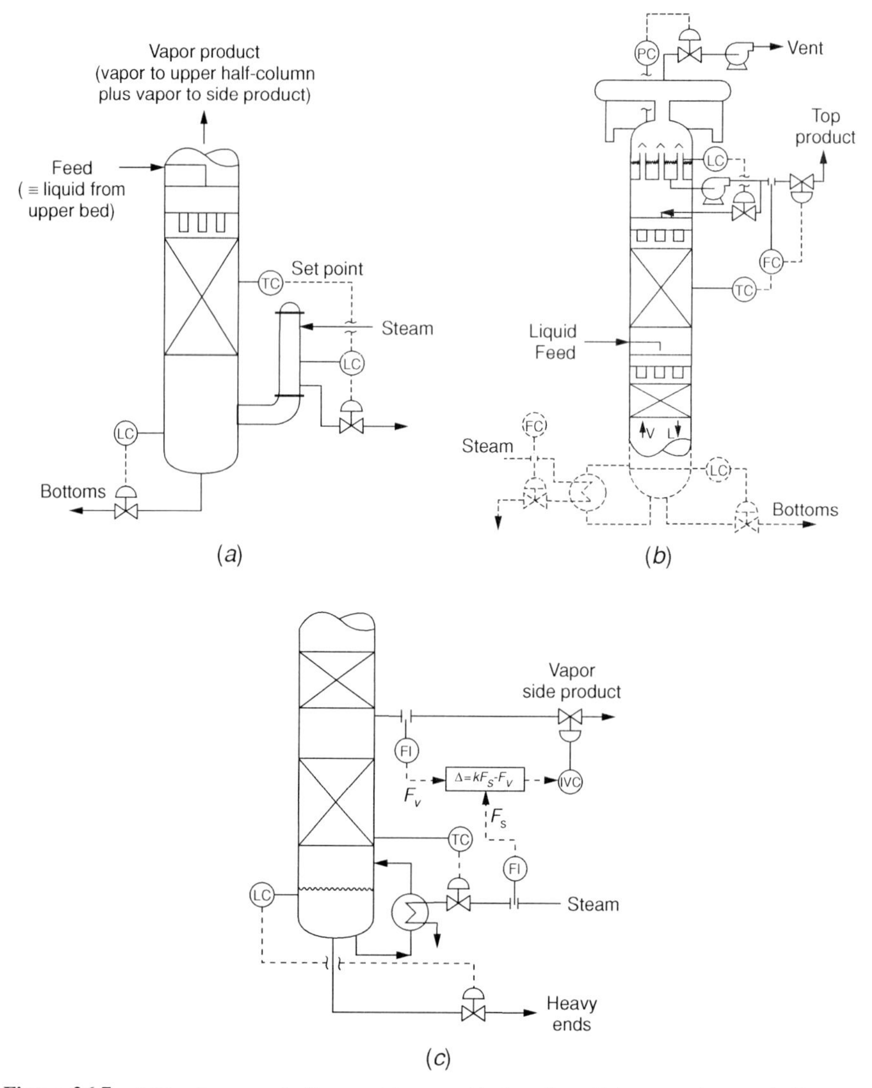

Figure 26.7 Addressing control of tower with vapor side draws by splitting tower into two half towers: (*a*) lower half-tower control; (*b*) upper half-tower control; (*c*) internal vapor controller for vapor side draws. (From Ref. 273. Reproduced with permission. Copyright © (1996) AIChE. All rights reserved.)

area, thus reducing heat transfer. In the side condenser, this would lower the amount of side product generated; in the reboiler, this would lower the reboiler heat input.

Epilogue Once the column instability (Case Study 24.1) was eliminated, this unique control system gave very stable and satisfactory control.

Chapter 27

Where Do Temperature and Composition Controls Go Wrong?

Three control malfunctions, each with around 30 case histories, hold the 14th, 16th, and 17th spots among distillation malfunctions (255). The three are control system assembly difficulties, temperature and composition control issues, and condenser and pressure control problems. Had the survey lumped them into one item, "control malfunctions," they would have featured prominently in the 3rd spot on the malfunction list. In the survey as well as in this book, it was preferred to split and itemize them due to the vast differences between the issues. However, it is important to recognize that, despite their low places, control issues feature very prominently on the malfunctions list.

Most of the temperature composition malfunctions come from chemical and olefins/gas towers, where splits are usually much tighter than between petroleum products in refinery towers. No clear trend of growth or decline was seen in these malfunctions (255).

There are three major composition control issues. The top issue is finding the best temperature control tray. This is followed by achieving successful analyzer control and obtaining adequate pressure compensation for temperature control. The search for a suitable control tray has been less of an issue in the last decade due to the publication of an excellent method by Tolliver and McCune (483). Still, there are some situations where no satisfactory temperature control can be found. With analyzers, the main problems have been measurement lags and on-line time. Recent advances in analyzer technology have improved both, but older analyzers are still extensively used. Modern analyzer controls are often associated with advanced controls and have grown in significance in the last decade.

Two approaches have been successful in curing temperature and composition control problems. The traditional approach uses solutions such as defining the best temperature control tray and cascading analyzers onto temperature controls. The

---

alternative approach, representing the more modern way of addressing the problems, is to use virtual analyzers based on model calculations from tower measurements, and using statistical process controls. These have overcome some of the inherent limitations of temperature and analyzer controls. Useful tricks, such as pressure correction to the temperature, or using an averaged temperature, have been incorporated with both approaches.

CASE STUDY 27.1 AMINE REGENERATOR TEMPERATURE CONTROL

Installation A refinery amine regenerator stripped a small amount of H_2S from rich amine. The H_2S concentration of the rich amine was severalfold lower that the design. The tower boil-up was controlled by a tray temperature in the lower part of the regenerator. The regenerator top pressure was about 5 psig.

Problem Temperature control was erratic. Due to the small concentration of H_2S, the control temperature was insensitive to the H_2S concentration. The control temperature was sensitive to pressure fluctuations. Pressure fluctuations of up to 2 psi were frequent. A rise in pressure would raise temperature and induce cutback in reboiler steam.

This tower is a good example of a situation where temperature control cannot be satisfactorily implemented.

Solution The temperature control was removed from "remote" and the steam was controlled on flow control. This eliminated the erratic behavior and stabilized the tower.

CASE STUDY 27.2 COMPOSITION CONTROL FROM THE NEXT TOWER

Henry Z. Kister and Tom C. Hower, reference 263. Reproduced with permission.

Installation An absorption–refrigeration gas plant. Rich absorption oil contained absorbed HCs from C_2 to gasoline. To regenerate the absorption oil, the deethanizer stripped out the C_1 and C_2 HCs. Gasoline and LPG (C_3–C_4) were recovered in the still overhead product. The C_2 impurities from the bottom of the deethanizer were recovered in the still overhead product (Fig. 27.1*a*). There was an economic incentive to recover as much C_2 as possible in the deethanizer bottoms (or still overheads) without exceeding the LPG purity specifications for C_2. As the deethanizer bottom stream was difficult to analyze, a C_2–C_3 analyzer was installed in the still overhead product line. The analyzer measurement for the C_2/C_3 ratio was used to adjust the set point of the deethanizer temperature controller.

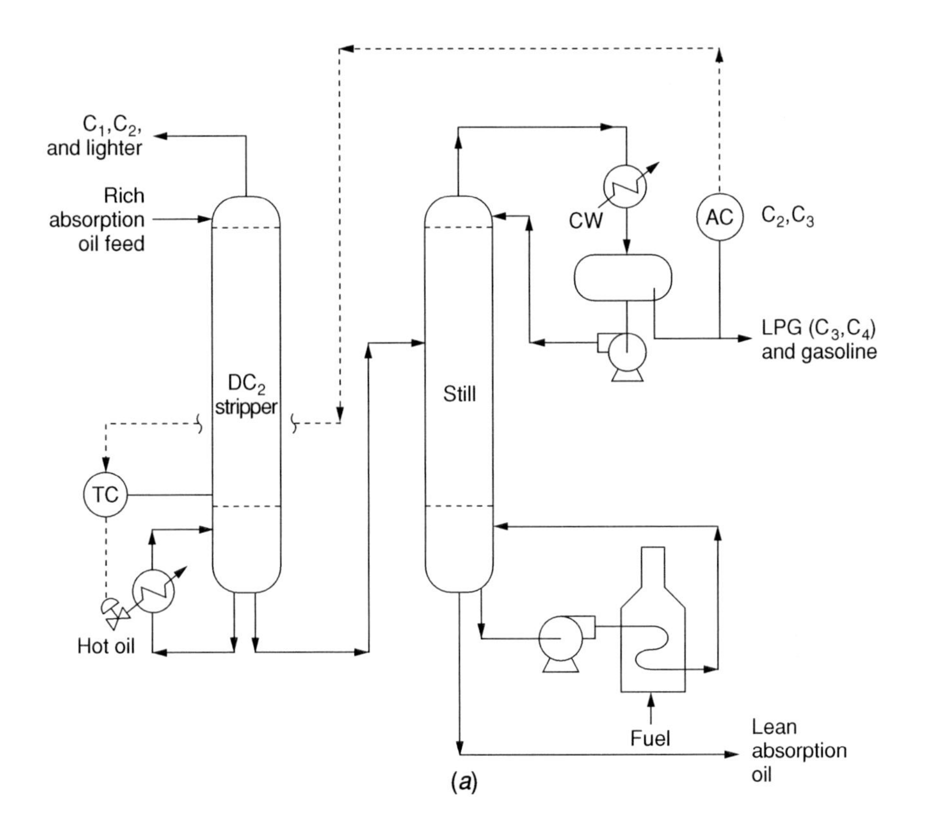

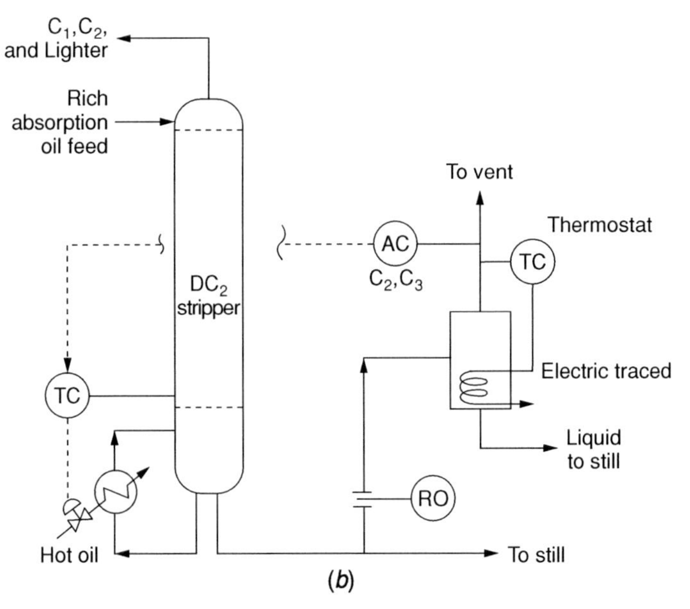

Figure 27.1 Analyzer from next tower gives poor control: (*a*) analyzer located in next tower product line, poor control; (*b*) modified system, with relocated DC_2 bottom analyzer, good control. (From Ref. 263. Reproduced with permission. Copyright © (1987) AIChE. All rights reserved.)

Problem The system did not work. Instead of getting better control of the C_2/C_3 ratio in the deethanizer bottom, the control was worse than without the analyzer.

Investigation System response was tested by taking the analyzer off control, then increasing the deethanizer temperature set point by a 5°F step. The analyzer reading changed steadily over a period of 5 hours following the change and only then stabilized.

Solution The analyzer was installed on the deethanizer bottom line (Fig. 27.1*b*) to eliminate the lag problem. A small liquid stream was sampled from the deethanizer bottom line at a fixed flow rate. The liquid flowed into a little flash pot to which a constant amount of heat was supplied. The analyzer was installed on the vent line from this pot. The correlation between the C_2/C_3 ratio in this stream and the deethanizer control temperature was determined experimentally. This system worked well with no further problems.

Chapter 28

Misbehaved Pressure, Condenser, Reboiler, and Preheater Controls

Three control malfunctions, each with around 30 case histories, hold the 14th, 16th, and 17th spots among distillation malfunctions (255). The three are control system assembly difficulties, temperature and composition control issues, and condenser and pressure control problems. In addition, reboiler and preheater controls were in the 25th spot. Had the survey lumped them up into one item, "control malfunctions," they would have featured prominently in the 3rd spot on the malfunction list. The survey as well as this book preferred to split and itemize them due to the vast differences between the issues. However, it is important to recognize that, despite their low place, control issues feature very prominently on the malfunctions list.

More pressure and condenser control case histories come from refinery than from chemical towers. One reason for this is refiners' extensive use of hot-vapor bypasses, which can be particularly troublesome (below). Like the other control issues, there is no apparent trend of growth or decline in pressure, condenser, and reboiler control issues.

About one-third of the pressure and condenser control case histories were problems with hot-vapor bypasses, practically all in refineries. There is little doubt that this is potentially the most troublesome pressure control method. Most of the problems are due to poor configuration of hot-vapor bypass piping, which evolves from poor understanding of its principles. These principles have been in the literature for more than 50 years, described in excellent papers in Whistler (533) and Hollander (217) and in some texts (250). When configured correctly, the author's experience is that hot-vapor bypasses are seldom troublesome. Cures have been to revert to the correct configurations.

Another troublesome pressure/condenser control is by cooling-water throttling. It has induced low cooling-water velocities and high outlet temperatures, leading to fouling, corrosion, and instability. Cures include switching to alternative methods or

Distillation Troubleshooting. By Henry Z. Kister

finding means (like inert blanketing or even aging) of keeping the valve open. A third troublesome issue is problems with vapor flow throttling resulting from low points that accumulate condensate in vapor product lines or from poor control configurations.

Reboiler and preheater controls have been troublesome in both refinery and olefins/gas plant towers. The reported case histories were equally split between reboilers and preheaters. Temperature control problems with preheaters were common, in most cases due to disturbances in the heating medium or due to vaporization in the feed lines. All the reboiler case histories reported involved a latent-heat heating medium. Hydraulic problems were common when the control valve was in the steam/vapor line to the reboiler while loss of reboiler condensate seal was common when the control valve was in the condensate lines out of the reboiler. The variety of solutions, well illustrated in Case Study 28.10, is a tribute to the ingenuity and resourcefulness of engineers, supervisors, and operators.

CASE STUDY 28.1 LIQUID LEG INTERFERES WITH PRESSURE CONTROL

Henry Z. Kister and Tom C. Hower, reference 263. Reproduced with permission. Copyright © (1987) AIChE. All rights reserved

Installation A potassium carbonate regenerator in a Benfield gas treating plant (Fig. 28.1).

Problem Column pressure varied erratically and could not be controlled properly. In addition, slugs of water entered the sulfur plant.

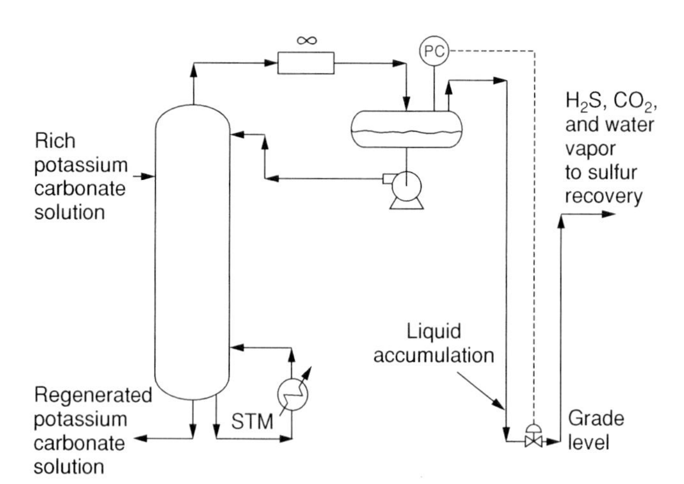

Figure 28.1 Liquid leg interferes with pressure control. (From Ref. 263. Reproduced with permission. Copyright © (1987) AIChE. All rights reserved.)

Cause The overhead vapor product pipe, after leaving the reflux drum, was lowered down to grade and then climbed up to pipe rack level. The backpressure valve was installed at grade to meet the maintenance requirement that all control valves be serviceable at grade. This created a low leg in the line. Water due to entrainment and atmospheric condensation accumulated in this leg and created a significant backpressure, which interfered with the pressure control loop.

Cure The backpressure valve was installed in the pipe rack, and the overhead product pipe was run directly from the reflux drum to the pipe rack. This eliminated the problem.

Alternative Cure In one refinery tower, an almost identical problem was experienced. This problem was eliminated by installing a liquid trap upstream of the control valve. The trap was a small drum below the valve, with an on/off level controller. The trap liquid was discharged to downstream of the control valve.

CASE STUDY 28.2 PRESSURE/ACCUMULATOR LEVEL CONTROLS INTERFERENCE

Installation Hydrogen chloride was recovered from halogenated HCs (Fig. 28.2). Tower overhead was condensed by boiling refrigerant in a kettle reboiler.

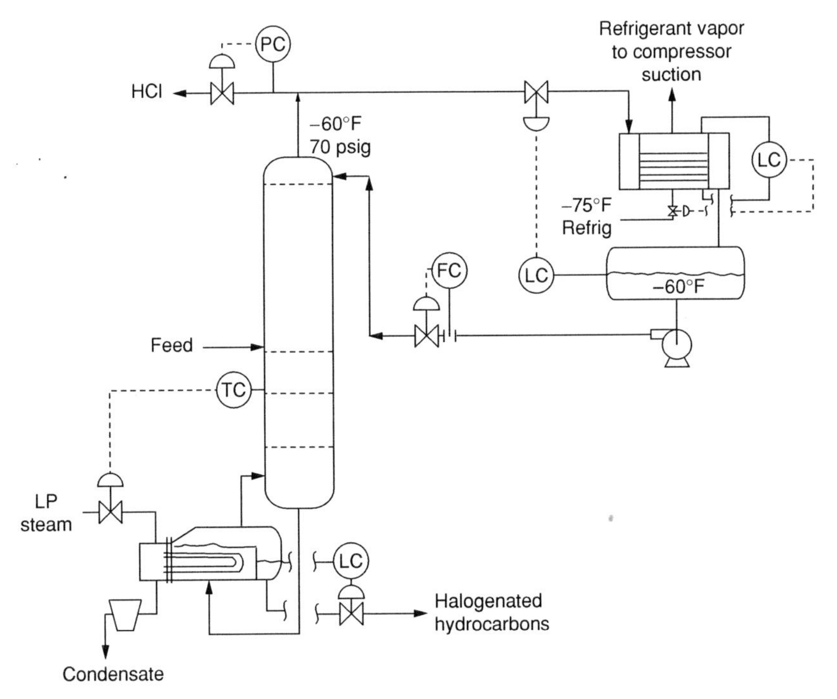

Figure 28.2 Control system on HCl recovery tower.

Problem The reflux drum-level control could not be operated satisfactorily. Every time it was placed in "automatic," it led to instability around the column overhead system.

Solution To avoid the instability, the level valve was operated fully open. The drum level was manually operated by adjusting the refrigerant level in the condenser.

Cascading the drum-level control onto the refrigerant level in the kettle was considered and rejected because it was not always possible to obtain a satisfactory signal from the kettle level transmitter. When the signal was unsatisfactory, there were problems with liquid entrainment into the refrigeration compressor suction.

Lesson Manipulating vapor product flow rate and manipulating vapor to the condenser are both very fast and very powerful controls. Tower pressure changes due to variations in vapor inventory, and these happen fast. Therefore, keeping steady pressure requires such fast and powerful control action. The converse is true for accumulator level. Level variations are much slower. The accumulator level therefore requires much slower control action. Having a fast, powerful manipulation of accumulator level is a common cause of interference and instability in tower overhead systems.

CASE STUDY 28.3 EQUALIZING LINE MAKES OR BREAKS FLOODED CONDENSER CONTROL

Installation Petrochemical tower with pressure control by flooded total condenser (Fig. 28.3*a*). The reflux drum pressure was controlled by manipulating a small valve in the 1-in. condenser bypass line. The 2-in. manual valve bypass was operated shut.

Problem Reflux drum pressure was unsteady. The bypass valve often fully opened, causing loss and fluctuation of drum pressure. These fluctuations destabilized the main tower pressure control.

Cause Liquid from a flooded condenser enters the reflux drum subcooled. The subcooled liquid had a much lower vapor pressure than the drum pressure. For proper operation, the vapor bypass flow rate must be high enough to keep the surface of liquid in the drum hot, that is, hot enough to produce a vapor pressure equal to the set point of the drum pressure controller. The valve was undersized to achieve this. Also, there could have been some interaction between the two pressure controllers.

Solution Opening the 2-in. control valve bypass increased the vapor flow rate to the drum and eliminated the instability.

Postmortem Opening the 2 inch control valve bypass reduced the control scheme of Figure 28.3a to the common classic scheme of Figure 28.3b, which works well with an adequately-sized bypass (87, 250).

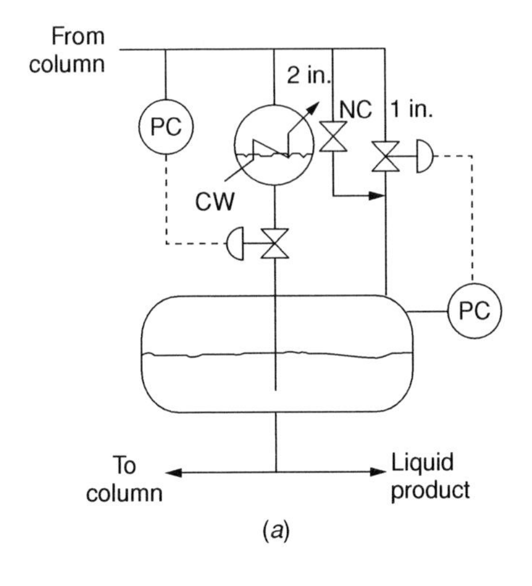

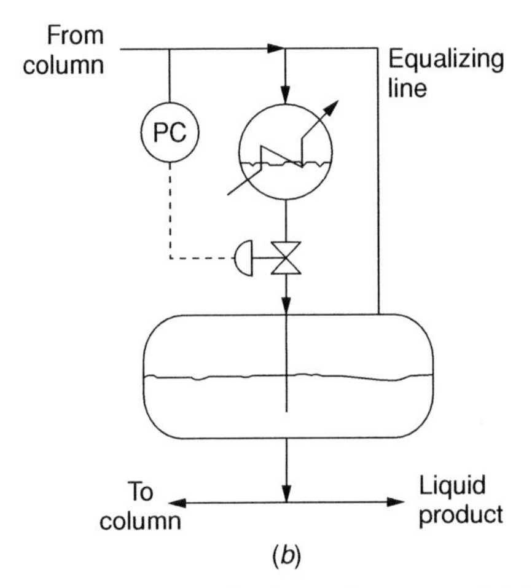

Figure 28.3 Vapor bypass makes or breaks flooded condenser control: (*a*) actual system; (*b*) classic control scheme that actual scheme was reduced to when opening 2-in. manual valve.

CASE STUDY 28.4 INERTS IN FLOODED REFLUX DRUM

Henry Z. Kister and Tom C. Hower, reference 263. Reproduced with permission.

Installation Absorption–refrigeration gas plant lean-oil still which separated LPG and gasolines as a top product from absorption oil. The still used a "flooded reflux drum" pressure control method (PC 1 in Fig. 28.4). Pressure was controlled by controlling liquid product leaving the drum. Since the reflux drum was full of liquid (or flooded), closing the valve in the product line would back up liquid into the condenser,

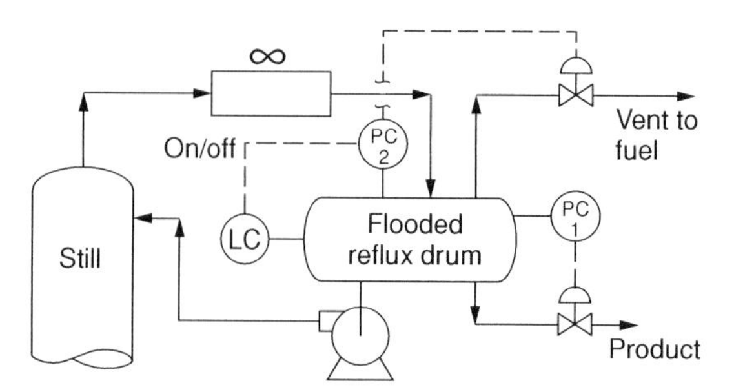

Figure 28.4 Automatic venting of flooded reflux drum. (From Ref. 263. Reproduced with permission. Copyright © (1987) AIChE. All rights reserved.)

flood condenser tubes, reduce the rate of condensation, and thus raise column pressure. Similarly, opening the product valve would lower liquid level in the condenser and act to decrease column pressure.

Problem Small quantities of C_1 and C_2 often entered the column and accumulated as vapor in the reflux drum, thus creating a vapor space near the top of the drum.

When the vapor space was formed, column pressure could not be controlled. Normally, manual venting would have been satisfactory to overcome the problem, but the plant was manned only 8 hours per day and was operated 24 hours a day.

Solution An automatic venting system (Fig. 28.4) was devised. A second pressure controller (PC 2), a level controller, and a control valve in the vent line were installed. The set point of PC 2 was lower than the set point of the normal pressure controller, PC 1. Normally, PC 2 was tripped off and did not operate, so that the vent valve was closed and PC 1 carried out the control action.

When inerts accumulated and a vapor gap formed, the level controller sensed a drop in level. The level controller then sent an air signal that activated PC 2. Pressure control 2 had a lower set point than PC 1 and therefore acted to open the vent valve. As the pressure would fall, PC 1 would close, thus helping liquid-level buildup. As soon as the inerts were vented and the liquid refilled the drum, the level controller stopped the air signal to PC 2, the vent valve closed, and operation returned to normal.

CASE STUDY 28.5 POOR HOOKUP OF HOT-VAPOR BYPASS PIPES

Henry Z. Kister and James F. Litchfield, Ref 260. Reprinted courtesy of *Chemical Engineering*

Installation A new debutanizer column separating C_4 and C_3 HCs from gasoline. Column overhead vapor was totally condensed by a battery of four submerged condensers (Fig. 28.5*a*). The reflux drum was elevated. The condensers were vented to the drum using 1-in. vent lines (not shown). Tower pressure was controlled by a hot-vapor bypass hooked up as shown in Figure 28.5*a*.

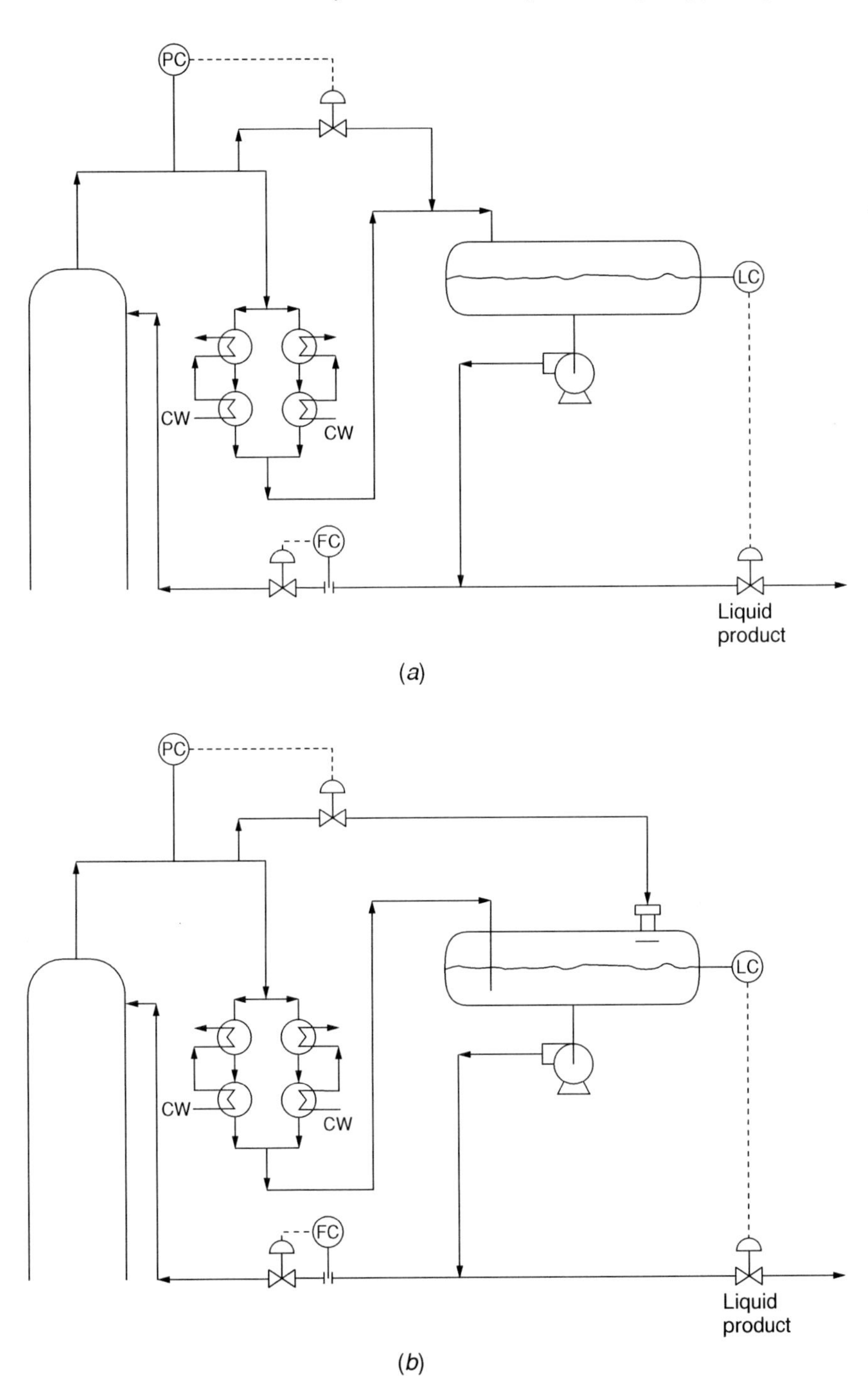

Figure 28.5 Hot-vapor bypass hookups: (*a*) incorrect, leads to pressure fluctuations; (*b*) modified, good pressure control. (From Ref. 260. Reprinted courtesy of *Chemical Engineering*.)

When the tower was put into service, it experienced severe pressure fluctuations. It was impossible to keep column pressure constant. This bottlenecked tower capacity.

Diagnosis Correct piping is mandatory for the success of the hot-vapor bypass control method. Bypass vapor must enter the vapor space of the reflux drum (Fig. 28.5*b*). The bypass should be free of pockets where liquid can accumulate; any horizontal runs should drain into the reflux drum. Most important, liquid from the condenser must enter the reflux drum well below the liquid surface. The bottom of the drum is the most suitable location, but extending the liquid line to near the bottom of the drum (Fig. 28.5*b*) is also acceptable. These recommendations were first published in the literature almost 50 years ago (217, 533), and have been strongly endorsed by key recent sources addressing column pressure control methods (87, 250).

Figure 28.5*a* shows a very poorly piped variation of this system. With this scheme, subcooled liquid mixes with dew point vapor. Collapse of vapor takes place at the point of mixing. The rate of vapor collapse varies with changes in subcooling, overhead temperature, and condensation rate. Variation of this collapse rate induces pressure fluctuations and control valve hunting. Similar problems were repeatedly described in the literature since the 1950s (217, 533), yet strangely enough, the incorrect hookups in Figure 28.5*a* keep reappearing in modern designs.

Solution The liquid and vapor lines were separated. The vapor line was modified so that it introduced the vapor into the top of the reflux drum. The liquid line was extended into the bottom of the reflux drum. Figure 28.5*b* shows the modified system. After this was implemented, the tower pressure no longer fluctuated, and the problem was completely solved. Following these modifications, one could feel the differences in temperature between the top part of the reflux drum (which contained hot vapor) and the bottom part (which contained subcooled liquid) simply by touching the drum.

Related Experience Another HC separation tower had a hot-vapor bypass system similar to that in Figure 28.5*a*. The tower experienced pressure fluctuations and inability to control pressure. To fix, the liquid and vapor lines were separated. The vapor line was modified so that it introduced the vapor into the top of the reflux drum. This was a major improvement but did not fully solve the problem. At a later time, the liquid line was extended into the bottom of the reflux drum, going to the Figure 28.5*b* arrangement. After this was implemented, the tower pressure no longer fluctuated.

CASE STUDY 28.6 PRESSURE CONTROL VALVE IN THE VAPOR LINE TO THE CONDENSER

In memory of Carl Unnuh (Retired), C. F. Braun Inc., Alhambra, Ca. Briefly described in Ref. 471.

Installation A total air condenser condensing column overhead. Condenser length was 32 ft and the tubes were sloped 5–6 ft. Condensate was backed up from the accumulator drum to ensure there was a liquid level inside the tubes. This level varied according to the capacity requirements. Column pressure was controlled by a control valve in the line from the column to the condenser (Fig. 28.6).

Problem A liquid hammer that shook the whole unit occurred during start-up.

Cause Under some conditions during start-up, the control valve would completely shut. When this occurred, the air cooler quickly condensed all the vapor available downstream of the valve. The condenser pressure dived. This caused liquid to be rapidly sucked from the accumulator drum, producing the liquid hammer.

Solution The valve was modified so it would not completely shut.

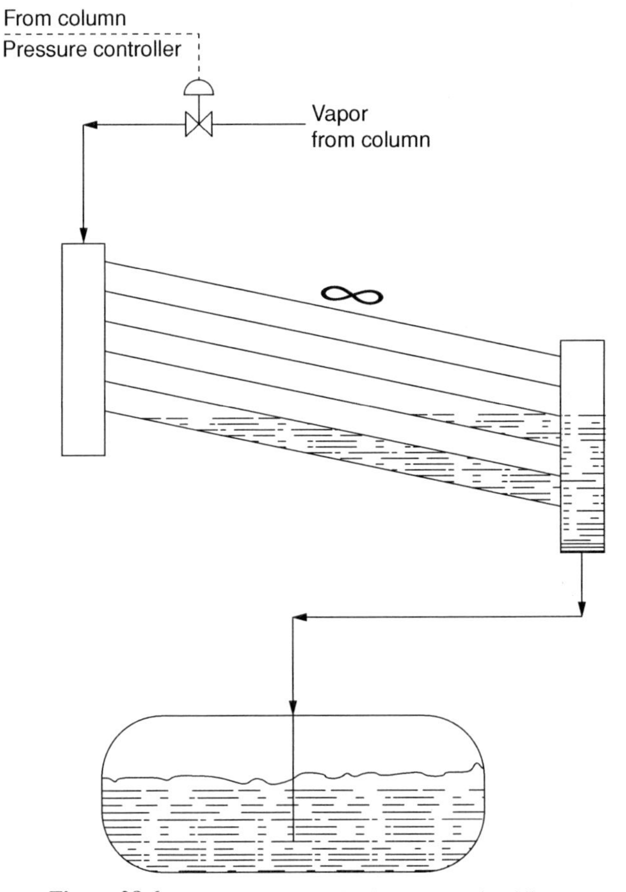

Figure 28.6 Pressure control valve in tower overhead line.

CASE STUDY 28.7 CAN CONDENSER FOULING BY COOLING-WATER THROTTLING BE BENEFICIAL?

Cooling-water throttling is one of the most troublesome condenser control methods (250). It is usually avoided because the throttling can lead to low cooling-water velocities and high outlet temperatures, both of which accelerate fouling and corrosion in the condenser (250). Nonetheless, there have been a few favorable experiences with cooling-water throttling controls, as described below.

Experience A Deep vacuum tower with a head pressure of 20 mm Hg. Initially, tower pressure was controlled by manipulating the cooling-water flow (Fig. 28.7*a*). The system led to excessive product losses in the ejector off gas. An inerts injection was added (Fig. 28.7*b*) to control tower pressure. The drum temperature controller was used to manipulate the coolant flow. This temperature was minimized to minimize product losses while kept high enough to avoid absorbing components from the off gas into the product.

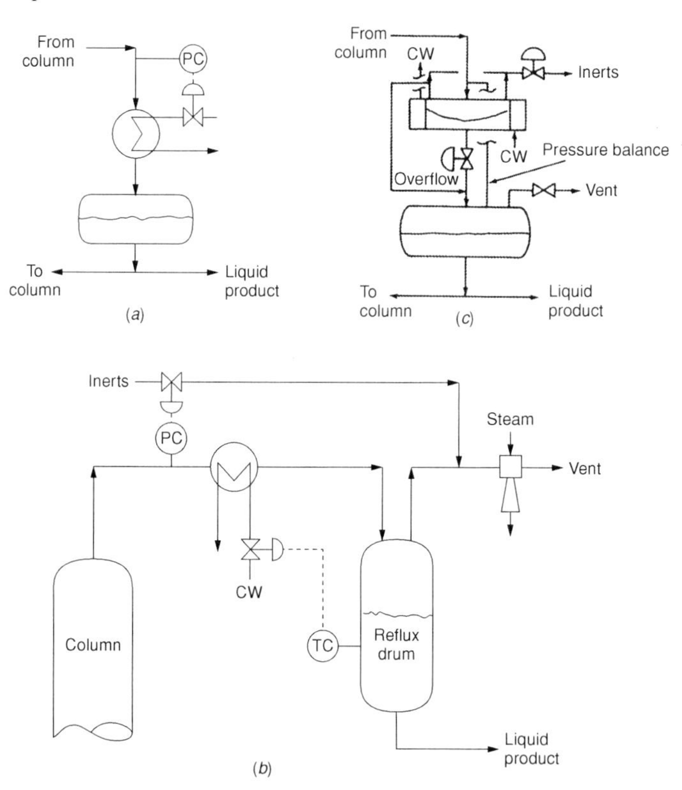

Figure 28.7 Throttling cooling water for control: (*a*) for tower pressure control; (*b*) using inert addition for pressure control and cooling-water throttling for reflux drum temperature control; (*c*) in flooded partial condenser. [(*a*, *c*) Reprinted with permission from Ref. 250. Copyright © 1990 by McGraw-Hill.]

Experience B Reflux drum temperature was controlled by a valve manipulating the cooling water (similar to Fig. 28.7*b*). Upon start-up, the valve was heavily throttled, cooling-water flow and velocity were low, and cooling-water outlet temperature was high. The low velocity and high temperature led to accelerated fouling. As the exchanger fouled, the heat transfer coefficient went down, and the controller opened the valve. This increased the cooling-water velocity and lowered the cooling-water outlet temperature. The fouling did not progress any further, and the condenser ran well to the next turnaround.

Experience C In another tower, a flooded partial condenser system was used (Fig. 28.7*c*). Upon turndown, liquid level rose in the condenser, reaching close to the vapor outlets. The condenser experienced entrainment and liquid surging. To alleviate the problem, cooling water to the condenser was manually throttled. This lowered the condenser ΔT, which in turn led to a lower liquid level in the condenser. At the lower liquid level, the entraining and surging did not occur. After some time, the condenser fouled enough, and the cooling-water throttling valve could be fully opened with no further problems.

CASE STUDY 28.8 CONTROL TO PREVENT FREEZING IN CONDENSERS

Background Figure 28.8*a* shows a tower pressure control scheme that is effective for preventing freeze-ups in the condenser (250). It is used with either cooling water or demineralized water. The scheme recirculates warm water from the cooler outlet to the cooler inlet to keep the inlet water temperature a safe margin above the freezing point of the chemicals on the process side. Column pressure control is achieved by manipulating the condenser inlet temperature within a desired range.

A popular variation of this scheme eliminates the pressure control in the cooling-water return line. The return pressure simply "rides" on the cooling-water return pressure. Another variation retains the control valve in the cooling water return, hooks the TC onto it, and eliminates the one in the cooling-water supply. Retaining the control valve in the cooling-water return reduces the likelihood of boiling the cooling water and counters the possibility of vacuum on the cooling-water side of elevated condensers.

Problem When the booster pump fails, cooling is lost to the condenser. Thus a booster pump failure is analogous to cooling-water failure.

Solution Figure 28.8*b* shows the modified system that overcame the problem (410). If the booster pump fails, the full cooling-water flow is still going through the condenser and can be automatically controlled by the pressure controller. The check valve in the booster pump discharge is very important and needs to be properly functioning. Without it, the water will bypass the condenser and go backward through the booster pump. The path of flow during booster pump failure is marked as dashed lines in Figure 28.8*b*.

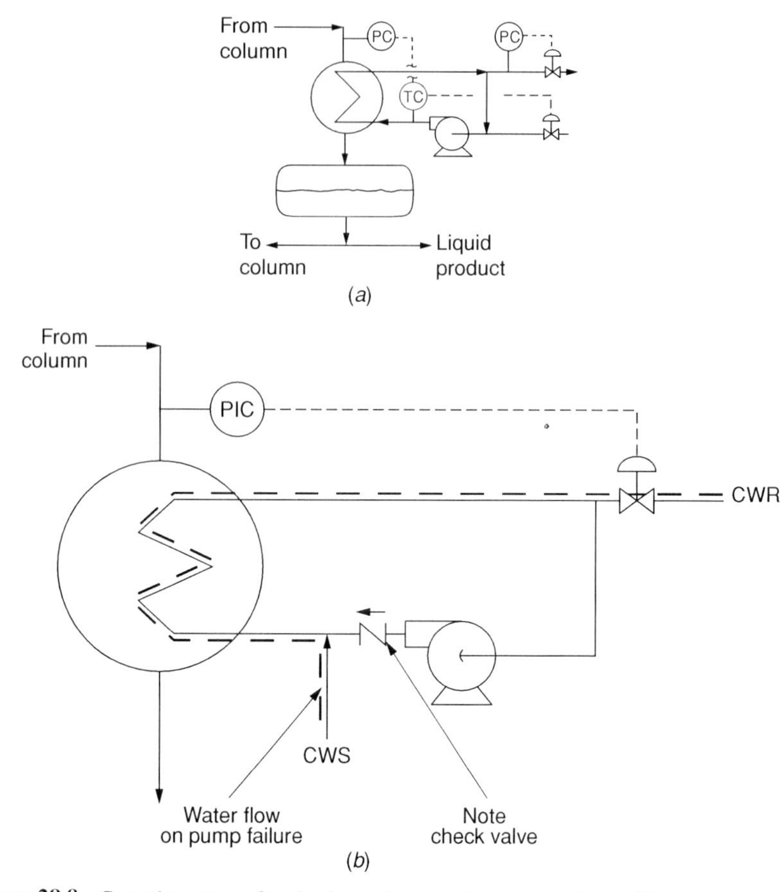

Figure 28.8 Control to prevent freezing in condensers: (*a*) common scheme; (*b*) scheme permitting continued operation during booster pump failure. [(*a*) Reprinted with permission from Ref. 250. Copyright © 1990 by McGraw-Hill.]

CASE STUDY 28.9 VALVE IN REBOILER STEAM INDUCES OSCILLATIONS DURING START-UP

Installation A solvent recovery tower. Overhead from the tower was an organic/water azeotrope. Bottom product from the tower was water. Column pressure was slightly higher than atmospheric. The tower vertical thermosiphon reboiler was heated by condensing 125 psig steam. The reboiler was oversized for the normal operation duty. The steam flow to the reboiler was controlled by a valve in the steam line to the reboiler (Fig. 28.9*a*). A tower tray temperature was cascaded onto the steam valve.

Problem At start-up, there were severe oscillations in the tower heat input and vapor rate. These oscillations were accompanied by hammering. The oscillations were severe enough to bring the start-up to a halt.

Cause Reboiler heat is supplied according to the equation

$$Q = UA\,\Delta T_{\ell m}$$

where Q is reboiler heat input (Btu/h), A is reboiler surface area (ft^2), U is the heat transfer coefficient (Btu/h ft^2 °F), and $\Delta T_{\ell m}$ is the log-mean temperature difference (°F).

When the control valve is in the steam supply, the total reboiler area is utilized, so A is the maximum available. At low-rate operation (e.g., during start-up), the heat duty is low, so Q is minimized. Also at start-up the tubes are clean, so the heat transfer coefficient U is high. At these conditions, the equation can be satisfied if $\Delta T_{\ell m}$ is minimized, that is, when the control valve closes and the condensing temperature approaches the boiling point of the process-side fluid.

Here the process boiling temperature was the boiling point of water at just above atmospheric pressure. At start-up, the steam would condense a few pounds of pressure above this. Thus the condensing pressure fell below the pressure in the condensate header, which was at 25 psig. The steam condensate was unable to flow forward, so it accumulated in the reboiler shell. Further, condensate would backflow from the higher pressure condensate header into the lower pressure reboiler. The reboiler steam contacted and collapsed onto this subcooled condensate, which produced the hammering.

The condensate built a level inside the reboiler until equilibrium was reached, with enough reboiler area covered by condensate to raise the steam chest pressure just above the condensate header. Flow forward into the condensate header resumed, and the system reached steady state. This new steady state was unstable and unable to survive even mild disturbances. Consider a disturbance that slightly increased the heat duty. The steam valve opened, raising the steam chest pressure, which raised reboiler ΔT, thus supplying more heat. At the same time, the higher pressure pushed the condensate level down, exposing more tube area for condensation. This action also increased the heat input, but only after a considerable time lag. The two actions are interactive and tend to chase each other. Together they plunge the heat duty into oscillations, cycling, and erratic behavior.

Solution The condensate valve to the steam trap (NO in Fig. 28.9*a*) was shut and the condensate valve to the deck (NC in Fig. 28.9*a*) was opened. With the reboiler condensing pressure above atmospheric, the condensate drained, so the reboiler heat input stabilized and no longer oscillated. The tower could operate in this mode only for a short time due to the environmental problem generated by sending hot condensate into the sewer. Plant rates were quickly raised. At the higher reboiler duty, steam chest pressure exceeded the condensate header, so the reboiler condensate could be returned to the condensate header.

Another Plant An identical problem was experienced in two light HC towers in a gas plant. Both were reboiled by oversized steam-heated kettle reboilers with the

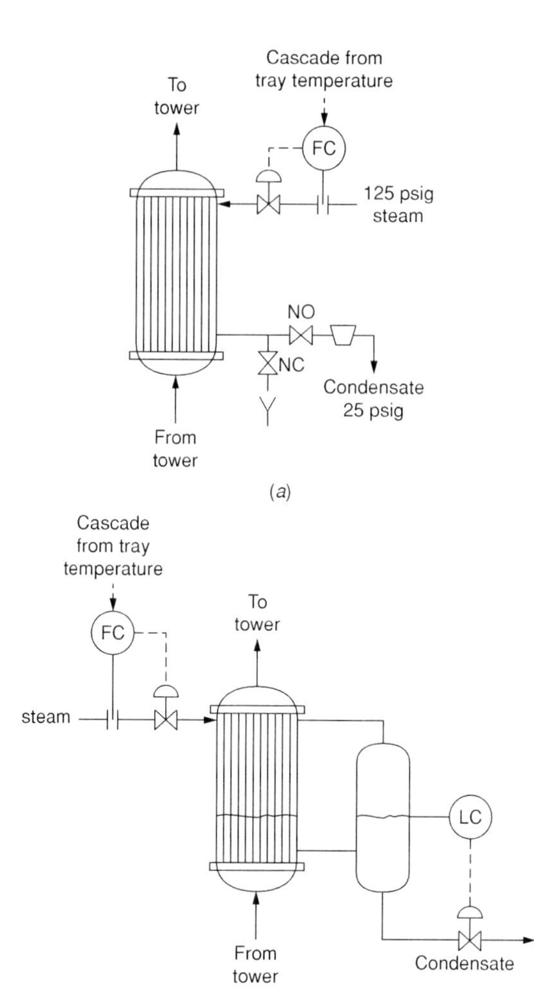

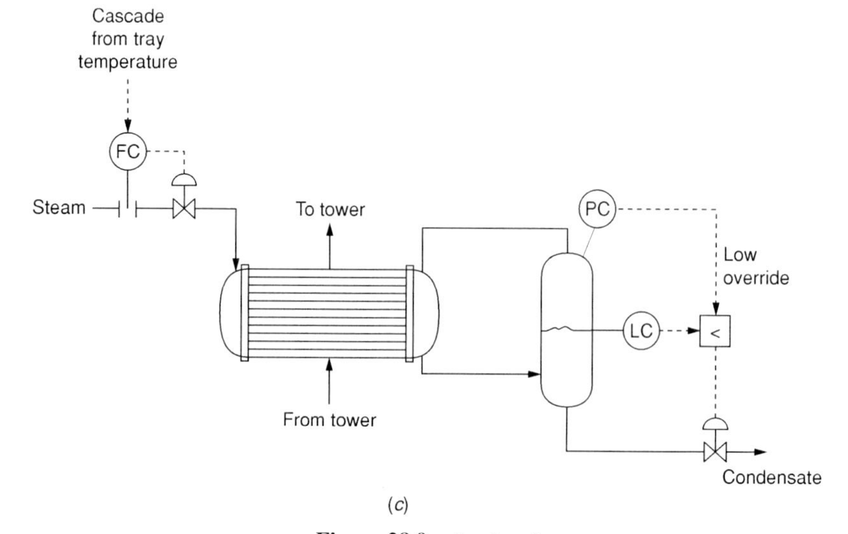

Figure 28.9 *Continued*

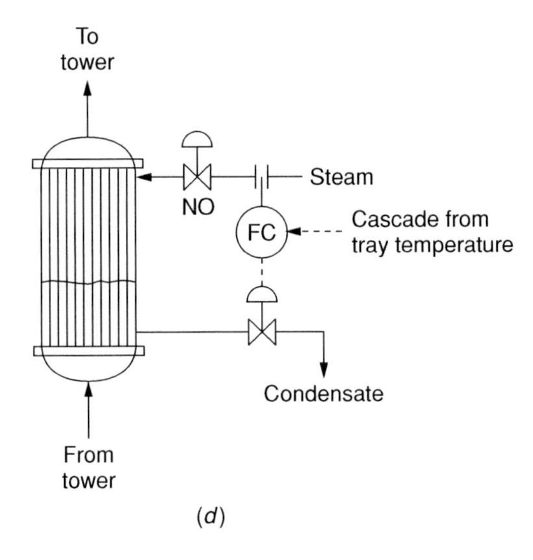

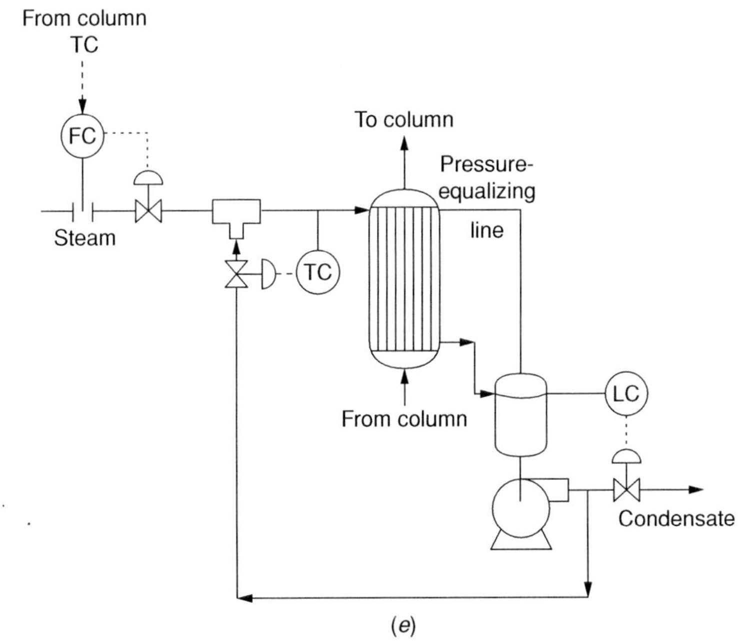

Figure 28.9 Control with valve in steam (or vapor) supply to reboiler: (*a*) original control scheme, led to severe oscillations at start-up; (*b*) adding condensate drum with level control to reduce effective reboiler area during periods of low-rate operation; (*c*) condensate drum and low-pressure override, which automatically reduces effective reboiler area at turndown; (*d*) switching to control by manipulation of valve in condensate line; (*e*) adding pump and desuperheating loop to eliminate oscillations and increase reboiler run length in fouling service.

control valves in the steam supply lines. In both, the instability was eliminated by opening the trap bypass to the deck.

Footnote Condensate draining to the deck needs to be avoided when there is a risk of the hot condensate causing vaporization of hazardous materials in the sewer system. See Case Study 1.3.

CASE STUDY 28.10 CONDENSATE DRUMS ELIMINATE REBOILER START-UP OSCILLATIONS

Installation A similar experience took place in many different towers in several different plants. In all, the tower had an oversized vertical thermosiphon reboiler with a control valve in the steam line to the reboiler. The arrangement was the same as in Figure 28.9*a*, except that the steam and condensate pressures varied.

Problem At low rates and during start-ups the reboiler heat input and vapor rate to the tower experienced severe oscillations.

Cause The cause was identical to that in Case Study 28.9, with the reboiler steam chest pressure falling below the condensate header pressure. Detailed explanation is in Case Study 28.9.

Solution The variety of solutions described below is a tribute to the ingenuity and resourcefulness of engineers, supervisors, and operators.

Testing In one of the cases, inerts accumulation was suspected. When the vent valve on the steam chest was open, air got sucked in, indicating a vacuum in the steam chest. In another case, a pressure gage indicated a steam chest pressure lower than that in the condensate header.

Towers A–C In two towers, a condensate drum with a level control was installed (Fig. 28.9*b*). In a third, the level control was mounted directly on the steam chest (no condensate drums). The condensate level was set high enough to ensure that the steam chest always exceeds the condensate header pressure. No more problems occurred after this.

Tower D In this tower, in addition to the drum and level control, a low-pressure override controller was added to automatically maintain the drum pressure above the condensate pressure (Fig. 28.9*c*). Normally, the drum pressure was higher than the condensate pressure, and the LC controlled the condensate valve. During start-up and turndown, when the pressure fell, the PC would take over and close the valve, backing condensate up the reboiler tubes.

Tower E This tower had a spare reboiler. For start-up, the fouled reboiler was used. Once the plant reached full rates, the tower was switched to the clean reboiler.

Towers F and G In these towers, inerts were injected into the condensing side of the reboiler. This reduced the heat transfer coefficient and raised the pressure on the condensing side. In one of the two, the oscillations were accompanied by hammering. Hearing hammering signaled the operators a need to increase the inerts injection rate.

Tower H In this tower, the steam trap in the condensate line of Figure 28.9*a* was replaced by a control valve. The valve in the steam supply line was kept fully open and the steam flow controller was hooked to the new condensate valve (Fig. 28.9*d*). This system, however, has its own drawbacks (250).

Towers I–K In these towers, pumping traps were installed in the condensate lines, in which steam pressure was used to pump the condensate into the condensate header.

Tower L In two different towers, both in polymerizing service, a condensate pump was added (Fig. 28.9*e*), permitting steam condensation at the lowest possible pressure and temperature. Condensate from the pump discharge was injected to desuperheat the reboiler inlet steam. These minimized reboiler temperatures and extended its run length.

Chapter 29

Miscellaneous Control Problems

This chapter includes the remaining "bits and pieces," that is, the control problems that did not fall into the previous classifications. These include interaction with the process, differential pressure control, flood control and indicators, batch distillation control, and problems in the control engineer's domain.

Of special interest is a new category of malfunctions: advanced control problems. One issue is updating multivariable controls (MVCs), which can be troublesome when the process train changes, especially if the MVC simultaneously optimizes an entire unit rather than individual towers. Another issue has been the response to bad measurements, with misleading measurements veering the control away from optimum. So far, the number of case histories of troublesome advanced controls has been low, which is a great tribute to the exciting technology of distillation tower advanced controls.

CASE STUDY 29.1 NATURAL FLOODING OR HYDRATES IN A C_2 SPLITTER?

Installation A C_2 splitter separating ethylene from ethane had 85 trays above the feed, 30 below. The tower was the main process bottleneck in an ethylene plant and was operated right at its flood limit. To watch for flood, the differential pressure (dP) between the top and bottom was monitored.

Problem The tower experienced severe flood about once every 2 months. A severe flood would generate off-specification ethylene, which often needed to be flared, and would induce large cuts in production rates for between a shift and a day in order to overcome.

Most floods were believed to have been caused by hydrates, that is, the deposition of icelike particles formed when small quantities of moisture (less than 1–2 ppm) enter the tower. The hydrates accumulate and plug trays; see Case Study 2.20 for

Distillation Troubleshooting. By Henry Z. Kister

more detail. Hydrates frequently occur in C_2 splitters. The corrective measure for hydrates was to inject methanol, which acts like antifreeze and dissolves the hydrates.

Analysis It was noticed that many times injecting methanol aggravated the flood instead of helping. This would have been the case had the tower flooded naturally (i.e., not due to a hydrate). For natural flood, the corrective action was to slightly back off rates.

Hydraulic calculations showed that the trays above the feed operated near their natural limit while the trays below the feed had some margin. On the other hand, the industry's experience has been that in most (but not all) high-pressure C_2 splitters tray plugging due to hydrates initiates below the feed. An idea postulated by an operator and endorsed by the staff was that recording the *dP* of the top and bottom sections separately would permit distinguishing hydrates from natural floods. These "flood recorders" would also permit catching a flood as it initiates, before there is a need for a drastic cut in rates.

Solution Separate *dP* recorders were installed across the top and bottom sections. The payout was a few months based on reducing the incidence of severe floods from six to five per year. In the years that followed, the number of severe flood incidents dropped to one per year. Almost always, it was the upper *dP* that rose first, suggesting natural flood and inducing an immediate and correct action of slight backing off in rates. Hydrates seldom occurred in that tower.

Distillation Troubleshooting Database of Published Case Histories

Chapter 1 Troubleshooting Distillation Simulations

Case	References	Plant/Column	Brief Description	Some Morals
			1.1 VLE	
1.1.1 Close-Boiling Systems				
118	464	Chemicals superfractionator	A laboratory error gave incorrect VLE, based on which a tower with 200 theoretical stages was built where over 300 stages were required. With the 200 stages, product purity could not be achieved. The plant was forced to rerun the purified material a second time through the tower, effectively halving plant capacity.	A VLE error can lead to a major failure.
122	250		A 2% difference in relative volatility in a low-relative-volatility ($\approx$1.1) system accounted for a difference of 50% in the tray efficiency. The designer's efficiency worked only with the designer's volatility; the operator's efficiency worked only with the operator's volatility.	
124	422	Aromatics ethylbenzene–styrene	Identical column simulations using major commercial simulators, all employing SRK VLE, calculated entirely different product purities. Reason was small differences in the critical temperature and pressure and in the acentric factor for styrene.	Beware of using equation-of-state predictions for close separations.
140	358	Butadiene	1,2-Butadiene is less volatile than 1,3-butadiene and leaves mostly in the bottom, but a commercial simulator predicted it would leave in the top. Problem was due to incorrect critical constants used in the equation of state.	Same as 124.
137	237	*n*-Heptane–toluene	Use of fraction composition data from batch distillation tests showed that the popular VLE choice for this system gave poorer simulation of plant data than alternative VLE procedures.	Batch distillation data can help select VLE options.
314, 315			*VLE error leads to mismatch between plant data and simulation, Section* 1.3.1.	
102, 147			*Due to extrapolation to pure product, nonideal system, Section* 1.1.4.	

1.1.2 Nonideal Systems (*See also Sections* 1.1.3–1.1.6)

123	469	Wastewater drum	DCM concentration in the wastewater was very low. Based on ideal behavior, the vapor vent to atmosphere from the wastewater storage tank would have contained little DCM. Measured DCM at the vent was 27 mol %.	Nonidealities are maximized near infinite dilution.
160	46	Olefins water quench/ C_2 splitter Many plants	Ammonia from treatment chemicals or feedstock impurities distilled up in the deethanizer and C_2 splitter due to a high activity coefficient, causing off-spec ethylene product. It also generated quench/process water quality and pH problems. Sufficient blowdown from, and improved pH control of, the quench water; feedstock specs; amine treaters; and absorption in dryers; have been (mostly successful) remedies.	
163	314	NGL deethanizer, depropanizer, debutanizer	Measured distribution of the more volatile sulfur compounds among fractionator products poorly matched predictions from the commercial simulations. Largest discrepancies were methyl mercaptan, CS_2 and DMS tending to go into lighter products then simulated. Least discrepancies were with COS and ethylmercaptan. Issues were also observed in some simulation predictions of component vapor pressures.	
164	314	NGL deethanizer, depropanizer, debutanizer	Measured distribution of the more volatile sulfur compounds among fractionator products was compared to predictions from three commercial simulators. One simulator gave poor predictions, especially for the splits of methyl and ethyl mercaptans. Predictions from the other two simulations were closer to measured, except for ethylmercaptan which they predicted to go into a lighter product than measured.	
1150			*C_4 vaporizes out of C_4/acetonitrite when dumped into sewer, Section 14.12.*	
DT1.1, DT1.3			*Nonideality enhances volatility of high-boiling components, inducing them into distillate.*	

(*Continued*)

Chapter 1 Troubleshooting Distillation Simulations (*Continued*)

Case	References	Plant/Column	Brief Description	Some Morals
DT1.2			*Caustic makes water less volatile, inducing it into debutanizer bottom.*	
DT1.4			*Presence of methanol enhances solubilities of* H_2S and CO_2 *in feed drum water.*	
158	398	Refinery HF alkylation main fractionator	Modeling tower overhead using NRTL, and bottom section with PR as well as using system-specific interaction parameters, gave a better match to plant data than using PR. (See also 338, Section 1.3.1.)	
157, DT1.5	524	Methanol–butanol–water	Simulation NRTL parameters regressed from binary data from a major data bank gave incorrect predictions for the heterogenous ternary.	
151	100, 101	*sec* butanol-SBE–water	NRTL parameters optimized for VLE data predicted azeotrope but overpredicted water solubility in SBE mixtures. Optimizing for LLE data predicted solubility but failed to predict butanol–SBE azeotrope. A hybrid approach worked best.	Hybrid approach may be the best way for a complex system.
153	101	Acetone–phenol	A data set taken from a major data bank turned out to be thermodynamically inconsistent and led to optimistic separation predictions.	Check thermodynamic consistency of VLE data.
109	413	Carbon tetrachloride, TCE-CTC tower	The concentration of TCE in CTC was higher than expected. A total reflux test showed that separation near the column bottom was worse than expected. Either VLE nonideality or decomposition of chlorinated ethanes at the reboiler temperature was the culprit.	Total reflux tests are invaluable. Watch out when extrapolating VLE data.
1.1.3 Nonideality Predicted in Ideal System (*See also Section* 1.1.6)				
148	91	EG-DEG	NRTL parameters were regressed from two independent data sources. Discrepancies between pure component boiling points of the two sources led to predicting a nonexistent azeotrope in this ideal system.	Check boiling point consistency.
152			*Due to nonideal VLE extrapolation, Section* 1.1.5.	
1.1.4 Nonideal VLE Extrapolated to Pure Products				
102	335	Acetylene solvent–water stripper	Solvent losses were far greater than design. Unsuccessful extrapolation of VLE data was one of the causes. Increasing number of trays and raising reflux helped reduce losses.	Caution is required when extrapolating to infinite dilution.

147	91	Ethanol–benzene	Wilson and Van Laar correlation of same VLE data gave different VLE at the low-volatility benzene-rich end, making a large difference to the number of stages.	Same as 102.
DT1.6			*Unusual hydrogen bonding reduces methanol volatility near infinite dilution in water.*	
152			*Due to poor extrapolation to different pressure, Section* 1.1.5.	
1.1.5 Nonideal VLE Extrapolated to Different Pressures				
129	182	AMS–phenol	Column pressure was lowered from 100 to 30 mm Hg to improve separation and valve trays replaced by screen trays to match the capacity. Separation did not improve. Extrapolating VLE from 100 to 30 mm Hg gave optimistic expectations that did not materialize.	Caution is required when extrapolating VLE data in pressure.
152	91, 100, 101	MEK–acetone–water	NRTL was used with binary parameters regressed from atmospheric data using two different options. Extrapolating one of the options to 6 atm predicted a nonexistent azeotrope for the ideal MEK–acetone binary. Cause was poor extrapolation to infinite dilution.	Same as 129.
154	101	Acetone–phenol	Simulation regression, based only on isobaric data, predicted more difficult acetone stripping at lower pressure. In-house regression of data from the same source, that also included isothermal data, gave an order-of-magnitude higher volatility at lower pressure.	Same as 129.
150	91	*sec*-Butanol-SBE	A three-parameter NRTL regression extrapolated to higher pressure much better than a five-parameter regression.	More does not always mean better.
1.1.6 Incorrect Accounting for Association Gives Wild Predictions				
155	100	Formic acid–acetic acid–water	NRTL parameters regressed from binary data gave incorrect azeotrope and VLE prediction when used with an ideal gas vapor. Correctly accounting for vapor-phase association gave good predictions.	Good VLE data cannot compensate for selection of incorrect vapor phase model.
156, DT1.7	524	Acetic acid–*n*-butylacetate–water	Same as 155.	Same as 155.

(*Continued*)

Chapter 1 Troubleshooting Distillation Simulations (*Continued*)

Case	References	Plant/Column	Brief Description	Some Morals
149	91	Acetone–DAA	Commercial simulation regression of limited data incorrectly accounted for dimerization of acetone to DAA. This resulted with nonideal behavior prediction for this ideal pair.	Beware of association when evaluating VLE.
1.1.7 Poor Characterization of Petroleum Fractions				
DT1.8, DT1.9			*Leads to simulations grossly underestimating residue yields.*	
117	169	Refinery five deep-cut vacuum towers	Design wash oil flow rate was too small, leading to drying, coking, high pressure drop, poor-quality gas oil, and short runs. Cause was simulations that underestimated the fraction of wash oil vaporized. In all cases, inaccurate TBP characterization of the heavy fractions of the crude led to the underestimates.	Incorrect TBP characterization of the heavy fractions breeds coking.
130	190	Refinery vacuum	Inaccurate TBP characterization of the heavy fractions of the crude led to a wash oil flow rate too small to prevent coking in a deep-cut wash bed. Coke plugged the level bridle and draw nozzle on the slop wax collector tray. Unable to drain, slop wax was reentrained into the wash bed.	Same as 117.
			1.2 Chemistry, Process Sequence	
110	413	Absorption of HF from HCl gas	HF absorption by wash with aqueous HCl was poor. HF escaping in the column overhead destroyed the downstream glass plant. The cause was that most of the "HF" in the feed was in the form of carbonyl fluoride. This component was sparingly soluble in water but hydrolyzed slowly to HF.	Examine the chemistry before choosing a separation process.

162	349	Natural gas hot-pot absorber	High CO_2, low H_2S gas was treated by a Benfield unit with internal heat recovery (by flashing the lean solvent and compressing the flashed steam with steam ejectors into the stripper). CO_2 removal was good, but only 35% COS was removed from the gas, instead of the desired 70%–90%. Downstream dehydration hydrolyzed COS, causing excessive H_2S in the product gas. Solved by changing dehydration desiccant and regeneration route.	
141	417	Phenol and reactant recovery, three-tower train mini plant	Distillate from first two towers was all the phenol. Distillate from third should have been phenol-free reactant but contained 1.5% phenol, formed by a previously unknown cracking reaction of the high boilers at the bottom. Solved by switching process sequence, so that high boilers are removed in the second tower and reactant is separated from phenol in the third. Easy to switch in a miniplant, almost impossible once a full-scale plant is built.	Miniplants are invaluable for finding unforeseen phenomena. Compare 128.
DT1.10			*Component analyzed as light impurity turns out a heavy component left over from previous campaign.*	
128			*Unexpected reaction near tower bottom contaminates overhead product, Section* 15.1	
112			*A previously unknown exotherm leads to nitro compound explosion, Section* 14.1.3.	
1291	185	Ammonia Benfield hot pot	The solvent was contaminated with organic acids, mainly acetic acid, leading to reduced CO_2 removal. Most acids were formed in the upstream shift converter from traces of methanol in its feed. Process condensate treatment is the planned cure.	
1296	226	Ammonia aMDEA absorber	Small changes in absorber pressure generated large variations in CO_2 slip, possibly because of the physical rather than chemical nature of the CO_2 in aMDEA absorption.	Solved by keeping the pressure steady.

(Continued)

Chapter 1 Troubleshooting Distillation Simulations (*Continued*)

Case	References	Plant/Column	Brief Description	Some Morals
134	536	Solvent–residue batch still, vacuum	Column separating reaction solvent and separation solvent from residue experienced excessive solvent losses to residue. Change in the upstream reactor, and using the same solvent for both reaction and separation, reduced feed inconsistencies, permitted semibatch operation, and reduced solvent losses.	Changes upstream of the column can eliminate column problems.
135			*Changing from single to multiple distillation process concentrates unstable chemicals, Section* 14.1.4.	
			1.3 Does Your Distillation Simulation Reflect the Real World?	
1.3.1 General				
DT1.11			*Incorrect characterization of feed components leads to impossible product specifications.*	
315	274	Aromatics	Diagnosis based on the initial simulation was control instability. A simulation reality check against plant data exposed needs for better energy balance data, a low reflux test, VLE review for one pair of components, and a surface temperature survey. Once adequately reflecting plant data, the simulation pointed to unexpected low tray efficiency. The corrective action became improving trays, not controls.	Ensure your distillation simulation reflects the real world.
311	275	Olefins demethanizer	Two simulation models gave a good match to plant data. Both suggested efficient packing in the lower sections. One model suggested efficient, the other inefficient packings, in the upper sections. Plant logs of the temperature-reflux dependence proved the model predicting poor upper efficiency to be correct. A revamp based on this model succeeded. Based on the high-efficiency model, the revamp would have failed. (See also 513, Section 4.8, and 940, Section 11.10.)	Same as 315.

312	272	Olefins water quench	A simulation based on a set of tower readings led to theories for explaining liquid carryover from the top. A detailed test invalidated the simulation and theories. A discrepancy between data and simulation, initially attributed to an incorrect temperature measurement, was proven in the tests to be due to error in flow measurement. This completely changed the explanation for the carryover. (See also 429, Section 7.4.2, and 738, Section 9.5.)	A set of readings does not constitute an adequate test.
338	398	Refinery HF alkylation main fractionator	While validating a simulation, laboratory analyzer measured much more C_6^+ in the recycle iC_4 side draw than the on-line analyzer and than could be conceived based on the reboiler duty. Joule–Thompson condensation and knockout of entrained heavy liquid during sample bomb purging were used to explain discrepancy. There was also a 16% discrepancy between the recycle iC_4 side draw and feed flow rate. (See also 158, Section 1.1.2, and 734, Section 10.8.)	
334	475	Refinery C_3 splitter	Misleading flow and temperature measurements frustrated reliable simulation and valid tray efficiency determination. Mass balance closure checks, surface temperature measurements, comparison to sales flowmeter, and calibration verification from basics helped rectify.	
314, DT1.12	254	Stabilizer	To develop a simulation for revamp, column was tested at high and low reflux. Low-reflux data matched the simulation well, high-reflux data gave poorer match. A Hengstebeck diagram led to an adequate explanation of the mismatch in terms of a VLE inaccuracy.	Troubleshoot simulations graphically.
329	267	Refinery depentanizer	Basing a simulation on matching simulated to measured bottom D86 distillation gave optimistic tray efficiency and a misleading simulation. Matching simulated to measured bottom-component analysis gave correct tray efficiency and good simulation.	Rely on component data, not ASTM D86, for this type of tower.
321			*Incorrect modeling of feed entry arrangement leads to purity problem, Section* 2.3.	
DT1.13			*Incorrect modeling of broad-boiling-range condenser fails to predict major bottleneck.*	

(*Continued*)

Chapter 1 Troubleshooting Distillation Simulations (*Continued*)

Case	References	Plant/Column	Brief Description	Some Morals
1.3.2 With Second Liquid Phase				
1426, 1279, 877			*Mismatch between measured and simulated temperature profiles shows unexpected presence of second liquid phase, Section* 2.5.	
1.3.3 Refinery Vacuum Tower Wash Sections				
218	175	Refinery vacuum	The wash oil flow was too small, leading to coking of the wash bed. This resulted from a single-tower simulation model predicting low wash dry-out ratios. Segmenting the simulation model into a number of flash units with recycles gave the correct dry-out ratio, requiring triple the previous wash rate. The revised simulation model correctly predicted plant data.	Coking can be prevented by a simulation that reflects the real world.
318	166	Refinery vacuum	Following replacement of trays by grid in the wash zone, the column experienced chronic coking leading to high pressure drop, reduced gas oil yield, and high metals content of gas oil. The design allowed for little vaporization in the wash bed. In reality, all the wash oil supplied vaporized and the bed dried up. Problem solved by redesigning the spray header for 3–4 times the original wash rate.	Same as 218.
117, 320, DT19.2			*Simulation underestimating the fraction of wash oil that vaporizes, Sections* 1.1.7 *and* 19.1.	
1.3.4 Modeling Tower Feed (*See also Section* 1.3.3)				
336	448	Refinery vacuum, wet	Feed to tower simulation was modeled as the sum of the product streams. The model therefore refed the overhead steam into the tower, inducing steam double counting and incorrect feed characterization.	Beware of the simulation basis.
1.3.5 Simulation/Plant Data Mismatch Can Be Due to an Unexpected Internal Leak				
DT20.2			*In a feed/bottoms interchanger.*	
DT9.4, DT12.5			*At a draw pan or chimney tray.*	
1353			*Across a preferential baffle at tower base, Section* 23.1.3.	

1.3.6 Simulation/Plant Data Mismatch Can Be Due to Liquid Entrainment in Vapor Draw				
755			*Due to reboiler return arrangement, Section* 8.4.4.	
734			*Due to low elevation of draw nozzle, Section* 10.8.	
1.3.7 Bug in Simulation				
326	294	Three chemical towers, high reflux ratio	Well-known commercial simulations with successful convergence and no error messages had erroneous energy balances on all three towers. Cause was a bug in the default convergence software. Repeating with an alternative convergence procedure gave valid mass and energy balances. Using the original simulation, all three reboilers would have been grossly undersized and tower feed grossly mislocated.	Verify the heat and mass balances for any simulations.
			1.4 Graphical Techniques to Troubleshoot Simulations	
1.4.1 McCabe–Thiele and Hengstebeck Diagrams				
313			*Detects a pinch caused by interreboiler addition, Section* 2.1.	
314			*Helps diagnose a VLE inaccuracy, Section* 1.3.1.	
1.4.2 Multicomponent Composition Profiles				
210			*Diagnoses separation problem, Section* 2.3.	
1.4.3 Residue Curve Maps				
212, 213, 214, 220			*Diagnoses unexpected product slate, Section* 2.6.1.	
			1.5 How Good Is Your Efficiency Estimate?	
307	413	Air stripper	Column using air to strip methanol, acetone, and ammonia from water failed to achieve design separation. Design efficiency was predicted from air humidification and oxygen-stripping studies in a single plate laboratory column. Wall and downpipe mass transfer enhanced efficiency in the laboratory column. This led to optimistic efficiency predictions.	Caution is required when predicting efficiency by adding laboratory-measured mass transfer resistances.

(*Continued*)

Chapter 1 Troubleshooting Distillation Simulations (*Continued*)

Case	References	Plant/Column	Brief Description	Some Morals
308	33	Pharmaceuticals IPA–water	The 2-ft ID azeotropic distillation column used benzene and IPE entrainer. HETP of the metal mesh packing was 12 in., considerably higher than the design HETP. Column HETP was scaled up from small-diameter columns that had good and frequent liquid distribution and redistribution.	Differences in distribution affect packed-tower scale-up.
523			*Low efficiency in small random packings, Section* 4.8.	
302	335	Acetylene Solvent-water stripper	Some causes of excessive solvent losses were an incorrect efficiency estimate and excessive entrainment. Performance was improved by adding trays and mist elimination pads under top section trays and raising reflux.	Mist elimination pads have effectively reduced entrainment.
304	95	Refinery vacuum	Tray efficiency was lower than expected. Column contained valve trays and operated at low liquid loads and with wide variations in vapor loads.	Beware of low-liquid-load issues.
315			*Unexpected low tray efficiency, Section* 1.3.1.	
337	448	Refinery vacuum lube	Using simulation, PAs were designed with three trays. In top PA, ΔT was so tight that heat transfer (not addressed by the simulation) controlled, requiring six trays. Error caught in time.	A heat transfer check is a must for cooling PAs.
342	86	Natural Gas MDEA and Sulfinol	An MDEA precontactor was followed by a sulfinol contactor. Due to the high partial pressures of H_2S and CO_2 in the feed, and to efficient trays, the acid gas absorption in the precontactor was tremendous and non-selective. This overloaded and overheated the MDEA solution. Cure was bypassing some gas directly to the sulfinol contactor. Manual bypassing was erratic, automatic bypassing worked.	

1.6 Simulator Hydraulic Predictions: To Trust or Not to Trust

1.6.1 Do Your Vapor and Liquid Loadings Correctly Reflect Subcool, Superheat, and Pumparounds?				
341	48	Olefins deemethanizer	Hydraulic loads predicted by simulation were insensitive to relatively large discrepancies between simulation and measurement.	
335	448	Refinery vacuum lube	Design of top trays in several PAs was based on the vapor flow rates leaving these trays, which were much lower (in the top PA by a factor of 6.6) than the vapor flow rates entering the trays. Error caught after trays were ordered but before fabrication.	Watch out when interpreting simulation output.
316			*Not properly accounting for subcooling, Section* 3.2.1.	
306			*Not properly accounting for superheat, Section* 3.3.	
411			*Not properly accounting for latent-heat variations, Section* 4.1.	
1.6.2 How Good Are the Simulation Hydraulic Prediction Correlations?				
514, DT1.14	254	Refinery vacuum	Two-inch Pall rings were replaced by 3-in. modern random packings. Expected capacity increase was 30%, but only 17% materialized. Both the default and supplier's options in a commercial computer simulation were optimistic, leading to the high expectation.	Do not trust packed-tower calculations on simulators.
DT1.15			*Several cases of optimistic vendor and simulation predictions for packing capacity.*	
515, DT1.16	254	High pressure	A commercial simulator gave optimistic prediction to packing capacity because it allowed extrapolation of a good correlation well beyond its applicability limits.	Same as 514.

Case	References	Plant/Column	Brief Description	Some Morals
			2.1 Insufficient Reflux or Stages; Pinches	
DT2.1			*No reflux, no separation.*	
309	464	Solvent recovery	Tower was piped up without reflux. This led to poor fractionation.	Do not overlook the obvious.
322			*Stripping steam valve blocked in, Section* 2.2.	
DT2.2, DT2.3, DT2.4			*Heavier mixtures require higher temperatures for good reboiled stripping.*	
1191	507	Gas glycerol dehydration	Condensation in the vent line and stack back-pressured the reboiler leading to excessive temperatures and poor stripping. Corrected by vapor being vented directly above the reboiler.	
201, 15119			*Bad meter leads to hard-to-detect reflux deficiency, Sections* 2.3 *and* 25.5.1.	
339	316	Oxygenated hydrocarbons	Utilizing the excess capacity of the purification column, the plant boosted reflux, top product recycle to reactor, and control temperature, to lower reactor conversion while keeping tower bottom on specification. This debottlenecked the reactor.	
340	191	Refinery crude fractionator	Diesel yield increased 5% and diesel in AGO declined from 40% to 20% by increasing number of bottom stripping trays and making them fouling-resistant (see Case 448, Section 18.4.1), and by raising bottom and AGO stripping steam rates, which in turn allowed higher reflux to the diesel-AGO fractionation section.	
204	306	Refinery debutanizer	The column feed was rich (72%) in butane. A few degrees extra preheat caused a large increase in feed vaporization accompanied by a large drop in stripping vapor rate. This increased butane in bottom product. The problem was solved by controlling the flow of steam to the preheater.	Excess preheating can cause fractionation difficulties.
317, 1559			*Excessive subcooling causes fractionation difficulties in rectifying, Sections* 3.2.1 *and* 3.2.2.	

313, DT2.5	254, 276	Olefins C$_2$ splitter	Addition of an interreboiler caused column design to approach a pinch. Pinch was undetected by a simulation; the simulation converged and worked well. Pinch detected by a McCabe–Thiele diagram.	Troubleshoot simulations graphically.
1547, 1598			*Control problems near a pinch, Sections 26.3.4 and 27.1.1. See also cases in Section 26.4.1.*	
314			*Pinch affects mismatch between plant data and simulation, Section* 1.3.1.	
226			*Excessive pumparound condensation below side draw limits product yield, Section* 3.1.5.	
142	476	Chemicals lights stripper	Stripper removed volatile component C from a reaction upstream. C needed to be immediately vaporized because in the liquid it reacted to an undesirable by-product. The stripper overhead was partially condensed and refluxed (no rectifying trays) in order to minimize drying downstream. The reflux had a higher concentration of C than the feed. This caused substantial product losses.	
342			*Excessive absorption overloads amine solution, Section* 1.5.	
1223	546	Natural gas amine regenerator	The large stripping heat supplied was insufficient to adequately strip H_2S from a rich amine solution due to excessive amine circulation rate. An eightfold cut in circulation permitted an eightfold cut in stripping heat simultaneous with a major improvement in H_2S stripping.	Good plant testing is invaluable for defining and overcoming plant problems.
1181	130	NGL glycol dehydration	Poor dehydration occurred in a grass-roots sour gas glycol contactor. Cause was poor regeneration due to excessive circulation. Cured by cutting circulation to half the design rates.	Good regeneration is essential.
230	416	Gas glycol regenerator several cases	Adding a flash drum upsteam of the regenerator gave two stages of BTX, VOC, and other component removal, reducing emissions. Also, removing the non-condensables ahead of the regenerator condenser improved heat transfer. In one case, reducing glycol circulation and flash drum pressure improved BTEX removal.	

(Continued)

Chapter 2 Where Fractionation Goes Wrong (*Continued*)

Case	References	Plant/Column	Brief Description	Some Morals
			2.2 No Stripping in Stripper	
322	306	Refinery gas oil side stripper	Jet fuel components were not stripped because stripping steam valve was blocked in. Flowmeter gave a misleading indication. No difference between inlet and outlet temperature led to correct diagnosis.	Do not overlook the obvious.
310	464	Inert gas stripper	Feed was subcooled so that the partial pressure of the components to be stripped was below the partial pressure obtained when injecting the inert gas. No stripping occurred.	
323	306	Refinery jet fuel side stripper	Jet fuel flash point was insensitive to steam rates, and stripping was poor. Reason was lack of insulation on stripper and its feed line. This subcooled the jet fuel so that the partial pressure of the components to be stripped remained below that required for vaporization. No stripping resulted.	Compare 310.
1028	308	Refinery diesel side stripper	Diesel product could not meet its flash point specification. Water entering with the stripping steam vaporized in the stripper and subcooled the diesel, making it difficult to vaporize. Adding a steam trap at the stripping steam line drain valve increased bottom temperature by 35°F and flash point by 55°F.	The observation that water drained from the steam supply line helped diagnose.
1274			*Due to high base level, Section* 8.1.2.	
			2.3 Unique Features of Multicomponent Distillation	
201, DT2.6	263	Natural gas lean-oil still	Column did not achieve required separation because of insufficient reflux induced by an undersized reflux orifice plate. Problem was difficult to diagnose because multiple steady states of the top temperature made it appear normal.	Temperatures may not tell full story in some multicomponent distillations.
205	304	Refinery vacuum	Top tower temperature was reduced by 130°F by cooling top pumparound. This led to an unexpected rise in column pressure, caused by the flash equilibrium behavior of the precondenser. When temperature was reduced, less liquid was condensed in the precondenser and less lights were absorbed.	Beware of absorption effects.

DT1.13			*Unconcensed gas was higher than expected due to lack of absorption effect.*	
DT22.4			*Absorption effect in condenser causes rapid fall in pressure at start-up and tray damage.*	
DT24.3			*Absorption effect in condenser interacts with flooding and controls.*	
DT2.7			*Product draw too close to overhead leads to excess light non-key in product.*	
227	316	Chlorinated hydrocarbon	Changing side-draw location from 13 to 2 stages from the top of a finishing column improved side-draw purity, which in turn permitted raising feed rate.	
210, DT2.8	254, 273	Chemicals	A vapor-side product impurity content was 10% (design 1%) due to a nonforgiving concentration profile. Over the eight design stages in the bottom bed, the concentration rose from 30% at the bottom to 50% four stages below the side draw, then dipped to 1% at the side draw. A miss by one to two stages would bring the concentration to 10%.	Use of concentration plot was invaluable for diagnosing problem.
321	159	Refinery FCC main fractionator, several towers	Following replacement of trays by structured packings, LCO/sponge oil draw temperature dropped 60°F, LCO product contained 5% more gasoline, and the LCO stripper stopped stripping. Reason was that the two trays between the sponge oil return and the LCO draw were eliminated. Gasoline-rich returned sponge oil mixed with tower liquid to form the LCO product. Problem minimized by minimizing sponge oil. In one case, solution was returning sponge oil below the LCO draw, generating a "pumpdown."	Computer simulation failed to predict due to incorrect modeling of the actual steps.
			2.4 Accumulation and Hiccups	
2.4.1 Intermediate Component, No Hiccups				
216	222	Ethanol extractive distillation	Heavy alcohols (fusel oils) were side drawn, cooled, then phase separated and decanted from the tower's ethanol–water mixture. Depending on the draw tray temperature and composition, the cooled side draw did not form two liquid phases. Solved by adding water to the decanter to ensure phase separation.	Problems diagnosed with help of a process simulation.
217	419	Alcohols–water	An intermediate component buildup caused the formation of a second liquid phase inside a column. Problem solved by decanting the organic phase and returning aqueous phase to column.	

(Continued)

Chapter 2 Where Fractionation Goes Wrong (*Continued*)

Case	References	Plant/Column	Brief Description	Some Morals
136	480	Olefins depropanizer, debutanizer	Small amounts of radon 222 (boiling point between propylene and ethane) contained in natural gas concentrated severalfold in the C_3 fraction. Decay into radioactive lead contaminated towers, auxiliaries, polymer deposits on trays, wastewater from reboiler cleaning, and generated problems of waste disposal and personnel entry into the towers.	Paper describes the contamination survey, how tackled and the lessons learnt.
1010	303	Refinery alkylation depropanizer	Column feed contained strongly acidic components, which dissolved in small quantities of water and caused a severe and recurring corrosion failure problem. The rate of corrosion failure was greatly reduced by adopting an effective dehydration procedure at start-up. To dehydrate, acid-free butane was total refluxed while drains were intermittently open until all water was removed.	
1040	88	Gas ethane recovery column extractive distillation	Tower feed gas contained excessive water (−2°F dew point, design −40°F) due to upstream limitation. Water peaked at side reboiler in the middle of stripping section, forming carbonic acid and corrosion. Increasing solvent ("additive") circulation, or surges in inlet gas, pushed water upwards, causing hydrates in the chilled condenser, with off-spec product up to a week. Hydrates removed by injecting large quantities of methanol into the reflux and thawing column.	
DT2.9			*Corrosion due to water accumulation in chlorinated hydrocarbon tower.*	
DT2.11			*Corrosion due to water accumulation in reboiled deethanizer absorber.*	
1042			*Excess water in stabilizer feed destabilizes temperature control, Section* 27.1.3.	
15143			*Purge minimization without frequent analysis leads to component accumulation, Section* 27.3.3.	

2.4.2 Intermediate Component, with Hiccups

Case	References	Plant/Column	Brief Description	Some Morals
1001, DT2.12	263	Natural gas deethanizer	Small quantities of water accumulated in a refluxed deethanizer and caused column to hiccup, or empty itself out either from top or bottom, every few hours. Cured by replacing reflux by oil absorption.	Water may accumulate in refrigerated columns.

DT2.10			*Hiccups in reboiled deethanizer absorber.*	
1038	42	Refinery deethanizer stripper	Tower flooded at feed drum temperatures below 100°F. To unflood, boilup was cut or drum warmed up at the expense of poorer C_3 recovery. Gamma scans show flood initiated in middle of tower, even though the highest hydraulic loadings were at the bottom. A retrofit with high capacity trays gave no improvement. Cause was water and C_2 accumulation. Replacement of the poorly designed draw and external water separation drum by an internal water removal chimney tray eliminated water accumulation.	
1039	42	Refinery reboiled deethanizer absorber	Tower had no vapor or liquid recycle to the feed drum. The absorber section had two external water intercooling loops with water removal. The stripper had a water draw with an external water removal drum. Water trapping during cold weather led to severe flooding and carryover. Warming the feed drum and cutting intercooler duty were cures at the expense of lower C_3 recovery. Later a properly-designed water draw tray was installed to eliminate water entrapment.	
751	165	Refinery coker deethanizer stripper	Residence time in downcomer-box water draw was too low to separate water, so no water was withdrawn. The water built up in the tower, periodically puking, disturbing pressure control and contributing to condenser corrosion and fouling in the downstream debutanizer. Replacement with a seal-welded chimney draw tray eliminated problem.	
DT2.13			*Hiccups in a debutanizer.*	
211, DT2.14	250	Azeotrope column	An intermediate component accumulated in the tower, causing regular cycling (hiccups). Cured by raising the top temperature. The additional product loss due to the higher top temperature was negligible.	
1213	437	Chemicals	A trapped component periodically built up in the upper section of a large distillation column. When it built up, the control temperature rose and increased reflux, eventually causing flooding. Gamma scans diagnosed the problem. Taking a purge side draw solved the problem.	Side draws are effective in preventing trapping of an intermediate component.

(*Continued*)

Chapter 2 Where Fractionation Goes Wrong (*Continued*)

Case	References	Plant/Column	Brief Description	Some Morals
228	120	Multicomponent packed tower	Column capacity dropped from design to almost zero in 3–4 days. After shutdown and restart, full capacity was reestablished and the cycle repeated. Cause was accumulation of a trace intermediate component in stripping section.	Cured by a vapor side draw between the two beds.
DT2.15			*Hiccups in methanol–water towers.*	
DT2.16			*Hiccups in ammonia stripper.*	
DT2.22, DT2.23			*Hiccups in azeotropic and extractive distillation towers.*	
DT22.16			*Accumulation of heavy alcohols induces foaming in solvent recovery tower.*	
DT15.2			*In-tower reaction products form an intermediate-boiling azeotrope that induces foaming.*	
2.4.3 Lights Accumulation				
1004	301	Refinery debutanizer	Column internals and reboiler tubes severely corroded after the water draw-off control valve on the reflux drum boot plugged. Manual draining was too inconsistent to prevent water (saturated with H_2S) refluxing the tower. Continuous flushing of water draw line with an external water source prevented recurrence.	Avoiding small-port control valves and continuous water flushing of water draw line can prevent blockage.
15157	454	Refinery FCC main fractionator	Plugging of a tap on the boot oil–water interface level transmitter locked its reading at about 50% when water filled the boot. Water refluxing to the fractionator over time cavitated and damaged the reflux pump and deposited salts that plugged top internals. Water in the naphtha destabilized the gas plant.	Problem eliminated by blowing the level tap.
1041			*Water in reflux leads to corrosion and plugging in crude fractionator, Section* 17.1.	

111	311	Refinery HF alkylation depropanizer and HF stripper	Depropanizer overhead went to an HF stripper. Stripper bottom was the propane product, while stripper overhead was recycled to the depropanizer overhead. When ethane entered the depropanizer due to an upstream unit upset, it entrapped in the overhead system and could not get out. Depropanizer pressure climbed and excessive venting was needed.	Cured by dropping stripper bottom temperature to allow ethane into the propane product.
DT2.7			*Water accumulates in overhead loop of glycol/residue tower.*	
1019			*Water accumulates in absorber–stripper recycle loop, Section* 2.4.5.	
2.4.4 Accumulation between Feed and Top or Feed and Bottom				
202, DT2.17	263	Natural gas lean-oil still	An added preheater which performed better than design caused column to "hiccup" and empty itself out every few hours from either the top or bottom. A bypass around the preheater eliminated the problem.	Oversized preheaters can cause accumulation. Bypasses are valuable.
229	42	Refinery deethanizer stripper	Chronic and severe flooding occurred at low (70–80°F) feed drum temperatures. Gamma scans showed the flood initiated above the internal water-removal chimney tray, eight trays from the top, even though hydraulic loadings were higher further down in the tower. C_2 accumulation and foaming were possible causes. Cured by adding a feed preheater.	
221	292	Refinery deethanizer stripper	Cold feed temperatures caused ethane condensation and accumulation, leading to hiccups once per week during winter. Solved by bypassing some of the feed around the feed cooler.	Oversized coolers can cause accumulation. Bypasses are valuable.
1036	171	Refinery deethanizer absorber	Premature flooding occurred in the absorber when feed temperature dropped below 100°F. Water and C_2 entrapment was the cause. A major contributor was problems in the feed separator boot interface level measurement, which led to feeding free water to the tower.	Solved by repairing level measurement, adding a water removal tray, and debottlenecking internals.
15123			*Warming stripper feed eliminates lights accumulation, Section* 28.3.1.	
1209	306	Refinery FCC absorber-stripper	Excessive condensation in the drum during winter repeatedly caused buildup of ethane until the stripper flooded. Keeping the recontact drum warm by cutting cooling to the condenser prevented recurrence.	Similar to 221.
DT2.13			*Preheater fouling aggravates hiccups.*	

(Continued)

Chapter 2 Where Fractionation Goes Wrong (*Continued*)

Case	References	Plant/Column	Brief Description	Some Morals
2.4.5 Accumulation by Recycling				
139	405	Olefins C_3 splitter	Frequency of propylene product going off-specification with methanol increased following installation of a system that enhanced BTX recovery, and also methanol recovery, out of wastewater. The recovered materials concentrated in the process. Corrective action was routing some water away from the process and stripping some with fuel gas into a waste gas burner.	Recycling can concentrate undesired components in the process.
160			*Ammonia accumulates in olefins water quench system, Section* 1.1.2.	
1019, DT2.18	224	Natural gas absorber and deethanizer (in series)	Modifications to recover the deethanizer overheads (previously sent to fuel) compressed, chilled, then recycled it to the absorber feed. Small quantities of water, previously going to fuel, returned to the absorber feed. The absorber top was too cold, and the deethanizer bottom too hot, to allow the water to escape. The water built up until freezing at the recycle chiller. For years the chiller was thawed to flare once per shift. Cured by adding a small package glycol dryer at the compressor discharge.	Always consider the interaction of a modification with the existing system.
DT2.19			*Recycling causes impurity buildup in ethanol tower.*	
DT4.4			*Lights accumulation due to bypassing a lights removal tower.*	
2.4.6 Hydrates, Freeze–Ups				
1020, DT2.20	276	Olefins C_2 splitter	Hydrates occurred two to three times per week. Stepping up methanol injection and regenerating secondary dryer gave only limited improvement. The hydrates formed between feed and interreboiler eight trays below. Methanol and dissolved hydrates got trapped in the kettle interreboiler, from which they slowly batch distilled back into the splitter. Mitigated by draining methanol/water from interreboiler.	An interreboiler can interact with and aggravate a hydrate problem.
1024	143	Natural gas demethanizer	Liquid could not be drawn from a chimney tray due to an ice plug. The plug and its location were identified using radioactive spot density measurements along the pipe. The ice plug was melted by external heat.	

DT29.1			*Separate top and bottom dP recorders distinguish floods from hydrates in C_2 splitter.*	
1019, DT2.18			*Freeze-up caused by water accumulation in gas plant deethanizer overhead loop, Section* 2.4.5.	
1040			*Hydrates promoted by high solvent rates in extractive distillation, Section* 2.4.1.	
1034, 1033			*Hydrates in debutanizer reflux pump lead to gas release, explosion, Sections* 14.4 *and* 14.11.	
1650			*Freeze-up in bottom valve of butylene wash tower leads to gas release, Section* 14.12.	
779			*Adding drain holes to demethanizer seal pan possibly helps against freeze-ups, Section* 9.2.	
			2.5 Two Liquid Phases	
224	352	Chemicals	Feed decanter removed most B (high boiler) from feed. Tower separated volatile A/B azeotrope from A bottom. Carryover of B-rich heavy phase into tower feed due to decanter interface level control problems shifted operation from one side of the azeotrope to the other, accompanied by heating up and flooding. Overcome by stopping feed, slumping column, reinventorying decanter and tower base, then restarting.	Azeotrope distillation problems can be initiated in upstream operations.
12111			*Excess water in batch tower feed drum forms unexpected second liquid phase, leads to explosion, Section* 14.1.3.	
101	113	Isopropyl acetate recovery	Poor decanter phase separation resulted from slow hydrolysis of isopropyl acetate to isopropanol, which was soluble in both organic and aqueous phases.	Beware of slow reactions when material is recycled.
DT2.21			*Siphoning in decanter outlet pipes.*	
220			*No split in decanter at certain compositions, Section* 2.6.1.	

(Continued)

Chapter 2 Where Fractionation Goes Wrong (*Continued*)

Case	References	Plant/Column	Brief Description	Some Morals
216			*Poor side-draw decanter phase separation, Section* 2.4.1.	
217, DT2.15			*Intermediate component buildup forms second liquid phase, Section* 2.4.1.	
898			*Poor mixing of two liquid phases in distributors of extractive distillation column, Section* 2.6.2.	
751			*Draw pan too small for phase separation, Section* 2.4.2.	
1004			*Plugged drain line from reflux drum boot, Section* 2.4.3.	
1426	383	Solvent dehydration by azeotropic distillation	Replacement of tower by a larger one caused instability, reduced capacity, and high solvent losses. The reason was refluxing of water in the cyclohexane entrainer. An undersized condensate line caused condensate buildup in the condenser all the way to its midpoint vent, from where it drained directly into the cyclohexane side of the decanter. Matching simulated temperature profile to plant data revealed excess water in the hydrocarbon reflux, pointing to a decanter malfunction.	Poor condensate removal bottlenecks condensers.
1279	383	Monomer and water separation from acid 15 ft ID	The column operated normally until suddenly it became unstable at high rates. Cause was a crack in the decanter baffle plate, which allowed water into the refluxed organic phase. The reflux water generated second liquid phase and therefore temperature instability on the trays. Matching simulated temperature profile to plant data revealed excess water in the organic reflux, pointing to a decanter malfunction.	
1545			*Malfunctioning decanter level control leads to poor phase separation and violent reaction, Section* 14.2.	
DT13.3			*Malfunctioning decanter level control leads to refluxing water into hot tower and tray damage.*	

877	383	Acid recovery from organics packed tower	Acid recovery was poor due to a malfunctioning collector that collected liquid from an internal condenser, splitting it into a heavy-phase reflux and a light-phase acid product. The phase separation overflow weir in the collector was too tall, installed in an incorrect location, and the product draw nozzle was undersized. All led to poor decanting. Once these were corrected, tower operated normally. Matching simulated temperature profile to plant data diagnosed presence of two liquid phases where only one was expected.	
15112	53	Esters batch column	Phase separation of the overhead condensate occurred upstream of the manually operated needle valve that performed the reflux/product split. The organic phase preferentially went to the reflux, which increased batch time. Solved by a new three-way valve reflux splitter operated by a timer which directs all the condensate either to product or to reflux.	
1418			*Presence of unexpected second liquid phase bottlenecks condenser, Section* 24.3.	
1177			*Phase separation in sump at shutdown leads to tray damage, Section* 22.4.	
1323			*Phase separation stops reboiler thermosiphoning, Section* 23.1.5.	

2.6 Azeotropic and Extractive Distillation

(*See also Temperature Control for Azeotropic and Extractive Distillation, Sections* 27.1.5 *and* 27.1.6, *and Two Liquid Phases, Section* 2.5)

2.6.1 Problems Unique to Azeotroping

212	445	Chemicals azeotropic column	Column separated a minimum-boiling AC azeotrope from a heavy C. Small amounts of light boiler B (lighter than the azeotrope) escaped in the top product. Changes in the reactor produced much more B in the feed. The light B was expected to go up, but much of it ended in the bottom. Reason was the formation of a much lighter AB azeotrope that distilled up, leaving a BC mix in the bottom.	Residue curve maps were invaluable in diagnosing.

(*Continued*)

Chapter 2 Where Fractionation Goes Wrong (*Continued*)

Case	References	Plant/Column	Brief Description	Some Morals
213	445	Chemicals azeotropic column	Light product B was separated from heavy reactants A and C. With a feed rich in A and lean in B, pure B could not be produced due to minimum-boiling A/C and A/B azeotropes, the latter boiling 1°C less than B. Problem solved by a column concentrating B (with some A and C) at the top, bottom being an A/C mix. C was added to the B concentrate en route to the next column, where it drew the A to the bottom (as an A/C mix), leaving B at the top.	Azeotropes can often be broken with in situ components. Also as 212.
214	445	Freon 22 (R22) reflux column	Column separated a light HF/HCl/R22 mix from heavy HF recycle to the reactor. Two simulations, identical except for initial-value differences of 0.15% HF, gave completely different results. This difference shifted the column across a distillation boundary, giving completely different end points.	Using residue curve maps, avoid crossing distillation boundaries.
220	385	Acetic acid–acetic anhydride–C_9 alkane	C_9 bottoms separated from ternary azeotrope that formed two liquid phases. Decanter acid/anhydride was distilled in second tower to remove C_9. Residue curve map showed that when the feed was low on anhydride, there is no phase split in the decanter and the process fails. Cure was diverting low-anhydride feeds away from the towers.	Similar to 212.
DT2.22			*Hiccups in azeotropic distillation tower.*	
1219	33	Pharmaceuticals solvent recovery IPA–water	Small amounts of methanol and acetone in the IPA reduced separation efficiency of an azeotropic distillation column using benzene and IPE entrainers. Problem was solved by removing all high volatiles before starting the azeotropic distillation.	Impurities can interfere with azeotroping.

1220	33	Pharmaceuticals solvent recovery IPA–water	Cold entrainer reflux reduced capacity of an azeotropic distillation column using benzene and IPE entrainers. Problem was solved by reheating the reflux.	In azeotropic distillation, subcooled reflux can lower column capacity.
2.6.2	**Problems Unique to Extractive Distillation**			
215	221, 222	Ethanol extractive distillation	A new plant could neither exceed 60% capacity nor produce fusel oils for 1 year. Cause was excessive boil-up preventing phase separation of fusel oil near the bottom of this tower plus a control problem (1585, Section 26.4.3) in a downstream (rectifying) column. Cure was reducing the steam rate by a factor of 2.	Problem diagnosed using process simulation.
898	288, 289	Aromatics BTX ED	A grass-roots tower had two beds of structured packings between the lean solvent entry and the feed entry. Poor premixing of reflux with the lean solvent, a hold-down blocking 8% of packing area and interfering with liquid distribution, no mixing in the interbed redistributor, nonsuitability of redistributor for two liquid phases, and packing disturbance due to lack of hold-down on lower bed, all contributed to poor benzene recovery and a capacity bottleneck.	After correcting deficiencies benzene recovery specifications were met and maximum capacity achieved.
1351			*Kettle reboiler maldistribution unique to extractive distillation, Section* 23.4.1.	
DT23.5			*Surging in extractive distillation tower with two reboilers in series.*	
216, 1040			*Accumulation problems in extractive distillation, Section* 2.4.1.	
DT2.23			*Hiccups in extractive distillation tower.*	

Case	References	Plant/Column	Brief Description	Some Morals
			3.1 Energy-Saving Designs and Operation	
3.1.1 Excess Preheat and Precool (*See also Preheater Controls, Section* 28.7 *and Accumulation between Feed and Top or Feed and Bottom, Section* 2.4.4.)				
225	434	Natural gas glycol contactor	Lean glycol entering contactor at 230°F led to excessive heat load on chilling train downstream and to poor NGL recovery. Solved by adding a lean glycol cooler.	
DT3.1, DT3.2			*Excess preheat raises reflux and energy consumption and bottlenecks capacity.*	
204			*Excess preheat causes fractionation problems, Section* 2.1.	
317			*Excess subcool causes fractionation problems, Section* 3.2.1.	
714			*Excess preheat causes flashing and bottleneck in feed tray downcomer, Section* 5.3.	
231			*Deficient crude preheat contributes to fractionator fouling, Section* 3.1.5.	
3.1.2 Side-Reboiler Problems				
313			*Pinch, Section* 2.1.	
1020			*Aggravating hydrates, Section* 2.4.6.	
1320, 1325			*Start-up difficulty, Section* 23.7.1.	
770, 767, 706			*Problems with side-reboiler draws and piping, Sections* 9.2, 10.1.1, *and* 10.1.2.	
3.1.3 Bypassing a Feed around the Tower				
1201, DT3.3	263	Olefins debutanizer	An end-of-run capacity limitation caused an off-specification product. A feed stream was bypassed around the column. This overcame the problem, with a surprising beneficial side effect of major energy savings.	Bypassing a feed can unload column and save energy.

1257	463	Refinery Alkylation DIB	Bottom contained 18% isobutane (design 5%) and column experienced a capacity restriction. One of the small column feeds contained 11% isobutane so the column degraded it instead of improving. Bypassing the stream around the column raised capacity and cut isobutane loss. (See also 317, Section 3.2.1.)	Never overlook the obvious.
450	293	Refinery FCC deethanizer stripper	Following revamp, tower flooded prematurely, restricting reactor severity. Alleviated by bypassing portion of feed to the bottom. To minimize bottom impurities, the bypass came from the main fractionator overhead receiver (low C_2), and went into the bottom sump so it gets stripped by the reboiler.	
1566	276	Olefins demethanizer	To overcome a capacity bottleneck, some feed was routed to a lower point. The bypass was difficult to operate and control because it was throttled manually and had no flowmeter.	Shortcuts can restrict benefits.
342			*To avoid excessive absorption in precontactor. Manual bypassing erratic, automatic worked, Section* 1.5.	
220			*To ensure two liquid phases are present in decanter, Section* 2.6.1.	
DT24.3			*Open nitrogen line to tower causes premature tray flood.*	
3.1.4 Reducing Recycle				
222	131	NGL stabilizer	Vapor make from the stabilizer overhead accumulator was excessive. This vapor was compressed and condensed. The condensate pump was overloaded by excess condensate and repeatedly failed, causing compressor trips. Solved by lowering accumulator temperature and raising stabilizer pressure as well as reducing compressor interstage cooling. The effect on stabilizer bottom was minimal because the condensate was recycled to the stabilizer feed.	
223	476	Chemicals lights	Tower removed lights from aqueous feed. Reflux was fresh water instead of condensed overhead. The fresh water reabsorbed the volatile components, countering the separation and increasing energy usage. Solved by modifying piping to reflux condensed overheads.	

(*Continued*)

Chapter 3 Energy Savings and Thermal Effects (*Continued*)

Case	References	Plant/Column	Brief Description	Some Morals
3.1.5 Heat Integration Imbalances				
12115	99	Gas NRU column	Tower had closed methane heat pump with a reboiler, condenser, subcooler, and side condenser. Keeping subcooler outlet temperature stable was key to balancing heat integration.	
12102	397	NGL demethanizer	Tower was a stripper fed by liquid flashed from an expander suction drum. Overhead from the tower chilled and condensed the liquid feed. Upon flooding, liquid entrainment in the overhead stepped up chilling, which in turn stepped up the feed rate, which in turn aggravated the flooding, and so on. This is termed a "cold spin," was initiated by ambient cooling, and led to outages and yield loss. Circumvented by flood control that would raise drum level and trim back side reboiler that was also used to chill feed.	Every heat integration system needs a flywheel.
DT3.4			*Cold heat integration spin in an olefin demethanizer.*	
1628			*Entrainment unbalances heat integration, leading to brittle failure, explosion, Section* 14.3.2.	
15156			*Cold heat integration imbalance due to misleading level measurement, Section* 25.7.3.	
226	171	Refinery FCC main fractionator	LCO product draw was just above the LCO PA, the only lower product being low-valued slurry oil. Combined heat duties of the LCO and slurry PA gave excessive loss of LCO to slurry oil. Solved by a supplementary LCO draw from the LCO PA.	

231	41	Refinery crude fractionator	A revamp to extra heavy crudes yielded excess atmospheric resid and AGO and poor kero/diesel yield. A HN side draw kept upper trays cool, inducing amine chloride salts, plugging and flooding. Amine originated from slops, chlorides from poor cold desalting. Weeping from TPA valve trays starved draw sump, leading to pump cavitation, restricting circulation. TPA return was cold, locally condensed water, causing corrosion and salting out. The low TPA heat removal and an undersized condenser led to operators cutting out stripping steam. Inadequate preheat, poor PA locations, and inability to cut heat removal at AGO PA were major contributors. Temporary fix was bypassing plugged trays and raising pressure. Cure was eliminating HN draw, eliminating AGO PA, replacing trays by packing in TPA and wash zone, retraying stripping section, revamping preheat system, relocating an upper PA down, and adding a condenser. Some salting out is still experienced, but can be effectively water-washed (see Case 12120, Section 12.7).	
DT3.5, DT3.6			*Heat integration imbalances in refinery fractionators cause flooding.*	
324	306	Refinery coker DC_2 absorber	Main fractionator PA reboiled tower. PA duty dropped during coke drum switchover, giving insufficient reboil. Problem solved by adding a supplementary steam trim reboiler.	Trim reboilers/ condensers make system robust.
219	164	Refinery crude fractionator	Overhead had two-stage condensation, first generated only reflux. Overhead from first condensed in second to make naphtha. Keeping interstage temperature high enough to prevent corrosion caused excessive heavies in naphtha. Operators overcame by raising tower pressure, but this lowered overall distillate recovery. Better solution by controlled addition of reflux from second stage and by adding controlled liquid transfer from first to second stage.	At a fixed pressure, temperature and composition cannot be independently specified.
1429, 1576			*Imbalances leading to reboiler/condenser issues, Sections* 24.7 *and* 23.9.2.	

(*Continued*)

Chapter 3 Energy Savings and Thermal Effects (*Continued*)

Case	References	Plant/Column	Brief Description	Some Morals
			3.2 Subcooling: How It Impacts Towers	
3.2.1	**Additional Internal Condensation and Reflux**			
316	143	Olefins demethanizer	High ethylene losses occurred after replacing trays by packing. Gamma scans showed poor liquid distribution in the upper two beds, flooding in the third, and poor vapor distribution in the bottom bed. No allowance for vapor condensation by the highly subcooled (70°F) feeds overloaded distributor capacities. Some improvement achieved by rerouting some of the cold main feed to an upper bed.	Do not neglect subcooling.
317	463	Refinery alkylation DIB, 8-ft-ID valve trays	Isobutane in bottom was 3–4 times the design, and column capacity was restricted. Excessive feed subcooling overloaded the bottom section. Also, a few trays above the feed had low liquid rates and could have been blowing. Several other problems were identified. Problem solved by adding a feed preheater, installing antijump baffles in the lower trays, and adding picket fence weirs to the low-liquid-rate trays. (See also 1257, Section 3.1.3.)	Same as 316.
330	465	Petrochemicals	Tower flooded 5–10% below design because additional vapor and liquid traffic induced by 100°F reflux subcooling was not accounted for in the internals design. Solved by using bubble point reflux.	Same as 316.
333	545		Subcooled feed led to flooding initiating at feed point. Tower returned to normal when feed preheater put into service.	
505	85	Multicomponent separation, two-feed column	Following replacement of trays by structured packing, bottom product could not meet specifications. Efficiency below the feed was half that expected. Raising reflux ratio did not improve separation. Problem was solved by preheating the highly subcooled feed to its bubble point.	Highly subcooled feed can cause a premature column bottleneck.
1210	306	Refinery naphtha reformer absorber	At low rates, lean solvent was subcooled and absorbed light components (ethane). This increased the internal circulation rates of lights and wasted energy. Cured by preheating the solvent with waste heat.	Preheating solvent prevents excessive subcooling.

3.2.2 Less Loadings above Feed				
1559	464	Chemicals	Feed to a two-column train fluctuated. Top purity of the first tower was held constant, which concentrated the disturbances in the feed to the second tower, destabilizing this tower. To overcome, a surge drum was added with 5–7 days residence time. Heat losses from the drum caused a 50°F drop in feed temperature to the second column, aggravating its reboiler limitation, which in turn necessitated a reduction in reflux ratio and therefore a lower purity of the top product.	
317			*Reduced loading above feed causes trays to dry, Section* 3.2.1.	
1129, DT3.7	194	Chemicals packed rectifier	Following operation with near-design boil-up, column would flood whenever feed was shut off. The subcooled feed entering the column base condensed 40% of the boil-up. When feed was shut off, the heat sink was eliminated, causing this uncondensed boil-up to enter the packing and flood it. Problem was solved by installing an override controller that reduced steam flow in response to excessive pressure drop.	Consider the effects of subcooling on capacity under start-up/shutdown conditions.
1280	308	Steam generation deaerator stripper, two cases	Stripper steam-stripped cold BFW and was mounted above the dearator drum. BFW rate exceeded design, while the steam supply line was restricting. Deaerator pressure fell, causing more vaporization, which flooded stripper. The flooding dropped deaerator level, which increased the cold BFW flow in, which further reduced the pressure and aggravated the flood. To unflood, operator manually restricted cold BFW flow. Permanent solution was by preheating the BFW.	Key to diagnosis was observing water in atmospheric vent.
3.2.3 Trapping Lights and Quenching				
1210			*In absorber, Section* 3.2.1.	
221			*Leading to accumulation and hiccups, Section* 2.4.4.	
206	173	Refinery vacuum	Cooled resid was returned to the flash zone. This increased heat recovery from bottom stream and increased column feed capacity. However, both effects were caused by degrading HVGO into resid due to quenching at the flash zone.	Diagnosed by a flash zone temperature 5°F colder than bottom.

(Continued)

Chapter 3 Energy Savings and Thermal Effects (*Continued*)

Case	References	Plant/Column	Brief Description	Some Morals
893			*Leak from small overflash collector tray quenches flash zone, Section* 9.3.	
872, 8112			*Entrainment from preflash drum quenches flash zone and maldistributes vapor, Section* 7.1.2.	
3.2.4 Others				
310			*Stop stripper action, Section* 2.2.	
1220			*Reflux undesired phase in azeotropic distillation, Section* 2.6.1.	
723	5	Refinery	Bubble caps located beneath an internal cold reflux pipe suffered chloride compound corrosion.	
208	168, 172	Refinery coker fractionator	Wash trays were replaced by grid, charge raised 10%, and recycle dropped from 1.2 to 1.1. Short runs, heavies carryover, and operational difficulties followed due to grid and collector plugging. Causes included (1) subcooled (rather than bubble-point) wash, which gave worse liquid distribution, increased spray nozzle plugging tendency, and may have led to a collapse of the spray angle; (2) a portion of the wash bypassed the grid bed to enter a shed deck section below, which reduced wash rate, enhancing coking tendency in the grid section. (See also 417, Section 4.4.2; 510, Section 19.4; 829, Section 7.1.2; 1236, Section 5.7.)	Subcooled wash can be troublesome in refinery fractionators. Do not bypass liquid around the wash section.
			3.3 Superheat: How It Impacts Towers	
306	306	Refinery FCC main fractionator	The design vapor rate in the slurry section did not allow for vaporization which occurs when a bottom feed with 300°F superheat contacts column liquid. Column therefore prematurely flooded. Solved by injecting subcooled quench liquid to desuperheat the feed. At a later stage, subcooled quench was replaced by a lighter liquid that vaporized, and premature flooding reoccurred.	Account for vaporization due to superheat in column loading calculation.

Chapter 4 Tower Sizing and Material Selection Affect Performance

Case	References	Plant/Column	Brief Description	Some Morals
			4.1 Undersizing Trays and Downcomers	
411	304	Pharmaceuticals methanol-water	Sieve trays flooded prematurely near the top of the bottom section due to low hole areas presumably because the trays were sized for the vapor loads at the tower base. Since water has a higher latent heat than methanol, the vapor load near the feed was higher.	Size trays for the maximum loads anticipated in the tower.
449	48	Olefins demethanizer	Tower bottleneck, flooding apparent above the swage, was caused by excessive downcomer back-up due to low hole areas on trays below the swage. Cure was new panels with more holes.	
444	288, 289	BTX ED	Stripping trays in a grass-roots tower flooded below design rates. Retrofitting with trays containing small slotted valves and larger downcomers, as well as modifying feed distributor, eliminated flood.	Properly size downcomers in ED tower.
431	250	Several pressure columns	Premature downcomer choke flood, at liquid rates as low as 40-50 percent of design, occurred when downcomer area was less than 5% of the column area, even though downcomer size appeared adequate.	
DT4.1, DT4.2			*Extremely small downcomers prematurely flood high-pressure alkaline absorbers.*	
706			*Insufficient downcomer area possibly induces flood, Section* 10.1.2.	
416	189	Refinery naphtha stabilizer	To gain stages, tray spacing in the bottom was reduced from 18 to 12 in. This caused flooding and loss of gasoline to the LPG. Remedy was raising the tray spacing to 18 to 21 in. in the heavily loaded bottom while lowering the tray spacing from 18 to 12 in. in the lightly loaded top.	
			4.2 Oversizing Trays (*See also valves popping out, Section* 22.5, *and leaking draw trays, Sections* 10.2 *and* 23.2.1)	
301, DT4.3	263	Olefins depropanizer	Instability and fluctuations occurred because of operation below the dumping loads in a section of the column just below the feed.	Review hydraulics above and below alternate feeds.
DT4.4			*Fractionator capacity appears limited when turndown impairs efficiency.*	

(*Continued*)

Chapter 4 Tower Sizing and Material Selection Affect Performance (*Continued*)

Case	References	Plant/Column	Brief Description	Some Morals
434	326	Natural gas MDEA regenerator	Steam consumption per gallon amine was 50% higher than design while residual acid gas loading in the lean amine was double the design. One explanation is poor efficiency due to the lower liquid load (500 gpm compared to 1200-gpm design).	
303	95	Steam distillation	Bottom section of an olefins naphtha splitter was oversized and operated at low rates. Separation was poor because of valve tray weeping. Increasing loadings solved it.	
410	308, 311	Refinery HF alkylation isostripper	Upon turndown, products went off-specification due to column efficiency loss, caused by weeping of the non-leak-resistant valve caps. Problem was overcome by a large reduction of column pressure.	Pressure reduction can help counter weeping.
401	95	Refinery vacuum	Weeping at low liquid loads in a PA section did not permit circulating liquid. Cured by replacing valves equipped with turned-down nibs (to prevent sticking) with valves that seat flush with the floor.	Avoid these nibs when liquid leakage is critical.
402	369	Refinery vacuum	Many valves on three PA trays were removed and their opening blanked. Ends of distribution pans and draw pans were seal welded. Leakage was reduced; separation and heat recovery improved.	Reducing valve density and seal-welding reduces leakage.
404	306	Refinery FCC DC_2 stripper	Bottom vapor load was triple that near the top, causing turndown problems. Successful resolution was achieved by different modifications to top 50% of trays, including (a) blanking half the valves; (b) using leak-resistant valve units, each containing a lightweight plate located below the normal disk; and (c) retraying with bubble-cap trays.	A number of useful techniques for improving valve tray turndown.
406	304	Refinery deethanizer	A dramatic increase in efficiency resulted from retraying column with leak-resistant valve units, each containing a lightweight plate located below the normal disk.	See 404.
412	13	Offshore gas amine absorber	At 25% of the design throughput, the lean-gas H_2S content was above specification. Blanking 60% of the valves on the trays solved the problem.	See 402, 404.

420	461	Refinery vacuum lube	Tower operated at reduced hydraulic loadings, resulting in poor separation between side cuts. Performance was improved by baffle installation to reduce tray open areas, blanking panels outside the baffles, blanking valves and caps, and adding picket-fence weirs.	
421	449	Petrochemicals superfractionator sieve trays	To permit operation at 12% of the design feed rate, all but the outlet 12 in. on every tray was blanked. This imposed a zig-zag vapor flow path, blowing opposite to the liquid flows, generating hydraulic gradients, inlet weep, poor efficiency, and off-specification product. Solved by reblanking so that the unblanked 12 in. was in the middle of the trays.	
			4.3 Tray Details Can Bottleneck Towers	
414	193	Chemicals 10 ft ID containing 72 trays below feed	Column flooded prematurely, with bottleneck just below feed. Bottleneck was identified by gamma scanning. Two trays below the feed were modified by reworking some unusual design features and lowering outlet weir. Column capacity did not change, but gamma scan showed improvement on modified trays. Modifying the next 10 trays below the feed improved capacity by 15%. Modifying the next 17 trays down gave another 15% capacity enhancement. Each step was guided by gamma scans.	Beware of unusual design features on trays. Gamma scans are effective for detecting tower bottlenecks.
405	304	Refinery reboiled absorber	Single-pass trays were replaced by two-pass trays in a capacity revamp. Higher capacity was achieved, but reducing flow path from 36 to 18 in. lowered efficiency to an extent that original trays had to be reinstalled.	Beware of efficiency loss when increasing tray passes.
DT4.5			*Lowering outlet weirs and redirecting liquid from side to center downcomers debottleneck four-pass trays in deethanizer.*	
426	141	High pressure	Trusses over weir restricted vapor disengagement from downcomer, causing premature flood.	
423	141		Inlet weir on tray blocked flow from tray above and caused capacity restriction.	
424	141		Radius tip pinched against inlet weir, causing capacity restriction.	

(*Continued*)

Chapter 4 Tower Sizing and Material Selection Affect Performance (*Continued*)

Case	References	Plant/Column	Brief Description	Some Morals
701			*Excessively tall outlet weir on bottom seal pan initiates premature flood, Section* 10.6.	
441	186, 450	Refinery FCC gas plant depropanizer	Following replacement of 40 conventional trays by 56 high-capacity trays with truncated downcomers, excessive propylene was lost in the bottom. Identified problems were a feed inlet unsuitable for the flashing feed, reboiler return and draw that constrained boil-up, and a tray design deficiency, but fixing these gave no improvement. A later diagnosis is that the short tray spacing lowered tray liquid flexibility, causing downcomers to unseal and the trays to operate like dual-flow trays. Reboiler fouling, which restricted boil-up and reflux rates, is also stated to contribute to the downcomer unsealing.	
743	250		A worker cut himself while climbing through a 16 × 20-in. manway during a false emergency.	
			4.4 Low Liquid Loads Can Be Troublesome (*see also small reflux below liquid draw, Section* 26.4.1)	
4.4.1 Loss of Downcomer Seal				
1101, DT4.6	256, 263	Olefins low temperature	Column could not be started up because vapor would not allow liquid to descend into downcomer. This caused excessive entrainment. Raising pressure at start-up permitted establishing downcomer seal.	Includes a method for modeling sealing problems.
1124	305	Gas glycol dehydrator	Downcomer liquid seal was lost on bubble-cap trays. Seal was reestablished by blocking in the gas flow to the tower and continuing glycol circulation for 30 min before reestablishing gas flow.	
435	8		Following a retray with conventional valve trays, capacity and product purity dropped. Reason was excessive downcomer clearances, which induced vapor up the downcomers and liquid dump through the trays. Diagnosed using tray and downcomer gamma scans. Solved by reducing downcomer clearances.	Beware of excessive downcomer clearances at low liquid rates.

436	182	AMS–phenol	Following replacement of valve trays by screen trays, both separation and capacity were poor. Causes were the VLE data (see 129, Section 1.1.5) as well as low liquid rates with possible vapor breakthrough into the downcomers. Problem alleviated by halving downcomer clearances, adding picket-fence weirs, increasing fractional hole areas, and diverting the feed from the downcomer to the tray.	
911	119	Light hydrocarbon	Premature flooding occurred on the third tray from the bottom because of variations on this tray that caused vapor leakage into the downcomer. Gamma scans identified trouble spot.	Abnormalities on only one tray can bottleneck a column.
943	250	Amine contactor	Valve tray panels were rotated 180°, placing the inlet weirs next to the outlet weirs. This left the downcomers unsealed in this low-liquid-rate service, resulting in entrainment. To cut entrainment, liquid rates were reduced; this gave poor wash and tray plugging.	
950	250		Downcomer clearance was 7–8 in. at the feed tray due to miscommunication. Premature flooding due to lack of downcomer seal resulted.	
439	495	Refinery debutanizer	Flooding, carryover, and high *dP* were experienced when rates were increased above 115% of original design. This was caused by several downcomers bowing out over the inlet weirs. The bowed downcomers cupped vapor and redirected its flow up the downcomers.	Unique patterns in the gamma scan helped diagnose this problem.
304			*Poor tray efficiency at low liquid loads, Section* 1.5.	
441			*Truncated downcomers losing seal, Section* 4.3	
4.4.2	**Tray Dryout**			
430	250		Stepped sieve trays at low liquid rates experienced entrainment and poor separation. Presumably, excessive weep occurred where liquid dropped over the intermediate weir, drying the downstream panel.	
903	61	Distillery whiskey still	Significant entrainment and weeping were experienced as a result of leaking bubble-cap trays at very low liquid rates. Gasketing solved problem.	Bubble-cap trays leak unless properly gasketed.

(*Continued*)

Chapter 4 Tower Sizing and Material Selection Affect Performance (*Continued*)

Case	References	Plant/Column	Brief Description	Some Morals
417	168	Refinery coker fractionator	Wash trays were replaced by grid, charge raised 10%, and recycle dropped from 1.2 to 1.1. Short run lengths, heavies carryover, and operational difficulties followed due to grid and collector plugging. A factor was use of six-pass perforated shed desks to handle a low liquid load (<0.3 gpm/in. weir length). The sheds were found distorted and heavily coked every shutdown. The dry sheds caused a seal loss in the downcomers bringing liquid from the collector above, leading to intermittent flooding at the collector. (See also 208, Section 3.2.4; 510, Section 19.4; 829, Section 7.1.2; 1236, Section 5.7.)	Shed decks require higher liquid rates (typically >2 gpm/in. weir) in plugging services.
443	534	Refinery heavy-oil vapor scrubber	Slurry recycle section used very small liquid flow rates (about 0.2 gpm/in.) in four-pass shed decks. This gave poor vapor–liquid contact, poor solid scrubbing, plugging, short runs, and fouling downstream.	See 417.
317			*Dry-out of rectifying trays due to excessive feed subcool, Section* 3.2.1.	
325			*Dry-out of fractionation trays due to excessive heat removal, Section* 10.2.	
			4.5 Special Bubble-Cap Tray Problems	
418	170	Refinery FCC main fractionator	Premature flooding took place either in the slurry section (equipped with baffle trays) or in the bubble-cap trayed sections above (wash or HCO pumparound). Calculations showed that these sections should have operated at 70–90% of flood.	
425	141		Bubble caps at outlet side blew liquid into downcomer and limited capacity. At high rates efficiency dropped prematurely due to entrainment.	
1124			*Loss of seal at low liquid rates, Section* 4.4.1.	
903			*Leakage and entrainment at low liquid rates, Section* 4.4.2.	

			4.6 Misting	
119	420	HCl scrubber	HCl content of the overhead vapor was much higher than equilibrium. Since the gas bulk temperature was lower than the dew point, misting (condensation of microscopic liquid drops in the bulk vapor) was likely and would explain the observation. This mist would not be washed by the column liquid.	In subcooled vapors, consider misting. Rate-based models can predict.
305	61	Whiskey distillery	Excessive entrainment was experienced with tunnel trough trays at 18-in. spacing. The problem was resolved by installing a 2-in. thick demister directly on top of the troughs.	A useful technique for minimizing entrainment.
302			*Mist elimination pads help eliminate entrainment, Section* 1.5.	
			4.7 Undersizing Packings	
501, DT4.7	263	Olefins water stripper	Unsuitable packings caused capacity restriction. Carbon steel packing corroded, ceramic packing chipped. Using undersized stainless steel packing did not solve the problem.	Ensure adequate choice of packing material and size.
			4.8 Systems Where Packings Perform Different from Expectations	
512	141	High pressure	Structured packings failed to meet design separation target. Many distillation towers.	
DT7.1			*In high-pressure, high-liquid-rate distillation, structured packings perform worse than trays they replaced even after maldistribution is corrected.*	
513, DT4.8	275	Olefins demethanizer	HETPs of random packings in the upper parts of the tower were about double those that would be expected for normal hydrocarbon distillation. The presence of hydrogen and the high relative volatilities are the likely explanation. (See also 311, Section 1.3.1, and 940, Section 11.10.)	Beware of low packing efficiencies in hydrogen-rich systems.
523	187	NGL depropanizer	Twelve valve trays were replaced by 22-ft bed of 1-in. rings. HETP was 30 in., much higher than expected. Possible causes included high ratio of tower to packing diameter and liquid maldistribution. Solved by high-capacity tray retrofit.	

(Continued)

Chapter 4 Tower Sizing and Material Selection Affect Performance (*Continued*)

Case	References	Plant/Column	Brief Description	Some Morals
511	460	Refinery vacuum	The 410 SS, 10-gage trays were replaced by 410 SS structured packing of 0.002 in. minimum thickness. After 4 months, packing fragments produced by naphthenic acid attack caused bottom pump cavitation. The life of the trays was 60 times longer due to the thicker metal.	Replacing the packing with 317L SS avoided recurrence.
319	68	Soapy water–polyalcohol oligomer (Poly-ol)	The 7-ft-ID random packed tower was unstable, experiencing unpredictable upsets, entrainment, and carryover. Below the feed the issues were high viscosity (~425 cP) and surface tension (~350 dyn/cm). Published methods for calculating capacity and pressure drop did not get close to predicting actual performance. (See also 633, Section 16.6.6, and 860, Section 6.8.)	Viscosity correction needed for high-viscosity flood/ΔP prediction.
525	215	Gas glycol contactor	High packing HETP (18-ft) led to excessive water in the overhead (52°F dew point, design 30°F). Shutdown inspection showed stains of streamlets about a drop wide on packing surfaces. Tests on a packing sheet sample showed that after acetone wash, TEG flowed in small drop-wide streamlets, while a wash with warm caustic/surfactant solution gave superior wetting. Replacing tower acetone wash by caustic/surfactant wash more than tripled staging.	Beware of surface effects at low liquid flows.
1227	84	Heavy-water distillation	Seal oil leaking into the feed from pumps covered column packing with a thin hydrophobic film. This reduced wettability and halved efficiency. A detergent wash was only partially effective for oil removal. Cured by injecting a low concentration of nonvolatile surfactant during several days of operation.	Solvent selection is critical and may require extensive off-line experimentation.
516, DT4.9	273	Chemicals vacuum	Packing flood was not accompanied by a sharp rise in ΔP. Flood could not be inferred from a high-rate gamma scan alone. A combination of detailed temperature profile, high- and low-rate gamma scans, and pressure drop measurements gave a conclusive diagnosis.	
			4.9 Packed Bed Too Long	
807	304	Refinery debutanizer	Two debutanizers operated in parallel and in identical service. One had an HETP of 39 in., the other 72 in. The only difference was that the latter did not have redistributors.	Redistribution is essential for good packing efficiency.

843	250		Poor separation was experienced in a 35-ft random packed bed that did not contain a redistributor. Addition of a redistributor significantly improved staging.	Same as 807.
DT6.1			*Splitting random packed bed and improving distribution quality improve separation.*	
507, 508, 510			*Excessive bed length induces excessive vaporization and coking in refinery wash beds, Sections* 19.1 *and* 19.4.	
			4.10 Packing Supports Can Bottleneck Towers	
801	343		Capacity was restricted by the 6-in. packing support bars which had the wide axis horizontal. If these had been vertical, they would have provided better structural strength and more open area.	Avoid restrictive packing supports.
804	304	Refinery crude fractionator	Pieces of packings squeezed through a grid bar support and got stuck in a product pump suction screen. A mesh screen was installed over the grid bars to prevent packing migration. The mesh reduced open area and caused premature flooding.	Same as 801.
836	141		Open area of a packing support was restricted to give good vapor distribution. The restriction caused a capacity bottleneck.	Same as 801.
929	141		Packing support had 60% open area versus normal 80–100%. Poor separation resulted.	Same as 801.
879	465	Atm stripper 2 ft ID	Bottom viscosity cycled by several centipoises. Cycling cause by home-made, low-open-area packing support plates. Solved by complete replacement of random packings and internals.	Same as 801.
809	45	Natural gas amine regenerator	Pieces of polypropylene saddles deformed at temperatures of about 250°F and passed through the support screen. Pieces were found in downstream equipment and blocked booster pump suction. When the still was opened, only 1 ft of the original 20 ft was found. Despite the loss, the amine was adequately regenerated. Repacking with ceramic saddles solved problem.	Watch out for plastic packing migrating through a support plate in hot services.
613			*Packing migration through support can eliminate foaming, Section* 16.4.3.	
840			*I-beam support interferes with vapor distribution, Section* 7.3.	

(Continued)

Chapter 4 Tower Sizing and Material Selection Affect Performance (*Continued*)

Case	References	Plant/Column	Brief Description	Some Morals
			4.11 Packing Hold-downs Are Sometimes Troublesome (*For packing carryover in absence of packing holddown, see Section* 22.12)	
1290	394	Refinery kerosene stripper	Tower flooding, induced by plugging of its Raschig rings, crushed the relief valve open, poured out, and fired. A layer of mud and rust about 100 mm thick was observed on the metal net above the packing, so not much kerosene was allowed to descend.	Beware of liquid in the relief. Nets above packings can be troublesome.
898			*Blocking 8% of packing area, Section* 2.6.2.	
828, 959			*Interference with distributor action, Sections* 6.10 *and* 11.10.	
			4.12 Internals Unique to Packed Towers	
838	418	3-ft and 1-ft, 3-in. columns	Separation efficiency was halved when a manhole nozzle was left uncovered inside a bed of structured packing. This problem was most pronounced under vacuum.	
885	103	CO_2 MEA stripper	Corrosion inhibitor in the MEA solution was ineffective at the manhole between packed beds. A 304L stainless steel lining prevented recurrence.	
			4.13 Empty (Spray) Sections	
524	519	Refinery vacuum	Test data and temperature surveys showed good heat transfer in an empty (no-packing) PA spray section when liquid and vapor distribution were good.	
DT4.13			*HCl absorption tails tower with sprays-only achieves better than the packing design absorption.*	
1195	286	Olefins gas cracker water quench	Tower had random packing in top, structured packing in bottom. A startup upset collapsed bed. With the plant on-line, spray nozzles were inserted through a ring of hot taps near the tower top. The plant could run at full design rates although overhead was hotter than design (at 115–120°F).	

Chapter 5 Feed Entry Pitfalls in Tray Towers

Case	References	Plant/Column	Brief Description	Some Morals
			5.1 Does the Feed Enter the Correct Tray?	
1283	267, 268	Refinery depentanizer	Feed was directed to either tray 8 or 12 from bottom by a three-way valve. It was thought that it entered tray 12, which was optimum, but surface temperature survey showed it entered tray 8.	Temperature surveys are invaluable.
301			*Oversized trays between alternate feed points induce dumping and instability, Section* 4.2.	
1654			*Incorrect entry point can cause overchilling, Section 14.3.2.*	
8112, 1270			*Crude entrainment in preflash vapor feed contaminates product, Sections 7.1.2, 17.7.*	
			5.2 Feed Pipes Obstructing Downcomer Entrance	
729	137, 138	Refinery hydrocarbon splitter, 78 four-pass trays	Valve trays retrofit by high-capacity trays raised capacity by 20%, short of the 35% target. Shortfall caused by feed pipes restricting downcomer entrance area. Also, tray spacing at the feed was not enlarged to accommodate the flashing feed. Reflux maldistribution could also have contributed. A further 9% capacity increase resulted from repiping feed and replacing feed and reflux trays by collector trays.	Beware of downcomer obstruction by feed pipes. Gamma scans helped identify problem.
769	187	NGL depropanizer	High-capacity tray retrofit fell short of expected capacity. The main feed pipe was parallel to and a short distance above the outlet downcomer, obstructing liquid descent and initiating premature flood. Diagnosed by neutron time studies and *dP* measurements. Solved by removing the feed tray.	Compare 729.
			5.3 Feed Flash Can Choke Downcomers	
739, DT5.1	276	Olefins demethanizer	For 16 years, column capacity was bottlenecked by a flashing feed entering a downcomer and choking it. Enlarging downcomers and adding a new rectifier did not help. Partial feed routing to a lower point helped. Solved by redesigning feed entry and replacing trays by packing in section below feed.	Flashing feeds should not enter downcomers.

(Continued)

Chapter 5 Tray Tower Inlet Pitfalls (*Continued*)

Case	References	Plant/Column	Brief Description	Some Morals
714	205	Refinery FCC gasoline debutanizer	Separation was poor and operation erratic. Column feed was 80°F hotter than tray liquid and entered the tray a short distance upstream of the downcomer. Insufficient mixing caused vaporization in the downcomer. Solved by feeding into a new chimney tray.	Do no enter hot feeds close to the outlet downcomer.
			5.4 Subcooled Feeds, Refluxes Are Not Always Trouble Free (*see Section* 3.2)	
			5.5 Liquid and Unsuitable Distributors Do Not Work with Flashing Feeds	
777	265	Refinery debutanizer	The tower flooded prematurely due to flashing feed entering at a huge velocity downward onto the tray floor. Flood eliminated by a well designed feed distributor. Downcomers of stripping section were enlarged and bottom chimney tray modified to eliminate secondary bottlenecks.	
740	250	Aromatics	A revamp partially vaporized an all-liquid column feed. The column feed distributor was not modified. The mixture issued at excessive velocities, causing premature flooding.	
779			*Changing feed tray to a chimney tray at higher spacing improves recovery, Section* 9.2.	
DT5.2, DT5.3			*Poor flashing feed entry bottlenecks towers.*	
778	80	Gas MDEA regenerator	Flashing feed issuing from an upward-pointing tapered slot in the feed inlet pipe impinged on the wash section seal pan, causing a portion of the seal pan to shear off. This led to hydraulic instability and reflux surges, but lean amine remained on-spec. Cure was a new seal pan with stiffener plates.	
773			*Crude tower flashing feed entraining seal pan liquid, Section* 8.4.6.	
748	382	Chemicals	Poor dual-flow tray efficiency was caused by feeding the tower flashing feed (95% vapor by volume) to a liquid distributor that was unable to handle vapor.	Dual-flow trays require good vapor and liquid distribution.

DT5.4			*Poor separation and premature flooding caused by high-velocity, downward-flashing feed entry to dual flow trays.*	
441			*Feed inlet unsuitable for flashing feed, Section* 4.3.	
			5.6 Flashing Feeds Require More Space	
736	141		Improper tray spacing at the feed location led to premature flooding.	
768	89	Refinery hydrocracker debutanizer	Tray spacing (24 in.) was not enlarged for introducing three 8-in. side-reboiler return nozzles, and it was not enlarged for a 15-in. deep sump that obstructed downcomer entry. Neither caused a bottleneck, presumably due to tower oversizing, and both were eliminated when the draw was replaced with a chimney tray (see 767, Section 10.1.1).	
729			*Among other problems, Section* 5.2.	
		5.7 Uneven or Restrictive Liquid Split to Multipass Trays at Feeds and Pass Transitions		
728	137	Refinery main fractionator 24 ft ID	Inlet pipe unevenly splitting PA return liquid to the four tray passes restricted PA circulation and bottlenecked tower capacity. Cured by adding restriction orifices to inlet pipe to equally split liquid to the passes. Side inlet weirs were replaced by false downcomers at the PA entry to improve distribution.	Feed must be equally split to passes of multipass trays.
720	304	Refinery	Plant test data on one PA section showed that the efficiency of the top tray was halved when its liquid distributor was removed.	Use liquid distributors for feeds into large columns.
756	267, 268	Refinery depentanizer	Switching from one- to two-pass trays was by running downcomer liquid to one of the two side-to-center panels. The other panel received no liquid from the tray above. This led to maldistribution, which was harmless because this tower section was overtrayed.	

(*Continued*)

Chapter 5 Tray Tower Inlet Pitfalls (*Continued*)

Case	References	Plant/Column	Brief Description	Some Morals
1236	168, 172	Refinery coker fractionator	Wash trays were replaced by grid, charge raised 10%, and recycle dropped from 1.2 to 1.1. Short runs, heavies carryover, and operational difficulties followed due to grid and collector plugging. A major factor was plugging of two out of three liquid pipes to the six-pass shed decks below, inducing severe per-pass maldistribution. Also, for the first short run wash rate was too low due to an incorrect meter factor. (See also 208, Section 3.2.4; 417, Section 4.4.2; 510, Section 19.4; 829, Section 7.1.2.)	A pipe pressure survey and a column radial temperature survey helped identify the plugged pipes.
702	7	Refinery	A restrictive design of a transition tray converting single-pass to two-pass flow caused premature flood.	
			5.8 Oversized Feed Pipes	
725, DT5.5	194	Chemicals	Column and its piping, operating at 30% of design rates, were shaken by water hammer at feed sparger. Due to excess orifice area, feed liquid probably ran out of upstream orifices, sucking vapor in via downstream orifices. This vapor collapsed onto the subcooled liquid. Hammer was eliminated by orienting sparger orifices upward to keep it full of liquid at low rates. A deflection bar was added above the orifices to prevent impingement onto the tray above.	Consider low-rate operation when designing spargers.
			5.9 Plugged Distributor Holes	
1262	190	Refinery vacuum	Quench pipe header continuously plugged, reducing circulation and generating hot spots in the bottom liquid. Problem solved by tripling liquid velocities to 6–11 ft/s.	Avoid low liquid line velocities in fouling service.
902			*Distributor pipe and holes plugged by debris, Section* 11.8.	

5.10 Low ΔP Trays Require Decent Distribution

(*see also Vapor Maldistribution Is Detrimental in Tray Towers, Section* 7.4)

765, DT5.6	262	Olefins quench towers (three towers)	Poor shed deck heat transfer was caused by poor liquid distribution and, to a lesser extent, by excessive shed deck width and marginal vapor distribution. Replacing perforated double-pipe liquid distributors with well-designed spray distributors, adding vapor inlet baffles, and replacing the wide shed decks by narrower ones greatly improved heat transfer.	Shed decks require good liquid distribution.
774	287	Olefins C_3 splitter, two towers in series	Dual-flow trays with sinusoidal corrugations gave 45% efficiency (design 70%) and high propylene losses. Gamma scans showed maldistribution after the first 30 trays of each tower. Tray supports were I-beams and U-channels along the centerlines that rose to 6 in. above the tray floor, splitting each tray into four quadrants with no remixing. Adding a redistributor every 30 trays and opening U-channels midway between redistributors raised efficiency 10%, which reduced propylene loss to just above the design.	
748, DT5.4			*Poor dual-flow tray efficiency, Section* 5.5.	

Chapter 6 Packed-Tower Liquid Distributors: Number 6 on the Top 10 Malfunctions (See also Section 11.10, Fabrication and Installation Mishaps in Packing Distributors)

Case	References	Plant/Column	Brief Description	Some Morals
			6.1 Better Quality Distributors Improve Performance	
6.1.1	**Original Distributor Orifice or Unspecified**			
812	353	Miscellaneous	Six columns: debutanizer; xylene tower; ethylene oxide absorber; Selexol towers, 3–14 ft ID. In each case, a standard liquid distributor was replaced by a high-performance distributor. In all cases, substantial improvements in separation efficiency resulted.	Well-designed high-performance distributors can improve efficiency.
811	367	Xylene fractionator	Reflux entered via a ladder pipe distributor, and the center bed was irrigated by an orifice distributor, both of standard construction. These were replaced by a lateral arm and orifice deck high-performance distributors. HETP in the beds was lowered by 20–40%.	Same as 812.
816	156	Styrene–ethylbenzene	The products did not meet design specifications Replacing the liquid distributor by a high-performance distributor solved the problem.	Same as 812.
894	188	Refinery vacuum	Replacing distributor in an LVGO/HVGO fractionation section by a high-performance distributor improved cut point by 36°F.	
839	344	Aromatics ethylbenzene–styrene 29.5 ft/24.5 ft ID top/bottom	With CS random packings, column developed 40 stages (design 76), improving to 50 after a good wash. Internal sampling showed liquid maldistribution originating at the reflux distributor. To improve, the orifice pipe reflux distributor was replaced by an orifice trough. Stripping section orifice pan liquid distributor was replaced by an orifice trough with liquid mixing. A vapor distributor was added between the sections. Hole density was reduced from 9 to 3 per square foot to enlarge hole diameters. Number of stages rose to 63. (See also 1135 and 1136, Section 12.6.)	Internal sampling, level measurements on distributors, and modeling by a two-parallel-column model proved invaluable here.
822	213	Ammonia Selexol	Unit did not meet design specifications. Problem solved by replacing standard distributors and redistributors in the absorber and regenerator towers with high-performance distributors.	Same as 812.

806	436	Ammonia hot-pot absorber	A maldistribution problem caused a CO_2 slip six times greater than design. A water test of the distributors at shutdown detected the problem. Distributor modification solved it.	Distributor water tests are invaluable.
899	485	Refinery FCC main fractionator	Gap between LCO and gasoline dropped from 24 to 0°C upon changing trays to structured packings due to poor initial liquid distribution.	
814, 863, DT6.1			*In plugging service, Section* 6.2.1.	
854			*Among other improvements, Section* 6.4.	
308, 523			*Inferior distribution leads to scale-up problem, Sections* 1.5 *and* 4.8.	
6.1.2 Original Distributor Weir Type				
DT6.2			*Replacing a notched-trough distributor by a fouling-resistant, high-quality distributor is a major contributor to better separation.*	
818	290	Test column C_6–C_7	Replacing a notched-trough distributor by a drip pan distributor effected a 30–40% reduction in HETP.	Same as 812.
870	163	Refinery FCC main fractionator	Replacing slurry section grid packing distributor with a better one increased slurry flash point from 57 to 91°C.	Same as 812.
6.1.3 Original Distributor Spray Type				
832	460	Petrochemicals superfractionator	Following replacement of trays by 10 beds of random packing column achieved 60% of the design efficiency, causing low-purity product. Replacement of the spray-type distributors by trough-type distributors reinstated on-specification products.	Distribution is key for successful packing performance.
897	534	Refinery heavy-oil vapor scrubber	Spray pattern did not cover the complete tower cross section and nozzles were oversized. This gave poor vapor–liquid contact in the grid PA bed leading to poor particulate scrubbing. Correcting flaws dramatically improved scrubbing.	Same as 832.
849	242	Refinery vacuum	Fractionation bed HETP with 1-in. random packing in a 25-ft-ID tower was 5–8 ft. Three generations of spray improvements did not help. Replacing sprays by MTS distributor halved HETP.	Same as 832.

(*Continued*)

Chapter 6 Packed-Tower Liquid Distributors: Number 6 on the Top 10 Malfunctions
(See also Section 11.10, Fabrication and Installation Mishaps in Packing Distributors) (*Continued*)

Case	References	Plant/Column	Brief Description	Some Morals
			6.2 Plugged Distributors Do Not Distribute Well	
6.2.1 Pan/Trough Orifice Distributors				
1278	261	Specialty chemicals, 2.5 ft ID, five beds	Salt plugging occurred just below the feed. Upsizing random packings did not go far enough to mitigate because the liquid distributor was also plugged. Replacing the high-irrigation-quality distributor by a lower quality, plugging-resistant one solved the problem.	A low-quality distributor that works beats a high-irrigation-quality distributor that plugs.
814	83	Specialty chemicals fouling service 6.5-ft-ID	2 in rings in tower using conventional pan-type distributors with large orifices could not achieve design efficiency. Solved by replacing all distributors with proprietary two-stage high-performance distributors.	Orifice distributors can plug in fouling services.
1198			*Distributor plugging contributes to packing fire, Section* 14.6.3.	
1298			*Orifice pan distributors repeatedly plug in dilution steam generator, Section* 18.3.	
863, DT6.1	382	Chemicals	Liquid maldistribution was caused by plugging of orifice pan feed distributor with solids that were presumed absent in the feed. Packing efficiency improved but still fell short of design after filtering the feed. Replacing poor-quality distributors by better ones further improved efficiency.	Closely examine solids in entering streams.
12117	215	Gas glycol contactor	Water dew point of outlet gas was high due to plugging of the structured packing liquid distributor holes. Cause was bypassing and infrequent changeover of the lean glycol filters. Cleaning the distributors and filters changeover on a regular schedule prevented recurrence.	
DT6.3, DT6.4, DT17.2			*Poor filtration leads to plugged distributors and poor performance.*	
1256	8	Formaldehyde stripper 6.5 ft ID	Formaldehyde removal was poor and random packing HETPs were high. Gamma scans showed that at constant operation liquid level built up in the distributor, rising above the distributor risers. The level then dumped. Maldistribution resulted. Distributor plugging was the likely cause.	

1275	465	Petrochemicals	Separation never reached its full potential and deteriorated with time due to plugging of peripheral openings of every orifice trough distributor. Out-of-levelness on every distributor and low peripheral drip point density on one distributor contributed. Cleaning, leveling, and replacing one distributor were the fix.	
8106	286	Olefins water quench	Run length was less than a year with a combination of spray nozzles, and random packings with pan distributors. Replacement by grid and V-notch distributors extended run length to more than 2-1/2 years with lower pressure drop and better heat transfer.	
12116	364	Gas hot-pot absorber	Gamma scans showed poor distribution in top bed and liquid stacking above top distributor. This area has fouled historically with carbonate buildup.	
12114	544		Fouling on the liquid distributor and top layers of packing caused distributor overflow and flooding	Gamma scans diagnosed.
1174	514	Refinery FCC main fractionator	Catalyst deposited in trough distributor that irrigated the slurry PA packing, causing liquid maldistribution. The flow imbalance caused coke buildup in the low-flow areas. Liquid and vapor maldistribution diagnosed by tracer tests.	Tracer tests are useful for diagnosing maldistribution.
DT19.4A			*Similar to* 1174.	
521			*Notched-trough distributor plugs in groundwater purification service, Section* 18.4.3.	
1226	194	Water scrubber	Fungus growth caused orifice pan distributor to plug and overflow.	
924			*Among other problems, Section* 11.10.	
1254			*Solved by on-line wash, Section* 12.7.	
6.2.2 Pipe Orifice Distributors				
808	426	Olefins water quench 13.5 ft ID	Ladder pipe distributor plugged after 3 days in service. Less than 1 lb of solids was sufficient to plug 80% of perforations. Problem eliminated by installing Y-strainers and enlarging distributor perforations. Good packing performance was achieved even though tray support rings were not removed.	Ensure a filter upstream of perforated distributors. Avoid small perforations.

(Continued)

Case	References	Plant/Column	Brief Description	Some Morals
831	460	Refinery naphtha splitter	Replacement of trays by packing led to off-specification top product and a 10% capacity loss. The cause was plugging of the reflux pipe orifice distributor with piping scale and oxidation products formed every turnaround and carried into the distributor upon restart. Problem mitigated by replacing the pipe orifice with a fouling-resistant trough-type distributor.	Study column history before specifying column internals.
852	209	50-in.-ID gauze packing	Plugging in ladder pipe distributor led to poor performance and head box overflow. The plugging was caused both by reactions and upstream corrosion. Replacement by MTS distributor eliminated the problem. (See also 851, Section 6.7.2.)	Small holes plug easily.
848	242	Chemicals, 3-ft-ID gauze packing	Column distilled propylene glycol/ethylene glycol and an agricultural chemical intermediate. Liquid flows were extremely low, 0.02–0.1 gpm/ft^2. A ladder pipe distributor plugged within 2 weeks. It was replaced by an MTS distributor. Worked well for more than 15 months.	Same as 852.
850	242	Chemicals vacuum 3 ft ID	Polymerization and plugging problems in a pipe lateral-type predistributor ($\frac{5}{16}$-in. holes) restricted column run length to 6-8 weeks. Replacing the predistributor with a slotted tube predistributor increased run length to 10 months.	Slots perform better than small holes in polymerizable services.
839			*Among other problems, Section* 6.1.1.	
6.2.3	**Spray Distributors**			
823	374	Refinery vacuum	Plugging of several spray nozzles caused packing fouling. Plugging resulted from relying on filter systems that were too distant from the column and using CS piping downstream of filters.	Rust particles from piping can plug distributors.
1228	172	Refinery vacuum	Wash section sprays plugged. The wash oil control valve bypass also bypassed the filters.	Avoid filter bypasses.

1234	377	Refinery vacuum	Spray header plugging caused excess metals in the gas oil of a deep-cut (1050°F) unit.	
8105	286	Refinery vacuum	Wash bed spray nozzles repeatedly plugged giving less than a year run length. Run length improved by replacing with a gravity distributor among other changes.	
DT6.4			*Plugging with debris caused plugging, coking.*	
1287	114	Refinery crude fractionator	Heat transfer in the top PA packed bed deteriorated within a year after start-up. Problem was spray distributor plugging at one end and damaged at the other end. Diagnosed by surface temperature surveys and gamma scans.	Paper gives detailed guidelines for surface temperature surveys.
12105	485	Refinery FCC main fractionator	Half the spray nozzles above a packed LCO PA bed plugged, causing 75°C variation in the temperature leaving the bed. The nonuniform vapor composition led to a 2.5-m HETP in the gasoline/LCO fractionation bed above.	
1153			*Turnaround hot work on plugged spray distributor causes a fire, Section* 14.6.1.	
924			*Among other problems, Section* 11.10.	
			6.3 Overflow in Gravity Distributors: Death to Distribution (*For overflow due to plugging, see Sections* 6.2.1 *and* 6.2.2)	
8107	429	Refinery crude fractionator	Jet fuel yield dropped from 12% to 6% following trays to packing retrofit. Cause was drain pipes from the chevron collector above the jet-fuel/diesel bed undersized for self-venting flow. This led to pulsating distributor level (measured), overflow of collector liquid onto the wall, and possibly also to distributor overflow. Liquid maldistribution, poor efficiency, and flooding in the jet fuel/diesel bed resulted. Cured by replacing collector and drain pipes by well-drained chimney tray and increasing bed length.	Diagnosed by through investigation involving field tests, gamma scans and simulation.

(*Continued*)

Case	References	Plant/Column	Brief Description	Some Morals
830	170	Refinery FCC main fractionator 24 ft ID	Replacing bubble-cap trays with structured packing led to a high gasoline end point, high losses of gasoline in the LCO product, and HETPs exceeding 10 ft. Severe maldistribution, verified by radial surface temperature surveys, and overflow of liquid above the risers were the causes. Remedy was a new distributor, replacing combination collector/redistributor by two separate pieces and repacking the tower.	Additional distribution height is often more beneficial than additional packed height.
842	250		Poor separation efficiency resulted from liquid spilling into the vapor risers of a packing redistributor. The spill was caused by undersizing of the total orifice area.	
1229	167	Refinery vacuum LVGO/HVGO fractionation	Following a tray to structured packing retrofit, LVGO/HVGO fractionation and LVGO product yield were poor. Internal reflux was 60% above design, overflowing the distributor. This reflux rate was not monitored and could only be inferred from the LVGO PA duty. Problem solved by cutting PA duty, which reduced internal reflux.	Problem diagnosed by an energy balance.
8101	485	Refinery crude fractionator	Separation between kerosene and light diesel was poor due to overflow of the orifice pan distributor irrigating a random packed bed. A heat balance showed that liquid flowed into the distributor from the kerosene draw tray above at more than twice the design rate. Poor liquid distribution from a ladder pipe distributor with excessive ratio of hole to pipe area to the PA bed below could have contributed.	Mass and energy balances and a temperature survey helped diagnose.
8104	291	Refinery FCC main fractionator	The orifice LCO pumparound (PA) distributor overflowed, causing 4–6 feet of liquid back-up to the fractionation bed above. Diagnosed by gamma scans and solved by a new trough distributor. The LCO PA retrun, containing rich sponge oil, entered a false downcomer, directing it into the distributor. This false downcomer was repeatedly found blown apart due to vaporization and insufficient strength and lying on the distributor. Solved by replacing with a new, strong flash-box.	Watch out for overflows and flashing in feeds.

857	133	Amine regenerator	At rates below design, liquid level on the feed distributor exceeded chimney height. At higher rates, this caused flooding of the top bed. Problem diagnosed by gamma scans.	
8102	400	Synthesis gas stripping column	Efficiency was low immediately after restart of 2.8-m-ID random-packed tower. Gamma scans showed maldistribution due to distributor overflows. Tower also had a history of fouling over a 4-year run.	
858	94	Olefins caustic scrubber	Caustic entrainment into the water section occurred when caustic circulation rates exceeded 75% of design. In situ water tests showed water levels exceeding riser height at 75% of design. Foaming could have been a contributor. Temporarily solved by restricting circulation rates at the cost of poorer caustic utilization.	In situ water tests are invaluable for distributors.
856	64	Chemicals caustic scrubber	Caustic was detected in a vent downstream of a pair of parallel caustic scrubbers; both supplied by the same caustic tank. Tracer tests found that only one of the two entrained. Entrainment eliminated by reducing caustic flow to that scrubber.	Overloading distributors can cause entrainment.
DT6.5			*Overflow explains poor performance of packing in heat transfer service.*	
882	545		Tower was not separating properly due to an overflowing parting box. A grid gamma scan showed the overflow but no maldistribution in the packing. A CAT scan confirmed the overflow and showed the maldistribution to be evenly spread among the grid chords.	
316			*Due to subcooling, Section* 3.2.1.	
1254, 1298			*Due to plugging, Sections* 12.7 and 18.3.	
1251			*Due to slug flow, Section* 6.5.	
955			*Detected in water test, Section* 11.10.	
845	250		Excessive horizontal velocity in a parting box pushed liquid against the narrow wall of the box. This induced liquid overflow; the overflowing liquid was entrained by the rising vapor, leading to an acid emission problem.	Avoid excessive velocities in parting boxes.
1249			*Due to damage, Section* 6.7.2.	

(Continued)

Case	References	Plant/Column	Brief Description	Some Morals
			6.4 Feed Pipe Entry and Predistributor Problems	
859	69		Liquid feed entered an orifice pan distributor at 5 ft/s pointed straight down. The kinetic head caused flow out of the pan in the vicinity of the feed to be much greater than away from that location.	Predistribution to the main distributor is important.
854	157, 428	Chemicals wastewater stripper, random packings	Improvement of benzene removal from 99.93 to 99.99% was achieved by extending the pipes feeding the liquid distributor below the liquid level. Later, liquid distributor was replaced by a high-performance distributor, at the same time when a high-capacity vapor-distributing tray was added (Case 853, Section 7.1.1). These increased benzene and toluene removal to better than 99.995%.	
864	382	Chemicals	A water test showed a feed pipe to a parting box caused excessive splashing, resulting in uneven flow to each of the distribution points. Problem eliminated by extending the feed pipe.	Distributor water tests can prevent disasters.
820, DT6.6	194	Acetic acid scrubber from off gas	Odor and excessive acetic acid emission resulted from poor scrubbing in a 2-ft-ID tower. Cause was an open vapor riser right under the liquid feed point to the distributor. The incoming liquid poured down the riser, bypassing the distributor. Removing and blanking the riser solved the problem.	Pay attention to feeding liquid onto distributors.
DT6.7			*Absorber feed pipe discharges reflux into a vapor riser.*	
895	434	Gas glycol regenerator	An alteration in the still feed location reduced overhead temperature from 250 to 190°F and drastically cut glycol losses.	
955			*Excessive liquid splashing, Section* 11.10.	
DT6.3			*Reflux issues with a horizontal momentum in the direction of reflux flow in header, but this was not the main problem.*	
850			*Feed pipe plugging, Section* 6.2.2.	
940			*Pan misoriented to liquid inlet, Section* 11.10.	

956			*Feed pipe terminating* $\frac{1}{16}$ *in. above floor, Section* 11.10.	
845, 882			*Overflowing parting boxes, Section* 6.3.	
1282			*Distributor damage due to excessive liquid velocity, Section* 6.7.2.	
898			*Poor mixing of solvent with reflux in extractive distillation feed distributor, Section* 2.6.2.	
			6.5 Poor Flashing Feed Entry Bottleneck Towers	
810	85	Aromatic hydrocarbon binary	Bubble point feed entered the column via a pipe distributor with underside perforations. When the feed became partially vaporized, lights moved into the bottom section, almost contaminating bottom product. Installing a chimney tray below the distributor eliminated the problem.	Liquid distributors do not work well with flashing feeds.
871	159	Refinery FCC main fractionator	Same as 321 (Section 2.3), except that the sponge oil return was via a spray distributor. The revamp failed because the sponge oil contained 30–50% volume vapor.	Spray distributors are unsuitable for flashing feeds.
8104			*Vaporization of sponge oil damages distributor, Section* 6.3.	
8110	444	NGL debutanizer random packing 15 inch ID	Inadequate separation of flashing feed led to maldistribution, flooding, terrible separation, and capacity loss. The pan liquid distributors had oversized holes and undersized risers. The riser pressure drop almost equaled distributor liquid head, so vapor competed with liquid for the distributor holes, causing frothing. Removing reflux distributor had no effect. Cured by retrofitting with well-designed flashing feed and reflux distributors.	
8111	444	NGL depropanizer	Same problem and solution as Case 8110, but here reflux distributor was not removed during the terrible-separation period. Tower was 23 in ID.	
940			*Flashing feed entering via a bare nozzle above orifice gravity distributor, Section* 11.10.	

(*Continued*)

Chapter 6 Packed-Tower Liquid Distributors: Number 6 on the Top 10 Malfunctions (*Continued*)

Case	References	Plant/Column	Brief Description	Some Morals
890	226	Ammonia aMDEA regenerator	Excessive velocities in the flashing feed gallery caused entrainment of aMDEA, excessive aMDEA losses, and severe corrosion repeatedly cutting holes in the tower wall. Cure was redesign of the flashing feed gallery, modifying its split-flow inlet to a tangential feed and replacing the resin-coated CS top section by an SS top section.	
1251, DT6.8	133	Debutanizer	The column was unable to achieve product specifications. The cause was slug flow of the column feed. Liquid collected at the pipe until a slug developed and was lifted into the column. There it instantly caused distributor overflows and uneven liquid-to-vapor ratios, leading to poor separation.	
DT6.9			*Slug flow in stripper feed pipe causes base-level oscillations, poor stripping.*	
15164	378	Natural gas Selexol H_2S stripper	Flashing feed control valve was 20 feet above grade. Severe line shaking was experienced. Cure was relocating valve to grade and anchoring it to a large concrete block (see also Case 15163, Section 29.1).	Compare Case 15103, Section 12.13.3.
			6.6 Oversized Weep Holes Generate Undesirable Distribution	
868	382	Water/EG, 30 in. ID, two beds Pall rings	Separation was very poor because most of the reflux passed through two $\frac{3}{4}$-in. weep holes right under the reflux feed pipe in the center of the weir riser distributor. Also, the weir risers were only force fitted through the deck and had gaps around them. Problem diagnosed using an in situ water test. It was fixed by plugging the weep holes and sealing the gaps.	In situ water tests are invaluable for distributors.
865	382	Chemicals	Water test of a V-notched trough distributor showed that at the lower flow rates almost all the liquid in each trough issued through a lone $\frac{3}{4}$-in. weep hole. Cured by reducing weep holes to $\frac{1}{4}$ in.	Distributor water tests can prevent disasters.

DT6.10			*Drain holes in a collector between two beds cause liquid to bypass distributor and poor separation.*	
821, DT6.11	194	Chemicals batch processing unit	A single-bed column did not achieve a simple separation because of a past modification of drilling 4-in. holes in the distributor floor. Liquid poured down the holes, bypassing the distribution orifices. Cured by equipping holes with risers.	
			6.7 Damaged Distributors Do Not Distribute Well	
6.7.1 Broken Flanges or Missing Spray Nozzles				
918	172	Refinery vacuum	Wash oil spray header flanges parted because the header arms were rigidly bolted to the vessel wall with no allowance for thermal expansion. This led to excessive asphaltanes in HVGO.	Diagnosed by zero pressure drop across sprays.
919	172	Refinery vacuum	Wash oil spray header flanges were made of 10-gauge metal with four bolts. These fell apart, dumping all liquid at inlet and causing gas oil quality to drop.	Use standard, raised-face pipe flanges.
880	458	Refinery vacuum	Spray nozzles had triple the capacity, and many of the flange gaskets were left out. This gave small spray nozzle pressure drop and deformed sprays and led to wash bed coking within 1–2 years.	
1287			*Spray distributor plugged at one end and damaged at other end, Section* 6.2.3.	
924			*Spray nozzles were missing their internals, Section* 11.10.	
6.7.2 Others				
1249	64	Packed	Liquid collector became dislodged, resulting in liquid overflow on one side and maldistribution.	
1282	216	Ammonia condensate stripper	Distributors were damaged by high-velocity liquid. The connection between the distributors and supports was broken. Poor liquid distribution to the random packings and poor efficiencies resulted.	Maldistribution diagnosed by gamma scans.
8104			*Vaporization damages distributor, Section 6.3.*	

(Continued)

Chapter 6 Packed-Tower Liquid Distributors: Number 6 on the Top 10 Malfunctions (*Continued*)

Case	References	Plant/Column	Brief Description	Some Morals
851	209	50-in.-ID gauze packing	Pilot column HETP was 10 in., prototype 16–24 in. Inspection showed warped thin-gauge metal collectors and distributors, evidence for gasket failure and collector leakage, and dry areas on the packings. Thermal stress was the prime cause. Welded sump rings for the liquid collectors and heavier metal gauge distributor prevented recurrence. (See also 852, Section 6.2.2.)	Seal welding is superior to gasketing in hot services.
DT6.12			*Large hole in distributor gives poor tower separation.*	
			6.8 Hole Pattern and Liquid Heads Determine Irrigation Quality	
876	306	Refinery FCC main fractionator 14 ft ID	Following replacement of trays by packing, LCO product went off specification. Cause was maldistributed reflux due to minimal liquid head in orifice pan distributor. Diagnosed by a temperature survey. Overcome by eliminating the LCO draw and taking all product from a lower point (combined LCO and HCO draw) in the tower.	
873	536	Pharmaceuticals batch	Column had variable reflux control. Product was off specification and column required much operator attention. Cause was liquid maldistribution to structured packings at lower reflux rates. Cured by converting from batch to semicontinuous and modified controller programming, sustaining good liquid distribution.	Distributor turndown could be a problem in variable–reflux operation.
8110, 8111			*Oversized holes and undersized risers lead to frothing on distributors, Section* 6.5.	
892	448	Water deoxygenator, vacuum	Liquid from a flash drum gravity-descended onto the tower packing via an orifice pipe distributor. Distributor pressure drop exceeded available head, restricting liquid flow to 70% of design. Solved by replacing pipe distributor by gravity distributor.	

867	382	Chemicals	The outer ring of drip points in a large orifice pan distributor had guide tubes to get liquid closer to the tower wall. A water test found that these guides actually put all the liquid from the tubes onto the wall. Shortening the tubes solved the problem.	Distributor water tests can prevent disasters.
860	68	Soapy water/ polyalcohol oligomer (Poly-ol)	A 7-ft-ID random-packed tower was unstable, experiencing unpredictable upsets, entrainment, and carryover. Below the feed, the issues were high viscosity (~425 cP) and surface tensions (~350 dyn/cm). Conventional orifice equation gave poor predictions to head–flow relationship, so distributor redesign was based on flow test with actual fluid. Distributor design also included features to mitigate foaming. (See also 319, Section 4.8, and 633, Section 16.6.6.)	
955, 934			*Mispunched holes, Section* 11.10.	
1275			*Low peripheral hole density, Section* 6.2.1.	
839, 8101			*Too many holes, Sections* 6.1.1 and 6.3.	
			6.9 Gravity Distributors Are Meant to Be Level	
953	496	Structured packings	Side draw above middle bed was off specification for heavies. Gamma scans showed maldistribution in middle and bottom beds. Both middle and bottom distributors were out of level; middle by 1 in. Packings below middle distributor were discolored in some locations due to drying. Leveling improved performance.	Ensure adequate levelness of distributors.
DT6.13			*Titled distributor leads to poor separation.*	
966	545	CO_2 absorber	Excessive CO_2 slip resulted from out-of-levelness of an orifice tray distributor. Maldistribution diagnosed by gamma scans. The fix was leveling and improved predistributor pipe.	
1275			*Among other problems, Section* 6.2.1.	
869, DT6.2			*Notched distributor, Section* 6.12.	

(*Continued*)

Chapter 6 Packed-Tower Liquid Distributors: Number 6 on the Top 10 Malfunctions (*Continued*)

Case	References	Plant/Column	Brief Description	Some Morals
			6.10 Hold-Down Can Interfere with Distribution	
828	309	Refinery vacuum 30 ft ID	Structured packing hold-down, specified to withstand 1-psi uplift, was a massive I-beam that interfered with 25% of the spray cones. This led to coke formation in the wash section and reduced heat transfer in PAs. Some packing damage resulted from heavy beams dragged across the top of the beds.	Heavy-duty design should not interfere with distribution.
959			*Due to incorrect assembly, Section* 11.10.	
898			*Among other problems, Section* 2.6.2.	
			6.11 Liquid Mixing Is Needed In Large-Diameter Distributors	
207	173	Refinery FCC main fractionator	Slurry section coked up 3 months after disc and donut trays were replaced by grid. The liquid distributor fed volatile wash liquid to one region of the packing and the much less volatile slurry PA liquid to another without premixing them. The volatile wash liquid vaporized, the packing in this region dried up, then coked.	A two-year plus run length was obtained following distributor modifications that mixed the liquids.
887	435	C_3 splitter six 6-m beds	HETP of 25-mm random packing was high because the redistributors did not homogeneously equalize any generated liquid maldistribution. Solved by improved redistributors and new random packings.	Liquid mixing is needed in large-diameter distributors.

898			*Poor mixing of two liquid phases in extractive distillation distributors, Section* 2.6.2.	
839			*Among other problems, Section* 6.1.1.	
830			*Among other problems, Section* 6.3.	
			6.12 Notched Distributors Have Unique Problems	
866	382	Chemicals	Water testing of a V-notched trough distributor showed that at the low rates the liquid wrapped around under the distributor and combined with several other liquid streams into a single-point downpour. At the high rates, about half the liquid still wrapped around. The other half flowed out, running into streams coming from an adjacent distributor trough, forming one downpour instead of two. Problem solved by adding drip guides that directed liquid to the design drip point layout.	Distributor water tests can prevent disasters.
869	382	Chemicals	Tower achieved less than one-third the expected efficiency due to liquid issuing from the V-notches flowing around underneath the troughs. This flow pattern was indicated by bathtub rings left behind by the liquid and is believed associated with the sensitivity of V-notches to levelness at low liquid rates.	This problem could have been identified by a water test before start-up.
868			*Gaps around weir risers, Section* 6.6.	
			6.13 Others	
807, 843			*Not enough redistribution, Section* 4.9.	
835, 940			*Distributor/redistributor parts interchanged, Section* 11.10.	

Case	References	Plant/Column	Brief Description	Some Morals
			7.1 Vapor Feed/Reboiler Return Maldistributes Vapor to Packing Above	
7.1.1 Chemical/Gas Plant Packed Towers				
817	235	CO_2 absorption from gas, three columns, 15 ft ID	Severe gas maldistribution was measured and caused uneven velocity and pressure profiles throughout the entire 50-ft-tall beds. The maldistribution was initiated at the bottom feed inlet. Efficiency in the column experiencing maldistribution was roughly half that measured in another similar column that had a specially designed gas inlet sparger.	Do not overlook vapor distribution.
813	83	Hydrogen peroxide 4 ft ID	The column performed poorly after its ceramic random packings were replaced with structured packing. Replacing the vapor distributor by an improved, higher pressure drop type solved the problem.	Same as 817.
8103	428	HCN steam stripper	Replacement of ceramic by high-capacity, high-efficiency random packings raised capacity but lowered efficiency. Cause was poor steam distribution better corrected by the higher pressure drop ceramic packing. Replacing the V-shaped baffle at steam inlet and the valve tray above by a steam sparger and a chimney tray with restricting orifices gave better stripping than ever.	Same as 817.
8109	515	LNG packed stripper	Vapor content of the feed was an order of magnitude greater than the stripping gas flow rate. Between 0.1% and 1.5% of the total liquid was entrained overhead. The carryover was proportional the feed rate. Gamma scans showed top of packing was a foot below design. Tracer tests showed gas maldistribution, with gas downflow in one quadrant of the 13-foot tall bed.	
815	83	Natural gas TEG dehydrator	Following replacement of trays with structured packing, the product failed to meet specifications and turndown was poor. Glycol rate had to be doubled to achieve dehydration. The most likely cause was gas maldistribution induced by an inlet vapor velocity head of 56 in. H_2O. The column achieved design after new structured packing as well as new vapor and liquid distributors were installed.	Same as 817.

853	157, 428	Wastewater stripper random packing	Tower designed to remove 99.97% of the benzene and 99.993% of the toluene from wastewater. It removed only 99.0% of the benzene and 99.5% of the toluene. Adding a steam diffuser raised benzene removal to 99.93% and toluene removal to 99.973%. Later, a high-capacity tray was added above the diffuser for further vapor distribution improvement. (See also 854, Section 6.4.)	Same as 817.
844, 855			*With inlet baffles, Section* 7.2.	
1242			*Channeling upon stacking random packings, Section* 18.5.	
7.1.2 Packed Refinery Main Fractionators				
824	374	Refinery FCC main fractionator	Disk and donut trays just above the feed and trays below the HCO draw were replaced by grid. Erratic temperatures and poor heat transfer resulted, caused by vapor maldistribution. A V-shaped wedge baffle was installed directly at the vapor inlet but did not help. Baffle and grid coked after 10 months operation, causing a capacity bottleneck.	Vapor maldistribution can be detrimental to grid performance. Vapor baffles coke inlets in this service.
803	306	Refinery FCC fractionator 14 ft ID	Cool slurry (700°F) did not evenly contact the entering hot (980°F) vapor. Spreads of up to 90°F were observed at the same elevation in the slurry PA grid bed. A spray distributor was used.	
1174, DT19.4			*Liquid and vapor maldistributiion in coked slurry section of FCC main fractionator, Section* 6.2.1.	
1259, 1260			*Grids weather high gas velocities in FCC main fractionators, Section 22.2.*	
862	542		Tower received a high-velocity feed below the packing with no vapor distributor. Grid gamma scan showed severe vapor maldistribution, with scan line opposite the vapor inlet far less dense than others, suggesting excessive vapor velocities displaced liquid traffic.	Follow recommended vapor distribution practices.
875	192	Refinery Flexicoker fractionator	Wash section coked 3 months after it was revamped with grid and a specialized vapor distributor mounted 9 in. below the grid. Shutdown inspection verified vapor maldistribution. Solution was relocating grid higher in the tower and adding a vapor-distributing tray between.	Same as 862.

(Continued)

Case	References	Plant/Column	Brief Description	Some Morals
829	168, 172	Refinery coker fractionator	Wash trays were replaced by grid, feed raised 10%, and recycle dropped from 1.2 to 1.1. Short runs, heavies carryover, and operational difficulties followed due to grid and collector plugging. Causes included the following: (1) Excessive weir lengths on downcomers from the collector tray giving low crests over the weirs and poor liquid split to the six-pass shed decks below. Thermal expansion on the collector tray led to out-of-levelness, which aggravated the problem. (2) Upon tray-by-grid replacement, vapor distribution was not improved. Grid is far more prone to vapor maldistribution than trays. This was aggravated by closeness of the top shed deck, a source of vapor maldistribution, to the collector tray. (3) The collector tray had excessive residence time and was not adequately sloped to remove solids. (4) Plugging of downpipe from the collector at low liquid rates. (See also 208, Section 3.2.4; 417, Section 4.4.2; 510, Section 19.4; 1236, Section 5.7.)	A high-pressure drop collector (here 6 in. w.g.) and adequate distance above the source of vapor maldistribution mitigated grid plugging. Reducing weir lengths also helped.
878	537	Refinery vacuum	Modification to the vapor horn halved the peak vapor velocities exiting the flash zone. This improved vapor distribution to the bed above and reduced entrainment from the flash zone, reducing the HVGO product tail by 120°F.	CFD modeling led to the improvement.
DT19.1			*Replacing simple 90° vapor horn by 360° state-of-the-art horn aggravates entrainment in refinery vacuum tower.*	
874	192	Refinery lube oil vacuum	Tower diameter was 6 ft on top, 10 ft at bottom. Feed entrainment caused a polymerization reaction that plugged the wash oil grid above. Entrainment was mitigated and run length increased sixfold by lowering feed nozzle from the 6-ft section to the swage and by improving vapor distribution.	Correct vapor feed is essential.
896			*Eliminating chimney tray liquid level gives more disengagement and vapor distribution height, Section* 9.8.	
872	162, 485	Refinery crude fractionator	Resid yield was 2% higher than expected after trays were replaced by packing. Crude entrainment in the preflash drum vapor quenched the flash zone near the vapor entry, making more resid. Directly above the quench zone, temperatures throughout the two beds above were 50°F colder than right across.	A highly subcooled liquid can initiate vapor maldistribution.

8112	43	Refinery crude fractionator	With pre-flash drum vapor entering the tower at AGO draw elevation, crude entrainment caused episodes of black, high-metals AGO. A revamp routed the pre-flash drum overhead to the flash zone, replaced trays by packing in the wash, diesel/AGO fractionation, and diesel PA zones, and raised rates. Following the revamp, diesel and AGO yields dropped 5%. Crude entrainment in the pre-flash drum quenched the region of the flash zone near the vapor entry making more resid. Directly above the quench, temperatures on top of the wash bed were 46°F colder than right across. The maldistribution gave poor fractionation. Raising heater outlet temperature from 700°F to 735°F to raise yield led to heater coking and stripping section fouling.	Field measurement identified problem. Entrainment diagnosed by color of condensed pre-flash overhead sample.
825			*Vapor inlet baffle grows coke, Section* 7.2.	

7.2 Experiences with Vapor Inlet Distribution Baffles

844	250		The addition of a "doghouse" baffle (a baffle parallel to the direction of fluid entry, in the shape of a doghouse, above a vapor feed) eliminated a vapor maldistribution problem.	
855	234		A deflector baffle was added at the inlet of the reboiler return. Off-specification product resulted. Tracer injection into the reboiler return showed that the baffle was forcing most of the vapor up to side of the column opposite the vapor inlet. The vapor maldistribution persisted through two packed beds. Solved by replacing deflector plate by a vapor sparger.	Poorly designed inlet baffles do not mitigate maldistribution.
DT6.1			*Arrangement of two V-baffles generates a vapor jet that rises up the packed bed.*	
825	374	Refinery FCC main fractionator	Vapor inlet had a baffle about midway in the vapor inlet zone. Coke grew on the baffle, starting on the back of the baffle up through the top of the grid packing above the feed, and needed to be dynamited out.	Inlet baffles coke in this service.
824, DT19.4			*Similar to* 825, *Section* 7.1.2.	

7.3 Packing Vapor Maldistribution at Intermediate Feeds and Chimney Trays

DT7.1			*Undersized, unsealed downpipes from a vapor distributor initiates flooding and poor separation in bed above.*	

(*Continued*)

Chapter 7 Vapor Maldistribution in Trays and Packings (*Continued*)

Case	References	Plant/Column	Brief Description	Some Morals
840	250	Refinery vacuum	I-beam interference with vapor issuing from accumulator tray chimneys generated severe vapor maldistribution. The maldistributed vapor profile was displayed as a carbon deposit on the surface of the bottom-packing layer above. (See also 841, Section 9.6.)	Beware of I-beam interference with vapor distribution.
839, 874			*At tower swage, Sections* 6.1.1 *and* 7.1.2.	
12105			*Maldistribution below bed leads to poor inlet composition profile and high HETP, Section* 6.2.3.	
DT6.2			*Vapor maldistribution due to damage and plugging of interbed demister.*	
			7.4 Vapor Maldistribution Is Detrimental in Tray Towers	
7.4.1 Vapor Cross-Flow Channeling				
427	271	Refinery crude fractionator	Upon a post-T/A restart, about 50–60% of tray liquid wept from sieve trays operating at 85% of flood. This degraded valuable furnace oil into gas oil. The same trays worked well before the T/A. Pre-T/A corrosion increased hole area from 13 to 16% of active area, but deposits kept it down. After T/A cleaning, the hole area became 16% and vapor cross-flow channeling set in, causing massive weep.	Keeping the hole area down mitigates vapor cross-flow channeling.
428	246	Chemicals absorber valve trays	Absorber had five cooling PAs. Pumps lost prime when gas rates fell even slightly below maximum. Absorption was mediocre at full gas rates and rapidly deteriorated upon turndown. Massive weep was observed at gas rate as high as 40–70% of flood. The cause was VCFC induced by the low dry pressure drop of the venturi (smooth orifice) valves.	Avoid venturi valves in services prone to VCFC.
442	195, 200	Aromatics toluene recovery	Premature flooding, instability, and a capacity bottleneck were caused by VCFC in the rectifying section. This is the first time VCFC was reported with sharp-orifice moving valves with standard (14%) slot areas. Other conditions, especially high ratio of flow path length to tray spacing, and tray trusses perpendicular to liquid flow, were conducive to VCFC. Problem eliminated by a retray with high-capacity trays designed to circumvent VCFC.	
DT7.2			*Valve trays with* 20% *open area and* 18-*in. spacing lose capacity due to VCFC.*	

7.4.2 Multipass Trays

DT7.3			*Center downcomer obstructs vapor bottom feed.*	
732	464	Refinery	The two reboiler return lines entered the column in the center section of four-pass trays (between the off-center downcomer). The entire vapor load was taken by this center section (which had about half the tray area), causing premature flood.	
429, DT7.4	272	Olefins water quench fixed valve trays	Vapor channeling through the center of two-pass trays led to liquid entrainment from the tower top. The channeling initiated at a poorly designed chimney tray (738, Section 9.5) and propagated through the trays above, due to their large slot area (18.5% of active area). Redesigning the chimney tray and blanking valves to reduce the slot area to 13–14% eliminated the problem. (See also 312, Section 1.3.1.)	

7.4.3 Others

432	250	Refinery vacuum 44 ft ID	Severe local corrosion occurred on the two-pass valve tray just above the overflash chimney tray. Vapor channeled through a small number of valves, which corroded. The tray operated at low vapor rates. About 40% of the valves should have been blanked.	Vapor channeling can be a severe problem at low vapor rates.
758	116, Case MS78	Refinery petnane isomerization caustic scrubber	The dry gas from the product stabilizer in a pentane isomerization unit contained HCl. The mixing was not instantaneous or completely effective. This led to corrosion in the bottom of the scrubber and its trays. To reduce the amount of expensive corrosion-resistant alloy, a static mixer was installed, with the caustic and gas flowing directly to the inlet. The mixing zone was only 12 in. long.	
433	250	Nitric acid absorber 10 ft ID	One-pass trays were at 12-in. spacing. Each tray was supported by two beams 6 in. deep oriented parallel to the liquid flow. The average froth height reached the bottom of the beams. Observations through viewing ports showed that the beams divided the tray into three cells, with most of the bubbling in the cell with lowest liquid level. The liquid would then violently move from the high-liquid-level cell to the low-liquid-level cell, generating violent back-and-forth liquid oscillation perpendicular to the beams.	Structural members should not extend deep enough to affect tray action.

Case	References	Plant/Column	Brief Description	Some Morals
			8.1 Causes of High Base Level	
8.1.1	**Faulty Level Measurement or Level Control** (*See also Level Instrument Fooled, Section* 25.7)			
1520	304	Refinery DC_2 absorber	Loss of bottom-level indication resulted in column flooding. Gasoline spilled over to the top knockout drum, thence to the fuel system, and ended spilling out of burners, causing several heater fires.	Ensure adequate level indication.
1168	12	Olefins demethanizer	Following introduction of liquid feed, it was not appreciated that the demethanizer level transmitter was disconnected. The apparent lack of level was attributed to having to control boil-up on the manual reboiler bypass because an isolation valve on the reboiler flow control set was broken. The tower flooded, filled the reflux drum, leading to excessive liquid drainage to flare. The level transmitter and alarm on the flare knockout drum were inadvertently isolated, so there was no indication that liquid was ascending the flare stack, which failed by low-temperature embrittlement.	Ensure all instrumentation is operational before introducing feed.
1611			*Failure of base-level controller causes liquid to pass out of relief valve and flash header cracking, Section* 14.3.2.	
1544	81	Olefins stripper	The base-level controller failed at start-up, and liquid level in the column rose to fill half the column. This caused excessive heavies in the top product. Diagnosed using gamma scans. Cutting feed rate was short-term solution. Using a gamma-ray absorption level indicator was a longer term cure.	
15133	543, 544	Deethanizer 26 trays	Column fully flooded due to liquid level exceeding reboiler return nozzle. There was no functional level gauge in the bottom. Diagnosed by gamma scan and cured by draining accumulated liquid while using a stationery gamma source/detector to monitor bottom level.	Same as 1520.
1560	141		Bottom liquid level rose above the bottom seal pan, causing excessive pressure drop and poor stripping. Level transmitter was improperly calibrated, and field level gage was neither blown nor checked.	Same as 1520.
15127	465	Petrochemicals	Out-of-calibration bottom-level controller caused liquid level to exceed reboiler return inlet, causing premature flood, high dP, and loss of product purity.	Same as 1520.

15160			*Steam condensation in level tubing of hydrocarbon tower leads to high base level and flooding, Section* 25.5.2.	
15109	299, 306	Refinery C_3/C_4 splitter	A level control tap was plugged, giving a false signal, which induced level rise above the reboiler return nozzle and tower flooding.	
DT8.1			*A construction blind in level controller piping leads to liquid level above vapor inlet.*	
15110	299, 306	Refinery C_3/C_4 splitter	Tower flooded after level float chamber was insulated. The insulation kept liquid hot, reducing its density and generating low signal when the level rose above the reboiler return. Recalibration eliminated problem.	Same as 1520.
1586	425	Refinery coker fractionator 14 ft ID	A faulty level indicator caused the column to fill up with liquid. Two days later, asphalt and tar were found in the upper products and the pressure drop across the bottom four sieve trays rose from 1–2 psi to 9 psi, indicating plugging. Plugging confirmed by gamma scans. (See also 1023, Section 13.1.)	Same as 1520.
1515	306	Refinery combination tower	Bottom liquid level rose above the vapor inlet nozzle because of a faulty level controller. The submergence backpressured the coke drum upstream. When the operator noticed this, he quickly lowered the bottom level. This caused foamover (a "champagne bottle" effect) in the coke drum.	Same as 1520. Avoid excessively rapid draining of column liquid.
1588, 12108, DT8.4, DT8.5			*Causing tray/packing damage, Section* 8.3.	
712, 1322, 1336			*Lack of level indication with kettle reboiler, Section* 23.4.1.	
8.1.2 Operation				
1589	97	Ammonia condensate stripper	High base level caused liquid carryover, which damaged reformer tubes downstream of the stripper. Afterward, a high-level trip with a voting system (two out of three) was added and linked to the plant shutdown system. At start-up, level control problems caused high levels in the stripper and these shut the plant down. To avoid recurrence, the trip was modified to only close the process condensate valve upon high level.	

(*Continued*)

Chapter 8 Tower Base Level and Reboiler Return: Number 2 on the top 10 Malfunctions (*Continued*)

Case	References	Plant/Column	Brief Description	Some Morals
1274	306	Diesel side stripper	Liquid-level rise above the stripping steam inlet caused flooding and poor stripping. Problem identified by comparing inlet to outlet temperature plus a pressure drop measurement.	Avoid base-level rise above the vapor inlet.
1269	302	Refinery DC_3 8 ft ID	Froth/foam at bottom of tower rose above reboiler return inlet, causing tower flooding. Particulates from corrosion could have induced foam. Problem solved by replacing tower by 11-ft-ID tower.	
1222			*Causes flooding and packing support damage, Section* 8.3.	
701			*Among other causes, Section* 10.6.	
1027			*Among other causes, Section* 13.5.	
8.1.3 Excess Reboiler Pressure Drop (*See also Excess* ΔP *in Circuit, Section* 23.4.1)				
1339			*In thermosiphon reboiler loops, Section* 23.1.1.	
1338			*In loop having a kettle and a thermosiphon reboiler in series, Section* 23.6.	
8.1.4 Undersized Bottom Draw Nozzle or Bottom Line				
1202			*Vapor in draw line, Section* 10.1.1.	
735, 1237			*Excess draw line pressure drop, Section* 10.1.2.	
8.1.5 Others				
1206			*Failure of base temperature controller, Section* 25.8.	
1224			*Loss of bottom pump, Section* 8.3.	
912, 954			*Debris in tower base, Section* 11.8.	
640, 644, 641, 1555			*Foaming, Sections* 16.1.2, 16.5.5, *and* 16.6.9.	
8.2 High Base Level Causes Premature Tower Flood (No Tray/Packing Damage) (*See also Faulty Level Measurement and Operation, Sections* 8.1.1 *and* 8.1.2, *Excess* ΔP, *in Circuit Section* 23.4.1, *and Level Instrument Fooled, Section* 25.7)				
701			*Operation problems, Section* 10.6.	
1202			*Vapor in draw line, Section* 10.1.1.	

735, 1237			*Excess draw line pressure drop, Section* 10.1.2.	
912, 954			*Debris in tower, Section* 11.8.	
640, 644, 641, 1555			*Foaming, Sections*, 16.1.2, 16.5.5, *and* 16.6.9.	
1339			*Excess pressure drop, thermosiphon reboiler, Section* 23.1.1.	
1338			*Excess pressure drop, kettle and thermosiphon reboilers in series, Section* 23.6.	
		8.3	**High Base Liquid Level Causes Tray/Packing Damage**	
1128, 15126	123	Chemicals, 18 incidents, 2.5–12 ft ID	In each incident, in various columns, one or many trays were damaged. Over half of the incidents were caused by base liquid level rising above the reboiler return nozzle. Other prime culprits were local vacuum in a column and poor installation.	Paper contains invaluable techniques for preventing excessive base level.
1203	369	Refinery vacuum	An automatic steam cutout on high level was installed in the stripper section of the tower. It saved the tower many times from tray damage.	
DT8.2, DT22.1			*High liquid level damages upper trays, but grid and structured packings remain undamaged.*	
1224, DT8.3	194	Chemicals	A temporary loss of bottoms pump caused base level to rise above the reboiler return nozzle. This caused bottom trays to collapse. Reflux was raised to meet purity with fewer trays, resulting in flooding.	Avoid base level above the reboiler return.
DT8.4			*Repeated tray damage in crude tower induced by misleading base level, aggravated by a small-diameter stripping can.*	
DT8.5			*Several cases, tray and packed towers.*	
1250	133	Quench column	The lower of two beds of random packings collapsed. Some of it ended at the column bottom, the rest in a downstream storage tank. Base level was well above the vapor inlet and may have caused the collapse.	
12106	307	Refinery vacuum	Stripping trays were dislodged by high base level, reducing HVGO yield. Pressure drop appeared normal due to liquid level above bottom pressure tap. Diagnosed by seeing no temperature change and experimenting with stopping steam.	

(*Continued*)

Chapter 8 Tower Base Level and Reboiler Return: Number 2 on the top 10 Malfunctions (*Continued*)

Case	References	Plant/Column	Brief Description	Some Morals
12108	127	Refinery vacuum	Stripping trays were dislodged by high base level, reducing HVGO yield. The base level was almost up to the feed inlet, although the level transmitter read 60%. Lowering the level, guided by static pressure measurement, reduced cracked gas production by 25%.	
1588	64	Refinery lube oil vacuum	Damage sustained due to level control problems led to off-specification products. Random packings were found in the bottom line. Gamma scans showed collapse of bottom three beds.	Ensure adequate level indication.
1165	114, 391	Refinery vacuum	High base liquid level led to displaced packing in the wash and HVGO sections of the tower. Uplifted wash section packing damaged the spray header distributor. Temperature surveys, plant tests, and gamma scans diagnosed problem and helped formulate a strategy for minimizing production losses.	
1222	304	Pharmaceuticals methanol stripper 3 ft ID	Column containing 16 trays in bottom and random packing in top, ran flooded with off-specification product for several months because base liquid level was above the reboiler return inlet. During this period, the packing support also collapsed, presumably due to pressure surges. The flooding was eliminated by lowering the liquid level below the reboiler return inlet.	
1027, DT22.5			*Due to abnormal operation, Section* 13.5.	
DT25.5, DT25.6			*Due to fooling of base-level transmitter by froth and by lights.*	
			8.4 Impingement by the Reboiler Return Inlet	
8.4.1 On Liquid Level				
744	455		A return line from a once-through thermosiphon reboiler was inclined downward. For 15 years, this caused tower instability, slugging problems, and cyclic level control.	Avoid bending down reboiler return lines.

DT8.6			*Transfers liquid from reboiler to base draw compartment, starves reboiler.*	
DT8.4			*Causes frothing of base level, a misleading level indication, and contributes to tray damage.*	
8.4.2 On Instruments				
709	320	Small-diameter steam stripper	Level float in a column bottom sump "bounced" and finally broke due to impingement of entering steam. Problem fixed by installing a shielding baffle over the level connection.	Avoid impingement on instrument connections.
DT25.4			*Impingement on level tap gives discrepancy between level transmitter and level glass.*	
8.4.3 On Tower Wall				
884	103	CO_2 MEA absorber	Corrosion inhibitor in the MEA solution was ineffective in the high-vapor-turbulence regions beneath the packing, and these corroded. Recurrence prevented by lining or replacing CS with 304L SS in the unwetted regions.	
802	285, 512	Ammonia hot-pot absorber	Severe localized corrosion of tower shell occurred due to impingement of liquid accumulated in a gas inlet distributor. Distributor was modified to eliminate accumulation. Corrosion at another spot was caused by gas issuing from the modified distributor impinging on tower wall. Poor wetting of the tower wall and gas maldistribution caused other localized corrosion incidents. Using SS shingles to protect wall areas was effective in checking corrosion but later let to stress corrosion.	Pay attention to inlet gas distributors. Maldistribution can lead to corrosion in corrosive services.
745	118	Ammonia amine contactor	Upon converting MEA to MDEA, base liquid level was raised from 4 ft below the inlet gas ladder pipe distributor to 6 in. above it. Impingement of inlet gas from the peripheral holes led to erosion corrosion on the vessel wall. Cured by welding shut the peripheral holes and by protecting tower wall.	

(*Continued*)

Chapter 8 Tower Base Level and Reboiler Return: Number 2 on the top 10 Malfunctions (*Continued*)

Case	References	Plant/Column	Brief Description	Some Morals
746	118	Ammonia MEA contactor	Following a catalyst failure, the feed gas contained high O_2 for 3–5 days. This led to extensive corrosion between the feed and the first tray, declining over the bottom five trays. Recurrence avoided by removing the bottom five trays and turning the gas distributor upside down so the exit holes pointed down. Liquid level in the tower was raised above the inlet gas distributor.	
8.4.4 Opposing Reboiler Return Lines				
755	440	Chemicals stripping column	Opposing return lines from two reboilers and associated internal hardware caused 0.5 lb entrainment/lb vapor all the way to the side-draw tray, several trays up. The heavies in the side draw led to excessive sewering of water and excessive water makeup consumption. Diagnosis was based on comparing simulation to plant measurement and on injecting nonvolatile tracer into reboiler process lines.	
760	115	Lube oil stripper	Bottom pump suction was difficult to maintain, and there was a heavy hydraulic hammer. Eliminating a high point in the pump suction gave no improvement. The problem was the opposing return lines from two reboilers, with the jet from one suppressing vaporization in the other, swinging the heat load alternately. Verified by shutting steam to one reboiler and fixed by adding an intervening baffle.	
8.4.5 On Trays				
1263			*Damage in cartridge trays, Section* 22.9.	
8.4.6 On Seal Pan Overflow				
759	196	Refinery sour water stripper 4 ft ID	Kettle reboiler vapor entering at 112 ft/s blew on the bottom seal pan overflow which was directly opposite the inlet and at the same elevation. This entrained liquid and contributed to flooding (1344, Section 23.4.1). Solved by removing bottom two trays and returning the vapor via a pipe distributor.	

766	195, 198, 200	Aromatics toluene recovery	A reboiler return nozzle was located directly underneath the center seal pan, resulting in entrainment of the liquid overflowing the seal pan. This did not appear to be the tower bottleneck.	
773	191	Refinery crude fractionator	Overflash liquid descending from the seal pan overflow of the bottom wash tray was entrained by 100-ft/s two-phase feed entering the tower, causing excessive metals in the AGO product. Cured by routing the overflash liquid away from the feed.	
780	338	Natural gas MEA absorber 2-pass valve trays	Inlet gas was deflected circumferentially to both sides of the tower, directly into the two chimney tray overflows. Changing gas inlet to enter beneath the center of the tray, where it did not contact the overflows, contributed to a 15% capacity increase (see also 664, Section 16.5.1).	
			8.5 Undersized Bottom Feed Line	
			(*See also Vapor Feed/Reboiler Return Maldistributes Vapor to Packing Above, Section* 7.1)	
1261	190	Refinery vacuum	Upon revamp to deep-cut operation, the stripping steam line was not enlarged. Critical flow in that line restricted stripping, and the deep cutpoint could not be achieved.	Do not overlook auxiliaries.
1280			*Initiating tower flood, Section* 3.2.2.	
			8.6 Low Base Liquid Level	
			(*See also Peroxide Towers, Section* 14.1.2)	
1510	284		Failure of a column bottom-level controller caused gas to enter the product storage tank and rupture it.	Watch bottom level.
DT8.7			*Insufficient surge in tower base causes instability and poor separation.*	
1134			*Contributing to ethylene oxide explosion, Section* 14.1.1.	
1562	202		Kettle reboiler bottom pump cavitated. Level indication was normal. Pump was shut down and checked but nothing found. Reason eventually traced to blockage in the level instrument yoke giving a false signal. The pump cavitated because there was no liquid to pump.	(1) Always cross check a level indicator. (2) Focus on problem, not on symptoms.

(*Continued*)

Chapter 8 Tower Base Level and Reboiler Return: Number 2 on the top 10 Malfunctions (*Continued*)

Case	References	Plant/Column	Brief Description	Some Morals
			8.7 Issues with Tower Base Baffles	
1318, 1353			*Liquid from bottom tray goes to wrong side of baffle, Section* 23.1.3.	
DT8.6			*Impingement from vapor return transfers liquid from one side of baffle to the other, starves reboiler.*	
DT8.8			*Gap between wall and unique angled baffle induces tray liquid to wrong side, starves recirculation pump.*	
DT11.4			*Fall of a poorly installed overflow weir on a reboiler draw pan starves reboiler.*	
905			*Unbolted hatchway in base baffle causes poor reboiler performance, Section* 11.6.	
DT26.6			*Base baffle leads to erratic control when sump level controls boil-up.*	
DT12.6			*Problems controlling liquid level during total reflux operation.*	
			8.8 Vortexing	
708	320		A carpenter's sawhorse was found in the bottom of the vessel. Following its removal, the bottom pump experienced loss of suction at low levels due to vortexing.	Vortex breakers should be routinely installed.
DT8.9			*A 7-ft vortex induces a premature reboiler limitation.*	

Chapter 9 Chimney Tray Malfunctions: Part of Number 7 on the Top 10 Malfunctions

Case	References	Plant/Column	Brief Description	Some Morals
			9.1 Leakage	
939	246	Absorber with cooling	Pumparound pumps lost prime even when gas rates were maximum. Gasketing sumps and tightening joints permitted pump operation at maximum gas rates.	
944	250		Gaskets on a total draw-off chimney tray were left out during a revamp. At start-up the tray leaked, making the tower inoperable.	
928	464	Refinery	Coke was removed from a gasketed total draw tray using a jackhammer. The tray was not regasketed. Upon restart, liquid leaked through the gaskets and could not be drawn.	
713	369	Refinery vacuum	Leakage from gasketed draw trays increased in service until no level could be maintained. Different gasketing materials and putty did not solve the problem. Seal welding significantly reduced leakage.	Gaskets may not prevent leakage in hot services.
920	167, 172	Refinery vacuum	HVGO yield was down and flash zone temperature was low due to leakage from HVGO and slop-wax collector trays. Inspection found a large hole in the HVGO collector.	Ensure adequate inspection.
827	172	Refinery vacuum	LVGO yield was down and HVGO yield up due to leakage from a gasketed LVGO chimney tray. The LVGO leak reduced the HVGO draw temperature, which caused heat removal problems and eventually resulted in high overhead temperatures to the ejectors.	See 713. Seal welding, not gasketing, should be used in this service.
DT9.1, DT9.2			*Chimney tray leaks overload vacuum system in refinery vacuum towers.*	
1258	166	Refinery vacuum	Following a revamp from trays/grid to structured packings and from damp to dry operation, cutpoint and product quality were both low. Overhead was hot and could not be cooled. Although the collector trays were all seal welded, the heat balance showed that about half of the LVGO produced downflowed, reducing the HVGO bubble point and limiting its heat removal.	Mass and energy balances are invaluable troubleshooting tools.

(*Continued*)

Chimney Tray Malfunctions: Part of Number 7 on the Top 10 Malfunctions (*Continued*)

Case	References	Plant/Column	Brief Description	Some Morals
881	458	Refinery vacuum	Upon replacement by packing, three bubble-cap trays were converted to total draw chimney trays with draw nozzles at the bottom of center downcomers from these trays. The downcomers were enclosed within deep seal pans. At high liquid rates, the seal pans overflowed. LVGO overflow cooled HVGO by 50°F, which reduced crude preheat and bottlenecked the crude heater. Cured by fully sealed chimney trays.	Problem diagnosed by mass and heat balances and simple plant tests. See 1258.
DT9.4			*Leak from damaged FCC main fractionator chimney tray leads to flooding, less steam generation.*	
924			*Among other problems, Section* 11.10.	
851			*Among other problems, Section* 6.7.2.	
DT9.5			*Flat hats can induce leaks.*	
DT23.5			*Dumping through a hatless chimney tray contributes to surging.*	
			9.2 Problems with Liquid Removal, Downcomers, or Overflows	
819, DT7.1	304	Refinery FCC C_3–C_4 splitter	Trays were replaced by structured packing in the top section. A chimney tray installed beneath the packing had undersized downpipes that were not liquid sealed. This led to chimney tray liquid overflowing the risers, causing maldistribution and possible local flooding. This led to a drop in efficiency as production rates were raised. Problem was solved by chimney tray modifications.	Beware of undersizing chimney tray downpipes. Ensure chimney tray downpipes are liquid sealed.
779	403	Natural gas ethane extraction demethanizer	Sequential retrofits of three parallel units fell sort of achieving design ethane recovery due to premature demethanizer floods at a consistent lower feed temperature of −52°C. In one unit, downpipes from the upper feed chimney tray were undersized. Corrected in the second unit with a 91.7% to 93.4% recovery improvement. In the third unit, raising tray spacing at the lower feed, changing the lower feed tray to a chimney tray, and adding drain holes to the seal pan above the lower feed further raised recovery to 95.1%.	

8107			*Collector tray overflows due to undersized drain pipes, Section* 6.3.	
DT9.3			*Chimney tray overflow reduces distillate yield in refinery vacuum tower.*	
DT5.2			*Narrow liquid exit slots at chimney tray downcomer back up liquid and flood tower.*	
707	63	Refinery FCC fractionator	Premature flooding resulted from absence of downpipes on a chimney tray, which forced liquid flow down vapor risers. Problem fixed by installing an external downpipe.	Avoid liquid downflow through vapor risers.
770	448	Natural gas demethanizer	Tower had two once-through side reboilers. Liquid to each side reboiler came from a total draw chimney tray, with the reboiler return entering below. When a side reboiler was out of service, liquid stacked above its draw chimney tray and flooded the tower.	Solved by providing internal overflows.
727	189	Refinery crude tower	The 10-in. kerosene draw pipe cleared the floor of the chimney tray by $\frac{1}{2}$ in. This restricted the kerosene draw rate. Increasing the clearance to 2.5 in. eliminated the problem.	
775	265	Refinery debutanizer	A 12-in. draw nozzle flush with the tray floor caused liquid level to reach 16 in. chimney height. Solved by raising chimney heights.	
DT23.5			*Liquid collector overflow starves reboiler draw.*	
1351, 1337			*Collector tray overflow due to excess pressure drop in reboiler circuit, Sections* 23.4.1 *and* 23.6.	
877			*Undersized outlet nozzle in internal condenser liquid collector, Section* 2.5.	
8100, 841, DT19.3			*Due to level measurement problems, Section* 9.6.	
973			*Loose chimney hat plugs draw, Section 11.8.*	
			9.3 Thermal Expansion Causing Warping, Out-of-Levelness	
829			*Out-of-levelness, Section* 7.1.2.	
893	188	Refinery vacuum	Leak from overflash collector tray quenched flash zone, reducing HVGO yield. The tray was damaged during each start-up because its refractory expanded at a different rate than the tray deck, causing weld cracking. Cured by leaving a gap between the refractory and tower wall.	

(*Continued*)

Chapter 9 Chimney Tray Malfunctions: Part of Number 7 on the Top 10 Malfunctions (*Continued*)

Case	References	Plant/Column	Brief Description	Some Morals
851			*Warping, Section* 6.7.2.	
921	309	Refinery vacuum 30 ft ID	Tower out-of-roundness caused I-beam support under the HVGO chimney tray to be several inches short. Slotted holes in the I-beam did not line up with predrilled holes in brackets fixed to wall. New holes were drilled, but these were not slotted and did not permit thermal expansion between the beam and tower wall.	Consider tower out-of-roundness.
			9.4 Chimneys Impeding Liquid Flow to Outlet	
826	310	Refinery vacuum	Downcomer boxes together with vapor risers on the LVGO chimney tray restricted horizontal liquid flow toward the draw nozzle, inducing excessive hydraulic heads upstream. At the high heads, much LVGO overflowed the risers, degrading into HVGO.	Avoid restriction in horizontal liquid path on collector trays.
737	141		Chimney tray risers impeded liquid flow to draw sump, forcing liquid to overflow prematurely. Trayed section below PA flooded as a result.	Same as 826.
883, DT9.6	264	Refinery vacuum	The long edges of the chimneys in the HVGO collector tray were perpendicular to the liquid flow toward the outlet nozzle, incurring high hydraulic gradient. At tray inlet, the liquid built up to the chimney height and overflowed into the bottom, causing product loss. Fixed by rotating tray by 90°.	See 826, 737.
8108	429	Refinery crude fractionator	Two years after the Case 8107 retrofit (Section 6.3), similar parallel unit retrofit from trays to packing also gave low jet fuel yield. This time the collector was a chimney tray with well-sized drain pipes and longer bed. Cause was collector overflow due to excessive hydraulic gradient due to chimneys and chimney reinforcements obstructing flow to outlet nozzle.	
			9.5 Vapor from Chimneys Interfering with Incoming Liquid	
726	189	Refinery crude fractionator	Entrainment of seal pan overflow by high-velocity vapor issuing from risers of the kerosene PA chimney tray caused flooding and a capacity restriction. Top of the risers was below the seal pan overflow.	Avoid vapor impingement on seal pan overflow.

837	141		Hats on chimney tray excessively restricted vapor, resulting in liquid entrainment.	
738, DT7.4	272	Olefins water quench	Liquid entering a chimney tray from side downcomers overflowed 8-in. inlet weirs. The waterfalls caused frothing at the tray inlets. The froth overflowed the 10-in. risers at tray inlet, channeling vapor via the central risers. The channeling propagated through 10 trays due to their excessive slot area (429, Section 7.4.2). The high vapor velocities near the tray center led to entrainment from the top of the tower.	Problem eliminated by a redesigned chimney tray and by blanking valves on the trays above.
762, DT9.7	264	Refinery FCC main fractionator	Seal-welded, total draw chimney trays would have failed to draw liquid totally because vapor from the outside chimneys would have blown liquid descending from the seal pans of the tray above into the chimney trays' overflow downcomers. Solved by closing the opening on the outside chimneys and adding 25-mm drip lips to the seal pans.	
DT23.5			*Reboiler return vapor blows liquid into chimney.*	
			9.6 Level Measurement Problems	
15106	330	Refinery vacuum	Overflash pumps experienced chronic cavitation because of erratic level control on overflash draw tray. Problem alleviated by using a pressure transmitter just above the pump suction to monitor liquid level in the suction line and to control pump flow.	
DT9.3			*Level indicator on refinery vacuum tower HVGO chimney tray reads* 40% *when tray overflows.*	
896			*Level control outside refinery vacuum tower eliminates chimney tray, Section* 9.8.	
8100	307	Refinery vacuum	Liquid overflowing the chimneys of the slop wax chimney tray at 60% level reading was entrained into the wash bed, causing black HVGO. Misleading level was caused by upper level tap being 10 in. below top of chimneys. Static pressure at upper tap read high due to static liquid head above.	Diagnosed by pressure survey. Cured by reducing level reading to 15%.

(Continued)

Case	References	Plant/Column	Brief Description	Some Morals
841	250	Refinery vacuum	Incorrect location of liquid-level taps induced liquid levels above the top of the chimneys of a slop wax accumulator tray. This generated entrainment and carbon deposits on the surface of the bottom packing layer. (See also 840, Section 7.3.)	Do not forget instrument taps.
DT19.3			*Level control problems on slop wax chimney tray lead to coking, overflow.*	
968, DT9.8	264	Refinery vacuum	The HVGO total draw chimney tray had an angle iron covering the support ring bolts in order to prevent joint leakage. The angle iron was installed so that it also covered the level measurement tap, giving a zero level reading. Overcome by measuring the head in the outlet pipe.	Same as 841.
833	449		Upon modification, level indication nozzles were left outside a collector tray.	Same as 841.
			9.7 Coking, Fouling, Freezing	
891			*Due to excess residence time of slop oil, Section* 19.2.	
829			*Among other problems, Section* 7.1.2.	
130			*Caused by inaccurate TBP characterization, Section* 1.1.7.	
1160	458	Refinery vacuum	HVGO chimney tray was damaged by using jackhammer to remove coke from it.	
1024			*Freezing up at outlet, Section* 2.4.6.	
			9.8 Other Chimney Tray Issues	
705	477	Natural gas crude oil stabilizer	Degassed liquid on a draw tray caused excessive backup of aerated liquid in the downcomer. The downcomer was submerged below the liquid level on the draw tray. The backup caused premature column flooding.	Allow for differences in aeration when using submerged downcomers.
761, DT9.9	264	Refinery FCC main fractionator	On a chimney tray, downcomer entrance was obstructed by the seal pan from above. On the tray below, downcomer entrance was obstructed by a draw sump. One or both caused entrainment, a separation problem, and a tower capacity bottleneck. Diagnosed using gamma scans and eliminated by eliminating obstructions in the next revamp.	

896	370	Refinery vacuum	Levels on the HVGO PA and the overflash are controlled in short sections of 24-in. line just outside the tower. The tower height previously occupied by liquid inventory was utilized to increase disengagement height under the wash bed. This improved HVGO quality.	
840			*I-beam interference maldistributes vapor to packing above, Section* 7.3.	
829			*Low crest over weirs gave maldistribution to section below, Section* 7.1.2.	
767, 776			*Hot and cold compartments on chimney trays from/to heat exchangers, Section* 10.1.1.	
877			*Poor decanting arrangement, Section* 2.5.	
972			*Collector dislodged due to poor clamping, Section* 22.3.	
888	35	Ammonia hot-pot regenerator	Flashing semilean solution entered a single-chimney tray just above the bottom sump. The flow directly hit the chimney, which broke off and rubbed against the tower shell, causing corrosion.	Avoid impingement at two-phase feed inlets.
889	35	Ammonia hot-pot regenerator	Flashing semilean solution entered a single-chimney tray just above the bottom sump. The SS shroud protecting the CS shell was poorly supported, its cleats were incorrectly fabricated from CS, and these failed. Flashing feed impingement cut a hole in the shell. Fixed by improved shroud and gas entry design.	
12118	80	Natural gas MDEA regenerator	Kettle reboiler experienced reduced steam flow, high outlet temperature, and high outlet amine concentration. A section of the chimney tray supplying liquid to the reboiler had opened out downwards, and the bottom tray collapsed, starving the reboiler of feed. Welding chimney tray sections prevented recurrence.	
780			*Inlet gas impinging on chimney tray overflows, Section* 8.4.6.	
DT9.4			*Upward force while filling tower with liquid can exceed uplift resistance of chimney trays with tall chimneys.*	

Chapter 10 Drawoff Malfunctions (Non–Chimney Tray): Part of Number 7 on the Top 10 Malfunctions

Case	References	Plant/Column	Brief Description	Some Morals
			10.1 Vapor Chokes Liquid Draw Lines	
10.1.1 Insufficient Degassing				
703	7	Refinery absorber	Choking of an outlet line from a downcomer trap-out limited absorber capacity. Increasing the height of the trap-out pan did not help. Degassing the draw-off liquid in a separate enlarged pan solved the problem.	Either avoid vapor in liquid outlets or design for it.
704	7	Refinery	Vapor choking of a long line from column to reboiler caused premature tower flooding. To solve the problem, the draw pan was converted to a degassing pan and the line was sloped and vented.	Same as 703.
711	95	Refinery DC_2 absorber	An undersized liquid draw-off line from a draw-off box caused vaporization in the line. This backed up liquid in the downcomer. The liquid overflowed into the section below. Poor separation resulted.	
DT10.1			*Vapor choking of side-draw line causes premature flooding in fractionator, larger vent does not help.*	
754	308	Refinery crude fractionator	Jet fuel draw was restricted by an undersized draw nozzle. Opening the flow control valve from 30 to 100% did not increase flow. Increasing level of liquid in draw sump increased flow.	
764	266	Refinery crude fractionator	An instability initiated shortly after a revamp that raised heat exchange surface in the top and mid PAs. The cause was a draw sump that mixed subcooled PA return with boiling tray liquid, had a flow restriction in the path to the draw nozzle, and could not handle a breakthrough of vapor bubbles in the downcomer liquid. Cured by draw sump modifications.	Diagnosed by surface temperature surveys. Paper gives detailed account on the troubleshooting.
DT10.2			*Vapor choking of undersized side-draw line produces instability, narrow operating range.*	
771	40	Refinery FCC main fractionator	Following a retrofit, tower experienced 70°F variation in gasoline end point, flooding, and capacity bottlenecks. Cause was cavitation of the LCO draw and PA pump due to aerated liquid in the draw box and a rundown line undersized for self-venting flow. A restrictive liquid inlet to the draw box impeded venting. Problem alleviated by adding a vent valve on the draw line.	Aerated liquid gravity rundown lines need to be sized for self-venting flow.

767	89	Refinery hydrocracker debutanizer	Draw to a forced-circulation side reboiler at 470°F was taken from sumps at the bottom of downcomers. Reboiler return at 530°F went to the active areas above the sumps, so the hot liquid could backflow to the sumps. The backflow vaporized the cold draw liquid, causing reboiler pump cavitation at two-thirds of the design flow. Solved by replacing draw by a chimney tray with separate cold and hot compartments.	
776	535	Refinery deethanizer absorber	A retrofitted side PA cooling loop never functioned reliably and was mothballed. Causes were a small draw box that did not permit degassing, and backmixing of PA return with the draw. Cured by replacing draw by a chimney tray which degassed liquid and was split into separate draw and return compartments to eliminate backmixing.	
1202	7	Refinery stabilizer reboiled by a fired heater	Premature tower flooding occurred when level in the bottom of the column rose above reboiler return nozzle and backpressured a uniquely designed bottom surge drum. Level rose either because vapor was present in liquid line to heater or liquid was entrained in drum vapor. Provisions which lowered surge drum level solved the problem.	Either avoid vapor in liquid lines or design for it. Always monitor tower bottom level.
1568			*Aerated draw pan backs up liquid to trays above, Section* 25.7.1.	
10.1.2 Excess Line Pressure Drop				
861	493	Refinery vacuum	Frequent head loss and cavitation of bottom pump was caused by eddies and back mixing in piping elbows located too close to pump inlet. Neutron scans confirmed tower level measurement and presence of vapor in pump suction line well upstream of elbows.	Alleviated by lowering the bottom temperature using circulating quench.
752	333	Refinery crude fractionator	Undersized PA draw nozzle (4 in., expanding into an 8-in. line) led to pump cavitation and poor heat transfer. This raised overhead condenser load and tower pressure, leading to a diesel yield loss.	
735	141		Bottom pump cavitated due to undersized bottom draw nozzle. Raising tower 10 ft did not help. Flooded bottom of the tower.	
1237	464	Chemicals	Following a tray to packing revamp, column capacity increased 10% (design was 30%). Reason was undersized bottom control valve causing a sump level rise and column flooding.	

(Continued)

Chapter 10 Drawoff Malfunctions (Non–Chimney Tray): Part of Number 7 on the Top 10 Malfunctions (*Continued*)

Case	References	Plant/Column	Brief Description	Some Morals
706	63	Refinery stripper	Unstable operation and premature flooding were experienced at 90% of design rates. Either block valves in a liquid draw-off line to a side reboiler or insufficient downcomer area was the culprit.	Gamma scans are useful in diagnosing problems.
1204	300	Refinery fractionators	In one case, it was thought that a draw-off tray was leaking, but the problem turned out to be a control valve stuck half open. In another case, PA performance was improved simply by opening two discharge valves that were pinched back for an unknown reason.	Do not overlook the obvious.
719	304	Refinery lube oil prefractionator	The amount of side product that could be withdrawn was restricted. The restriction was caused by an overflowing draw pan. The pan overflowed because the side product control valve was located on the horizontal pipe between the draw nozzle and the first elbow turning down.	Locate control valves downstream of long vertical runs of outlet pipes.
10.1.3 Vortexing *(for tower base vortexing, see Section* 8.8)				
747	228	Chemicals	A pump with 60 ft of vertical suction from a draw box was erratic and almost uncontrollable. Reason was high velocities (5 ft/s) at the outlet nozzle and low (6 in.) submergence above the top of the nozzle. These led to vortex formation, carrying gas downward. Problem solved by installing a vortex-breaking baffle above the nozzle.	
10.2 Leak at Draw Tray Starves Draw *(for leak at draw pan to a once-through thermosiphon reboiler starving draw, see Section* 23.2.1)				
716	300	Refinery	A valve tray used in trap-out service excessively leaked. The leakage was eliminated by seal welding tray sections and by welding a strip onto the periphery of the tray a few inches from the support ring.	Successful techniques for minimizing leakage.
717	300	Refinery crude fractionator	A leaking valve tray in trap-out service was replaced by an all-welded chimney tray which was seal welded to the tray ring. Leakage was eliminated.	An all-welded chimney tray can eliminate leakage.

763, DT10.3	264	Refinery lube oil feed preparation	It was impossible to withdraw heavy-intermediate cut from the draw sump of a conventional valve tray. This cut leaked through the valves and ended in the bottom. Solved by replacing the valve tray with a seal-welded chimney tray.	See 717.
DT10.4			*Sealing gaps in draw pan eliminates leaks and product loss that persisted more than a decade.*	
724	304	Refinery lube oil vacuum	Lube oil was drawn as a side cut from a trap-out tray (total draw-off). When all tower baffle trays were replaced by valve trays, lube oil rate declined by 12% due to leakage at the trap-out tray. To restore the original draw rate, the trap-out tray was replaced first by a bubble-cap tray, then by a valve tray with venturi openings; finally, the outlet nozzles were expanded. Each of these steps progressively further lowered the lube oil rate. Installation of a seal-welded chimney trap-out tray solved the problem and achieved a lube oil rate 19% above original.	A seal-welded chimney tray should be used for total draw-offs.
325	333	Refinery crude fractionator	Diesel PA was between LD and HD draws. During crude switches, the high heat removal rate in the PA plus leakage from the tray dried the LD draw, causing the LD stripper to lose level and the pump to cavitate. Drying also caused low reflux and poor separation of LD and HD. Cure was merging LD and PA draws, taking both from a seal-welded total draw tray.	Total draws should be used when the stream drawn is most of the tray liquid.
321			*Weeping from TPA valve trays cavitates pump, reducing circulation, Section* 3.1.5.	
730	160	Refinery coker fractionator	Side-draw product quality could not be maintained through coke drum switches. System improved by replacing the total draw valve tray by a seal-welded chimney tray (to prevent weep at the low vapor rates), using bubble-point instead of subcooled refluxes, and monitoring pumpback rates.	Good internal design practices can improve column controls.
718	304	Refinery wax fractionator	Upon retray, the new trays were rotated 90° to their original orientation. Internal piping was required from the new intermediate product draw sump to its draw nozzle and level gage. A flange on the internal draw line leaked and starved the line of liquid. Poor separation resulted. The internal level gage lines plugged.	Internal shop flanges often spread apart and leak. Avoid internal instrument lines.
DT12.5			*Overheating during outage causes damage and leak at draw pan.*	
401, 402			*Affecting PA circulation, Section* 4.2.	

(Continued)

Chapter 10 Drawoff Malfunctions (Non–Chimney Tray): Part of Number 7 on the Top 10 Malfunctions (*Continued*)

Case	References	Plant/Column	Brief Description	Some Morals
			10.3 Draw Pans and Draw Lines Plug Up	
1272	192	Visbreaker fractionator	Coke breaking off the outside of a downcomer restricted and prevented flow into the draw pan below, forcing a unit shutdown.	
1277	178		Sidestream withdrawal was restricted because bubble caps had dislodged and blocked the draw-off nozzle.	
1241			*Accumulating unstable component, Section* 15.3.	
DT18.9			*With debris from damaged tray.*	
1111			*Infrequently used line, Section* 14.11.	
			10.4 Draw Tray Damage Affects Draw Rates	
1267	164	Refinery crude fractionator	Following modification to the diesel PA section and a pressure surge at start-up, no diesel PA could be drawn and there was poor separation between diesel and gas oil. Cause was tray damage. An on-line fix was to draw diesel a few trays above the PA draw and pass some of it through the PA exchangers as a pumpdown reflux.	Temperature and pressure surveys diagnosed problem. Gamma scans and simulations did not.
DT22.1			*Diesel make drops following damage to draw pan during upset.*	
1281	178		Bottom flow was restricted because bottom downcomer had fallen.	
1341			*Damaged tray feeding a once-through thermosiphon reboiler, Section* 23.2.1.	
			10.5 Undersized Side-Stripper Overhead Lines Restrict Draw Rates	
721	304	Refinery asphaltic crude fractionator	Column efficiency severely dropped following a capacity revamp. Problem was caused by the bottom downcomer being converted into a draw-off box with no overflow, coupled with an undersized side-stripper overhead line. Excessive pressure drop in this overhead line backed liquid into the draw-off box, and this liquid flooded the column.	Ensure draw trays are equipped with overflows. Beware of undersized lines.

753	308	Refinery jet fuel stripper	Naphtha could not be properly stripped from jet fuel due to an undersized stripper overhead line. Upon steam addition, stripper pressure rose, impeding feed entry and causing a loss in bottom level.	A large stripper overhead line solved the problem.
			10.6 Degassed Draw Pan Liquid Initiates Downcomer Backup Flood	
749	159	Refinery FCC primary absorber	Degassed liquid on a draw pan backed up aerated liquid in a downcomer submerged below the pan liquid level, causing premature column flooding. A fix that cut the number of valves on the tray reduced weeping and aggravated problem. Solved by redesigning draw pan to reduce submergence.	Same as 705, Section 9.8.
701	7	Refinery	Liquid carryover from the top of the tower occurred at less than design rates. Believed causes were an excessively tall outlet weir on the bottom seal pan and a submerged reboiler return nozzle.	
			10.7 Other Problems with Tower Liquid Draws	
772	40	Refinery FCC main fractionator	Premature flooding occurred because LCO product and PA draw box obstructed entrance area to the downcomer below. Despite the large draw box, tray spacing was not extended to the tray below.	
DT10.5			*Downcomer seal lost at product draw when reflux minimized, causing major tower disturbance.*	
733	451	VCM	One-pass trays were replaced by two-pass trays. A side-draw downcomer trap-out nozzle was not modified. Following the revamp, it drew liquid from only one tray half, causing dry-out of this half tray and vapor bypassing that flooded the tray above.	Do not overlook side-draw connections.
771			*Restrictive liquid inlet to draw box impedes venting, Section* 10.1.1.	
721			*Draw-off box with no overflow on tray, Section* 10.5.	
715			*Water accumulation in dead pocket below draw nozzle, Section* 13.3.	
751			*Insufficient residence time for hydrocarbon–water separation, Section* 2.4.2.	

(Continued)

Chapter 10 Drawoff Malfunctions (Non–Chimney Tray): Part of Number 7 on the Top 10 Malfunctions (*Continued*)

Case	References	Plant/Column	Brief Description	Some Morals
			10.8 Liquid Entrainment in Vapor Side Draws	
734	398	Refinery HF alkylation main fractionator	A high C_6 concentration was measured in the iC_4 vapor side draw. This would coincide with liquid carryover making up 20% of the side product. Believed cause is low elevation (338 mm) of the bottom of the vapor draw nozzle above the tray floor and high nozzle velocities. (See also 338, Section 1.3.1.)	
DT10.6			*Proximity of draw nozzle to tray floor induces liquid into vapor side draw.*	
DT10.7			*Liquid weep from collector weep hole into vapor side draw degrades product purity.*	
757	450, 465	UDEX aromatic unit stripper Tetra solvent	Following replacement of bubble caps by truncated-downcomer high-capacity valve trays, capacity increased, loss of aromatics to the solvent was eliminated, but there was much more solvent in the aromatics vapor side draw. Believed to be caused by liquid weeping from tray above into the vapor draw pipe that had upward-directed holes, as well as enhanced agitation and entrainment from the trays.	
			10.9 Reflux Drum Malfunctions	
10.9.1 Reflux Drum Level Problems				
1610			*High level, carryover following pump failure, Section* 21.8.	
1521, 15156			*High and low level following loss of or incorrect level indication, Sections* 21.7. *and* 25.7.3.	
1520, 1168, 1586			*High level due to tower filled with liquid from base, Section* 8.1.1.	
10.9.2 Undersized or Plugged Product Lines				
750	165	Refinery coker debutanizer	Undersized reflux/product draw nozzle from the reflux drum caused excessive head loss and vaporization resulting in cavitation of the high-head product/reflux pump. Corrosion products could have played a role.	
DT10.8			*Excessive aeration due to a liquid waterfall, with outlet too small to handle aeration, causes instability.*	
1004, 15157			*Plugged water draw-off level tap and control valve, Section* 2.4.3.	
10.9.3 Two Liquid Phases (*for decanter problems, see Section* 2.5)				

Chapter 11 Tower Assembly Mishaps: Number 5 on the Top 10 Malfunctions

Case	References	Plant/Column	Brief Description	Some Morals
			11.1 Incorrect Tray Assembly	
907	360	Drying column	Tray perforations varied from one column section to another. During construction, tray sections were mixed up. Resolving problem was costly.	
931	141		Tray from one tower section installed in another. This led to a capacity restriction.	
942	250		One tray with a grossly diminished hole area caused premature flood.	
949	250		Bubble caps were installed under the tray panels. The column flooded at 30–40% of design.	
917	310	Refinery crude fractionator	Poor installation of one of the four stripping section trays caused a tray to dry. This resulted in excessive lights in the tower bottom.	
930	141		Tray panels installed with valves beneath downcomer of tray above. Instability, limited capacity resulted.	
DT11.1			*Blanking strips installed over valve floats bulge, allowing valves to open.*	
DT22.7B			*Valve legs were bent, not permitting valves to open.*	
DT11.2			*Backward installation of directional valves.*	
932	141	Absorber	Sieve trays installed with holes behind false downcomer. Caused entrainment and limited reflux flow.	
946			*Home-made valves lasting a short time, Section* 22.5.	
			11.2 Downcomer Clearance and Weir Malinstallation	
916	194	Refinery naphtha splitter	Column flooded prematurely after valve trays were replaced by sieve trays. Flooding was caused by large pieces of scale and debris restricting rectifying section downcomer clearances. Design clearances were 1 in.; installed were $\frac{5}{8}$–$\frac{3}{4}$ in. because scale left on tray support rings raised the new panels.	Properly inspect downcomer clearances following installation.

(*Continued*)

Chapter 11 Tower Assembly Mishaps: Number 5 on the Top 10 Malfunctions (*Continued*)

Case	References	Plant/Column	Brief Description	Some Morals
958	306	Refinery crude fractionator	A stripping tray was installed with zero clearance under the downcomer. This flooded the stripping section, propagating into the upper section and causing the bottom two side cuts to be black (off specification). Flooding was diagnosed by a pressure survey and persisted 8 years.	Same as 916.
DT11.3			*Picket fence weir installed on the wrong tray induces premature flood.*	
911, 943, 950			*Large gaps cause seal loss, Section* 4.4.1.	
		11.3 Flow Passage Obstruction and Internals Misorientation at Tray Tower Feeds and Draws		
901	263, DT4.7	Olefins water stripper	A bottom downcomer installed backward caused a restriction between the downcomer bottom and the seal pan wall. Cyclic flooding resulted.	Adequately inspect even when hard to get at.
904	7	Refinery stripping tower	Bottom seal pan was inadvertently blocked off during a revamp in which a reboiler was replaced by steam injection. Column operated, but flooded prematurely.	Closely inspect modification areas following revamps.
DT11.4			*Poorly installed overflow weir on a reboiler draw pan falls off, starving reboiler.*	
963	197	Refinery TAME depentanizer	C_5 product was drawn from sumps at bottom of side downcomers of two-pass trays. The installer welded the downcomer to draw sumps so that no liquid could enter the tray below. Tower flooded. Inspection found installation was per drawings and failed to detect problem.	Critically examine details even when per drawings.
964	465	Petrochemicals	The internal liquid pipe from a tray sump to a once-through thermosiphon reboiler was removed during a revamp. This led to insufficient thermosiphon driving head, insufficient reboil, and off-specification product. Fixed by converting the draw pan into a seal pan and raising liquid level.	
913	248	Caustic absorber	An internal pipe on a bottoms draw-off side nozzle was bent upward instead of downward. This caused vapor rather than liquid to escape out of the bottom.	Ensure adequate inspection.

926	189	Refinery naphtha stabilizer	Reflux pipe was directed into outlet downcomer instead of inlet seal area. Downcomer bolting plate was installed horizontally instead of vertically, blocking two-thirds of the downcomer entrance area. This, plus design errors, caused an efficiency loss.	Same as 913.
727			*Draw pipe clearing chimney tray flow by* $\frac{1}{2}$ *in., Section* 9.2.	
			11.4 Leaking Trays and Accumulator Trays (*for poor gasketing/seal welding at chimney trays, see Section* 9.1)	
716			*Seal welding at draw-offs, Section* 10.2.	
903			*Drying of bubble cap trays, Section* 4.4.2.	
935	141		Tray gasketing material blocked downcomer, causing a capacity restriction.	
			11.5 Bolts, Nuts, Clamps (*See also Uplift Due to Poor Tightening during Assembly, Section* 22.3)	
906			*Leakage of once-through thermosiphon draw pan due to loose bolts, Section* 23.2.1.	
921			*Chimney tray bolting not permitting thermal expansion, Section* 9.3.	
918			*No allowance for thermal expansion causes spray header to part, Section* 6.7.1.	
922	309	Refinery vacuum 16 ft ID	Tray supplier placed 1.5-in. slots, instead of bolt holes, around the edge of the stripping trays to allow thermal expansion. The installers burned (instead of drilled) holes in the tray support rings with the trays in place, at random locations (instead of at the outlet edge of the slot).	Good design can be negated by poor installation.
			11.6 Manways/Hatchways Left Unbolted	
945	250	Several cases	Manways were left unbolted sitting on tray decks, leaving large gaps in the tray floor. In one case, column still functioned; in others, poor separation and entrainment resulted.	Ensure adequate inspection.

(*Continued*)

Case	References	Plant/Column	Brief Description	Some Morals
933	141		Manways left off trays led to a poor separation. Gamma scans did not identify problem. In fact, the trays with properly installed manways were thought to be entraining.	Gamma scans need process cross checks.
974	363	Gas Sulfinol-M regenerator	Tray manways were resting on the support rings, waiting to be installed in one of the two passes. This raised steam consumption by 50% over the next five years. Diagnosed by gamma scans.	Same as 945.
DT11.5			*Good simulation leads to open manways, explains poor separation.*	
962	268	Refinery depentanizer	Manways on four trays were left sitting on the trays. Problem was not detected until next turnaround due to misleading simulation.	Same as 945.
DT11.6			*An internal manway left open in a sleeve containing packed bed caused vapor to bypass bed and poor product quality.*	
905	320		A hatchway was left unbolted in a preferential baffle separating column bottom draw-off and reboiler compartments. This caused poor reboiler performance.	Same as 945.
			11.7 Materials of Construction Inferior to Those Specified	
908	360		Trays were specified to be 316SS, but four of the column trays installed and many nuts and bolts were 304SS. These would have failed in this service.	Ensure adequate material inspection.
923	309	Refinery vacuum	Stripping trays and bolts specified as 410SS. Installed was CS.	Same as 908.
889			*Shroud cleats incorrectly fabricated from CS and failed, Section* 9.8.	
919			*Spray header flange made from sheet metal fell apart, Section* 6.7.1.	
			11.8 Debris Left in Tower or Piping	
954	498	C_3 splitter	A fire blanket left in the tower caused high liquid levels, which in turn flooded the entire column.	Gamma scans diagnosed.
960	308	Coal gasification	A plastic bag left inside a packed bed of Pall rings caused premature flooding in an off-gas scrubber.	Inspect for debris.

902, DT11.7	363	Olefins water quench	A welding rag left in the column reflux line found its way to and partially blocked the reflux distributor to two-pass trays. Excessive entrainment resulted.	Inspect for debris in lines connected to the column.
973	286	Olefins water quench	A loose vapor riser hat from a draw pan plugged the circulation draw forcing a shutdown. Hats were welded in-pace to prevent reoccurrence.	
909	446	LPG	During commissioning, reboiler pump strainers were broken due to blockage by debris. Pieces of strainer casings damaged the pumps. Strainer casings with extra support bars avoided recurrence.	It is best to keep debris out of outlet lines.
912	305	Gas-processing debutanizer	The carcass of a dead rat lodged in the kettle reboiler inlet nozzle and backed up liquid into the tower. When level reached the reboiler return nozzle, the column flooded prematurely.	Keep manholes closed when no one is in the column.
DT6.4			*Plugged packing distributors.*	

11.9 Packing Assembly Mishaps

11.9.1 Random

910	50	Soda ash recovery ammonia still, 10 ft ID	Ceramic packings were dumped through a chute installed at the manhole. This caused pieces of packing to stratify in layers on an inclined plane ("hill" formation) as well as breakage. This resulted in poor liquid distribution, low efficiency, and low capacity.	Avoid hill formation while packing a column. Plan to avoid ceramic breakage.
DT6.1			*"Hill" formation during installation causes maldistribution, poor separation.*	
DT11.8			*Leaving no space between bed and distributors leads to premature flooding.*	
DT11.9			*Packing handling issues include torn cardboard boxes spilling packing, need to clean mud, and particles falling during loading.*	
947	250		About 20% of the polypropylene rings were damaged when being loaded into the tower. The rings were dropped from a height of 23 feet onto a steel support at about -15°F.	Avoid excessive fall for plastic at low temperatures.
914	290	Test column	Screening chips from breakage of ceramic saddles during shipment was troublesome. It was necessary to resort to picking out chips by hand over 144 ft^3 of packing.	
936	141	Several columns	Tray support rings left in the tower revamped with packings led to poor separation and premature flooding.	

(*Continued*)

Chapter 11 Tower Assembly Mishaps: Number 5 on the Top 10 Malfunctions (*Continued*)

Case	References	Plant/Column	Brief Description	Some Morals
886	435	Ethylene oxide Benfield absorber	Trays were replaced by a 10-m bed of modern random packing in the top and 6 m of same in bottom without cutting out the tray support rings. Works well.	
808, 940			*Support rings left in tower, no ill effects, Sections* 6.2.2 *and* 11.10.	
11.9.2 Structured				
961	308		Structured packing flooded at 50% of design because workers stepped on and crushed intermediate layers of packings.	Ensure adequate construction supervision.
941	250		Poor efficiency resulted from a wide annular open gap left between a structured packing and a column wall.	
965	465	Aromatics BTX ED	Tower did not meet benzene recovery specifications (see 898, Section 2.6.2). Contributing factors seen by inspection included packing up to 1/2 in. from wall and up to 1/2 in. opening in middle.	Same as 961.
DT11.10			*Gaps between packing and wall, between packed bricks, bed swelling, hammering bricks, sloping bricks occurred in different towers.*	
828			*Packing damage due to heavy beams dragged on top of bed, Section* 6.10.	
11.9.3 Grid				
925, 951			*Poor assembly causes grid bed to disintegrate, Section* 22.3.	
			11.10 Fabrication and Installation Mishaps in Packing Distributors	
940	275	Olefins demethanizer random packings	Turnaround inspection detected problems with existing internals. In the upper sections, a flashing feed entered via a bare nozzle above an orifice pan distributor. An orifice pan distributor was misoriented to the liquid inlet nozzle. In the lower sections, side panels of a redistributor were interchanged with those of another with larger holes. Old tray support rings were not removed (tower ID was 5.5 ft). Despite these, the packings in the lower sections operated efficiently. The top sections were inefficient but were not greatly improved by correcting the faults. (See also 311, Section 1.3.1, and 513, Section 4.8.)	Inspection is invaluable for detecting flaws.

971	350		A 20-ft-ID, 150-ft-tall packed tower was unstable and produced off-specification products following a retrofit with new structured packings and distributors. Cause was incorrect feed distributor installation that channeled liquid to the center of the tower. CAT scan led to diagnosis and cure.	
924	309	Refinery vacuum	A water test after installation of structured packings in a 30-ft-ID tower revealed plugged spray nozzles, spray nozzles missing their internals, plugged distributors, and leaking chimney trays.	Water tests can be invaluable.
835	141		Feed distributor, put in reflux service due to drawing error, gave liquid maldistribution and poor separation.	
959	308	Formic/acetic acid	Holddown plate for $1^1/_2$-in. Teflon rings was installed immediately below the feed distributor instead of 15 in. below. This restricted the open area at the interface of the distributor, the holddown, and the rings. Flooding initiated at 50% of the design loads at that interface.	Installers tried to use the same support for holddown and distributor.
955	382	Chemicals	Several essentially identical liquid distributors for the same packed tower gave uneven liquid flows in a water test because the original holes were not deburred. The higher friction also caused liquid level to rise above the vapor risers. Also, undersized feed pipe holes caused water splashing outside of the distributor.	Distributor water tests can prevent disasters.
934	141		Holes were punched in opposite directions in a distributor. Caught during a water test in the shop. Levels in some troughs were about a third deeper than others and would have overflown before reaching design rates.	Same as 955.
956	382	Chemicals	A distributor feed pipe that should have terminated $1^1/_2$ in. above the floor of a parting box actually terminated $\frac{1}{16}$ in. above the floor.	Ensure adequate inspection.
DT11.8, DT11.11			*Distributor rotated* 90° *to correct orientation causes poor separation.*	
DT11.12			*Inverted chimney hat installation leads to capacity loss.*	
DT11.13			*Holes mispunched, undersized liquid equalizing, horizontal momentum, liquid splashing, jump over sides of parting box, missing bolts, and poor drainage occurred in different distributors.*	
953, 966, 1275			*Out-of-levelness, Sections* 6.9 *and* 6.2.1.	

(*Continued*)

Chapter 11 Tower Assembly Mishaps: Number 5 on the Top 10 Malfunctions (*Continued*)

Case	References	Plant/Column	Brief Description	Some Morals
880			*Missing gaskets, Section* 6.7.1.	
918			*Rigid bolting of spray header to tower hall, Section* 6.7.1.	
919			*Thin sheet metal flanges, Section* 6.7.1.	
			11.11 Parts Not Fitting through Manholes	
915	408	C_3 splitter	The column was being retrayed. Existing tray panels would not fit through manholes and had to be hot-cut first. This added 4 days to the retray. Column sway during a violent storm caused a further delay. Despite the difficulties, the retray was completed within the time available.	Paper contains excellent data on timing and labor for retray field work.
927	460	Refinery crude fractionator	A 17.25-in. manhole was specified as 18 in. All major collector and distributor parts had to be cut in two, then rewelded inside, adding two extra days to the turnaround.	
			11.12 Auxiliary Heat Exchanger Fabrication and Assembly Mishaps	
975	502	Ethyl acetate low-pressure column	Condenser capacity fell below design due to a construction fault that had the exchanger inlet on the same side of the exchanger as the outlet. Capacity achieved after fault corrected. CFD denied alternative theory of non-condensables collection near exchanger bottom.	
937, DT11.14	224	Natural gas absorber and deethanizer	An interchanger heating the absorber bottom (deethanizer feed) by cooling the absorber lean oil was built as a cocurrent instead of countercurrent exchanger. This gave a cold deethanizer feed and a hot absorber lean oil, giving poor separation in both towers. The solution was to repipe one side of the interchanger.	Checking design and installation details can save headaches at start-up.
			11.13 Auxiliary Piping Assembly Mishaps (*for incorrect meter installation, see Section* 25.5)	
967			*Missing vent hole leads to vacuum, Section* 22.1.	
DT11.15			*Seal leg in vent line from storage to tower destabilizes column.*	
DT11.16			*Cooling water pumped backward through condenser.*	

Chapter 12 Difficulties During Start-Up, Shutdown, Commissioning, and Abnormal Operation: Number 4 on the Top of 10 Malfunctions

Case	References	Plant/Column	Brief Description	Some Morals
			12.1 Blinding/Unblinding Lines	
1137, DT12.1	225	Natural gas lean-oil still	At commissioning, lean oil was circulated through the fired reboiler, its temperature gradually raised to 540°F. Upon reaching 350°F, the rise stopped. The cause was no flow due to a blind left over from hydrotesting. The dead-headed pump was damaged. The plant was lucky not to rupture a heater tube. Key instruments were not operational, impeding correct diagnosis, and obvious interpretations turned out misleading.	Blinds should have long handles and tags. Instrumentation needs to be operational for commissioning.
DT8.1			*Construction blind in level transmitter piping causes misleading indication and tower flooding.*	
1116	6	Refinery crude fractionator	Crude tower gases were released to the atmosphere and detonated. This followed unblinding and valve removal from a line which contained the gases. Blind removal followed a breakdown in communication.	Blind removal should require a written permit.
11101	540	Refinery coker main fractionator 26 ft ID	For start-up preparation, lines were deblinded while tower contained air. Upon attempting to remove a blind on the low-pressure natural gas circuit, valve passing was observed. The next morning the blind was removed with only a minimal amount of gas passing noted. Six hours later, an explosion dislodged trays downwards, broke support beams, lifted relief valves and damaged the overhead accumulator. The explosion is believed to have been initiated by a relatively small volume of flammable gases collected in the tower upper head space. Paper has detailed analysis.	A small quantity of gas can cause an explosion in a tower. Properly service valves. Follow good blinding/deblinding procedures.
1167	152	Cryogenic	The bottom half of a tower partitioned with an internal head was blinded and cleared for personnel entry with the upper half on-line. Hearing a hissing noise, a superintendent discovered that the insulated *dP* lines had not been isolated and leaked hydrocarbons in.	Blinding schedules need to be concise, comprehensive, with all blinds signed off.

(Continued)

Chapter 12 Difficulties During Start-Up, Shutdown, Commissioning, and Abnormal Operation: Number 4 on the Top of 10 Malfunctions (*Continued*)

Case	References	Plant/Column	Brief Description	Some Morals
1172	140	Refinery FCC main fractionator	A flange was opened to remove an 8-in. spade from a line from the overhead receiver to the flare during start-up. An H_2S-containing gas was released and overcame the worker and others who came to help. The line contained gas because the block valve between the overhead receiver and the flare was already open.	Permits for removing spades should require that the lines are gas free and depressured.
1108, 1109			*Backflow of chemicals into tower while open, Section* 14.10.	
1114, 1119, 1197			*Chemical releases and fire due to unblinded lines, Sections* 14.10 and 14.8.	
1034, 11103			*Explosions due to unblinded lines, Sections* 14.4 *and* 14.1.3.	
1110, 1125			*Leading to release of trapped chemical, Section* 14.4.	
1113			*Absorber–regenerator not properly blinded during shutdown wash, Section* 12.6.	
1122			*Purging no substitute for blinding, Section* 12.4.	
1151			*Air entered tower containing combustibles, Section* 14.7.	
1641			*Explosion when hydrocarbons routed via a line open to atmosphere, Section* 14.5.	
			12.2 Backflow	
			(*See also Fires Caused by Backflow, Section* 14.8 *and Chemical Releases by Backflow, Section* 14.10.)	
1126	6	Refinery coker sponge absorber	The light distillate product pump, which supplied lean oil to the absorber, lost suction at start-up. Gas from the absorber backed through the lean-oil line and traveled into the hydrotreater charge pump, causing it to gas up and lose suction. This resulted in hydrotreater catalyst damage.	Beware of reverse flow.
1172			*Backflow of chemicals into tower while open, Section* 12.1.	
1292			*Backflow initiates upward gas surge, Section* 22.4.	
DT12.2			*Backflow during an outage causes water backflow into a tower containing dry hydrogen chloride and aggressive corrosion.*	
114			*Backflow at commissioning leads to undesirable reaction, Section* 15.4.	

DT22.11			*Backflow during commissioning causes tray damage.*	
1133, 644			*Storage contamination at shutdown leads to reaction, foaming, Sections* 15.4 *and* 16.1.2.	
			12.3 Dead-Pocket Accumulation and Release of Trapped Materials	
715			*Water accumulation in tower dead pocket, Section* 13.3.	
1247			*Freezing of water accumulated in dead leg, Section* 14.3.3.	
1595			*Water accumulation in an infrequently used line, Section* 28.1.	
1110, 1125			*Release of hydrocarbons trapped in valves, Section* 14.4.	
1111			*Release of chemicals trapped in plugged, infrequently used draw, Section* 14.11.	
1627			*Hydrocarbon release when blocked drain clears, Section* 14.11.	
133			*Accumulated nitro sludge overheated, Section* 14.1.3.	
			12.4 Purging	
1115	304	Refinery	High-point vent on a distillation column was left open, causing product loss for an entire week.	
1117	6	Refinery FCC columns	A supplier error caused the unit to be purged with a gas containing 93% oxygen. Several explosions and fires resulted.	Always test purge gas before use.
1122	223	Chemicals	A shutdown column contained flammable gas. Work was performed on a ground-level exchanger in the column product line. The line was purged by inert gas pumped backward into the column and out through the top of the column. Air managed to get into the column, causing an explosion that damaged trays.	Purging is no substitute for blinding.
11104	356	Phenylethylamine batch column 3-ft ID × 20 ft tall	Hastelloy still, mounted over a still pot, contained Hastelloy mesh packing. At batch completion, nitrogen was injected at column base, exiting to an atmospheric vent via a scrubber, and still pot was drained. Several hours after draining, an internal fire damaged shell and packing. Nitrogen flow was too small, allowing air to enter via the drain. A contributing cause was shutdown of instrumentation.	Ensure adequate inerting.

(*Continued*)

Chapter 12 Difficulties During Start-Up, Shutdown, Commissioning, and Abnormal Operation: Number 4 on the Top of 10 Malfunctions (*Continued*)

Case	References	Plant/Column	Brief Description	Some Morals
1148	481	Olefins C_3 splitter	At shutdown, residual hydrocarbons in column and piping were purged using nitrogen supplied by vaporizing liquid nitrogen from a tanker truck. Twenty-four hours after purging started, a 16-in. CS pipe spool section from the tower into one of the two reboilers ruptured. The cause is believed to be inadvertent introduction of liquid nitrogen into the piping system, which overchilled and overstressed the metal. The liquid nitrogen vaporizer is believed to have malfunctioned without an alarm.	To prevent recurrence, a low-temperature alarm and automatic shutdown of liquid nitrogen were added and procedure modified.
			12.5 Pressuring and Depressuring	
1103, 1130, 1158			*Pressuring from top down damages valve trays, Section* 22.6.	
1156			*Rapid flashing of liquid pool at base dislodges trays, Section* 22.4.	
1135			*To dehydrate aromatics vacuum tower, Section* 12.6.	
969	51	Specialty chemicals batch	Shortly after relocation, a small glass still receiver exploded when high-pressure N_2 was used to relieve the vacuum following an otherwise successful distillation. The still was hooked into the incorrect N_2 system during relocation.	
			12.6 Washing (*See also shutdown wash to prevent packing fires, Section* 14.6.2.)	
1104	360		Water used in pre-start-up wash was heavy in solids and laid a thick mud deposit on trays and exchangers.	Check source of wash water.
1139	250	Amine	Insufficient quantity of wash solution was charged into the system for shutdown wash. Pump suction was lost due to lack of inventory.	
1120	180	Ammonia Benfield hot pot	Frequent plugging occurred at the lean-solution pump suction strainer. Problem persisted despite frequent strainer cleaning. The plugging was caused by particulate matter, including rust, which remained in the system after washing. Unit operated well following a rewash.	It pays to check for particulates at the completion of a wash.

1113	328	Ammonia hot-pot regenerator	Several start-up/shutdown accidents in hot-pot regenerators are described. In all, the system was being water flushed, with the absorber under pressure of insoluble gas. The water absorbed small amounts of gas (natural gas, hydrogen, nitrogen) in the absorber and desorbed it in the regenerator or its piping. When the gas was combustible, explosions occurred once hot work was performed inside the regenerator or on its vent line. When the gas was nitrogen, a suffocating atmosphere resulted inside the regenerator.	Blind column connections and maintain good ventilation during hot work. Watch out for absorption of gases in wash water.
11109	378	Natural gas Selexol H_2S absorber-regenerator	During commissioning, the absorber was pressured up with sweet gas, and Selexol circulated. Due to pump problems, this continued for several days. Small amounts of gas were absorbed in the absorber and desorbed in the stripper. The desorbed gas accumulated in a downstream unit, leading to an explosion that caused minor damage.	See Case 1113.
1118	6	Refinery caustic scrubber	Hydrogen sulfide was liberated to atmosphere when a caustic scrubbing system was acid washed.	Consider reaction of wash with tower deposits.
1135 (also 1136)	344	Aromatics EB–styrene 29.5 ft/24.5 ft ID top/bottom	At a shutdown, the column was water washed, then opened. The random CS packings rusted; upon chemical cleaning, the iron removed was equivalent to 1.4% of the packing weight. Following the wash, the column is now steam heated, then steamed downward, then nitrogen pressured and vacuumed several times, to an exit gas dewpoint of −20°F. (See also 839, Section 6.1.1)	A special wash procedure developed and described in paper.
1136 (also 1135)	344	Aromatics EB–styrene 29.5 ft/24.5 ft ID top/bottom	A water wash followed by a chemical wash could not remove insoluble material brought in with the feed and lodged in the reflux distributor and the bed below. Based on a published theory that water-wetted rust particles (hydrophilic) agglomerate to form paste in the presence of hydrocarbons, a new wash procedure was developed. A boiling hydrocarbon wash was followed by a boiling water wash to strip hydrocarbons, finally a boiling water wash at maximum rate to disperse plugs and wash down rust. This procedure alleviated plugging and enhanced separation stages from 40–44 to 50. (See also 839, Section 6.1.1)	Same as 1135

(*Continued*)

Chapter 12 Difficulties During Start-Up, Shutdown, Commissioning, and Abnormal Operation: Number 4 on the Top of 10 Malfunctions (*Continued*)

Case	References	Plant/Column	Brief Description	Some Morals
1131	376	Refinery vacuum	Packing plugging was caused by initial operating problems with the HVGO flush system. Improving flush system increased run length from 3 to 10 years.	Start-up procedure can affect run length.
106			*Caustic from wash drawn into vacuum tower during operation, Section* 14.2.	
1119			*Caustic backflowed into steam system, Section* 14.10.	
1138			*Chlorides in water, Section* 12.8.	
DT12.3			*Caustic wash effectively dissolves salts, polymer on packing and internal reboiler tubes.*	
			12.7 On-Line Washes	
1221	304	Pharmaceuticals methanol dehydrator	Excessive caustic injected into feed was entrained into the methanol-rich top section and precipitated there due to water vaporization. The deposits plugged the $\frac{3}{16}$-in. sieve tray holes. This induced premature flooding. Problem solved by on-line water wash, effected by raising boil-up and cutting reflux for a few hours, thus inducing water up the column. Longer term solution was cutting caustic injection.	A useful on-line washing technique as devised. Small perforations plug.
DT12.3			*Caustic on-line wash removes formaldehyde dust plugging.*	
1232	376	Refinery FCC main fractionator	With some feedstocks, salting out is experienced on some upper trays. Symptoms include a high gasoline end point, a reduction in LCO draw, and an increase in flash zone pressure. Alleviated by a 15-min. on-line boiler feedwater wash injected into the top reflux. This may need repeating until improved.	
1284	547	Refinery FCC main fractionator	Due to a high chloride content of the AGO, salting out of ammonium chloride plugged trays near the top of the tower about once per month. Plugging was removed by an on-line water wash, with water removed a few trays down. A better solution implemented now is AGO desalting.	Some useful guidelines for on-line water wash included.
1273	199	Refinery FCC main fractionator	Salt laydown on top PA trays caused high pressure drop. An on-line water wash system successfully removed deposits.	

DT12.4			*Procedure that successfully water washed salt deposits in FCC main fractionator and reformer stabilizer.*	
1265	499	Refinery hydrotreater fractionator	Poor separation and high naphtha end point were caused by salting out. Gamma scans showed that the plugging moved between the fifth tray down (out of six) and the PA bed underneath, supporting a salting-out diagnosis. A water wash restored normal operation.	
12120	41	Refinery crude fractionator	Salting out occurred at the packed TPA, with severe fouling in the trayed upper fractionation zone below. Problem was most severe when running light crudes that cooled tower overhead and TPA return by about 20°F and raised vapor loads. Water washing at outages was effective. Since the product from the upper fractionation zone was drawn from a seal-welded chimney tray, the water and foulant could be totally removed without flowing down the tower.	
1295	244	Refinery coker debutanizer	Plugging with chloride salts occurred 3–10 trays above the feed. Gamma scans in this region showed liquid-full downcomers in the center and on one side, while those on the other side contained highly aerated froth. A water wash during a short shutdown eliminated problem.	
1254	94	Olefins caustic wash random packings	Distributor holes plugged after 2 years in service, causing flooding, high dP, and poor absorption. On-line wash with a hydrocarbon liquid cleared the plugs, increasing column run length to 5 years. Each bed was washed separately, with the gas bypassed around it during the wash.	Clever design and on-line washing can extend column life.
1176	117	Chemical solvent stripper	Grid scans and high dP indicated plugged packings. Tower was taken off-line and condensate washed. Repeat scan showed the plug cleared, and opening the tower was not needed.	
12101	399	25 ft ID	Grid scan showed nonuniform fouling on the lowest dual-flow tray in a section with flooding above. Repeat scan after a chemical wash showed the fouling moved around but was not eliminated.	
1227			*Injecting surfactant removed oil from packing surfaces, Section* 4.8.	

(*Continued*)

Chapter 12 Difficulties During Start-Up, Shutdown, Commissioning, and Abnormal Operation: Number 4 on the Top of 10 Malfunctions (*Continued*)

Case	References	Plant/Column	Brief Description	Some Morals
			12.8 Steam and Water Operations	
1112, 1147, 1614			*Condensation causes vacuum and implosion, Section* 22.1.	
1008			*Condensation leads to pressure surge, Section* 13.6.	
1140, DT22.3, DT22.14, DT22.15			*Cold water step-up generates local vacuum that damages trays, Section* 22.7.	
1105, 1106			*Overheating by the steam, Section* 12.9.	
1138	250		Steam–water operations were discontinued in SS columns after it was found that chlorides in the water caused metal deterioration.	
1193			*Frequent steaming of SS tower with high-chloride steam leads to stress corrosion cracking, leaks, Section* 20.4.1.	
1144			*Rapid drainage of water caused downcomer damage, Section* 22.8.	
1149			*Overchilling by cold water at hydrotest, Section* 12.11.	
			12.9 Overheating	
1106	284		A large distillation column was made in two halves in series connected by a large vapor line containing a bellows. Steaming the line during shutdown excessively raised one end of the bellows above the other.	Beware of overheating by steaming.
1107	284		A reflux line was fixed rigidly to brackets welded to the shell. At start-up, differential expansion of the hot tower and the cold line tore one of the brackets off the tower, causing a leak of flammable vapor.	Check supports of auxiliary lines as the tower heats up.
11102	109	Refinery vacuum	During start-up, a heavy leak of oil developed from the inlet flange of one of the tower exchangers, resulting in a fire. Pipework did not have adequate allowance for thermal expansion at start-up.	
502 DT16.1	263	Gas hot-pot absorber	Plastic packing melted upon start-up because of reaction and possibly also hot spots.	

503	436	Ammonia hot-pot absorber	On many occasions, plastic packing melted upon solution circulation (power) failure. Absorber feed was cooled by the regenerator reboiler. This cooling was interrupted when circulation ceased, causing hot feed to enter the column.	Avoid plastic packing where hot feed can enter.
1105	548	Chemicals stripper aqueous stream	Polypropylene packings melted because of evaporation of a water seal which desuperheated stripping steam issuing from a submerged bottom feed distributor. The incident occurred during a brief maintenance shutdown. Resolidified plastic which oozed through the packing support later caused pump damage.	
506	290		High temperature during start-up caused aluminum packing to lose strength and become compressed. This incurred excessive pressure drop.	Consider abnormal operation.
DT12.5			*Draw pan damage due to overheating at outage.*	
1651, 1137			*Pump explosion, damage, due to dead-head overheating, Sections* 14.4, 12.1.	
			12.10 Cooling	
1012	6, 36	Refinery crude fractionator	Tray damage occurred at shutdown due to premature exposure of column internals to cold water and air. Steam cooling of the tower was not adequate prior to air and water introduction.	Adequately cool before introducing air or water.
1112			*Vacuum and tower implosion upon cooling, Section* 22.1.	
1123			*Fire when tower opened while containing combustibles, Section* 14.7.	
1142	250		Upon cooling, residual liquid in the tower became highly viscous, forming a hard, solid, difficult-to-remove mass at the bottom sump.	Compare 1194, Section 12.13.1
			12.11 Overchilling	
1611, 1616, 1628, 1168			*Flashing of liquefied hydrocarbon liquids overchills metal, Sections* 14.3.2 *and* 8.1.1.	
1148			*Due to liquid nitrogen entering tower during purging, Section* 12.4.	

(*Continued*)

Chapter 12 Difficulties During Start-Up, Shutdown, Commissioning, and Abnormal Operation: Number 4 on the Top of 10 Malfunctions (*Continued*)

Case	References	Plant/Column	Brief Description	Some Morals
1149	105	Refinery	Following modifications the tower was hydrotested. Instead of using water at not less than 20°C as specified, water at around 10°C was used. Brittle fracture occurred, resulting in 2 months lost production.	
			12.12 Water Removal	
12.12.1 Draining at Low Points (*See also Undrained stripping steam line, Section* 13.5)				
1028			*Undrained stripping steam line leads to poor stripper performance, Section* 2.2.	
1006, 1007			*Undrained transfer line leads to pressure surge, Section* 13.2.	
1021, 1030			*Undrained accumulator drum leads to pressure surge, Section* 13.7.	
12.12.2 Oil Circulation				
1005, 1015			*Water pockets at pump suction lead to pressure surges, Section* 13.4.	
1011			*Skipping the oil circulation step leads to pressure surges, Section* 13.3.	
1016			*Water entering tower during oil circulation, Section* 13.2.	
1132	201	Refinery FCC main fractinator	During liquid circulation and dry-out, hot liquid from the slurry PA was routed to the fractionator upper sections in order to speed up heating. These lines were not shut prior to catalyst circulation in the reactor and regenerator. A reactor upset caused catalyst carryover, which reached the tower upper sections via the open lines, fouling distributors, packings, and collector trays.	Start-up procedure can affect run length.
DT9.4			*Inventorying tower with oil exerts excessive upward force, damaging chimney tray with tall chimneys.*	
12.12.3 Condensation of Steam Purges				
1008, 1013			*Lead to pressure surge, Section* 13.6.	
12.12.4 Dehydration by Other Procedures				
1010			*Total refluxing hydrocarbon, Section* 2.4.1.	
1135			*Pressuring and depressuring, vacuum tower, Section* 12.6.	

12.13 Start-Up and Initial Operation

12.13.1 Total-Reflux Operation				
1194	286	Olefins oil quench	Tower was placed on circulation without the addition of fresh feed with volatile components. Vaporization of lights caused a bottom viscosity runaway and solidification that forced plant shut down.	Compare 1142, Section 12.10.
103			*Concentration of unstable component leading to explosion, Section* 14.1.4.	
120			*Long residence time leads to undesirable reaction, Section* 15.6.	
DT12.6			*Baffle separating draw and reboiler compartment gives problem during total reflux start-up.*	
1010			*For water removal, Section* 2.4.1.	
109			*For testing separation, Section* 1.1.2.	
12.13.2 Adding Components That Smooth Start-Up				
1175	121	Chemicals dilute HNO_3 concentration	There was a low-ppm specification on nitrates in distillate and a minimum specification on bottom acid concentration. To speed start-up, low-nitrate water was added to reflux drum, thus preventing contamination, while bottom concentration was slowly raised.	
12.13.3 Siphoning				
DT2.21			*Due to undersized pressure balance lines.*	
1416			*Due to high point in condenser drain lines, Section* 24.2.1.	
1197, DT22.11			*Siphon formation with one end pressured, other open to atmosphere, causes sudden emptying of line, Section 14.8.*	
15103	339		Temperature control manipulating a valve in the vertical leg of the cooling-water supply line to an elevated condenser was erratic. The valve was destroyed by cavitation. Plant operation was at 50% of design. Siphoning from the return line and low pipe and condenser friction losses created vacuum immediately downstream of the valve. The valve pressure drop exceeded the maximum pressure drop to avoid valve cavitation. Release of dissolved air bubbles under vacuum and switch of downflow between self-venting and siphoning could also have been the cause. Problem solved by eliminating valve and control.	Consider worst-case scenario when specifying cooling-water throttling valves to elevated condensers.

(Continued)

Chapter 12 Difficulties During Start-Up, Shutdown, Commissioning, and Abnormal Operation: Number 4 on the Top of 10 Malfunctions (*Continued*)

Case	References	Plant/Column	Brief Description	Some Morals
15164			*Severe shaking in stripper feed line with control valve 20 ft above grade, Section 6.5.*	
12.13.4 Pressure Control at Start-Up				
1159	479	Refinery FCC main fractionator	Start-up pressure control was by steam injection into the reactor and throttling a control valve in overhead line to condenser. The condenser was far oversurfaced for start-up duty, giving pressure fluctuation of ± 25 kPa. Instability, catalyst loss, and high-utility consumption resulted. Solved by pressuring tower with nitrogen and pressure controlling bleed from reflux drum to flare.	At start-up, noncondensing gases can help pressure control.
			12.14 Confined Space and Manhole Hazards	
1171	108	Olefins depropanizer spare base section	Following turnaround cleaning, a hot sodium nitrite solution used for passivating tower internals was prepared inside the tower. Concentrated nitrite was diluted with water charged by a hose into the top manhole, with nitrogen blown through the solution to mix. A supervisor went to check whether the water was flowing in and was found with his head in the top manhole.	Beware of asphyxiation hazard at manholes. Use outside standby.
1182	15, 281		At turnaround, a tower was emptied, washed, and nitrogen purged. A manhole cover at the base was removed. One of the two men removing the top manhole cover was overcome but was pulled to safety by the other. Apparently, due to a chimney effect air entered at the base and displaced the lighter nitrogen.	See 1171.

1113			*Suffocating atmosphere formed during absorber–regenerator wash, Section* 12.6.	
1169	401	Waste gas quench tower	Workers were overcome when working inside the tower skirt. The tower contained gas under positive pressure, and the valve that vents the tower to atmosphere via a water seal was closed. The tower skirt had four openings and therefore was not considered confined.	Skirts of vessels should be considered confined spaces.
1192	31	Formic acid	To reduce fumes while modifying a manhole cover, a temporary 3 mm metal cover was placed over the open manhole, held by 3 bolts. A slight vacuum was pulled on the column. When the modified cover returned, two of the 3 bolts were removed. The temporary cover was suddenly sucked into the column along with two workers, killing them.	
1170	108	Solvent recovery	A worker was deprived of air when an air supply hose on a portable breathing apparatus became detached while in use inside tower. Prompt action by others restored air supply quickly.	Ensure outside standby and adequate hose attachment.
1633			*Flange leak inside a fractionator skirt, Section* 20.4.1.	

Case	References	Plant/Column	Brief Description	Some Morals
			13.1 Water in Feed and Slop	
1035	443	Refinery coker fractionator	During start-up, LCGO circulated in the LCGO and HCGO PAs. After introducing hot feed, an HCGO pump-out from storage to the HCGO PA was started. The pump-out came from the bottom of a tank and contained water. Flashing of the water upon tower entry severely damaged several valve trays, which included no heavy-duty features. Strangely, trays immediately above were damaged downward. Paper has enlightening details and photos.	"Heavy-duty" internals should be used in towers prone to pressure surges. Tank pump-outs should be from an elevated position.
1023	425	Refinery coker fractionator 14 ft ID	Bottom four sieve trays were plugged with 9 psi pressure drop across them. A new batch of slop was fed, entering tower between trays 2 and 3. Pressure drop dived to 3 psi. Gamma scans showed tray displacement, probably caused by a water pocket in the slop. (See also 1586, Section 8.1.1.)	Ensure slops are water free before they enter a hot fractionator.
1029	459	Refinery crude fractionator	Top 15 trays were damaged due to water entry during operation. The water most likely entered in a naphtha feed 11 trays below the top, some of which was imported from another unit. It is believed that a seal leak generated a water leg in the standby naphtha transfer pump, and the leg was sent into the tower when the standby pump switched into service.	Check for water before connecting spare pumps.
1018	442	Aromatics raffinate stripper	Column had a long history of upsets from pressure surges caused by introduction of free water. These severely damaged most of the trays. Replacement by heavy-duty fixed-valve trays, 10 ga. 410 SS decks, shear clips, and explosion doors permitted trays to weather explosions well.	
403	301	Refinery main fractionator	This column was prone to pressure surges because of accidental introduction of water. Valve trays needed replacing approximately once per year. Cast iron bubble-cap trays were used in a very similar unit and could weather such surges.	

DT13.2B, D			*Water entering with pumpback of flare drum hydrocarbon into olefins oil quench tower and from leaking valve in hot organic vacuum service causes tray damage.*	
			13.2 Accumulated Water in Transfer Line to Tower and in Heater Passes	
1006	3	Refinery combination tower	Nearly all trays were damaged by a pressure surge. A low point was formed in a long horizontal line from a coke drum to the tower. Condensate collected at the low point. When the coke drum was heated, the low point was lifted, dumping water into the tower and causing a pressure surge.	Ensure drainage of lines connecting two vessels to avoid water trapping.
1007	3, 4	Refinery FCC main fractionator	A pressure surge severely damaged trays and support beams. At start-up, water was trapped above a block valve in the vertical (downflow) fractionator feed line. A drain just above the block valve was plugged. When the block valve was opened, the water was dumped into hot oil, causing a pressure surge.	Ensure complete draining of feed lines before opening.
1016	6	Refinery vacuum	Most trays and some tray supports were damaged by a pressure surge during restart following an outage. Oil was circulated at 280°F through the tower and heater, and the column was under full vacuum. Source of water was condensed steam that was accumulated in one heater pass. Recommendations included avoiding starting up under full vacuum; pumping the tower out when temperature falls below 300°F during an outage; monitoring coil outlet temperatures; and preventing water accumulation in heater coils.	
DT13.2 A			*Generates pressure surge that rips trays in olefins oil quench tower.*	
			13.3 Water Accumulation in Dead Pockets	
715	3	Refinery combination tower	Column consisted of two sections separated by an upward bulging internal head which served as a draw pan. The liquid outlet was 3 in. above the lowest point; water accumulated below that. When hot oil later filled the pan, a pressure surge occurred and damaged trays.	Ensure adequate drainage of trap-out pans.

(*Continued*)

Chapter 13 Water-Induced Pressure Surges: Part of Number 3 on the Top 10 Malfunctions (*Continued*)

Case	References	Plant/Column	Brief Description	Some Morals
1002	369	Refinery crude fractionator	Weep holes in the bottom seal pan plugged and trapped water. The water vaporized at start-up, causing a pressure surge that lifted trays off their supports. Problem was solved by installing downpipes (extending below bottom liquid level) to drain the pan.	Ensure weep holes operate properly.
1011	6, 36	Refinery vacuum	Pressure surges occurred upon feed introduction and caused tray damage in several cases. The start-up procedure did not use oil circulation to flush out water. The surges resulted from pockets of water remaining in draw pans and PA circuits.	In hot fractionators, oil circulation is essential for flushing out water.
1595			*Water accumulation in an infrequently used line, Section* 28.1.1.	
		13.4	**Water Pockets in Pump or Spare Pump Lines**	
1005	3	Refinery vacuum	All trays were bumped by a pressure surge at start-up. A block valve was opened to establish flow of hot circulating oil to a pump. A pocket of water trapped between the block valve and a second block valve at the pump suction flashed on contact with the hot oil, resulting in a pressure surge.	Even small quantities of water can cause major damage in vacuum towers.
1015	6	Refinery vacuum	A side-stream accumulator was lifted off its foundation by a pressure surge at start-up. During hot-oil recirculation, back flush of the spare pump was being attempted. The pump discharge valve was cracked open before the suction valve was closed. A pocket of water in the pump or its piping was sucked into the hot oil, causing the surge.	Ensure adequate drainage of spare pumps before connecting to a hot-oil system.
DT13.2C, E, H			*Cause pressure surge when spare pump connected to system in crude tower, hot aromatics vacuum tower, but not in a refinery vacuum tower.*	
1029			*Water leg in standby feed pump, Section* 13.1.	

1217	6	Refinery vacuum	Most of the trays were torn off their supports following a pressure surge. The surge was caused by vaporization of a slug of flushing oil. The slug entered the tower from a spare bottoms pump that was inadequately isolated.	Ensure proper isolation of spare pumps in service prone to pressure surges.
			13.5 Undrained Stripping Steam Lines	
1003	369	Refinery	Undrained water in stripping steam line entered tower upon start-up, causing a pressure surge that dislodged trays. A valve right at the column flange with a blowdown drain just upstream can eliminate this problem.	
1037	127	Refinery crude fractionator	Adding an isolation valve at the column flange on the stripping steam line and a properly sized start-up vent just upstream of the valve and purging the steam line via this vent at start-up have minimized pressure surges from slugs of water entering the fractionator.	
1025	456	Refinery vacuum	Upstream piping problems on the stripping steam, improper start-up procedures, and mechanically ineffective tray design combined to destroy stripping trays at start-up. This resulted in bitumen being off specification (poor penetration). Problem diagnosed by temperature and pressure survey.	
1026	306	Refinery crude fractionator	Vacuum distillate rate was insensitive to crude fractionator stripping steam flow rate. A pressure survey confirmed the stripping trays had been damaged, presumably due to water entry at start-up.	As 1003, 1037.
1027	333	Refinery crude fractionator	Diesel product yield was low due to damaged stripping trays. The damage was due to either wet stripping steam or high liquid levels during start-ups, shutdowns, and upset conditions. Diagnosed by a pressure survey. Cured by a heavy-duty tray design retray.	As 1003, 1037.
DT13.1			*Wet stripping steam in side stripper causes damage to preflash vacuum fractionator.*	

(Continued)

Case	References	Plant/Column	Brief Description	Some Morals
1017	201	Refinery vacuum	Stripping trays were damaged by a water-induced pressure surge.	
1031, DT10.3	264	Lube oil feed preparation	Stripping trays were repeatedly dislodged by pressure surges caused by water entry with the wet stripping steam. Solved by eliminating the steam injection and the stripping.	
DT13.2F, G			*Wet stripping steam causes loss of trays, pressure spikes in refinery vacuum and crude towers.*	
		13.6	**Condensed Steam or Refluxed Water Reaching Hot Section**	
1008	3	Refinery combination tower	A pressure surge severely damaged trays at start-up. Steam bleeds were used to keep the upper part of the tower free of air. Some condensed steam drained into hot oil that was introduced near the bottom, resulting in a pressure surge.	Avoid steam for keeping column free of air. Start with warm rather than hot oil.
1013	306	Refinery combination tower	Tower trays were repeatedly upset due to pressure surges resulting from water accumulating in the tower during short unit outages. The source of water was condensation of purge steam used under the column relief valves to prevent their inlets from plugging.	Beware of relief valve steam purges.
DT13.3			*Due to malfunctioning interface controller, causes major damage in olefins oil quench tower.*	
1032	71	Heavy naphtha vacuum distillation	Following replacement of trays by structured packing, tower experienced intermittent flooding and poor efficiency and could not run even at half the previous rates. Cause was violent flashing of free water in the reflux as the two-liquid phase mixture entered the packing well above its boiling point. Because packings have low liquid holdup, the water quickly descends to the hotter sections. Problem eliminated by reinstalling the top two trays.	Problem diagnosed using pilot tests which are described in detail.

1014	439	Organic chemicals vacuum column	Severe tray damage occurred due to a pressure surge in a large-diameter tower separating high-molecular-weight (106+) water-insoluble organics. Cooling water leaking from the condenser found its way down because of liquid maldistribution. The pressure surge occurred when the water reached the reboiler.	
			13.7 Oil Entering Water-Filled Region	
1009	2	Refinery vacuum	A pressure surge severely damaged trays at shutdown. The surge occurred when hot oil leaked into the tower and contacted condensate from steaming out the column feed furnace.	Prevent hot oil leaks into the column while steaming.
1021	16	Refinery FCC main fractionator	During start-up, fractionator was steamed. Condensate collected at the bottom from where it was pumped into a drum. The water is normally drained through a valve at the bottom of the drum. On the accident day, the valve was shut, leaving water in the drum. As the tower was brought on-line, hot oil entered the drum and caused instant vaporization and pressure surge, which ruptured the drum causing an explosion and fire with five fatalities.	Ensure complete water drainage before introducing hot oil.
1030	116, Case MS72	Refinery FCC main fractionator	During a short shutdown of an FCC unit, the hot-feed drum cooled to near ambient temperature. Probably due to steaming out of a fractionator connected to the feed drum, some 5 barrels of steam condensate got in. Apparently, this was not realized as no one drained if off at start-up. There were about 25 barrels of oil below the feed drum suction nozzle, so it took some time for the heat from the 250°C oil charged through the feed drum to reach the water. When it did, the water vaporized instantly and the vessel exploded.	As 1021.

Case	Reference	Plant/Column	Brief Description	Some Morals (Note 1)

* Note 1. Many of the lessons learned from explosions, fires, and chemical releases go well beyond distillation, and address the management and the specific precautions required to handle certain chemicals. Being a distillation book, the morals included in this Data base are limited to those directly pertaining to distillation, and exclude those that focus on the management and handling of the chemicals. Many of the cited references contain detailed discussion of additional lessons learned.

14.1 Explosions Due to Decomposition Reactions

14.1.1 Ethylene Oxide Towers

Case	Reference	Plant/Column	Brief Description	Some Morals (Note 1)
115	279	EO	Polymerization of EO in a distillation column caused overheating, which caused a decomposition reaction. The polymerization may have been catalyzed by iron carried over from an upstream column. Several similar incidents are said to have occurred.	Avoid contamination of EO.
1134	516	EO redistillation still	ORS 1 exploded. Blast and ensuing fire caused one fatality and extensive plant damage. The accident occurred because of a series of coinciding circumstances: reboiler circulation was reduced; dry-out occurred near the top of the reboiler tubes; EO vapor became stagnant locally near the top of the reboiler tubes; a highly exothermic reaction catalyzed by iron oxides generated a localized hot spot; dry, stagnant EO decomposed. Following accident, ORS towers were modified to maintain base levels at or above top reboiler tube sheets at all times with automatic shutdowns just below that; avoid condensate backup in reboiler shells; positively purge inerts from reboiler shells; and minimize heating media temperatures.	Thermosiphon circulation failure may occur without warning. It is promoted by inerts and by low base levels. Paper describes actions to avoid recurrence.
1612	279	EO	Five separate incidents have been described in which external fires caused overheating, which in turn led to decomposition reactions and explosions in EO distillation columns or their auxiliaries. At least one involved a fatality; in some, the column was destroyed.	Pay attention to equipment layout and fireproof insulation.
131	283	EO	Rust that accumulated in a dead-end spot may have catalyzed an explosive decomposition.	Avoid pockets that can collect rust in this service.

125	181, 298, 346	EO	A leak from a manhole flange caused EO accumulation in the insulation. The EO slowly reacted and eventually overheated the column until reaching the EO decomposition temperature. The column exploded violently, with glass damage 7 miles away.	
126	19, 25, 181, 280, 298	EO	A hairline crack in the level indicator allowed EO to accumulate in insulation reacting to form polyethylene glycol. Metal insulation sheathing was later removed for maintenance and fire ignited. Decomposition of EO inside the column was initiated by the external heat. Two explosions occurred.	
127	298	EO	Tower exploded following a decomposition of EO. The center of explosion was 100–200 ft above grade. Substantial damage occurred in neighboring units.	
14.1.2	**Peroxide Towers**			
116	279	Cumene oxidation	An explosion occurred in the base of a distillation column containing 65% CHP in cumene. An interruption in bottom takeoff may have caused overheating or air leakage into the column may have caused an explosion. There were fatalities, and the column blew 600 ft into the air, then fell on other equipment.	Prefer low-inventory techniques to distillation for concentrating CHP.
1316	279	Cumene oxidation	CHP concentration reboiler exploded as a result of a late change in the design: The low-level alarm was set too low.	Watch out for low liquid levels when concentrating unstable substances.
1512	112	Cumene oxidation	A "duplicate" column vaporizing cumene from CHP was installed. In the duplication process, the reboiler was deepened. The setting of the low-level alarm did not take the deepening into account. This resulted in the reboiler exploding.	Be alert to differences when duplicating from an "identical" or "similar" unit.
1513	112	Hydrogen peroxide–water	A low-level signal at the reboiler served as a safety device. The level float, which was located in the reboiler boiling liquid, failed to detect a low-level condition. This caused an explosion.	Locate level devices on a bridle, not in boiling liquid.

(*Continued*)

Chapter 14 Explosions, Fires, and Chemical Releases: Number 10 on the Top 10 Malfunctions (*Continued*)

Case	References	Plant/Column	Brief Description	Some Morals (Note 1, p. 518)
1207	327	Anthraquinone	A feed tank to a still separating hydrogen peroxide from organics was switched. At the switch, the feed filter appeared to block. The liquid level dropped at the column vaporizer, resulting in hydrogen peroxide concentration and an explosion.	Columns should be designed so that feed failure is not hazardous.
138	325	Chemicals	An organic system with an overall hydrogen peroxide concentration of 1% was distilled. A small amount of emulsion containing organics and concentrated peroxide separated at the still base and exploded.	Beware of concentrating peroxide.
14.1.3	**Nitro Compound Towers**			
113	26	Nitrotoluene distillation	An alternative still was used for the first time for this product. Blockages in the condenser led to excessive pressure drop and overheating in the still. This led to a runaway decomposition and an explosion.	Adequately monitor temperature and pressure.
112	26	O-Nitrotoluene recovery vacuum column (batch)	The residues were held at 150°F and air admitted. A previously unknown exotherm set in, causing an explosion.	Recommendations were to modify plant and process.
108	111	Nitro chemicals	Nitrocellulose precipitated out of the solvent in a solvent recovery steam distillation still and exploded.	
1606	111	Chloro–nitro distillation	The vacuum system failed. This permitted a temperature rise to the self-accelerating decomposition level. An explosion resulted.	
12111	282	DFNB, DMAC batch recovery tower	Water was distilled at the cold start of the batch cycle. Then, DMAC, a reaction solvent, came out and was recycled to the reactor. On this occasion, a water leak in storage induced much more water into the column feed drum, forming an unexpected second, light liquid phase in the drum. This water entered the column when hot, initiating rapid hydrolysis of DMAC into acetic acid, which distilled off and was recycled to the reactor with the DMAC. The acetic acid set an explosive decomposition reaction that blew the reactor and wrecked the plant, causing injuries and a fatality.	

11103	504	Aniline/ nitrotoluenes MNT tower 7 ft ID 145 ft tall	During an outage, 1,200 gallons of MNT were left at the tower base and vacuum was broken. Isolation valves on reboiler steam were shut but not blinded. They leaked, heating tower base (normally at 350°F) to over 400°F. A runaway decomposition set in, rupturing the tower, hurling large pieces more than 1,500 feet, injuring three, setting fires, and narrowly missing oil storage. Instruments showed steam flow and high base temperatures for almost a week before the explosion, but there were no alarms and the instruments were not actively monitored. Possible contributors include reactive monomer near the top, presence of residues on the structured packing, air introduction, and breaking tower vacuum at the outage.	Many morals in report. Beware of nitrotoluene decomposition and protect tower. Monitor and alarm conditions leading to decomposition. Properly isolate during outages.
133	18, 541	Mononitrotoluene	Organic sludge built over 30 years to the depth of 13–14 in. in the base of the tower was softened by steaming for removal. The sludge overheated to 165°C, initiating runaway reaction and explosion. There were five fatalities. The still base thermometer did not contact the sludge and read air temperature, misleading operators.	Beware of hazards of nonroutine operations. Other morals are in the paper.
14.1.4	**Other Unstable-Chemical Towers**			
103	231	Butadiene refining column	A detonation demolished the column and caused widespread damage. Cause was vinylacetylene doubling in concentration and detonating in the absence of air. The rise in concentration was due to leakage of butadiene (light key) out of the tower during total reflux operation. A thorough analysis of the accident is presented. Recommendations include avoiding total reflux operation in such services; keeping vinylacetylene concentrations low and continuously monitored in this type of column; and others.	
143	183	Pilot plant DMSO	An old 88% DMSO/7% water/5% BMD mixture was batch distilled in a jacketed reactor. The desired vacuum of 40 mm Hg was not achieved and kept deteriorating due to off gas from rapid degradation of DMSO catalyzed by hydrogen bromide, which was apparently formed by long-term BMD oxidation in storage. Stopping the heating and blowing the rupture disk did not contain the pressure rise, and the reactor separated, releasing a vapor cloud that exploded.	

(*Continued*)

Chapter 14 Explosions, Fires, and Chemical Releases: Number 10 on the Top 10 Malfunctions (*Continued*)

Case	References	Plant/Column	Brief Description	Some Morals (Note 1, p. 518)
144	183	Indole derivative DMSO recovery	DMSO was fractionated under vacuum using steam. Outage caused shutting off the steam and cooling water. Six hours later, a blowout occurred due to autocatalytic decomposition of impure DMSO.	
145	183	Indole derivative DMSO recovery	The last 100 gal of a 450-gal batch was left in still recovering DMSO from waste. At 130°C and 60–80 mm Hg, the pressure increased and blew out the bottom manhole flange.	
107	111	Propargyl bromide	An explosion occurred when this material was distilled under pressure. Tests showed that, although stable at atmospheric distillation, it detonates when distilled under pressure.	Pressure (or temperature) affects reactions in stills.
1511	112	Recovery of epichlorohydrin from tars	The thermowell used for controlling heat input into the column was located at the reboiler outlet. It fouled up. The operators tried to control heat input by watching the column top temperature. This was unsuccessful: The reboiler overheated, resulting in an explosion.	In heat-sensitive services, provide alternative temperature indications.
1208	111	Insecticide	In tower recovering hexane from residue, excessive concentration of residue caused an explosion.	
135	23	Pharmaceuticals batch distillation	Changing plant operation from a single- to a multiple-distillation process rendered the residue more thermally unstable. This led to an unstable mixture in a reactor, which exothermically decomposed, fired, and exploded, injuring one person.	
1642	25	Aniline removal from by-products, batch still	Still was shut down to repair a leaking reboiler valve. Reboiler and feed were isolated, but condenser and reflux pump continued to operate. 8–9 hours later, high level was seen in the reflux drum, and still base temperature was high. Operators attempted to reduce level and temperature, but pressure increased, relief valve lifted, and vent on top of accumulator vented liquid. Shortly after, the still exploded.	

165	504	meta-chloroaniline	Batch tower was destroyed by a runaway reaction and overpressurization with debris propelled off-side. Tower had no provision to mitigate a thermal runaway.	
166	505	Hydroxylamine (HA)	A first of a kind batch distillation concentrated 30% wt aqueous solution of HA and potassium sulfate to 50% HA distillate. Water removal in the fore runs concentrated HA in the charge drum liquid to 86%, well above the MSDS-referenced 70% above which HA may decompose explosively. After shutdown, the HA in the charge drum and piping explosively decomposed due to high concentration and temperature destroying the facility, killing 4, injuring 14, and damaging surrounding businesses.	Process design and operation procedures must keep below decomposition concentrations. Hazard analysis is a must.
167	478, 505	Hydroxylamine (HA)	HA distillation was being restarted after 5-hour outage. HA concentration reached 80–85% wt, well above the HA decomposition concentration of 70%. An explosion destroyed the HA distillation tower, killing 4 and injuring 58.	
168	210	Methyl isothiocyanate product recovery from residue	Residue was transferred from storage to batch still. Transfer line was then flushed with methyl, isothiocyanate, and isolated at two block valves. During the weekend shutdown, the isolated pipe section remained steam traced. The steam pressure regulator failed, heating the pipe to about 140°C. The pipe ruptured, causing little damage. Normally, the residue was thermally stable to 230°C, but a sample taken decomposed exothermically at 140°C due to a small (3%) presence of water. The effect of water on residue stability was not previously known.	Contaminants can promote hazardous reactions. Several invaluable generic lessons are presented in paper.
			14.2 Explosions Due to Violent Reactions	
106	327, Vol. 1, Case 363	Vitamin A intermediate 1′ Pentol still (high vacuum)	A small amount of caustic left over in the piping from a shutdown cleaning operation was drawn into the still. The mixture and still exploded, causing fatalities, injuries, and extensive damage.	Avoid washes with chemicals whose entry to the column during operation is hazardous.

(Continued)

Chapter 14 Explosions, Fires, and Chemical Releases: Number 10 on the Top 10 Malfunctions (*Continued*)

Case	References	Plant/Column	Brief Description	Some Morals (Note 1, p. 518)
1307	111	Air separation column reboiler	Poor venting of noncondensable from the condensing side of the reboiler interrupted thermosiphon action. This in turn reduced the effectiveness of removal of hydrocarbon impurities from the reboiler liquid. Hydrocarbon accumulation caused an explosion.	Ensure adequate venting. Avoid buildup of hazardous impurities in reboiler.
1545	6	Refinery HF alkylation depropanizer	Both the level indicator and level controller failed on the overhead receiver, which separated liquid HF from liquid HCs. HF overflowed into the HC product route, which included a bed of solid KOH. Violent reaction between KOH and HF overpressured the vessel, causing multiple explosions and rupture of the vessel.	Ensure adequate level indication.
			14.3 Explosions and Fires Due to Line Fracture	
14.3.1	**C_3–C_4 Hydrocarbons**			
1617	298	Gas concentration C_3 splitter	Fracture of overhead line from column led to formation of cloud , which ignited, causing heavy blast damage.	
1618	298	Refinery FCC absorber–stripper	A 12-in. line from stripper to absorber developed a C_3–C_4 leak, which ignited and exploded. Some windows broken up to about 1 mile away.	
1619	298	Refinery reforming DC_3	The 6-in. overhead line failed, releasing C_3 hydrocarbons. The vapor ignited and an explosion occurred.	
1620	25, 298	Refinery FCC DC_3	An 8-in. elbow in the overhead line ruptured due to internal corrosion. A severe blast followed, destroying the control room and toppling the 26-ft-ID main fractionator.	
1637	17	Refinery naphtha DC_3	An elbow at the base of the vertical section of the feed line failed by external corrosion, spilling HCs that ignited.	
1639	25	Refinery H_2SO_4 alkylation DC_3	A 12-in. elbow in a line from the reflux accumulator failed, discharging 4000–5000 gal that formed a large vapor cloud and ignited, severely damaging the alkylation unit, FCC unit, and control building.	

1643	25	Olefins deethanizer	During an acetylene converter upset, a fire was noticed at the base of the deethanizer, believed to have been initiated by high-pressure propylene leaking from a reboiler flange. The fire spread to other distillation columns and to the storage area, causing explosions and major destruction.	
1625	402, 520	Refinery FCC debutanizer	Following an electrical storm, debutanizer distillate valve was shut but indicated open. To relieve pressure buildup, operator opened distillate route to compressor interstage drum and from there drained to flare. Deficiencies in flare system did not permit adequate liquid removal, overfilling flare drum and forcing liquid into a corroded discharge pipe that broke, releasing a vapor cloud that exploded, injuring 26.	Beware of false instrument signals.
1640	25, 298	Cat polymerization stabilizer	A pipe from the stabilizer reboiler failed 2 weeks after initial start-up of this new unit, releasing 2.4–3-m-deep and 150-m-long propane cloud. Blast destroyed the cat polymerization unit and heavily damaged others.	
1646	530	Refinery DC_4	Failure of a tee-piece connection at the tower base at start-up caused a HC leak that ignited.	
14.3.2	**Overchilling**			
1611	26	Olefins	Failure of base-level controller caused cold liquid to pass out of relief valve and into the CS flare header. This overchilled and cracked the header. A vapor cloud formed and ignited, causing fatalities.	Adequate liquid-level monitoring is imperative.
1168			*Similar to* 1611, *Section* 8.1.1.	
1616	298	Olefins demethanizer	Failure of blowdown line from top of column released methane at -263°F. A vapor cloud exploded, destroying olefins plant and damaging buildings within 1000 ft.	
1654	347	Olefins Demethanizer	A study of eight >20 years old existing bimetallic demethanizers (3.5 Ni alloy on top, killed carbon steel at bottom) showed that in some cases, loss of reboiler with continuing reflux and feeds may lead to unacceptable risk of overchilling in the bottom section. One demethanizer was replaced by 304 SS tower. Other measures to prevent, mitigate, or reduce the risk included relocating the bottom (warmer) feed to an upper feed point, above the bimetallic transition, installing low temperature cutouts on reflux or upper feed, low temperature override controls, cold safety training, and improved procedures.	A thorough report on the risk assessment and findings is included.

(Continued)

Chapter 14 Explosions, Fires, and Chemical Releases: Number 10 on the Top 10 Malfunctions (*Continued*)

Case	References	Plant/Column	Brief Description	Some Morals (Note 1, p. 518)
1628	11, 241	Natural gas deethanizer	Two parallel oil absorbers removed C_2+ hydrocarbons from natural gas. The rich oil was flashed, then regenerated in a deethanizer, then in a hot fractionator. Lean oil fractionator bottom was cooled by preheating fractionator and deethanizer feeds and reboiling the deethanizer before returning to the absorbers. The absorbers bottom compartments were used for knocking out and removing liquid condensate. Rich oil was drawn from a chimney tray above.	Beware of brittle failure. Hazop impact of loss of circulation on heat-integrated system. Avoid high liquid levels.
			At the time, the condensate level in one absorber exceeded the upper level tap and probably the gas inlet and was probably entrained into the rich-oil draw. This raised the level in the rich-oil flash drum. The high-level cut lean-oil circulation and a low-flow switch tripped the lean-oil pump. The deethanizer reboiler lost heating and chilled from 85 to -48°C. When icing and a leak were noticed in the fractionator preheater, lean-oil circulation was restarted to counter the leak. Upon reheat, the thermal stress ruptured the reboiler by brittle failure, causing an explosion that killed two, injured eight, and interrupted gas supplies for 9–19 days.	Provide emergency isolation from suppliers of flammable liquids. The paper contains many more lessons.
14.3.3	**Water Freeze**			
1247	298	Olefins	Water accumulation froze inside a dead leg in a reflux system. Ice cracked steel pipe, releasing a vapor cloud that exploded. Blast caused severe damage inside plant and surrounding communities.	
12119	67	Refinery vacuum	Freeze up and subsequent failure of a 2-inch pipe released a high-pressure spray of naphtha that ignited and exploded. The failed line was out of service for 20 years, but had not been fully decommissioned, forming a dead leg that accumulated water.	

14.3.4	**Other**			
1645	25	Natural gas absorber 10 ft ID	Failure of a threaded $1\frac{1}{2}$ in. drain connection on a rich-oil line at the base of the absorber released rich oil and gas at 850 psi and $-40°F$. The vapor cloud ignited, eventually causing the 75-ft-tall tower to collapse on other equipment.	
1638	25	Refinery FCC main fractionator	A 12-in. recycle slurry line in a pipe rack ruptured. The 600–700°F slurry ignited. The FCC reactor, regenerator, fractionator, and piping sustained severe damage.	
1621	28, 298	VCM quench tower	Erosion due to carbon entrainment caused the rupture of the inner side of a 4-in. elbow on the quench tower feed. A vapor cloud released and ignited. There was limited blast damage but major damage to instrumentation and electrical cabling.	
1648	30	Synthetic fuels quench tower	A CS oil return line to the tower failed due to corrosion/erosion. Iron particles in the oil contributed. Released hydrocarbons were ignited.	
1636	511	Ammonia MDEA absorber/ regenerator	A false absorber bottom low-level signal tripped the rich-solution control valves at the inlets to the two parallel regenerators. One could reopen, but not the other. Flashing in both pipes set strong vibrations that dislodged a section of line just upstream of the closed control valve, spilling MDEA. The absorber bottom could not be blocked in, so when MDEA ran out, hydrogen came out of the broken pipe and fired.	Ensure adequate support, mechanical strength, and resistance to vibrations.
1649			*Joint fails due to inadequate relief, Section* 21.5.	
		14.4	**Explosions Due to Trapped Hydrocarbon or Chemical Release**	
1110	327, Vol. 2, Case 838	C_4 hydrocarbons	A reboiler outlet isolation valve was blinded on the reboiler side at shutdown. Liquid remained trapped in the valve bonnet. After the reboiler was cleaned, it was water washed, then drained. To facilitate draining, the blind was removed and the reboiler valve opened. The trapped HCs were released and exploded in the column, killing a worker and lifting trays.	Always blind on the column side before entry. Watch out for gas release from rubbery deposits. Open valves during purging and flushing to remove trapped materials.

(*Continued*)

Chapter 14 Explosions, Fires, and Chemical Releases: Number 10 on the Top 10 Malfunctions (*Continued*)

Case	References	Plant/Column	Brief Description	Some Morals (Note 1, p. 518)
1125	6	Refinery fractionator	An explosion and fire were caused by a butane release during a start-up following an interruption. After bolts in the reboiler pump suction screen housing were loosened, heavy deposits trapped in the plugged suction isolation valve broke loose, releasing 150 psi butane into the atmosphere.	Ensure proper blinding. Watch out when cracking flanges in plugging services.
1034	20	Refinery hydrocracker debutanizer	The reflux pump stopped pumping due to hydrate formation. It was isolated but not blinded. The pump and suction line were full of ice. The pump was removed and a steam hose was pushed into the suction line to melt the ice. This released a vapor cloud that exploded and fired. It was later established that the suction valve was one turn open but appeared closed due to a hydrate in its seat.	Beware of hydrates. Follow good blinding procedures. Compare 1033, Section 14.11.
1651	151	Organic acids	Tower bottom containing organic acids and viscous sludge was pumped by a high-head pump. Casing of dead-headed spare pump exploded, with parts up to 35 ft away, some time after pump was remotely started with suction and discharge isolation valves shut. Pump had no running light in the control room.	HAZOP remote start capabilities.
133			*Heating trapped sludge in nitrotoluene tower, Section* 14.1.3.	
			14.5 Explosions Induced by Commissioning Operations	
1641	25	Olefins depropanizer	During a restart after an outage, the reflux drum was vented. The vent was inadvertently routed to the cracked gas compressor via a line that was open for maintenance. A vapor cloud formed and exploded.	
1117			*Purging HC towers with oxygen, Section* 12.4.	
1122			*Inert gas purge unable to prevent explosive mixture, Section* 12.4.	
1113			*Water wash of absorber–regenerator system formed explosive mixture, Section* 12.6.	
1116, 11101			*Blind removed from gas-containing line, Section* 12.1.	

14.6 Packing Fires

14.6.1 Initiated by Hot Work Above Steel Packing

11107	139	Ethylbenzene recycle tower	Tower was open for 3 days when carbon steel structured packing caught fire. Initial fire was extinguished with water, but recurred and moved from bed to bed. Water and N_2 did not put fire out. Hours later, column fell over. Hot slag or molten metal could have ignited combustibles on the packing below. It was not known if combustibles were in the bed.	Believed to be a metal fire with CS structured packing as fuel. Avoid hot work above packings.
1152	62	Refinery vacuum	Hot work dropped embers into a packed bed that had sufficient coke to ignite, causing a fire.	Use fire blankets.
1153	62	Refinery vacuum	Plugged wash distributor was being removed by torch-cutting bolts off the flanges. There was tar in the header, which melted, dripped into the packing, and ignited from the torch.	Use fire blankets. Assume coke/tar is present.
1184	139, 392	Refinery vacuum	LVGO SS structured packings were steam cleaned outside tower, then reinstalled. Overhead nozzle work after packing reinstalled produced sparks that ignited residual HCs in bed despite fire blanket protection.	Steam cleaning, fire blankets do not provide enough protection.
11110	125	Refinery coker main fractionator	Following steaming and chemical cleaning, cutting by torch was performed above the naphtha fractionation bed. Hot slag ignited the structured packing, destroying the packing and damaging the shell. Quenching the fire by water-filling the tower took several hours and delayed extinguishing it.	Avoid hot work above packing. Have plans to quickly put out fire if it breaks.
1187	139, 392		A clip was welded to tower wall above packing. Despite fire blanket protection, hot slag ignited a small fire within the bed.	See 1184.
1188	139, 392		At the turnaround, a 2-in. thermocouple nozzle was erroneously cut into a hydrocarbon-laden bed of structured packing, igniting a large fire.	Follow proper hot-work procedures.
970	124		New packing coated with a thin film of lubricating oil from the manufacturing process caught fire after hot work was performed above the packing.	

(Continued)

Chapter 14 Explosions, Fires, and Chemical Releases: Number 10 on the Top 10 Malfunctions (*Continued*)

Case	References	Plant/Column	Brief Description	Some Morals (Note 1, p. 518)
14.6.2	**Pyrophoric Deposits Played a Major Role, Steel Packing**			
509	376	Refinery fractionators	Autoignition occurred in packed beds during turnarounds in seven different fractionators, even though towers appeared to have been properly steamed out. In all cases, packings were destroyed. In one case, the column shell was damaged. Pyrophoric iron sulfides, possibly accelerated by residual HCs, are believed to be the cause. Remedial actions taken include installing wash water capability above packing and monitoring bed dP for coking.	
11111	468	Refinery FCC main fractionator 16 ft ID	Turnaround steaming was followed by chemical wash that circulated neutralizers and oxidizers at rates exceeding the design liquid rates through each of the structured packed beds. 12–18 hours after manholes were opened, the upper two packed beds caught fire. Column temperatures reached 1300°F within minutes, damaging the vessel and the upper two packed beds. The fire was caused by pyrophoric deposits of iron sulfide, and was water-quenched from the top two manholes. It was determined that the chemical wash failed to remove all pyrophoric material from the packing.	Washing/steaming does not guarantee deposit removal. Packing removal or continuous wetting and good monitoring are necessary. More lessons in paper.
1154	62, 139	Refinery crude fractionator	Light gas oil PA bed caught fire at the interface between two grades of structured packings. At that location, unusual scale deposits can form due to distribution disturbances. The fire was rapidly put out, but the tower bulged at the fire points. The bed was not water washed before the shutdown.	Shutdown wash, keeping packing wet, monitoring bed temperatures for hot spots, and CO monitoring at manways help avoid fires.
DT14.1, DT14.2			*Recurrence of structured packing fire in crude towers prevented by improved metallurgy, washing, shutdown cooling, and spraying water when open.*	

1178	124, 125	Refinery vacuum	Tower was retrofitted with 410S SS structured packing. Design was 1.7% sulfur crude, but later 3.0% sulfur crude was processed. At the next turnaround, the tower was steamed and cooled for 2 days. Upon opening manholes, the fractionation bed below the top PA caught fire, its packing destroyed. Quick manhole closure prevented injury and shell damage.	Steaming and water wash are insufficient for removing pyrophoric sulfur from packing.
1179	124, 125	Refinery vacuum	Tower processing high-sulfur crudes was retrofitted with 316L SS structured packing. Pyrophoric sulfur compounds from corrosion of PA exchangers entered in the reflux and PA return, coated the packing, and ignited when manholes were opened in the next turnaround.	Same as 1178. Also, beware of external formation of pyrophoric sulfur.
1180	124	Refinery	At turnaround, a packed bed ignited after vapor-phase treatment with potassium permanganate. The iron sulfide was not completely oxidized and ignited during tower ventilation. Proper monitoring and quick manhole closure prevented shell and minimized internal damage.	Vapor-phase treatment has not always been fully effective.
1155	62	Refinery	A shutdown packing fire took place after the packing was washed by an oxidizing amine to oxidize pyrophoric sulfide deposits. The quality and condition of the wash were not known.	Even the best procedure may not be perfect.
1189	423	Refinery vacuum	At turnaround, tower was deoiled and steamed, and a water wash connection was made at the LVGO PA. A flange was opened in the tower overhead line for removing a spool piece. Air ingress autoignited pyrophoric iron sulfide inside the tower, causing an explosion that damaged internals.	Properly blind and wash before removing spool pieces.
1141	250		A bed of CS structured packings burnt after air was allowed into a column shortly after steaming was completed. Pyrophoric deposits are common in this service.	

(Continued)

Chapter 14 Explosions, Fires, and Chemical Releases: Number 10 on the Top 10 Malfunctions (*Continued*)

Case	References	Plant/Column	Brief Description	Some Morals (Note 1, p. 518)
14.6.3	**Tower Manholes Opened While Packing Hot, Steel Packing**			
1198	126	Glycol recovery from TEA (triethanolamine)	Spontaneous ignition of organics on structured packings at 120°C occurred while the tower was open for maintenance. Partial plugging in packing and distributor retarded cooling and retained organics. Distributor was plugged directly above burned area. White smoke was seen, 304 SS packing melted, and a portion of the shell glowed red and buckled. Fire extinguished by nitrogen purge and water-filling.	Consider water-filling. Reconsider structured packings in plugging service. Monitor, provide early warning.
11100	126		At shutdown, nitrogen was disconnected with packings at 140°C–150°C. All but two bolts were removed from the upper and lower manholes and a flange was opened to drain. Half an hour later, sludge on the packing ignited.	Ensure adequate cooling.
14.6.4	**Other, Steel Packings Fires**			
1199	126	Paint	After cleaning and opening, it was observed that CS structured packing temperature rose 25°C and oxygen concentration decreased, indicating rapid oxidation.	
11108	139	Styrene Benzene-toluene 14-foot ID 105 feet tall	Top bed had 316SS wire mesh structured packings. Tower was open for a week before commencing hot work. A fire occurred 2 hours after work crew left, burning a big hole midway through the top bed deforming the shell. Traces of polystyrene on packing surfaces fed fire. Firewater applied could have released H_2, causing explosion. Many precautions such as solvent washing, covering top bed with plywood covered with fire-resistant tarps and a wet, fire-resistant blanket above were in effect, but failed to prevent fire.	Structured packing may ignite despite regorous protection. Avoid hot work above packings and plywood isolation. Monitor temperature. Ensure adequate firewater pressure at top manhole.
11104			*Packing fire due to air entry during N_2 page, Section* 12.4.	

14.6.5	**Titanium, Zirconium Packing Fires**			
11105	139, 336	Chemicals Multiple bed Ti structured packing	Sparks, formed when a battery-operated grinder touched stainless steel or titanium, are believed to have initiated a thermite reaction (an exothermic reaction between a metal oxide and another metal that has a greater affinity for oxygen) between the thin Ti packing sheets and a dried iron oxide layer accumulated on the packing surfaces. Two beds ignited and were consumed, temperatures exceeded 600°C, with molten Ti melting through the tower bottom head. Fire put out by water addition from top. Explosion noises were heard at the initiation of water addition.	Beware of the thermite reaction and of high fire potential of Ti packing coated with combustibles and/or iron oxide. Water addition may help or hurt.
11106	24, 139, 337	Chemicals 42-inch ID 73 feet tall	Top and middle beds were Ti Pall rings. 316SS tower was shut down after an upset caused by a plug in the tower. Ti packing in top bed was found corroded to 70% TiH_2, and broken into small pieces that migrated through the packing support. Upon manhole opening, a flash fire occurred, followed by Ti fires that burnt through the shell in two places. Recurrence prevented by replacing Ti with Inconel packing.	See Case 11105. Metal oxides and small pieces contribute to Ti packing fires. Water and CO_2 may not extinguish Ti fire.
1186	139, 392		1-in. titanium random packing ignited at the turnaround, most likely due to pyrophoric residue on the packing. Switched to thicker wall Inconel 625 to prevent recurrence.	
1185	392		During manufacturing, a single element of zirconium (Zr) structured packing ignited. When burning element pushed out, intense heat deformed cement driveway.	Handling reactive metals requires special precautions.
			14.7 Fires Due to Opening Tower before Cooling or Combustible Removal (*See also Tower Manholes Opened While Packing Hot, Section* 14.6.3)	
1123	223	Fatty acid vacuum column	Immediately after shutdown, the bottom manhole was opened while column contained 400°F liquid. Air was sucked in due to condensing vapor. A violent combustion resulted, causing injury and widespread tray damage.	Drain liquid and adequately cool column before introducing air.

(*Continued*)

Chapter 14 Explosions, Fires, and Chemical Releases: Number 10 on the Top 10 Malfunctions (*Continued*)

Case	References	Plant/Column	Brief Description	Some Morals (Note 1, p. 518)
1151	62	Refinery vacuum	Overhead system opened for maintenance with no proper blinding before tower was washed and cooled. Air entered through open overhead line, igniting combustibles inside the tower.	
			14.8 Fires Caused by Backflow	
1197	219, 506	Refinery crude fractionator	Following discovery of leaks and corrosion-thinning in the 6 in. naphtha draw line to the stripper (Case 1041, Section 17.1), 100-ft of the line were to be isolated and replaced. A leaking corroded shut-off valve on the control valve bypass repeatedly prevented isolation and drainage for 2 weeks. On the accident day, the pipe was cut 8-feet below the top, and the section between the tower isolation valve and the cut removed. A second cut, 26-ft. below the first, leaked naphtha, so a flange 38-ft below the second cut was opened to drain. When enough naphtha was drained, the pressure from the stripper exceeded the naphtha head in the line, and pushed back the naphtha trapped in the line, suddenly releasing it out of the open pipe. The release ignited, probably by a hot fractionator surface, killing four.	Lines should not be cut unless fully isolated, blinded, drained, depressured and flushed. Non-routine operations should be closely audited, hazard-evaluated and closely supervised.
1162	116, Case MS47	Refinery debutanizer	At shutdown, tower was steamed out and drained, but the valves isolating the debutanizer from the butane storage were not closed, which allowed a reverse flow into the column. In addition, the debutanizer vent to flare was only partly open. As a result, the column was under some pressure. When an operator opened the drain valve, it was partially plugged so he opened it further. A gush of condensate and butane was released and the operator had to escape the butane cloud, so could not reclose the valve. The cloud was ignited. To prevent recurrence, procedures were rewritten, operators were retrained, and critical drains were fitted with two valves, one of which was well away from the final drainpipe outlet.	

1173	395	Ammonia MDEA absorber–regenerator	Regenerator was being shut down for maintenance, while absorber was kept at process gas pressure (400 psig). MDEA in the regenerator was pumped out to the absorber and flash drum. Once emptied, pump lost suction and tripped. When the deinventory line from the pump discharge was opened to MDEA storage, reverse flow of process gas from the absorber to storage set in and blew the liquid out of the line. The MDEA storage was not N_2 purged, so an explosive mixture formed and exploded, lifting the storage tank off its base. There were injuries and more damage.	
1218	147	Ammonia MDEA absorber	The main pump delivering semilean amine from the desorber to the absorber was shut down for maintenance and the spare pump was started. The spare pump delivered one-third of the normal flow, causing poor absorption. This could not be tolerated downstream, so the plant was shut down. Shortly after the spare pump was switched off, hydrogen escaped through the seals of the main pump and ignited. The incident was caused by backflow through the main pump, induced by failure of the main-pump nonreturn valve.	Beware of reverse flow. Nonreturn valves cannot be relied upon. The article contains a comprehensive description and measures to avoid recurrence.
1145	250	Olefins caustic wash	At start-up, caustic backflowed into the depressured cold box and attacked an aluminum heat exchanger, resulting in a fire. Piping was later changed to prevent recurrence.	
			14.9 Fires by Other Causes	
1644	25	Phenol stripper	The benzene recycle pump plugged. Pressure built up in the stripper. The relief valve lifted, discharging benzene vapor that ignited and exploded.	
1190	423	Refinery naphtha stabilizer	After steaming, reboiler floating head cover was opened so tubes can be cleaned. Two days later, fire and smoke rose from the head cover. Cause was pyrophoric iron sulfide ignition of residual HCs.	Properly wash, clean to remove deposits, residues.
1520			*Liquid carryover from tower overhead, Section* 8.1.1.	
1290			*Liquid pouring out of relief valve, Section* 4.11.	

(*Continued*)

Chapter 14 Explosions, Fires, and Chemical Releases: Number 10 on the Top 10 Malfunctions (*Continued*)

Case	References	Plant/Column	Brief Description	Some Morals (Note 1, p. 518)
1286, 1349			*Tube failure in fired reboiler, Section* 20.2.1.	
11102			*Pipework not adequate for start-up thermal expansion, Section* 12.9.	
			14.10 Chemical Releases by Backflow	
1108	284	Chemicals	Ammonia flowed from storage backward through a leaking valve into the reflux drum of a column that was shut down. From there, it flowed into the column and out of an open end in the bottom line.	Properly blind to avoid reverse flow.
1109	284	Chemicals	Toxic gas leaked from a blowdown header back into a shutdown column through a closed valve and killed an operator who was draining the column.	Same as 1108.
1127	27	Ammonia recovery absorber	Gas overpressure lifted three storage tanks off their plinths. One split at the base, releasing ammonia liquor. The tanks received liquid feed from the bottom of a high-pressure absorber. The rundown line branched off a PA circuit at an elevated position. The pump failed during a start-up, and gas from the column backflowed through the upper leg of the PA into the tank's rundown line.	Beware of reverse flow. Rundown lines should branch off near the bottom of a PA circuit.
1114	304	Refinery	Propane leaked into the steam system via a steam purge connection. This resulted in propane issuing from a fire-suppressant steam purge nozzle.	Ensure proper blinding.
1119	6	Refinery crude and vacuum	Caustic backflowed into column steam lines. This resulted in caustic being sprayed over a wide area via an atmospheric steam vent.	Pay attention to blinding before a chemical wash.
			14.11 Trapped Chemicals Released	
1111	327, Vol. 2, Case 924	Chemicals stripper column	An infrequently used intermediate draw-off line was plugged just above an isolation valve, which was at grade. When the isolation valve was being removed for maintenance, the plug suddenly cleared, spraying the worker with water and sludge.	Isolate infrequently used draw-off lines at the column to avoid dead legs.

1627	29	Refinery depropanizer	While attempting to clear a blocked reboiler drain line, a blocked 2-in. gate valve was cleared, causing high-pressure C_4 hydrocarbons to escape to atmosphere. The valve could not be closed. To reduce the escape, the tower was depressured to the flare via a line that was blocked by debris, collecting a large volume of liquid. When hit by the depressuring gas, a slug formed and dislodged the flare line. It fell 10 m to the ground and buckled on impact. Pure luck prevented a potentially disastrous line rupture.	Beware when clearing blocked lines.
1033	20	Refinery hydrocracker debutanizer	The casing drain of the reflux pump suddenly released a large volume of LPG. The valve was not completely closed due to hydrate in its seat. The hydrate melted after the pump had been on-line for several hours.	Beware of hydrates. Compare 1034, Section 14.4.

14.12 Relief, Venting, Draining, Blowdown to Atmosphere

(*See also Chemical Leaks to Atmosphere, Section* 20.4.1)

1605	149		A runaway reaction occurred in column and could only be stopped by water flooding. The flooding blew the relief valve, which was not yet connected to the quench/flare system, causing an atmospheric discharge of noxious fumes.	Relief devices must be properly connected to vent header before start-up.
1168			*High bottom levels induce flooding and HC release, Section* 8.1.1.	
1653	67	Butadiene Final purification column	A butadiene vapor cloud was released from a 1 m split rupture on the reboiler return pipe. The split was caused by the tremendous forces during the formation of popcorn polymer. Fortunately, the cloud did not ignite. Liquid pooling in an undrained dead-leg line to the safety valve was a contributor.	

(*Continued*)

Chapter 14 Explosions, Fires, and Chemical Releases: Number 10 on the Top 10 Malfunctions (*Continued*)

Case	References	Plant/Column	Brief Description	Some Morals (Note, p. 518)
1150	431	Butadiene heavy-end recovery	A total power failure caused C_4-containing acetonitrile to reach the heavy-ends accumulator. Thinking it was water, operators drained the high accumulator level to sewer. When joined with other aqueous streams, the solubility of C_4 diminished, and it vaporized, causing a 10-ft geyser from on oily water sewer manhole, an odor, and a benzene release.	Analyze "water" before sewering it.
1650	109	Butylene wash tower	Oversized tower level control valve failed wide open due to a freezing problem, discharging much liquid butylene into the water disengaging drum. The vent to blowdown was undersized, so butylene came out of the drum water seal leg. Quick action prevented explosion.	Adequately size bottom valves and vent lines.
DT21.1			*Atmospheric discharge and rain of HCs caused by condenser hardware changes.*	
1626	14	Pharmaceuticals 4-amino antipyrine sulfonic acid	A vessel boiled over during distillation of an intermediate, releasing chemicals into the biological wastewater treatment plant. The wastewater plant was not informed, and the chemicals discharged to the river.	
1514	327	Vacuum batch distillation of a toluene cut	Foaming occurred at the kettle. The level indicator therefore failed to detect the low-level condition. Because of the low level, the still temperature indicator showed a vapor temperature, which was lower than the liquid temperature. The low temperature increased the heat input. An exothermic reaction took place, causing eruption of residue.	Watch out for fooling of level instruments by foam and for temperature indicators below the bottom tray.
1635			*Incorrectly set relief valve and high-pressure trip, Section* 21.5.	
1115			*High-point vent left open, Section* 12.4.	
1118			*Hydrogen sulfide liberated during acid wash, Section* 12.6.	
1505			*Control system inducing cooling water to boil, Section* 28.3.1.	

Case	Reference	Plant/Column	Brief Description	Some Morals
			15.1 Excessive Bottom Temperature/Pressure	
			(For reactions leading to explosions, see Sections 14.1.2–14.1.4)	
128, DT15.1	250	Glycols product column	A reaction near the bottom produced a light that contaminated the top product. To mitigate, bottom temperature was lowered 30–40°F, but this lowered top-product recovery. Running the cold bottom into a new column, whose overheads were recycled upstream, recovered both the light and the lost product.	Lowering bottom temperature can stop undesirable reactions.
141, DT2.7			*Decomposition reaction near bottom contaminates top product, Section* 1.2.	
DT2.9			*Decomposition reaction near bottom yields corrosive compound that accumulates in tower.*	
1193			*Reaction near bottom yields corrosive chemical, causes leaks, Section 20.4.1*	
109			*Decomposition reaction at the reboiler possibly contaminates bottom product, Section* 1.1.2.	
1514			*Leading to eruption of residue, Section* 14.12.	
			15.2 Hot Spots	
			(For decomposition reactions in ethylene oxide towers, see Section 14.1.1)	
DT15.1			*Eliminating reboiler hot spots eliminates undesirable reaction.*	
133			*Leading to explosion in nitro compound towers, Section* 14.1.3.	
			15.3 Concentration or Entry of Reactive Chemical	
			(For reactions leading to explosions, see Sections 14.1.2–14.1.4)	
1307			*Hydrocarbon impurities in air separation towers, Section* 14.2.	
1241	486	Chemicals	Column purged a tar stream at extremely low flow rates through a $^1/_4$ in. aperture. The flow was made higher to prevent line plugging. This caused product loss. The large purge also helped prevent accumulation of an unstable reaction by-product. Purge rate was reduced by simultaneous improvement of flow control and modifying reactor to produce less undesirable by-product.	

(Continued)

Chapter 15 Undesired Reactions in Towers (*Continued*)

Case	Reference	Plant/Column	Brief Description	Some Morals
142			*Due to refluxing in stripper, Section* 2.1.	
161	46	Olefins caustic scrubber	Vinyl acetate in vents from polyethylene plant led to polymerization problems in caustic towers from two plants.	
			15.4 Chemicals from Commissioning	
106			*Caustic wash leftovers causing violent reaction, Section* 14.2.	
1118			*Acid wash liberating toxic gas, Section* 12.6.	
1145			*Caustic backflow attacking aluminum heat exchanger, Section* 14.8.	
1605			*Runaway reaction leading to atmospheric discharge of chemicals, Section* 14.12.	
114	13	Offshore gas amine	H_2S in the absorber lean gas was well above specification. Cause was aldehyde inadvertently backing up from a storage tank into the amine charge. The aldehyde reacted more strongly with H_2S than the amine, and H_2S could not be properly stripped in the regenerator. Cure was a new solvent charge.	Piping was modified to prevent recurrence.
1133	141		Unexpected component generated in storage tank during shutdown. It led to poor separation. Problem solved by bringing in new feedstock.	
644			*Contamination of amine inventory at shutdown causes foaming. Section* 16.1.2.	
		15.5	**Catalyst Fines, Rust, Tower Materials Promote Reaction**	
115, 1134, 131			*Leading to decomposition reaction in ethylene oxide towers, Section* 14.1.1.	
104			*Leading to polymerization, Section* 15.8.	
DT15.2			*Formic acid catalyzes reactions that lead to foaming.*	
146	239	Refinery DEA absorption	Continuous removal of HSS suppressed a reaction that converted DEA into DEA–formamide. This reduced DEA makeup.	
DT15.3			*Reaction that destroys ceramic packings also removes an undesirable impurity from product.*	

			15.6 Long Residence Times
101			*Reaction products interfering with phase separation, Section* 2.5.
120	250		Operation of column at total reflux for a lengthy period of time led to a reaction that formed an undesirable component that was very difficult to get rid of.
1653			*Popcorn polymer forming in an undrained leg of a relief valve leads to line rupture, Section* 14.12.
143, 1133			*Component generated in storage, Sections* 14.1.4 *and* 15.4.
12108, 132, DT11.6			*Crude cracking due to excess residence time, Sections* 8.3 *and* 24.1.2.
			15.7 Inhibitor Problems
121	250	Chemicals	An additive was injected into the tower feed to eliminate small quantities of aldehydes. At turndown, the additive also attacked ketones, causing product losses. Problem solved by continuous operation at high rates, shutting down when product stocks built up.
159	416	Natural Gas Glycol dehydration	Ammonium carbonate deposits plugged water lines from the regenerator condenser to storage. Sodium nitrite was injected to scavenge H_2S from the dehydrator gas feed, and the reaction formed ammonia. Some of the ammonia was absorbed by the glycol in the contactor, released in the regenerator, and together with CO_2 dissolved in the condenser water. When cooled, ammonium carbonate precipitated.
1561	486	Chemicals	To prevent hazardous reactions of thermally unstable chemicals, a solid stabilizer was added batchwise to the reboiler. The stabilizer, however, generated a waste disposal problem and also entrapped recoverable product. Excess stabilizer was required because stabilizer levels were monitored by laboratory sampling with a turnaround time of 4 hours. By installing on-line laboratory instrumentation, the excess was eliminated.

(*Continued*)

Chapter 15 Undesired Reactions in Towers (*Continued*)

Case	Reference	Plant/Column	Brief Description	Some Morals
			15.8 Air Leaks Promote Tower Reactions	
116			*Leading to explosion in CHP towers, Section* 14.1.2.	
112			*Leading to explosion in nitro compound towers, Section* 14.1.3.	
104	149	First-of-a-kind process	Lower trays in vacuum refining column plugged and later buckled due to excessive pressure drop. Plugging was caused by polymerization of the product. Product polymerization required the presence both reactor catalyst and air. Traces of the former were carried over; air entered due to a substantial leak. Repairing the air leak cured the problem.	Air leakage and/or catalyst carryover into a column can induce an undesirable reaction.
			15.9 Impurity in Product Causes Reaction Downstream	
1545, 12111			*Causing violent reaction, Sections* 14.2, 14.1.3.	
110			*Attacking downstream equipment, Section* 1.2.	
162			*Generating product impurity, Section* 1.2.	

Case	References	Plant/Column	Brief Description	Some Morals
			16.1 What Causes or Promotes Foaming?	
16.1.1	**Solids, Corrosion Products** *(See also Improving Filtration, Section 16.5.1 and Other Contaminant Removal Techniques, Section 16.5.5.)*			
615	150, 233	Heavy water using GS process "cold" tower H_2S-water contactor	Foaming occurred when the suspended solids content of the feed water was high. Antifoam injection suppressed the foaming. Some tray designs coped better with foaming than others. Pressure drop measurements and gamma scans were useful for diagnosing foam flood.	Suspended solids can promote foaming.
618	123	Chemicals	Small amounts of sodium chloride impurity precipitated out of solution and caused foaming. Problem solved by antifoam injection.	Particulates in liquid can cause foaming.
647	297	Ammonia hot-pot absorber	Heavy foaming occurred in winter nights shortly after a pilot-scale trial absorber was added in parallel with the plant absorber. The long rich-solution line from the trial absorber was uninsulated, and night temperatures of 12–15°C were sufficient to precipitate potassium bicarbonate particles that promoted foaming, erosion, and fouling.	Precipitation can promote foam.
639	492	Soluble polymer	Tower has been in service for over 30 years and never been entered. Gamma scans ordered to investigate sudden performance deterioration diagnosed foaming. "Shake tests" in the laboratory did not show foaming. Solution polymerization, fouling, and suspended polymer particulates could have contributed.	
637	432	NGL fractionation amine contactors	Three contactors in parallel received lean amine from a common regeneration. The piping of two was greatly oversized, and the resulting low velocities caused settling of solids along pipe bottom. The buildup continued until constriction-created velocities became large enough to reentrain the solids, resulting in sudden fouling and foaming of the two amine contactors.	Bigger is not always better. Avoid places where solids can deposit in amine systems.
646			*Foaming catalyzed by activated carbon particles from damaged filter, Section 16.1.3.*	

(Continued)

Case	References	Plant/Column	Brief Description	Some Morals
620, 628, 662			*Foaming mitigated by switching to less corrosive solvent, Section* 16.5.4.	
1269			*Particulate-catalyzed foaming mitigated by a large tower, Section* 8.1.2.	
16.1.2 Corrosion and Fouling Inhibitors, Additives, and Impurities (*See also Other Contaminant Removal Techniques, Section 16.5.5.*)				
601, DT16.1	263	Natural gas hot-pot absorber	Foaming occurred in packed absorber because a corrosion inhibitor injected into the gas well ended up in the column. Laboratory tests did not identify the problem. Tests under actual operating conditions did. Antifoam injection at the correct concentration solved it.	Tests for foaming are best carried out under actual plant operating conditions.
614	466	Natural gas treating	Foaming was caused by a corrosion inhibitor used in the boiler feed water. Steam condensate used for solvent makeup contained the inhibitor.	External corrosion inhibitors can cause foaming.
654	56	Olefins caustic absorber	Polymerization inhibitor effectively mitigated polymerization and foaming in tower.	
653			*Similar to* 654, *but with certain feedstocks, Section* 16.2.1.	
635	359	Olefins caustic scrubber	Use of antifouling dispersant in this tray tower needed to be supplemented by antifoam injection to prevent foaming. The antifoam injection had to be increased when the dispersant concentration was raised.	Additives can induce foaming.
634			*Similar to* 635, *but foamed severely only with structured packings, Section* 16.6.5.	
640	312	Refinery crude preflash tower	Foaming in base of tower was caused by flow improvers added by the pipeline to reduce pressure drop. Once the foam level rose above the feed inlet, the trays flooded and the distillate became black.	Additives can induce foaming.
629			*Well-treating fluids in natural gas plant, Section* 16.5.1.	
660			*Foam promoter in makeup water, Section* 16.5.1.	
644	465	Amine regenerator	Unit worked well between start-up and first shutdown and foamed afterward at the tower base. The reflux drum would fill with amine. Contamination of amine inventory during the shutdown is believed to be the cause. Keeping base level low was used as short-term fix. (See also 15126, Section 25.7.1.)	

649			*Degradation products promoted foaming, Section* 16.4.3.	
610	509	Butadiene (acetonitrile solvent)	Extractive distillation pilot column simulated plant column by using pentane/isoprene and acetonitrile. Foaming occurred and was promoted by increased water content. Oldershaw column tests were in good agreement with pilot column results. Foaming was suppressed by antifoam injection.	Variations in water content may affect foaming.
16.1.3	**Hydrocarbon Condensation into Aqueous Solutions**			
621	376	Refinery FCC amine absorber	At high rates foaming was experienced, attributed to presence of liquid HCs, possibly due to entrainment of LCO sponge oil. Amine is clean and sediment free. Keeping amine temperature 10 degrees hotter than inlet gas and antifoam injection were remedies.	
656	230	Gas Sulfinol M and MDEA	Severe foaming due to condensation of heavy hydrocarbons into the treating solution limited capacity, generated excessive solution losses, and destabilized the sulfur plant. Adding a silica gel heavy-hydrocarbon extraction unit upstream improved, but did not mitigate. Foaming was mitigated within an hour by adding carbon beds that filtered 15% of the lean solution and also improved particulate filtration. A new analytical technique that determines hydrocarbons in a treating solution guided changeouts of carbon beds.	
646	404	Natural gas Sulfinol	After several months on-line, foaming set in and bottlenecked the gas plant. Causes were condensation of heavy HCs in the feed gas and breakthrough of fine sulfur-impregnated activated carbon from damaged filter elements upstream of the unit. Cures were skimming oil, adding antifoam, and replacing fitter elements.	
659	363	Natural gas Sulfinol-M contactor	Foaming was reduced from frequent to rare over six years. Major cause was cutting hydrocarbon absorption by reducing sulfolane concentration from 25% to 6%, routine skimming, and addition of a coalescing catalyst to inlet gas. Periodic reclamation by vacuum distillation or ion exchange were effective. Silicone antifoam worked well, but required reclaiming to prevent foam-aggravating accumulation. Alcohol-based antifoam evaporated out in regenerator overhead. Increasing bottom three trays spacing from 24 to 30 inches also helped.	

(*Continued*)

Chapter 16 Foaming (*Continued*)

Case	References	Plant/Column	Brief Description	Some Morals
629, 643			*Among other factors, Section* 16.5.1.	
630	527	Natural CO_2 production glycol contactor	Diesel originating from drilling mud was carried over by high gas rates into the glycol contactor, where it caused foaming and high glycol losses. Initially, a defoamer mitigated foaming. Later, the rich glycol was sent to an existing low pressure flash drum with 3–5 hours retention time for glycol–diesel separation.	Foaming can be caused by hydrocarbon entering a glycol system.
661, DT16.2			*Major contributor to foaming, Section* 16.5.5.	
16.1.4 Wrong Filter Elements				
622	376	Refinery coker amine absorber	A filter specified 100% cotton contained polyester and other materials. These dissolved and catalyzed foaming. Filter replacement eliminated problem.	Beware of filter materials.
617, 627, 629			*Wrong filter types/sizes, Section* 16.5.1.	
16.1.5 Rapid Pressure Reduction				
1515			*Foamover due to rapid depressuring, Section* 8.1.1.	
16.1.6 Proximity to Solution Plait Point				
DT16.7			*Foaming in HC tower just below water removal tray.*	
DT2.23			*Causing hiccups and cycling in extractive distillation.*	
DT15.2, DT22.14			*Accumulation of intermediate boiling components or azeotropes of in-tower reaction products.*	
			16.2 What Are Foams Sensitive To?	
16.2.1 Feedstock				
604	95	Refinery crude fractionator	Serious foaming occurred when processing one type of crude. In this case, the residue was retained in the system at elevated temperatures for a relatively long time.	Foaming tendency is sensitive to the crude.
605	306	Refinery visbreaker fractionator	At high visbreaker conversions, the column bottom would foam and carry over into the distillate. Injecting 10 ppm of silicone defoamer into the vapor space above the fractionator bottom solved problem.	

653	56	Olefins caustic absorber	Foaming and polymerization occurred when a recycle stream from the polyethylene plant was reprocessed in tower. Mitigated by stepping up programs inhibiting polymerization and reaction.	
610			*In extractive distillation, Section* 16.1.2.	
16.2.2	**Temperature**			
612			*Extensive testing in DMF absorber, Section* 16.3.4.	
607			*Experiences in sponge absorber, Section* 16.5.4.	
16.2.3	**Pressure**			
607			*Experiences in sponge absorber, Section* 16.5.4.	
			16.3 Laboratory Tests	
16.3.1	**Sample Shake, Air Bubbling**			
601			*Unsuccessful in hot-pot absorber, Section* 16.1.2.	
609			*Unsuccessful in sulfolane extractive stripper, Section* 16.3.2.	
612			*Unsuccessful in DMF absorber, Section* 16.3.4.	
639			*Unsuccessful in soluble polymer tower, Section* 16.1.1.	
648			*Depend on sample location, Section* 16.4.2.	
16.3.2	**Oldershaw Column**			
609	509, 510	Sulpholane extraction extractive stripper	Foaming occurred near the top of the column. Bubbling air through a mixture sample failed to detect foaming, but Oldershaw and pilot column tests indicated foaming. Injecting antifoam or a small quantity of kerosene effectively suppressed foaming.	
DT2.23			*Foaming shows up in extractive distillation Oldershaw tests.*	
652	132	Chemicals	Foaming was the suspected cause of premature flooding. Oldershaw column tests showed no foaming whatsoever. The flood was found to be due to misfitting tray members.	"No foam" in Oldershaw tests means "no foam" in tower.
610			*Good in extractive distillation service, Section* 16.1.2.	
642			*Good for viscous residue foaming, Section* 16.6.7.	

(Continued)

Case	References	Plant/Column	Brief Description	Some Morals
16.3.3 Foam Test Apparatus				
611			*Good for natural gasoline absorber, Section* 16.4.3.	
16.3.4 At Plant Conditions				
612	59	DMF absorber	Foaming occurred in a reboiled absorber using DMF separating light mono- and diolefins, but bubbling air through solvent samples failed to detect it. Initial antifoam addition was unsuccessful because of poor dispersal. Effective dispersal of antifoam solved the problem. The investigation preceding the cure showed that foaming is temperature sensitive. Foaming was not observed in a similar but oversized column. A thorough investigation and analysis is described.	Foam testing under other than actual plant conditions can mislead. Foaming can be sensitive to temperature and downcomer size.
601			*Only successful method for hot-pot absorber, Section* 16.1.2.	
602, 648			*Successfully diagnosed foaming, Sections* 16.4.3 *and* 16.4.2.	
16.4 Antifoam Injection				
16.4.1 Effective Only at the Correct Quantity/Concentration				
625, DT16.3	224	Natural gas amine contactor	Excess antifoam generated massive foaming, liquid carryover, sour gas, and loss of bottom level. It took 24 hours of remixing and circulation with no gas to reinstate good operation.	Excess antifoam can aggravate foaming.
650	239	Refinery amine	A 2-in. layer of antifoam broke out on the surface of the lean amine sample. Despite this, the system experienced foaming.	As 625.
603	95		Too much antifoam was added to an absorber–stripper system removing CO_2 from H_2-rich gas to overcome a foaming problem. This interfered with tray froth action, causing poor absorption.	Too much antifoam can be as detrimental as too little.
DT22.10			*Too much antifoam plugs trays.*	

626	234	Refinery oil absorber	Foaming caused liquid entrainment. Removing the top two trays did not help. Antifoam injection solved problem. Gamma scans were effective in monitoring addition and determining optimum quantity.	Removing trays seldom mitigates foaming.
631	133	Amine absorber	Antifoam mitigated a foaming problem. The plant tried two different antifoams and three different concentrations of each, using gamma scans to evaluate their effectiveness.	
601, 649			*In hot-pot and aMDEA absorbers, Sections* 16.1.2 *and* 16.4.3.	
635			*Depending on the concentration of the antifouling additive, Section* 16.1.2.	
16.4.2 Some Antifoams Are More Effective Than Others				
648	352		Trial of a new antifoam produced premature flooding that initially did not look like foaming, gave different symptoms than previous foaming, and did not show up in shake tests on the regular test samples. A later sample from just below the feed showed foaming when tested at actual process conditions. Return to previous antifoam stabilized tower.	As 601, Section 16.1.2.
631, 629, 660			*In amine absorbers, Sections* 16.4.1 *and* 16.5.1.	
611			*In natural gasoline absorber, Section* 16.4.3.	
659			*In Sulfinol-M towers, Section* 16.1.3.	
16.4.3 Batch Injection Often Works, But Continuous Can Be Better				
602	95		In selective absorption of a light component from a gas stream using a nonvolatile solvent, batchwise addition of antifoam did not improve performance, so it was initially concluded that no foaming occurred. Laboratory tests at plant conditions finally verified foaming. Continuous antifoam injection solved the problem.	Batchwise antifoam addition may be ineffective.
649	226	Ammonia aMDEA absorber	Foaming was a major problem in the first month in operation after switching from hot-pot to aMDEA. Consequences were increased in CO_2 slip and hydrogen peaks in the regenerator off gas. Cured by adjusting the amount of antifoam and switching from batch to continuous dosing. Too many antifoam and degradation products were found to promote foaming.	

(Continued)

Case	References	Plant/Column	Brief Description	Some Morals
613	45	Natural gas amine regenerator	Foaming in this packed column was eliminated by batchwise antifoam injection. Previously, when bed height was low because of packing migration through the supports, foaming was not observed.	Batchwise antifoam addition is sometimes effective.
611	154	Natural gas absorber	Foaming occurred in this absorber, which used crude oil as solvent. Foaming and antifoam effectiveness was successfully tested in a foam test apparatus. Problem was solved by intermittent antifoam injection.	Same as 613.
16.4.4 Correct Dispersal Is Important, Too				
DT16.4			*Use of static mixer can make or break injection effectiveness.*	
612			*Initial antifoam injection unsuccessful because of poor dispersal, Section* 16.3.4.	
DT22.10			*Antifoam settling in mixing drum.*	
16.4.5 Antifoam Is Sometimes Adsorbed on Carbon Beds				
627, 629			*In amine systems, Section* 16.5.1.	
16.4.6 Other Successful Antifoam Experiences				
608	372	Refinery preflash	Several columns experienced foaming problems. Problems successfully solved by antifoam injection.	
623	368	Refinery	Foaming in the solvent deasphalting asphaltene stripper caused plugging of overhead condenser (due to asphaltene carryover). Eliminated by antifoam injection into feed.	
624	368	Refinery	Upsets due to foaming in the solvent deasphalting deasphalted oil tower were eliminated by antifoam.	
638	492	Amine contactor	Problems with CO_2 and H_2S removal, temperature excursions, and *dP* swings were caused by foaming. Gamma scans diagnosed and antifoam injection solved the problem.	
645	545	Solvent wash column	At high solvent rates foaming initiated at the feed tray and flooded trays above. Gamma scans diagnosed and antifoam mitigated.	

More are listed in order of appearance in all sections of Chapter 16 *except those already listed above in Section* 16.4: 615, 618, 610, 621, 646, 630, 605, 609, 664, 662, 661, 658, 606, 616. *Also DT*6.1, *DT*15.2, *DT*16.5.

16.4.7 Sometimes Antifoam Is Less Effective

617			*MEA absorber, Section* 16.5.1.	
634			*Mitigated with trays, but not with structured packings, Section* 16.6.5.	
607			*Antifoam injection only partially effective, Section* 16.5.4.	
DT16.7			*Possibly due to removal in water draw tray above foaming zone in hydrocarbon tower.*	

16.5 System Cleanup Mitigates Foaming

16.5.1 Improving Filtration

617	388	Natural gas MEA absorber	A foaming problem was treated with limited success by antifoam injection, frequent activated carbon regeneration, and reduced MEA circulation. Problem was finally solved by using finer filter elements and adding a filter to the lean-amine circuit.	Improving filtration can effectively cure some foaming problems.
627	531	Natural gas amine regenerator	Foaming was experienced. One cause was ethane carryover from flash tank on ethane liquid treater upstream. Replacement of the cartridge filter by a 5-μm bag filter, cleaned once per week, helped. Constant antifoam injection also helped, but some was absorbed by the activated carbon bed.	Good filtration and adsorption and adequate defoamers can mitigate amine foaming.
629	389	Natural gas MDEA absorber	Foaming, due to liquid HCs and well-treating fluids, caused capacity loss, plant shutdowns, and amine losses. Changing amine filters from 10 to 2 μm did not help; tests showed particles as large as 50 μm slipping through. Laboratory tests showed that carbon filtration promoted foaming by adsorbing the antifoam but not the foam-generating contaminants. Mitigated by 10-μm absolute filter elements, changing activated carbon specifications, adding a coaleser in the gas feed downstream of the plant inlet filter–separator, and changing antifoam.	

(*Continued*)

Chapter 16 Foaming (*Continued*)

Case	References	Plant/Column	Brief Description	Some Morals
663	74	Coal seam gas amine	Plant used rotating amine filtration for three parallel trains with 10μ pleated cartridge filters containing cellulose/polyolefin media. Switching to 10μ absolute rated rigid depth filtration cartridges tripled filter life and removed contaminants, changing amine from dirty black to clean light purple. This reduced fouling and foaming.	
643	411	Refinery DEA absorbers and regenerators	Cleaning up the system mitigated foaming, corrosion, fouling, and absorption problems and reduced upsets. The major improvements were adding cartridge filters and modifying flash drum internals to adequately remove HCs. Article contains details on how to clean up and what are the benefits.	See 617, 627, and 629.
660	90	Natural gas MEA 6 contactors	Severe foaming persisted after seven years of modifications including installing carbon beds, reclaiming, reducing regenerator pressure, improving solution and gas filtration, changing type and concentration of corrosion inhibitor, desuperheating regenerator steam, reducing MEA strength, increasing lean amine H_2S loading to help protective film, raising amine temperature, and changing trays to fixed valve trays. Root Cause Failure Analysis led to a threefold reduction in foaming severity. Key modifications were modifying coalescing filters and installing cyclones on feed gas; improving solution filtration; changing antifoam; eliminating the foam promoter in the makeup water; more frequent and better contactor acid washes; and improvements to operation.	
664	338	Natural gas MEA absorber/ regenerator	Severe foaming, corrosion, and reboiler plugging problems were mitigated by cleanup that included improved solution filtration, hydrocarbon skimming, inlet gas filtration and coalescing, modifications to minimize hydrocarbon condensate carry-over into the absorber, and an amine heater. Regenerator steam to solution ratio was reduced from 1.6 to 1.0 lb/US gpm, allowing lower bottom temperature, lower steam pressure and less degradation. Antifoam successfully used to suppress remaining foaming (see also 780, Section 8.4.6).	
622			*Using correct materials for filter elements, Section* 16.1.4.	

646			*Replacing damaged filter elements, Section* 16.1.3.	
636, 651, 656			*Improved filtration among other factors, Sections* 16.5.5, 16.1.3.	
16.5.2 Carbon Beds Mitigate Foaming But Can Adsorb Antifoam				
619	375	Refinery amine	Foaming was mitigated by installation of a carbon filter on the lean-amine stream.	
656			*Carbon beds adsorb hydrocarbons, mitigate foam, Section* 16.1.3.	
651			*Improved adsorption among other factors, Section* 16.5.5.	
627, 629			*Carbon beds adsorbed antifoam, Section* 16.5.1.	
617, 660			*Limited success with carbon beds, Section* 16.5.1.	
16.5.3 Removing Hydrocarbons from Aqueous Solvents (*See also Hydrocarbon Condensation, Section 16.1.3.*)				
643			*Improving flash drum internals to remove HCs, Section* 16.5.1.	
629, 664			*Improving skimming & coalescing, Section* 16.5.1.	
636, 651			*Improved HC removal among other factors, Section* 16.5.5.	
16.5.4 Changing Absorber Solvent				
607	373	Refinery FCC secondary absorber	At least 15 instances of foaming occurred when LCO was used as the lean oil. Increasing tray spacing, increasing downcomer area, reducing pressure drop, increasing temperature, and injecting antifoam were only partially successful remedies. Replacing LCO by naphtha effectively cured problems. In other cases, use of antifoam, injecting naphtha into LCO, and raising pressure were effective remedies.	
620	376	Refinery amine unit FCC gas	Amine discoloration and foaming were experienced at high H_2S loadings. The maximum acid gas/solution loading could have been exceeded, causing excessive iron sulfide to form and catalyze foaming. Problem initially controlled by corrosion inhibitor. Final solution by switch to MDEA, which rejected CO_2; this reduced the acid gas/solution loading.	Keep amine clean. Avoid overloading the amine solution.
628	326	Natural gas MEA amine unit	Unit experienced severe corrosion and foaming. The rich-amine filter was plugged with iron sulfide and needed changing once or twice per day. Mitigated by switching to MDEA. The lower corrosion rate of MDEA reduced generation of foam-stabilizing corrosion products.	Corrosion products catalyze amine foaming.

(*Continued*)

Case	References	Plant/Column	Brief Description	Some Morals
662	208	Gas MEA amine	Severe corrosion in an MEA system (absorber and regenerators contained bubble cap trays) caused leaks and foaming. Problems persisted following swap of MEA to formulated MDEA, improving corrosion inhibition, upgrading pipe metallurgy in key sections, and keeping CO_2 loadings below 0.35 mole/mole solution. Excessive antifoam was required. Problems eliminated by switch to aMDEA after thorough system cleanout.	
659			*Reducing concentration of Sulfinol-M solution, Section* 16.1.3.	
660			*Limited success with reducing MEA strength, Section* 16.5.1.	
16.5.5 Other Contaminant Removal Techniques				
636	92, 239	Refinery MDEA absorption/ regeneration	A large rate (as much as 1000 lb/day) of HSS buildup caused severe corrosion, fouling, and constant foaming in absorbers and regenerators, bottlenecking HC throughput. Caustic addition and a caustic plus electrodialysis program did not help. An ion exchange HSS removal dropped HSS concentration fivefold, eliminating fouling, foaming, amine losses, and plant bottleneck. Later improvements in filtration and HC removal gave further benefits.	Removal of impurities helps an amine system.
651	239	Refinery MDEA	Removing HSS and improving filtration, adsorption, and HC removal dramatically reduced foaming upsets and the resulting flaring and excessive wastewater treatment as well as corrosion and fouling.	Amine system cleanup mitigates foaming.
661	86	Gas MDEA and Sulfinol contactors & stills	High corrosion rates produced suspended solids and fouling, limiting solution circulation rates and run length. Solids plus hydrocarbon condensation led to severe foaming in the contactors and stills with solution carry-over to the gas dehydration and sulfur plant. Filters were overwhelmed. Silicone antifoam and corrosion inhibition helped but did not mitigate. Procedure changes, selective metallurgy upgrades, an enhanced inhibitor program, heat exchange modifications, reducing condensed hydrocarbons, and amine reclaiming, effectively alleviated foaming.	

655	348	Gas Sulfinol M	Foaming led to excessive circulation and reduced throughput. Reclaiming the solution by vacuum distillation after mixing with caustic removed HSS and sodium contaminant, eliminating foaming and permitting lower circulation and higher throughputs.	
657	517	Gas amine	Daily foaming upsets prevented increase in amine concentration and restricted capacity. Amine degradation products and solids below the removal range of the plant filter were prime causes. Adding a device that concentrates and removes contaminants eliminated the upsets and permitted higher capacity and amine concentration.	
658	517	Amine	Frequency of antifoam addition was reduced from multiple times per day to once per day after adding a device that concentrates and removes contaminants. Solids and long-chain carboxylic acids were identified to play major role in foam development.	
641	302	Refinery hydrocracker depropanizer	Foam/froth in the tower base rose above the reboiler return inlet, initiating flooding in the tower. Foaming due to corrosion products formed in the debutanizer overhead upstream is believed responsible. To stop the foaming, the pH of the wash water in the upstream hydrogen off-gas drum was raised, removing the corrosive component (HCl) upstream of the debutanizer.	
16.6 Hardware Changes Can Debottleneck Foaming Towers				
16.6.1 Larger Downcomers				
606, DT16.5	49	Aldehyde column	Premature flooding occurred due to foaming in downcomers from just below the feed up. This was detected by gamma-ray scans. Adding antifoam improved performance but was undesirable in the process. Downcomer enlargement in the trouble area solved the problem.	Enlarging downcomers and adding antifoam are effective against foaming.
DT15.2			*Enlarging downcomers in formaldehyde tower raises capacity.*	

(Continued)

Chapter 16 Foaming (*Continued*)

Case	References	Plant/Column	Brief Description	Some Morals
409	278	Olefins HP condensate stripper	Column flooded prematurely, presumably due to foaming. Capacity was increased by 50% by replacing original sieve trays by valve trays that had half the weir height and 50% more downcomer top area. Extensive performance data are presented.	Pay attention to downcomer performance in systems with high foaming tendencies.
607			*Increasing downcomer area partially successful remedy for foam, Section* 16.5.4.	
DT16.6			*Flooding due to excessive downcomer velocities in stripping trays of crude preflash tower eliminated by removing trays.*	
16.6.2	**Smaller Downcomer Backup (Lower Pressure Drop, Larger Clearances)**			
DT16.7			*Increasing downcomer clearances debottlenecks foaming tower.*	
DT16.8			*Removing blanking strips in small towers debottlenecks foaming.*	
409			*Used among other changes in successful fix for foam, Section* 16.6.1.	
607			*Partially successful remedy for foam, Section* 16.5.4.	
16.6.3	**More Tray Spacing**			
607			*Partially successful remedy for foam, Section* 16.5.4.	
659			*For bottom 3 trays helps, Section* 16.1.3.	
16.6.4	**Removing Top Two Trays Does Not Help**			
616	81	Refinery oil absorber	Foaming occurred, primarily near the top of the column. Removing the top two trays to provide more disengagement space did not solve the problem. Gamma scans identified the foaming, and antifoam injection cured it.	Gamma scans are useful for diagnosing foaming problems.
626			*Almost identical to* 616, *Section* 16.4.1.	
16.6.5	**Trays Versus Packings**			
634	359	Olefins caustic scrubber	Trial addition of antifouling dispersant to the top loop of the tower led to severe foaming that could not be mitigated by antifoam. The structured packings are believed responsible for the severity because tray towers using the same dispersant program have not experienced severe foaming. (See 635, Section 16.1.2.)	Additives and column hardware affect foaming.

632	218	Steam stripper	Foaming occurred in stripper removing NH_3 from aqueous stream. After-the-fact pilot tests showed significant foaming with sieve trays, but foaming minimized with structured packings.	Column hardware affects foaming.
16.6.6 Larger Packings, High-Open-Area Distributors Help				
633	68	Soapy water/ polyalcohol oligomer (Poly-ol)	Seven-foot-ID random packed tower was unstable, experiencing unpredictable upsets, entrainment, and carryover. Above feed, foaming took place because of the soapy water. Foaming was mitigated by upsizing packings, adding collector between beds, and changing distributors to minimize vapor–liquid contact while increasing open area. (See also 319, Section 4.8, and 860, Section 6.8.)	
DT16.2			*Haphazardly removing every other trough from liquid distributor does not help with foaming.*	
16.6.7 Increased Agitation				
642	536	Solvent residue batch still, vacuum	A stagnant mixture of the highly viscous residue foamed. Laboratory distillation proved that enhanced agitation helped, so the kettle single-blade propeller agitator was retrofitted with a dual axial impeller. This permitted the residue to be pumped.	
16.6.8 Larger Tower				
1269			*Larger tower replacement mitigates foaming in bottom sump, Section* 8.1.2.	
612			*Foaming observed in tower but not in another, oversized tower, Section* 16.3.4.	
16.6.9 Reducing Base Level				
1555	368	Refinery solvent deasphalting	Level indication in asphalt tower was problematic due to a foam in the tower. To solve, tower bottom was emptied to get rid of the foam. Also, the level float was replaced by a *dP* indicator calibrated by finding the actual level with tricocks.	
1268			*Reducing base level alleviates foaming, Section* 25.7.1.	

Case	References	Plant/Column	Brief Description	Some Morals
			17.1 Piping Scale/Corrosion Products *(for commissioning debris, see Section* 11.8)	
1248	496	Refinery alkylation depropanizer	The upper trays (1–28) suffered severe corrosion damage: Trays 19–28 completely collapsed and their debris plugged tray 29. This initiated flooding over 40 ft of tower height.	Diagnosed by gamma scans.
1041	506	Refinery crude fractionator	Excessive water in the crude and water in the reflux led to corrosive salts in the fractionator and naphtha side draw. Corrosion products and salts plugged low piping and the naphtha side draw control valve. The block valve on the control valve bypass was operated partially open for 10 months and experienced corrosion-erosion so it could not be isolated. Piping sections experienced thinning and leaks. These were major contributors to the accident in Case 1197, Section 14.8.	Use water washes and other measures to eliminate corrosive salts.
518	501	Refinery crude fractionator	Upper section of this all-CS tower/internals experienced severe corrosion, causing fouling in the kero PA section and premature flooding. Gamma scans and *dP* diagnosed.	Ensure correct materials.
DT17.1		•	*Structured packings corrode and plug by corrosion products in refinery vacuum tower.*	
1297	239	Refinery amine	In three MDEA, one DEA, and one MEA systems, effective removal of HSS, typically from >4% to <1% amine weight, reduced corrosion rates (mpy) typically by a factor of 4 or more as well as fouling. Caustic addition in two of the five gave mixed results.	HSS removal effectively removes solids from ethanolamine systems.
DT18.3, DT18.4			*Effect of hole size, valve type.*	
1214			*On top trays, Section* 18.5.	
1288			*Plugged downcomers, Section* 18.2.	

413			*Sieve versus valve trays, Section* 18.4.1.	
916, DT18.2			*In downcomer seal area, Section* 11.2.	
1135, 1120			*Following shutdown wash, Section* 12.6.	
501, 1290			*Plugged packed bed, Sections* 4.7 *and* 4.11.	
808, 831, 852, 823, DT6.4			*Plugged packing distributors, Sections* 6.2.2 *and* 6.2.3.	
1627			*Plugged reboiler drain line, Section* 14.11.	
DT12.3			*Water wash removed from fouled internal reboiler.*	
643, 620, 628, 636, 522			*Plugged amine absorption/regeneration systems, Sections* 16.5.1, 16.5.4, 16.5.5, *and* 18.4.3.	
1607			*Plugged relief valve inlet, Section* 21.5.	
750			*Contributed to cavitation of reflux/product pump, Section* 10.9.2.	
			17.2 Salting Out/Precipitation (*See also Section* 12.7)	
1264	490	Refinery sour-water stripper	Flooding and high ΔP at high rates were caused by heavy deposits of calcium salts about three to four trays from the bottom. The salting out is believed to be due to a decrease in calcium solubility with increasing pH near the bottom. There was heavy scale on the reboiler presumably due to further decrease of calcium solubility at higher temperature. The refinery plans dilute acid wash if problem reoccurs.	Gamma scans identified fouling and the fouled region.
1164	70	Groundwater purification air stripper	Plastic random packings heavily fouled with iron deposits and biological growth in service were fully cleaned by ozone injection to control biological growth and inorganic polyphosphate to prevent deposition of iron, calcium, and manganese ions.	Procedure fully described in paper.
1211, 1216, 12100, DT18.3, DT18.4, DT22.10			*Plugged trays, Sections* 18.1, 18.4.1.	

(*Continued*)

Chapter 17 The Tower as a Filter: Part A. Causes of Plugging—Number 1 on the Top 10 Malfunctions (*Continued*)

Case	References	Plant/Column	Brief Description	Some Morals
1288			*Plugged downcomers, Section* 18.2.	
1278, 12116			*Plugged packing and distributors, Section* 6.2.1.	
105, 520, 521			*Plugged packing, Sections* 18.4.2 *and* 18.4.3.	
15157			*Due to water refluxing in HC tower, Section* 2.4.3.	
231, 1041			*Due to cold temperatures near top of crude tower, Sections* 3.1.5, 17.1.	
159			*Ammonium carbonate deposits plug outlet lines from regenerator, Section* 15.7.	
1166	329	Refinery FCC main fractionator	Top three trays fouled up and flooded at start-up. Cause was ammonium chloride deposition due to rapid cooling of the column and shifting deposits from the overhead system back into the tower. Cured by salt dispersant injection.	Gamma scans helped diagnosis.
12103	371	Refinery HF alkylation isostripper	Iron fluoride fouling was experienced below the feed. An isostripper upset that lifted the relief valves resulted in redistribution of the fluorides within the column, saving a shutdown.	
636, 651, 647			*Fouled amine absorbers and regenerators, Sections* 16.5.5 *and* 16.1.1.	
			17.3 Polymer/Reaction Products	
1252	134	Olefins DC_3 stripper	Polymer fouling on the stripper's top three trays caused flooding near the top of the stripper and near the bottom of the rectifier. Chemical treatment removed the polymer and eliminated flooding.	Diagnosed by gamma scans and *dP* measurement.
1183	204	Refinery deoiling column	Thick coats of polymeric and iron sulfide deposits in the bottom sections required days to remove using hammers, chisels, and whisk brooms. Laboratory testing led to developing a solvent combining surfactants, enzymes, and oxidizer to dissolve deposits.	
12109			*Popcorn polymer on valve trays in butadiene column, Section* 18.4.1.	
DT18.3, DT18.4			*Plugged trays, downcomers.*	
1289			*Plugged bottom downcomers, Section* 18.2.	
DT18.2			*Accumulated behind "interrupter bar" inlet weirs.*	
DT19.5C			*Blocked windows of baffle trays and bottom-liquid offtake.*	

1242			*Formed at bottom of packed bed, Section* 18.5.
850, 852, 1254			*Plugged packing distributors, Sections* 6.2.2 *and* 12.7.
1241			*Plugged bottom line, Section* 15.3.
DT12.3			*Removed by soaking packings.*
161			*Due to reactive component in feed, Section* 15.3.
104			*Due to air leak, Section* 15.8.
DT23.5			*Foul reboiler, leading to surging.*
1321			*Due to high reboiler metal temperature, Section* 23.8.2.
			17.4 Solids/Entrainment in the Feed
1205	113		Feed tank was emptied for the first time, stirring up settled solids from the bottom. These plugged column.
1225, 1235			*Entrainment of iron, carbonates, and mud in vapor entering tower, Section* 18.1.
1270, 443			*Entrainment of tars in vapor feed, Sections* 17.7 *and* 4.4.2.
519			*Alumina solids in feed, Section* 18.4.2.
863, 1174, DT6.1, DT6.2, DT6.3, DT6.4, DT17.2, DT19.4			*Solids in feed plugged packing distributors, Section* 6.2.1.
1207			*Explosion in peroxide tower resulting from feed filter plug, Section* 14.1.2.
1221			*Entrainment and deposition of caustic from feed to upper trays, Section* 12.7.
1586, 623			*Entrainment of solids from tower bottom plugs trays, condenser, Sections* 8.1.1 *and* 16.4.6.
			17.5 Oil Leak
1271	427	Urea strippers	Tars originating from compressor lubricant degradation entered strippers and deposited, reducing run length. Replacing lubricant by one of greater thermal and oxidation stability eliminated the problem.

(*Continued*)

Chapter 17 The Tower as a Filter: Part A. Causes of Plugging—Number 1 on the Top 10 Malfunctions (*Continued*)

Case	References	Plant/Column	Brief Description	Some Morals
1121	318	Natural gas NRU LP stripper	At start-up, turboexpander lube oil leaked into the stripper feed and plugged the column. The NRU was shut down, and a hydrocarbon solvent was used to clean the system.	
			17.6 Poor Shutdown Wash/Flush	
1104, 1120, 1131			*Plugged trays, packings, and pump strainers, Section* 12.6.	
1132			*Poor liquid circulation leads to plugged packing and distributors, Section* 12.12.2.	
			17.7 Entrainment or Drying at Low Liquid Rates	
1270	162	Refinery crude fractionator	Preflash drum vapor entered crude tower three trays above the flash zone and one tray under the gas oil draw. Entrainment of crude in the drum vapor caused tar-like fouling on the trays below and black gas oil with a large metal concentration.	
8112			*Crude entrainment in preflash drum vapor contaminates product, Section* 7.1.2.	
1586			*High bottom level in a coker fractionator bottom, Section* 8.1.1.	
943, 15105			*At low and/or drying liquid rates, Sections* 4.4.1 *and* 26.4.1.	
			17.8 Others	
1233	377	Refinery crude fractionator	The top five trays plugged up. Likely cause is corrosion inhibitor injected into overhead system coupled with operating conditions.	
DT22.10			*Deposition of excess defoamer.*	
1020, 1024, 1019			*Ice, hydrates, Sections* 2.4.6.	
1244			*Sticky chemicals, Section* 18.1.	
1142, 1194			*Viscosity runaways, Sections* 12.10 and 12.13.1.	
1226			*Fungus growth, Section* 6.2.1.	
1164			*Biological growth, Section* 17.2.	

Chapter 18 The Tower as a Filter: Part B. Location of Plugging—Number 1 on the Top 10 Malfunctions

Case	References	Plant/Column	Brief Description	Some Morals
			18.1 Trays (*See also Sections* 12.7, 17.1–17.3, *and* 18.4)	
1211	306	Refinery FCC C_3 splitter	A small amount of KOH solution from an upstream dryer was carried over into the column. In the column, KOH precipitated and plugged sieve decks with $\frac{3}{16}$-in. holes. Acid wash removed deposits.	Small perforations plug.
1216	81	Phenol	A salt precipitated out in the lower section of the column. The deposits restricted vapor upflow and caused poor fractionation. Gamma scans identified the problem, tray changes solved it.	
1225	194	Natural gas amine regenerator	Premature flooding occurred after several months in service. Iron and other metal carbonates formed deposits 1 in. thick on some of the top valve trays. The solids originated in the natural gas stream. Problem was solved by cleaning, and recurrence prevented by annual acid wash.	
1235	189	Refinery crude fractionator	Wash valve trays plugged. Likely reason is entrainment of mud from flash zone below.	
104			*By polymerization due to air leak, Section* 15.8.	
1321			*By polymerization due to high reboiler metal temperature, Section* 23.8.2.	
943, 443			*Due to low liquid flow rates, Sections* 4.4.1 *and* 4.4.2.	
1270			*Due to fouling liquid entrainment into low liquid flow rate region, Section* 17.7.	
15105			*Due to drying in low liquid flow rate region, Section* 26.4.1.	
1586			*Due to high bottom level causing carryover of fouling materials, Section* 8.1.1.	
1201			*End of run, Section* 3.1.3.	
1233			*Plugging of trays by corrosion inhibitors, Section* 17.8.	
1020			*Due to hydrates, Section* 2.4.6.	

(*Continued*)

Case	References	Plant/Column	Brief Description	Some Morals
1244, DT18.1	250		Valve trays handled extremely sticky chemicals without sticking where high throughput was constantly sustained. The column run length was restricted by downcomer plugging. In another case where throughput varied, valve sticking occurred while handling mildly sticky chemicals.	
1104			*Due to solids in shutdown wash water, Section* 12.6.	
1236			*Plugged pipes that distribute liquid to multipass trays, Section* 5.7.	
1002			*Plugging of weep holes, Section* 13.3.	
1255			*Bubble cap trays at top of fractionator, Section* 22.8.	
DT19.5			*Buildup on closely spaced baffle trays.*	
			18.2 Downcomers	
1288	44	Refinery crude fractionator	Downcomers from the collector tray beneath the top PA packed bed, and/or from the top three trays in the fractionation section below, plugged by salts, scale, and/or corrosion products. This caused the collector tray to overflow, the packing to flood, and restricted tower rates. Diagnosed by a detailed pressure survey and solved by bypassing top PA return liquid to a hot tap below the plugged region.	Pressure surveys provide a powerful diagnostic tool.
1299	514, 515	Refinery FCC main fractionator	Downcomer restriction just below the HCO draw caused high dP and rate restriction. Diagnosed by gamma scans. Solved by liquid bypass around the restricted trays. Initial bypass was undersized, so a second bypass was needed to return to full rates.	
1289	543	Depropanizer 60 trays	After a new reboiler was put into service, tower experienced flooding and instability, probably due to polymer buildup. Gamma scans showed that the plugging occurred somewhere in the bottom four downcomers. Solved by a hot tap into tray 56 downcomer and bypassing the liquid into the bottom.	
DT18.2			*Solids accumulate behind interrupter bars, inlet weirs.*	
1244			*Handling sticky chemicals, Section* 18.1.	

1255			*In a coker fractionator, Section* 22.8.	
916			*Clearance under downcomer plugged, Section* 11.2.	
1214			*Top downcomer plugged by corrosion products, Section* 18.5.	
1143			*Played a possible role in vapor gap problem, Section* 22.6.	

18.3 Packings

(*See Section* 6.2 *for plugged distributors and Section* 18.4 *for additional plugged packing cases*)

1298	286, 552	Ethylene dilution steam generator	Random packings and orifice pan distributors repeatedly plugged within 4 months of start-up, giving distributor overflows, high pressure drops, and short runs. Fouling eliminated by retrofit with grid and V-notch distributor and redistributor. There was no efficiency loss.	See 1278.
501, 1290, DT17.1			*Corrosion products plug bed, Sections* 4.7 *and* 4.11.	
DT17.2			*Fitter removal causes plugging in bed, distributors.*	
1242			*Polymeric compounds plug bottom of bed, Section* 18.5.	
1135			*Oxidation at shutdown, Section* 12.6.	
1164			*Successful prevention of fouling in groundwater purification, Section* 17.2.	
8102			*History of packing fouling in synthesis gas-stripping column, Section* 6.3.	
1198, 11106			*Plugging of packing/distributors leads to fires, Sections* 14.6.3 *and* 14.6.5.	
1154			*Deposits between two grades of structured packing cause fire, Section* 14.6.2.	
1131			*Poor flush procedure at shutdown, Section* 12.6.	
1132			*Poor liquid circulation at start-up, Section* 12.12.2.	
DT6.2			*Dust plugs interbed demister, causing vapor maldistribution.*	

18.4 How Packings and Trays Compare on Plugging Resistance

18.4.1 Trays versus Trays

413	194	Refinery naphtha splitter	Replacement of CS valve trays by SS sieve trays reduced frequency of required tray cleaning to remove scale.	
DT18.3, DT18.4			*Several experiences where changing hole size or valve type made a major difference to plugging and valve sticking.*	

(*Continued*)

Chapter 18 The Tower as a Filter: Part B. Location of Plugging—Number 1 on the Top 10 Malfunctions (*Continued*)

Case	References	Plant/Column	Brief Description	Some Morals
12100	207	Refinery two parallel sour-water strippers	$^{1}/_{2}$-in. hole sieve trays suffered chronic fouling with salts and insoluble organics, with typical 5-month runs. Retrofitting with large, directional fixed valves greatly improved fouling resistance and run length.	
12109	396	Butadiene	Valve trays containing moving round valves experienced severe popcorn polymerization. Replacement by fouling-resistant trays containing fixed valves, uniform liquid flow devices, and special downcomers mitigated fouling. Unspecified process changes contributed too.	
12110	396	VCM EDC heavy ends	Runs were short due to fouling of sieve trays below the feed. Replacement by fouling-resistant trays as in 12109 and adding a transition tray that mixed feed and reflux led to longer runs and higher capacities.	
448	191	Refinery crude fractionator	Replacing four conventional stripping trays by seven fouling-resistant sieve trays helped minimize fouling with heavy Venezuelan crudes and improved stripping, enhancing diesel yield (see Case 340, Section 2.1). Fouling-resistant features included large (0.75-1-inch) holes, rectangular tray design, high weir loadings, low special weirs, high downcomer clearances, and minimum downcomer cross section area.	
18.4.2 Trays versus Packings				
519	21	Air stripping of benzene from wastewater	Alumina solids saturated with alkylbenzene oils built up in the tower packing, causing flood and high pressure drops. The solids and oils also released absorbed benzene under stripping conditions. Problem solved by replacing packings by specialty baffle trays, "Froth Scrubber."	For fouling services, fouling-resistant internals are needed.
520	206	Chemicals steam stripping	Tars and salts accumulating in the tower bottom caused recurring plugging. The weight of tar deposits crushed the ETFE packing. Problem eliminated by fouling-resistant fixed-valve trays.	Same as 519.
105	113	Aqueous feed/solvent	Removal of the volatile solvent from an aqueous feed solution caused dissolved solids to precipitate in the packing below the feed, plugging the column. The solution was replacing the packing with sieve trays.	

DT18.5			*A structured packing retrofit of an olefins oil quench tower does not survive a viscosity runaway incident.*	
18.4.3 Packings versus Packings				
517	240		Packed-bed plugging was monitored by logging the ratio of clear vapor gamma transmission to transmission at the fouled spot ("absorption ratio"). The highest ratios were right under the feed. Metal structured packings reached a ratio of 150 within 2 months. Ceramic structured packings took over 18 months to reach that level. A further replacement with 2-in. metal rings produced ratios less than 7.5 over 2 years.	Gamma transmission log is effective in monitoring packing fouling.
521	70	Groundwater purification air strippers, several towers	Iron oxide precipitation from high-iron groundwater severely fouled Pall rings, "nonfouling" random packings, and structured packings. The packing type (all plastic) made little difference. In one case, deposition of iron oxide from entrained groundwater even plugged a notched-trough distributor. The entrained water appears to have evaporated from the distributor walls, leaving the deposits behind.	
522	362	Ammonia CO_2 absorber arsenic-activated Vetracoke	1.5-in. ceramic saddles were leached by the solution, causing turbidity, sludge, plugging and flood. Replacement by 1-in. SS Pall rings did not eliminate plugging. Replacing the 10 m^3 at the bottom of the 80-m^3 bottom bed by 2-in. open modern random packing, while keeping 1-in. Pall rings above, all in SS, alleviated the plugging. (See also 1294, Section 20.2.3.)	
DT18.6			*Packing size and shape make a difference to plugging.*	
1278			*Large random packings mitigate salting out in packing, Section* 6.2.1.	
8106			*Runlength increases after random packing replaced by grid, Section* 6.2.1.	
			18.5 Limited Zone Only	
1214	305	Natural gas depropanizer	Flooding caused by top downcomer plugging with corrosion products occurred in the rectifying section. Measured pressure drop was low because only the top tray was flooded.	Flooding may occur even at low-column-pressure drop.

(Continued)

Chapter 18 The Tower as a Filter: Part B. Location of Plugging—Number 1 on the Top 10 Malfunctions (*Continued*)

Case	References	Plant/Column	Brief Description	Some Morals
1242, DT18.7	224	Acetylene compressor discharge aftercooler	Heavy components formed polymeric compounds that plugged the bottom foot or so of the random packings. On turnaround, the plant hand stacked all the rings in a staggered arrangement. Upon restart, the column failed to cool the gas due to channeling. The packings then were removed and dumped randomly, with only the bottom foot stacked. Column run length quadrupled.	Stacking random packings can promote washing and suppress solid buildup.
522			*Packing plugged at bottom of amine absorber, Section* 18.4.3.	
1252			*Top three trays of olefins depropanizer stripper plugged by polymer, Section* 17.3.	
1289			*Bottom few downcomers of depropanizer plugged by polymer, Section* 18.2.	
231, 12120, 1233, 1288, DT18.8			*Top few trays and downcomers of refinery crude fractionator plug, Sections* 3.1.5, 12.7, 17.8 *and* 18.2.	
518			*Kerosene PA of refinery crude fractionator plugged by corrosion, Section* 17.1.	
1235, 1270, 15105			*Wash trays of refinery crude fractionator plugged, Sections* 18.1, 17.7, *and* 26.4.1.	
1255			*Top few trays of refinery coker fractionator, Section* 22.8.	
1166, 1273, 1284			*Top of refinery FCC fractionator, Sections* 17.2 *and* 12.7.	
1264			*Bottom few trays in sour-water stripper due to salting out with increased pH, Section* 17.2.	
1265			*Trays just above PA in hydrotreater fractionator, Section* 12.7.	
1586			*Bottom few trays of a coker fractionator, Section* 8.1.1.	
1278			*Salt plugging just below the feed, specialty chemicals, Section* 6.2.1.	

			18.6 Draw, Exchanger, and Vent Lines
909, 912, DT18.9, DT23.6			*Reboiler liquid line, Section* 11.8.
DT23.5			*Fouled reboiler leads to surging in tower.*
1286			*Reboiler heater charge pump strainers, Section* 20.2.1.
1644			*Product recycle pump, Section* 14.9.
1142			*Bottom line, Section* 12.10.
1241			*Bottom purge line, Section* 15.3.
1041			*Naphtha draw line from a crude fractionator, Section* 17.1.
1024			*Liquid draw from a chimney tray, Section* 2.4.6.
1111			*Infrequently used intermediate draw line, Section* 14.11.
1424			*Ejector lines, Section* 24.7.
1607			*Relief valve inlet line, Section* 21.5.
1652			*Relief valve tail pipe, Section* 21.5.
1350			*Reboiler vent line, Section* 23.9.1.
1627			*Vent line to flare and reboiler drain line, Section* 14.11.
1120			*Lean-amine solution pump suction strainer, Section* 12.6.
1004			*Reflux drum boot drain line, Section* 2.4.3.
750			*Reflux/product pump suction line, Section* 10.9.2.
159			*Water draw from regenerator condenser, Section* 15.7.
12104	521	Refinery FCC main fractionator	Adding a filter upstream of the steam generators in the slurry PA circuit drastically reduced exchanger fouling. Fouling reoccurred after holes developed in filter baskets.

(*Continued*)

Chapter 18 The Tower as a Filter: Part B. Location of Plugging—Number 1 on the Top 10 Malfunctions (*Continued*)

Case	References	Plant/Column	Brief Description	Some Morals
			18.7 Feed and Inlet Lines	
902			*Debris in reflux pipe distributor, Section* 11.8.	
1262			*Debris in quench pipe distributor, Section* 5.9.	
637			*Solids settled in oversized lean-amine pipes, Section* 16.1.1.	
1007			*Drain on inlet transfer line, Section* 13.1.	
1207			*Feed filter blockage leads to explosion in peroxide tower, Section* 14.1.2.	
DT16.4			*Antifoam plugs static mixer in suction of feed pump.*	
			18.8 Instrument Lines	
15135, 15121			*Temperature indicators used for level indication to overcome tap plugging, Section* 25.2.	
1569	250	Refinery vacuum	Gas used for instrument purge contained solids, and these plugged the purge gas restriction orifice. The purge flow was lost.	
718			*Internal instrument line, Section* 10.2.	
1529			*Plugged tap of steam flowmeter fools advanced controls, Section* 29.6.2.	
1521			*Reflux drum level taps plug, causing liquid discharge to flare and pump damage, Section* 21.7.	
15109			*Plugged level control tap causes high liquid level and flood, Section* 8.1.1.	
1511			*Fouled thermowell leads to explosion in tower handling thermally unstable chemicals, Section* 14.1.4.	

Chapter 19 Coking: Part of Number 1 on Tower Top 10 Malfunctions

Case	References	Plant/Column	Brief Description	Some Morals
			19.1 Insufficient Wash Flow Rate, Refinery Vacuum Towers (*See also Spray Distributors, Section* 6.2.3)	
12107	161	Refinery vacuum	In most vacuum towers revamped for deep cut between 1988 and 2003, wash beds coked up due to design wash rates being too low to prevent coking, by as much as a factor of 4.	Ensure adequate wash.
117, 130			*Due to inaccurate TBP characterization of heavy fractions, Section* 1.1.7.	
218, 318			*Due to conventional simulation that predicts optimistic dry-out ratios, Section* 1.3.3.	
209	315	Refinery vacuum deep cut	Wash section coked up 3 months after trays were replaced by structured packing. This was accompanied by a pressure drop rise, a high metal content in the gas oil, and a low cut point. Wash rate was low and the spray nozzles had a high pressure drop (high entrainment), so that little liquid got to the lower part of the wash bed. Improvement achieved by doubling wash rate, replacing packing by larger, less efficient type, and redesigning sprays.	
320	174	Refinery vacuum	Wash bed coked within 6 months following replacement of low-efficiency by high-efficiency packings. Asphaltene balance showed that practically all the overflash was entrainment. The efficient packing vaporized more wash liquid, causing unwetting and coking in the lower packings.	Efficient packing promotes dryout in this service.
DT19.1, D19.2			*Tall wash beds and efficient packing lead to excessive vaporization and chronic coking.*	
331	458	Refinery vacuum	Spray header replacement to reduce wash oil rates and recover more gas oil caused wash bed coking and high pressure drop 1–2 years later. Pressure drop went further up after grid was cleaned by high-pressure lance from the top. New grid and original wash rates reinstated good operation.	

(*Continued*)

Case	References	Plant/Column	Brief Description	Some Morals
332	458	Refinery visbreaker vacuum	Coked wash beds and high pressure drops were caused by low wash rate. Metered slop wax (including entrainment from flash zone) was 0.1 gpm/ft^2.	
507	172	Refinery vacuum	Rapid coking was experienced in a 9-ft-deep wash bed, even though it operated at relatively low (980°F) cut point. The reason was dry-out due to excessive evaporation of the wash oil. Shortening the bed to 5 ft deep increased run length.	Excess bed height in this service promotes dry-out.
508	172	Refinery vacuum	Coking took place 6 months after a 2.5-ft-deep bed of Pall rings was replaced by a 5-ft-deep bed of grid and structured packings in the wash zone. At the same time, wash oil rates were reduced, giving almost zero overflash. The column was cutting deep (~1125°F).	Same as 507.
1231	375	Refinery vacuum	Many units operating in deep-cut mode experienced coking in the wash zone due to drying in the middle. This happened with grids, rings, and structured packings.	Ensure adequate wash.
DT19.3B			*An operating procedure that started wash oil about a week after start-up.*	
			19.2 Other Causes, Refinery Vacuum Towers	
504	304	Refinery fractionators	Grid showed high capacity and excellent resistance to plugging and coking in two fractionator wash sections. Coked sections, however, could not be cleaned at turnaround and needed replacement.	
415	375	Refinery vacuum	Several inches of coke was found deposited on the wash zone trays each turnaround. Following structured packing replacement, the wash zone was found clean.	
891	484	Refinery vacuum	The wash zone had an upper packed reflux section and a lower trayed slop oil PA section. Coking persisted despite many changes of trays, packings, and distributors over 20 years. Replacing the slop oil chimney tray by a draw pan mitigated coking, improved HVGO yield, but made it impossible to maintain the slop oil PA without pump freeze-ups.	Excess residence time of slop oil promotes coking.

1160			*Coke on HVGO chimney tray, Section* 9.7.	
828			*Due to I-beam interference with liquid distribution, Section* 6.10.	
841, DT19.3A			*Due to poor level measurement and excess liquid level on accumulator tray, Section* 9.6.	
1552	172	Refinery vacuum	Severe coking of tower internals near the flash zone was caused by excessive firing of the feed heater during rainstorms. The heater temperature control thermocouple was inserted only 1 in. into the transfer line. It cooled during rainstorms, leading to excess firing.	Watch out for thermocouples that are too short.
DT11.6			*Throttled gate valve at heater outlet causes excess ΔT, forcing yield cut to prevent cracking and coke.*	
15138	453	Refinery visbreaker vacuum	Tower suffered severe coking every 12 months. Contributing factor was a short heater outlet control thermocouple that read 8°C low and caused excessive firing. Also, the controlled fuel oil flow to the heater was not measured directly but was calculated as a small difference between two large numbers, so it fluctuated, causing the coil outlet temperature to fluctuate.	
DT19.5B			*Coke accumulates behind support ledges of stripping trays, causes pump trips.*	
			19.3 Slurry Sections, FCC Fractionators	
DT19.4			*Several cases with grid packing.*	
824, 825			*Due to vapor maldistribution at grid bed and vapor inlet baffles, Sections* 7.1.2 and 7.2.	
207			*Due to inadequate mixing of slurry and wash in distributor, Section* 6.11.	
1174, DT19.4A			*Due to plugging in slurry liquid distributor, Section* 6.2.1.	
DT19.5A			*Coke accumulates behind support ledges of segmental baffle trays.*	

(*Continued*)

Chapter 19 Coking: Part of Number 1 on Tower Top 10 Malfunctions (*Continued*)

Case	References	Plant/Column	Brief Description	Some Morals
			19.4 Other Refinery Fractionators	
510	168	Refinery coker fractionator	Wash trays were replaced by grid, charge raised 10%, and recycle dropped from 1.2 to 1.1. Short runs, heavies carryover, and operational difficulties followed due to grid and collector plugging. Too efficient a grid bed, leading to excessive vaporization and drying, was a contributor. Shortening the bed from a 7.5-ft high-efficiency grid to a 2-ft low-efficiency grid plus a 2-ft high-efficiency grid helped alleviate the problem. (See also 208, Section 3.2.4; 417, Section 4.4.2; 829, Section 7.1.2; 1236, Section 5.7.)	Same as 507, Section 19.1.
417			*Tray dry-out at low liquid loads, Section* 4.4.2.	
208			*Subcooled liquid increases nozzle-plugging tendency, Section* 3.2.4.	
829			*Vapor maldistribution, Section* 7.1.2.	
1236			*Plugging of liquid pipe to shed decks, Section* 5.7.	
440	332	Refinery crude fractionator	Coking in the stripping section blocked downcomers and trays reducing run length to 18 months and/or requiring reducing the stripping steam to near zero. Trays were modified for fouling-resistant design.	
8112			*Excessive temperature foul crude tower stripping trays, Section 7.1.2.*	
1272			*Plugging of draw line in visbreaker fractionator, Section* 10.3.	
			19.5 Nonrefinery Fractionators	
419	462	PVC VCM steam stripper, atmospheric	Within 24 hours from start-up, bottom-product color went off specification due to coking of PVC in stagnant hot spots on the sieve trays. Most deposits were in downcomers opposite the outlet, along the upstream of inlet and outlet weirs, and on the weir edges near the shell. Acceptable on-specification run lengths were achieved by removing outlet weirs, installing modified arc downcomers, leaving flushing gaps at the edges of the inlet weirs, and recycling product to feed to keep up the liquid velocities.	Keeping the solids moving can be an effective strategy in fouling services.
DT18.5			*Coke fines plug structured packing in an olefin oil quench tower.*	

Case	References	Plant/Column	Brief Description	Some Morals
			20.1 Pump, Compressor	
1246	250	Depropanizer	Excessive propane concentration was measured in the bottom. The cause was a seal failure on the reboiler pump; the pump used propane as seal gas.	
1271, 1121			*Oil leak plugs towers, Section* 17.5.	
1218			*Seal leak causing fire, Section* 14.8.	
1029			*Seal leak leads to water pressure surge, Section* 13.1.	
1227			*Seal oil leak reduces packing wettability, Section* 4.8.	
			20.2 Heat Exchanger	
20.2.1 Reboiler Tube				
1102	306	Refinery debutanizer	Excessive tube leakage in a reboiler caused reboil, reflux and column instability, and inability to run at design pressure.	A complex-looking problem may have a simple cause.
1348	407	Ammonia Benfield hot-pot regenerators	The plant had two parallel process trains. A once-through horizontal thermosiphon reboiler was heated by hot absorber feed gas. Following a number of hot and cold cycles induced by frequent start-ups and shutdowns, the floating head gaskets failed in both trains. Gas leaked into the shell side, causing corrosion, atmospheric leaks, and product contamination.	Shell was changed from CS to SS to prevent recurrence.
1215	81	Batch distillation	Leakage of steam from the reboiler contaminated overhead product with water. A radiotracer technique diagnosed the source of leak and measured the rate of leakage.	Tracers are useful for diagnosing exchanger leaks.
1022	234	Batch distillation	A small steam leak at the reboiler contaminated overhead product. Source and quantity of leak determined by radioactive tracer.	As 1215.
1253	134	C_2 splitter	Tracer studies showed that 1.5% of the high-pressure reboiler flow leaked into the low-pressure side.	As 1215.

(Continued)

Case	References	Plant/Column	Brief Description	Some Morals
DT20.1			*A reboiler leaking about* 1.5% *of the steam flow in aniline plant tower diagnosed by tracer tests.*	
1286	116, Case MS12	Refinery naphtha fractionator	Rapid plugging of the reboiler heater charge pump strainers caused difficulty in maintaining flow to the furnace. Inadequate safety instrumentation permitted the operator to continue firing the heater in the hope of restoring the flow. The heater tubes overheated and ruptured in 4 minutes.	
1349	142	Offshore gas glycol dehydration regenerator firetube reboiler	Thermal expansion led to fireside bowing, which caused tension in the welds attaching the firetubes to the tube plates. A weld failed. Glycol entered the firetube and burned. Operators shut off the gas, but glycol in the shell continued to burn, exhausting to a vent via a PSV/vapor breaker. Thermal expansion of the piping venting the combustion products sheared the bolts of the PSV/vapor breaker, releasing vaporized glycol that ignited and produced an external fire and damage.	Ensure system components adequately accommodate thermal expansion.
1628			*Leading to overchilling and explosion, Section* 14.3.2.	
1421			*Contributing to product impurity, Section* 24.4.	
20.2.2 Condenser Tube				
1014			*Leading to water pressure surge, Section* 13.6.	
20.2.3 Auxiliary Heat Exchanger (Preheater, Pumparound)				
1212	305	Natural gas glycol contactor	Poor dehydration was caused by a leaking feed-effluent exchanger that leaked cold, wet glycol into the heated dry glycol flowing to the top of the contactor.	
1266	500	Refinery stabilizer	A 0.5% preheater leak, diagnosed by tracer tests, caused liquid buildup on the feed tray, temperature profile anomalies, and poor performance.	
1276	178	Refinery crude fractionator	Kerosene showed color due to a tube leak in the crude/mid-PA exchanger just above the kero draw-off.	

DT20.2			*Feed-bottom interchanger leak causes poor product quality and premature flood.*
1285	458	Refinery vacuum	Leaking crude-HVGO PA exchanger caused overhead pressure to suddenly rise by 10 mm Hg. Identified by slop oil rate near doubling showing a large amount of naphtha, kerosene, and diesel.
DT20.3			*Several exchanger leaks in olefin oil quench tower PA.*
1294	362	Ammonia CO_2 absorber	The lean and semilean arsenic-activated Vetracoke solution coolers developed leaks due to underdeposit corrosion. This caused solution and cooling-water contamination. Eventually solved by going to less aggressive solvent. (See also 522, Section 18.4.3.)
12112	351	Refinery HF alkylation DC_3	About 4% of tower bottom stream leaked into the HF-rich recycle via one of the two bottom coolers, causing poor product quality. Diagnosed by tracer tests. Cured by bypassing the exchanger bank.
1308			*Steam preheater leak sends noncondensables that gas blanket the reboiler, Section* 23.9.1.
1649			*Steam leak into isolated preheaters leads to overpressure, Section* 21.5.
			20.3 Chemicals to/from Other Equipment
20.3.1 Leaking from Tower			
103			*Leak of overhead valve during total reflux operation leads to explosion, Section* 14.1.4.
1114			*Leak into steam system via purge connection, Section* 14.10.
20.3.2 Leaking into Tower			
1108, 1109, 11101			*Through unblinded valve at shutdown causes accidents, Sections* 14.10, 12.1.
1009			*Hot oil leaking into tower at shutdown causes water pressure surge, Section* 13.7.
1614			*Leaking recycle water valve fills tower with water, which generated vacuum upon draining, Section* 22.1.

(*Continued*)

Case	References	Plant/Column	Brief Description	Some Morals
20.3.3 Product to Product				
DT8.1			*Kerosene product off specification for heavies due to leakage of valve on kero/diesel cross connection.*	
1197			*Leaking bypass valve prevents isolation, leading to fatal fire, Section* 14.8.	
			20.4 Atmospheric (*See also explosions and fires due to line fracture, Section* 14.3)	
20.4.1 Chemicals to Atmosphere				
125, 126			*Causing ethylene oxide accumulation in insulation, which led to explosion, Section* 14.1.1.	
1193	324	DMF-triethlyamine	After switching from acetone separation to new service, 316L SS base of tower experienced recurrent leaks, causing hazardous and toxic chemical releases. Reboiler pipe leak was due to reaction yielding formic acid at the tower base, together with a reboiler pipe fabricated of 304L instead of 316L SS, and poor weld quality. Shell leaks and collapsed trays were due to frequent steaming and drying between campaigns using high-chloride steam causing stress corrosion cracking. Recurrence prevented by upgrading base material, and keeping column bottom wet.	
1647	30	MEA stripper	Following a number of leak incidents, the SS MEA feed line and flange separated from the CS column nozzle. Caused by galvanic corrosion of nozzle neck.	
1629	390	Refinery HF alkylation depropanizer	Failures in one flange and one weld in the depropanizer charge CS pipe forced shutdown of alkylation unit. The main factor that accelerated the corrosion was the operating temperature (190°F), being well above the maximum recommended 150°F for CS in HF service.	
1632			*Flange leak initiates fire that causes overhead receiver failure, Section* 21.1.	

1163	116, Case MS60	Refinery FCC main fractionator	A valve on the inlet to the main fractionator was closed during start-up and then had to be opened to bring the unit into operation. The procedure was reversed during shutdown. The thermal shock of the changeover caused persistent flange leaks. The position was improved by better gasketing, but it was only finally resolved by a change in operating procedures, which avoided the need to use this valve.	
1633	116, Case MS49	Refinery crude fractionator	The bottom flange inside the tower skirt leaked residue. The leak got progressively worse and autoignition followed. The cramped conditions inside the skirt could have been the reason that the joint was badly made up in the first place and made it impossible to tighten up the bolts once the leak had started.	Compare 1169. Such flanges should be located outside the vessel skirt.
1169			*Leak into the tower skirt, Section* 12.14.	
1107			*Caused by rigid support stretched by thermal expansion, Section* 12.9.	
1161			*At a flange of stab-in reboilers due to piping thermal expansion, Section* 23.5.	
1622	283	Glass column	Water was sprayed to disperse leak of inflammable vapor after glass cracked. The water drops were electrically charged, and the charge was collected on the metal insulation cover, which was not grounded. A spark was seen to jump from the insulation cover to the water line. Fortunately, it did not ignite the leak.	

20.4.2 Air into Tower

104			*Causing polymerization reaction, Section* 15.8.	
116			*Causing explosion in peroxide tower, Section* 14.1.2.	
1623	179	Hydrocarbons	A vent condenser on overhead line to vacuum system captured escaping volatile hydrocarbons. An inerts flowmeter on the vent line (normally no flow) was invaluable in detecting air leaks that induced escape of volatile HCs.	Monitoring "inerts" vent flow reduces emissions.
DT20.4			*Leak through a corroded steam line nipple disrupts PA flow in the tower upper cooling loop.*	

Case	References	Plant/Column	Brief Description	Some Morals
			21.1 Relief Requirements	
1601	522	10-ft-ID column	Distillation column relief requirement was based on failure of the steam flow controller. The relief requirement was more than halved by adding a restriction orifice in the steam supply line.	A method for reducing relief discharge requirements.
1602	522	High-boiling hydrocarbons 15 ft ID	Column relief requirement was based on the loss of cooling to the condenser and full steam on the reboiler. Adding additional (redundant) controls to shut off steam to the reboiler on high temperature or pressure effected a severalfold reduction in relief requirement.	A method for reducing relief discharge requirements.
1632	116, Case MS8	Refinery depropanizer	A depropanizer overhead receiver failed when exposed to a fierce fire caused by a flange leak. The relief valve design was not adequate for the fire case. However, heating the surface of the vessel above the liquid level compounded the problem.	
1634	115	Aromatics extraction glycol stripper 16 ft ID	Recommended design standard called for vacuum breaking using a total replacement of the volume of vapor entering the overhead condensers. This flow rate equaled the refinery's total fuel gas consumption; so if installed, vacuum breaking would have shut the refinery down. Revising the vacuum breaker system to supply enough fuel gas to gas blanket the condenser provided effective vacuum breaking while reducing the fuel gas consumption 300 times.	
DT21.1			*Top PSV not designed for blockage of overhead resulting from bowing of exchanger bundles and rotation of impingement plate.*	
			21.2 Controls That Affect Relief Requirements and Frequency	
1603	522	8-ft-ID column	Top section of column was destroyed when cooling-water valve to condenser was inadvertently shut. Steam supply was controlled by column *dP*; the loss of cooling caused the controller to open. Two failures thus occurred simultaneously; relief capacity was designed only for one.	Carefully examine control behavior when determining relief requirements.

1604	522		Column top head was torn loose and tray remnants blown out as a result of inadequate relief capacity. The reason for inadequate relief capacity was identical to 1603.	As for 1603.
1624	93	Natural gas demethanizer	Relief pressure was set too close to normal operating pressure, so slight pressure swings opened the relief valve. The problem was compounded by lack of automation on the inlet and outlet compressors and instability in the tower temperature controls. The number of alarm trips was drastically reduced by APC, which acted on disturbances in the gas volume balance due to inlet supply or compression.	
DT28.8			*Control of cooling-water temperature to condenser configured to prevent cooling loss when booster pump fails.*	
			21.3 Relief Causes Tower Damage, Shifts Deposits	
DT21.2			*Oversized relief devices lead to rapid depressuring and tray uplift in one case, downbending in other.*	
1613	442	Ammonia stripper 6 ft ID	Trays and supports were damaged due to reversal of vapor flow upon lifting of a relief valve located below the bottom tray. Recurrence prevented by replacing moving-valve trays with fixed-valve trays with truss lugs and resetting the relief valve to higher pressure.	Reverse flow through valve trays can lead to tray damage.
1013			*Condensing of steam purges to relief valve during outage initiates water-induced pressure surge, Section* 13.6.	
1653			*Liquid pooling in an undrained dead-leg line to a relief valve leads to popcorn polymer, line rupture, Section* 14.12.	
1603			*Controls induce double failure, Section* 21.2.	
12103			*Relief moves deposits up and clears plugging, Section* 17.2.	
			21.4 Overpressure Due to Component Entry	
1609	306	Refinery iC_4–nC_4 splitter	Total condenser was close to maximum capacity, and no adequate venting was available. Each time propane in the column overhead would rise, the relief valve would lift.	Relief valves are not vent valves for light nonkeys.

(Continued)

Chapter 21 Relief and Failure (*Continued*)

Case	References	Plant/Column	Brief Description	Some Morals
1146	250	Refinery HF alkylation depropanizer	Problems with interface level in the reflux drum caused carryover of hydrofluoric acid into the propane product route. This overpressured downstream equipment.	
1545			*Same as* 1146, *causing an explosion, Section* 14.2.	
111			*Accumulation of lights, Section* 2.4.3.	
1595			*Water entering hot HC tower, Section* 28.1.	
1510			*Base-level controller failure overpressures downstream storage tank, Section* 8.6.	
913			*Incorrectly installed internal pipe on bottom draw-off overpressures downstream storage tank, Section* 11.3.	
			21.5 Relief Protection Absent or Inadequate (*See also Sections* 21.1–21.4)	
1607	306	Refinery depropanizer	Column pressure reached 450 psig. The inlet line to the relief valve (set at 300 psig) was plugged by corrosion products. Both the pressure controller and high-pressure alarm came off the same transmitter and gave no indication of high pressure. Problem was only discovered when the feed pump could not maintain flow to tower.	In fouling services, a sensing element for an alarm or trip must be separate from that used for control.
1649	32	Azeotrope still	Water was trapped in the tower feed line between shut valves upstream and downstream of a preheater. Steam entered the preheater via a shut, but passing control valve. Hydraulic pressure of the heated water rose above 800 psig, causing the gasket in the main preheater joint to fail. When organics were fed to the still the joint leaked. A large fire resulted.	Relief devices are needed on heat exchangers.
1644			*Fire and explosion due to relief to atmosphere, Section* 14.9.	

1290			*Liquid discharge due to plugged packing, Section* 4.11.	
1652	67	Deethanizer	Steam was bled into the tail pipe of an atmospheric relief valve to prevent leak ignition. The drain hole was plugged, accumulating condensate in the tail pipe. Up to 1 ton of condensate could have accumulated and caused scalds upon discharge. The hole was rodded out.	Keep tail pipe drain holes clear.
1631	116 Case MS3	Refinery coker unit	In checking the relief valve logic on a unit in connection with proposed modifications, an existing fault was found. A previous modification had installed a valve which could block off the overhead receiver from the relief valve on the fractionator column, which was intended to protect it.	
1605			*Chemical discharge to atmosphere because relief valve not connected to flare, Section* 14.12.	
1635	136	Benzene concentration	Due to incorrect low setting of the tailings column relief pressure, the valve lifted "light" at start-up. The high-pressure trip was also incorrectly set. A benzene/gasoline mix was released to atmosphere.	Check relief valve and trip setting.
1615	482	Chemicals	Condenser was a steam generator. At start-up, it cooled excessively. The steam outlet control valve was throttled, raising steam pressure. A miscalibrated relief valve limited the pressure rise. The cooling remained excessive, which limited the vapor distillate rate. Problem solved by shutting column down and resetting relief valve pressure.	
969			*Glass still explodes when pressured by incorrect N_2 source, Section* 12.5.	

21.6 Line Ruptures

(*See also Section* 14.3 *for explosions and fires due to line fracture*)

1653			*Reboiler line rupture due to popcorn polymer, Section* 14.12.	
1627			*Flare line dislodged, releasing C_4 to atmosphere, Section* 14.11.	

(*Continued*)

Chapter 21 Relief and Failure (*Continued*)

Case	References	Plant/Column	Brief Description	Some Morals
			21.7 All Indication Lost When Instrument Tap Plugged	
1521	304	Refinery depropanizer	Reflux drum level indicator, level gage, and level alarm were connected to the same taps. The upper tap plugged, and all became erratic. This caused liquid to flow into the flare because of excessive level. Later, the reflux pump blew a seal because of cavitation resulting from low level.	Level indicator and level gage should not share the same tapping.
1607			*Contributing to tower overpressure, Section* 21.5.	
			21.8 Trips Not Activating or Incorrectly Set	
1608	306	Ethylbenzene fractionator	Column was reboiled by a fired heater. Heater fuel was controlled by a tray temperature, and there was a high-temperature trip at the process-side heater outlet. When the circulation pump briefly failed, the column cooled, and the controller increased heating rate. The trip failed to function. When circulation was reestablished, an extremely high vaporization rate resulted, producing a pressure surge that dislodged trays.	Where trips are critical, use a high-reliability trip system. Test it regularly.
1610	6	Refinery FCC main fractionator	Massive carryover of liquid from the reflux drum destroyed internals of the overhead compressor and damaged its turbine driver. The incident followed a fire at the product pump, which made it and its spare inoperable. The accumulator level rose past the compressor trip level, but the trip failed to activate.	Trips cannot always be counted on.
1147			*Tower imploded because nitrogen blanketing trip fails to activate, Section* 22.1.	
1635			*Incorrectly set trip contributes to release of HCs, Section* 21.5.	

			21.9 Pump Failure
1610			*Product pump failure leads to damage to overhead compressor, Section* 21.8.
1608			*Fired heater circulation pump failure leads to tray damage, Section* 21.8.
1218			*Absorber–regenerator circulation pump failure leads to seal leak that fired, Section* 14.8.
503			*Absorber–regenerator circulation pump failure leads to melting of plastic packings, Section* 12.9.
			21.10 Loss of Vacuum
1606			*Leading to explosion in nitro compound tower, Section* 14.1.3.
1130, 1158			*Causing vapor backflow and tray collapse, Section* 22.6.
			21.11 Power Loss
			(*See also Section* 21.9 *for pump failure*)
1630	465	Extractive distillation	A power failure during commissioning caused a unit upset that dislodged trays.

See also:
Section 6.7, Damaged Distributors Do Not Distribute Well.
Section 8.3, High Base Liquid Level Causes Tray/Packing Damage.
Chapter 13, Water-Induced Pressure Surges, all sections.
Chapter 14, Explosions, Fires, and Chemical Releases, Sections 14.1–14.9.
Chapter 21, Relief and Failure, Sections 21.2–21.4, 21.6–21.11.

Case	References	Plant/Column	Brief Description	Some Morals
			22.1 Vacuum	
1112	327, Vol. 2, Case 1088	Chemicals solvent recovery still	At shutdown, the still was being steamed. After steaming, cold water was applied to cool the column. The sudden cooling caused a partial vacuum, and the still imploded. The open vent did not have sufficient capacity to relieve the vacuum.	Either adequately design the vessel for vacuum or avoid rapid cooling after steaming.
1147	523	Ammonia amine regenerators	At an outage, the top sections of the two in-parallel regenerators collapsed due to vacuum. During the outage, amine circulation and reboiler steam were cut to conserve energy. The steam reduction led to condensation and vacuum. The towers were isolated from vents. The nitrogen-blanketing trip failed to properly activate and prevent the vacuum.	Same as 1112, plus ensure critical trips are always operational.
1614	424	Stripper: 9 ft ID bottom, 5.5 ft ID top, atmospheric	Stripper overhead was vented into absorber feed. Plant was being commissioned with water circulation. A leaking recycle water valve caused both absorber and stripper to fill up. The vent line from the stripper could not provide vacuum relief because it was flooded at the absorber feed. Upon draining, the stripper collapsed just below the swage.	Recurrence prevented by installing vacuum breakers and rerouting the stripper vent.
967	116, Case MS9	Refinery naphtha reforming	During construction, a vent hole was omitted from the inlet pipe on a fractionator overhead receiver. As a result, the column ran under vacuum when started up. Luckily, it did not collapse.	

22.2 Insufficient Uplift Resistance

1238	442	Chemicals steam stripper 9 ft ID	Bottom two to three trays suffered repeated valve loss, were damaged, and needed replacement on a 2–3-year cycle. Recurrence prevented by replacing those with heavy-duty fixed-valve trays with cross-channel braces, shear clips, and double nuts.	Heavy-duty tray designs can prevent damage.
1239	442	Natural gas amine regenerator 5 ft ID	Column suffered chronic loss of valves and trays in bottom section due to corrosion and frequent process surges. The bottom 25% of trays were replaced with fixed-valve trays with trusses, extra strength, and upgraded metallurgy. This eliminated problem.	Same as 1238.
1240	442	Refinery vacuum 11 ft ID	Repeated loss of stripping trays was eliminated by replacing trays with grid, heavy-duty supports, and a steam "H" pipe distributor. The retrofitted internals weathered upsets without damage over 7 years.	Same as 1238.
1260	412	Refinery FCC main fractionator	Disk and donut trays in the slurry section were dislodged during upsets. These were replaced by grid that weathered upsets well.	
1259	291	Refinery FCC main fractionator	In two turnarounds, the grid at the bottom of the slurry pumparound bed was found slightly damaged. 180-ft/s feed vapor was ripping the bed apart. Despite the damage, performance remained good. Cure was reinforcing damaged area without removing the bed.	
DT8.2			*Grid weathers upsets better than trays.*	
DT22.1			*Larger holddown clips prevent recurrence of tray uplift.*	
1025, 1027, 1035			*Having insufficient uplift resistance contributed to tray uplift, Sections* 13.5, 13.1.	
1018, DT8.5 I, J, DT22.5, DT25.6B			*Heavy-duty tray design improves uplift resistance, Section* 13.1.	
422			*Heavy-duty tray design and tray replacement by packing improve upset resistance, Section* 22.7.	
403			*Cast iron bubble caps weathering pressure surges better than valve trays, Section* 13.1.	

(*Continued*)

Chapter 22 Tray, Packing, and Tower Damage: Part of Number 3 on the Top 10 Malfunctions (*Continued*)

Case	References	Plant/Column	Brief Description	Some Morals
1177			*Enhancing tray strength can help even with heavy-duty trays, Section* 22.4.	
1230	167	Refinery lube vacuum	"Explosion-resistant" trays were damaged. Gamma scans suggested trays were intact and operating properly. Pressure survey and lack of separation confirmed the damage.	Gamma scans need process cross checks.
805	304	Refinery crude fractionator	A pressure surge caused breakage of clamps holding together a bed limiter. Sections of the limiter were dislodged; random packings were damaged and carried over. This resulted in poor separation.	As 1238.
DT9.4			*Upward liquid force while filling tower with liquid exceeds uplift resistance of chimney tray with tall chimneys.*	
889, 12118			*Stronger chimney trays prevent damage recurrence, Section 9.8.*	
			22.3 Uplift Due to Poor Tightening During Assembly	
938, DT22.2	224	Acetylene absorber	Unbolted sieve trays were dislodged by a major upset, which occurred occasionally. Good bolting prevented recurrence.	
957	491	Highly corrosive service	Manways were secured by wedge-shaped clips, which were supposed to be wire tied in place. The clips were installed but not tied down. During operation, the manways were displaced, leading to severe weeping and poor separation. Gamma scans diagnosed.	
972	286	Olefins water quench	A start-up upset dislodged the drain pan that collected liquid from the top bed to feed the sprays to the bed below. Caused by hold-down clamps improperly installed and bolts missing. Temporary cure was hot tapping a set of spray nozzles similar to those used in the same tower on an earlier occasion (Case 1195, Section 4.13).	
925	376	Refinery vacuum	Improper grid installation plus a unit upset led to a loss of the wash oil grid. The bed was unified with through bolts to avoid recurrence.	
951	250		Poor tightening caused an entire grid bed to fall to the bottom of the column.	
952	250		Poor tightening caused loud banging noise from an operating column.	

889			*Poor support of protective shroud, Section* 9.8.	
1292			*Low mechanical strength a factor in uplift, Section* 22.4.	
			22.4 Uplift Due to Rapid Upward Gas Surge	
1292	35	Ammonia Benfield hot-pot absorber	The bottom bed was uplifted 1.6 m, and the redistributors of the bottom two beds were also uplifted due to a sudden upward gas surge. The most likely cause was gas backflow through the semilean solution line, pump, and check valve from the low-pressure regenerator. Low mechanical strength was also a factor.	Beware of reverse flows. See 1218, Section 14.8.
1196	80	Natural gas MDEA absorber	14 valve trays were damaged upwards and collapsed with downcomer bent inwards, causing excess H_2S in the product gas. Probable cause was sudden thrust of high-pressure gas when tower was depressured.	
1156	194, 539	Chemicals	Rapid flashing of a pool of liquid at the base of a 10 ft-ID tower dislodged several trays.	
1177	184	Volatile HC/water/ chlorinated HC tower, atmospheric	The 20–30 lowest "heavy-duty" Bayercap trays were repeatedly buckled or split upward during tower shutdowns. Upon shutdown, stripping trays liquid drained and phase separated in sump into heavy chlorinated HC layer and lighter aqueous layer. Drained-rectifying tray liquid then formed a third light HC layer on top. A 40°C temperature difference initiated vigorous boiling at the interface. This reduced the hydrostatic head in the sump, initiating boiling at the water-chlorinated HC interface, which mixed the phases. The enhanced contact violently boiled the HCs, causing the pressure surge. Cure was reducing sump level, cooling sump liquid, and enhancing tray strength.	This new "delayed-boiling" phenomenon was diagnosed and discovered by thorough laboratory tests.
DT21.2			*Oversized rupture disk leads to tray uplift.*	
1217			*A slug of flushing oil entering a refinery vacuum tower, Section* 13.4.	
DT22.3			*Reinstating PA circulation to hot packing induces rapid vaporization and packing uplift in quench tower.*	

(*Continued*)

Case	References	Plant/Column	Brief Description	Some Morals
DT8.5 B, C			*Rapid vaporization intensifies slugging action due to high liquid level.*	
DT22.3			*Stepping up cold near tower top rapidly drops pressure, uplifting packings.*	
DT22.4			*Absorption of light component in condenser rapidly drops top pressure and uplift trays.*	
1608			*A sudden heat step-up uplifts trays, Section* 21.8.	
DT22.5			*Compressor surge uplifts trays in olefin caustic wash tower.*	
DT22.6			*Trays scream after vacuum pump returns from outage to hot tower, rapidly dropping top pressure.*	
			22.5 Valves Popping Out	
1243	250	Several cases, valve trays	Valves popped out of their seats. This problem was not realized until the column was shut down for inspection or until the popped-out valves damaged the column bottom pump. In all these cases, operation was at high throughputs; in some, the fraction of popped-out valves was high.	
DT22.7			*Valve pop-out, numerous experiences.*	
1238, 1239			*Repeated valve losses, Section* 22.2.	
446			*Due to flow-induced vibrations, Section* 22.10.	
946	250	Valve tray	Home-made valves lasted a very short time before popping out.	
			22.6 Downward Force on Trays	
1103	446	LPG lean-oil stripper	Column was pressured up through a connection in the overhead system while liquid circulated through its valve trays. The gas could not travel downward, causing mechanical damage to top 12 trays. This later resulted in premature flooding.	Pressure columns from the bottom up, especially with valve trays.
1130	194	Chemicals vacuum tower	A sudden loss of vacuum from the top of the tower caused vapor backflow through the column. This exerted a downward force on the trays, which in turn caused the trays to collapse.	

1158	220	Refinery vacuum in mild hydrocracking 21 ft ID	Over the years, tower had experienced several upsets that damaged its valve trays. The upsets were caused by loss of vacuum and pressure surges emanating from the top of the column. These forced valves shut and exerted tremendous downward forces on the trays, which either caused trays to collapse or, in some cases, completely destroyed them.	Trays were replaced by structured packings. These survived all upsets during the first 10 months.
1143, DT22.8	250	Vacuum 10-ft-ID valve trays	Feed entered 20 trays above bottom. The 16 lower trays distorted downward, with major support beams bent and twisted into V shapes up to 1 in. deep. Below the feed, the column was full of liquid when the reboiler was started. The boiling of some liquid left a vapor gap under the bottom tray, which caused the tray to fail downward and shifted the gap upward. Downcomer plugging could have played a role.	
1613, DT21.2			*Relief valve at bottom of tower leads to valve tray damage by backflow, Section* 21.3.	
DT22.9			*Sudden loss of vacuum generates downflow that bends down bubble cap trays.*	
DT22.10, DT22.12			*Rapid condensation by cold water entering tower full of steam generates downward force that damages trays.*	
DT22.11			*Reflux drum liquid, forced back into the tower via overhead line by pressuring drum, bends trays down.*	
11101			*Explosion in tower upper head, Section* 12.1.	
		22.7	**Trays below Feed Bent Up, above Bent Down and Vice Versa**	
1140	250	Valve trays	During commissioning, column contained steam and was fed with cold water. When the water rate was stepped up, rapid condensation took place at the feed tray, generating a local vacuum. Trays above were bent downward; those below were bent upward.	
DT22.13			*Similar to* 1140, *but during the run, after preheater fouled up and feed to sour water stripper was cold.*	

(*Continued*)

Case	References	Plant/Column	Brief Description	Some Morals
422	442	Chemicals stripper, 7-ft-ID valve trays	Gas inlet was located several trays from the bottom. During operation, plant experienced sudden bursts of feed gas velocity that collapsed trays below and displaced trays above the feed. Recurrence prevented by retrofitting a section of packing (that could resist flow reversal) below the feed and several heavy-duty trays with explosion doors above the feed.	Avoid reverse flow with valve trays. Heavy-duty designs can prevent damage.
			22.8 Downcomers Compressed, Bowed, Fallen	
1144	250	Refinery stabilizer	At shutdown, column was filled with water, then the bottom manhole was opened. Sieve tray drainage was much faster than downcomer drainage. Water remaining in the downcomers in locations of wide tray spacing (e.g., near manholes) exerted a hydrostatic force that bent these downcomers toward the tray, pulling them out of their braces every shutdown.	
1255	8	Refinery coker fractionator 9 ft, 6 in ID, two-pass	Instability, heavy entrainment, and high *dP* occurred during portions of the coker cycle. Inspection showed a lower center downcomer to be compressed so that its bottom width was 2 in. instead of 8 in. This induced local flooding. The five top trays were fouled to the tops of the bubble-cap risers. Gamma scans suggested that some of the lower side downcomers were also plugged.	
439			*Downcomers bowing over inlet weirs induce flooding, Section* 4.4.1.	
1196			*Downcomers bent inwards following gas upsurge, Section* 22.4.	
1281			*Bottom flow restricted because of fallen downcomer, Section* 10.4.	
DT11.4			*Poorly installed overflow weir on a reboiler draw pan falls off, starving reboiler.*	
778			*Seal pan damage due to flashing feed impingement, Section* 5.5.	

			22.9 Uplift of Cartridge Trays	
741	250		A bundle of cartridge trays uplifted and separated from another, leaving an unsealed downcomer between. This caused premature flood.	Ensure adequate holddown.
742	250		A bundle of cartridge trays uplifted and separated from another. Several gasket pieces ended in the bottom of the tower.	Same as 741.
1263	489	Chemicals batch 2.5 ft ID	The 43 trays were in nine cartridges, interconnected by ten 1.5-in-diameter tie rods, secured with interlocking hardware at the column base. The bottom two trays partially corroded/eroded during operation, causing the tie rods to be dislodged, which ultimately caused trays to slip from 2 in. near the top to 14 in. near the bottom. Further drop was stopped by the tower thermowells.	Gamma scans clearly showed the slipping of trays.
			22.10 Flow-Induced Vibration	
408	72	Chemicals, five columns, pressure, vacuum, 5–25 ft ID	Flow-induced vibrations at gas rates close to the weep point caused damage to sieve and valve trays, support beams, and beam-to-column supports, at times within hours of operation at the damaging vapor rates. In one case, total internal collapse resulted; in another, shell cracking occurred. Successful cures included reducing fractional hole area, stiffening support beams, and avoiding low operating rates.	Article proposes effective techniques for preventing flow-induced vibrations.
407	129	Oxidation reaction effluent absorbers	Heavy flow-induced vibrations in two 14-ft-ID towers in a train of four, while operating at low rates, caused cracking of beam to support welds and of some sieve trays. Cure was raising gas flow rates.	Same as 408.
445	474	Acetic acid–water	Flow-induced vibrations caused cracking in two-pass, 12-ft-ID fixed-valve trays that periodically operated at low hydraulic loads. Recurrence prevented by avoiding low loads.	
446	474	Chemicals	Flow-induced vibrations caused cracking and popping out of a large number of round valves in one-pass, 8-ft-ID trays while operating at hydraulic loads so low that valves were fully closed.	

(*Continued*)

Chapter 22 Tray, Packing, and Tower Damage: Part of Number 3 on the Top 10 Malfunctions (*Continued*)

Case	References	Plant/Column	Brief Description	Some Morals
447	474	Chemicals	Flow-induced vibrations repeatedly caused cracking in one-pass, 13.5-ft-ID 10-gage sieve trays. The trays operated at 14% of jet flood, which was below the harmonic vibration region. Vibrations were observed during higher load operation at steamout.	
DT22.14			*Flow-induced vibrations mitigated by replacing sieve by valve trays.*	
437	539	Chemicals, 11-ft-ID valve trays	As tower reached 25% of its capacity upon initial start-up, efficiency dropped dramatically. Cause was flow-induced vibrations and damage to the top-section one-pass trays and supports; the two-pass trays in the bottom remained intact. This recurred following two sets of supplier-recommended modifications. Problem solved by increasing tray natural frequencies.	An effective technique for preventing vibration damage.
DT22.14			*Double-locking nuts prevent loosening of tray nuts during operation.*	
438	539	Chemicals	Violent action and resulting large deflections wedged washers in the cracks between tray panels, shook loose hardware lying on the tray decks, and led to fatigue failure.	
710			*Possibly affected by reboiler return velocities, Section* 23.3.	
			22.11 Compressor Surge	
DT22.5			*Trays uplift during compressor surge.*	
1245	250	Several	Demisters and trays were dislodged or damaged by backpressure during a compressor surge.	
1157	144	Olefins caustic scrubber	Surge of the cracked gas compressor at start-up caused uplift of packing support in one section, sending packings to the recirculating pump filters.	

			22.12 Packing Carryover	
834	464	Heat transfer	Fluidization of random packings led to a loss of heat transfer and excessive condenser load that limited plant capacity. No bed limiter was used.	Do not forget the bed limiter.
846	250		Metal random packings settled unevenly, some in distributor troughs, following an upset.	Same as 834.
847	250	Several	Carryover of plastic random packages occurred from beds that did not have bed limiters.	Same as 834.
898			*Lack of holddown causing packing disturbance, Section* 2.6.2.	
			22.13 Melting, Breakage of Plastic Packing	
502, 503, 1105			*Plastic packings melt due to overheating or outage, Section* 12.9.	
809			*Plastic packings soften and pass through support, Section* 4.10.	
947			*Plastic packings break during loading at cold temperatures, Section* 11.9.1.	
			22.14 Damage to Ceramic Packing	
910, 914			*Breakage during loading into tower, Section* 11.9.1.	
501			*Breakage during operation, Section* 4.7.	
522			*Leaching by solution, Section* 18.4.3.	
DT15.3			*Reaction that chews ceramic packing keeps product on specification.*	
			22.15 Damage to Other Packings	
12113	544		Gamma scan showed a density gradient in a 12-ft-tall packed bed immediately after turnaround. Gamma scan before next turnaround showed much steeper gradient, flooding at the bottom, and bed 2 ft shorter. Crushed packing was the most likely cause.	
1195			*Hot tapping spray nozzles reinstates performance after packed bed collapsed in startup upset, Section* 4.13.	
506			*Softening of aluminum packings due to overheating at start-up, Section* 12.9.	
520			*Tar deposits crush ETFE packings, Section* 18.4.2.	

Case	References	Plant/Column	Brief Description	Some Morals
			23.1 Circulating Thermosiphon Reboilers	
23.1.1 Excess Circulation				
1309	304	Refinery toluene column	Restricting liquid circulation through a thermosiphon reboiler almost tripled heat transfer coefficient. The high rate of circulation evidently interfered with nucleation.	Watch out for excessive circulation rates.
1339	312	Refinery C_3/C_4 splitter	Reducing circulation by throttling thermosiphon reboiler effluent reduced base level below the reboiler return and eliminated tower premature flood.	
23.1.2 Insufficient Circulation				
1134			*Leading to dry-out near top of tubes, Section* 14.1.1.	
23.1.3 Insufficient ΔT, Pinching				
1318	433	Refinery sulfuric acid alkylation DIB	Fifty percent of the tray liquid bypassed the preferential baffle in the bottom sump, ending in the product side. This rendered reboiler process inlet temperature 10°F hotter than the bottom temperature, leading to reboiler heat transfer and product purity bottlenecks. A detailed reboiler simulation helped diagnose. Baffle/seal pan changes solved problem.	Reboiler inlet temperature hotter than bottom suggests baffle malfunction.
1353	171	Refinery depentanizer	To raise reboilers LMTD, all bottom-tray liquid was to be fed to the reboiler-draw side of a preferential baffle in the tower base. Test run data showed that some of the bottom-tray liquid entered the product side of the baffle.	Cured by converting the bottom tray to a chimney tray.
1328	250		Heavy residue accumulated at the base of a vertical thermosiphon reboiler, causing a temperature pinch.	
23.1.4 Surging				
1329, DT23.1	250		Small quantities of water caused surging in a vertical thermosiphon reboiler. Problem eliminated by elevating the bottom-liquid offtake about a foot, making the volume below the offtake a reservoir which constantly supplied a small amount of water to the reboiler.	

1330	250		Reboiler surging occurred in a vertical thermosiphon reboiler whose ΔT was small and the vapor product was not properly vented. Due to the poor venting, bottom pressure and bottom temperature rose, ΔT dipped, boiling stopped, pressure and bottom temperature dove, ΔT rose, and the reboiler took off again. Dips in steam pressure triggered the surging.	
23.1.5 Velocities Too Low in Vertical Thermosiphons				
1323, DT23.2	225	Gasoline stripper 17 in. ID	Vertical thermosiphon reboiler ID was larger than column ID. Liquid velocities at the reboiler base were about 200 ft/h, which permitted settling of small quantities of water. When these accumulated (about once a week), the thermosiphon stopped. Manual reboiler draining yielded water. Once drained, the reboiler returned to normal. In a related experience, problem was solved by installing a bucket trap at the reboiler drain valve with a float set for oil–water separation.	
23.1.6 Problems Unique to Horizontal Thermosiphons				
1314	98	Refinery 200–240°F hydrocarbons	Column product failed to meet specification because of puffing in a once-through horizontal thermosiphon reboiler. The puffing caused some liquid to bypass the trap-out pan. The puffing was caused by vapor binding at the distribution baffles. Drilling vent holes in the baffles improved operation.	Ensure distribution baffles in horizontal thermosiphon reboilers are vented.
			23.2 Once-Through Thermosiphon Reboilers	
23.2.1 Leaking Draw Tray or Draw Pan				
1327	250	Refinery coker debutanizer	A trap-out pan collected bottom-tray liquid and fed it to a once-through reboiler. At low rates, the bottom valve tray leaked, and the reboiler was starved of liquid. The bottom-tray 16-gauge circular valves with turned-down nibs were replaced by 12-gauge valves that seat flush with the tray floor. This mitigated leakage and permitted reboiler start-up.	Compare 1311.
1311	304	Refinery depropanizer	A once-through thermosiphon reboiler could not be started up because tray weeping at start-up starved the reboiler of liquid. Solved by adding a valved dump line connecting reboiler liquid and bottom sump.	Provide valved dump lines for once-through reboilers.

(Continued)

Chapter 23 Reboilers that did not Work: Number 9 On The Top 10 Malfunctions (*Continued*)

Case	References	Plant/Column	Brief Description	Some Morals
DT23.3			*Weeping of venturi valves in bottom tray makes start-up of tower and once-through thermosiphon difficult.*	
1340	165	Refinery coker deethanizer stripper	A trap-out pan collected liquid from the bottom valve tray, feeding it to a once-through reboiler. During coke drum switches (low vapor loads) liquid dumped through the valves, starving the reboiler of liquid and inducing lights into the bottoms. The fix was making the collector tray a seal-welded chimney tray.	Compare 1311, 1327.
1341	165	Refinery coker debutanizer	A trap-out pan collected liquid from bottom valve tray, feeding it to a once-through thermosiphon reboiler. Tray damage caused liquid bypassing, which dramatically lowered the bottom temperature. Fix was making collector tray a seal-welded chimney tray.	Compare 1340.
906	95	Extractive distillation	Bolts at the flanges of a sectionalized draw-off pan feeding a once-through thermosiphon reboiler were left hand tight. Flange leakage caused excessive reboiler outlet temperature.	Ensure adequate tightening of bolts on draw pan joints.
722	304	Refinery depropanizer	A once-through thermosiphon reboiler was starved of liquid because the overflow weir of the trap-out pan feeding the reboiler was level with a seal pan weir. Liquid bypassed the trap-out pan. Problem was solved by raising trap-out pan weir by 6 in.	Trap-out pan weir should be higher than seal pan weir.
D23.4			*Low weir and narrow opening cause liquid to miss draw pan to once-through thermosiphon.*	
23.2.2	**No Vaporization/Thermosiphon**			
1333	387	Natural gas demethanizer	Three vertical once-through thermosiphon reboilers only achieved 30% of the design heat transfer, limiting boil-up and causing ethane to be off specification. No vaporization appeared to take place. Design liquid velocity in the tubes was 0.3 ft/s. Gas injection to gas lift the liquid worked well, but the available gas contained CO_2 which froze on the upper tower trays. Recycling ethane liquid to boost velocities also worked well, but cost capacity and energy. Problem solved by installing a small rod inside each tube, which reduced flow area without lowering heat transfer area.	A clever technique for enhancing heat transfer.

964			*Pipe from draw pan to reboiler removed during a revamp, Section* 11.3.	
DT23.5			*Reboiler fouling leads to loss of thermosiphon and overflowing draw pan.*	
23.2.3 Slug Flow in Outlet Line				
1312	304	Refinery depropanizer	Slug flow in an oversized outlet line from a once-through thermosiphon reboiler caused fluctuations in column pressure and bottom level.	Beware of oversized reboiler outlet lines.
			23.3 Forced-Circulation Reboilers	
1313	153		The performance of a forced-circulation reboiler was poor and was much the same whether power to the pump was on or off. Problem was caused by NPSH required exceeding NPSH available.	Ensure pump system compatibility.
1331	250		A restriction was placed in the vapor line downstream of the reboiler and sized to prevent vaporization in the reboiler. The restriction experienced erosion at an intolerable rate.	
710	113	Chemicals	Vapor from a forced-circulation reboiler caused vibration. This resulted in loosening of tray fasteners and tray failure. Increasing nozzle size and stiffening the support beams solved the problem.	Do not undersize reboiler lines and nozzles.
1268			*Excessive bottom liquid/froth level, Section* 25.7.1.	
1286			*Plugging of reboiler heater charge pump strainers, Section* 20.2.1.	
			23.4 Kettle Reboilers	
23.4.1 Excess ΔP in Circuit				
1310	304	Refinery depropanizer	High reboiler circuit pressure drop backed liquid up above the reboiler return nozzle, flooding the column. Caused by introducing kettle reboiler feed at a point from which liquid could not easily spread, failure to allow for head over the kettle overflow baffle, and low liquid driving head from the column.	Carefully review kettle circuit ΔP.
1317	355	NGL fractionation depropanizer	A retray failed to raise capacity because liquid line from base to kettle reboiler was undersized. As rates increased, base liquid backed up, covered the reboiler return inlet, and got entrained into the trays, causing premature flooding. Diagnosed using gamma scan time studies and solved by increasing line size.	Beware of undersized kettle reboiler lines (compare 1310).

(Continued)

Chapter 23 Reboilers that did not Work: Number 9 On The Top 10 Malfunctions (*Continued*)

Case	References	Plant/Column	Brief Description	Some Morals
1322	202	Light HCs 350 psia 107 high-capacity trays	Reboiler inlet and outlet lines were undersized. As reflux and boil-up rates were raised, liquid backed up at the tower base and covered the reboiler return inlet, causing flooding beyond 65% of design rates. There was no base-level indication; neutron backscatter helped diagnose. To solve, lines were enlarged and vapor was returned to tower at higher elevation.	Base-level indicator is a prime troubleshooting instrument.
1326	250		High pressure drop in a vapor return sparger from a kettle reboiler caused liquid backup above the reboiler return inlet. Premature flooding resulted. Problem solved by chopping off sparger to reduce pressure drop and removing the bottom tray to promote vapor distribution.	Compare 1310, 1317.
1335	494	Refinery isomerization prefractionator	High packing *dP* and poor separation were caused by base liquid level rising above the reboiler return nozzle and flooding the bottom two beds. Lowering the reboiler eliminated the flooding and established good operation.	Avoid liquid level exceeding the reboiler return nozzle.
1336	164	Refinery FCC DC_2 stripper	High pressure drop in the kettle reboiler return line, coupled with use of a level indicator on the column rather than the reboiler, resulted in a flooded reboiler.	
1344	196, 197	Refinery sour water stripper 4 ft ID	Base level exceeded reboiler return inlet, causing premature flood. Top of kettle baffle was only 2 ft below the reboiler return elevation, and the piping plus reboiler pressure drop exceeded 2 ft of liquid. Problem diagnosed using gamma scans and solved by removing two bottom trays and discharging reboiler return at higher elevation. (See also 759, Section 8.4.6.)	Compare 1310, 1317.
712	467	Refinery	Tower flooded prematurely after being switched to a new service. Cause was an excessive pressure drop in a kettle reboiler circuit backing up liquid to above the reboiler return nozzle. Elevating vapor inlet above liquid level solved the problem. Lack of level indication made diagnosis difficult.	Compare 1310, 1317.

1346	196	Refinery gasoline fractionator	Following replacement of trays by packings, tower was bottlenecked at 105% of revamp design capacity. Bottleneck was caused by excessive kettle pressure drop raising sump level above the reboiler return. Capacity raised to 118% of design by converting the bottom collector tray into a total draw tray, which directly fed the kettle at greater head.	
1347	196	NGL fractionation DIB	Three parallel kettle reboilers had a common liquid feed header and a common vapor return header. Liquid levels in kettles were uneven due to nonsymmetrical pipings, different surface areas and heating media, and different baffle elevations. Excessive liquid level set in one draw compartment caused all three draw compartments to overflow, entrain liquid, and cause excess pressure drops in the reboiler circuit. The tower base level rose above the reboiler return, and the tower flooded prematurely.	Cured by lowering the draw compartment level set point.
1351	415	Chemicals extractive distillation	Excessive pressure drop in a kettle reboiler circuit backed up liquid on the kettle draw chimney tray. Liquid overflowed the chimneys into the bottom product, leading to poor lights recovery. Fixed by raising chimney height. A kettle maldistribution pattern unique to extractive distillation and similar systems was a contributor.	As 1310, 1317.
1304	153		Kettle reboiler supplied insufficient heat. Reason was excessive pressure drop in the vapor line from the reboiler causing low liquid level in the reboiler shell.	As 1310, 1317.
DT23.6			*Partially blocked liquid line to kettle reboiler causes excessive liquid in tower base and flooding.*	
DT18.9			*Partially plugged kettle draw does not impair operation of an amine regenerator.*	
23.4.2 Poor Liquid Spread				
1332	250		Channeling was experienced in a kettle reboiler. The channeling was evidenced by the shell surface being much warmer in the center (above the inlet) than near the shell ends. The problem was eliminated, and heat transfer greatly improved, after a horizontal baffle which directed liquid toward the sides was installed above the inlet nozzle.	Watch out for maldistribution in horizontal reboilers.
1351, 1310			*Causing or contributing to excess pressure drop, Section* 23.4.1.	

(*Continued*)

Case	References	Plant/Column	Brief Description	Some Morals
23.4.3	**Liquid Level above Overflow Baffle**			
1526			*Due to fooling of level gage by an oil layer above glycol, Section* 25.7.2.	
			23.5 Internal Reboilers	
1334	497	Refinery H_2S stripper	Increasing heat duty caused flooding in the bottom 17 trays. Gamma scans showed flood initiation at the internal reboiler, with froth propagating into the trays. Lowering the base liquid level from flush with the top of the tubes to halfway up the tubes caused flooding to recede to the bottom 5 trays.	Frothing at an internal reboiler often causes premature tower flood.
DT26.4			*Froth generated by boiling possibly affects base-level control.*	
1319	189	Refinery alkylation HF stripper	An oversized "bathtub" housing an internal reboiler left little escape area for vapor disengaging from the sump. At high rates, this vapor upflow restricted liquid descent from the bathtub, initiating tower flood. Eliminated by replacing the internal reboiler by a kettle.	Beware of restrictive designs where two-phase flow is present.
1345	465	Petrochemicals DIB	Overflow weirs on the chimney tray housing tube bundles of internal reboilers were only 3 in. tall, which wetted only two rows of reboiler tubes. Raising bottom liquid level to top of the tube bundles permitted reboil but lost level indication. Unstable control resulted.	Internal reboilers need liquid pools.
DT12.3			*Salts deposit on tube outside in bathtub internal reboiler.*	
1161	116, Case MS23	Refinery	A heavy leak developed on a flange at two stab-in heat exchangers on a fractionator during start-up. This was due to inadequate provision for thermal expansion in the pipework joining these exchangers at the start-up conditions.	
			23.6 Kettle and Thermosiphon Reboilers in Series	
1337	164	Refinery FCC deethanizer stripper	A collector tray fed a thermosiphon reboiler and a kettle reboiler in series. Bottom liquid product was drawn from the kettle overflow compartment. Undersized kettle vapor return line and excessive pressure drop from the collector tray to the kettle caused overflow of C_2-rich liquid from the collector tray into the bottom product. Cured by eliminating the hydraulic bottlenecks.	Diagnosed by temperature and pressure measurements. See also 1338.

1338	159	Refinery FCC deethanizer stripper	The bottom sump fed a thermosiphon reboiler and a kettle reboiler in series. When charge rates exceeded design, excessive pressure drop in the reboilers raised sump liquid level above the reboiler return inlet, which initiated tower flooding. Solution achieved by decoupling the reboilers, with thermosiphon receiving feed from a new collector tray and discharging into sump while kettle receiving feed from sump.	Thermosiphon-kettle in-series coupling can lead to complex hydraulics that can bottleneck towers.
			23.7 Side Reboilers	
23.7.1 Inability to Start				
1320, DT23.7	276	Natural gas demethanizer	At start-up, the tall liquid legs from/to the thermosiphon side reboiler were full of process liquid. The liquid head suppressed boiling. Further, during the shutdown, lights batch distilled out, leaving high boilers in the side-reboiler and liquid legs. Upon heating, boiling and thermosiphoning did not initiate. Problem solved by connecting methane gas to the side-reboiler outlet process line. This gas lifted the liquid and started thermosiphon circulation.	Gas injection can help initiate thermosiphons.
1325	449	Light hydrocarbon	Commissioning a new thermosiphon side reboiler with half the duty of the main reboiler incurred excessive sudden vapor generation, which induced tray flooding. At start-up, the tall liquid legs to/from the reboiler were full of process liquid, which suppressed boiling. Heat input needed to be high to get boiling (and therefore thermosiphoning) started, causing the excessive sudden vapor generation. Problem solved by injecting overhead gas as a lift gas to the reboiler inlet to start thermosiphon circulation.	See 1320.
23.7.2 Liquid Draw and Vapor Return Problems				
706			*Due to excess valve pressure drop, Section* 10.1.2.	
767			*Mixing of side reboiler draw and return liquids chokes draw, Section* 10.1.1.	
768			*Insufficient tray spacing at the side reboiler return, Section* 5.6.	
770			*Tower flooding at side reboiler outage, Section* 9.2.	
23.7.3 Hydrates				
1020, 1040			*Side reboiler helps keep hydrates in tower, Sections* 2.4.6, 2.4.1.	

(*Continued*)

Chapter 23 Reboilers that did not Work: Number 9 On The Top 10 Malfunctions (*Continued*)

Case	References	Plant/Column	Brief Description	Some Morals
23.7.4 Pinching				
313			*Due to excess heat removal, Section* 2.1.	
23.7.5 Control issues				
1578			*Side reboiler aggravates instability due to two composition control loops, Section* 26.2.	
			23.8 All Reboilers, Boiling Side (*See also Reboiler Tube, Section* 20.2.1)	
23.8.1 Debris/Deposits in Reboiler Lines				
1302, DT23.8	263	Natural gas demethanizer	Erratic vertical thermosiphon reboiler action resulted from a piece of masking tape stuck in a reboiler flange.	Do not use masking tape as flange covers.
1303	153		Heat transfer from a vertical thermosiphon reboiler declined with time. Cause was accumulation of residue in liquid line to reboiler.	Draw off residue from base of reboiler.
912			*Debris in liquid line to reboiler, Section* 11.8.	
23.8.2 Undersizing				
1321	464		Due to reboiler undersizing, higher pressure steam was used. Higher metal temperature and polymer formation resulted. The polymer plugged tower trays and reduced run length.	
23.8.3 Film Boiling				
1315, DT23.9	354	Chemicals	Poor heat transfer occurred in a vertical thermosiphon reboiler heating 440°F column bottom (expected outlet temperature was 550°F) by liquid Dowtherm which entered at 725°F. Film boiling caused the problem. Reversing the Dowtherm flow from coocurrent to countercurrent solved the problem.	Reversing flow direction can overcome film boiling problems.

23.9 All Reboilers, Condensing Side

23.9.1 Non condensables in Heating Medium				
1305	47		In a tower separating recycle reactants in the C_6 range, vertical thermosiphon reboiler performed poorly because of inerts accumulation on its condensate side. Venting was inadequate.	Ensure adequate venting on reboiler condensate side.
1343	551	Benzene–toluene/styrene	Poor pre-start-up venting of noncondensables from the steam side of a vertical thermosiphon reboiler limited heat transfer. Fixed by venting.	Same as 1305.
1350	352	Chemicals	Vertical thermosiphon reboiler was heated by flash steam with shell kept under vacuum. Partial plugging of vent line from shell to ejector led to tower cycling. Adding 50 psig steam helped, but problem came back when vent line fully plugged.	Reboiler vents are necessary.
1308	306	Refinery depropanizer	A newly installed preheater had several tube leaks. Column feed backflowed into the steam supply and from there into the steam side of the reboiler. The volatile feed gas blanketed the reboiler, causing erratic behavior in the column.	What may appear to be a reboiler tube leak can be caused by a leak elsewhere.
1306	310	Refinery	Inability to vent accumulated CO_2 from the steam (tube) side of a horizontal reboiler caused corrosion and tube leakage near the floating head. Problem was solved by extending an upper tube to make up a vent tube from the floating head to a vent valve located at the channel head.	A novel technique was developed to solve problem.
1307			*Poor venting of noncondensable led to an explosion in air separation tower, Section* 14.2.	
1134			*Inerts in reboiler heating side contributed to explosion in ethylene oxide tower, Section* 14.1.1.	
23.9.2 Loss of Condensate Seal				
1301, DT23.10	263	Olefins DC_1	Condensate seal was lost on a vertical thermosiphon reboiler causing loss of heat transfer.	Beware of reboiler seal loss.
DT23.11			*Similar to* 1301, *but in a steam reboiler.*	

(Continued)

Chapter 23 Reboilers that did not Work: Number 9 On The Top 10 Malfunctions (*Continued*)

Case	References	Plant/Column	Brief Description	Some Morals
1576	296	Olefins C_2 splitter heat pumped	Reflux control on the reboiler/condenser measured flow of the heating vapor and throttled condensate (reflux). When the compressor discharge pressure dropped, condenser ΔT declined, condensation dropped, and condensate level in the reboiler dropped rapidly, causing gas breakthrough with no reflux to the column. High column pressure and off-specification product resulted. A low-level override controller on the reboiler condensate eliminated problem. (See also 1575, Section 26.2.)	Watch out for loss of level when controlling a reboiler by a condensate valve.
1342	308		Blowing a horizontal thermosiphon condensate seal due to a faulty steam trap caused a loss in reboiler duty of about 50%.	As 1301.
23.9.3	**Condensate Draining Problems** (*See also Throttling Steam/Vapor to Reboiler or Preheater, Section* 28.5.)			
DT23.12			*Reboiler limitation and instability due to inability to drain condensate.*	
1516			*Twenty percent of reboiler area waterlogged, Section* 28.6.	
23.9.4	**Vapor/Steam Supply Bottleneck**			
1324	270	Olefins refrigerated fractionator	The fractionator capacity was limited by heat transfer in the vertical thermosiphon reboiler. Heat input was controlled by partial flooding of the reboiler tubes. Tests using neutron backscatter monitored the partial flooding and showed that the bottleneck was vapor supply to the reboiler, not shortage of heat transfer area.	Neutron backscatter is effective for detecting condensate level.

Case	References	Plant/Column	Brief Description	Some Morals
			24.1 Inerts Blanketing	
24.1.1 Inadequate Venting				
1402	153		Capacity of a horizontal, in-shell condenser was well below design. Inlet vapor entered in the middle; ends were inert blanketed. Vents solved problem.	Ensure adequate venting.
1423	456, 457	Refinery naphtha stabilizer	Poor heat transfer was caused by lack of a vent line in the ground-level condenser (using hot-vapor bypass control). Once or twice per week the hot-vapor bypass would close, and the operators vented from the reflux drum to the flare. This induced vapor blowby from the exchanger to the drum that purged out the non condensables. After 15–30 min of venting, normal operation resumed.	Same as 1402. Vent line is planned for next turnaround.
1406	306	Refinery debutanizer	The ability to condense the overhead product was lost because of vapor blanketing in the condenser shell. Venting solved the problem. A newly installed, nitrogen-purged instrument caused the problem.	
1404	471		Vapor entering a vertical downflow in-shell condenser contained high-molecular-weight condensables and a low-molecular-weight inert. Poor condensation was caused by channeling that led to inert blanketing.	Use sealing strips; ensure adequate sweeping.
1412	229	Refinery	Tube bundle of a water-cooled in-shell horizontal downflow condenser experienced corrosion at the end due to accumulation and lack of venting of corrosive gases. The addition of a vent line plus a HC gas purge into exchanger inlet more than doubled tube life and lowered corrosion inhibitor consumption.	A valuable technique for alleviating accumulation of undesirable components.
1419	482	Chemicals vacuum	Column condenser was a steam generator. During start-up, the condenser did not cool properly, generating too much vapor product. Manual opening of the steam valve vented all the inerts from the generator into the steam system. This caused problems with steam users but resolved the condenser problem.	Same as 1402.
DT24.3			*Open nitrogen line to the tower overloads condensing system; also, inerts build up due to condenser draining problem.*	

(Continued)

Chapter 24 Condensers That Did Not Work (*Continued*)

Case	References	Plant/Column	Brief Description	Some Morals
24.1.2	**Excess Lights in Feed**			
132	331	Refinery vacuum	Noncondensable cracking products raised tower pressure from 30 to 56 mm Hg absolute, inducing loss of distillate to residue. The cracking was due to excessive temperature and oil residence time in the atmospheric crude tower heater. Cured by reducing crude tower heater temperature and changing crude.	
1433	75	Natural gas debutanizer	Excess lights in the feed limited condenser capacity in summer, leading to high pressure, instability, and flaring. Cured by stripping out more of the lights in the upstream tower (demethanizer) during summer.	
1425	332	Refinery crude fractionator	Capacity limitation of the overhead compressor caused tower pressure going off control during the heat of the day, spiking to 25 psig. Pressure returned to 11 psig by installing a line from the compressor suction to the suction of the vacuum unit vent gas compressor that was unloaded.	
1413	334	Refinery vacuum	Higher pressure reduced distillate yield. Entrainment of LVGO from the top sprays, causing waxing on the condenser tubes, was partially responsible and was mitigated by lowering the LVGO PA flow. Poor stripping due to capacity limitation in the upstream crude fractionator was the main cause. The stripping was improved by tray modification.	Laboratory analysis of vacuum tower overhead and tests at varying stripping rates diagnosed problem
111			*Lights build up in overhead system, Section* 2.4.3.	
1609			*Lights entering tower cause relief valve to lift, Section* 21.4.	
			24.2 Inadequate Condensate Removal	
24.2.1	**Undersized Condensate Lines**			
1401	153	HC gas condensation	Low heat transfer in a horizontal, two-pass, in-tube condenser was caused by an undersized condensate line using gravity flow.	Ensure adequate condensate removal.

1416, DT24.1	273	Chemicals vacuum	Violent surging of the column pressure and accumulator level occurred at above 50% of the design rates. The cause was undersized condensate drain lines with high points. These led to intermittent accumulation and siphoning of liquid in the overhead condenser. Problem eliminated by enlarging drain lines and eliminating the high points.	Avoid undersizing condenser drain lines. Avoid high points in these lines.
1417, DT24.2	250	Olefins C_2 splitter	Low heat transfer in a horizontal, two-pass, in-tube condenser was caused by an undersized condensate drain line.	See 1401.
1426			*Undersized condensate line interferes with decanter action, Section* 2.5.	
DT24.3			*Traps noncondensables in condenser.*	
24.2.2 Exchanger Design				
1403	153		A horizontal, in-tube condenser with axial inlets and outlets did not achieve design capacity. Axial outlet did not permit condensate drainage.	Ensure adequate condensate removal.
			24.3 Unexpected Condensation Heat Curve	
DT1.13			*With wide-boiling-range mixtures, condensation path depends on hardware and may differ from simulation.*	
205, DT22.4			*With wide-boiling-range mixtures, absorption effect is important, Section* 2.3.	
1410	533	Absorption plant stripper	Rayleigh fractionation occurred with a wide-boiling mixture. Some vapor, which left the condenser uncondensed, mixed with condensate in the condensate outlet pipe, causing a sudden 10°F temperature rise due to vapor condensation. The system still worked, but not much leeway was left. Situation could have been remedied by venting or by injecting liquid near the back of the condenser.	
1418	250	Aromatics	Following a revamp, column received much more water in the feed. The water distilled up, forming a second liquid phase upon condensation. The condenser was unable to achieve its duty when the two liquid phases separated, causing excessive product loss.	

(*Continued*)

Chapter 24 Condensers That Did Not Work (*Continued*)

Case	References	Plant/Column	Brief Description	Some Morals
			24.4 Problems with Condenser Hardware	
1409	319	Chemicals	Vacuum jets overloaded with uncondensed vapor. A downflow in-tube condenser was used with large baffle windows. Reducing baffle windows solved the problem.	
1430	534	Refinery FCC main fractionator	Replacing high-pressure-drop trim condensers by larger, low-pressure-drop unbaffled H-type exchangers permitted raising receiver pressure but gave poor heat transfer, higher receiver temperature, and higher wet gas rate. Bundle redesign with double segmental baffles was the cure.	Carefully balance heat transfer versus pressure drop.
1431	416	Gas glycol regenerator	Installation of a forced draft air condenser reduced condensation, back-pressuring the reboiler and forcing the relief valve to vent almost continuously. Cause was preferential air flow along the outside condenser tubes duct with little cooling on the inside tubes. Cured by installing baffles on the air side.	
1432	102	Refinery FCC main fractionator	Changing air condenser bundles from two pass to one pass, and changing 15 to 35 hp motor to increase heat transfer, lowered total pressure drop through the air and water condensers from 14 psi to 5.5 psi. This allowed a higher overhead receiver pressure, lowering wet gas production by 35%, and eliminating a wet gas compressor bottleneck.	
1411	6	Refinery crude fractionator	Overhead to crude exchangers equipped with impingement plates on both sides for 180° rotation. The lower impingement plates in three exchangers collapsed, blocking their outlets. A sudden pressure rise followed and lifted several atmospheric relief valves. Recurrence was prevented by removing the lower impingement plates.	
1421	365	Methanol methanol–water	Methanol contained excessive water, especially in summer. Cured by cleaning finned tubes with detergent and high-pressure water jets, increasing fan blade angles, and plugging two leaky reboiler tubes.	
DT21.1			*Bowing of tubes and rotation of inlet impingement plate to outlet lead to blocked overhead and atmospheric discharge.*	

			24.5 Maldistribution between Parallel Condensers	
1407	304	Refinery	A new set of condensers was added in parallel to an existing set to increase condensation capacity. Instead of increasing, condensation capacity decreased. Vapor maldistribution was the cause.	Beware of maldistribution in parallel condensers.
1422	432	Debutanizer	A new pair of in-parallel condensers was added upstream of the existing condenser. One of the two new condensers was 30 ft above the other. Liquid filled the lower condenser, giving zero flow/condensation through it. Flow was reestablished by pinching a block valve on the upper condenser outlet.	Same as 1407.
1427	476	Chemicals	A second condenser was added in parallel to an existing one. Due to nonsymmetrical manifold piping, the new condenser operated 30% below its rated design capacity.	
			24.6 Flooding/Entrainment in Partial Condensers	
1415, DT24.4	225	Olefins C_3 splitter	Excessive opening of the control valve venting a knockback condenser induced excessive vapor velocities through the condenser, causing liquid entrainment. The liquid flashed in the valve and overchilled the metal downstream. Problem eliminated by a valve limiter.	Avoid excess vapor flows in knockback condensers.
DT24.5			*Entrainment from a water-cooled vent condenser.*	
1405	471		Horizontal in-shell condenser with vapor upflow and condensate downflow did not reach design capacity. This was caused by liquid entrainment at excessive vapor velocities.	Avoid excessive velocities in vapor upflow condensers.
1408	79	Refinery crude fractionator	Column pressure and product gas rate sharply fluctuated during low-rate operation in this and other units. The condenser was a partial condenser located at ground level, with an elevated reflux drum. Problem was caused by slug flow in the riser from the condenser to the drum.	Size risers to avoid slug flow at low rates.
1506	471		Condensation in a horizontal in-shell partial condenser with liquid outlet at the bottom and vapor outlets at the top was controlled by varying liquid level in the condenser. Excessive entrainment was caused by condenser pressure drop building a large hydraulic gradient.	Avoid high levels and high pressure drops in such condensers.

(Continued)

Chapter 24 Condensers That Did Not Work (*Continued*)

Case	References	Plant/Column	Brief Description	Some Morals
			24.7 Interaction with Vacuum and Recompression Equipment	
1428	361	Refinery vacuum, wet tower	Vacuum was pulled by a three-stage ejector, with a condenser after each stage. In winter, cold cooling water permitted lower first-stage discharge pressure, but this did not lead to highly desirable suction pressure reduction. Achieved by adding a small (10% capacity) ejector parallel to the first stage.	
1429	430	Refinery DIB heat pumped	One summer, after years in service, recompressor repeatedly tripped on high discharge temperature. DIB was partially reboiled by DC_4 overhead, partly by condensing its compressed overhead. That summer, DC_4 pressure was raised to overcome fouling of its air condenser. This shifted condensing duty away from the DIB overhead, which in turn raised recompressor outlet pressure and temperature. Solved by cleaning and overhauling system.	Beware of heat duty shifts in heat integration.
1424	331	Refinery vacuum	One of the two parallel second-stage ejectors was plugged, resulting in tower pressure rise, which reduced distillate recovery. Blocking in both the process and steam sides to this ejector unloaded the intercooler and third-stage ejector and permitted lower pressure in the tower.	
DT25.1			*Misleading pressure measurement appears like poor steam ejector performance.*	
			24.8 Others	
1414	202	C_{10} HCs and alcohols, 100 trays	Top pressure was 500 mm Hg absolute (normal 200 mm Hg), temperatures were up, and products were off specification with boil-up rates 10–20% of normal. Reason was that the condenser water supply valve was one-quarter open. It was throttled on a previous run and was not reopened.	Never overlook the obvious.

Case	References	Plant/Column	Brief Description	Some Morals
			25.1 Incorrect Readings	
			(*for incorrect base-level reading leads to tower flooding and tray/packing damage, see Sections* 8.1.1 *and* 8.3; *and for incorrect flow and temperature measurements and analyses that mislead simulation, see Section* 1.3.1)	
1611			*Incorrect base-level reading leads to overchilling flare header and explosion, Section* 14.3.2.	
1636			*False low-level trip signal leads to a fire, Section* 14.3.4.	
1206			*Base temperature controller malfunction leads to flammable liquid spill from reflux drum, Section* 25.8.	
1562			*Low tower base-level and pump cavitation due to false level signal, Section* 8.6.	
15106, DT19.3			*Erratic level on overflash draw tray, Section* 9.6.	
1042			*Faulty feed drum oil-water interface indicators destabilize temperature control, Section* 27.1.3.	
322			*No stripping due to misleading flow measurement, Section* 2.2.	
1558			*Incorrect dP measurement fools advanced controls, Section* 29.6.2.	
DT25.1			*Misleading pressure measurements appear like poor steam ejector performance.*	
DT24.3, DT25.3			*Pressure transmitter readings too high induce premature flood.*	
1553	172	Refinery vacuum	Loss of vacuum (5 mm Hg) was observed during hot weather. Reason was use of an ordinary manometer, which picked up a fall in barometric pressure in hot weather.	At deep vacuum, use vacuum manometers.
1554	172	Refinery vacuum	Pressure drop measured across the tower was 4 mm Hg when 7 mm Hg was the correct reading. The measurement was performed using an ordinary manometer and did not account for the static head difference of air between the column bottom and top.	Same as 1553.

(*Continued*)

Chapter 25 Misleading Measurements: Number 8 On Top 10 Malfunctions (*Continued*)

Case	References	Plant/Column	Brief Description	Some Morals
203	343		Separation efficiency of a component appeared extremely poor, while that of other components was OK. This was caused by analyzer error and was discovered by calculating a component balance.	Use mass and component balances to verify analyses.
1540, 1574, DT1.10			*Incorrect analysis misleads troubleshooters and controls, Section* 27.3.1.	
			25.2 Meter or Taps Fouled or Plugged	
15135	177	Vacuum tower	Plugging of level taps was frequent. The operators then used TIs at three different heights to control bottom level: hot TI is covered with liquid, cold uncovered. A fuzzy control mimicked operator action, giving multiplication factor depending on TI readings. The controller was set to cycle the level around the middle TI.	An excellent control method for fouling services.
15121	308		In plugging service, level indication by a battery of "ram horn" level indicators succeeded where all else failed. Each indicator is a thermowell extending to the vessel wall only and is poorly insulated, with a curved pipe draining liquid into the tower. A sudden temperature increase indicates presence of liquid.	
DT8.1			*Construction blind looks like gasket in level transmitter piping.*	
1511			*Fouling of reboiler outlet thermowell leads to explosion with unstable chemicals, Section* 14.1.4.	
1607, 1521			*All indication lost when instrument tap plugs, Sections* 21.5 *and* 21.7.	
15109			*Level tap plugging causes excessive base level and flooding, Section* 8.1.1.	
15157			*Level tap plugging on reflux drum boot prevents water removal, Section* 2.4.3.	
130			*Coked bridle nozzles on slop wax collector tray, Section* 1.1.7.	
1529			*Plugged tap on steam flowmeter fools advanced controls, Section* 29.6.2.	
718			*Internal level gage lines plug, Section* 10.2.	
DT13.3			*Fouling leads to incorrect interface level measurement, inducing water into hot-oil tower and tray damage.*	

15114	22		Flow rate of viscous residue was measured by a wedge flowmeter with remote chemical seal diaphragm elements. The meter suffered an offset error or total failure periodically, inflicted either from steaming out to clear plugging or from plugging. Diaphragm damage during steam operation, fill liquid leak, fill liquid problems, and air leakage could be causes.	

25.3 Missing Meter

(*for lack of functional bottom-level indication, see Sections* 8.1.1 *and* 23.4.1)

1137			*Uninstalled flowmeter prevents detecting no flow through pump and heater tubes, Section* 12.1.	
15138			*Lack of flow measurement promotes coking, Section* 19.2.	
1566			*Tower feed bypass difficult to operate and control due to absence of flow indication, Section* 3.1.3.	
1345			*Level indication forfeited to get sufficient boil-up, internal reboiler, Section* 23.5.	
1229			*Unmonitored internal reflux lead to packing distributor overflow, Section* 6.3.	
11104			*Instrumentation shutdown contributes to explosion, Section* 12.4.	

25.4 Incorrect Meter Location

(*for level taps on chimney trays, see Section* 9.6)

133			*Temperature indication not contacting fluid led to overheating and explosion in nitro compound service, Section* 14.1.3.	
1512, 1513			*Reboiler in peroxide service explodes because level monitoring is mislocated, Section* 14.1.2.	
1552, 15138			*Coking results from thermocouple not inserted deep enough, Section* 19.2	
731	449		Pressure connection in the vapor space of one-pass trays ended in the center downcomer upon revamp into two-pass trays.	Do not forget instrument connections.
15152	307	Refinery vacuum	Measured pressure drop across a coked wash bed was a low 3 mm Hg because the lower pressure tap was inside the vapor horn. A later measurement outside the horn gave 8 mm Hg higher pressure than inside.	

(Continued)

Chapter 25 Misleading Measurements: Number 8 On Top 10 Malfunctions (*Continued*)

Case	References	Plant/Column	Brief Description	Some Morals
			25.5 Problems with Meter and Meter Tubing Installation	
25.5.1 Incorrect Meter Installation				
15119, DT25.2	405	Olefins C_3 splitter	New tower was touchy and separation was poor. The orifice run of the reflux flowmeter had seamed (spiral-wound) pipe, with spiral weld beads on the interior protruding 3–4 mm, leading to a false high reading. Energy balances helped identify. Flow factor recalibration, based on the pump curves, rectified the problem. The tower was unnecessarily shut down several times during troubleshooting.	Excellent lessons were drawn and presented in the paper.
1587	496		Following a retrofit, poor separation, excessive pressure drop, and liquid carryover occurred. Cause was poorly installed orifice plates, leading to understatement of reflux and therefore overrefluxing. Gamma scans helped diagnose. Cutting reflux reinstated on-specification products.	Good troubleshooting can avert shutdown.
DT8.4			*Poor location of level taps leads to incorrect level indication and high-level tray damage.*	
201			*Undersized orifice plate causing poor separation, Section* 2.3.	
709			*Impingement of reboiler return on level float, Section* 8.4.2.	
25.5.2 Instrument Tubing Problems (*for taps plugging, see Section* 25.2)				
15161	507	Gas glycerol dehydration	Incomplete insulation and poor heat tracing of level bridles led to false levels with separators overfilling or emptying. Solved by temperature-controlled electric heat tracing.	
15142	475	Refinery C_3 splitter	Measured tower *dP* almost doubled during warm weather due to liquid condensation and accumulation in a low spot of the uninsulated $\frac{1}{4}$-in. tubing. Tower product flowmeters read incorrectly due to disturbances to propylene glycol fill of their tubing, such as loss of fluid and air bubble trapping	*dP* problem eliminated by larger, free-draining tubing.

15160	400	Refinery crude topping column	Base level became erratic and uncontrollable when stripping steam was added to tower, causing flooding due to high base levels. Cause was steam condensing, then flashing in the legs of the level transmitter. Gamma scans showed no flooding and steady base level over time, pointing to an instrument problem. Cured by a gas seal on level transmitter legs.	Gamma scans provide a useful level check.
DT25.3			*Pressure transmitters below their taps lead to misleading readings, premature flood.*	
15116			*Instrument air tubing leakage reduces product make, Section* 29.5.	
DT25.4			*Impingement by reboiler return on upper level tap.*	
			25.6 Incorrect Meter Calibration, Meter Factor	
1236			*Incorrect meter factor led to a wash rate that was too low and contributed to coking, Section* 5.7.	
15127, 15110, 1560, 15156			*Out-of-calibration base level causes liquid rise above reboiler return and flood, Sections* 8.1.1 *and* 25.7.3.	
1577			*Lack of pressure/temperature compensation leads to poor control, Section* 27.2.2.	
			25.7 Level Instrument Fooled	
25.7.1 By Froth or Foam				
1568	250	Refinery	Troubleshooters were misled by a normal level indication in a downcomer trap-out pan when the pan was filled with aerated liquid that backed up to the trays above. The level glass measured the liquid head between its own tappings, giving a lower reading than the height of aerated liquid in the pan.	Watch out when interpreting level measurements when liquid is aerated.
15139			*Unsuccessful attempt to measure and control level at a partial draw, Section* 29.6.2.	
1268	312		Fractionation was poor and pressure drop was high in a tower with forced-circulation reboiler. Reducing the indicated liquid level from 58 to 32% improved both. Froth or foam in the base exceeded the reboiler return and initiated tower flooding.	Excessive base liquid/froth level causes flood. See also 1568.

(*Continued*)

Chapter 25 Misleading Measurements: Number 8 On Top 10 Malfunctions (*Continued*)

Case	References	Plant/Column	Brief Description	Some Morals
DT25.5			*Frothing fools base-level measurement when upper level tap is above reboiler return inlet.*	
DT8.4			*Frothing at tower base due to impingement and entry of steam issuing from sparger.*	
15126	465	Amine regenerator	Bottom sump foamed and flooded tower, but level indication was normal. The lower foam density led to a low-level indication. There was liquid in a sample from the upper level tap. Manually changing bottom flow rate showed no change in bottom level. (See also 644, Section 16.1.2.)	Same as 1568.
1555			*Level indication problematic due to foam in tower, Section* 16.6.9.	
1514			*Level indication fooled by foam in kettle, Section* 14.12.	
DT26.4			*Level indication possibly fooled by froth generated by internal reboiler.*	
25.7.2 By Oil Accumulation above Aqueous Level				
1527	305	Natural gas amine absorber	A HC phase settled above the amine in the bottom sump. Due to the lower density of this phase, the level indicator read low. Liquid level rose above the vapor inlet nozzle, while level indication was still normal. This prematurely flooded the tower. The flooding flushed the HC layer overhead. Once the layer was flushed, the flooding stopped by itself.	Ensure adequate skimming to avoid settling of oil above an aqueous phase.
15115	306	Refinery amine absorber, several towers	Hydrocarbon liquid (SG = 0.6) settled above the amine (SG = 1.0), fooling the level transmitter into underestimating the bottom level. The transmitter read normal level when the liquid level rose above the gas inlet. The gas entrained the liquid causing tower flood.	Separate level indicators with lower taps at different elevations are not fooled by the hydrocarbon phase.
1526	305	Natural gas glycol regenerator	The regenerator (kettle reboiler) draw compartment level exceeded the overflow baffle. The high level was undetected because of oil accumulation in the compartment. The level gage was fooled by the low density of the oil.	Check if oil can be skimmed before trusting level indication in this service.
DT8.5 E			*A layer of low-gravity insoluble organics above water gives misleading level indication, leads to tray damage.*	

25.7.3 By Lights				
DT25.6			*Level transmitters calibrated for normal process liquid in base are fooled by light liquid in base during start-up.*	
15156	454	Natural gas demethanizer	Upon shifting from ethane rejection to ethane recovery operation, lower liquid density fooled level transmitters into underestimating liquid level. Base level rose above reboiler return, inducing intermittent flooding. High accumulator level induced entrainment, which disturbed control of heat integration.	Level transmitter recalibration eliminated problems.
25.7.4 By Radioactivity (Nucleonic Meter)				
1509	284	Chlorine heavy-ends column	Radioactive bromine used as a tracer in a brine stream which was electrolyzed to make chlorine ended up in the column. It concentrated at the base and interfered with the action of the nucleonic level controller, eventually flooding the column.	Nucleonic level devices are affected by radioactive materials.
25.7.5 Interface-Level Metering Problems				
1545			*Leads to overpressure of downstream vessel and an explosion, Section* 14.2.	
1146			*Leads to overpressure of downstream vessel, Section* 21.4.	
DT13.3			*Leads to a water-induced pressure surge.*	
1036, 1042			*Leads to water accumulation, Sections* 2.4.4 and 27.1.3.	
224			*Changes azeotroping patterns, Section* 2.5.	
			25.8 Meter Readings Ignored	
1206	284		A temperature controller at the column base went out of order. Seven hours later, flammable liquid spilled out of the reflux drum. Several abnormal instrument readings were overlooked during this period.	Do not ignore abnormal instrument readings.
11103			*During outage, leading to explosion in nitro tower, Section* 14.1.3.	
DT8.5 A, G			*Leading to high liquid-level damage, in one case due to ambiguous legend.*	
			25.9 Electric Storm Causes Signal Failure	
1625			*Leading to an explosion, Section* 14.3.1.	

Case	References	Plant/Column	Brief Description	Some Morals
			26.1 No Material Balance Control	
1508	343		Column was unable to meet bottom design purity even at higher than design reflux rates. Top purity was on specification. Problem was caused by control system setting too low at top product rate. The remainder of the light component was forced to leave out of the bottom.	Ensure proper material balance control.
1570			*Energy balance controls with temperature manipulating boil-up better than material balance control in C_3 splitter, Section* 26.3.2.	
1581	110	Methanol–water	Off gas that was 90% N_2–10% MeOH was scrubbed by water, the MeOH–water mix separated by a distillation column, and the water was recycled to the scrubber. The column used a conventional material balance control. The scrubber bottom was on level control, with water makeup on flow control. This scheme had no mechanism for retaining the water inventory. Having the scrubber level control the water makeup solved the problem.	Consider component inventory when assembling controls.
1293	297	Ammonia Benfield hot pot	Liquid levels in the absorber, regenerator, and flash drum were high after a restart. After the required (30%) solution concentration was achieved, the levels were brought down by allowing water to evaporate. This increased the concentration to 38–40%, which accelerated corrosion, erosion, and fouling, forcing a shutdown.	Closely watch the water balance in these systems.
1597	295	HCl absorbers	Hydrogen chloride was absorbed from a feedgas by weak acid in a cooled "acid tower" absorber. The remaining unabsorbed HCl is absorbed from the gas by freshwater in a "primary tower," making the weak acid. Both absorbers are cocurrent with partial circulation of bottom liquid to the top. Controlling primary tower level by freshwater flow, acid tower level by recirculation/bottom flow split, and acidity by primary tower flow split gave extremely steady control even during upsets and start-up.	Dynamic simulation using commercial software was instrumental in developing the control system.
1559			*Destabilizing downstream tower, Section* 3.2.2.	

26.2 Controlling Two Temperatures/Compositions Simultaneously Produces Interaction				
1564, DT26.1	225	Olefins C_2 splitter, 115 trays	This is a sequel to 1563, Section 26.3.2. At a later date, the ΔT controller was hooked to the reflux while the tray 10-temperature controller was hooked to the boil-up. When a slight upset occurred, the two temperature controllers started chasing each other, giving erratic reflux, reboil, and temperature control.	Interaction of two composition controllers leads to poor control.
1578	296	Olefins C_2 splitter conventional	Top composition was controlled by reflux manipulation and bottom composition by boil-up manipulation, simultaneously. Despite a feed-forward correction of feed variations using ratio controls, an interaction between the composition controls occurred, aggravated by an interreboiler. Controlling the interreboiler with a heat/feed rate ratio controller and speeding up the boil-up composition control via a ΔT controller improved product purity, stability, and energy consumption.	Same as 1564. See also 1563, Section 26.3.2.
1575	296	Olefins C_2 splitter heat pumped	Top purity was controlled by manipulating boil-up, bottom purity by manipulating reflux. Interaction between the controllers, magnified by large time lags, caused erratic responses. Control improved by cascading the bottom purity to a temperature control that manipulated boil-up and cascading the top purity to a reflux/overhead ratio controller. The ratio controller gives some decoupling of the interaction while the temperature control sped the bottom purity control compared to the top. (See also 1576, Section 23.9.2, and 1577, Section 27.2.2.)	Same as 1578.
15166	1	Refinery C_3 splitter heat pumped	Propylene product purity and tower mass balance were well-controlled. Bottoms purity was erratic with propylene concentration at times rising well above the 1.5% spec. Excessive lags, sensitivity to ambient disturbances, insensitivity of tray temperatures to composition, and analyzer lags, were the major issues. Cure was MISO strategy in the DCS that controlled bottoms purity by the trim condenser (minor) reflux, with feed forward compensation from feed rate and atmospheric temperature.	

(Continued)

Chapter 26 Control System Assembly Difficulties (*Continued*)

Case	References	Plant/Column	Brief Description	Some Morals
1565, DT26.2	225	Natural gas lean-oil still	Fired reboiler outlet temperature was controlled by manipulating fuel flow, while the column top temperature manipulated the air condenser louvres. Control was unstable and operation erratic. Problem solved by disconnecting the louvre control.	Same as 1564.
1594	421	Xylene C_8–C_9+ separation, 30 trays	Unsatisfactory performance resulted when bottom temperature controlled steam to reboiler while an upper temperature controlled reflux flow. To overcome, upper temperature loop was opened, leaving reflux on flow control. Later integration of a dynamic simulation model with a neural network model gave optimum set point selection and further improvement.	Same as 1564.
26.3 Problems with the Common Control Schemes, No Side Draws				
26.3.1 Boil-Up on TC/AC, Reflux on FC (Figure 26.4*a*)				
1534			*Poor performance with very small distillate flows, Section* 26.3.3.	
1563, 1570, 1575			*Fast response in large, trayed towers, Sections* 26.3.2 *and* 26.2.	
15144	340	NGL depropanizer	Bottoms were <3% of feed. Distillate and bottoms were level controlled from the receiver and base, respectively, reflux was flow controlled, and boilup was composition controlled with feed-forward control from feed to reflux and boil-up. Bottom composition fluctuated widely. Model predictive control stabilized product composition.	Do not control a level on a small stream.
1547			*Problem with ambient disturbances, Section* 26.3.4.	
1541			*Problem with feed temperature disturbances, Section* 26.3.4.	
DT26.5			*Tuning solves problem with reboiler swell.*	

1593	93	Natural gas demethanizer	Jacketed water heated glycol, which in turn reboiled tower. Bottom tray temperature regulated a valve in the water supply. Temperature control was poor because the two-step heat exchange introduced large, multiple lags. Also, the water supply temperature varied, so ΔT from water to tower bottom swung 30–80°F in a day. Temperature swings were drastically reduced, permitting a 15°F reduction in bottom temperature, by implementing APC, which compensated for variations in jacket water temperature and on tray 4 of the tower.	
26.3.2 Boil-Up on FC, Reflux on TC/AC (Figure 26.4*b*)				
1563, DT26.1	225	Olefins C_2 splitter, 115 trays	Top-section ΔT controller cascaded to the reflux. The main control tray was 50 trays below the top. Control was slow and sluggish due to the hydraulic lags over the 50 trays, causing excessive ethylene losses. It was replaced by temperature control 10 trays above the bottom cascading to the boil-up, with reflux on flow control. Both top and bottom compositions became far more stable, and the ethylene losses greatly reduced.	In large-trayed fractionators, the fast response of the boil-up manipulation is advantageous.
1575			*Fast response of reboiler manipulation advantageous in large trayed tower, Section* 26.2.	
1570	250	Olefins C_3 splitter, 120 trays	Material balance control, with near-bottom temperature manipulating reflux, gave slow and very sluggish response. A switch to an energy balance control with the same tray temperature manipulating boil-up was a major improvement. The energy balance control (normally not recommended!) was successful here because the column was not close to a limit and could tolerate slow cycles in reflux and boil-up.	
1533			*Sensitivity to ambient disturbances, Section* 26.3.3.	

(*Continued*)

Chapter 26 Control System Assembly Difficulties (*Continued*)

Case	References	Plant/Column	Brief Description	Some Morals
26.3.3	**Boil-Up on FC, Reflux on LC** (Figure 26.4*d*)			
1531	57	Chemicals bubble-cap column	Column acted like a stripper; reflux-to-distillate ratio was 0.43. When reflux flow was on accumulator level control, a small change in heat input led to large changes in reflux flow. Reflux flow at times fell below the minimum required for tray wetting.	Avoid controlling reflux by accumulator level when reflux ratios are low.
1533	366	Refinery FCC debutanizer	The column was highly sensitive to changes in ambient conditions. Reflux was controlled by a tray temperature and distillate was on accumulator level control. Interchanging these controls desensitized the column and improved its stability.	The modified system minimizes upsets due to disturbances in the coolant.
1571	250	Solvents	Cooling-water temperature periodically rose or fell by 20°F over a short time interval. Despite this, the column was barely affected. The column had a total condenser, with tray temperature manipulating the product and accumulator level manipulating reflux.	Same as 1533.
1584			*Accumulator level destabilized by feed and heat input disturbances, Section* 28.3.2.	
15148	122		Changing accumulator level control from product to reflux alleviated upsets due to rainstorms in air-condensed tower.	Same as 1533, 1571.
1534	345	Ethylbenzene–xylene splitter	Reflux-to-distillate ratio was 70:1. Boil-up was controlled by bottom composition, distillate by accumulator level, and reflux by distillate composition. Control was unstable and extremely slow. System was modified to control reflux by accumulator level, boil-up by steam flow, and distillate by set flow adjusted manually for distillate composition. This was better but still slow. Tight product quality was achieved when manual setting of distillate flow was replaced by an on–off control.	An extensive analysis of the system and its dynamics is presented.

15141	121	Microelectric material, batch	To achieve high purity, tower was operated at very high reflux ratio and small draw rates. Once sufficient material concentrated in the top section, it was switched to no reflux and full distillate takeoff.	
26.3.4 Boil-Up on LC, Bottoms on TC/AC (Figure 26.4*e*)				
1547	203	Pharmaceuticals (minimum boiling toluene azeotrope)–toluene	Distillate and bottoms were controlled by accumulator and base levels, respectively, feed and reflux on flow control and boil-up on temperature control. Pall rings were replaced by higher capacity rings (bottom) and wire-mesh structured packing (top) to increase capacity and reduce reflux. The column was sensitive to ambient disturbances (e.g., rainstorms). The reflux reductions escalated this sensitivity to an extent that annulled the revamp benefits. The temperature control was ineffective due to its narrow range of variation. Problems solved by controlling boil-up on base level and bottom product on flow control.	Frequent ambient disturbances destabilize indirect MB control. A non-MB control can do well where changes in feed rate and composition are minor.
1541	380	Debutanizer	Column feed was heated by a feed–bottom interchanger, then by a steam preheater. Feed temperature was controlled by adjusting preheater steam. Boil-up was controlled by a tray analyzer and bottom flow by the base level. A disturbance in steam pressure at times rendered the feed temperature control inoperative, leading to analyzer control cycling. Interchanging the level and analyzer controls eliminated the problem.	Good control can prevent disturbance amplification via a feed-bottom interchanger.
1530	76, 77	Chemicals 12 ft ID, 100-valve trays	Control of bottom level by manipulating boil-up was unstable due to inverse response. Stepping up reboiler steam displaced tray liquid into the column base so that bottom level rose instead of falling. Problem was solved by manipulating reflux flow to control bottom level. An extensive analysis of inverse response, including predictive equations, is presented.	Column inverse response can be troublesome when boil-up is manipulated by bottom level.
DT26.3			*Similar to* 1530 *in a xylene splitter.*	

(*Continued*)

Case	References	Plant/Column	Brief Description	Some Morals
1532	57	Chemicals bubble-cap column	Control of bottom level by manipulating boil-up was sluggish and unsatisfactory due to reboiler inverse response. Increasing reboiler heat input backed up liquid into the column base so that bottom level rose instead of falling ("reboiler swell"). Problem was solved by controlling bottom level by manipulating bottom flow and bottom composition by manipulating boil-up. (See also 1531, Section 26.3.3.)	Inverse response of the reboiler may be troublesome when boil-up is manipulated by bottom level.
DT26.4			*Inverse response in a tower with an internal condenser and no reflux drum.*	
DT26.5			*Reboiler swell causes reboiler to stop working.*	
DT26.6			*Preferential baffle in tower base makes this control scheme erratic.*	
26.3.5 Reflux on Base LC, Bottoms on TC/AC				
1530, DT26.3, DT26.4			*Used to solve an inverse-response problem, Section* 26.3.4.	
			26.4 Problems with Side-Draw Controls	
26.4.1 Small Reflux below Liquid Draw Should Not Be on Level or Difference Control				
1556	160	Refinery crude fractionator	AGO (bottom side-draw) quality fluctuated due to use of calculation to control reflux to the wash section below. The calculated reflux was a difference between two large numbers and was erratic. Problem alleviated by a flow control of this reflux.	A difference between two large numbers is not a suitable control variable.
15105	332	Refinery crude fractionator	Wash tray internal reflux was the small, uncontrolled (overflow) difference between tower liquid and a large AGO draw above. This reflux swung erratically, inducing tray drying, fouling, black AGO, and excessive reflux (overflash). Solved by converting to a total AGO draw, making wash reflux a flow-controlled pumpback. The trays were also replaced by structured packings.	Same as 1556, 1557.

15118	333	Refinery crude fractionator	Wash section reflux (overflash) was controlled by changing the heavy diesel product draw rate above, making it the difference between two large numbers. This was difficult to operate and caused loss of diesel to the resid and poor heavy diesel quality. Solved by adding a total draw tray with flow-controlled reflux plus replacing trays by packings.	Same as 1556, 1557.
DT26.7			*Similar to* 1556, 15105, *and* 15118, *except that it led to excess overflash, and cured solely by a total draw tray and flow-controlled reflux (no packing).*	
1557	160	Refinery FCC main fractionator	Reflux to the section below the LCO side draw was by internal overflow. This stream, much smaller than the LCO draw above, fluctuated, causing tray dry-out and heavy ends in the LCO. Problem solved by making the LCO draw a total draw and flow controlling the reflux.	When reflux is small, it must not be controlled by a level or internal overflow.
15117	128	Refinery FCC main fractionator	Section below LCO draw was unstable, with fluctuations in temperatures and in LCO stripper level. The instability was reduced when the LCO draw removed all available liquid, drying up the LCO/HCO fractionation section below. No HCO was drawn, and the HCO PA provided enough reflux for LCO/DO separation. Lack of flow control on the DO product heating the LCO stripper augmented the instability. Fluctuations mitigated by DMC implementation (which presumably always removed all liquid available at LCO draw). This permitted running hotter DO and better LCO yield.	See 1557. A small internal overflow reflux is the difference between two large streams. As it fluctuates, trays dry and downcomers unseal, causing instability.

(*Continued*)

Chapter 26 Control System Assembly Difficulties (*Continued*)

Case	References	Plant/Column	Brief Description	Some Morals
26.4.2	**Incomplete Material Balance Control with Liquid Draw**			
15158	538	Aromatics BTX prefractionator	BTX liquid side draw was on flow control, operator adjusted using an on-line product analyzer. Top product was on flow control, boil-up on tray temperature control, reflux on accumulator level control, and bottoms on base-level control. Adjustment of reboiler temperature led to tower loading and unloading, instability, long delays, and off-specification products. Improved by integrated plant MVC, which included an inferential product analyzer and better pressure/temperature controls in the prefractionator subcontroller.	
26.4.3	**Steam Spikes with Liquid Draw**			
1585	221	Ethanol rectifier	A new plant could neither exceed 60% capacity nor produce fusel oils for 1 year. Cause was excessive boil-up in the extraction column (Case 215, Section 2.6.2) plus erratic steam flow in the rectifier. To draw fusel oils, low temperature was required near the rectifier bottom, but it rose sharply with steam spikes. Problem overcome by operators manually countering the spikes.	
26.4.4	**Internal Vapor Control Makes or Breaks Vapor Draw Control**			
1567, DT26.8	273	Chemicals vacuum	Vapor side-draw control problems were eliminated after the side-draw rate was controlled by an IVC. It kept a constant ratio of *dP* above the side draw to the *dP* below the side draw.	
15159	538	Aromatics BTX stripper	BTX vapor side draw was on HV control in vapor line to total air condenser. Tower pressure was controlled by valve in overhead line to another total air condenser. Stripper pressure fluctuated with ambient conditions. Adjustments to air coolers disturbed the stripper. Improved by MVC setting overhead pressure and temperature as minimum-move variables.	
26.4.5	**Others**			
15143			*Infrequent analysis leads to insufficient side-draw removal, Section* 27.3.3.	

Case	References	Plant/Column	Brief Description	Some Morals
			27.1 Temperature Control	
27.1.1	**No Good Temperature Control Tray**			
1504	10	A mixed alcohol–ether column	Excessive alcohol losses occurred because a temperature control point sensitive to the key products and at the same time insensitive to other components could not be found. The column separated volatile azeotropes from mixed alcohols. Analyzer control solved the problem.	In some multicomponent towers, good temperature control cannot be achieved.
1599	38	Refinery gasoline stabilizer	Temperature control manipulating boil-up failed to keep iC_5 impurity in distillate on specification. The major source of disturbance was composition variation of a secondary (slop) feed. Distillate brought on specification by an on-line neural estimator that inferred iC_5 in the product and manipulated boil-up.	Same as 1504.
1543			*Temperature control having problem with feed fluctuations, Section* 27.3.4.	
1582	409	Solvent–water	Water concentration of the bottom product was successfully controlled via a stripping section temperature, but it was difficult to control the solvent concentration at the top. Problem solved using an observer model and adaptive multivariable control. One temperature in the top section and one in the bottom, both pressure compensated, were key inputs.	Finding a best control tray can be difficult.
1583	409	Isomeric column	Bottom purity was successfully controlled via a stripping section temperature, but it was difficult to control distillate purity. Problem solved using an observer model and adaptive multivariable control. Two temperatures in the top section and one at the bottom, all pressure compensated, were key inputs. The reason for two top temperatures is that the distillate contained two components plus heavy impurities.	Same as 1582.
DT27.1			*Adequate temperature control cannot be achieved in amine regenerator stripping little* H_2S.	

(*Continued*)

Case	References	Plant/Column	Brief Description	Some Morals
15166			*Temperatures insensitive to composition in C_3 splitter, Section* 26.2.	
1547			*Temperature control becomes ineffective when reflux is reduced, Section* 26.3.4.	
1598	148	Methanol–water	Temperature sensing for control was unreliable below the feed because feed contained variable amount of salts. Above the feed, temperatures were sensitive to raising steam rate above the steady state, but not to lowering steam rate below it. Increasing reflux and boil-up by 5% reinstated temperature sensitivity.	A useful trick to enhance temperature sensitivity if capacity permits.
27.1.2 Best Control Tray				
15137	453	Refinery deethanizer stripper	Ethane in the bottom varied from near zero to 2000 ppm. Boil-up was controlled by bottom temperature, which was totally insensitive to the small ethane concentration. Adding advanced control did not improve.	Poor control tray location gives poor control.
1549	304	Pharmaceuticals methanol stripper from water	Control temperature was located at the reboiler outlet, where the mixture was almost pure water and temperature was insensitive to composition. This, combined with the operator's natural reaction, promoted flooding.	Manual operation of the boil-up rate was better than temperature control.
1563			*Better temperature control tray contributes to better control, Section* 26.3.2.	
27.1.3 Fooling by Nonkeys				
1518	306	Refinery iC_4–nC_4 splitter	A sudden rise in propane (light nonkey) occurred. The top temperature controller counteracted the falling temperature by increasing *n*-butane (heavy key) in the top product.	Temperature controllers can be fooled by nonkeys.
1538			*Similar to* 1518, *Section* 27.3.5.	
1042	75	Natural gas stabilizers	Faulty oil-water interface indicators in the feed drums resulted in excess water entering the tower. This destabilized the column and affected its temperature control and bottom product quality. Cured by repairing faulty indicators.	
1598			*Salts affecting control temperature, Section* 27.1.1.	

27.1.4	**Averaging (Including Double Differential)**			
1536	65	Aromatics benzene column	Double-differential temperature control gave stable control of top and bottom product purities. Both product purities were in the parts-per-million range. A conventional temperature control was unable to accomplish this.	An effective technique for high-purity splits. A thorough analysis presented.
1582, 1583			*Similar principles, in advanced controls, Section* 27.1.1.	
27.1.5	**Azeotropic Distillation**			
1550	66	Azeotropic alcohol/ether/ water/heavies column	Heavies, alcohol/ether/water azeotrope, and dry alcohol product were the bottoms, distillate, and bottom-section side-draw products, respectively. Azeotrope was split into an aqueous purge and an organic distillate/reflux stream in top decanter. Column experienced erratic operation and excessive alcohol losses and steam consumption. Problems were mitigated by changing boil-up control from a single temperature to a two-temperature average near the water/organic break point, adding a second temperature control on bottom tray to replace bottom flow control, and changing pneumatics to DCS.	Average temperature control can be advantageous near sharp composition breaks. With four products, two composition controls are better than one.
1551	66	Azeotropic alcohol/ether/ water/heavies column (same as in 1550)	Problems in 1550 were completely eliminated by further changes, including (a) implementing break-point position control, a technique that subtracted adjacent temperature readings, identified the largest difference as the break-point interval, and used its position as the control signal; (b) adding analyzer control on the aqueous/organic decanter split; (c) adding pressure compensation to temperature controls; and (d) adding feed-forward controls.	Break-point position control is beneficial for azeotropic distillation. In columns with four products, three composition controls are often better than two.
27.1.6	**Extractive Distillation**			
1580	9	Extractive formic/acetic acids	The control tray temperature was sensitive to the ratio of reflux to solvent. When the high-boiling solvent rate increased, the temperature rose, cutting back on boil-up, which led to more formic acid (light) in the bottom. Problem solved by compensating control temperature for variations in solvent-to-reflux ratio.	Control temperatures can be sensitive to heavy nonkey concentration.

(*Continued*)

Chapter 27 Where do Temperature and Composition Controls go Wrong? (*Continued*)

Case	References	Plant/Column	Brief Description	Some Morals
15129	55	Aromatics benzene–extractive distillation	The nonaromatics content in the benzene (bottom) was kept lower than necessary because it was difficult to maintain a set target. A peak sensitive temperature was used to indicate nonaromatics content. A drop of temperature often led to over-reboiling or excessive solvent rates, inducing benzene losses and instability. Improved by multivariable control based on a neural network and virtual on-line analyzer.	
15149			*Slow analyzer response, Section* 27.3.2.	
15150	60	Butadiene	Tower-to-tower level shifts between large extractive distillation and stripper towers led to swings and excessive steam consumption. Mitigated by a MVC that adjusted the two levels simultaneously.	
27.1.7 Other				
15140	352	Chemicals	Temperature controller set point was set higher than the boiling point of the pure component in the base. To increase boiling point, the sump pressure had to be increased by large addition of steam, initiating loading and unloading cycles.	Temperature control does not work if the set point exceeds the boiling point.
			27.2 Pressure-Compensated Temperature Controls	
27.2.1 ΔT Control				
1535	528	Refinery alkylation unit DIB, several columns	Boil-up was controlled by bottom-section ΔT. This worked well when the bottom contained no components heavier than C_4. Occasionally, the control became unstable when lights content in the bottom was high. At a later stage, a feed containing heavy nonkeys was added to the column, and the ΔT control became unstable. Changing to straight temperature control improved stability. A thorough analysis is presented.	ΔT control may be unsatisfactory when product nonkey impurities are relatively high.
1542	513 •	Refinery alkylation unit DIB	Boil-up was controlled by the bottom-section ΔT controller. By a careful choice of the ΔT measurement locations and operation to the right of the maximum in the curve of ΔT versus bottom composition, the system was made to work even in the presence of significant nonkeys. A thorough analysis is presented.	ΔT control can be made to work even with nonkeys in the product (compare 1535).

1539	441	Ethanol–water	Reflux was controlled by a top-section differential temperature controller. A low differential temperature signaled excess reflux but also occurred without any reflux at all or when the column was cold.	Beware of ΔT control issues during severe disturbances.
1536			*Double-differential temperature control, Section* 27.1.4.	
27.2.2 Other Pressure Compensation				
1590	182	AMS–phenol	Column control improved after a pressure compensation was added to the temperature controller set point. The column operated at 30 mm Hg absolute.	
1551, 1582, 1583			*Helpful in azeotropic distillation and in advanced control, Sections* 27.1.5 *and* 27.1.1.	
1577	296	Olefins C_2 splitter heat pumped	Reflux and overhead flowmeters measured vapor with no pressure compensation. A rise in pressure induced the controller to raise the reflux flow. The increase was magnified by use of the reflux-to-overhead ratio controller. Eventually, top composition got purer and cut back reflux, but after an amplified upset. (See also 1575, Section 26.2.)	Pressure compensation of vapor flow measurements can prevent upsets.
			27.3 Analyzer Control	
			(*See also Experiences with Composition Predictors in Multivariable Controls, Section* 29.6.5)	
27.3.1 Obtaining a Valid Analysis for Control				
1540	380	Depropanizer	Analyzer control was troublesome when sample point was located in the overhead vapor line. Isobutane concentration was twice at the center of the line than near the wall, and the concentration gradient was unsteady.	Beware of nonreproducibility of samples drawn from the column overhead line.
1574	296	Olefins C_2 splitter	Boil-up was controlled by an IR analyzer measuring ethylene in the bottom. The IR analyzer actually measured ethylene plus propylene. When propylene reached splitter bottom, the controller added heat. This made no sense. Problem solved by replacing IR by GC analyzer.	Beware of IR analyzer selectivity.
27.3.2 Long Lags and High Off-Line Times				
15132	146	Refinery FCC and alkylation units, several columns	Compositions inferred from temperature, pressure, and reflux measurements and a shortcut model tracked analyzer measurements very well. Their use in APC overcame slow analyzer responses (in one case, a 2-h dead time) and problems with on-line availability (in one case, a full month off-line).	

(*Continued*)

Chapter 27 Where do Temperature and Composition Controls go Wrong? (*Continued*)

Case	References	Plant/Column	Brief Description	Some Morals
15149	60	Butadiene extractive distillation	Delayed signal from analyzer measuring butanes in bottoms led to product quality giveaways and excess solvent circulation. An inferential on-line measurement of the butane content closely matched analyzer and was incorporated in continuous DCS to eliminate the time lags.	
15130	73	Oils, vacuum tower	A viscometer in APC controlled the bottom viscosity at a specified value. The viscometer read incorrectly 10% of the times, leaving the tower without its main control. A predictive model tuned to track the analyzer when reading correctly was added as backup. The model takes over when the analyzer output is flagged invalid. This enhanced lights recovery without violating bottoms specifications.	
1538			*Accumulator sample point gives poor control, sampling tray near top was better, Section* 27.3.5.	
15158, 15166			*Among other problems, Sections* 26.4.2, 26.2.	
27.3.3 Intermittent Analysis				
15146	397	NGL demethanizer	Use of multistream GC, with sample system issues, impeded C_1/C_2 ratio control on boil-up. Solved by using feed rate and tower pressure to compensate analyzer signal between analyzer updates.	
1591	384	Chemicals steam stripper	Total pollutant concentration in bottom was reduced by a factor of 3 due to more frequent analytical measurements and an advisory software that recommended feed-forward control actions to operators.	Analytical improvements can improve purity.
15143	145	Refinery alkylation DIB	Distillate was C_3/iC_4, bottoms C_5–C_8, and side purge nC_4. Purge rate, adjusted per daily laboratory analysis, was minimized to minimize iC_4 loss. Insufficient purge led to nC_4 buildup. This and variable C_5 in feed impeded temperature control and destabilized unit. Stabilized by inferential model control.	
1561			*Frequent analysis reduces product loss in bottoms, Section* 15.7.	
15168			*Among other problems, Section* 29.5.	

27.3.4 Handling Feed Fluctuations				
1543	503	Refinery naphtha splitter and DIB	Column temperature controls were unable to prevent periodic off-specification product resulting from feed fluctuations. Replacing temperature controls by analyzer controls eliminated problem and gave smooth, tight composition control.	Analyzer control can give superior performance to temperature control.
15100	106	Chemicals heavy/light isomers of C_5 diolefins	Large feed composition variations, the need to use analyzers with long time cycles, a long dead time in composition response, a history of flooding, and upset recovery periods as long as 3 days generated a chronic control problem. An adaptive multivariable predictive controller, with controller gains predicted from steady-state simulations and with a flood predictor, solved the problem.	
1537, 1541			*Poor analyzer control due to fluctuations in feed preheat, Sections* 28.7 *and* 26.3.4.	
27.3.5 Analyzer–Temperature Control Cascade				
1538	470	DIB	Top-section temperature controller responded well to feed changes but produced an offset when lights were present. Installing IR analyzer control with a sampling point at accumulator outlet gave poor control. Relocating the sampling point to a tray near the column top was better, but control was destabilized by rapid feed disturbances. A cascade system using a chromatograph sampling at the accumulator outlet to adjust the set point of the temperature controller gave good response, eliminated the offset, and handled feed disturbances well.	An analyzer temperature cascade can give better control than temperature or direct analyzer control.
1502			*Another successful application, Section* 27.3.6.	
27.3.6 Analyzer on Next Tower				
1502, DT27.2	263	Natural gas deethanizer	Controlling bottom impurity by an analyzer located in the next column overhead did not work because of excessive dynamic lags. A simple sampling system was developed to obtain an adequate sample from the deethanizer bottom stream.	
15131	146	Refinery FCC deethanizer	The C_2 in the LPG (C_3/C_4 product) was kept on specification using APC. The C_2 analyzer was on the distillate from the next tower (debutanizer), and a dead time of 4 hours was observed. An inferential shortcut model closely tracked the analyzer steady-state value and overcame the dynamics problem.	See 1502.

Case	References	Plant/Column	Brief Description	Some Morals
			28.1 Pressure Controls by Vapor Flow Variations	
1503, DT28.1	263	Natural gas	A low leg in a hot-pot regenerator overhead line filled with liquid and backpressured the column, causing control instability.	Avoid low legs in column overhead lines.
1191			*Condensation in vent line backpressures glycerol regeneration reboiler, Section* 2.1.	
1595	107	Refinery FCC main fractionator	Responding to a faulty low-pressure signal, the tower pressure controller slowed the wet gas compressor and opened the spillback valve from the compressor discharge to the top of the fractionator. The valve was at ground level, forming a dead leg that accumulated 2–10 m^3 of condensate. Upon opening, that liquid dumped into the fractionator. The tower pressure surged, lifting the relief valve within 20 seconds, and the compressor surged.	See 1503. Best solution is to eliminate dead leg.
1596 (see 1595)	107	Refinery FCC main fractionator	The wet gas compressor antisurge control was by spillback from the discharge to tower overhead. There was an automatic pressure vent to flare from the reflux drum. Upon surge, the spillback valve opened. This raised reflux drum pressure. The pressure controller vented the extra spillback to flare, and the compressor could not recover from surge.	The pressure control on the vent from reflux drum to flare was eliminated.
15136	452	Vacuum tower	Tower pressure was controlled by manipulating the steam valve to the ejector. Sudden swings from 10 to 40 mm Hg vacuum and back resulted from minute changes in motive steam pressure.	This pressure control is not recommended.
15167	75	Natural gas demethanizer	Column pressure swung due to large feed gas flow swings, affecting bottoms purity. Operations were reluctant to control pressure by adjusting recycle compressor due to its sensitivity. Alleviated by adding maximum and minimum constraint limits to the pressure control.	
1330			*Poor atmospheric tower venting leads to reboiler surging, Section* 23.1.4.	
DT28.2			*This control system destabilized by a drum level control manipulating tower overhead to condenser.*	

28.2 Flooded Condenser Pressure Controls

Ref.	Page	Service	Description	Comments
28.2.1 Valve in Condensate, Unflooded Drum				
15120	308	Refinery C_5–C_6 splitter	Top pressure was swinging between 12 and 20 psig. Cause was an oversized flooded condenser pressure control valve that operated between 5 and 15% opening. Pressure swings were aggravated by alternating flooding and dumping on the trays resulting from the pressure changes.	
DT28.3			*Too small an equalizing line destabilizes this control system.*	
1506, DT28.7C			*Contributing to entrainment in partial condenser, Section* 24.6.	
28.2.2 Flooded Drum				
1501, DT28.4	263	Natural gas lean-oil still	Inerts accumulation in flooded reflux drum caused unflooding of the drum and poor control. Manual venting could not solve problem because plant was not continuously attended.	A simple automatic venting control solved the problem.
15108	165	Refinery coker debutanizer	Flooded drum pressure control worked poorly. Tower pressure swings induced erratic distillate purity. Cause was high and variable noncondensables, with reflux drum purging three times per shift. Also, a small change in fin-fan outer heeder box level made big changes in flooded surface area. Pressure swings were eliminated in a revamp that added a trim condenser, eliminated the noncondensables, and changed to hot-vapor bypass pressure control.	Compare 1501.
28.2.3 Hot-Vapor Bypass				
1524	217		Condenser controlled using a hot-vapor bypass. Subcooled liquid leaving the condenser was mixed with the hot bypass vapor prior to entering the reflux drum. Severe shock condensation occurred. Problem was solved by entering the vapor and liquid separately into the drum.	With hot-vapor bypass, vapor must enter the vapor space and liquid must enter below the liquid surface.

(*Continued*)

Chapter 28 Misbehaved Pressure, Condenser, Reboiler, and Preheater Controls (*Continued*)

Case	References	Plant/Column	Brief Description	Some Morals
1572	250		Column pressure was controlled by a hot-vapor bypass with a reflux drum elevated above the condenser. The hot vapor was connected to the subcooled liquid before entering the drum. This resulted in poor pressure control. The problem was completely eliminated by separating the liquid from the vapor and extending the liquid line well below the liquid surface. The vapor line entered onto the drum vapor space.	See 1524.
DT28.5			*Violating moral of Case* 1524 *(above) gives severe pressure fluctuations with hot-vapor bypass.*	
1592	432	Depropanizer	A hot-vapor bypass control did not work because subcooled condensate and hot vapor joined upstream of the reflux drum. Also, the air condenser was elevated above the drum without throttling and the pressure controller took signal from the reflux drum. To permit operation, the bypass was blocked, the condensate outlet was manually throttled, and tower pressure was controlled by adjusting the air condenser louvers.	See 1524, 1572.
15128	116, Case MS16	Refinery naphtha fractionation	An improperly designed hot-vapor bypass around a fin-fan condenser resulted in hydraulic hammer in the pipework. The severity can be judged by the fact that lightweight fireproofing was dislodged from the supporting structure. The unit was shut down before anything failed.	See 1524, 1572.
1525	217		Condenser controlled using a hot-vapor bypass. Subcooled liquid entered the reflux drum vapor space (presumably due to unflooding the liquid inlet) and contacted drum vapor that was 100°F hotter. The rapid condensation sucked the liquid leg between the condenser and drum in seconds.	With hot-vapor bypass, the liquid must always enter below the liquid surface.
1522	87	Narrow-boiling-range distillate	Column pressure was controlled using a hot-vapor bypass scheme. Severe pressure and reflux drum level upsets occurred whenever the reflux drum surface was inadvertently agitated.	Avoid surface agitation with hot-vapor bypass.

15125	292	Refinery DIB	Pressure fluctuations lowered product purity and caused periodic premature flooding. Cause was severe accumulator level fluctuations interfering with the hot-vapor bypass pressure control. Improving feed controls stabilized accumulator level, which in turn alleviated pressure fluctuations.	Accumulator level fluctuations can interfere with hot-vapor bypass.
1517	306	Refinery large debutanizer	Column was limited by overhead condensing capacity during summer. A new condenser was purchased but never used because just before its installation it was discovered that the control valve in the condenser vapor bypass (process side) leaked. Blocking in the bypass increased condenser capacity by 50%.	Never overlook the obvious.
15145	198	Refinery FCC depropanizer	Tower had hot-vapor bypass control, but with the pressure controller mounted on reflux drum. Response was nonlinear. Valve undersizing caused it to be operated open. A hand valve installed in the condenser outlet to increase *dP* lowered condenser capacity.	With this method, pressure transmitter should be on tower, not on drum.
1423			*Poor condenser heat transfer caused by lack of condenser vent with a hot-vapor bypass scheme, Section* 24.1.1.	
1523	87		Column pressure was controlled by a valve located in the condenser bypass. The total condenser was located above the reflux drum and drained freely (no liquid held in the condenser). This method did not work.	Beware of condenser bypass control when the condenser is not partially flooded.
28.2.4 Valve in the Vapor to the Condenser				
1507, DT28.6	471		Closure of a control valve in column overhead line to an air condenser caused rapid condensation and a severe liquid hammer downstream of the valve. Control valve was modified so that it would not shut.	
1159			*Pressure fluctuation at start-up, Section* 12.13.4.	
DT22.4			*Helps mitigate damage due to rapid absorption in overhead condenser.*	

(*Continued*)

Chapter 28 Misbehaved Pressure, Condenser, Reboiler, and Preheater Controls (*Continued*)

Case	References	Plant/Column	Brief Description	Some Morals
			28.3 Coolant Throttling Pressure Controls	
28.3.1 Cooling-Water Throttling				
1573	250	Several	Accelerated fouling occurred in condensers whose controllers throttled cooling-water flows. Water outlet temperatures were as high as 180–200°F, and this caused the fouling.	Throttling cooling water can cause accelerated fouling.
DT28.7			*Several experiences, good and bad, with cooling-water throttling.*	
15123	292	Refinery DC_2 stripper	Throttling cooling water to feed cooler eliminated ethane condensation and accumulation in stripper top but generated low velocities and high water temperatures in the cooler, which in turn fouled it.	Throttling cooling water fouls exchangers.
1505	78	Chemicals	Controlling reflux drum temperature by throttling cooling water to condenser caused boiling of cooling water when control valve closed. This resulted in atmospheric product release.	Beware of issues of throttling cooling water.
DT24.5			*Cooling-water throttling causes instability, boiling of cooling water, and corrosion.*	
15103			*Siphoning due to control valve in cooling-water line to condenser, Section* 12.13.3.	
28.3.2 Manipulating Airflow				
1584	96	Refinery hydrocracker preflash	Overhead product was vapor, overhead temperature was controlled by manipulating airfin blade pitch, and drum level was controlled by reflux flow. The tower experienced instability. A particular problem was control of the overhead drum level, which was destabilized by major feed and heat input disturbances. Problem overcome by DMC.	Accumulator control by reflux manipulation is troublesome at low reflux rates.
28.3.3 Steam Generator Overhead Condenser				
1615			*Miscalibrated steam relief valve bottlenecks throughput, Section* 21.5.	
28.3.4 Controlling Cooling-Water Supply Temperature				
DT28.8			*Avoiding loss of condenser coolant when booster pump fails.*	

28.4 Pressure Control Signal				
28.4.1 From Tower or from Reflux Drum?				
15124	292	Refinery hydrotreater fractionator	Pressure control on tower was poor because pressure transmitter was mounted on reflux drum, and condenser pressure drop fluctuated. Taking signal from tower top pressure eliminated instability and improved light-ends yield.	Tower pressure control transmitter should be located upstream of condenser.
1592, 15145, DT24.3			*Signal from reflux drum causes or contributes to instability, Section* 28.2.3.	
28.4.2 Controlling Pressure via Condensate Temperature				
15107	165	Refinery coker debutanizer	Tower pressure control changed set point of condensate temperature, which in turn manipulated the pitch of the air cooler fan blades. This system did not work. Large variations in noncondensable concentration could have been a factor.	Controlling pressure via condensate temperature often gives poor results.
28.5 Throttling Steam/Vapor to Reboiler or Preheater				
DT28.9, DT28.10			*Multitude of experiences and solutions with oscillations when condensing pressure in reboiler falls below condensate header.*	
1519	306	Refinery	The heat input control valve was located in the 30-psig steam line to the reboiler. The condensate was at 20 psig. When the valve was throttled, condensate would back up into the reboiler and waterlog tubes. The column would call for more heat and the valve would reopen until the condensate drained. It would then throttle, and the cycle was repeated.	Problem was overcome by relocating the valve to the condensate line.
1548	386	Stripper	At a capacity-boosting revamp, a preheater was added to supplement boil-up requirements. The preheater heat input control valve was in the 100-psig steam line to the preheater. At turndown to near the pre-revamp rates, the preheater heat duty fell. The inlet valve closed, dropping the preheater pressure below the condensate header, and the preheater would stop operation.	Cure was shifting heat load from the reboiler to the preheater, thus keeping the preheater loaded.

(Continued)

Chapter 28 Misbehaved Pressure, Condenser, Reboiler, and Preheater Controls (*Continued*)

Case	References	Plant/Column	Brief Description	Some Morals
15104	457	Refinery depropanizer	Condensate pot was vented to the steam inlet line, not to the reboiler. Pressure drop in the reboiler and its entrance backed up condensate, partially flooding reboiler tubes. To compensate heat transfer loss, condensate pot level valve was run wide open blowing steam into the condensate header. Temperature control by manipulating steam inlet valve gave wide variations in bottom composition.	Condensate pots should be vented to the reboiler, not to inlet header.
1352	198	Refinery debutanizer	A new steam horizontal thermosiphon reboiler failed to achieve design heat duty, limiting tower capacity. It was thought that excessive reboiler/piping pressure drop backed condensate pot liquid into the reboiler, but relocating the pot pressure-equalizing line to bypass the pressure drop did not eliminate the problem.	
DT3.2			*Cooling condensate pot liquid to prevent NPSH problems.*	
1342			*Faulty steam trap causes blowing of condensate seal and loss of heat transfer, Section* 23.9.2.	
15138			*Flow control signal, calculated as a small difference between two large numbers, gives poor control and coking in fired heater, Section* 19.2.	
			28.6 Throttling Condensate from Reboiler	
1301, DT23.11, 1576			*Loss of condensate seal causing vapor breakthrough, loss of heat transfer, Section* 23.9.2.	
1516	306	Refinery	Rust layer formed on the inside of the channel head of a reboiler to the level where the steam condensate normally ran. This indicated that 20% of the heat transfer area was waterlogged and ineffective.	
15169	532	Refinery FCC deethanizer stripper	Steam to start-up reboiler was flow-controlled. Condensate from the reboiler was level-controlled, with level transmitter spanning the condensate pot below and a portion of the reboiler tubes. At low steam loads, condensate level exceeded the upper transmitter tap. The condensate valve was 100% open with the level exceeding 100%.	Condensate level transmitters should span the entire tube field.

1537	321	NGL debutanizer	The column feed was preheated by the bottoms, then by a steam preheater. Preheater steam was controlled by the feed temperature downstream. The feed enthalpy fluctuated with fluctuations in column bottom flow. This interfered with the column product analyzer control. Problem was cured by a feed enthalpy controller, which regulated preheater steam flow.	A feed enthalpy control is needed if heat input to the feed fluctuates.
1541			*Preheater control instability destabilizes tower, Section* 26.3.4.	
15162	99	Gas NRU column	Fluctuations occurred when base level controlled bottom flow to preheater with preheater bypass held constant. Cured by holding liquid flow to preheater constant and controlling base level on the bypass.	
204			*Problems with preheat temperature control leads to insufficient reflux, Section* 2.1.	
1546	441		Column bottoms, drawn from the weir compartment of a kettle reboiler, preheated column feed. Preheat was not controlled. A rising bottom level increased bottom flow and feed preheat. The greater preheat reduced column downflow, bottom level, and bottom flow. This in turn reduced preheat and raised bottom level. A cycle developed.	Controlling preheat could have avoided the problem.
15113	357	NGL deethanizer	Feed was throttled to tower pressure, then preheated. Feed temperature, controlled by manipulating preheater bypass, was difficult to control due to vaporization. Tower instability resulted. Solved by relocating throttling valve to downstream of preheater and limiting preheat to minimize flashing.	
15134	518	Refinery crude fractionator	When the preflash tower and its bottom pump were bypassed, vaporization occurred upstream of the control valves splitting the flow to the heater passes. This gave uneven split and instability, with the outlet temperature from one pass 40°C hotter than the others with its valve wide open. Solved by reducing feed preheat by shifting PA duties from feed preheat to reboilers and to the crude overhead condenser.	It is difficult to evenly split two-phase flow.
1548			*Turndown problem with valve in steam supply to preheater, Section* 28.5.	

Case	References	Plant/Column	Brief Description	Some Morals
			29.1 Interaction with the Process	
1547, 1598			*Reducing reflux ratio destabilizes controls, Sections* 26.3.4 *and* 27.1.1.	
1559			*Disturbance amplification in tower train, Section* 3.2.2.	
15122, 15168			*Tight level controls allow feed swings to destabilize tower, Section* 29.5.	
15163	378	Natural gas Selexol H_2S stripper, packed	In two parallel trains, feed was on level control from HP flash drum. Occasional drum level disturbances destabilized tower, giving swings in steam demand and inadequate stripping. In third train, swings were mitigated by adding MP flash drum, with a level control with large dead band resetting flow control (see also Case 15164, Section 6.5).	
15111	176	Refinery crude fractionator	Crude switches disrupted tower operation for up to 6 hours, causing off-specification products, yield losses, and bottom-level fluctuation. A multivariable predictive controller based on measurements and model responses of the feed preheat train as well as the tower mitigated these problems.	
15150			*Column-to-column level swings in extractive distillation, Section* 27.1.6.	
1589			*Tower high-level switch trips entire plant, Section* 8.1.2.	
			29.2 ΔP Control	
15147	122		Despite tuning, using tower ΔP to control boil-up was unsuccessful because boil-up had little effect on ΔP. Solved by using stripping section temperature to control boil-up.	Beware of ΔP control issues.
1603, 1604			*Causing two relief failures simultaneously, Section* 21.2.	
			29.3 Flood Controls and Indicators	
15153	316		Flooding episodes were reduced from 12 to 2 per year by an incipient flooding indicator. Oscillations of tray 10 temperature and bottom sump level repeatedly preceded flood episodes. Their onset would switch the flood indicator on, signaling to operators to back off rates.	

DT29.1			*Separate dP recorders below feed and above feed distinguish floods from hydrates and dramatically cut flood episodes.*	
12102			*Flood control circumvents cold spins, Section* 3.1.5.	
			29.4 Batch Distillation Control	
15141			*Switching off reflux for high-purity product, Section* 26.3.3.	
873			*Modified controller programming improves turndown liquid distribution, Section* 6.8.	
15112			*Formation of second liquid phase interferes with product/reflux split, Section* 2.5.	
			29.5 Problems in the Control Engineer's Domain	
15165	378	Natural gas Selexol CO_2 absorber	Selexol flow control valve from LP flash drum began rapid uncontrolled cycling, causing a hydraulic hammer that moved the line and damaged bolts and gasket on the absorber inlet flange. Increasing actuator size did not cure. Changing valve plug and internal actuator settings eliminated problem.	
1579	482	Chemicals	Gas flow to fired reboiler, and column temperatures, cycled due to 8% hysteresis in the gas flow control valve. Problem eliminated by installing a valve positioner.	
15122	292	Refinery DC_2 stripper	Tower flooded prematurely due to fluctuations in feed rate. Cause of fluctuations was extremely tight tuning of liquid-level control in the feed surge drum.	Surge drum level controls should be loosely tuned.
15168	75	Natural gas Stabilizer and debutanizer	Swings in feed flows and compositions and tight controls of reflux drum and tower levels destabilized tower. Lack of on-line analyzers and slow, infrequent lab analyses added to give poor condensate RVP control. Cured by non-linear level controllers that mitigated swings and by inferential model MVC.	
15151	60	Butadiene stripping from acetonitrile	Vinylacetylene and ethylacelytene content of the raw heavy tail stream fluctuated. Using a gain-scheduling strategy together with a process model smoothed fluctuation.	

(*Continued*)

Chapter 29 Miscellaneous Control Problems (*Continued*)

Case	References	Plant/Column	Brief Description	Some Morals
15116	306	Refinery crude fractionator	Kerosene production declined due to a loose instrument air-tubing connection to the kerosene draw control valve. The control valve did not get enough air to open fully. This took several months to diagnose.	Look for the simple cause.
			29.6 Advanced Controls Problems	
29.6.1	**Updating Multivariable Controls**			
15101	238	Refinery FCC main fractionator	Over 9 years in service, the MVC model had not been updated and lost accuracy due to operational changes. The predictor for the light cycle oil 90% point became useless. A software-based predictor using neural network was successfully used to correctly predict that point.	Control models may deteriorate due to operational changes.
15154	406	Refinery FCC	The FCC plant was configured as one large MVC covering the reactor/regenerator, main fractionator, and gasoline towers to account for interactions. A revamp which updated reactor and gasoline handling technology rendered the DMC models nonrepresentative and unusable. Solution was separating the MVC into reactor, fractionator, and debutanizer models, transferring fractionator levels from the DMC to the DCS, and reengineering the MVC using plant data but no step testing.	
29.6.2	**Advanced Controls Fooled by Bad Measurements**			
1529	306	Refinery debutanizer	An advanced feed-forward control system caused steam flow, reflux, pressure, and temperature to drop. Switching the steam flow to manual resuscitated the column. Problem was caused by a plugged tap of a steam flowmeter. The malfunctioning meter misled the controller.	
1558	449, 453		The advanced control constrained production rate using the *dP* across the top packed bed. The upper *dP* tap was in the overhead line, so the measurement included the outlet nozzle pressure drop. This inflated *dP* measurement restricted production.	

15139	453	Refinery FCC main fractionator	LCO was originally drawn on flow control with internal overflow of reflux. An advanced control added a level indicator for the advanced controller. The advanced controller then reset the LCO draw rate. Unappreciated was the fact that on a partial draw tray the overflow weir sets the liquid level, so the level input degrades from the advanced control.

29.6.3 Issues with Model Inaccuracies

15155	406	Refinery HN stripper	The level control valve periodically wound up to 100%, so operators had to intervene, taking the HN product flow out of DMC and cutting it back. This problem lowered HN yield. Cause was a linear DMC model representing the response of the very nonlinear valve. Cure was transforming the level controller output to better represent the nonlinear characteristic.

29.6.4 Effect of Power Dips

1528	306	Refinery iC_4–nC_4 splitter	An advanced feed-forward control system was installed. Each time a power outlet was used in the control room, the feed-forward system was affected, just like running an electric appliance interferes with TV reception. This caused erratic reflux and reboil behavior.

29.6.5 Experiences with Composition Predictors in Multivariable Controls

(*for deficiencies of temperature and analyzer controls, see Sections* 27.1.1 *and* 27.3.2.–27.3.4)

15102	232	Refinery dehexanizer (two towers)	In absence of on-line analyses, it was extremely difficult to optimize towers and respond to process changes. Virtual analyzer, based on neural network model, gave good composition predictions except when process conditions were outside the training range. Remodeling for the extended data range rendered the virtual analyzer signals reliable enough for inclusion as input to a multivariable controller.

References

1. Alsop, N., and J. M. Ferrer, "One Step Up: Control Simulation Gets Dynamic," *The Chem. Engnr.*, July 2004, p. 37.
2. American Oil Company (AMOCO), *Hazard of Steam*, 2d ed., AMOCO Chicago, IL, 1984.
3. American Oil Company (AMOCO), *Hazard of Water*, 6th ed., AMOCO Chicago, IL, 1984.
4. American Oil Company (AMOCO), *Safe Ups and Downs*, 3d ed., AMOCO Chicago, IL, 1984.
5. American Petroleum Institute, "Pressure Vessels," in *Guide for Inspection of Refinery Equipment*, 4th ed., Chapter VI, API, Washington, DC, Dec. 1982.
6. American Petroleum Institute, *Safety Digest of Lessons Learnt*, Publication 758, Sections 2–4, API, Washington, DC, 1979–1981.
7. Andersen, A. E., and J. C. Jubin, "Case Histories of the Distillation Practitioner," *Chem. Eng. Prog.*, *60*(10), 1964, p. 60.
8. Anderson, C. F., "Simultaneous Scanning Techniques for More Accurate Analysis," Paper presented at the AIChE Spring National Meeting, New Orleans, LA, Mar. 29–Apr. 2, 1992.
9. Anderson, J. E., "Control by Tray Temperature of Extractive Distillation," in W. L. Luyben (ed.), *Practical Distillation Control*, p. 405, Van Nostrand Reinhold, New York, 1992.
10. Anderson, J. S., and J. McMillan, "Problems in the Control of Distillation Columns," *IChemE Symp. Ser.*, *32*, 1969, p. 6:7.
11. Anonymous, "A Lesson Too Late," *The Chem. Engnr.*, Aug. 19, 1999, p. 18.
12. Anonymous, "A Major Incident During Startup," Loss *Prevention Bull.*, *156*, IChemE, Dec. 2000, p. 3.
13. Anonymous, "Absorber Changes Solve Offshore High -H_2S Problems," *Oil Gas* J., May 23, 1988, p. 40.
14. Anonymous, "Accidents Renew Hoechst's P. R. Woes," *Chem. Eng.*, Mar. 1996, p. 45.
15. Anonymous, "Chimney Effect," *Loss Prevention Bull.*, *110*, IChemE, Apr. 1993, p. 8.
16. Anonymous, "Citgo Explosion Explained," *The Chem. Engnr.*, Aug. 15, 1991, p. 12.
17. Anonymous, "Corrosion under Insulation Causes Refinery Fire," *Loss Prevention Bull.*, *87*, IChemE, June 1989, p. 5.
18. Anonymous, "Exothermic Runaway Caused Hickson Deaths," *The Chem. Engnr.*, Dec. 10, 1992, p. 7; "Hickson and Welch Questions Continue," Oct. 15, 1992, p. 6.
19. Anonymous, "Explosion at the BASF Antwerp Ethylene Oxide/Glycol Plant," *Loss Prevention Bull.*, *100*, IChemE, Aug. 1991, p. 1.
20. Anonymous, "Formation of Butane Hydrate Leads to Hydrocracker Fire," *Loss Prevention Bull.*, *98*, IChemE, Apr. 1991, p. 9.
21. Anonymous, "Froth Scrubber Traps Vapors, Solids; Quenches Hot Gases," *Chem. Proc.*, Nov. 1990, p. 96.
22. Anonymous, "How Can We Measure Flow of High Freezing-Point Residue," *Control*, July 1998, p. 72.
23. Anonymous, "HSA Criticizes Hickson," *The Chem. Engnr.*, Nov. 10, 1994, p. 5.
24. Anonymous, "Incident at a Chemical Plant," *Loss Prevention Bull.*, *161*, IChemE, October 2001, p. 6.
25. Anonymous, "Introduction to Large Property Damage Losses in the Hydrocarbon-Chemical Industries: A Thirty-Year Review," *Loss Prevention Bull.*, *99*, IChemE, June 1991, p. 1.

26. Anonymous, "Miscellaneous Case Histories," in C. H. Vervalin (ed.), *Fire Protection Manual*, Vol. 2, p. 29, Gulf Publishing, Houston, TX, 1981.
27. Anonymous, "Over Pressuring of a Storage Tank," *Loss Prevention Bull.*, *75*, June 1987, p. 19.
28. Anonymous, "Pipe Rupture and Fire in a Vinyl Chloride Monomer (VCM) Plant," *Loss Prevention Bull.*, *97*, p. 12, IChemE, Feb. 1991.
29. Anonymous, "Shell's Near-Disaster Lands a Fine of £100,000," *The Chem. Engnr.*, Dec. 12, 1991, p. 8; "Shell—The Facts," Jan. 16, 1992, p. 4.
30. Anonymous, "Some Corrosion Incidents from Past LPBs," *Loss Prevention Bull.*, *163*, IChemE, February 2002, p. 24.
31. Anonymous, "The Problems of Temporary Manhole Covers," *Loss Prevention Bull.*, *167*, IChemE, October 2002, p. 23.
32. Anonymous, "Unrevealed Hazards," *Loss Prevention Bull.*, *164*, IChemE, April 2002, p. 23.
33. APV DH-682, *Distillation Handbook*, 2nd ed., Chicago, IL.
34. Aspen Technology, *Distil v. 6.0*, 2003, http://www.aspentech.com/.
35. Bali, V. K., and A. K. Maheshwari, "Case Study of CO_2 Removal System Problems/Failures in Ammonia Plant," *Ammonia Plant Safety*, *39*, AIChE, 1999, p. 285.
36. Ballmar, R. W., "Towers Are Touchy," in *API Proc., Section III—Refining*, 40, 1960, p. 279.
37. Banik, S., "Weeping from Valve Trays," paper presented at the AIChE National Meeting, Houston, TX, Apr. 1989.
38. Baratt, R., G. Vacca, and A. Servida, "Neural Network Modeling of Distillation Columns," *Hydrocarbon Processing*, June 1995, p. 35.
39. Barber, A. D., and E. F. Wijn, "Foaming in Crude distillation Units," *IChemE Symp. Ser.*, *56*, 1979, p. 3.1/5.
40. Barletta, T., "Pump Cavitation Caused by Entrained Gas," *Hydrocarbon Processing*, Nov. 2003, p. 69.
41. Barletta, T., and K. J. Kurzym, "Consider Retrofits to Handle High-Viscosity Crudes", *Hydrocarbon Processing*, September 2004, p. 49.
42. Barletta, T., and S. Fulton, "Maximizing Gas Plant Capacity," *PTQ*, Spring 2004, p. 105.
43. Barletta, T., E. Hartman, and D. J. Leake, "Foam Control in Crude Units", *Petroleum Technology Quarterly*, Autumn 2004, p. 117.
44. Barletta, T., J. Nigg, S. Ruoss, J. Mayfield, and W. Landry, "Diagnose Flooding Columns Efficiently," *Hydrocarbon Processing*, July 2001, p. 71.
45. Baumer, J. A., "DEA Treats High-Volume Fractionation Plant Feed," *Oil Gas J.*, Mar. 15, 1982, p. 63.
46. Baumgartner, A. J., M. W. Blaschke, S. T. Coleman, R. Kohler, and T. E. Paxson, "Feedstock Contaminants in Ethylene Plants—an Update," *Proc. 16th Ethylene Producers Conference*, New Orleans, Louisiana, 2004.
47. Bell, K. J., "Coping with an Improperly Vented Condenser," *Chem. Eng. Prog.*, *79*(7), 1983, p. 54.
48. Bellner, S. P., W. Ege, and H. Z. Kister, "Hydraulic Analysis is Key to Effective, Low Cost Demethanizer Debottleneck," *Oil Gas J.*, Nov 22, 2004, p. 56.
49. Betts, B. W., and H. N. Rose, "Radioactive Scanning of Distillation Columns," the Joint Symposium on Distillation, The University of Sydney/The University of NSW (Australia), 1974, p. F1.
50. Biales, G. A., "How Not to Pack a Packed Column," *Chem. Eng. Prog.*, *60*(10), 1964, p. 71.
51. Bickerton, J., "Still Explosion," *Loss Prevention Bull.*, *158*, IChemE, 2001, p. 15.
52. Biddulph, M. W., "Tray Efficiency Is Not Constant," *Hydrocarbon Processing*, *56*(10), 1977, p. 145.
53. Bidrawn, S., "Glass Process Systems Perform Trouble Free," *Chem. Proc.*, Oct. 1990, p. 110.
54. Billet, R., *Distillation Engineering*, Chemical Publishing Company, New York, 1979.
55. Bischof, H., T. Pelster, T. Morrison, K. Robler, and M. Sugars, "Advanced Control Improves German Aromatics Operation," *Oil Gas J.*, May 8, 2000, p. 68.
56. Blaschke, M., "Causes and Remedies in Caustic Tower Fouling," *PTQ*, Summer 2003, p. 133.
57. Bojnowski, J. J., R. M. Groghan, Jr., and R. M. Hoffman, "Direct and Indirect Material Balance Control," *Chem. Eng. Prog.*, *72*(9), 1976, p. 54.
58. Bolles, W. L., "Estimating Valve Tray Performance," *Chem. Eng. Prog.*, *72*(9), 1976, p. 43.
59. Bolles, W. L., "The Solution of a Foam Problem," *Chem. Eng. Prog.*, *63*(9), 1967, p. 48.

60. Bonavita, N., R. Martini, and T. Grosso, "A Step by Step Approach to Advanced Process Control," *Hydrocarbon Processing*, Oct. 2003, p. 69.
61. Bosworth, C. M., "Alcohol Rectification," *Chem. Eng. Prog.*, *61*(9), 1965, p. 82.
62. Bouck, D. S., "Vacuum Tower Packing Fires," paper presented at the API Operating Practices Symposium, Apr. 27, 1999, with Addendum/Revisions, Feb. 21, 2000.
63. Bouck, D. S., and C. J. Erickson, "Gamma Scans—A Look into Trouble Towers," paper presented at the AIChE Annual Meeting, Miami Beach, FL, Nov. 1986.
64. Bowman, J. D., "Troubleshoot Packed Towers with Radioisotopes," *Chem. Eng. Prog.*, Sept. 1993, p. 34.
65. Boyd, D. M., "Fractionation Column Control," *Chem. Eng. Prog.*, *71*(6), 1975, p. 55.
66. Bozenhardt, H. F., "Modern Control Tricks Solve Distillation Problems," *Hydrocarbon Processing*, *67*(6), 1988, p. 47.
67. BP "Hazards of Trapped Pressure and Vacuum," Process Safety Booklet II, Sunbury on Thames, United Kingdom, 2003.
68. Bravo, J. L., "Design and Application of Packed Towers for Foaming, Viscous and High Surface Tension Systems," paper presented at the AIChE National Spring Meeting, Houston, TX, 1995.
69. Bravo, J. L., "Design and Application Pitfalls for Packed Gas-Liquid Towers," paper presented at the AIChE Annual Meeting, Miami Beach, FL, Nov. 2–6, 1992
70. Bravo, J. L., "Effectively Fight Fouling of Packing," *Chem. Eng. Prog.*, Apr. 1993, p. 72.
71. Bravo, J. L., J. R. Fair, and A. F. Seibert, "The Effect of Free Water on the Performance of Packed Towers in Vacuum Service," paper presented at the AIChE Annual Meeting, Los Angeles, CA, Nov. 2000.
72. Brierley, R. J. P., P. J. M. Whyman, and J. B. Erskine, "Flow Induced Vibration of Distillation and Absorption Column Trays," *IChemE Symp. Ser.*, *56*, 1979, p. 2.4/45.
73. Brooks, K. S., R. Dreyer, and A. Burston, "Improve APC Uptime Using Analyzers with Predictive Models," *Hydrocarbon Processing*, Feb. 2001, p. 87.
74. Brower, G., S. Smith, and R. Repetti, "Milagro Amine System Costs Reduced," Lawrence Reid Gas Conditioning Conference, March 1997, p. 386.
75. Bu Al Rougha, S. Y., H. Ni, S. Viswanathan, and S. Gejji, "Multivariable Control of Atheer's Habshan Gas Plant," 79th Annual Convention of the Gas Processors Association, Atlanta, Georgia, March 12–15, 2000.
76. Buckley, P. S., R. K. Cox, and D. L. Rollins, "Inverse Response in a Distillation Column," *Chem. Eng. Prog.*, *71*(6), 1975, p. 83.
77. Buckley, P. S., R. K. Cox, and D. L. Rollins, "Inverse Response in Distillation Columns," paper presented at the AIChE Annual Meeting, Houston, TX, Mar. 16–20, 1975.
78. Buckley, P. S., W. L. Luyben, and J. P. Shunta, *Design of Distillation Column Control Systems*, Instrument Society of America, Research Triangle Park, NC, 1985.
79. Cady, P. D., "How to Stop Slug Flow in Condenser Outlet Piping," *Hydrocarbon Processing*, *42*(9), 1963, p. 192.
80. Chakraborty, A., and A. Bagde, "Case Studies on Gas Sweetening Process," 52nd Laurence Reid Gas Conditioning Conference, Norman, Oklahoma, February 24–27, 2002.
81. Charlton, J. S. (ed.), *Radioisotope Techniques for Problem Solving in Industrial Process Plants*, Gulf Publishing, Houston, TX, 1986.
82. Chen, G. K., "Packed Column Internals," *Chem. Eng.*, Mar. 5, 1984, p. 40.
83. Chen, G. K., "Troubleshooting Distribution Problems in Packed Columns," *The Chem. Engnr.*, Sept. 1987 (Suppl.), p. 10.
84. Chen, G. K., and K. T. Chuang, "Recent Developments in Distillation," *Hydrocarbon Processing*, *68*(2), 1989, p. 37.
85. Chen, G. K., T. L. Holmes, and J. H. Shieh, "Effects of Subcooled or Flashing Feed on Packed Column Performance," *IChemE Symp. Ser.*, *94*, 1985, p. 185.
86. Cheng, N., M. Bignold, and G. Hanlon, "Shell Canada's Caroline Gas Plant Reaches Smooth Operation," 47th Laurence Reid Gas Conditioning Conference, p. 1, Norman, Oklahoma, March 2–5, 1997.

87. Chin, T. G., "Guide to Distillation Pressure Control Methods," *Hydrocarbon Processing*, *58*(10), 1979, p. 145.
88. Chokkarapu, K., R. Eguren, and R.N. MacCallum, "Selection and Evaluation of Upgrading a Glycol Dehydration System in High CO_2 Gas Service," 54th Laurence Reid Gas Conditioning Conference, Norman, Oklahoma, February 22–25, 2004, p. 43.
89. Chung, W., and D. Nielsen, "A Dual Chimney Tray—A Highly Effective Technique for Adding a Side Reboiler to an Existing Column," in *Distillation 2001: Frontiers in a New Millennium, Proceedings of Topical Conference*, p. 587, AIChE Spring National Meeting, Houston, TX, Apr. 22–26, 2001.
90. Cosma, G., C. Duncan, B. Gillies, M. Green, and A. Laundry "Use of a Task Team Approach for Amine System Problem Solving," 49th Lawrence Reid Gas Conditioning Conference, Norman, Oklahoma, February 1999.
91. Cox, K. R., R. N. French, and G. J. Koplos, "Limiting Lemmas and Lemons: Some Common Pitfalls of Modeling Gibbs Excess Energy Data," in *Distillation Tools for the Practicing Engineer, Topical Conference Proceedings*, AIChE Spring National Meeting, New Orleans, LA, March 10–14, 2002, p. 212.
92. Cummings, A. L., and S. M. Mecum, "Remove Heat Stable Salts for Better Amine Plant Performance," *Hydrocarbon Processing*, Aug. 1998, p. 63.
93. Cummings, D., and D. Tobias, "Advanced Controls Refine Yields," *Control*, May 1997, p. 60.
94. Cunha, J. A. de C., and P. R. de M. Freitas, "Acid Gases Absorber Improvements in Copene's Ethylene Plant," paper presented at the AIChE Spring National Meeting, Houston, TX, March 9–13, 1997.
95. Custer, R. S., "Case Histories of Distillation Columns," *Chem. Eng. Prog.*, *61*(9), 1965, p. 89.
96. Cutler, C. R., and S. G. Finlayson, "Design Considerations for a Hydrocracker Preflash Column Multivariable Constraint Controller," paper presented at IFAC Conference, Atlanta, GA, June 1988.
97. Dajaeger, J., and E. Das, "Commissioning and Operation of 1800 MTPD Ammonia Plant," *Ammonia Plant Safety*, *34*, AIChE, 1994, p. 38.
98. Davies, J. A., "Trouble with Transients," *Chem. Eng. Prog.*, *61*(9), 1965, p. 74.
99. Davis, R. A., G. N. Gottier, G. A. Blackburn, and C. R. Root, "Plant Design Integrates NGL Recovery, N_2 Rejection," *Oil Gas J.*, Nov. 6, 1989, p. 33.
100. de Villiers, W. E., R. N. French, and G. J. Koplos, "Navigate Phase Equilibria via Residue Curve Maps," *Chem. Eng. Prog.*, Nov. 2002, p. 66.
101. de Villiers, W. E., R. N. French, and G. J. Koplos, "Navigating Phase Equilibria via Residue Curve Maps," in *Distillation Tools for the Practicing Engineer, Topical Conference Proceedings*, AIChE Spring National Meeting, New Orleans, LA, March 10–14, 2002, p. 225.
102. Dean, C. E., S. W. Golden, and D. W. Hanson, "Understanding Unit Pressure Balance Key to Cost-Effective FCC Revamps," *Oil Gas J.*, May 10, 2004.
103. DeHart, T. R., D. A. Hansen, C. L. Mariz, and J. G. McCullough, "Solving Corrosion Problems at the MEA Bellingham Massachusetts Carbon Dioxide Recovery Plant," paper presented at the NACE Conference, San Antonio, TX, Apr. 1999.
104. Diehl, J. E., and C. R. Koppany, "Flooding Velocity Correlation for Gas-Liquid Counterflow in Vertical Tubes," *Chem. Eng. Prog. Symp. Ser.*, *65*(92), 1969, p. 77.
105. DiGesso, J., "Business Interruption Insurance," *The Chem. Engnr.*, Oct. 1989, p. 52.
106. Dollar, R., L. L. Melton, A. M. Morshed, D. T. Glasgow, and K. W. Repsher, "Consider Adaptive Multivariable Predictive Controllers," *Hydrocarbon Processing*, Mar. 1993, p. 103.
107. Dolph, G. A., "Dynamic Simulation for Emergency Control Strategies," *HTI Quarterly*, Summer 1995, p. 51.
108. Donaldson, T., "Confined Space Incidents—a Review 1980–1999," *Loss Prevention Bull.*, *154*, IChemE, Aug. 2000, p. 3.
109. Donaldson, T., "Design Faults—a Review," *Loss Prevention Bull.*, *160*, IChemE, August 2001, p. 4.
110. Downs, J. J., "Distillation Control in a Plantwide Control Environment," in W. L. Luyben (ed.), *Practical Distillation Control*, Van Nostrand Reinhold, New York, 1992, p. 413.
111. Doyle, W. H., "Industrial Explosions and Insurance," *Loss Prevention*, *3*, 1969, p. 11.
112. Doyle, W. H., "Instrument-Connected Losses in the CPI," *Instrum. Technol.*, *19*(10), 1972, p. 38.
113. Drew, J. W., "Distillation Column Startup," *Chem. Eng.*, Nov. 14, 1983, p. 221.

114. Duarte, R., M. Perez, and H. Z. Kister, "Combine Temperature Surveys, Field Tests and Gamma Scans for Effective Troubleshooting," *Hydrocarbon Processing*, April 2003, p. 69.
115. Duguid, I., "Avoid Process Engineering Mistakes," *Chem. Eng.*, Nov. 2001, p. 97.
116. Duguid, I., "Take This Safety Database to Heart," *Chem. Eng.*, July 2001, p. 80, (includes database at www.che.com/CEEXTRA).
117. Dumas, B., "Gamma Scan and CAT-Scan Assist in Maintaining a Proper Δp," *Tru-News*, Vol. 10, Ed. 3, Tru-Tec Services Inc., La Porte, TX, 2002, p. 1.
118. DuPart, M. S., T. R. Bacon, and D. J. Edwards, "Understanding Corrosion in Alkanolamine Gas Treating Plants," *Hydrocarbon Processing*, May 1993, p. 89.
119. Eagle, R. S., "Trouble-shooting with Gamma Radiation," *Chem. Eng. Prog.*, *60*(10), 1964, p. 69.
120. Eckert, J. S., "Selecting the Proper Distillation Column Packing," *Chem. Eng. Prog.*, *66*(3), 1970, p. 39.
121. Eckles, A. J., "Control Strategies for Ultra-High Purities or Extremely Challenging Separations," in *Distillation 2003: On the Path to High Capacity, Efficient Splits, Topical Conference Proceedings*, p. 370, AIChE Spring National Meeting, New Orleans, LA, Mar. 31–Apr. 3, 2003.
122. Eder, H., "For Good Process Control, Understand the Process," *Chem. Eng.*, June 2003, p. 63.
123. Ellingsen, W. R., "Diagnosing and Preventing Tray Damage in Distillation Columns," *DYCORD 86, IFAC Proceedings of International Symposium on Dynamics and Control of Chemical Reactors and Distillation Columns*, Bournemouth, U.K., Dec. 8–10, 1986.
124. Ender, C., and D. Laird, "Minimize the Risk of Fire During Column Maintenance," *Chem. Eng. Prog.*, Sept. 2003, p. 54.
125. Ender, C., and D. Laird, "Reduce the Risk of Fire During Distillation Column Maintenance," *World Refining*, November 2002, p. 30.
126. Englund, S. M., "Inherently Safer Plants: Practical Applications," *Process Safety Progress*, *14*(1), January 1995, p. 63.
127. Erasmus, N. M., and T. Barletta, "Crude Unit Start-Ups: The Results of a High Liquid Level," *PTQ Revamps & Operations Suppl.*, 2003, p. 38.
128. Eriksson, P. O., A. Tomlins, and S. K. Dash, "FCCU Advanced Control System Achieves 2-Month Payout," *Oil Gas J.*, Mar. 23, 1992, p. 62.
129. Erskine, J. B., and W. Waddington, "Investigation into the Vibration Damage to Large Diameter Sieve Tray Absorber Towers," Paper No. 211, the International Symposium on Vibration Problems in Industry, UKAEA and NPL, Keswick, England, 1973.
130. Eskaros, M. G., "Glycol Dehydration," *Hydrocarbon Processing*, July 2003, p. 80.
131. Eskaros, M. G., "Improve NGL Plant Performance", *Hydrocarbon Processing*, Aug. 2001, p. 84-A.
132. Fair, J. R., B. R. Reeves, and F. Seibert, "The Oldershaw Column: Useful for Solving Distillation Problems," in *Distillation Tools for the Practicing Engineer, Topical Conference Proceedings*, p. 27, AIChE Spring National Meeting, New Orleans, LA, March 10–14, 2002.
133. Ferguson, D., "Radiation Scanning of On-Line Absorption and Stripping Columns," Paper presented at the AIChE Spring National Meeting, Houston, TX, Mar. 1993.
134. Ferguson, D., "Radioisotope Techniques for Troubleshooting Olefins Plants," in *Procedings of the seventh Ethylene Producers Conference*, Houston, TX, AIChE, 1995.
135. Fishwick, T., "A Hydrogen Sulphide Release Affects Four Workers," *Loss Prevention Bull.*, *155*, IChemE, Oct. 2000, p. 5.
136. Fishwick, T., "Pressure Relief Problems," *Loss Prevention Bull.*, *160*, IChemE, 2001, p. 12.
137. Fleming, B., E. L. Hartman, and G. R. Martin, "Don't Underestimate the Importance of Reflux/Feed Entry Design for Trayed Fractionators," paper presented at the AIChE Spring National Meeting, Houston, TX, Mar. 20–24, 1995.
138. Fleming, B., G. R. Martin, and E. L. Hartman, "Pay Attention to Reflux/Feed Entry Design," *Chem. Eng. Prog.*, Jan. 1996, p. 56.
139. Fractionation Research Inc. (FRI) Design Practices Committee, "Causes and Prevention of Packing Fires," paper presented at the AIChE Spring Meeting, New Orleans, Louisiana, April 25–29, 2004.
140. Fractionation Research Inc. (FRI), "The Performance of Trays with Downcomers," Motion Picture A, FRI, Stillwater, OK, 1985.

141. France, J. J., "Troubleshooting Distillation Columns," paper presented at the AIChE Spring Meeting, Houston, TX, Mar. 1993.
142. Freeman, C., "Failure of a Glycol Reboiler," *Loss Prevention Bull., 156*, IChemE, Dec. 2000, p. 21.
143. Freeman, L., and J. D. Bowman, "Use of Column Scanning to Troubleshoot Demethanizer Operation," paper presented at the AIChE Spring National Meeting, Houston, TX, Mar. 31, 1993.
144. Freitas, P. R. de M., R. G. Mattos, G. Heck/Ph. Roth, and V. Kaiser "New Gas Separation Technology at Work in Copene," paper presented at the AIChE Spring National Meeting, Atlanta, GA, Apr. 17–21, 1994.
145. Friedman, Y. Z., "First Principles Inference Model Improves Deisobutanizer Column Control," *Hydrocarbon Processing*, Mar. 2003, p. 43.
146. Friedman, Y. Z., and G. T. Reedy, "An Inferential Model for Distillation Columns," *PTQ*, Autumn 2001, p. 97.
147. Fromm, D., and W. Rall, "Fire at Semi-Lean Pump by Reverse Motion," *Plant/Operations Prog., 6*(3), 1987, p. 162.
148. Fruehauf, P. S., and D. P. Mahoney, "Improve Distillation Column Control Design," *Chem. Eng. Prog.*, Mar. 1994, p. 75.
149. Gans, M., S. A. Kiorpes, and F. A. Fitzgerald, "Plant Startup—Step by Step," *Chem. Eng.*, Oct. 3, 1983, p. 74.
150. Garvin, R. G., and E. R. Norton, "Sieve Tray Performance under GS Process Conditions," *Chem. Eng. Prog., 64*(3), 1968, p. 99.
151. Giles, D. S., and P. N. Lodal, "Case Histories of Pump Explosions while Running Isolated," *Process Safety Progress 20*(2), June 2001, p. 152.
152. Gillard, T., "Entry into Vessels—a Near Miss," *Loss Prevention Bull. 143*, IChemE, Oct. 1998, p. 18.
153. Gilmour, C. H., "Troubleshooting Heat Exchanger Design," *Chem. Eng.*, Jun. 19, 1967, p. 221.
154. Glausser, W. E., "Foaming in a Natural Gasoline Absorber," *Chem. Eng. Prog., 60*(10), 1964, p. 67.
155. Glitsch, Inc., *BALLAST® Tray Bulletin 4900*, 6th ed., Glitsch, 1993.
156. Glitsch, Inc., *High Performance Distributors*, The Glitsch Column, No. 262A, Dallas, Texas.
157. Glitsch, Inc., "Novel Column Internals Boost Stripping Efficiency," *Chem. Eng.*, Jan. 1994, p. 129.
158. Gmehling, J., U. Onken, and W. Arlt, *Vapor-Liquid Equilibrium Data Collection; DECHEMA*, Frankfurt am Main, Germany, 1981.
159. Golden, S. W., "Case Studies Reveal Common Design, Equipment Errors in Revamps," *Oil Gas J.*, Apr. 7, 1997, and April 14, 1997, p. 62.
160. Golden, S. W., "Improved Control Strategies Correct Main Fractionator Operating Problems," *Oil Gas J.*, Aug. 21, 1995, p. 58.
161. Golden, S. W., "Managing Vanadium from High Metals Crude Oils," *PTQ Revamps Operations Suppl.*, 2003, p. 26.
162. Golden, S. W., "Prevent Preflash Drum Foaming," *Hydrocarbon Processing*, May 1997, p. 141.
163. Golden, S. W., "Revamping FCC's—Process and Reliability," *Pet. Tech. Quarterly*, Summer 1996, p. 85.
164. Golden, S. W., "Temperature, Pressure Measurements Solve Column Operating Problems," *Oil Gas J.*, Dec. 25, 1995, p. 75.
165. Golden, S. W., "Troubleshooting a Debutanizer Condenser System," paper presented at the AIChE Spring Meeting, New Orleans, LA, Mar. 8–12, 1998.
166. Golden, S. W., A. W. Sloley, and B. Fleming, "Refinery Vacuum Column Troubleshooting," paper presented at the AIChE Spring National Meeting, New Orleans, LA, Mar. 31, 1993.
167. Golden, S. W., and A. W. Sloley, "Simple Methods Solve Vacuum Column Problems Using Plant Data," *Oil Gas J.*, Sept. 14, 1992, p. 74.
168. Golden, S. W., and G. R. Martin, "Analysis of a Delayed Coker Packed Column Failure," paper presented at the AIChE Spring National Meeting, Houston, TX, Mar. 20–24, 1995.
169. Golden, S. W., D. C. Villalanti, and G. R. Martin, "Feed Characterization and Deepcut Vacuum Columns: Simulation and Design," paper presented at the AIChE Spring National Meeting, Houston, TX, Mar. 20–24, 1995.

170. Golden, S. W., G. R. Martin, and K. D. Schmidt, "Field Data, New Design Correct Faulty FCC Tower Revamp," *Oil Gas J.*, May 31, 1993, p. 54.
171. Golden, S. W., J. Moore, and J. Nigg, "Optimize Revamp Projects with a Logic-Based Approach," *Hydrocarbon Processing*, Sept. 2003, p. 75.
172. Golden, S. W., N. P. Lieberman, and E. T. Lieberman, "Troubleshooting Vacuum Columns with Low-Capital Methods," *Hydrocarbon Processing*, July 1993, p. 81.
173. Golden, S. W., N. P. Lieberman, and G. R. Martin, "Correcting Design Errors Can Prevent Coking in Main Fractionators," *Oil Gas J.*, Nov. 21, 1994, p. 72.
174. Golden, S. W., S. Craft, and D.C. Villalanti, "Refinery Analytical Techniques Optimize Unit Performance," *Hydrocarbon Processing*, Nov. 1995, p. 85.
175. Golden, S. W., V. B. Shah, and J. W. Kovach III, "Improved Flow Topology for Petroleum Refinery Crude Vacuum Distillation Simulation," paper presented at the 44th Canadian Chemical Engineering Conference, Calgary, Alberta, Canada, Oct. 2–5, 1995.
176. Gonzalez-Martin, R., M. Solar, and D. W. Hoffman, "Minimizing the Impact of Crude Feedstock Changes," *PTQ*, Winter 1999/2000, p. 119.
177. Gous, G., "Control Level Under Fouling Conditions," *Hydrocarbon Processing*, Nov. 2000, p. 71.
178. Goyal, O. P., "Guidelines Aid Troubleshooting," *Hydrocarbon Processing*, Jan. 2000, p. 69.
179. Goyal, O. P., "Reduce HC Losses Plant-Wide—Part 1", *Hydrocarbon Processing*, Aug. 1999, p. 97.
180. Grover, B. S., and E. S. Holmes, "The Benfield Process for High Efficiency and Reliability in Ammonia Plant Acid Gas Removal—Four Case Studies," in *Nitrogen 1986*, p. 101, British Sulphur Corp. Ltd., Amsterdam, Apr. 20–23, 1986.
181. Gustin, J.-L., "Safety of Ethoxylation Reactions," *Loss Prevention Bull.*, *157*, IChemE, 2001, p. 11.
182. Guy, J. L., and J. A. Bonilla, "Case History of a Retrayed Column Troubleshooting Techniques and Methods," paper presented at the AIChE Spring National Meeting, New Orleans, LA, Mar. 29–Apr. 2, 1992.
183. Hall, J., "Hazards Involved in the Handling and Use of Dimethyl Sulphoxide (DMSO)," *Loss Prevention Bull.*, *114*, IChemE, Dec. 1993, p. 9.
184. Hallenberger, K., and M. Vetter, "Plate Damage as a Result of Delayed Boiling," in *Proceedings of International Conference on Distillation and Absorption*, Baden-Baden, Germany, European Federation of Chemical Engineers, Sept. 30–Oct. 2, 2002.
185. Hanekom, P., and P. Gibson, "Contamination of Benfield CO_2 Removal System by Carboxylic Acid Salts," *Ammonia Plant Safety*, *37*, AIChE, 1997, p. 281.
186. Hanson, D. W., and E. L. Hartman, "High Capacity Distillation Revamps," *PTQ*, Autumn 2001, p. 53.
187. Hanson, D. W., and I. Buttridge, "Revamp, Troubleshooting Optimize NGL Plant Depropanizer Operations," *Oil Gas J.*, Aug. 25, 2003, p. 88.
188. Hanson, D. W., and M. Martin, "Low Capital Revamp Increases Vacuum Gas Oil Yield," *Oil Gas J.*, Mar. 18, 2002.
189. Hanson, D. W., and N. P. Lieberman, "Refinery Distillation Troubleshooting," paper presented at the AIChE Annual Meeting, Miami Beach, FL, November 12–15, 1995.
190. Hanson, D. W., E. L. Hartman, and S. Costanzo, "Modify Crude Units to Deepcut Operation," *Hydrocarbon Eng.*, Jan./Feb. 1997, p. 2; supplemented by private communication, Jan. 1998.
191. Hanson, D. W., J. Langston, J. Keen and C. Johnson, "Low-Capital Crude Unit Revamp Increases Product Yield," *Oil Gas J.*, March 24, 2003.
192. Hanson, D. W., N. P. Lieberman, and E. T. Lieberman, "De-entrainment and Washing of Flash-Zone Vapors in Heavy Oil Fractionators," *Hydrocarbon Processing*, July 1999, p. 55.
193. Harrison, M. E., "Gamma Scan Evaluation for Distillation Column Debottlenecking," *Chem. Eng. Prog.*, Mar. 1990, p. 37.
194. Harrison, M. E., and J. J. France, "Distillation Column Troubleshooting," *Chem. Eng.*, Mar. 1989, p. 116; Apr. 1989, p. 121; May 1989, p. 126; and Jun. 1989, p. 139.
195. Hartman, E. L., "New Millennium, Old Problems: Vapor Cross Flow Channeling on Valve Trays," in *Distillation 2001: Frontiers in a New Millennium, Proceedings of Topical Conference*, p. 108, AIChE Spring National Meeting, Houston, TX, Apr. 22–26, 2001.

196. Hartman, E. L., and D. W. Hanson, "Reboilers Behind Distillation Tower Problems," *PTQ*, Spring 2001, p. 111.
197. Hartman, E. L., and E. Menzes, "Successful Column Revamps Require Careful Design," *Oil Gas J.*, Jan. 24, 2000, p. 42.
198. Hartman, E. L., and T. Barletta, "Reboiler and Condenser Operating Problems," *PTQ*, Summer 2003, p. 47.
199. Hartman, E. L., D. W. Hanson, and B. Weber, "FCCU Main Fractionator Revamp for CARB Gasoline Production," *Hydrocarbon Processing*, Feb. 1998, p. 44.
200. Hartman, E. L., I. Buttridge, and D. F. Wilson, "Increase Aromatics Complex Profitability," *Hydrocarbon Processing*, Apr. 2001, p. 97.
201. Hartman, E. L., S. D. Williams, and D. W. Hanson, "Startup, Shutdown and Control of Refinery Packed Fractionators," *Hydrocarbon Processing*, Nov. 1995, p. 67.
202. Hasbrouck, J. F., J. G. Kunesh, and V. C. Smith, "Successfully Troubleshoot Distillation Towers," *Chem. Eng. Prog.*, Mar. 1993, p. 63.
203. Hatfield, J. A. (Merck & Company), "High Efficiency Tower Packings and Responsive Control Schemes," *Chem. Proc.*, Sept. 1988, p. 130; supplemented by private communication, Oct. 1988.
204. Hatton, A., D. Conley, and B. Collins, "Texas Refiner Achieves Fast Column Turnaround with Combination Treatment," *Oil Gas J.*, July 7, 2003, p. 58.
205. Hausch, D. C., "How Flooding Can Affect Tower Operation," *Chem. Eng. Prog., 60*(10), 1964, p. 55.
206. Hauser, R., and J. Richardson, "Prevent Plugging in Stripping Columns," *Hydrocarbon Processing*, Sept. 2000, p. 95.
207. Hauser, R., and R. T. Kirkey, "Refinery Tests Demonstrate Fixed Valve Trays Improve Performance in Sour Water Stripper," in *Distillation 2003: On the Path to High Capacity, Efficient Splits, Topical Conference Proceedings*, p. 163, AIChE Spring National Meeting, New Orleans, LA, Mar. 31–Apr. 3, 2003.
208. Hearn, W. J. and R. Wagner, "Solvent Conversion of the Gas Treatment System at the Badak LNG Plant," 79th Annual Convention of the Gas Processors Association, Atlanta, GA, March 12–15, 2000.
209. Helling, R. K., and M. A. Des-Jardin, "Get the Best Performance from Structured Packing," *Chem. Eng. Prog.*, Oct. 1994, p. 62.
210. Hendershot, D. C., A. G. Keiter, J. Kacmar, J. W. Magee, P. C. Morton, and W. Duncan, "Connections: How a Pipe Failure Resulted in Resizing Vessel Emergency Relief Systems," *Process Safety Progress, 22*(1), March 2003.
211. Hengstebeck, R. J., "An Improved Shortcut for Calculating Difficult Multicomponent Distillations," *Chem. Eng.* Jan. 13, 1969, p. 115.
212. Hepp, P. S., "Internal Column Reboilers—Liquid Level Measurement," *Chem. Eng. Prog., 59*(2), 1963, p. 66.
213. Hernandez, R. J., and T. L. Huurdeman, "Solvent Unit Clean Synthesis Gas," *Chem. Eng.*, Feb. 1989, p. 154.
214. Hills, P. D., "Designing Piping for Gravity Flow," *Chem. Eng.*, Sept. 5, 1983, p. 111.
215. Holder, M. R., "Performance Troubleshooting on a TEG Dehydration Unit with Structured Packing," 46th Laurence Reid Gas Conditioning Conference, Norman, Oklahoma, March 3–6, 1996, p. 100.
216. Holiday, A., and D. Meyer, "Novel Distillation Process for Condensate Stripping," in *Distillation: Horizons for the New Millennium, Topical Conference Preprints*, p. 186, AIChE Spring National Meeting, Houston, TX, Mar. 14–18, 1999.
217. Hollander, L., "Pressure Control of Light-Ends Fractionators," *ISA J., 4*(5), 1957, p. 185.
218. Holmes, T. L., "To Generate or Not to Generate," *Chem. Eng.*, Aug. 1995, p. 8.
219. Holmstrom, D., S. Selk, S. Wallace, and I. Rosenthal, "A Multiple Fatality Incident at the Tosco Avon Refinery, Martinez, California," *Loss Prevention Bull., 167*, October 2002, p. 4.
220. Hood, A. R., A Stander, G. Lilburne, and S. Soydaner, "Vacuum Column Revamp—A Fast-Track Turnkey Project," *Hydrocarbon Processing*, Nov. 1999, p. 63.
221. Horwitz, B. A., "Don't Let Startup or Debugging Problems Bug You," *Chem. Eng. Prog.*, Nov. 1994, p. 62.
222. Horwitz, B. A., "Hardware, Software, Nowhere," *Chem. Eng. Prog.*, Sept. 1998, p. 69.

223. Howard, W. B., "Hazards with Flammable Mixtures," *Chem. Eng. Prog.*, *66*(9), 1970, p. 59.
224. Hower, T. C. Jr., and H. Z. Kister, "Solve Column Process Problems," Part 1, *Hydrocarbon Processing*, May 1991, p. 89.
225. Hower, T. C. Jr., and H. Z. Kister, "Solve Column Process Problems," Part 2, *Hydrocarbon Processing*, June 1991, p. 83.
226. Huurdeman, T. L., and S. Goole, "Operating Experience with the Change-Over from Hot Potassium Carbonate Solution to aMDEA in a CO_2 Removal System," *Ammonia Plant Safety*, *43*, AIChE, 2002, p. 90.
227. Hyprotech Ltd., Distil v. 5.0, 2001, http://www.hyprotech.com/.
228. Ingram, J. H., "Suction Side Problems—Gas Entrainment," *Pumps Syst. Mag., Sept.* 1994.
229. Irhayem, A. Y. N., "Purging Prevents Condenser Corrosion," *Chem. Eng.*, Aug. 15, 1988, p. 178.
230. Iyengar, J. N., P. W. Sibal, and D.S. Clarke, "Operations and Recovery Improvement Via Heavy Hydrocarbon Extraction," 48th Lawrence Reid Gas Conditioning Conference, Norman, Oklahoma, March 1–4, 1998, p. 161.
231. Jarvis, H. C., "Butadiene Explosion at Texas City," *Loss Prevention, 5*, 1971, p. 57; R. H. Freeman and M. P. McCready, ibid., p. 61; R. G. Keister, B. I. Pesetsky, and S. W. Clark, ibid., p. 67.
232. Johnston, M., K. Kimura, R. Hinkle, and S. Melville, "Virtual Analyzers Help Refiners Meet RFG Specifications," *Hydrocarbon Processing*, June 1999, p. 57.
233. Jones, D. W., and J. B. Jones, "Tray Performance Evaluation," *Chem. Eng. Prog.*, *71*(6), 1975, p. 65.
234. Jones, T. L., "Investigation of Problems in Distillation Column Operation Using Radioisotope Technology," *IChemE Symp. Ser.*, *128*, 1992, p. B213.
235. Kabakov, M. I., and A. M. Rozen, "Hydrodynamic Inhomogeneities in Large-Diameter Packed Columns and Ways to Eliminate Them," *Khim. Prom.*, No. 8, 1984, p. 496; *Soviet Chem. Ind.*, *16*(8), 1984, p. 1059.
236. Kaes, G. L., *Refinery Process Modeling—A Practical Guide to Steady State Modeling of Petroleum Processes Using Commercial Simulators*, Athens Printing Company, Athens, GA, 2000.
237. Kalthod, V. G., G. Joglekar, S. M. Clark, and J. R. Fair, "Distillation Column Performance Testing: Continuous and Batch Approaches," in *Preprints of the Topical Conference on Separation Science and Technologies, Part I*, p. 225, American Institute of Chemical Engineers, The AIChE Annual Meeting, Los Angeles, CA, Nov. 17–19, 1997.
238. Keaton, M., M. Keenan, and J. Keenan, "Online Soft Analyzers Benefit Refining," *Hydrocarbon Processing*, June 1998, p. 105.
239. Keller, A. E., A. L. Cummings, and D. K. Nelsen, "Amine Purification System to Increase Crude Processing," *PTQ*, Spring 2003, p. 49.
240. Kennedy, G., "A Proactive Approach to Distillation Maintenance," *Tru-News*, Vol. 5, Ed. 2, Tru-Tec Services Inc., La Porte, TX, 1997, p. 2.
241. Kenney, G. D., M. Boult, and R. M. Pitblado, "Lessons for Seveso II from Longford Australia," *Loss Prevention Bull.*, *158*, IChemE, 2001, p. 17.
242. Killat, G. R., and D. Perry, "High Performance Distributor for Low Liquid Rates," paper presented at the AIChE Annual Meeting, Los Angeles, CA, Nov. 17–22, 1991.
243. King, C. J., *Separation Processes*, 2nd. ed., McGraw-Hill, New York, 1980.
244. Kirmse, B., K. Krase, T. Votaw, and D. Ferguson, "Determine the Cause of Flooding to Regain Column Efficiency," in *Distillation 2003: on the Path to High Capacity, Efficient Splits, Topical Conference Proceedings*, p. 52, AIChE Spring National Meeting, New Orleans, LA, Mar. 31–Apr. 3, 2003.
245. Kister, H. Z., "Are Column Malfunctions Becoming Extinct—or Will They Persist in the 21st Century?" *Trans. IChemE*, *75*, Part A, Sept. 1997, p. 563.
246. Kister, H. Z., "Can Valve Trays Experience Vapor Cross Flow Channeling?" *The Chem. Engnr.*, June 10, 1993, p. 18.
247. Kister, H. Z., "Can We Believe the Simulation Results?" *Chem. Eng. Progr.*, Oct. 2002, p. 52.
248. Kister, H. Z., "Column Internals," *Chem. Eng.*, May 19, 1980, p. 138; July 28, 1980, p. 79; Sept. 8, 1980, p. 119; Nov. 17, 1980, p. 283; Dec. 29, 1980, p. 55; Feb. 9, 1981, p. 107; Apr. 6, 1981, p. 97.
249. Kister, H. Z., "Component Trapping in Distillation Towers: Causes, Symptoms, and Cures," *Chem. Eng. Progr.*, August 2004, p. 22.

250. Kister, H. Z., *Distillation Operation*, McGraw-Hill, New York, 1990
251. Kister, H. Z., *Distillation Design*, McGraw-Hill, New York, 1992.
252. Kister, H. Z., "Recent Trends in Distillation Tower Malfunctions," in *Distillation 2003: On the Path to High Capacity, Efficient Splits, Topical Conference Proceedings*, AIChE Spring National Meeting, New Orleans, LA, Mar. 31–Apr. 3, 2003, p. 3.
253. Kister, H. Z., "Trouble-Free Design of Refinery Fractionators," *PTQ*, Autumn 2003, p. 109.
254. Kister, H. Z., "Troubleshoot Distillation Simulations," *Chem. Eng. Prog.*, June 1995, p. 63.
255. Kister, H. Z., "What Caused Tower Malfunctions in the Last 50 Years?" *Trans. IChemE*, *81*, Part A, Jan. 2003, p. 5.
256. Kister, H. Z., "When Tower Startup Has Problems," *Hydrocarbon Processing*, *58*(2), 1979, p. 89.
257. Kister, H. Z., and D. R. Gill, "Flooding and Pressure Drop Prediction for Structured Packings," *IChemE Symp. Ser.*, *128*, 1992, p. A109.
258. Kister, H. Z., and D. R. Gill, "Packed Tower Capacity: Scaleup and Prediction Traps," *Proceedings, Chemeca 92*, Canberra, Australia, p. 185-2, 1992.
259. Kister, H. Z., and D. R. Gill, "Predict Flood Point and Pressure Drop for Modern Random Packings," *Chem. Eng. Prog.*, *87*(2), Feb. 1991, p. 32.
260. Kister, H. Z., and J. F. Litchfield, "Diagnosing Instabilities in the Column Overhead" *Chem. Eng.*, Sept. 2004, p. 56.
261. Kister, H. Z., and S. Chen, "Solving a Tower's Salt Plugging Problem," *Chem. Eng.*, Aug. 2000, p. 129.
262. Kister, H. Z., and S. Schwartz, "Olefins Producer Uses Shed Decks to Improve Quench Tower Operation," *Oil Gas J.*, May 20, 2002, p. 50.
263. Kister, H. Z., and T. C. Hower, Jr., "Unusual Case Histories of Gas Processing and Olefins Plant Columns," *Plant/Operations Prog.*, *6*(3), 1987, p. 151.
264. Kister, H. Z., B. Blum, and T. Rosenzweig, "Troubleshoot Chimney Trays Effectively," *Hydrocarbon Processing*, Apr. 2001, p. 101.
265. Kister, H. Z., D. E. Grich, and R. Yeley, "Better Feed Entry Ups Debutanizer Capacity," *PTQ Revamps Operations Suppl.*, 2003, p. 31.
266. Kister, H. Z., D. W. Hanson, and T. Morrison, "California Refiner Identifies Crude Tower Instability Using Root Cause Analysis," *Oil Gas J.*, Feb. 18, 2002, p. 42.
267. Kister, H. Z., E. Brown, and K. Sorensen, "Sensitivity Analysis Is Key to Successful DC_5 Simulation," *Hydrocarbon Processing*, Oct. 1998, p. 124.
268. Kister, H. Z., E. Brown, T. Yanagi, and K. Sorensen, "Testing a Depentanizer Containing Nye Trays," paper presented at the AIChE Annual Meeting, Los Angeles, CA, Nov. 16–20, 1997.
269. Kister, H. Z., G. Balekjian, J. F. Litchfield, J. Damm, and D. Merchant, "Absorber Troubleshooting: Systematic Investigation Pays Off," *Chem. Eng. Prog.*, *88*(6), 1992, p. 41.
270. Kister, H. Z., H. Pathak, M. Korst, D. Strangmeier, and R. Carlson, "Troubleshoot Reboilers by Neutron Backscatter," *Chem. Eng.*, Sept. 1995, p. 145.
271. Kister, H. Z., K. F. Larson, and P. E. Madsen, "Vapor Cross Flow Channeling on Sieve Trays: Fact or Myth?" *Chem. Eng. Prog.*, Nov. 1992, p. 86.
272. Kister, H. Z., K. F. Larson, J. M. Burke, R. J. Callejas, and F. Dunbar, "Troubleshooting a Water Quench Tower," *Proceedings 7th Ethylene Producers Conference*, Houston, TX, AIChE, 1995.
273. Kister, H. Z., R. Rhoad, and K. A. Hoyt, "Improve Vacuum-Tower Performance," *Chem. Eng. Prog.*, Sept. 1996, p. 36.
274. Kister, H. Z., S. Bello Neves, R.C. Siles, and R. da Costa Lima, "Does Your Distillation Simulation Reflect the Real World?" *Hydrocarbon Processing*, Aug. 1997, p. 103.
275. Kister, H. Z., S. G. Chellappan, and C. E. Spivey, "Debottleneck and Performance of a Packed Demethanizer," *Proc. 4th Ethylene Producers Conf.*, New Orleans, LA, p. 283, AIChE, 1992.
276. Kister, H. Z., T. C. Hower, Jr., P. R. de M. Freitas, and J. Nery, "Problems and Solutions in Demethanizers with Interreboilers," *Proc. 8th Ethylene Producers Conf.*, New Orleans, LA, AIChE, 1996.
277. Kitterman, L., "Tower Internals and Accessories," paper presented at Congresso Brasileiro de Petroquimica, Rio de Janeiro, Nov. 8–12, 1976.

278. Kler, S. C., R. J. P. Brierley, and M. C. G. Del Cerro, "Downcomer Performance at High Pressure, High Liquid Load, Analysis of Industrial Data on Sieve and Valve Trays," *IChemE Symp. Ser., 104*, 1987, p. B391.
279. Kletz, T. A., "Fires and Explosions of Hydrocarbon Oxidation Plants," *Plant/Operations Prog., 7*(4), 1988, p. 226.
280. Kletz, T. A., "Lessons of Another Ethylene Oxide Explosion," *The Chem. Engnr.*, Feb. 1990, p. 15.
281. Kletz, T. A., "Some Loss Prevention Case Histories," *Process Safety Prog.*, Oct. 1995, p. 271.
282. Kletz, T. A., "The 1990 Shell Explosion," *Loss Prevention Bull., 100*, Aug. 1991, p. 21.
283. Kletz, T. A., "Unexpected Equipment Involved in Process Plant Accidents," *Hydrocarbon Processing*, Oct. 1999, p. 85.
284. Kletz, T. A., *What Went Wrong*, 2nd ed., Gulf Publishing, Houston, TX, 1988.
285. Kolff, S. W., "Corrosion of a CO_2 Absorber Tower," *Plant/Operations Prog.*, *5*(2), 1986, p. 65.
286. Kolmetz K., W. K. Ng, P. W. Faessler, A. W. Sloley, and T. M. Zygula, "Case Studies Demonstrate Guidelines for Reducing Fouling in Distillation Columns," *Oil Gas J.*, August 16, p. 60; and August 23, 2004, p. 43.
287. Kolmetz, K., and C. M. Tham, "Flow Phenomenon in Staged and Non Staged Distillation Equipment," paper presented at the Regional Symposium on Chemical Engineering and 16th Symposium of Malaysian Chemical Engineers, Kuala Lumpur, Malaysia, Oct. 29, 2002.
288. Kolmetz, K., J. Gray, M. Chua, and R. Desai, "BTX Extractive Distillation Capacity Increased by Enhanced Packing Distributors," paper presented at the AIChE Spring Meeting, New Orleans, LA, Mar. 10–14, 2002.
289. Kolmetz, K., M. Chua, R. Desai, J. Gray, and A. W. Sloley, "Staged Modifications Improve BTX Extractive Distillation Unit Capacity," *Oil Gas J.*, Oct. 13, 2003, p. 60.
290. Kunesh, J. G., "Practical Tips on Tower Packing," *Chem. Eng.*, Dec. 7, 1987, p. 101.
291. Laird, D. "Packed FCCU Main Fractionator Upgrades for Performance and Reliability," paper presented at the NPRA Annual Meeting, San Antonio, TX, March 21–23, 2004.
292. Laird, D., and J. Cornelisen, "Control-System Improvements Expand Refinery Processes," *Oil Gas J.*, Sept. 25, 2000, p. 71.
293. Langdon, D., T. Barletta, and S. Fulton, "FCC Gas Plant Stripper Capacity," *PTQ, Revamp Operation Suppl.*, Autumn 2004, p. 3.
294. Le, N. D., B. J. Sel, and V. H. Edwards, "Doublecheck Your Process Simulations," *Chem. Eng. Prog.*, May 2000, p. 51.
295. Lear, J., M. Thatcher, and S. Newman, "Saving Cost through Simulating the Control of a Hydrochloric Acid Plant," *Chem. Eng. Australia*, Sept.–Nov. 1997, p. 12.
296. Leegwater, H., "Industrial Experience with Double Quality Control," in W. L. Luyben (ed.), *Practical Distillation Control*, p. 331, Van Nostrand Reinhold, New York, 1992.
297. Lele, G. S., P. H. Ghate, and B. S. Chaknayet, "Heavy Corrosion Problems in Benfield—CO_2 Removal System of Ammonia Plant," *Ammonia Plant Safety, 35*, AIChE, 1995, p. 216.
298. Lenoir, E. M., and J. A. Davenport, "A Survey of Vapor Cloud Explosions: Second Update," *Process Safety Prog., 12*(1), 1993, p. 12.
299. Lieberman, N. P., "Basic Field Observations Reveal Tower Flooding," *Oil Gas J.*, May 16, 1988, p. 39.
300. Lieberman, N. P., "Common Crude Unit Problems, Remedies," *Oil Gas J.*, Aug. 11, 1980, p. 115.
301. Lieberman, N. P., "Design Processes for Reduced Maintenance," *Hydrocarbon Processing, 58*(1), 1979, p. 89.
302. Lieberman, N. P., "Diagnosis Key to Flooding Problem Solution," *Oil Gas J.*, July 16, 1990, p. 62.
303. Lieberman, N. P., "Drying Light End Towers Is Critical for Preventing Problems," *Oil Gas J.*, Feb. 16, 1981, p. 100.
304. Lieberman, N. P., *Process Design for Reliable Operation*, 2nd ed., *Gulf Publishing*, Houston, TX, 1988.
305. Lieberman, N. P., *Troubleshooting Natural Gas Processing*, PennWell Publishing, Tulsa, OK, 1987.
306. Lieberman, N. P., *Troubleshooting Process Operations*, 2nd ed., PennWell Publishing, Tulsa, OK, 1985; 3rd ed., 1991.

307. Lieberman, N. P., and D. W. Hanson, "Vacuum Unit Troubleshooting," *PTQ Revamps Operations Suppl.*, 2003, p. 19.
308. Lieberman, N. P., and E.T. Lieberman, *A Working Guide to Process Equipment*, McGraw-Hill, New York, 1997.
309. Lieberman, N. P., and E. T. Lieberman, "Design, Installation Pitfalls Appear in Vac Tower Retrofit," *Oil Gas J.*, Aug. 26, 1991, p. 57.
310. Lieberman, N. P., and E. T. Lieberman, "Inadequate Inspection Cause of Flawed Vac Tower Revamp," *Oil Gas J.*, Dec. 14, 1992, p. 33.
311. Lieberman, N. P., and G. Liolios, "HF Alky Unit Operations Improved by On-Site Troubleshooting to Boost Capacity, Profit," *Oil Gas J.*, June 20, 1988, p. 66.
312. Lieberman, N. P., and S. W. Golden, "Foaming Is Leading Cause of Tower Flooding," *Oil Gas J.*, Aug. 14, 1989, p. 45.
313. Lieberman, N. P., private communication, Aug. 1986; described in Ref. 250.
314. Likins, W., and M. Hix, "Sulfur Distribution Prediction with Commercial Simulators," 46th Laurence Reid Gas Conditioning Conference, Norman, Oklahoma, March 3–6, 1996, p. 254.
315. Lin, D., K. Wu, A. Yanoma, and S. Costanzo, "Entrainment Limits and Operating Capacity of Large-Size Structured Packing with Sprayed-Type Distributor," paper presented at the AIChE Spring National Meeting, Houston, TX, Mar. 9–13, 1997.
316. Litzen, D., and J. L. Bravo, "Uncover Low-Cost Debottlenecking Opportunities," *Chem. Eng. Prog.*, Mar. 1999, p. 25.
317. Lockett, M. J., *Distillation Tray Fundamentals*, Cambridge University Press, Cambridge, 1986.
318. Looney, S. K., B. C. Price, and C. A. Wilson, "Integrated Nitrogen Rejection Facility Produces Fuel and Recovers NGL's," *Energy Prog.*, *4*(4), 1984, p. 214.
319. Lord, R. C., P. E. Minton, and R. P. Slusser, "Guide to Trouble-Free Heat Exchangers," *Chem. Eng.*, June 1, 1970, p. 153.
320. Love, F. S., "Troubleshooting Distillation Problems," *Chem. Eng. Prog.*, *71*(6), 1975, p. 61.
321. Lupfer, D. E., and M. W. Oglesby, "Automatic Control of Distillation Columns," *Ind. Eng. Chem.*, *53*(12), 1961, p. 963.
322. Luyben, W. L., "10 Schemes to Control Distillation Columns with Sidestream Drawoffs," *ISA J.*, *13*(7), 1966, p. 37.
323. Luyben, W. L., "Control of Columns with Side Stream Draw-Off," in J. T. Ward (ed.), *Instrumentation in the Chemical and Petrochemical Industries*, Vol. 3, Plenum, New York, 1967.
324. Lyon, S., "Suspected Weld Failures in Process Plan Equipment," *Loss Prevention Bull.*, *163*, IChemE, February 2002, p. 7.
325. MacKenzie, J., "Hydrogen Peroxide without Accidents," *Chem. Eng.*, June 1990, p. 84.
326. Mak, H. Y., "Gas Plant Converts Amine Unit to MDEA-Based Solvent", *Hydrocarbon Processing*, Oct. 1992, p. 91.
327. Manufacturing Chemists' Association (MCA), *Case Histories of Accidents in the Chemical Industry*, MCA, Washington, DC, Vol. 1, 1962; Vol. 2, 1966; Vol. 3, 1970.
328. Markham, R. S., and R. W. Honse, "Carbon Dioxide Stripper Explosion," *Ammonia Plant Safety*, *20*, 1978, p. 131.
329. Martin, D. O., and R. O. Allen, "Preventing Salt Fouling in FCC Main Fractionators," *PTQ*, Spring 2001, p. 41.
330. Martin, G. R., "Refinery Experience Provides Guidelines for Centrifugal Pump Selection," *Oil Gas J.*, Mar. 11, 1996.
331. Martin, G. R., "Understand Real-World Problems of Vacuum Ejector Performance," *Hydrocarbon Processing*, Nov. 1997, p. 63.
332. Martin, G. R., and B. E. Cheatham, "Keeping Down the Cost of Revamp Investment," *PTQ*, Summer 1999, p. 99.
333. Martin, G. R., E. Luque, and R. Rodriguez, "Revamping Crude Unit Increases Reliability and Operability," *Hydrocarbon Processing*, June 2000, p. 45.
334. Martin, G. R., J. R. Lines, and S. W. Golden, "Understand Vacuum-System Fundamentals," *Hydrocarbon Processing*, Oct. 1994, p. 91.

335. Martin, H. W., "Scale-Up Problems in a Solvent-Water Fractionator," *Chem. Eng. Prog.*, *60*(10), 1964, p. 50.
336. Mary Kay O'Connor Process Safety Center, "A Fire in Titanium Structured Packing Involving Thermite Reactions," Department of Chemical Engineering, Texas A&M University, College Station, Texas, 2004, www.mkopsc.tamu.edu/safety_alert/08_03_01.htm.
337. Mary Kay O'Connor Process Safety Center, "Safety Alert: Incident at a Chemical Plant," Department of Chemical Engineering, Texas A&M University, College Station, Texas, 2004, www.mkopsc.tamu.edu/safety_alert/04_16_01.htm.
338. Masson, W. B., "Westcoast McMahon Plant Gas Treating Experiences," Lawrence Reid Gas Conditioning Conference, Norman, Oklahoma, February 27–March 2, 1994, p. 37.
339. Mathur, J., "Draw Some Help from Hydraulic Grade Lines," *Chem. Eng. Prog.*, Oct. 1990, p. 50.
340. Mathur, U., and R. J. Conroy, "Successful Multivariable Control without Plant Tests," *Hydrocarbon Processing*, June 2003, p. 55.
341. McCabe, W. L., and E. W. Thiele, "Graphical Design of Fractionating Columns," *Ind. Eng. Chem.*, *17*, 1925, p. 605.
342. McCune, L. C., and P. W. Gallier, "Digital Simulation: A Tool for Analysis and Design of Distillation Columns," *ISA Trans.*, *12*(3), 1973, p. 193.
343. McLaren, D. B., and J. C. Upchurch, "Guide to Trouble-Free Distillation," *Chem. Eng.*, Jun. 1, 1970, p. 139.
344. McMullan, B. D., A. E. Ravicz, and S. Y. J. Wei, "Troubleshooting a Packed Vacuum Column—A Success Story," *Chem. Eng. Prog.*, July 1991, p. 69.
345. McNeill, G. A., and J. D. Sacks, "High Performance Column Control," *Chem. Eng. Prog.*, *65*(3), 1969, p. 33.
346. Mellin, B. E., "Ethylene Oxide Plant Explosion, 3 July 1987, BP Chemicals, Antwerp, Belgium," *Loss Prevention Bull.*, *100*, IChemE, Aug. 1991, p. 13.
347. Messinger, J. R., and M. Findlay "Cold Safety Risk Assessment of Existing Bimetallic Demethanizers" Proc. 13th Ethylene Producers Conference, Houston, Texas, AIChE, April 22–26, 2001.
348. Millard, M., and T. Beasley, "Contamination Consequences and Purification of Gas Treating Chemicals Using Vacuum Distillation," paper presented at the 43rd Annual Lawrence Reid Gas Conditioning Conference, University of Oklahoma, Norman, OK, March 1–3, 1993.
349. Miller, E. W., S. J. Soychak, A. E. Reed, R. K. Bartoo, and R. Ackman "Brady Plant Treating Project," 49th Laurence Reid Gas Conditioning Conference, Norman, Oklahoma, 1999, p. 48.
350. Mixon, W., "CAT-Scan Technology Identifies the Cause of Liquid Maldistribution in a Revamped Distillation Tower," *Tru-News*, Vol. 9, Ed. 4, Tru-Tec Services Inc., La Porte, TX, 2001, p. 1.
351. Mixon, W., "Tracing Your Process Flows," *Tru-News*, Vol 10, Ed. 1, Tru-Tec Services Inc., LaPorte, TX, 2002, p. 1.
352. Moffatt, S., V. Monical, and S. Ramchandran, "Take a Second Look at the Reasons for Your Distillation Column Upsets," in *Distillation 2003: On the Path to High Capacity, Efficient Splits, Topical Conference Proceedings*, p. 76, AIChE Spring National Meeting, New Orleans, LA, Mar. 31–Apr. 3, 2003.
353. Moore, F., and F. Rukovena, "Liquid and Gas Distribution in Commercial Packed Towers," paper presented at the 36th Canadian Chemical Engineering Conference, Oct. 5–8, 1986; same paper published in *Chemical Plants and Processing* (European edition), Aug. 1987, p. 11.
354. Moore, J. A., comments in Rubin, F. L. (chairman), and P. Minton, "Heat Exchangers That Did Not Work," Panel Discussion, AIChE Meeting, Anaheim, CA, June 6–10, 1982.
355. Morgan, R. D., "Ultra-Frac High Capacity Trays for Heavily Loaded Distillation Columns," paper presented at the IChemE Fluid Separation Process Group Debottlenecking Seminar, London, England, Mar. 16, 1994.
356. Morrison, D. R. III, A. R. Carpenter, and R. A. Ogle, "Common Causes and Corrections for Explosions and Fires in Improperly Inerted Vessels," *Process Safety Progress*, *21*(2), June 2002, p.142.
357. Mortko, R. A., "Optimized NGL Fractionator Expansion," *HTI Quarterly*, Autumn 1994, p. 87.
358. Moura, C. A. D., and H. P. Carneiro, "Common Difficulties in the Use of Process Simulators," *B. Tech. Petrobras*, *34*(3/4), Jul./Dec. 1991; quoted in R. Agrawal, Y.-K. Li, O. Santollani, M. A. Satyro, and A. Vieler, "Uncovering the Realities of Simulation," *Chem. Eng. Prog.*, May 2001, p. 42.

359. Mullenix, D., M. Jordan, H. Bundick, F. Martin, and V. Lewis, "Control of Carbonyl Polymer Fouling in Caustic Towers," in *Proceedings of the 8th Ethylene Producers Conference*, AIChE, 1996, p. 89.
360. Murray, R. M., and J. E. Wright, "Trouble-Free Startup of Distillation Columns," *Chem. Eng. Prog.*, *63*(12), 1967, p. 40.
361. Musial, T., "Addition of Winter-Ejector Allows Deep-Cut Operation at Lower than Design Pressure," in *Distillation 2003: On the Path to High Capacity, Efficient Splits, Topical Conference Proceedings*, p. 157, AIChE Spring National Meeting, New Orleans, LA, Mar. 31–Apr. 3, 2003.
362. Muthumanoharan, R., S. Muruganandam, and K. Manikandan, "Corrosion in CO_2 Removal Section of Ammonia Plant," *Ammonia Plant Safety*, *40*, AIChE, 2000, p. 259.
363. Mykitta, R. S. "Operating Experiences of Shell's Yellowhammer Gas Plant," 49th Laurence Reid Gas Conditioning Conference, Norman, Oklahoma, February 1999.
364. Naklie, M. M., L. Pless, T. P. Gurning, and M. Ilyasak, "Radiation Scanning Aids Tower Diagnosis at Arun LNG Plant," *Oil Gas J.*, Mar. 26, 1990.
365. Nath, B. K., "Methanol Plant Applies SPC: A Case Study," *Hydrocarbon Processing*, Jan. 1996, p. 116-B.
366. Nisenfeld, A. E., "Reflux or Distillate—Which to Control?" *Chem. Eng.*, Oct. 6, 1969, p. 169.
367. Norton Company, *Intalox High Performance Separation Systems*, Bulletin IHP-1 UK, Akron, OH, 1987.
368. NPRA, Panel Discussion, "Auxiliary Equipment, Corrosion Focus of Refining Meeting," *Oil Gas J*, Apr. 4, 1994, p. 62.
369. NPRA, Panel Discussion, "Refiners Respond to Distillation Queries," *Oil Gas J.*, Jul. 28, 1980, p. 189.
370. NPRA, *Q & A Session on General Processing*, p. 65, 54th Annual NPRA Meeting, Chicago, IL, October 10–12, 2001.
371. NPRA, *Q & A Session on Light Oil Processing*, p. 190, 54th Annual NPRA Meeting, Chicago, IL, October 10–12, 2001.
372. NPRA, *Q & A Session on Refining and Petrochemical Technology*, 1981, p. 21.
373. NPRA, *Q & A Session on Refining and Petrochemical Technology*, 1983, p. 52.
374. NPRA, *Q & A Session on Refining and Petrochemical Technology*, 1988, pp. 17, 59, 60.
375. NPRA, *Q & A Session on Refining and Petrochemical Technology*, 1993, pp. 21, 22, 144.
376. NPRA, *Q & A Session on Refining and Petrochemical Technology*, 1994, pp. 25, 26, 69, 177, 178.
377. NPRA, *Q & A Session on Refining and Petrochemical Technology*, 1995, pp. 20, 24.
378. Oberding, W., R. Goff, M. Townsend, D. Chapin, D. Naulty, and V. Worah, "Lessons Learned and Technology Improvements at the Lost Cabins Gas Plant," Lawrence Reid Gas Conditioning Conference, February 22–25, 2004, p. 273.
379. O'Connell, H. E., "Plate Efficiency of Fractionating Columns and Absorbers," *Trans. AIChE*, *42*, 1946, p. 741.
380. Oglesby, M. W., and J. W. Hobbs, "Chromatograph Analyzers for Distillation Control," *Oil Gas J.*, Jan. 10, 1966, p. 80.
381. Ognisty, T. P., and M. Sakata, "Multicomponent Diffusion: Theory vs. Industrial Data," *Chem. Eng. Prog.*, *83*(3), Mar. 1987, p. 60.
382. Olsson, F. R., "Detect Distributor Defects Before They Cripple Columns," *Chem. Eng. Prog.*, Oct. 1999, p. 57.
383. Opong, S., and D. R. Short, "Troubleshooting Columns Using Steady State Models," in *Distillation: Horizons for the New Millennium*, Topical Conference Preprints, p. 129, AIChE Spring National Meeting, Houston, TX, Mar. 14–18,1999.
384. O'Rouke, B., "Experimental Design Goes on Line," *Control*, Sept. 1995, p. 82.
385. Partin, L. R., "Use Graphical Techniques to Improve Process Analysis," *Chem. Eng. Prog.*, Jan. 1993, p. 43.
386. Pathak, V. K., and I. S. Rattan, "Turndown Limit Sets Heater Control," *Chem. Eng.*, Jul. 18, 1988, p. 103.
387. Pattinson, S., "Changes in Demethanizer Reboil Solve Efficiency Problems," *Oil Gas J.*, May 1, 1989, p. 102.
388. Pauley, C. R., and B. A. Perlmutter, "Texas Plant Solves Foam Problems with Modified MEA System," *Oil Gas J.*, Feb. 29, 1988, p. 67.

389. Pauley, C. R., D. G. Langston, and F. C. Betts, "Redesigned Filters Solve Foaming, Amine-Loss Problems at Louisiana Gas Plant," *Oil Gas J.*, Feb. 4, 1991, p. 50.
390. Penuela, L., and J. Chirinos, "Inspection Program Evaluates HF-Alkylation Carbon Steel Piping," *Oil Gas J.*, May 31, 1999, p. 55.
391. Perez, M., R. Duarte, and L. Pless, "Temperature and Gamma Scans Identify Problems in Crude Tower," *Oil Gas J.*, Feb. 1, 1999, p. 44.
392. Perry, D., T. Poole, and M. Manifould, Nor-Pro Saint Gobain Corp. (now Koch-Glitsch LP), "Multi-Company Review of Distillation Column Incidents," paper presented in One Day Conference on Packing Fires, Houston, TX, Nov. 15, 2001.
393. Perry, R. H., and D. W. Green (ed.), *Chemical Engineers' Handbook*, 7th ed., McGraw-Hill, New York, 1997.
394. Perunicic, M., and M. Skotovic, "Case Studies of Some Fire Accidents in Yugoslavia's Oil Refineries," *Loss Prevention Bull.*, *96*, IChemE, Dec. 1990, p. 23.
395. Peterson, R. R., D. C. Haring, T. G. Johnson, and J. M. Senn, "Explosion of MDEA Storage Tank," *Ammonia Plant Safety*, *41*, AIChE, 2001, p. 118.
396. Pham, L. V., M. Binkley, T. M. Zygula, J. Y. Yang, and R. M. Garner, "High-Performance Anti-Fouling Tray Technology and Its Application in the Chemical Industry," paper presented at the AIChE Spring Meeting, Houston, TX, Mar. 12, 1997.
397. Phillips, J. R., and B. D. Payne, cited by Anonymous, "Low Maintenance Control Systems Improve NGL Recovery at Texas Gas Plant," *Oil Gas J.*, May 19, 2003, p. 56.
398. Pilling, M., and P. Mannion, "Process Simulation Aids Expansion of Alkylation Unit Main Fractionator," paper presented at the AIChE Spring National Meeting, Houston, TX, Mar. 9–13, 1997.
399. Pless, L., "Gamma Scans Confirm the Presence and Severity of Fouling," in *Tru-News*, Vol. 10, Ed. 3, Tru-Tec Services Inc., La Porte, TX, 2002, p. 1.
400. Pless, L., and B. Asseln, "Using Gamma Scans to Plan Maintenance of Columns," *PTQ*, Spring 2002, p. 115.
401. Porter, S. R., and P. J. Mullins, "Waste Gas Leads to Near Fatality," *Loss Prevention Bull.*, *154*, IChemE, Aug. 2000, p. 15.
402. Powell-Price, M., "The Explosion and Fires at the Texaco Refinery, 24 July 1994," *Loss Prevention Bull.*, *138*, IChemE, Dec. 1997, p. 3.
403. Rahman, A. A., A. A. Yusof, J. D. Wilkinson, and L. D. Tyler, "Improving Ethane Extraction at the Petronas Gas GPP-A Facilities in Malaysia," paper presented at the 83rd Annual Convention of the Gas Processors Association, March 15, 2004.
404. Rahman, A. R. A., N. H. K. Al Thani, M. Ishikura, and Y. Kikkawa, "First LNG from North Field Overcomes Feed, Start-up Problems," *Oil Gas J.*, Aug. 24, 1998, p. 41.
405. Reid, J. A., and C. Wallsgrove, "Odd But True Stories of Troubleshooting Olefins Units," in *Distillation: Horizons for the New Millennium*, Topical Conference Preprints, p. 141, AIChE Spring National Meeting, Houston, TX , Mar. 14–18, 1999.
406. Rejek, U., B. Tookey, S. Park, S. Goodhart, and S. Finlayson, "Advanced Process Control Re-Engineering," *PTQ*, Winter 2004, p. 95.
407. Rengarajan, T., and J. J. Patel, "Experience with Severe Leakage in CO_2 Stripper Reboiler," *Ammonia Plant Safety*, *35*, AIChE, 1995, p. 227.
408. Resetarits, M. R., J. Agnello, M. J. Lockett, and H. L. Kirkpatrick, "Retraying Increases C_3 Splitter Column Capacity," *Oil Gas J.*, June 6, 1988, p. 54.
409. Rhiel, F. F., "Model-Based Control," in W. L. Luyben (ed.), *Practical Distillation Control*, p. 440, Van Nostrand Reinhold, New York, 1992.
410. Richardson, R. E., Union Carbide, private communication, 1993.
411. Richert, J., and P. Gilbert, "Tune-Up of Amine System Avoids Costly New System," *Oil Gas J.*, Mar., 15, 1999, p. 33.
412. Robinson, B. A., and A. E. Hodel, "Internals Increase Capacity of Fractionation Column," *Chem. Proc.*, April 1988.
413. Rose, L. M., *Distillation Design in Practice*, Elsevier, Amsterdam, 1985.
414. Ross, S., and G. Nishioka, "Foaminess of Binary and Ternary Solutions," *J. Phys. Chem.*, *79*(15), 1975, p. 1561.

415. Rubbers, E., K. Green, T. Fowler, H. Z. Kister, and W. J. Stupin, "Distillation Reboiler Startup Can Pose Challenges," *Chem. Eng.*, Feb. 2004, p. 55.
416. Rueter, C., C. Beitler, C.R. Sivalls and J. M. Evans, "Design and Operation of Glycol Dehydrators and Condensers" 47th Laurence Reid Gas Conditioning Conference, Norman, Oklahoma, March 2–5, 1997, p. 135.
417. Ruffert, D. I., "The Significance of Experiments for the Design of New Distillation Column Sequences," in *Distillation 2001: Frontiers in a New Millennium, Proceedings of Topical Conference*, p. 133, AIChE Spring National Meeting, Houston, TX, Apr. 22–26, 2001.
418. Rukovena, F. Jr., and T. D. Koshy, "Packed Distillation Tower Hydraulic Design Method and Mechanical Considerations," *Ind. Eng. Chem. Res.*, *32*, 1993, p. 2400.
419. Russel, R., "Column Simulation," discussion of Horwitz (Ref. 222), *Chem. Eng. Prog.*, Feb. 1999, p. 7.
420. Ryan, J. M., C. L. Hsieh, and M. S. Sivasubramanian, "Predict Misting and Bubbling in Towers," *Chem. Eng. Prog.*, Aug. 1994, p. 83.
421. Sabharwal, A., N. V. Bhat, and T. Wada, "Integrate Empirical and Physical Modeling," *Hydrocarbon Processing*, Oct. 1997, p. 105.
422. Sadeq, J., H. A. Duarte, and R. W. Serth, "Anomalous Results from Process Simulations," paper presented at the AIChE Annual Meeting, Miami Beach, FL, Nov. 1995.
423. Sahdev, M., "Pyrophoric Iron Fires," Chemical Engineers Resource Page, www.cheresources.com.
424. Sanders, R. E., "Don't Become Another Victim of Vacuum," *Chem. Eng. Prog.*, Sept. 1993, p. 54.
425. Sattler, F. J., "Nondestructive Testing Methods Can Aid Plant Operation," *Chem. Eng.*, Oct. 1990, p. 177.
426. Sauter, J. R., and W. E. Younts III, "Tower Packings Cut Olefin-Plant Energy Needs," *Oil Gas J.*, Sept. 1, 1986, p. 45.
427. Savell, L., "Polyglycol Lubricant Reduces Stripper Tar Build-Up," *Chem. Proc.*, Nov. 1991, p. 152.
428. Schafer, T. A., Y. S. Lam, and S. Vragolic, "Organic Strippers: Vapor Maldistribution Problems and Solutions," paper presented at the AIChE Spring National Meeting, New Orleans, LA, Mar. 29–Apr. 2, 1992.
429. Schnaibel, G., and M. E. P. L. Marsiglia "The Recurring Jet Fuel Problem (and What it Took to Solve It), paper presented at the AIChE Spring National Meeting, New Orleans, Louisiana, April 25–29, 2004.
430. Schneider, D. F., "Heat Integration Complicates Heat Pump Troubleshooting," *Hydrocarbon Processing*, May 2002, p. 53.
431. Schneider, D. F., "Plant Power Failure and Its Indirect Effects: A Case Study," *PTQ*, Winter 1998/1999, p. 125.
432. Schneider, D. F., and M. C. Hoover, "Practical Process Hydraulic Considerations," *Hydrocarbon Processing*, Aug. 1999, p. 47.
433. Schneider, D. F., J. Musumeci, and R. Chavez, "Analysis of Alky Unit DIB Exposes Design, Operating Considerations," *Oil Gas J.*, Sept. 30, 1996, p. 41.
434. Schroeder, A. J. Jr., "Studies of the Performance of Dehydration Systems Pay Off," in *Sour Gas Processing and Sulfur Recovery*, Petroleum Publishing Co., Tulsa, OK, 1979.
435. Schultes, M., "Raschig Super-Ring: A New Fourth Generation Random Packing," in *Distillation 2001: Frontiers in a New Millennium, Proceedings of Topical Conference*, p. 498, AIChE Spring National Meeting, Houston, TX, Apr. 22–26, 2001.
436. Seidel, R. O., "Experience in the Operation of Activated Hot Potassium Carbonate Acid Gas Removal Plants (U.S.)," paper presented at the Seminar on Raising Productivity in Fertilizer Plants, Baghdad, Iraq, Mar. 23–25, 1978.
437. Severance, W. A. N., "Advances in Radiation Scanning of Distillation Columns," *Chem. Eng. Prog.*, *77*(9), 1981, p. 38.
438. Sewell, A., "Practical Aspects of Distillation Column Design," *The Chem. Engnr.*, *299/300*, 1975, p. 442.
439. Shah, G. C., "Troubleshooting Reboiler Systems," *Chem. Eng. Prog.*, *75*(7), 1979, p. 53.
440. Shankwitz, G. P., and T. P. Stauffer, "Improve Your Existing Process," *Chem. Eng. Prog.*, June 2000, p. 35.

441. Shinskey, F. G., *Distillation Control for Productivity and Energy Conservation*, 2nd ed., McGraw-Hill, New York, 1984.
442. Shiveler, G. H., "Use Heavy-Duty Trays for Severe Services," *Chem. Eng. Prog.*, Aug. 1995, p. 72.
443. Shiveler, G. H., and H. Wandke, "Tray Damage in a Coker Main Fractionator," paper presented at the AIChE Annual Meeting, Reno, NV, Nov. 2001.
444. Shiveler, G., D. Love, and D. Pierce, "Retrofit of Depropanizer and Debutanizer Packed Column," paper presented at the AIChE Spring Meeting, New Orleans, Louisiana, April 26–28, 2004.
445. Short, D. G. R., "Using Residue Maps for Solving Separation Problems," paper presented at the AIChE Spring National Meeting, Houston, TX, Mar. 9–13, 1997.
446. Shtayieh, S., C. A. Durr, J. C. McMillan, and C. Collins, "Successful Operation of a Large LPG Plant," *Oil Gas J.*, Mar. 1, 1982, p. 79.
447. Simpson, L. L., "Sizing Piping for Process Plants," *Chem. Eng.*, June 17, 1968, p. 192.
448. Sloley, A. W., "Avoid Common Distillation Equipment Pitfalls," in *Distillation Tools for the Practicing Engineer, Topical Conference Proceedings*, p. 105, AIChE Spring National Meeting, New Orleans, LA, Mar. 10–14, 2002.
449. Sloley, A. W., "Avoid Problems During Distillation Column Startups," *Chem. Eng. Prog.*, July 1996, p. 30.
450. Sloley, A. W., "Customized Tower Revamps—Success and Failures," *Fuel Technol. Manag.*, Jan. 1998, p. 48.
451. Sloley, A. W., "Drawing Distillation Products," paper presented at the AIChE Spring National Meeting, Houston, TX, Mar. 9–13, 1997.
452. Sloley, A. W., "Effectively Control Column Pressure," *Chem. Eng. Prog.*, Jan. 2001, p. 38.
453. Sloley, A. W., "Instrumentation: The Key to Revamp Benefits," *PTQ*, Spring 2001, p. 127.
454. Sloley, A. W., "Liquid Levels and Density," Chemical Engineers Resource Page, www.cheresources.com.
455. Sloley, A. W., "Properly Design Thermosyphon Reboilers," *Chem. Eng. Prog.*, Mar. 1997, p. 52.
456. Sloley, A. W., "Reducing the Danger of Maintenance Exposure," *Petrol. Quarterly*, Spring 1998, p. 59.
457. Sloley, A. W., "Simple Methods Solve Exchanger Problems," *Oil Gas J.*, Apr. 20, 1998, p. 85.
458. Sloley, A. W., "Troubleshooting Refinery Vacuum Towers," in *Distillation 2001: Frontiers in a New Millennium, Proceedings of Topical Conference*, p. 251, AIChE Spring National Meeting, Houston, TX, Apr. 22–26, 2001.
459. Sloley, A. W., "Water Damage to a Refinery Atmospheric Crude Fractionator," in *Distillation: Horizons for the New Millennium*, Topical Conference Preprints, p. 234, AIChE Spring National Meeting, Houston, TX, Mar. 14–18, 1999.
460. Sloley, A. W., and B. Fleming, "Mechanical Specification of Mass Transfer Equipment," paper presented at the Seventeenth Annual Energy Sources Technology and Exhibition, New Orleans, LA, Jan. 24, 1994.
461. Sloley, A. W., and B. Fleming, "Successfully Downsize Trayed Columns," *Chem. Eng. Prog.*, Mar. 1994, p. 39.
462. Sloley, A. W., and G. R. Martin, "Subdue Solids in Towers," *Chem. Eng. Prog.*, Jan. 1995, p. 64.
463. Sloley, A. W., and S. W. Golden, "Analysis Key to Correcting Debutanizer Design Flaws," *Oil Gas J.*, Feb. 8, 1993, p. 50.
464. Sloley, A. W., S. W. Golden, and E.L. Hartman, "Why Towers Do Not Work," *Nat. Eng.*, Part 1, Aug. 1995, p. 19; Part 2, Sept. 1995, p. 16.
465. Sloley, A. W., T. M. Zygula, and K. Kolmetz, "Troubleshooting Practice in the Refinery," paper presented at the AIChE Spring National Meeting, Houston, TX, Apr. 2001.
466. Smith, R. F., "Curing Foam Problems in Gas Processing," *Oil Gas J.*, July 30, 1979, p. 186.
467. Snow, A. I., and W. S. Dickinson, "Analysis of Tower Flooding," *Chem. Eng. Prog.*, *60*(10), 1964, p. 64.
468. St. Laurent, R., "Structured Packing Fire in FCC Main Fractionation Column," paper presented at the API Operating Practices Symposium, May 18, 2004.
469. Staggs, D. W., "The Impact of Non-Ideal Vapor/Liquid Behavior on Solvent Emissions," paper presented at the AIChE Spring National Meeting, Houston, TX, Mar. 20–24, 1995.

470. Stanton, B. D., and A. Bremer, "Controlling Composition of Column Product," *Control Eng.*, *9*(7), 1962, p. 104.
471. Steinmeyer, D. E., and A. C. Mueller, "Why Condensers Don't Operate as They Are Supposed To," Panel Discussion, *Chem. Eng. Prog.*, *70*(7), 1974, p. 78.
472. Stichlmair, J. G., and J. R. Fair, *Distillation Principles and Practice*, Wiley, New York, 1998.
473. Strigle, R. F. Jr., *Packed Tower Design and Application*, 2nd ed., Gulf, Houston, TX, 1994.
474. Summers, D. R., "Harmonic Vibrations Cause Tray Damage," Paper 307g, presented at the AIChE Annual Meeting, San Francisco, CA, Nov. 18, 2003.
475. Summers, D. R., R. Alario, and J. Broz, "High Performance Trays Increase Column Efficiency and Capacity," in *Distillation 2003: On the Path to High Capacity, Efficient Splits, Topical Conference Proceedings*, p. 467, AIChE Spring National Meeting, New Orleans, LA, Mar. 31–Apr. 3, 2003.
476. Swisher, C., "On the Trail of Hidden Design Flaws," *Chem. Eng.*, Feb. 1995, p. 133.
477. Talley, D. L., "Startup of a Sour Gas Plant," *Hydrocarbon Processing*, *55*(4), 1976, p. 92.
478. Tamura, M., "Hydroxylamine Explosion at a Chemical Plant," *Loss Prevention Bull.*, *172*, August 2003, p. 25.
479. Tao, Z., "Use of Dry Gas, Nitrogen Improves FCCU Reactor Pressure Control," *Oil Gas J.*, Nov. 8, 1999, p. 68.
480. Taylor, G. L., and M. Cleghorn, "Naturally Occurring Radioactive Materials: Contamination and Control in a World-Scale Ethylene Manufacturing Complex," paper presented at the AIChE Spring National Meeting, New Orleans, LA, Feb. 28, 1996.
481. Thielsch, P. E., and F. M. Cone, "Texas Facility Investigates Catastrophic Piping Failure," *Chem. Proc.*, May 1997, p. 50.
482. Tolliver, T. L., "Control of Distillation Columns via Distributed Control Systems," in W. L. Luyben (ed.), *Practical Distillation Control*, p. 351, Van Nostrand Reinhold, New York, 1992.
483. Tolliver, T. L., and L. C. McCune, "Finding the Optimum Temperature Control Trays for Distillation Columns," *InTech*, *27*(9), 1980, p. 75.
484. Torres, G. A. da Silva, C. Ney da Fonesca, N. Morais Pinto, and S. Lage de Araujo, "How a Simple Modification Resulted in a Great and Unexpected Change in the Operation of a Vacuum Tower Wash Section," in *Distillation 2003: On the Path to High Capacity, Efficient Splits, Topical Conference Proceedings*, p. 284, AIChE Spring National Meeting, New Orleans, LA, Mar. 31–Apr. 3, 2003.
485. Torres, G. A. da Silva, S. Waintraub, and E. Hartman, "Crude Column Revamp Using Radial Temperature Profiles," *PTQ Revamps Operations Suppl.*, 2003, p. 12.
486. Trebilcock, R. W., J. T. Finkle, and T. DiJulia, "Reduce Distillation Waste Streams," *Chem. Eng. Prog.*, June 1994, p. 50.
487. Trompiz, C. J., Petroleos de Venezuela SA, personal communication, 1994.
488. Trompiz, C. J., and J. R. Fair, "Entrainment from Spray Distributors for Packed Columns," *Ind. Eng. Chem. Res.*, *39*, 2000, p. 1797.
489. Tru-Tec Division of Koch Engineering Company Inc., "Before and after Scan Results Identify Structural Failure of Cartridge Trays," *Tru-News*, Vol. 3, Ed. 2, La Porte, TX, 1995, p. 3.
490. Tru-Tec Division of Koch Engineering Company Inc., "Column Scan Identifies Fouling in a Sour Water Stripper," *Tru-News*, Vol. 4, Ed. 1, La Porte, TX, 1996, p. 1.
491. Tru-Tec Division of Koch Engineering Company Inc., "How to Detect Partial Damage to Trays and Packings," *Tru-News*, Vol. 3, Ed. 1, La Porte, TX, 1995, p. 5.
492. Tru-Tec Division of Koch Engineering Company Inc., "Identifying Process Related Problems—Foaming," *Tru-News*, Vol. 4, Ed. 4, La Porte, TX, 1996, p. 1.
493. Tru-Tec Division of Koch Engineering Company Inc., "Neutron Survey Identifies Flashing Resid in Vacuum Bottom Suction Piping," *Tru-News*, Vol. 4, Ed. 1, La Porte, TX, 1996, p. 3.
494. Tru-Tec Division of Koch Engineering Company Inc., "Scanning Before a Turnaround," *Tru-News*, Vol. 5, Ed. 3, La Porte, TX, 1997, p. 2.
495. Tru-Tec Division of Koch Engineering Company Inc., "Scans Identify Vapor Bypassing in a Debutanizer," *Tru-News*, Vol. 4, Ed. 3, La Porte, TX, 1996, p. 1.
496. Tru-Tec Division of Koch Engineering Company Inc., "TRU-SCAN Case Studies," *Bull. TCS-1*, 1990.
497. Tru-Tec Division of Koch Engineering Company Inc., "What Is Limiting This Column's Capacity, Column Design or Operation?" *Tru-News*, Vol. 4, Ed. 4, La Porte, TX, 1996, p. 1.

498. Tru-Tec Services Inc., "Troubleshooting Multi-Downcomer Trays," *Tru-News*, Vol. 6, Ed. 3, La Porte, TX, 1998, p. 1.
499. Tru-Tec Services (TTS) Inc., *Tru-Scan Case Study 2*, TTS, La Porte, TX, 1998.
500. Tru-Tec Services (TTS) Inc., *Tru-Trace Case Study 1*, TTS, La Porte, TX, 1998.
501. Tru-Tec Services Inc., "You Make the Call," *Tru-News*, Vol. 7, Ed. 1, La Porte, TX, 1999, p. 1.
502. Tuck, M. and M. Ashley, "New Technology to Produce Ethyl Acetate," *PTQ*, Spring 2004, p.131.
503. Tyler, C. M., "Process Analyzers for Control," *Chem. Eng. Prog.*, *58*(9), 1962, p. 51.
504. U.S. Chemical Safety and Hazard Investigation Board, "Investigation Report, Explosion and Fire, First Chemical Corporation, Pascagoula, Mississippi," Report No. 2003-01-I-MS, October 2003.
505. U.S. Chemical Safety and Hazard Investigation Board, "The Explosion at Concept Sciences: Hazards of Hydroxylamine," Report No. 1999-13-C-PA, March 2002.
506. U.S. Chemical Safety and Hazard Investigation Board, "Investigation Report, Refinery Fire Incident, Tosco Avon Refinery," Report No. 99-014-1-CA, March 2001.
507. Udvari G., L. Gerecs, Y. Ouchi, F. Nagakura, E. A. Thoes, and C. B. Wallace, "CO_2 Dehydration Scheme Aids Hungarian EOR Project," *Oil Gas J.*, Oct. 22, 1990, p. 74.
508. University of Massachusetts, *Mayflower—Software for Distillation*, Version 10, 1995.
509. Van der Meer, D., "Foam Stabilisation in Small and Large Bubble Columns," *VDI Berichte*, *182*, 1972, p. 99.
510. Van der Meer, D., F. J. Zuiderweg, and H. J. Scheffer, "Foam Suppression in Extract Purification and Recovery Trains," in *Proceedings of the International Solvent Extraction Conference (ISEC 71)*, Soc. Chem. Ind., London, 1971, p. 350.
511. Velayudhan, V., and D. G. Inamdar, "Failure of a MDEA Line," *Ammonia Plant Safety*, *40*, AIChE, 2000, p. 57.
512. Verduijn, W. D., "Corrosion of a CO_2-Absorber Tower Wall," *Plant/Operations Prog.*, *2*(3), 1983, p. 153.
513. Vermilion, W. L., "Precise Control of Alky-Unit Deisobutanizers," *Oil Gas J.*, Aug. 21, 1961, p. 98.
514. Vidrine, S., "Radioisotope Technology—Benefits and Limitations in Packed Bed Tower Diagnostics," in *Distillation 2003: On the Path to High Capacity, Efficient Splits, Topical Conference Proceedings*, p. 589, AIChE Spring National Meeting, New Orleans, LA, Mar. 31–Apr. 3, 2003.
515. Vidrine, S., and P. Hewitt, "Radioisotope Technology—Benefits and Limitations in Packed Bed Tower Diagnostics," paper presented at the AIChE Spring Meeting, New Orleans, Louisiana, April 25–29, 2004.
516. Viera, G. A., L. L. Simpson, and B. C. Ream, "Lessons Learned from the Ethylene Oxide Explosion at Seadrift, Texas," *Chem. Eng. Prog.*, Aug. 1993, p. 66.
517. Von Phul, S.A., "Sweetening Process Foaming and Abatement Part II: Case Studies," 52nd Laurence Reid Gas Conditioning Conference, Norman, Oklahoma, February 24–27, 2002, p. 9.
518. Waintraub, S., A. J. F. Lemos, L. C. M. Paschoal, and M. dos R. M. Mainenti, "How to Circumvent the Shutdown of a Preflash Tower in an Atmospheric Distillation Unit: An Abnormal Operation," in *Distillation 2001: Frontiers in a New Millennium, Proceedings of Topical Conference*, p. 300, AIChE Spring National Meeting, Houston, TX, Apr. 22–26, 2001.
519. Waintraub, S., G. A. da Silva Torres, F. Martins de Queiroz Guimaraes, and B. de Almeida Barbabela, "Is It Necessary to Use Internals in Heat Transfer Sections of Refinery Columns," paper presented at the AIChE Annual Meeting, San Francisco, CA, Nov. 16–21, 2003.
520. Walker, A., "Do Plant Pilots Need Cockpit Training?" *The Chemical Engineer*, Sept. 25, 1997, p. 10.
521. Walker, G., and R. Sadeghbeigi, "Coke Trap Reduces FCC Slurry Exchanger Fouling for Texas Refiner," *Oil Gas J.*, Sept. 8, 2003, p. 52.
522. Walker, J. J., "Sizing Relief Areas for Distillation Columns," *Chem. Eng. Prog.*, *66*(9), 1970, p. 38.
523. Wallace, D. P., "Failure of Two CO_2 Regenerator Towers," *Ammonia Plant Safety*, *32*, AIChE, 1992, p. 158.
524. Wasylkiewicz, S. K., and H. Shethna, "VLE Data for Synthesis of Separation Systems for Azeotropic Mixtures," in *Distillation Tools for the Practicing Engineer, Topical Conference Proceedings*, p. 242, AIChE Spring National Meeting, New Orleans, LA, Mar. 10–14, 2002.
525. Wasylkiewicz, S. K., L. C. Kobylka, and F. J. L. Castillo, "Optimal Design of Complex Azeotropic Distillation Columns," *Chem. Eng. J.*, *79*, 2000, pp. 219–227.

526. Wasylkiewicz, S. K., L. N. Sridhar, M. F. Doherty, and M. F. Malone, "Global Stability Analysis and Calculation of Liquid-Liquid Equilibrium in Multicomponent Mixtures," *Ind. Eng. Chem. Res.*, *35*, 1996, pp. 1395–1408.
527. Watson, K. D., and D. K. Siekkinen, "Sheep Mountain CO_2 Project Development and Operation," *Energy Prog.*, *6*(3), 1986, p. 155.
528. Webber, W. O., "Control by Temperature Difference?" *Petrol. Ref.*, *38*(5), 1959, p. 187.
529. Weiland, R. H., University of Newcastle, N.S.W., Australia, now with Koch-Glitsch LP, USA, private communication, Apr. 1992.
530. Weirauch, W., "Report Reveals Causes of UK Refinery Incidents," HPI Impact, *Hydrocarbon Processing*, Oct. 2003, p. 25.
531. Wesch, I. M., "Canadian Gas Plant Uses Amine Unit to Sweeten Liquid Ethane," *Oil Gas J.*, Feb. 24, 1992, p. 58.
532. Weston, K., and S. White, "Steam Reboiler System Operations," *PTQ, Revamp & Operation Suppl.*, Autumn 2004, p. 27.
533. Whistler, A. M., "Locate Condensers at Ground Level," *Petrol. Ref.*, *33*(3), 1954, p. 173.
534. White, S., and S. Fulton, "Specifications—Importance of Getting Them Right," *PTQ*, Autumn 2003, p. 117.
535. Whitt, K., and C. Herron, "FCCU Gas Plant Revamp Boosts C3 Recovery," *PTQ Revamps Operations Suppl.*, 2003, p. 19.
536. Williams, J. A., "Optimize Distillation System Revamps," *Chem. Eng. Prog.*, Mar. 1998, p. 23.
537. Williamson, J. C., "Improve Vapor and Mixed-Phase Feed Distribution," World Refining, p. 22, May 2000.
538. Wilson, B., and B. Das Biswas, "Advanced Control for Solvent Extraction," *PTQ*, Winter 2004, p. 105; and B. Das Biswas, private communication, Jan. 2004.
539. Winter, J. R., "Avoid Vibration Damage to Distilaltion Trays," *Chem. Eng. Prog.*, May 1993, p. 42.
540. Woodward, J. L., J. K. Thomas, and B. D. Kelly, "Lessons Learned from an Explosion in a Large Fractionator," *Process Safety Progress*, *22*(1), 2003, p. 57.
541. Woolf, G., "Painful Lessons," *Chem. Engnr.*, June 30, 1994, p. 11.
542. Xu, S. X., "Identifying Packed Tower Maldistribution," *Tru-News*, Vol. 5, Ed. 4, Tru-Tec Services Inc., La Porte, TX, 1997, p. 1.
543. Xu, S. X., and L. Martos, "Flooding Phenomenon in Distillation Columns and its Diagnosis, Part I, Trayed Columns," in *Distillation 2001: Frontiers in a New Millennium, Proceedings of Topical Conference*, p. 123, AIChE Spring National Meeting, Houston, TX, Apr. 22–26, 2001.
544. Xu, S. X., and L. Pless, "Distillation Tower Flooding—More Complex Than You Think," *Chem. Eng.*, June 2002, p. 60.
545. Xu, S. X., and L. Pless, "Understand More Fundamentals of Distillation Column Operation from Gamma Scans," in *Distillation 2001: Frontiers in a New Millennium, Proceedings of Topical Conference*, p. 605, AIChE National Meeting, Houston, TX, Apr. 22–26, 2001.
546. Yarborough, L., L. E. Petty, and R. H. Wilson, "Úsing Performance Data to Improve Plant Operations," in *Proc. 59th Annual Convention of the Gas Processors Associations*, Houston, TX, Mar. 17–19, 1980, p. 86.
547. Ye, X., "Gas Oil Desalting Reduces Chlorides in Crude," *Oil Gas J.*, Oct. 16, 2000, p. 76.
548. Zilka, M. I., "The Chemical Engineer as a Super Sleuth," *Chem. Eng.*, May 17, 1982, p. 121.
549. Zuiderweg, F. J., J. G. Kunesh, and D. W. King, "A Model for the Calculation of the Effect of Maldistribution on the Efficiency of a Packed Column," paper presented at the AIChE National Meeting, Pittsburgh, PA, Aug. 1991.
550. Zuiderweg, F. J., P. J. Hoek, and L. Lahm, Jr., "The Effect of Liquid Distribution and Redistribution on the Separating Efficiency of Packed Columns," *I. Chem. E. Symp. Ser.*, *104*, 1987, p. A217.
551. Zygula, T. M., and O. Barkat, "Troubleshoot Vertical Thermosiphon Reboilers," *Chem. Eng. Prog.*, July 2000, p. 71.
552. Zygula, T. M., K. Kolmetz, and R. Sommerfeldt, "Troubleshooting an Ethylene Feed Saturator Column," *Oil Gas J.*, Aug. 25, 2003, p. 88.

Index

(Bold page numbers signify major references on indexed item)

Abnormal operation (*see also* Startup; Shutdown) 215–223, 291–292, 499–511, 586
Absorber 466, 491, 586
- acetylene 294–295, 588
- alkaline 145, 241, 459, 462, 544, 548
- amine (*see* Amine absorber)
- ammonia recovery 536
- backflow 215, 500, 535–536
- Benfield hot pot (*see* Hot pot)
- caustic (*see* Caustic absorber)
- control (*see* Control)
- deethanizer (*see* Deethanizer)
- ethylene oxide 446
- HCl 620
- HF 402
- hydrocarbon (*see also* Lean oil absorber, deethanizer, sponge absorber) 241, 484, 549, 550, 556
- methanol 620
- naphtha reformer 428
- nitric acid 467
- oxidation reaction effluent 593
- Selexol (*see* Selexol)
- sponge 500, 553

Absorption
- and desorption of sparingly soluble gas 503, 541
- effect **18–20**, 25, 166, 295–296, 342, 412, 425, 609
- excessive 295–296, 408
- heat of- 42, 209–210, 609
- light components 295–296, 342, 386, 428, 503, 609
- oil 30–33, 40–42, 47–50, 414, 418
- -refrigeration gas plant 30–33, 40–42, 47–50, 209–211, 374–376, 381–382, 418, 526–527
- vs. reflux 40–42, 425

Accident (*see also* Blinding; Chemical release; Confined space; Explosion; Fatalities; Fire; Injuries) 215, 233, 347, 503, 517–538
Accumulation
- corrosion 25, 37–38, 40, 219, 414, 416, 427, 430, 558
- control-induced 360–362, 634
- cycling (*see* Hiccups)
- dead pocket (*see* Dead pocket)
- foaming 60, 240, 313, 545
- heavies 27, 596
- hiccups (*see* Hiccups)
- intermediate component 4, 25, 37–55, 57–61, 240, 313, 414–419, 521, 539, 634
- light component 4, 42, 49–51, 415–417, 428, 605, 640
- liquid (*see* Flooding)
- recycle-induced 49–52, 79–80, 417, **418**, 620
- simulation 46–47
- symptoms & cures 37–55, 57–60, 414–419
- Water (*see* water accumulation)

Accumulator (*see* Reflux drum)
Acentric factor 398
Acetals 239
Acetates (*see also* Butyl acetate; Ester; Isopropyl acetate) 43–46
Acetic acid 401, 403, 422, 497, 520, 631
- dehydration 9–11, 401
- scrubber 124–125, 454

Acetic anhydride 422
Acetone (*see also* Ketone) 59, 294, 400–402, 422
- stripping 407

Acetonitrile 538, 545, 645
Acetylene plant 264–265, 294–295, 400, 408, 568, 588

Distillation Troubleshooting. By Henry Z. Kister

Acetylenic compounds (*see also* Ethyl acetylene; vinyl acetylene) 322
Acetylides 294
Acid gases (*see also* Amine; Carbon dioxide; Hydrogen sulfide) 76, 552–554
Acid recovery tower 421
Acid tower (*see also* Acetic acid; Alkylation; Fatty acid; Nitric acid) 620
Acid wash 503, 552, 559, 563
Acids, organic (*see also* Acetic acid; Formic acid) 528, 555
Acrylate 220, 261
Activated carbon (*see* Foam, carbon beds)
Activity coefficient 1–11, 38–42, 45–47, **399–402**, 538
Additive (*see also* Corrosion inhibitor; Extractive distillation, Foam; Inhibitor)
Aeration (*see* Base level; Chimney tray; Condenser liquid; Distributor, liquid; Downcomer; Draw-off; Foam; Reflux drum)
Aftercooler tower 264–265, 568
AGO 369–370, 410, 427, 465, 475, 626
Agricultural chemicals 450
Air (*see also* Leak; Stripper)
  separation 524
  supply (inside tower) 511
Alarm 155, 519, 521
Alcohol (*see also* amyl alcohol; aromatic alcohol; butanol; DAA; ethanol; IPA; methanol; phenol; propanol) 25, 43–47, 59–60, 239, 303–305, 313, 413, 545, 612, 629, 631
Aldehyde (*see also* formaldehyde) 51, 59–60, 239, 241, 245–246, 303–305, 540–541, 555
Alpha-methyl styrene (*see* AMS)
Alkylation
  butylenes H_2SO_4 79
  depropanizer 79, 414, 417, 524, 558, 577, 582, 633
  DIB (*see also* Splitter, iC_4-nC_4) 425, 428, 596, 632, 634
  HF stripper 417, 602
  isostripper 187, 259, 432, 560
  main fractionator 188, 400, 405, 490
Alternative feed point 77–79, 431, 441
Amine:
  absorber 126, 140, 241, 243–244, 347, 432, 435, 473–475, 535, 540, 543, 545, 548–555, 618
  acid gas loading 552–553
  aMDEA 403, 456
  circulation 411, 551, 555
  clean-up 543–546, 551–555
  DEA 76–77, 540, 552, 558
  excessive absorption 408
  HSS 540, 554–555, 558
  losses 456, 540, 545, 551, 554
  MDEA 408, 432, 442, 473, 483, 527, 535, 545, 549, 551, 553–554, 558, 589
  MEA 440, 473–475, 551–554, 558, 578
  oxidizing 531
  non-selective 408
  reclaiming 545, 552, 554–555
  regenerator 241, 263, 266–270, 374, 411, 439–440, 442, 453, 456, 483, 494, 535, 540, 544, 550–552, 554, 563, 578, 586–587, 618
  regenerator steam rate, pressure 552
  strength 545, 552, 555
  Sulfinol 241, 408, 494, 545, 554–555
  TEA 532
  triethylamine 578
4-Amino antipyrine sulfonic acid 538
Ammonia 399, 536, 541
  plant 145, 205, 403, 446–447, 456–457, 469, 473–474, 483, 502–503, 507, 527, 535, 543, 549, 567, 575, 577, 586, 589, 620
  recovery absorber 536
  still 495
  stripping 47, 241, 310–311, 407, 557, 581
Ammonium carbonate 541
Ammonium chloride 504–505, 560
AMOCO safety series (*see* BP safety series)
AMS 401, 435, 633
Amyl alcohol 51
Analysis (*see also* Control, composition) 281
  cold liquid compositions 86
  laboratory 12–14, 19, 86, 353, 405, 541, 608, 614, 645
  on-line 405, 541, 633–635, 645, 647
Analyzer (*see* Analysis; Control, composition)
Angle iron (*see also* Shed decks) 110, 175–176, 482
Aniline 282, 521–522
Anthraquinone 520
Antifoam (*see* Foam)
Antifoulant 544, 547, 556, 560
Antifreeze 3, 52, 396, 414, 418
Antijump baffles 134, 428
Aqueous (*see* Water)
Aromatic alcohol (*see also* Phenol) 58
Aromatic derivatives 258
Aromatic extraction 13, 580
Aromatics (*see also* Benzene; Ethylbenzene; Splitter; Toluene; Xylene) 226, 236, 404, 423, 442, 446, 455, 466, 475, 490, 496, 503, 512, 609, 628, 632
Arsenic – activated Vertacoke 567, 577

Asphalt tower 557
Asphaltenes 167, 271, 457, 550
Asphyxiation 503, 510–511
Assembly (*see also* Distributor, liquid; Draw-off; Packing) 193–213, 291
baffles 198
blanking strips 194–195
column sway 498
debris (*see* Debris)
directional valves 195–196
downcomer 85, 492–493, 615
downcomer clearances 194, 435, **491–492**
drawings error 492, 497
gaps 119, 160, 204–205
grid 588
heat exchanger 209–211
instrument taps obstructed 175–176, 482
instruments 616–617
interchanging parts 491, 496–497
leakage 193, 435
manways 194, 196, 198–200, **493–494**, 588
materials of construction 83–85, 193, 437, 494
misorientation of internals 492–493
nuts, bolts tightening 131–132, 193, 294–295, 435, 471, 477, 480, 493–494, 579, **587–588**, 598
obstructions of flow passages 193, 493
parts fitting through manholes 498
piping 211–213, 493
seal pans 85, 492
tray panels 193–196, 435, **491**, 547, 588
weirs 196, 198, 435, 491
Association (of molecules) 9–11, **401–402**
ASTM D86 26, 176–178, 405
Atmospheric crude tower (*see* Crude tower)
Azeotrope 2, 6–11, 43–46, 51, 240, 388–389, **400–401**, 415, 419, 421–422, 520, 582, 625, 628, 631
Shift (from one to another) 419, 421, 422, 520
Azeotropic distillation
acetic acid – n butyl acetate – water 9–11, 401
Formaldehyde – acetone – water 59
methanol – n butanol – water 6–8, 400
organics – benzene – water (dehydration) 57–58
organics – hydrocarbon – water (dehydration) 46, 59, 420
sec butanol – SBE - water 400
temperature control 631

Back flow (*see* Reverse flow)
Backpressure 211–212, 379, 636
Backscatter (*see* Neutron backscatter)
Baffle:
angled baffle (diverts seal pan overflow to a preferred sump compartment) 159–160
antijump 134, 428
condenser, cross flow 610
distribution (horizontal reboilers) 597, 599, 601
division (*see* Baffle, overflow)
doghouse 465
downsizing baffles (fencing-off blanked tray active area) 433
feed round deflector (located above feed pipe and deflects feed downwards) 104, 107, 444
gap in 160
horizontal deflector baffle 191
impingement (at bottom sump) 151, 473–474
impingement (in condenser) 288, 610
inlet gas/reboiler return deflector (*see also* Baffle, v-baffle) 110, 133, 155–157, 272, 277–278, 445, 465, 474–476
instrument tap shielding 473
kettle reboiler overflow 266, 599, 601
overflow baffle (divides sump to a drawoff compartment and a circulation suction compartment) 159–160
reboiler preferential baffle (divides tower base to separate bottoms and reboiler draw compartments) 146, **155–157**, 197–198, 222–223, 368, 476, 494, 596
tangential feed 149–151, 272, 464
trays (*see also* Shed decks) 278–279, 487
v-baffle at reboiler return/gas inlet 113–114, 355, 463
Balance (*see* Material balance; Component; Control assembly; Heat)
Banging (*see also* Hammering) 588
Base level 145–161, 355–356, 468–476, 589
aeration 149–152, 161, 347, 355–356, 519, 538, 557, 589, 617–618
cooling 589
damage (*see* Base level, uplift)
foaming 470, 544, 557, 617
frothing 602, 617
gamma scan (*see* Gamma scans, high base level)
high 55, 85, 145–155, 198, 222–223, 266–270, 292, 297, 301–302, 304, 315, 326, 355–356, **468–472**, 485, 489, 494–495, 515, 526, 544, 555, 557, 591, 596, **599–603**, 616–620
high level leads to discharge from column top 468–469, 619
high level leads to vapor gap damage 300–302, 591

Base level (*Continued*)
impingement 145, 151, **155–157**, 355, **472–475**
kettle circuit pressure drop (*see* Kettle reboiler)
loss 489, 548
low 146, 233, 475, 518–519, 538, 544, 616
outlet line restriction 145, 488
rapid draining 301, 305, 469
raising to overcome reboiler limitation 155–157, 198, 266–270, 322, 602
reliability switch 154, 469, 471
startup/shutdown incidents 147–149, 152, 154, 222–223, 292, 297, 356, 468–469, 507, 526, 589
sudden rise 295, 307
swings (*see also* Base level control) 77, 85, 128, 223, 596–597, 599
uplift of trays/packing due to high level 145–155, 270, 291–293, 296–297, 355–356, **471–472**, 515
vapor entrainment 146, 485
Base level control
absorber 620
faulty 145, 468–469, 475, 525, 527, 538, 602, 617, 619
fuzzy 614
MVC 646–647
loss due to baffle issues 155–157, 222–223
loss at startup 222–223, 310–311, 356, 525
small bottom stream 622
swings 368, 472, 614, 617, 622, 632, 643–644, 647
tower-to-tower shifts 632, 646
windup 647
Base level measurement
alarm incorrectly set or misunderstood 155, 519
calibration 468–469, 557
false indication 145–154, 326, 347, 355–356, **468–469**, 475, 519, 525, 527, 538, 557, **614–619**
fooled by aeration 149–152, 347, 355–356, 519, 538, 557, 617–618
fooled by impingement 355
fooled by lighter liquid 347, 356, 469, 619
fooled by radioactivity 619
fooled by second liquid phase 152, 347, 618
heat tracing/insulation issues 616–617
impingement on 473
no indication 600
nucleonic 326, 468, 619
from static head 472
tap location 149–152, 355
tap plugging 469, 614
from temperature measurement 614
from tower pressure drop 148–149, 269
unavailable due to baffles 156, 222–223
upper-lower transmitters 151–152, 355–356
Basic HETP 88
Batch distillation 12–13, 241, 404, 421, 457–458, 501–502, 520–523, 538, 557, 575, 625, 645
of component(s) in continuous distillation 45, 54–55, 295–296, 328, 418, 521
control 458, 625, 645
converted to semicontinuous 458
data 398
Bathtub 220, 602
Bayercap trays 589
Bed length (*see* Packing, bed length)
Bed limiter (*see* Holddown)
Bellows 506
Bench tests (*see also* Pilot; Tests) 407–408, 545, 547
Benfield hot pot (*see* Hot pot)
Benzene (*see also* BTX) 4, 57–59, 401, 422–423, 454, 463, 532, 538, 566, 583, 605, 632
Bimetallic 525
Biological growth 559
Blanking (*see also* Valve trays) 143, 179, **432–433**, 467
Blanking strips 194–195, 251
Blinding 146–147, 215–218, **499–501**, 521, 527–528, 534, 536
Blowdown 55, 399, 418
Blowing (lines) 285–286
Blowing (trays) 428
Blown condensate seal 331–333, 378, 605–606
Boiler feedwater 285, 313, 429, 504, 544
Boiling
cooling water 344, 387, 640
delayed (two liquid phases) 517, 589
film 330, 604
nucleation 596
point elevation 328, 589, 603
range (*see* Wide boiling; Narrow boiling)
wash 503
Boilover 234, 538
Boilup
control (*see* Control, reboiler)
excessive 423, 494, 628
rate change (*see* Control assembly)
swings 316–317, 388–393, 474, 575, 596–597, 621, 623, 647

Bolts (*see* Assembly, nuts, bolts)
Boot (on Reflux drum) 4, 38, 42, 221, 416–417
Boroscope 258
Bottle shake (*see* Foam, testing)
Bottom
  draw 470, 485, 492, 494
  -feed interchanger (*see* Feed-bottom interchanger)
  level (*see* Base level)
  line 145, 470, 485, 495, 525, 596
  loss/reduction of 147–149, 488
  pump loss 147–149, 471, 485
  temperature (*see* Temperature, bottom)
BP safety series of booklets 225
Braces
  cross-channel 587
  downcomer 592
Break point (azeotrope) 631
Breakage (*see* Packing)
Breathing (near-atmospheric column) 96, 313
Brittle failure (*see* Overchilling)
Broad boiling range (*see* Wide boiling)
Bromo organic compounds 521–522
BTX, BTEX 411, 418, 423, 431, 496, 628
Bubble cap trays 258, 294, 301, 303, 430, 432, 434–**435**, 437, 452, 478, 487–488, 490–491, 512, 554, 592, 624, 626
Buckling (*see* Thermal expansion)
Bumping (*see* Damage)
Burping (*see* Hiccups)
Butadiene 60, 66–68, 261, 398, 521, 537–538, 545, 566, 632, 645
  extractive distillation 60, 545, 632
  final purification column 537
  heavy ends recovery 538
  refining column 521
  Stripper 632, 645
Butanol 6–8, 51, 400–401
Butenes 60, 66, 538
Butyl acetate 9–11, 401
Butyl ether (*see* SBE)
By-gas 95
Bypass:
  the column 67–68, 77, 79–83, 408, **424–425**
  control valve bypass 468, 558, 643
  filters 121, 255, 448, 450
  heat exchanger 48–49, 284, 417, 577, 643
  hot vapor (*see* Control, pressure)
  manual vs. auto 408, 425
  plugged trays 258, 266, 427
  preflash crude tower 643
  section of column 98–99, 425, 430, 441, 505, 564
C-factor 21, 91–93, 106, 314
C_2 Splitter 28–30, 52–55, 61–63, 339–340, 358–360, 395–396, 399, 411, 418, 575, 606, 609, 621, 623, 633
C_3-C_4 Splitter 71, 134–138, 469, 478, 596
C_3 Splitter 2–4, 212–213, 343–344, 351–354, 405, 418, 445, 460, 494, 498, 502, 524, 563, 611, 616, 621, 623
C_4 separation (*see* Debutanizer, Stabilizer, Splitter)
C_5 separation (*see* Depentanizer)
C_6-C_7 separation 447
C_8-C_9 separation 622
C_{10} HC's and alcohols separation 612
Caged valve trays 262, 300
Calcium salts 559
Calibration (Instrument) 347–350, 353, 405, 616–617
Campaign (*see also* Multipurpose plant) 236, 337, 578
Cap (*see* Chimney tray hat; Valve trays)
Capacity (*see also* Entrainment; Flooding; Pressure drop; Turndown)
  compressor bottleneck 108
  debottleneck 14, 20–21, 28–30, 62–63, 68, 85–90, 97–100, 122–124, 157, 284
  limitation 14, 18, 62–71, 79–81, 84, 97–104, 176–178, 201–202, 281, 354–355, 424, 431, 433–443, 482, 485, 491–496, 545, 551, 555, 595, 608, 628, 646
Carbon beds (*see* Foam)
Carbon dioxide (*see also* Amine; Caustic; Hot-pot) 5, 74, 159, 171, 205, 241–242, 266, 294, 378, 440, 447, 459, 462, 541, 546, 550, 554, 598, 605, 645
  high content in natural gas 403, 598
Carbon monoxide 294
Carbon tetrachloride 400
Carbonyl fluoride 402
Carboxylic acids (*see* Organic acids)
Cartridge trays 593
Carryover (*see* Flood; Entrainment; Reflux drum)
Cassette (*see* Cartridge trays)
CAT scan 453, 497
Catalyst carryover (into tower) 121, 233, 237, 276–277, **540**, 542
Catalytic reformer 13
Caustic
  absorber, C_5 isomerization 218–219, 467
  absorber, chemicals 241, 453
  absorber, olefins 74–75, 159–160, 171–172, 241, 248, 261, 296–297, 453, 505, 535, 540, 544, 547, 556, 594
  absorber, refinery 503

Caustic (*Continued*)
backflow 218–219, 535–536
C_3 Splitter 563
circulation rate restriction 453
consumption 160, 261, 453
deposits 4, 504, 563
injection into feed, tower 467, 504, 554, 558
in piping 523
for reclaiming 555
solid bed 524
spent 160, 297
wash 219–220, 438, 523, 536
Cavitation (*see also* Damage) 146, 160, 163, 169, 171, 179, 182, 228, 285–286, 347, 427, 438, 466, 474–477, 481, 484–485, 487, 500, 502, 535, 584
CFD modeling 464, 498
Channeling (*see* Condenser maldistribution; Distribution; Valve trays; Vapor cross flow channeling; Vapor maldistribution)
Characterization (of petroleum fractions) 1, 11–12, 271, **402**
Check valve 219, 297, 387–388, 535
Chemical reaction (*see also* Coking; Degradation; Polymerization)
air-leak promotes- 519–520, 542
bottom temperature hot 33, 38, 233, 237, **239**, 400, 403, 518–523, 538, **539**, 578
catalysis of 233, 237, 240, 518, 521, 523, **540**, 542
chemicals from commissioning 503, 523, 535, 537, **540**, 544
concentration of reactive component 233, 237, 411, 519–523, **539–540**
condensation 33, 239
corrosive 38
cracking 199–200, 403, 472, 574, 608
decomposition 2, 38, 233, 237, 400, 518–523
degradation 74, 521
dehydration 237, **240**
hot spot 233, 237, 239, 518–519, **539**, 574
hydrolysis 2, 402, 403, 419, 520
inhibitor problems 541
long residence times 200, 237, 419, 472, 521, 540, **541**, 608
oxidation 236, 503, 521
polymerization (*see* Polymerization)
precipitation of reactive component 233
pressure rise induces temperature rise 520
reactive feed or product impurities 237, 402–403, 520, 524, 538, 540, **542**
relief 537
rise in pressure 233, 520–523
runaway 518–528, 537
simulation 1, 402–404
slow 518–519, 521
thermite 533
violent 25, 234, **523–524**, 538
Chemical release
by backflow 215, 233–234, **536**
leaks 281, 506, 527, 578–579, 582, 602
relief, venting, draining to atmosphere 233–234, 287–288, 347, 440, 503, 522, **537–538**, 580–583, 619, 640
trapped chemicals released 233–234, **536–537**
Chemical system VLE
NRTL 6–9, 38, **400–401**
UNIQUAC 6, 8, 38
Wilson 6, 8, 401
Van Laar 401
Chemical wash (*see* Washing)
Chemistry of a process 1, 2, 237–240, **402–404**, 539–542
Chevron collector 188–189, 451
drain hole 189
drain pipe 451
overflow 451, 457
Chimney effect 510
Chimney tray
aeration of liquid 141, 178, 481–482
coking 477, 482
damage 170–171, 477, 482–483, 495, 588
drain downcomer undersized 104, 137–138, 163, 478–479
drain downpipe absent 479
draw line undersized 163, 479
draw nozzle location 479
freezing 163
hat 171–172, 324, 481, 495
hot and cold compartments 485
hydraulic gradients 141, 163, 172–173, 480
impingement 480–481, 483
inadequate degassing 163
interference with supports 466
interference vapor/liquid 163, 173–174, 325, 475, 480–481, 483
internal reboiler 602
leakage 163–176, 193, 311–312, **477–478**, 482, 486–487, 497
level 141, 163, 166–171, 175–176, 272, 276, 311–312, 477, 481–482
level measurement 163, 175–176, 402, 481–482
liquid collection in packed tower 117
liquid entrance into- 141, 173–174, 176–178, 480–482
obstruction of downcomer entrance 176–178, 482

out-of-levelness 464
overflow 101–104, 133, 137–138, 141, 163–170, 172–173, 248–250, 276, 322–325, 402, 477–482, 564, 601–603
overflow downcomer/downpipe 173–174, 276, 475, 478–479, 481
plugging, coking, fouling 163, 402, 418, 430, 436, 464, 477, 482, 573–574
pressure drop 137, 464
reboiler draw 322–328, 479
reentrainment from 137, 141, 402, 480–482
residence time 464
risers open area 104, 113
sloping 464
startup problem 170–173, 175–176
submerged inlet downcomer 482
supports 175, 480, 483
tests to check for leak, overflow 166–171, 497
thermal expansion 166, 175–176, 464, 479–480
unsealed downpipes from- 137, 436, 478
vapor distributor (*see* Distributor, vapor)
vapor entry into 325
water removal from HC's 101–104, 248–250, 415, 417, 504–505
water test 497
Chlorides 220–221, 504–506, 543, 560, 578
Chlorinated HC's (*see also* Halogenated HC's; TCE; Vinyl chloride) 37–38, 400, 413, 589
Chlorine 619
m-chloroaniline 523
Choking
downcomer (*see* Downcomer choke)
lines (*see* Self-venting flow)
CHP 519
Chute and sock 113, 115, 117, 204
Clamps 132, 292–293, 307, 310–311, 313, **587–588**
Cleaning (*see also* Reboiler; Packing)
Clean-up (*see* Amine)
Clearance, downcomer 194, 250, **434–435**, **491–492**, 566
Clips (*see* Clamps)
Close-boiling systems (*see also* Splitter) 1, 3, 398
Coal gasification 494, 552
Coalescer 4, 5, 40, 545, 551–552
Cocurrent heat exchange 209–211, 330, 498, 604
Coil (*see* Fired heater; Temperature, coil outlet)
Coker
coke drum foamover 469
coke drum switchover 427, 487, 598
debutanizer 42–43, 490, 505, 597–598, 637, 641
deethanizer 415, 427, 598
main fractionator 147, 225, 241, 272, 300, 430, 436, 444, 463–464, 469, 487, 499, 512–513, 516, 529, 574, 583, 592
sponge absorber 500
Coking (*see also* Plugging; Vacuum refinery tower; FCC main fractionator) 215, 253, 257, 262, **271–279**, 402, 430, 436, 449, 460, 463, 477, 482, 488, 529–530, **571–574**
Collapse (*see* Implosion; Steam-water operation; Vapor, collapse)
Cold spin (*see* Heat integration spin) 61, **68–69**, 426
Cold water H_2S contactor 241
Collector (*see also* Chevron collector; Chimney tray) 128, 457–458, 464, 508
Color 369, 574
of product or reflux 59
Coloration 115–116
Commissioning 171–172, 212–213, 215–223, 234, 291–292, 305, 310–311, 495, **499–511**, 585, 586, 591
Compensation, pressure (*see* Control, temperature)
Component (*see also* Accumulation):
addition 313, 509
balance 1–2, 19, 86, 283, 353, 361, 614, 620
concentration 5, 33–55, 57–60, 240, 411, 413–419, 425, 519–521, 539, 619–620, 634
high boiling 26–27, 70, 316, 325, 328, 630–632
intermediate key 4, 25, 35–55, 57–61, 240, 313, 335–339, 413–419, 521, 539, 619, 634
key 15–18, 33–37, 629–630
low-boiling 13–14, 33–34, 42, 50, 316, 325, 328, 630, 635, 640
low concentration (*see* Low concentration, Infinite dilution) 3–5, 8–9, 400–401, 625, 631
pinch 9, 15–18, 28–30, 66, 411
trapping (*see* Accumulation)
Composition profile (*see also* Multicomponent) 2, 6–8, 33–37, 131, 407, 413
Compressor
limitation 608, 610, 612
start-up 296–297
surge 221, 292, 594, 636
trip 50, 425, 612
Computer (*see* Control, advanced; Simulation)
Condensate
cooling 66
draining to deck 334, 389, 392
pot (drum) 390–392, 642
pump 66

Condensate (*Continued*)
  removal, from condenser 335–340, 420, 586, 608–609
  removal, from reboiler 316, 333–334, 518, 606, 641–642
  seal 331–333, 378, **605–606**
  stripper 241, 247–250, 457
Condensation
  direct contact (*see* Pumparounds; Water quench)
  hydrocarbon into aqueous solution (*see* Base level measurement; Foam)
  in instrument lines 352, 354–355, 616
  multicomponent 18–20, 166, 295–296, 311, 412, 609
  rapid 292, 295–296, 305, 310–313, 586, 591–592, 638
  steam in pipe 349
  steam purges 508, 586
  two-stage 427
  in vapor product lines 378–379
  in vent line and stack 410
  water near top of dry tower 427
  wide boiling range **18–20**, 166, 295–296, 335, 412, 609
Condenser (*see also* Control, pressure and condenser; Tube leak; Venting; Vibrations) 335–346, 607–612
  axial outlet 609
  baffles 610
  blockage 288, 520, 610
  capacity limitation 339–340, 427, 498, 607–612, 639
  control (*see* Control, pressure)
  cooling water supply 212–213, 509, 612
  cooling water outlet temperature 344, 386–387, 640
  corrosion 335, 344, 377, 607
  damage 610
  direct contact (*see* Pumparound; Water quench)
  double split-flow 18, 288, 610
  draining 335–340, 420, 586, 608–609
  entrainment 335, 343–344, 387, 611
  equilibrium (vapor-liquid) 18–20
  exchanger design 609–610
  fabrication/assembly mishaps 209–211, 498
  flooded 335, 338–340, 379–385, 611, 637–639
  fouling 19, 335, 344, 377, 386–387, 610, 612, 640
  heat transfer 607–610, 639
  hydraulic gradient 611
  impingement baffle 288, 610
  inert blanketing 297, 335, 340–343, 346, 378, 580, 607–608, 641
  internal 68–69, 344–346, 363–367, 421
  knockback 68–69, 335–339, **343–346**, 421, 579, **611**
  liquid aeration 338
  maldistribution, single condenser 335, 607, 610, 611
  non-equilibrium (or local equilibrium) 18–20, 412, 609
  outlet line 335–340, 420, 586, 609, 611
  seal 331–333, 378, **605–606**
  shells in series 18, 288
  side draw 337
  Smith's statement 335
  spray 340–343
  steam generator 583, 607, 640
  tube bowing 288
  vent (*see* Vent condenser)
  venting 297, 335, 340 343, 382, 386, 581, 597, 607–609
  vertical upflow (*see* Vent condenser)
Confined space 510–511
Construction debris (*see* Debris)
Contamination
  product 13, 237, 281, 455, 523, 575, 577
  lean solution 403, 540, 544, 551–555, 577
  utility 281
Control (*see also* Lag; Oscillations)
  absorber system 620
  ambient disturbances 621, 624–625, 628
  batch distillation 458, 625, 645
  composition swings 359, 621–622, 629, 635, 642, 645
  condenser (*see* Control, pressure)
  differential pressure 72, 371–372, 395, **580–581**, 628, 644, 646
  differential pressure ratio 371–372, 628
  effect on relief 580–581
  energy consumption 621, 631–632
  feed enthalpy 643
  feed-forward 367, 621–622, 631, 634, 646–647
  flood 72, 395–396, 426, 635, 644–645
  fuzzy 614, 631
  heat pumped column 621, 633
  instability 344–346, 363–369, 374–380, 620–647
  interaction with process 395, 644
  interreboiler 621
  off-spec product 363, 620–621, 626–628, 637, 644
  packed vs. tray column 131
  preheat 61, 377–378, 410, 625, 641, **643**
  product recovery low 359, 369, 623, 626–627, 629, 631–635, 641, 644

pressure swings 360–362, 377–387, 636–639
ratio 621, 633
reboiler (*see* Control, reboiler)
reflux instability 366–367, 620–621, 624–627, 647
split range 344–346
variable reflux (batch) 458, 625
variables 364–365
Control, advanced
Advisory software 631, 634, 644
APC 581, 623, 630, 633–635, 646–647
decoupling 357, 621
dynamic simulation model 622
DMC 157, 627, 640, **646–647**
fooled by bad measurements 646–647
gain scheduling strategy 645
MISO 621
models 357, 373, 622, 633–635, **644–647**
multivariable (MVC) 395, 628–629, 632, 635, **644–647**
neural model 622, 629, 632, 646–647
observer model 629
power dip 647
statistical process control (SPC) 357, 373
training range 647
updating MVC models 646
virtual analyzer 365, 373, 628–629, 632–634, 647
viscometer 634
Control, assembly of system
azeotrope break point 631
difference control 626–627
direct MB (*see* scheme 26.4d, 26.4e)
dynamic response 357–360, 623–624
energy balance 623
extractive distillation 628, 631–632, 634
high-purity 625, 631
indirect MB (*see* scheme 26.4a, 26.4b, 26.4c)
interaction of two temperature controllers 357–362, 621–622
internal reflux control, refinery fractionator 369, 387, 626–627
internal vapor controller (IVC) 371–372, 628
inverse response 357, 362–367, 625–626
material balance 357, **364–365**, 371, **620–628**
on-off 624–625
reboiler swell 367–368, 626
Richardson's law 357, 369–370, 622, 624, 626–627, 640
scheme 26.3 362–363, 365–367, 625–626
scheme 26.4a 359–360, 364, 368, **621–623**, 625–626, 628
scheme 26.4b 357–360, 364, 369–370, **623**
scheme 26.4c 364
scheme 26.4d 364, 369–370, **624–625**, 640
scheme 26.4e 357, 362–368, **625–626**
side draw, liquid 357, 626–628
side draw, vapor 369, 371–372, 628, 634
side stripper 369–370, 626–627
small bottoms flow 362–363, 368, 622, 625
stripping column 371, 624
two composition 357–362, 621–622
Control, composition
analyzer 365, 373–376, 625, 628–629, 631, **633–635**, 643, 645
analyzer on next tower 635
analyzer/temperature cascade 373–376, 621, 635
measurement lags 373–376, 628, **633–635**
on-line availability 373, 633
sample location 633–635
two composition 357–362, 621–622
vapor pressure controller 365
virtual analyzer 365, 373, 628–629, 632–634, 647
Control, condenser (*see* Control, pressure)
Control, level
absorber 620
bottom (*see* Base level)
feed drum 644
interaction between base and reflux drum level controls, scheme 26.3 367
non-linear 645, 647
override 332–333, 606
reflux drum 379–380, 584, 624
refrigerant in kettle 380
by a small stream 357, 369–370, 622, 624, 626–627
at start-up 222–223, 310–311, 347–348, 356, 468
tower-to-tower 632, 646
tuning 644–645, 647
Control, override
low feed 72, 429
low-level 332–333, 606
low pressure 390, 392
low temperature 525
Control, pressure and condenser 377–388
compressor manipulation 581, 636
by condensate temperature 641
constraints 636
coolant recirculation 387–388, 640
cooling water throttling 344–346, 365, 377, **386–387**, 509, **640**
cooling water throttling on reflux drum TC, inerts injection/venting on PC 386–387, 640
fan blade pitch 640–641

Corrosion
condenser 335, 344, 377, 607
due to corrosive alkaline solution 552–554, 558, 587, 620
due to corrosive gases 605, 607
due to corrosive salts 558
due to excessive temperature 254, 344, 377
due to impingement on column wall 145, 456, 473–474
due to vapor channeling 467
due to water in hydrocarbon column with acidic components 25, 37–38, 40, 219, 414, 416, 427, 430, 558
-erosion 473, 527, 543, 558, 593
external 524
exposure to acids 84, 285–286, 438, 473, 578
inhibitor 241, 243, 440, 473, 544, 552–554, 562, 607
inhibitor does not reach 440
products (*see* Plugging)
reboiler 575, 605, 642
stress- 473, 578
valve trays 298–300
Corrosive chemical service 300, 588, 593
COS 399, 403
Crack (*also see* Gap) 593
Cracking 199–200, 403, 472, 574, 608
Critical temperature, pressure 398
Crude
assay 11–12
preheat 70, 427
solvent 550
stabilizer 101–104
switch 261, 487, 644
Crude fractionator
assembly problem 196, 292–293, 491–492, 498, 646
base level 149–152, 617, 644
black distillate 492, 562
blinding 536
bottoms 226
can 149–152
chimney tray problem 479–480
coking 272, 465, 574
condensation of overhead 18–20, 287–288, 608, 610–611
cooling procedure at shutdown 235, 507
corrosion 299–300, 558
corrosion inhibitor 562
crude switches 487, 644
cut points 68–70, 427, 626–627
damage 149–152, 225, 227–228, 292–293, 451, 488, 512, 514–515, 588
downcomer unsealing 186, 626
draw-off restriction 479, 484–485, 488
fire 235, 530, 579
flange leak 579
flash zone entrainment 475, 563
foaming 546
heater outlet temperature 465, 608, 643
hydrocarbon release/explosion 287–288, 499
instability 484
instrument tubing 646
leakage at draw-off 184–185, 486–487
off-spec products 465, 475, 492, 562, 576, 626–627, 644
overflash excess 369–370, 626–627
PA exchanger leak 576
PA heat transfer 479, 485, 487, 643
packing maldistribution 451–452
packing supports/holddown 439, 588
plugging 265–266, 451, 465, 514, 558, 562–564, 566, 574, 626
preflash drum entrainment 562
preflash tower bypassing 643
premature flood 488, 558
pressure, excessive 485, 608
pressure, fluctuation 611
product yields 26, 68–70, 184–185, 410, 427, 451, 464–466, 480, 485, 488, 515, 626, 644, 646
quenching at feed 464–465
relief 287–288, 610
residence time, hot resid 546, 608
salting out 505, 558, 564
separation between products 452, 465, 487–488, 562, 588, 626–627
simulation 18–20
short runs 574
stripping section 149–152, 241, 465, 491–492, 546, 566, 574, 608, 617
undrained/wet stripping steam 515
VCFC 466
venting 608, 610–611
wash rate control 369–370, 626–627
wash section drying 626
Cryogenic 81–83, 105–106, 152, 327–328, 434, 499
CS_2 399
CTC 400
Cumene hydroperoxide 519
Cumene oxidation 519
Cut point 68–70, 275, 446
Cutoff, Cutout (*see* Trips)
Cycle oil (*see* HCO, LCO)
Cycling (*see also* Hiccups; Oscillations) 113, 605, 623, 625, 632, 643, 645
Cyclohexane 59, 420

DAA 402
Damage (*see also* Base level, uplift; Condensation, rapid; Corrosion; Depressuring, rapid; Distributor, liquid; Downward; Explosions; Failures; Fires; Mechanical strength; Packing; Relief; Upflow; Valve trays; Vaporization, rapid; Water-induced pressure surges) 95–96, 215, 281, 287, **291–314**, 347, 435, 440, 442, 488, 542, 578, 584–585, **586–595**
  condenser 335, 344, 377, 607
  decanter 420
  demister 119, 594
  filter 545–546
  pump 218, 299–300, 495, 499, 507, 528, 590
  scratch marks 307
  storage tank 475, 536
  trays vs. packings 591–592
  two liquid phases 25, 517, 589
Data validation 1, 14–18, 90–95, 404–407
DCM 399
DEA (*see* amine)
Dead-headed pump 218, 499, 528, 582
Dead leg (liquid) **211–212**, 307–308, 328, **378–379**, 384, 512, 526, 536–537, 586, 616, 636
Dead pocket 225, 501, 513, 518, 526
Deaerator 429
Deasphalted oil tower 550
Debris 122, 193, 200–201, 203, 254, 269, 329, **494–495**, 537, 604
Debutanizer (*see also* Stabilizer) 4, 446, 456, 611, 625
  gas plant 127–128, 399, 455, 495, 608, 643, 645
  olefins 66–68, 299, 414, 424
  petrochemical 4
  refinery 3, 4, 38–39, **42–43**, 382–384, 410, 415–416, 435, 438, 442–443, 479, 485, 490, 505, 525, 528, 534, 537, 555, 575, 597–598, 607, 612, 624, 635, 637, 639, 641–642, 646
Decant oil 71, 221–222, 627
Decanter 9, 25, 51–52, 55–59, 413, **419–422**, 524, 538, 631
Decomposition (*see* Chemical reaction, decomposition)
Decoupling (*see* Control, advanced)
Deethanizer 468, 583
  gas plant 40–42, 50, 209–211, 374–375, 399, 414, 418, 498, 526, 635, 643
  olefins 53, 399, 525
  refinery 38–40, 80–81, **415**, **417**, 425, 427, 431, 468, 484–485, 489, 598, 600, 602–603, 630, 635, 640, 642, 645
Deep-cut (*see* Vacuum refinery tower)
Deflagmator (*see* Vent condenser)
Defoamer (*see* Foam)
DEG 400
Degassing 179, 183, 191, **335–340**, **484–485**, **489**
  time 180
Degradation (*see* Chemical reaction)
Dehexanizer 647
Dehydration
  acetic acid 9–11
  gas plant 50, 198–199, 241, 410–411, 424, 434, 438, 448, 454, 462, 541, 546, 576, 610, 616, 618
  organics 57–59, 631
  procedure (to dry tower prior to startup) 414, 512–517
  solvent 43–46, 420
Deinventorying 234, 535
Deisobutanizer (*see also* Alkylation; Splitter, iC_4-nC_4) 79
Delayed boiling (two liquid phases) 517, 589
Demethanizer
  gas plant 105–106, 327–330, 418, 426, 478–479, 581, 598, 603–604, 608, 619, 623, 634, 636
  olefins 68, 85–90, 97–100, 331–332, 404, 409, 425, 428, 431, 437, 441, 468, 496, 525, 605
Demister 118–119, 408, 437, 594
Densitometer (*see* Gamma scans time studies)
Density (*see* Specific gravity)
Deoiling column 560
Depentanizer •13–14, 261, 405, 441, 443, 492, 494, 596, 637
Deposit dissolving (*see* Washing)
Depressuring 503
  sudden (*see also* Implosion; Steam-water operation) 216, 292, **295–298**, 469, 589
Depropanizer (*see also* Alkylation) 564, 575, 638
  FCC 434, 524, 639
  gas plant 3, 42, 399, 437, 441, 455, 567, 599, 622
  hydrocracker 241, 555
  naphtha 524
  olefins 77–79, 196–197, 299, 414, 431, 510, 528, 560
  refinery 241, 434, 470, 524, 537, 580, 597–599, 605, 639, 642
  reforming 524
Desalter 427, 504
Desorption 503, 566
Desuperhearing 393, 430, 507, 552
Detonation (*see* Explosion)

Dew point
  of mixture 86, 348, 350, 427, 437
  water in gas 198–199, 414, 438, 448, 503, 576
Dewaxing 26
Diacetone alcohol 402
Diacetyl 59
DIB (*see* Alkylation; Splitter, iC_4-nC_4)
Dichloromethane 399
Diehl and Koppany correlation 338, 344
Diethyleneglycol 400
Diesel 68, 147
  cloud point 185
  flash point 26–27, 412
  in glycol dehydrator 546
  pumparound 288, 487
  separation from other products 452, 487, 627
  stripper 26–27, 470, 487
  from vacuum tower 199, 226
  yield 184–185, 292–293, 410, 427, 464, 485, 488, 627
Differential pressure (*see* Control; Pressure drop)
Differential temperature (*see* Control)
2,3 diketonebutane 59
Dilution steam generator 565
Dimerization 9–11, **401–402**
Dimethyl acetamide 520
Dimethyl formamide 241, 548, 578
Dimethyl sulphoxide (*see* DMSO) 521–522
Dip pipe (reflux drum) 307–308
Direct contact cooling (*see* Pumparound; Water quench)
Discharge (*see* Chemical release)
Disengagement (vapor from liquid) 179, 183, 191, **335–340**, **484–485**, **489**
Disk and donut trays 229, 272, **276–278**, 460, 463, 587
Dislodging (*see* Damage)
Dissolved water (in hydrocarbons) 38
Distillation boundary (*see* Azeotrope shift, Residue curve map)
Distillation Region Diagram (*see also* Residue Curve Maps) 9–11
Distillery 51–52, 435, 437
Distributed component (*see* Component, intermediate)
Distribution:
  baffles, in reboiler 597
  degraded by bed height 73, 88, 117, 438–439
  degraded by flood 94
  dual flow trays 97, 107, 442, 445
  gamma scans (*see* Gamma scans)
  from internal sampling 446
  modeling maldistribution 85–90, 446, 464
  multipass trays 81, 97, 134, 201, 441, **443–444**, 464, 489
  pressure surge due to poor- 517
  -quality index 109–110, 116–117, 119
  shed decks 108–110, 436, **444–445**, 464
  from surface temperature survey 112–113, 131–132, 446, 451–452, 458, 472
  vapor (*see* Distributor, vapor; Vapor maldistribution; Condenser)
Distributor, flashing feeds 455–456
  excessive velocities 456
  gallery type 127, 456
Distributor, liquid: 111–132
  aeration 123, 453, 455, 459
  assembly 111, **126**, 193, 202, **205–209**, **496–498**
  attachment 131–132
  bypassing 128–129, 454
  center-peripheral unevenness 137–138, 449, 497
  compartmentalized design 119
  complex 277
  damage 126, 131, 452, **457–458**, 472
  distribution quality 109–110, 111, 116–117, 137, **446–448**, 497
  drip point density 119, 446, 461
  entrainment 122, 202
  fabrication 111, 496–498
  feed entry 111–112, **124–126**, 205–206, 450, 452, **454**, 497
  feed pipe clearance 497
  feed pipe orientation 112, 124–126, 202, 205–206, 454, 497
  feed velocity 112, **454**, **457**, 497
  flashing feed 112, 442, 452, **455–456**, 496
  flow tubes 276–277, 459, 461
  flow testing (*see* Distributor, water testing)
  flow testing with actual fluid 459
  foam in distributor 453, 459, 557
  fouling-resistant 255, 257
  gasketing 113, 117, 202, 458
  hats 125–126, 206–207
  head, low 458
  head-flow relationship 120, 458–459, 497
  high performance 110, 116–117, 119, 206, **446–448**, 454
  high viscosity, surface tension 459
  hole not deburred 497
  hole diameter 122, 446, 448–450, 496
  hole pattern 112, 117, 449, **458–459**, 496
  hole punching direction 497
  hole/pipe area ratio (ladder pipe distributors) 452
  horizontal momentum 120, 453

Distributor, liquid (*Continued*)
inspection 112, 202, 205–206, 243, 496–498
installation 111, **126**, 193, 202, **205–209**, **496–498**
interchanging distributor panels 496–497
intermediate quality 117, 119
irrigation quality 109–110, 111, 116–117, 137, **446–448**, 497
ladder pipe (orifice pipe) 446, 449–450, 452, 458
leakage 113, 202, 456
level measurement 128, 446, 451
level oscillations 127–128, 451
liquid depth 451
misorientation 202, 205–206, 496
mixing 112, 117, 423, 446, **460**
MTS 447, 450
notched trough (*see also* Distributor, v-notch) 276–277, **447**, 567
open area (for vapor flow) 207
orifice area 111, 129, 452, 455, 457
orifice pan 112–117, 123–126, 129, 131–132, 138 206, **446–448**, 455, 457–459, 496, 565
orifice trough 120–124, 138, 243–244, **446–449**, 452, 497
out-of-levelness 112, 119, **131–132**, 449, **459**, 461
overflow 111, 115, 122–124, 127–128, **448–453**, 456, 497, 565
overflow tubes 129
parting box 205–206, 453–454
perforated pipe 121, 459
plugging 111, 497, 115–116, **121–122**, **255**, 257, 276–277, **448–451**, 497, 503, 508, 529, 532, 565, 567, 595
pressure drop 207, 455, 458
redistribution (*see* Redistributors)
removing distributor or part 243, 455
risers 122–126, 129, 454–455, 457
slug flow 127–128, 456
splashing 112, **454**, 497
spray (*see also* Spray) 95–96, 110, 121, 153–154, 272, 440, **447**, **450–451**, 455, **457**, 497, 571
standard 446
subcooling effects 428
trough (*see* Distributor, orifice trough; Distributor, v-notch) 450, 595
trough covers 122
troughs removed 243
turndown 458
two liquid phases 423
v-notch 117–120, 205–206, 447, 449, 456, **461**, 565
wall flow 459
water test (*see* Distributor, Water test)
weep holes 456–457
weir riser (*see also* Drip points; Distribution; Leveling) 456
Distributor, vapor (*see also* Vapor maldistribution; Vapor horn)
addition 446, 462–463, 474
chimney tray 137–138, 462
damage 155
diffuser 463
impingement on wall 473–474
initiating flooding 137–138, 478
inlet deflector baffle 110, 133, 277–278, 463, **465**
liquid-covered sparger 473–474, 507
liquid in sparger 473
mounted close to packing 463
pressure drop 113, 462, 464
sparger 107, 149–152, 266, 462, 465, 473–474, 507, 587, 600
tray 463
V-baffle 113–114, 462–463
vane distributor 106, 277–278
vapor horn 149–151, 168, 272–274, **464**, 615
Distributor, water test
at vendor shop 35, 112, 117, 120, **122–124**, 137, 454, 459, 461, 497
in-situ 112, 113, 117, 120–123, 205, 447, 453, 456
sprays 96, 110, 122, 272, 497
Diving bell effect 151
DMC (*see* Control, advanced)
DMF 241, 548, 578
DMSO 521–522
Double-locking nuts (*see* Nuts) 291, 314, 587–588
Downcomer (*see also* Assembly; Plugging):
aeration 250, 482, 489
area 73–77, 99, 245–247, 431, 442, 486, 493, 553, 555, 592
backup 80–81, 112, 250, 431, 482, 484, 489, 556
blocks 81
bowing 292, 309, 435, 592
braces 592
capacity prediction (*see also* Trays, hydraulic predictions) 73–77, 431
chimney tray 104, 137–138, 163, 478–479
choke 74, 76, 80–81, 100, 137, 245–247, 250, 431, 442, 493, 553, 555, 592
clearance 194, 250, **434–435**, **491–492**, 566
clearance, bottom downcomer 85, 492
damage 488, 592

false 101–104, 443, 452, 491
feed into 97–100, 435, 441, 493
fouling-resistant 566
gamma scan 240, 245–246, 266, 434–435, 441, 555
maldistributing inlet vapor 140
modified arc 574
obstruction of inlet 97, 176–178, 250, 433, **441**, 443, 489, 492–493
plugging 257–259, 261, 301, 322–324, 488, 491, 493, **564–565**, 574, 591
radius 433
residence time 76
sealing 73, 79, 81–83, 186, 266, 369, **434–435**, 491, 626–627
sizing 73–77, 431
sloped 85, 492
submerged 482, 489
trapout (*see* Draw-off)
trousers downcomer 140, 266–270
truncated 434
unsealed 73, 79, 81–83, 137, 266, 369, **434–435**, 491, 593
velocity 73–77, **245–247**, 250
water-testing 35
width (*see also* Damager; Foaming; Residence time; Seal pan; Tolerance 74–77, 592

Downpipe 74–75, 104, 135–138, 155–156, 451, 514
entrance head loss 137
missing 492
plugging 444, 464
undersizing 135–138, 451, 478–479
unsealing 137, 156, 436, 478

Downsizing 433

Downward tray damage 228, 289, 292, **300–313**, 512, 581, **590–592**, 594

Dowtherm 330, 604

Drain holes 128, 189, 456–457, 478, 514, 583

Drain lip 174

Draining:
condenser 335–340, 420, 586, 608–609
instrument lines 354
piping 211–212, 513, 534
rapid (base level) 301, 305, 469
reboiler (condensing side) 316, 333–334, 518, 606, 641–642
startup, shutdown, commissioning 261, 301, 512–517, 527, 592

Draw-off, liquid side-draw (non-chimney tray; *also see* Side draw):
aeration 179–183, 484–486
assembly 179, 492
damage 179, 221–222, 292–293, 488, 513
dead pocket 513–514
hydraulic restriction in sump 484, 488
interaction with PA return 181–183
internal piping 487
leakage 179, 184–185, 221–222, 292–293, 427, 432, **486–488**
level 184–185, 221, 348, 617, 647
line, from decanter 421
obstructing downcomer entrance 443, 489, 492
from one panel only, multipass trays 489
pan overflow 486
plugging 179, 488
poor design (no details) 179, 415, 489
poor venting 484
quenching of vapor bubbles 182
restriction 486
unsealing 186
vapor choke 179–183, 484–486
vortex 486

Draw-off, liquid to reboiler
to interreboiler 316, 485
leakage, draw to once-through thermosiphon reboiler 315–316, 319–322, 492, 597–598
mixing of hot and cold liquids 485
overflow, kettle draw 601
vapor choke 484–485
weir problem 598

Draw-off, vapor (*see also* Chimney tray; Control-side draw) 187–189, 490
draw box 187
liquid entrainment 187–189, 490
weeping into- 188–189, 490

Drip lip 173–174, 481

Drip points (*see* Distributor, liquid)

Dry, Drying (*also see* Dehydration)
internals 220
shed decks 436
trays 73, 428, **435–436**, 487, 489, 491, 562, 624, 626–627
after water wash 261, 503

Drying column 491

Dual flow trays 63, 106–107, 261, 442, 445, 505

Dump line (reboiler) 223, 320–321, 597

Dumping 27, 77–79, 82–83, 85, 316–317, 434

Duplicate column 77, 519

Dust 117–120, 220

Dynamic Matrix Control (*see* Control, advanced)

Dynamic simulation 620

Economizer (reboiler) 63, 322, 328

EDC 138, 239, 566

Eddies 278–279

Efficiency
  apparently poor 80, 614
  differences between binary pairs 17, 614
  effect of VLE errors (*see* VLE inaccuracies)
  estimate in simulation 1, 37, 86–89, 407–408
  Extrapolation to different process conditions 18
  measurement (from plant data) 14–18, **85–90**, 404–405
  measurement (from plant data): allowing for different internals 85–90
  packing (*see* HETP)
  prediction 2, 407–408, 438
  scaleup 407, 408
  tray, poor 80, 106, 404, 431, 435–436, 493–494
  turndown 80
Effluent minimization 253
EG 131, 400, 450
  Water-EG 456
EGEE (Ethylene glycol monoethyl ether) 43–46
Ejector 348–350, 610, 612, 636
Electrodialysis 554
Emission (*see also* Chemical release) 117–120, 124–125, 281, 411, 453–454
Empty (spray) section 95–96, **440**
Emulsion separation 520
End point 146, 284, 452, 646
Energy balance (*see* Heat Balance)
Energy savings 30, 61, 63–68, 424, 432, 621
Entrainer 9–11, 57–59, 422–423
Entrainment (*see also* Flood; Foam)
  bubble caps 435–437
  from distributor (*see* Distributor, liquid)
  flash zone, in refinery vacuum tower 2, 200
  rate measurement 2, 201
  reboiler return 474–475
  from reflux drum (*see* Reflux drum)
  from top of tower (*see also* Flood) 68, 141–143, 200–201, 405, 434–435, 453, 456, 459, 462, 481, 491, 493, 495
  tray to tray 408
  vapor in liquid outlet 146, 179–181, 191, 335–340, 342, 484–486, 492
  in vapor draw (*see* Draw-off, vapor)
  from vent condenser 335, 343–346, 611
Epichlorohydrin 522
Equalizing line (*see* Pressure balance line) 55–57, 380–381, 642
Equation of state 3, 38
  Peng-Robinson (PR) 5, 400
  Soave-Redlich-Kwong (SRK) 398
Equilibrium (*see* Condenser; VLE)
Erosion (*see also* Corrosion/erosion) 473, 527, 543, 599, 620
Erratic (*see* Instability)
Esters (*also see* Acetates; Butyl acetate; Ethyl acetate; Isopropyl acetate) 51, 421
Ethane recovery column 414
Ethanol (*see also* Alcohol) 43–46, 51–52, 59, 413, 423, 628, 633
Ethanolamine (*see* Amine)
Ether (*see also* EGEE; IPE; SBE) 629, 631
Ethyl acetate 498
Ethyl acetylene 645
Ethylbenzene 398, 584
  recycle tower 529
  -styrene 152, 446, 503, 624
Ethylene
  dichloride 133, 239, 566
  fractionator 28–30, 52–55, 61–63, 339–340, 358–360, 395–396, 399, 411, 418, 575, 606, 609, 621, 623, 633
  glycol 131, 400, 450, 456
  glycol monoethyl ether 43–46
  hydration 52
  oxide 233, 299, 446, 496, **518–519**
Ethyl mercaptan 399
Event timing analysis 147–149
Explosion (*see also* Water-induced pressure surges) 25, 233, 237, 281, 347–348, **518–528**
  acetylenes 521
  air/oxygen introduced 501, 519–520, 531
  air separation towers 524
  backflow 215
  butadiene 521
  C_1-C_4 hydrocarbon releases 216, 233–234, **524–528**
  C_5+ (heavier) hydrocarbon releases 499, 517, 527
  catalyst/metal fines 233, 518, 521, 523
  commissioning 215–216, 234, 499, 501, 503, 525–527
  concentration of chemical/hydrocarbon 4, 233, 503, 519–524
  dead-headed pump 528
  DMSO 521–522
  ethylene oxide towers 518–519
  excess temperature 233, 518–523
  hydroxylamine 523
  leak 499, 519, 521
  low base level 518–520
  nitro compound towers 520–521
  overchilling 216, 525–527
  overheating 518–523

packing fire-fighting releasing hydrogen 532–533
peroxide towers 519–520
pressure rise increases temperature 520–523
in storage 535
trapped hydrocarbons released 527–528
tray/packing damage (only) 499, 501, 531
unstable chemicals 518–523
violent reaction 523–524
water freeze 526
Explosion doors 512, 592
Explosion proof trays (*see* Heavy-duty design)
Explosive limit 4, 535
Extractive distillation 51, 59–60, 241, 303–305, 322–325, 413–414, **423**, 431, 496, 545, 547, 585, 598, 601, 628, 631–632, 634
foaming 60, 545, 547
kettle maldistribution 601
simulation 60, 413, 423
solvent to feed ratio 59–60, 423, 631–632, 634
solvent/reflux mixing 423
Extrapolation of
Crude oil distillation assay 11–12
flood correlations to high pressure 23–24, 409
test data (to other process conditions) 18
VLE (*see* VLE data extrapolation)

F-factor, hole 314
Failure: 287
brittle (*see* Overchilling)
condenser blockage 288, 520, 610
control valve 525, 538, 580
double- 580–581
exchanger tube 218, 281, 499, 526, **576**
external fire 580
feed 520
instrumentation 501, 619
level measurement (*see* Base level measurement; Measurement, misleading; Reflux drum)
loss of boilup 518, 524–526
loss of coolant 216, 234, 288, 522, 580, 610, 640
loss of electric power 218, 221–222, 287, 507, 538, 585
loss of instrument air 287
loss of steam 154, 304–308, 522
pump 215, 287–288, 295, 304–305, 387–388, 425, 471, 500, 507, 535, 584
reflux drum 580
relief 287–288, 537, 580
rotating equipment 215, 287
steam pressure regulator 523
vacuum generation 287, 297–298, 301, 303, 520, 590–591
Falling-film
absorber 95
evaporator 335–339
False downcomer 101–104, 443, 452, 491
Fastening trays, supports 131–132, 193, 294–295, 435, 471, 477, 480, 493–494, 579, **587–588**, 598
Fatalities 4, 215–216, 233, 281, 287, 347, 510–511, 517–521, 523, 525–527, 534, 536
Fatigue 314
Fatty acid 285, 533
FCC (Fluid Catalytic Cracking): 501
absorber 417, 489, 524
amine absorber 545, 553
debutanizer 43, 442, 525, 624, 635
depropanizer (*see also* Splitter, C_3-C_4) 434, 524, 639
sponge absorber 553
stripper 261, 417, 425, 432, 524, 600, 602–603, 635, 642
FCC main fractionator
blinding/unblinding 500
catalyst carryover 276–277, 449, 508
chimney tray bottleneck 173–174, 176–178, 479, 482
coking 133, 272, **276–279**, 449, 460, 463, 465, 564, **573**
condenser/overhead system issues 610, 636
damage 170–171, 221–222, 225, 227, 279, 452, 513, 517, 584, 587
distributor plugging 276–277, 449, 451, 508
draw-off bottleneck 181–183, 484, 489
drying below LCO draw 627
feed line flange leak 579
fire 530, 584
gas inlet velocity 277, 587
heat exchanger fouling 569
heat duty imbalance 70–71, 170–171, 426
hot oil enters water-filled drum 517
inlet baffle 133, 272, 277–278, 463, 465
LCO yield 627
LCO-DO separation 627, 646
lights in LCO 413, 455
liquid maldistribution 276–277, 447, 449, 451–452, 458, 460
on-line wash 220–221, 504–505
MVC updating 646–647
poor gasoline-LCO separation 176, 452, 484
premature flooding 170–171, 176–178, 430, 436, 452, 479, 482, 489

FCC main fractionator (*Continued*)
pressure control 510, 636
reflux drum carryover 584
reflux to section below HCO draw 627
salting out 221, 504–505, 560
shed decks/disc and donut replacement by grid 272, 277–278, 460, 463, 587
startup 510, 579
superheat 430
temperature spreads, slurry PA 463
vapor maldistribution 277–278, 463, 465
water in naphtha 416
wet gas rate 610
Feed (*see also* Distributors; Feed/reflux entry; Feedstock; Subcooling)
alternative 77–79, 431, 441
-bottom interchanger 47–49, 210, 310, 417, 498, 526, 576, 625, 643
bypassing the tower 67–68, 77, 79–80, 408, **424–425**
components (*see also* Component) 5, 11–14, 26–27, 35–55, 57–60, 240, 413–419, 425, 520, 609, 629–630, 634–635
condenser 62–66, 77–79, 429, 640
cyclone 552
decanter (or liquid phase separator) 5, 46, 55–57, 229–231, 417, 419
entry impeded 489
entry simulation 271, 406
filter (*see* Filter)
fluctuations 127–128, 294–295, 404, 429, 588, 592, 635–636, 643–645
impurity 35–55, 57–60, 414–419
introduction, startup 295–296, 310–313, 468, 579
interruption 71–72, 294, 429, 520
line 127–128, 513, 524, 527
multiple feeds 30, 66–68, 97–100, 283–284, 425, 441, 525, 629, 632
-overhead interchanger 68–69, 426
point location 30, 45, **79**, 138, 425, **441**, 454, 525
preheat (*see also* Crude preheat) 4, 40, 42–43, **47–49**, **61–66**, 81, 210, 281, 283–284, 301, 410, **417**, **424**, 428, 498, 526, **576–577**, 582, 605, 625, 641, **643**
quenching at feed zone 61, 133, 295, **429–430**, 464, 479
stopping, shutdown 71–72, 579
superheat 152, 272, 276–279, 409, 430
temperature 40, 42–43, 47–49, 63, 81, 284, 313, 415, **417**, 424, 426, **428–429**, 507, 578, 625, 640, 643
water to hot oil fractionator (*see* Water-induced pressure surges)
Feed/reflux entry to tray towers 97–110, 430–431
baffle **104**, 107, 108
changing to chimney tray 442, 478
discharge angle 106–107, 442
discharge velocity 101, 107, 201, 442, 444, 475
into downcomer 97–100, 435, 441–442, 493
flashing **97–106**, 108, 313, 434, 441–442, 475, 483, 643
free area at feed 106
hot 442
impingement on other internals 442, 475
inlet weirs at feed 104, 443
interaction with seal pan above 101, 475
maldistribution 108–110, 201, **443–444**, 643
mixing feed and reflux 423, 566
multipass trays 97, 101, **441–444**, 495
obstructing downcomer entrance 97, 441
pipe hole area 101, 107, 444, 495
pipe 101–108, 493, 495
plugging 108
routing to different location 77–79, 97–100, 431, 441, 493
tray removal or replacement by CT 441–442
tray spacing at flashing feed entrance 441, 443, 478
vapor-liquid segregation 107
Feedstock 14, 253
heavier 26–27
Fermentation 51–52
Field observations, tests (*see* Tests)
Film boiling 330, 604
Filter (*also see* Strainer) 110, 115–116, 120–121, 254–255, 448–450, 520, 543, 545–546, **551–555**, 569, 594
Finishing column 413
Fire 233, 287–288, 347
backflow 215
blanket 494, 529, 532
external, causing decomposition reaction inside tower 518–519
heating the column 580
inerting 501
leaks 236, 281, 519, 524–527, 579–580, 582
line rupture 524–527
liquid discharge to fuel gas 468
opening tower, no wash 534
opening tower while hot 533–534
oxidants presence 236, 501
release of trapped hydrocarbons 527–528, 534

relief 580
relief, atmospheric 287–288, 440
snuffing 236
tube rupture, fired heater 576
Fire, packing
air ingress 236, 501
hot work 236, **529**, 532
insufficient wash 235, 530–531
manhole opened while packing hot 532
pyrophoric deposits 235–236, **529–531**
random, steel 236
snuffing 236
structured, steel 234–236, **529–532**
temperature monitoring at outage 236, 530–532
titanium/zirconium 533
wire mesh 236, 501
Fired heater (*see also* Temperature, coil oulet) 26, 31–32, 48, 216–218, 225, 272, 360–363, 485, 499, 513, 573, 576, 584, 622, 645
Firewater 310–312, 532
First-of-a-kind process 2, 523, 542
Fixed valve (*see* Valve trays)
Flange: 218
cover 329–330
internal 457, 487
leak (*see* Leak)
Flash point 26, 284, 412, 447
Flash zone (*see* Crude tower; Vacuum refinery tower)
Flashing (*see also* Feed/reflux entry to tray towers; Vaporization)
Flood
control 72, 395–396, 426, 635, 644–645
correlation 2, 20–24, **409**, 438
cyclic (*also see* Hiccups) 85, 492
definition 90
downcomer backup 112, 250, 431
downcomer choke 74, 76, 80–81, 100, 137, 245–247, 250, 431, 442, 493, 553, 555, 592
due to assembly mishap 491–492, 497
due to bent valve legs 299
due to chimney tray 101–104, 134–138, 163, 176–178, 478–482
due to collector leakage 170–171
due to control issues 628, 630, 632, 637, 639, 644–645
due to cut point change 68–70
due to distributor overflow 111, 115, 122–124, 127–128, 448–449, **451–453**, 455
due to downcomer backup at draw pan 482, 489
due to downcomer inlet obstruction 176–178, 433, 489
due to draw vapor choke, restriction 179–183, 484–485, 488–489
due to entrainment of seal pan liquid 474–475, 480
due to feed entry 97–106, 441–443
due to feed interruption 71–72, 429
due to high base level 145, 147–149, 315, 355–356, **468–471**, 485, 494–495, 544, 555, 596, **599–603**, 617–619
due to inlet weir issues 433
due to incorrect pressure measurement 341–342, 354–355
due to incorrect reflux measurement 616
due to internal reboiler frothing or choked downflow 316, 602
due to leak 283–284
due to packing overfill 202
due to packing support/holddown 439, 497
due to poor assembly 85, 196, 202, 495–496, 547
due to preheater leak 283–284, 576
due to subcooling 428–429
due to tray supports left in tower upon packing 495
due to violent flashing 516
due to unsealed downcomer 73, 83, 186, 266, 369, 434–435, 491, 593
due to VCFC 138–139, 466
due to vapor maldistribution 467
effect of hole area 431, 466, 491
effect of seal pan 85
foam (*see* Foam)
heat integration spin 61, 68–69, 426
hydrates 52–55, 395–396, 418–419
massive entrainment from top 74, 76, 438
packed tower 90–95, 438
periodic (*see* Hiccups)
prediction 2, 20–24, **409**, 438
pressure drop 20–24, 90–95
in pressure services 23–24, 77, 409, 431, 437
system limit 338
temperature profile 90–95
testing 90–95
tray spacing effect 431
undersized downcomers 73–77, 245–247, 431, 555–556
vapor-sensitivity 250
vent condenser 335, 343–346, 611
wetted-wall column 338
Flooded reflux drum (*see* Control, pressure)

Flow:
improvers 544
induced vibrations (*see* Vibrations)
inerts flow rate 579
measurement 32, 86, 212, 351–354, 404–405, 412, 444, 573, 579, 615–616, 626–627
measurement from control valve opening 165
measurement from pump curves 165, 353, 616
measurement from spray nozzle pressure drop 165
negative 212
Flow path length
Long 139, 178, 466
Short 433
Zig-zag 433
Fluctuations (*see* Oscillations)
Fluidization of packing 595
Flushing (*see also* Washing) 121, 127, 230, 234, 504, 514, 527, 562
Flushing gaps (in inlet weir) 574
Flushing, line 416
Foam 237, **241–251**, 417
acid gas loading 552–554
aeration of downcomer liquid 250
agitation 557
antifoam 76, 112, 240, 242–245, 250–251, 303, **543–556**
antifoam, batch vs continuous 549–550
antifoam concentration 243–244, 304, 544, 548–549
antifoam dispersal 244–245, 303–304, 548, 550
antifoam mixing 245, 303–304
antifoam type 545, 549, 551–552
base 470, 544, 555, 557, 618
batch kettle 538
carbon beds 545, 550–554
contaminant concentration and removal 555
corrosion inhibitor promoting- 241, 243, 544, 552
corrosion products catalyzing 552–555
cyclones on feed gas 552
dirty aqueous solution 243–244, 552–554
downcomer 76, 112, 240, 245–250, 548, 553, **555–556**
downward velocity 240, 247, 250, 555–556
due to component accumulation 60, 240, 313, 545
effect of temperature 545, 547, 548, 552–553
extractive distillation 60, 545, 547
feedstock 543, **546–547**
filtration **543**, 545–546, **551–555**
gamma scans 245–246, 249–250, 543, 549, 550, 555–556
height 247, 470
HC condensation into aqueous solution promoting- 241, 243–244, **545–546**, 551–554
impurity 543–547, 549, 551–556
ion exchange 545, 554
make-up water 544, 552
operation 552, 554
organic acids catalyze- 555
packed towers 242–244, 544, 550, 556–557
packing distributor 453, 557
packing size 557
precipitation 543
reclamation 545, 552, 554–555
removing top trays 549, 556
Ross type (near plait point) 60, 240, 249–250, 417, 545, 547
service 112, 161, 241, 470, 538, 543–557
sight glass watching 313
solids catalyzing- 241, 243–244, 470, 543, 552–555
solution circulation rate 551
solution concentration 545, 552, 555
solution losses 545, 551, 554
solvent change 244, 553–554
testing 76, 243–244, 543–545, 547–550, 618
testing with level glass 243, 544, 548
testing in pilot, Oldershaw columns 543, 545, 547, 550
tray design 543, 545, 548–549, 553, 555–557
trays vs. packing 556–557
wash 552
Foamover (coke drum) 469
Fooling of level instrument (*see* Base level measurement)
Forced circulation reboiler 158, 316, 318–319, 495, 575, **599**, 617
outlet time restriction 599
pump seal leak 575
Formaldehyde 59, 220, 239–240, 448
Formic acid 240, 401, 497, 511, 578, 631
Fouling (*see also* Condenser; Packing; Plugging; Preheater; Reboiler; Thermowell) 237, 552, 554, 558, 620
cooling water system 285
inhibitor 544, 556
steam system 285
Four-pass trays (*see* Multipass trays)
Fractional hole area 73, 79, 262, 299, 303, 314, **431**, 435, 466, 491, 593–594
Fractionation issues 25–60
Fractionator (*see also* Crude fractionator; Vacuum refinery tower; FCC main fractionator; Alky main fractionator; Coker main fractionator;

Visbreaker main fractionator) 163, 180–181, 215–216, 236, 262, 299, 443, 512, 515, 528, 531
Free water (in hydrocarbons or in water-insoluble organics) 38, 40, 230–231, 249–250, 316, 414–417, 516, 609, 630
Freezing (*see also* Hydrates) 49–50, 387–388, 418, 526, 538, 579, 598
  point (*see* Melting point)
  in dead leg 526
  in valve or pipe 215, 234, 418
Freon 422
Froth 363
  height 363, 467, 470, 548
  regime 363
  scrubber 566
Frothing (*see also* Base level aeration; Distributor, liquid, aeration; Internal reboiler)
Fuel oil 262
Fungus 449
Furnace (*see* Fired heater)
Fusel oil 51–52, 413, 423, 628

Gamma scans:
  absorption ratio 567
  CAT scan 453, 497
  downcomers 240, 245–246, 266, 434–435, 441, 555
  downcomer unsealing 434–435
  draw-off bottleneck 486
  dual flow trays 106
  entrainment 176–177, 482
  flood initiation location **90–95**, 100–101, 105, 106, 141, **176–177**, 240, 249, 266, 435, 441, 482, 486, 505, 550, 595, 602, 616–617
  flood, packings 90–95, 438, 595
  flood, trays 101,105, 106, 176, 249, 266, 354
  foaming 245–246, 249–250, 543, 549, 550, 555–556
  frothing from internal reboiler 602
  guide tray modifications 433
  hiccups 415, 417
  high base level 149–151, 156, 269, 326, 468, 494, 599–600, 602, 617
  hole in packing 254
  incorrect diagnostics 199, 221, 262, 488, 494, 588
  instability 78, 127–128
  large diameter tower 129
  liquid distributor overflow 115, 452–453
  liquid level on chimney tray 141, 169, 177–178
  liquid maldistribution, trays 141, 201
  liquid maldistribution, packing 90–95, 115–116, 120, 126, 449, 451–453, 457, 459, 497
  liquid maldistribution, packing, seen only in CAT scan 453, 497
  missing manways 494, 588
  packed bed density 127, 254, 595
  packing missing 295, 472, 595
  pipeline 326
  plugging location 258, 469, 505, 558, 560, 563, 592
  plugging monitoring 567
  time studies 78, 127–128, 326, 468, 599, 617
  tray damage 266, 488, 512, 588
  vapor maldistribution 463
  vortex 161
Gap (mechanical) 119, 160, 184–185, 204–205, 456, 496
Gas (*see* Vapor)
Gas blanketing (*see* Inert blanketing)
Gas chromatograph 12–13, 19, 633–635
Gas cloud (*see* Vapor cloud)
Gas-lifting
  liquid leg 211–212, 305–308, 342, 534, 583
  reboiler liquid 320–321, 326–329, 598
  reflux drum liquid 305–308
Gas oil (*see also* AGO; HCGO; HVGO; LCGO; LVGO) 226
Gas solubility 5, 503, 541
Gasketing 113, 117, 184, 202, 218, 311–312, 435, 458, 477, 493, 575, 579, 582, 593
Gasoline (*see also* Pyrolysis gasoline; Naphtha; Stripper) 30–33, 40–42, 47–50, 176–178, 217, 241, 328, 360, 374, 381, 431, 451–452, 484, 505
Geyser 538
Glass plant 402, 502, 579
Glycerol 410, 616
Glycol (*also see* DEG; Propylene glycol; TEG)
  dehydrator 50, 198–199, 241, 410–411, 424, 434, 438, 448, 462, 541, 546
  distillation from other organics 25–26, 33–35, 237–239, 532, 539, 580
  heating medium 623
  losses 454
  regenerator 411, 454, 576, 610, 618
Graphical techniques (*see* Simulation, graphical troubleshooting techniques)
Gravity:
  high point in line 338, 474, 609
  line size (*see* Self-venting flow)
  settling (*see* Residence time)
Grid packing 147, 193, 271–278, 430, 436, 444, 447, 449, 460, 463–465, 477, 565, 571–572, 574, 587–588

Groundwater purification 559, 567
GS heavy water process 543
Guarantee 13–14, 124

Halogenated HC's (*see also* Chlorinated HC's) 379
Hammering 107, 146, 332, 385, 388–389, 393, 444, 474, 637–639, 645
Handhole 191, 204
Harmonic vibration region 594
Hat:
- chimney tray (*see* Chimney tray, hat)
- distributors/redistributors (*see* Distributor, hat)

Hatchway (bottom baffle) 494
Hazard (*see* Chemical release; Confined space; Damage; Explosions; Failure; Fire; Fatalities; Injuries; Leaks; Overchilling; Steaming; Washing; Water-induced pressure surges)
Hazop, hazard analysis 215, 288, 347, 523, 526, 528, 534
HBr 521
HCl 37, 95–96, 219, 239, 379–380, 402, 422, 437, 467, 555, 620
HCGO 512
HCO 71, 170–171, 436, 564, 627
Head (*see* Static head)
Heads 51–52
Heat:
- of absorption (*see* Absorption)
- balance (*see also* Vacuum refinery tower) 1, 68–72, 86, 163–171, 273, 281, 353, 404–405, 407, 452, 477–478, 616
- duty shift (*see* Pumparound)
- integration (*see also* Heat pumped columns, Multi-effect) 61, 63–70, 81, 216, 326–328, 507, 526, 575
- integration imbalance 61, 71, 216, **426–427**, 612
- integration spin 61, **68–69**, 426
- losses from column, auxiliaries 412, 429
- pocket (*see* Hot spot)
- -pumped columns 77–79, 606, 612, 621
- stable salts (*see* Amine, HSS)
- tracing 616
- transfer (*see* Condenser; Reboiler)
- transfer coefficient, direct contact 122–124, 595

Heater (*see* Fired heater; Temperature, coil outlet)
Heating the column 221–222
Heating by external fire (*see* Fire)
Heavier feedstock (*see* Feedstock, heavier)
Heavy boiler (*see* Component)
Heavy-duty design 225, 291, 294, 297, 309–313, 356, 457–458, 460, 512, 515, **587–589**, 592
Heavy ends column 368, 566, 619
Heavy water 241, 438
Hengstebeck diagram 2, 15–18, 405, 407
Henry's Law (*see also* Infinite dilution) 5
Heptane 398
HETP (*see also* Efficiency) 35, 37, 112–128, 131–132, 134–138, 201, 205, 254–255, 404, 408, 437–440, 495–497, 571–572, 574
- basic 88
- effect of reasonable degree of maldistribution 88
- effect of uncovered manhole 440
- too efficient 571–572, 574
- high in hydrogen-rich systems 73, 85–90, 437
- high in high tower to packing diameter ratio 437
- increase with reflux 136
- prediction 85–90, 437–438

Hexane 307, 522
HF 240, 402, 422, 582
HF alky MF (*see* Alkylation)
Hiccups (*see also* Accumulation) 25, 38–49, 57–60, 414–419
- effect of feed composition 49, 52
- effect of feed preheat 40, 42–43, 47–49, 415, 417
- gamma scans 415, 417
- simulation 46–47
- solution by changing feed temperature 42, **47–49**, 415, **417**
- solution by eliminating component 42, 50, 52, 59, 417
- solution by side draw 46, 55, 59, 313, 415–416
- solution by smaller temperature difference 38, 42–47, 414–417
- solution (failed) by tray retrofit 415
- temperature behavior 38–40, 43–49, 54–55, 415

High-capacity trays (*see* Trays)
High level (*see* Base level; Reflux drum)
High point (gravity lines) 338, 474, 609
High-pressure water jets 203, 220, 279
History of event 147–149
Holddown:
- cartridge trays 593
- packing 73, 134, 423, **440**, **460**, 497, 588, 595

Hole:
- area, distributor (packing) 111, 129, 452, 455, 457
- area, feed pipe (trays) 101, 107, 444, 495
- diameter/size (trays) 79, 257–261, 504, 563, 566
- F-factor (trays) 314

fractional hole area (trays) 73, 79, 262, 299, 303, 314, **431**, 435, 466, 491, 593–594
weep 128, 189, 456–457, 478, 514, 583
Horizontal momentum (distributor) 120, 453
Horizontal thermosiphon (*see* Thermosiphon)
Hot:
pot absorber 205, 241–243, 403, 447, 449, 473, 496, 502, 507, 543–544, 589, 620
pot regenerator 241, 378–379, 483, 503, 575, 620, 636
pot solvent contamination 403
spot 233, 236–237, 239, 518–519, 530, 574
tap 258, 266, 440, 564, 588
vapor bypass (*see* Control, pressure)
work 4, 234, 236, 503, **529**, 532
HSS (*see* Amine)
HVGO (*see* Vacuum refinery tower)
Hydrate 3, **42**, **49–55**, 395–396, 414, **418–419**, 528, 537
Hydraulic
gradient, chimney trays (*see* Chimney tray)
gradient, condensers 611
gradient, trays 133, 433
hammer 107, 146, 332, 385, 388–389, 393, 444, 474, 637–639, 645
loadings 2, 22, 26, 28–30, 54, 61, 70–72, 79–81, 99–100, 249, 314, 409, 428–431, 593–594
predictions (*see* Packing; Simulations; Trays)
Hydroblasting
heat exchanger tubes (inside/outside) 220
random packing 203
trays 279
Hydrocracker
debutanizer 443, 485, 528, 555
depropanizer 241, 555
preflash 640
vacuum 591
Hydrogen
bonding 8
chloride 37, 95–96, 219, 239, 379–380, 402, 422, 437, 467, 555, 620
component 294
fluoride (*see also* Alkylation) 240, 402, 422, 582
high hydrogen service, packed tower 73, 85–90, 437
peroxide 462, 519–520
sulfide (*see also* Amine; Caustic; Hot-pot; Stripper) 5, 27, 74, 126, 159, 171–172, 241–242, 266, 310–311, 374, 378, 411, 416, 432, 456, 500, 503, 540–541, 543, 550, 552–553, 589
Hydrolysis 2, 402, 403, 419, 520
Hydrostatic force (*see* Static head)
Hydrotest 198, 218, 261, 499, 508
Hydrotreater main fractionator 146–147, 283–284, 505, 641
Hydroxylamine 523
Hysteresis (*see* Control, valve)

I-beam (*see* Support packing; Support, trays)
IBP (*see also* Flash point) 32
Ice (*see* Freezing; Hydrates)
ICO 170–171
Immiscible (*see* Two liquid phases)
Impingement
in chimney tray 480–481, 483
in condenser 288, 610
at feed 442, 475
tower base 145–146, 151, **155–157**, 355, **472–475**
Implosion 216, 586
Impulse line (*see* Instrument connection)
Impurity (*see* Component; Foam; Off-specification)
Indole derivative 522
Inert(s): (*see also* Start-up)
blanketing, condenser 297, 335, 340–343, 346, 378, 580, 607–608, 641
blanketing, at failure/outage 297, 586
blanketing, reboiler 518, 524, 605
condenser (*see* Vent condenser)
control (*see* Control, pressure)
flooded reflux drum 381–382, 637
flow measurement 579
gas, commissioning 305–309, 501–502, 510, 605
gas, for fire snuffing 236
generation by reaction 233
injection 346, 378, 386, 393
instrument purge (*see* Instrument purge)
mixing solution 510
product 33–35
pressuring up 287
purge 340–343, 346, 501–502, 518
in steam 605, 607
storage 535
venting 287, 297, 335, 340–346, 381–382, 386, 501, 518, 521, 524, 605, 607–608, 637
Infinite dilution 3–5, 8–9, **400–401**
Infrared analyzer 633, 635
Inhibitor: (*see* Chemical reaction; Corrosion; Foam; Fouling; Polymerization) 541
Initial boiling point (*see also* Flash point) 32
Injuries 203–204, 215–216, 233, 281, 287, 347, 434, 510–511, 517, 520–523, 525–526, 533, 535–536, 583

Inlet (*see* Baffle; Feed; Feed/reflux entry; Distributor; Reboiler; Reflux; Vapor maldistribution)
  weep 258, 433, 466
  weir 73, 259, 433, 435, 574
Insecticide 522
Inspection (*see also* Liquid distributor) 193–213, 291, 477, 491–496, 592
  bottom seal pans 198
  packed tower 113, 119, 121, 126, 128, 132, 193, **202–207**
  trays 193–196, 299
Instability (*see also* Base level; Flooding; Gamma scans; Hiccups; Oscillation; Pressure; Reboiler; Slug flow; Vibrations) 95–96, 179–181, 281, 340–346, 362–368, 377–380, 472, 486, 491, 497, 545, 575, 581, 592, 602, 605, 620–647
  touchy column 351, 616
Installation (*see* Assembly)
Instrument (*see also* Base level measurement; Measurement, lack of; Measurement, misleading; Orifice plates; Start-up):
  calibration 347–350, 353, 405, 616–617
  connections 481–482, 487, 614–617, 646
  flushing 230
  meter tubing problems 347, 616–617
  mislocation 347, 354, 519, 521, 573, 615, 646
  plugged taps 152, 230, 257, 347, 416, 469, 522, 570, 582, 584, 614–615, 646
  purge 340–343, 348, 530–532, 570, 607, 617
  reading disbelieved, ignored, or not monitored 212, 218, 348–350, 354–355, 521, 619
Insulation 352, 412, 469, 519, 616
Interaction (*see* Control)
Interaction parameters 6–8
Interchanging (*see* Assembly; Distributor, liquid)
Intercooler 108, 415
Interface level 4, 229–231, 347, 416, 417, 419, 524, 582, 589, 619, 630
Intermediate draws (*see* Chimney tray; Draw-off)
Intermediate key component 4, 25, 35–55, 57–61, 240, 313, 335–339, 413–419, 521, 539, 619, 634
Internal:
  condenser 68–69, 344–346, 363–367, 421
  head 513
Internal reboiler: 220, 316, 363–367, **602**
  "bathtub" arrangement 220, 602
  flange leak 602
  fouling 220
  frothing 316, 365, 602
  overflow weir 602
Interreboiler (*see also* Draw-off; Thermosiphon) 28–30, 52–55, 61, 97–100, 316, 326–329, 411, 418, **424**, 443, 479, 485–486, **603–604**
  control issues 604, 621
  heated by tower feed 327–328
  hydrates intensification 52–55, 414, 418, 603
  inability to start 316, 603
  pinching 28–30, 411, 604
  sudden vapor generation 603
  vapor return from 316, 603
Interrupter bars (on valve trays) 258
Inventorying (start-up) 170–171, 295–296, 502
Ion exchange 545, 554
IPA 47, 408, 419, 422–423
IPE 408, 422–423
Iron fluoride 560
Iron sulfide 530–531, 535, 553, 560
Irrigation (*see* Distributor, liquid)
Isomerization prefractionator 600
Isoprene 545
Isopropyl acetate 419
Isopropyl alcohol 47, 408, 419, 422–423
Isopropyl ether 408, 422–423
Isostripper 187, 259, 432, 560

Jack hammer 277, 477, 482
Jet fuel
  draw tray damage 147
  flash point 412
  yield 147, 451, 480, 484
Joint (*see* Fastening; Flange; Gasketing; Leak; Seal welding)

Kerosene
  cut point 68–70
  off-spec 146–147, 576
  PA failure 287–288
  separation from diesel 452
  suppressing foam 547
  yield 427, 479, 646
Ketone (*see also* Acetone; MEK) 51, 59–60, 303–305, 541
Kettle reboiler
  backflow from tower to draw sump 269
  condensing tower overhead 339–340, 379–380
  draw compartment 269, 339, 618
  excessive circuit pressure drop 145, 266–270, 315, 325–326, 470, 495, **599–603**
  in series with thermosiphon 602–603
  liquid inlet to 599
  liquid supply to 269, 483
  maldistribution 601

overflow baffle 599
-return impingement 474
-return sparger 600
Key component 15–18, 33–37, 629–630
Kister and Gill equation 21–22, 90–95
Knockback condenser (*see* Vent condenser)

Laboratory (*see also* Tests):
analysis 12–14, 19, 86, 353, 405, 541, 608, 614, 645
column 60, 219, 240, 407–408, 545, 547
foam test (*see* Foam)
Ladder pipe (packing distributor) 446, 449–450, 452, 458
Lag 324, 359, 373–**376**, 389, 621, 623, 628, **633–635**
LCGO 512
LCO 176–178, 181–183, 221–222, 413, 426, 451, 458, 484, 545, 553, 627, 646–647
Leaching, ceramic packing 567
Leak, external (*see also* Tube leak): 281–286, 554, 575–579
accumulation in insulation 519
air into tower 233, 236, 237, 281, 285–286, 521, 531, 542, **579**
chemicals to atmosphere 281, 506, 511, 519, 524–526, 534–536, 558, 575, **578–579**, 580, 602
exchanger flange 506, 575, 602
in/out of tower from/to other equipment 228, 281–283, 499, 517, 520–521, 534, 536, **575–577**, 586, 605
oil (from machinery) 281, 438, 512, 535, **575**, 561, 562
piping 506
product to another product 147
pump seal 438, 512, 535, 575, 584
rate measurement 282–284, 575–577
reboiler/preheater valve 521–522, 582
at startup/shutdown 499, 506, 521, 535, 586
Leak, inside tower
chimney tray 163–176, 193, 311–312, **477–478**, 482, 486–487, 497
draw-off (non chimney tray), side draw 179, 184–185, 221–222, 292–293, 427, 432, **486–488**
draw-off to once-through reboiler 315–316, 319–322, 492, 597–598
flange 110, 457, 487
symptoms 169
testing (internal leak) 166, 168–169, 194, 312, 497
tray (*see* Weeping)
Lean oil (in natural gas plant):
absorber 32, 50, 209–211, 418, 498, 526
still 30–33, 47–49, 216–218, 360–362, 374–376, 381–382, 412, 417, 499, 526, 622, 637
stripper 590
Level (*see also* Base level; Chimney tray; Distributor, liquid; Gamma scans; Reboiler; Reflux drum):
gage 128, 487
glass 146, 243
interface 4, 230–231, 347, 416, 417, 419, 524, 582, 589, 619, 630
kettle reboiler 475, 618
measurement (*see also* Base level measurement; Distributor, liquid; Measurement, misleading) 128, 176, 347, 472, 481, 614–619, 647
switch 154
taps 149–152, 481
Levelness
distributor, packing 112, 119, **131–132**, 449, **459**, 461
trays 352
Light-boiler 13–14, 33–34, 42, 50, 316, 325, 328, 630, 635, 640
Lights depletion 316–317, 325
Line (*also see* Pipe; Reflux; Self-venting):
blowing 285–286
steaming 285–286
rupture, fracture (*see also* Tube leak) 233–234, 502, **523–528**, 537, 578, 583
Liquid (*see also* Base level; Collector; Condensate; Distributor; Downcomer; Draining; Draw-off; Level; Low liquid rate; Reboiler; Redistributor; Residence time; Thermosiphon; Washing):
carryover (*see* Entrainment; Flooding)
circulation 110, 216–218, 414, **512–517**, 526, 586, 590
density (*see* Specific gravity)
hammer (*see* Hammering)
leg **211–212**, 307–308, 328, **378–379**, 384, 512, 526, 536–537, 586, 616, 636
leg, gas lifted 211–212, 307–308, 328, 342, 379, 534
leg, pulling vacuum 342, 586
trap 379
trapped in valve 527–528
LLE (*see also* VLLE) 400
Local equilibrium 18–20
Long bed (*see* Packing bed length)
Long flow path 139, 178, 466

Louvers (*see* Control, pressure)
Low
  boiling point 13–14, 33–34, 42, 50, 316, 325, 328, 630, 635, 640
  concentration 3–5, 8–9, 400–401, 625, 631
  feed, level, pressure, temperature override (*see* Control, override)
  point (*see also* Liquid leg) 211–212, 378–379, 384, 513, 616, 636
  rate operation (*see* Dumping; Turndown; Weeping)
  reflux test 14–18, 404–405, 472
  tray spacing 251, 313, 431, 434, 441, 443, 467, 478, 489, 545, 553, 556
Low liquid rate
  dry trays 73, 428, **435–436**, 487, 489, 491, 562, 624, 626–627
  loss of downcomer seal 73, 81–83, 266, 269–270, **434–436**, 491, 626–627
  low tray efficiency 408, 432, 435–436
  packing (*also see* Vacuum refinery tower, wash section coking/dryout) 128, 262, 430, 450, 458
  recycle product to increase- 574
LPG 3–4, 30–33, 40–42, 47–50, 209, 217, 241, 374, 381, 431, 495, 537
Lube cuts
  color 199–200
  yield 199–200, 486
Lube oil prefractionator
  feed preparation 183–184, 487
  prefractionator 486
Lube vacuum tower (*see* Vacuum refinery tower)
Lubricant (on packing) 561
LVGO (*see* Vacuum refinery tower)

Maintenance 255, 299–300, 379
Major support beams (*see* Support)
Maldistribution (*see* Condenser; Distribution; Distributor; Vapor maldistribution;)
Management of change 288, 354
Manganese ion 559
Manhole 73, 277, 440, 495, **497**, **510–511**, 519, 522, 531–533, 592
Manometer 350, 613
Manway 194, 196, 198–200, 434, 493, 588
Mass balance (*see* Material balance)
Mass spectrometer 13
Material balance (*see also* Control) 1, 86, 163, 283, 353, 361, 405, 407, 452, 620
  plant overall 285
Materials of construction (*see also* Corrosion) 83–85, 184–185, 193, 234, 254, 437, 473, 483, 494, 554, 565, 578, 587
McCabe-Thiele diagram 2, **28–30**, **63–66**, 407, 411
MDEA (*see* amine)
MEA (*see* amine)
Measurement, lack of: 347–348, 452, 501, 521, 576, 600, 615
  base level 600
  at commissioning 216–218, 351, 468, 499, 576
  differential pressure 202
  flow 217–218, 452, 521, 627
  interface level 524
  level, reboiler side 222–223
  temperature 217–218, 521–522
Measurement, misleading: 212, 347–356, 444, 613–619
  composition 2, 12–13, 353, 405
  control valve opening 525
  flow 2, 32, 86, 351–354, 404–405, 412, 444, 573, 616, 633, 646
  interface level 229–231, 347, 416, 419, 524, 582, 619, 630
  level (*see also* Base level measurement) 149–152, 347, 475, 481–482, 616–617, 647
  poor location/positioning 347, 354, 519, 521, 573, 615, 647
  pressure 12, 341–342, 347–350, 354–355, 613
  pressure drop 613, 616, 646
  temperature 347, 405, 521, 538, 573
Mechanical strength (*see also* Support) 291–297, 310, 442, 452, 457–458, 515
MEK 313, 401
Melting 242–243, 292, 506–507, 532–533
  point 237, 300–301, 387–388
Mercaptan 399
Mercury compounds 3, 294
Metals in petroleum fractions (*see* Vacuum refinery tower, metals)
Methanol 2–8, 400
  absorption from gas 620
  -EG 131
  impurity 47, 422
  injection 3, 52, 396, 414, 418
  -n butanol – water 6–8
  reacting in column 240
  recovery from wastewater 418
  stripping 407, 472, 630
  -water 5, 8–9, 46–47, 431, 504, 610, 620, 630
Methylisocyanate 523
Metylethylketone 313, 401
Methyl mercaptan 399
Microelectric 625
Migration (packing through support) 73, 439, 533, 550

Minimum stripping 28–30
Mini-plant (*see* Pilot plant)
Misleading measurement (*see* Measurement, misleading)
Mist eliminator 594
 intermediate between packed beds 118–119
 intertray (*see also* Damage) 408, 437
Misting 437
Mixing (in redistributor) 112, 117, 423, 446, **460**
MOC 288, 354
Modified arc downcomers 574
Monomer and water separation from acid 420
Monoolefins/diolefins 548
Moore and Rukovena method 109–110, 116–117, 119
MSDS 523
MTS distributor 447, 450
Multicomponent distillation (*see also* Component; Composition profile) 30–37, 311, 407, **412–413**, 428, 614, 629–630
Multi-effect 63–66
Multi-feed 30, 66–68, 97–100, 283–284, 425, 441, 525, 629, 632
Multipass trays 80–81, 134, 187, 352, 436, 441, 443, 467
Multiplicity, temperature 30–33, 412
Multipurpose plant (*see also* Campaign) 12–13

Naphtha 19, 38, 42–43, 68–70, 181, 235, 246, 416, 431, 493, 512, 526, 529, 534, 553, 558
 feed preheaters 285
 heavy 177–178, 427, 516, 647
 reforming 586
 splitter 432, 450, 491, 565, 576, 601, 635
 sponge absorber solvent 553
 stabilizer 431, 493, 535, 607, 629
 stripper 27, 534, 647
Neural model (*see* Control, advanced)
Neutron backscatter
 base level 600
 draw sump level 269, 485
 liquid in pipes 269, 485
 reboiler condensate level 606
 time studies 441
Nickel in petroleum fractions (*see* Vacuum refinery tower, metals)
Nipple 285–286
Nitric acid
 absorber 467
 concentration 509
Nitrite 510, 541
Nitro compounds 233, 520–521
Nitrogen (*see* Inerts; Purge; Suffocation)
Nitrogen, rejection unit 426, 562, 643
Noise (*see* Hammering; Sounds)
Noncondensables (*see* Inerts)
Nonideality, VLE 1–11, 38–42, 45–47, **399–402**, 538
Nonkey (*see* Component)
Notch (*see* Distributor, liquid; Weir, picket fence)
NPSH 66, 316, 599
NRTL 6–9, 38, **400–401**
NRU 426, 562, 643
Nucleation (boiling) 596
Nucleonic (base level measurement) 326, 468, 619
Nuts 131–132, 193, 294–295, 314, 435, 471, 477, 480, 493–494, 579, **587–588**, 598
 double-locking 291, 314, 587–588

Obstruction of flow passages (*see* Assembly; Chimney tray; Downcomer; Draw-off, liquid; Seal pan)
Odor 124, 454, 537–538
Offshore 432, 540, 576
Off-specification product 3, 8, 12–13, 35–37, 57, 67–68, 90, 134, 146–149, 163, 179, 184, 187–189, 197, 210, 240, 284, 351, 354, 363, 395–396, 398–399, 418, 424, 427, 437, 447, 450, 456, 458–459, 462, 465, 468, 472, 487, 491–492, 497, 540, 544, 574–575, 589, 596–598, 602, 606, 610, 612, 616, 620–621, 626–628, 637, 644
Oil (*see* Absorber, hydrocarbons; Leak; Lean oil; Liquid circulation; Lubricant; Water-induced pressure surges)
 quench tower 226, 229–231, 259, 262, 284–285, 509
 rain 288
 -water separation trap 319, 597
 -water separator 5, 46–47, 229–231, 630
Oldershaw column 60, 545, 547
On-line analyzer (*see also* Control, composition) 405, 541, 633–635, 645, 647
Once-through thermosiphon 315, 319–322, 472, 492, 575, 597–599
Open area (*see* Chimney tray; Distributor, liquid; Hole; Support, packing)
Organic acids (*see also* Acetic acid; Formic acid) 528, 555
Organo-metallic compounds (*see* Vacuum refinery tower, metals)
Orifice (*see* Distributor; Restriction orifice)
Orifice plates
 condensation in line 352
 inadequate pipe in orifice run 351–354, 616
 incorrectly sized/installed 32, 412, 616

Oscillations (*see also* Base level; Boilup; Feed; Hiccups; Pressure, swings; Pressure, surges; Reflux, instability; Reflux drum; Thermosiphon; Vibrations):
- froth on tray 467
- preheat 625, 643
- product flow 55–57
- VCFC 139

OSHA 288
Out of levelness
- distributor, packing 112, 119, **131–132**, 449, **459**, 461
- tray 352

Out of roundness 160, 480
Outage (*see* Failure; Startup; Shutdown)
Overchilling 216, 234, 281, 468, 502, **507–508**, **525–527**, 611
Overcome by toxic gas 500
Overflow (*see* Baffle, reboiler; Chevron collector; Chimney tray; Distributor, liquid; Draw-off, liquid; draw-off, liquid to reboiler; Vacuum refinery tower)
Overhead vapor line 335–339, 488–489, 524
Overheating (*see also* Reboiler) 216, 221–222, 262, 292, 347, **506–507**, 518–523
Overpressure (*also see* Relief) 288, 475, 492, 502, 536, 581–582
Over-reboiling 32, 332, 628, 632
Over-refluxing 60, 66, 186, 351–354, 369, 411, 423, 452, 616
Override (*see* Control, override)
Oxidizing agents (*see* Amine, oxidizing; Potassium permanganate)
Oxygen deficiency 236, 532
Oxygen stripping 407
Ozone injection 559
Oxygenerated hydrocarbons 47, 240, 410

Packing (*see also* Distributor; HETP):
- aluminum 507
- assembly 113, 115, 117, 193, 202–205, 264, **494–496**
- bed length 73, 88, 117, 119, **438–439**, **571–574**
- bed limiter (*see* Packing, holddown)
- breakage 84, 193, 292, 437, 495, 533, 595
- carryover 588, 595
- ceramic 84, 193, 240, 264, 292, 437, 439, 462, 495, 567, 595
- chute and sock 113, 115, 117, 204
- cleaning 203–204, 220, 255, 529–530, 572
- collapsed uplifted bed 95–96, 589, 594
- compression 126, 152, 507, 566, 595
- construction supervision 496
- deformation 203, 240, 263, 495–496, 588, 595
- ETFE 566
- factor 21, 22, 93
- fire-resistant metallurgy 234–235, 533
- fires (*see* Fires)
- foaming 242–244, 544, 550, 556–557
- fouling (*see also* Distributor, liquid, plugging; Plugging, packing) 126, 202–204, 220, 254–255, 257, **263–265**, 440, 448–449, 453, 521, 530–532, **565–568**
- gaps in packing 119, 204–205, 496
- grid 147, 193, 271–278, 430, 436, 444, 447, 449, 460, 463–465, 477, 565, 571–572, 574, 587–588
- handling 202–204
- hills 113, 115, 205, 495
- high hydrogen service 73, **85–90,** 437
- high surface tension service 73, 438, 459, 557
- high viscosity service 73, 438, 459, 557
- holddown 73, 134, 423, **440**, **460**, 497, 588, 595
- hydraulic prediction 2, 20–24, 90–95, 409, 438
- hydroblasting 203
- inspection 113, 119, 121, 126, 128, 132, 193, **202–207**
- installation 113, 115, 117, 193, 202–205, 264, **494–496**
- installation supervision 193, 202, **204–205**
- melting 242–243, 292, 506–507, 532–533, 595
- migration 73, 439, 533, 550
- oil layer 202–204, 438, 529
- overfilling beds 202, 204, 497
- oxidation, rapid 236, 503, 532
- plastic 95, 242–243, 264, 292, 439, 495, 506–507, 559, 595
- PVDF 295
- raking 113, 117
- random 20–22, 84–90, 95, 112–124, 134–138, 149–155, 201–204, 220, 236, 240, 242–244, 263–265, 295, 327, 438–440, 446–463, 471–472, 494–497, 503, 505–507, 533, 557, 559, 565–568, 572, 588–589, 594–595, 625
- random, replacement by structured 134–138, 157
- replacement by sprays 95–96, 437
- replacement by trays 85, 126, 257, 262, 566
- removal 529–530

safety, handling and loading 202–204
screening 495
size change 20–21, 88–90, 263–264, 437, 448, 557, 567, 571
stacking random packings 264–265
shipping 495
storage 202–204
structured 73, 90–95, 122, 134–138, 147, 154, 188–189, 193, 208–209, 254, 335–339, 413, 438, 440, 446–454, 458–460, 462, 477, 496, 516, 521, 529–533, 556–557, 567, 571–572
supports (*see* Support, packing)
surface area 119, 122
titanium 533
tower to diameter ratio 437
type change 20–21, 84, 240, 439, 452, 462, 533, 567, 625
wet packing 264
wetting to improve efficiency 438
wetting to prevent fires 234–235, 530–532
wire mesh 22, 35–37, 204, 236, 254, 450, 458, 501, 532, 625
zirconium 533
Paint manufacturing 532
Pall rings (*see* Packing, random)
Pan (*see* Chimney tray; Distributor; Draw-off; Seal pan)
Parting box (*see* Distributor, liquid)
Passes, change of number (*see also* Multipass) 73, 444
Peng-Robinson (equation of state) 5, 400
Pentane/isoprene 545
Pentol 523
Perforation (*see* Distributor, liquid; Hole)
Performance testing (*see* Test)
Peroxides 233, 519–520
cumene hydroperoxide (CHP) 519
hydrogen 462, 519–520
pH 220, 399, 555, 559
Pharmaceuticals 236, 408, 422–423, 431, 458, 472, 504, 522, 538, 625, 630
Phase
Phase diagrams 2, **9–11**, 407, **421–422**
Phenylethylamine 501
Phenol 400–401, 403, 435, 535, 563, 633
Picket fence weirs 196, 428, 433, 435
Pilot plant 2, 12, 59–60, 247, 403, 516, 521, 543, 545, 547, 557, 589
Pinch
composition 9, 15–18, 28–30, 66, 411
thermosiphon reboiler temperature **315–317**, 326–328, 332, 597, 606
Pipe, Piping (*see also* Distributor; Feed/reflux entry to tray towers; Flange; Leak; Line; Plugging; Self-venting flow; Thermal expansion; Threaded connection)
cutting 534
for hot vapor bypass 384
misconnected 213
oversized 543
shaking 456, 509, 527
spiral-wound 353–354, 616
supports 457, 506
thinning 286, 558
underground 212–213, 349
velocity, fouling service 344, 386, 444, 543
Plait point 60, 240, 241, 249–250, 545, 547
Plugging (*see also* Coking; Washing) 145, 215, 253–279, 558–574
antifoam 304–305
biological growth 559
bottom outlet 279
bubble-cap trays 258, 592
catalyst carryover 276–277, 449, 508
chimney tray 163, 402, 418, 430, 436, 464, 477, 482, 573–574
condenser 288, 550, 610
control valve (*see* Plugging, draw-off line)
corrosion inhibitor 562
corrosion products 84, 122, 253–254, 259, 262, 437, 440, 450, 490, 502, 530–532, 552–553, **558–560**, 564, 567
debris 84, 96, 122, 200–201, 254, 269, 329, 437, 491, 493–495, 537, 558, 604
demister 119
distributor 111, 497, 115–116, **121–122**, **255**, 257, 276–277, **448–451**, 497, 503, 508, 529, 532, 565, 567, 595
downcomer 257–259, 261, 301, 322–324, 488, 491, 493, **564–565**, 574, 591
downpipes 444, 464
drain 513, 534, 537
draw-off line/control valve 257, 416, 418, 475, 487, 494, 536, 558, 568
dust 119, 220
ejector 612
entrainment into towers 562
feed line 257, 570, 495
filter 110, 121, 255, 450, 520, 553
fungus 449
heat exchanger 50, 502, 569
freezing/hydrates 50, 52–55, 234, 301, 395–396, 418
head baffle 324
inlet weirs 258–259, 574

Plugging (*see also* Coking; Washing) (*Continued*)
- instrument connections 152, 230, 257, 347, 416, 469, 522, 570, 582, 584, 614–615, 646
- internal pipe 444, 495
- limited zone 257, 261, 264–265, **568**
- line 297, 418, 537, 539, 541, 543
- mist eliminators 119
- moving deposits 543, 560, 581
- mud 440, 502, 563
- oil 561
- packing 84, 93, 121, 127–128, 220, 236, 242–243, 254–255, 257, **263–265**, 435, 440, 448, 450, 494, 503–505, 508, 530–533, 559, **565–568**, 572, 574
- pipe distributor holes 108, 495
- polymer 220, 230, 253, 258–259, 261, 264, 322, 414, 450, 543–544, **560–561**, 564, 604
- precipitation (*see* Plugging, salt)
- reboiler inlet/inlet line 269, 300, 326, 527–528, 537, 596, 604
- reboiler tubes, head 220, 324, 535, 552, 559
- relief device 582
- salt 122, 220–221, 259–261, 416, 427, 448–449, 504–505, 520, 541, 543, **558–560**, 563–566
- scale (*see also* Plugging, corrosion products) 253, 259, 450, 491, **558–559**, 564–565
- shed decks 279, 436, 444
- sludge 261–262, 521, 532, 567
- solid agglomeration 122, 543
- solidification 507, 509
- solids in feed 115–116, 121–122, 253, 543, **561**
- spray nozzles 110, 122, 235, 430, **449–451**, 497
- stagnant hot spots 574
- static mixer 245
- strainer 279, 438–439, 495, 502, 538, 576
- tars 561–562, 566
- trays 66, 395, 427, 435, 469, 502, 504–505, 512, 558–563, **565–566**, 574, **581–582**
- trays active areas 257–261, 304–305, 466, 563
- valve 215, 234, 416
- valve trays 52–55, **257–262**, 322–324, 564
- vent line 537, 569, 605
- weep holes 514, 583
- weirs 574

Pockets (*see* Liquid legs)

Poly-ol (polyalcohol oligomer) 241, 438, 459

Polymerization 2, 220, 230, 258, 264, 322–325, 393, 450, 518, 537, 540, 542–544, 547, 560–561, 566, 604
- inhibitor 544, 547, 560

Polyphosphate 559

Popcorn polymer 537, 566

Pop-out (floats out of valve trays) 258, **298–300**, 587, **590**, 593

Potassium
- bicarbonate 543
- carbonate (*see* Hot pot)
- hydroxide 563
- permanganate 531
- sulfate 523

Precipitation (*see* Plugging)

Precooling 62–63, **424**

Preflash
- drum 464–465, 562
- tower 241, 246–247, 544, 550, 643
- vacuum tower 226

Preheater (*also see* Feed preheat; Tube leak) 40, 42–43, 47–49, 61, 209–211, 281, 283–284, 309, 410, **417**, **424**, **428**, 526, **576–577**, 582, 605, 625, 641, 643
- bypass 47–49, 284
- control 61, 377–378, 410, 625, 641, **643**
- fouling 310
- leak 281, 283–284, 526, **576–577**, 582, 605

Pressure (*see also* Control, pressure; Control, temperature; Relief; Vacuum)
- back- 211–212, 379
- balance line 55–57, 380–381, 642
- flooding in pressure services 23–24, 77, 409, 431, 437
- influence on reaction 233
- low 196, 575, 583
- measurement 12, 347–350, 354–355, 613, 615
- over- 288, 475, 492, 502, 536, 581–582
- partial (in stripping) 412
- raising 81–83, 427, 434, 553, 608, 610
- rapid fall (*see also* Depressuring) 216, 292, **295–298**, 307, 469, 589
- rapid rise (*see also* Pressuring) 221, 292, 297, 417, 502, 577, 580, 606, 610
- reduction 432
- spikes 228
- structured packing in pressure services 73, 138
- surges, non-water-induced (*see also* Downward; Reboiler; Upflow) 335–339, 587, 589, 594, 609
- surges, unspecified cause 587–588
- surges, water-induced 215–216, **225–231**, 291, **512–517**, 636
- survey 21, 258, 341, 481, 488, 492, 515, 564, 588, 602

swings (*see also* Control, pressure) 95–96, 211, 335–339, 340–343, 360–362, 377–387, 581, 596–597, 599, 609, 611, 636–639
transmitter below tap 354–355
Pressure drop
backward 303, 354
condenser circuit 610
correction for static vapor head 134
excessive (*also see* Vacuum refinery tower) 90, 106, 147–149, 176, 182, 220, 243, 261, 341, 441, 468–470, 488–489, 505, 507, 512, 520, 530, 542–543, 558–560, 563–567, 592, 615–617
fall of 295, 512
flood – (packing) 20–24, 90–95
fluctuations 77, 146, 218, 325–326, 431, 550
kettle reboiler circuit 145, 266–270, 315, 325–326, 470, 495, **599–603**
vs. load relationship **90–95**, **136**, 249, 371, 438
low 149, 196, 307, 352
measurement 613, 615–616, 646
overhead vapor line 337
packed columns 90–95, 134, 136, 438
reduction 262, 553
rise of 44, 52, 68, 90, 99, 134–138, 147–149, 230, 249, 262, 304, 324, 571, 616
side stripper overhead line 488–489
specification 139
survey 21, 258, 341, 481, 488, 492, 515, 564, 588, 602
Pressuring 216, 305–308, 311, 503, 590
sudden 221, 292, 297, 502
Process water stripper (*see* Stripper)
Product column 33–35, 237–239, 539
Product contamination 147
Product loss 215, 501
Product recovery, low 38, 40, 43, 99, 163, 167–170, 172–173, 176–179, 184–185, 239, 271–276, 343, 359, 369, 386, 426–428, 432, 434, 445, 452, 478, 480, 487, 496, 539, 541, 609, 623, 626–627, 629, 631–635, 641, 644
Propanol (*see also* Alcohol) 43–46, 51
Propargyl bromide 522
Propylene fractionator 2–4, 212–213, 343–344, 351–354, 405, 418, 445, 460, 494, 498, 502, 524, 563, 611, 616, 621, 623
Propylene glycol 450
Protective clothing 258
PSV (*see* Relief)
Puking (*see* Hiccups)
Pulse (*see* Tracer)
Pump
cavitation 146, 160, 163, 169, 171, 179, 182, 228, 285–286, 347, 427, 438, 466, 474–477, 481, 484–485, 487, 500, 502, 535, 584
curves 353
damage 218, 299–300, 495, 499, 507, 528, 590
dead-headed 218, 499, **528**, 582
failure 215, 287–288, 295, 304–305, 387–388, 425, 471, 500, 507, 535–536, 584, 589
hydrate 537
power measurement 353
seal leak 438, 512, 535, 575, 584
strainer (*see also* Filter) 279, 438–439, 449, 495, 502, 576
suction loss (*see* Pump cavitation)
trip 279, 535
valve floats, packing, in suction 258, 299, 438–439, 590, 594
water pocket 225, 228, 512, **514–515**
Pumparound
chemical/petrochemical towers (*also see* Caustic; Oil quench; Water quench) 117–120, 286, 466, 477, 536
heat duty maximization 186, 426, 484, 487
heat duty shifts between pumparounds 68–71, 164–166, 170–171, 285–286, 426, 452, 643
heat duty loss 182, 427, 451, 485
poor location 427, 487
refinery fractionators **68–71**, 164–171, 181–183, 186, 196, 261, 288, 409, 426–427, 452, 477–478, 480, 484–487, 504–505, 512, 514, 531, 558, 576–577, 627, 643
refinery, other towers 310–313, 485
restricted circulation 427, 443, 485–486, 495
return, interaction with draw-off 181–183, 484
side draw location relative to PA 426, 484, 487
Pumpback, Pumpdown 369, 487–488, 626–627
Pumping trap 393
Punching (holes in packing distributor) 497
Purge (*see also* Venting) 51–52
at commissioning 216, 234, **501–502**, 516
gas interchanged 501
instruments 340–343, 348, 530–532, 570, 607, 617
insufficient 501
steam 516
Push (for tray liquid) 195
PVC 574
Pyrolysis gasoline 13, 66–68, 229–231, 262, 285
Pyrometer (*see* Surface temperature survey)
Pyrophoric deposits 234–236, 529–533, 535

Quench 471
- desuperheat 430
- at feed zone 61, 133, 295, **429–430**, 464, 479
- oil quench tower 226, 229–231, 259, 262, 284–285, 509
- synthetic fuels 527
- VCM 527
- waste gas 511
- water quench tower 108–110, 122–124, 140–143, 200–201, 229–231, 295, 399, 405, 440, 445, 449, 481, 495, 588

Radial temperature survey (*see* Surface temperature survey)
Radioactive
- contamination 414
- decay 414
- tracer (*see* Tracer)

Radioisotope (*see* Tracer)
Radon 414
Raffinate 13, 512
Rain of oil 288
Rapid
- condensation 292, 295–296, 305, 310–313, 586, 591–592, 638
- depressuring 216, 292, **295–298**, 469, 589
- pressuring 221, 292, 297, 502
- reflux drum emptying/filling 295–296, 305–308, 335–339, 341, 584, 609, 638
- upflow 291, 294, **589**
- vaporization 152, 289, 295, 297–298, 313, 452, 516–517, 584, 589, 603, 636

Rayleigh condensation **18–20**, 166, 295–296, 335, 412, 609
Reaction (*see* Chemical reaction)
Reboiled deethanizer absorber (*see* Deethanizer, refinery)
Reboiler (*see also* Baffle, reboiler; Control, reboiler; Draw-off, liquid to reboiler; Falling film; Fired heater; Forced circulation; Internal; Kettle; Thermosiphon; Tube leak) 315–334, 596–606
- aluminum plate 327
- bottom product off-take 596
- cleaning 322–324, 527, 535
- condensate removal 316, 333–334, 518, 606, 641–642
- condensate subcooling 332, 334
- distribution baffle (horizontal reboilers) 597, 601
- drainage (condensing side) 316, 642
- draining (boiling side) 328
- dump line 223, 320–321, 597
- economizer 63, 322, 328
- film boiling 330, 604
- fire 535
- fouling 310, **322–325**, 391–393, 434
- gas injection (*see* Thermosiphon)
- heat transfer 596, 599, 601, 604–606, 642
- heated by bottoms 322
- heated by feed 63–66, 326–330, 507, 575
- hot spot 239, 518, 598
- inert blanketing 518, 524, 605
- inerts injection (steam side) 393
- inlet line blockage 269, 300, 326 596
- inlet temperature 596
- isolation at outage 521
- limitation 160, 196, 321–325, 330, 334, 429, 434, 492, 494, 596–598, 602, 604, 606, 642
- liquid level (condensing side) 331–333, 378, 605–606, 641–642
- LMTD 222, 315, 330, 597, 604, 606, 623
- loss of condensate seal 331–333, 378, 605–606
- nucleation 596
- opposing return lines 474
- puffing (thermosiphon) 597
- return inlet (*see also* Base level) 133, 145–157, 325–326, 467, 472–475, 599–601
- return line 158, 325, 434, 472, 474, 489, 525, 537, 599–601
- startup 152–155, 301, 311, **320–321**, 326–329, 591, 597–598, 602–603
- starving of liquid 156–157, 198, 316, **319–325**, 483, 596–598, 602
- surge (*see* Thermosiphon)
- swell 367–368, 626
- swinging 474, 596–597, 623, 642
- temperature difference, too large 330, 604
- temperature difference, too small 597, 606
- temperature pinch (*see* Thermosiphon)
- vapor supply to- 66, 429, 606
- venting (condensing side) 316, **605**, 642
- vibrations 599

Reclaimer 545, 552, 554–555
Rectifier 71–72, 152
Recovery (*see* Product recovery, low)
Recycle 61
- effect on accumulation 49–51, 399, **418**
- product to tower 574, 598
- product to reactor 79, 410
- promotes slow reaction 419
- reduction 425

Redistributor, Redistribution (*see also* Distributor, liquid)
- -collector combination 141–143, 452
- dual flow trays 445

frequency 73, 112, 117, 137–138, **438–439**, 445
mixing 112, 117, 423, 446, **460**
Reflux (*see also* Distributor, liquid; Feed/reflux entry)
balancing in two-stage condensation 427
difference between two large numbers 369, 487, 626–627
excess 60, 66, 186, 351–354, 369, 411, 423, 452, 616
gravity 189–191
high/low test 14–18, 404–405, 472
insufficient 25–27, 30–33, 353, 367, 369, **410–411**, 412, 428–429, 435, 458, 491, 606, 624–627, 636
instability 189–191, 366–367, 575, 606, 620–621, 624–627, 647
line 189–191, 524, 528
minimization 173
sensitivity to 351–354
splitter 421
-temperature dependence 30–33, 87–89, 404
total (*see* Total reflux)
vs. absorption 40–42, 425
water into HC tower 42, 225, **229–231**, 416, 420, 427, **516–517**, 558, 636
Reflux drum
aeration 189–191
agitation of surface 638
boot 4, 38, 42, 221, 416–417
boot, level 221, 416–417
carryover from 221, 468, 584, 619
chimney tray 336–337, 344–346
dip pipe 307–308
elevated 382–384, 637–639
filling/emptying fast 295–296, 305–308, 335–339, 341, 584, 609, 638
fire 580
flooded (*see* Control, pressure, flooded drum)
gas back-lifting drum liquid 305–308
glass 502
inlet pipe 586
interface level measurement failure 524, 582
level 179, 189–191, 229–231, 335–339, 468, 490, 522, 524, 584, 609, 619, 638
level control 379–380, 584, 624
level swings 55–57, 189–191, 335–339, 638
plugged outlet line 179, 490
relief 580, 583, 586
reverse flow 305–308, 500
temperature control 386–387
temperature difference 384
undersized outlet line 191, 490
venting 381–382, 607, 637
volume 295–296
Regenerator (*see also* Amine; Hot pot)
Reid vapor pressure 15, 645
Relief (*also see* Depressuring; Failure)
atmospheric 287–288, 440, 535, 537, 583, 610
block-off 583
capacity 288, 580–581
commissioning 502, 537
control behavior 580–581, 636
double failure 580–581
downstream unit 475, 492, 524, 582
due to unexpected lights 287, 581
due to unexpected second liquid phase 287, 636
due to plugged packing 440
frequency 581
inert blanketing 580
instrument action 580
line to valve 537, 582
liquid discharges 287–288, 440, 525, 537
minimizing 580
moving deposits 560, 581
oversizing 289
plugging 582
pressure 581
rates 580
reboiler, preheater 582
requirement 287, 580
setting 581, 583
sizing 287–288
steam purges on relief valve 516, 583
tower overpressured 288, 502, 536, 580–582
tray damage 289, 581
vacuum 305, 308, 580, 586
valve incorrectly set 287
valve lifting 221, 288, 440, 537, 583, 610, 636
Residence time (*see also* Vacuum refinery tower)
chemical reaction 237
chimney tray 464
for degassing 338–340
downcomer 76
excessive, causing coking, foaming 464, 546
two liquid phase separation (*see also* Chimney tray, water removal from HC) 55, 101, 319, 415, 546
Residue, organics **27–28**, 33–35, 237–239, 522–523, 538, 596, 604, 615
Residue, (vacuum "Resid"), refinery 11–12, 167–170, 172–173, 175, 271–272, 406, 608
flash point 199–200
Residue curve map 2, **9–11**, 407, **421–422**
Residue yield (refinery) 11–12, 406
Restriction orifice 66, 443, 462, 570, 580

Retray (*see* Tray)
Reverse diffusion 90
Reverse flow
  through condenser 212–213
  process lines 215–216, 218–219, 233–234, **500**, **534–536**, 605
  steam condensate to reboiler 389–393, 641–642
  through pump 387, 500, 535, 589
  reflux drum liquid 305–308
  in trays 216, 289, 292, 303–313, 512, 581, 590–592
Rings (*see* Packing, random)
Riser (*see* Chimney tray; Distributor, liquid)
Root Cause Failure Analysis 552
Rosin 285
Ross type foaming 60, 240, 249–250, 417, 545, 547
Rumble (*see* Sounds)
Runaway reaction 518–528, 537
Rundown lines, gravity (*see* Self-venting flow)
Rupture
  disc 289
  heater/exchanger tube 218, 281, 499, 526, **576**
  line (*see also* Tube leak) 233–234, 502, **523–528**, 537, 578, 583
  storage 475, 536
RVP 15, 645

Saddles (*see* Packing, random)
Salting out (*see* Plugging)
Salt dispersant injection 560
Sampling (*see also* Control, composition)
  bomb purging 405
  from inside tower 45, 59, 88, 147, 446
  Joule-Thompson condensation 405
  reproducibility 633
  water ex-packed bed (*see* Distributor Water test)
SBE 400–401
Scaleup
  from laboratory column 407–408
Scream (*see* Sounds)
Screen trays 401, 435
Screens (in filter) 121, 438–439, 528
Scrubber 117–121, 124–125, 220, 436, 447, 449, 453–454, 494, 620
Seal
  condensate (reboiler) 331–333, 378, **605–606**
  downcomer 73, 79, 81–83, 186, 266, 369, **434–435**, 491, 626–627
  loop 55–57, 95–96, 336–339, 511, 538
Seal pan
  below bottom tray 85, 140, 321, 474–475, 489, 492, 514, 598
  above chimney tray 173–174, 480–481
  common with reboiler draw 197–198
  at downcomer trapout 186, 478, 598
  obstructing downcomer inlet 177–178, 250
  at tower feed 101–104, 442, 478
Seal welding 166, 169, 173–176, 179, 184–185, 311–312, 432, 477, 481, 486–487, 598
Sealant 185
Sealing (*see* Seal, downcomer)
Seamed (*see* Pipe, spiral wound)
Sec-butanol 400–401
Sec-butyl ether 400–401
Selexol 446, 456, 503, 644–645
  absorber 645
  hydrogen sulfide stripper 456, 644
Self-locking nuts (*see* Nuts)
Self-venting
  downpipes 137, 451
  flow, correlation 137, 179–181, 191, **335–340**, 342, 451
  lines 179–181, 191, 335–340, 342, 484–485, 509, 608
Settling time 229–231, 319, 415
Sewer **4**, 43–44, 51, 334, 389, 392, 474, 538
Shear clips 512, 587
Shed decks (*see also* Angle irons, Baffle trays) 108–110, 262, 272, **276–278**, 436, 444–445, 464, 566
  replacement by grid 272, 436, 444, 587
Shell-side condensation 18–20
Shutdown 215–223, 285, 291–292, 300–302, 313, 351–354, 426, 469, 488, 495, 499–511, 521, 535, 551, 560, 575, 578, 586, 589, 592, 620
  lean solvent pump 215, 500, 535–536, 589
  taking feed out 71–72, 579
  unnecessary 351–354, 616
Side draw (*see also* Control assembly)
  elimination of (in refinery fractionator) 427, 458
  fusel oil 51–52, 413
  liquid 5, 33–35, 47, 51–52, 59, 179–186, 285, 354, 413, 415
  location 38–40, 42, 413
  location, refinery main fractionator (*see* Pumparound)
  vapor 35–37, 187–189, 335–339, 413
  water 39–40, 42, 101–104, 415
Side reboiler (*see also* Draw-off; Thermosiphon) 28–30, 52–55, 61, 97–100, 316, 326–329, 411, 418, **424**, 443, 479, 485–486, **603–604**
Side stripper (*see* Stripper)
Sight glass 120, 189, 313–314, 467

Sieve trays 81, 171–172, 187, 218–219, 294, 313–314, 433, 466, 491, 504, 512, 556–557, 563, 574, 592–594
Silica gel 545
Silicone 545, 554
Simpson's rule 179–181, 191, 335–339
Simulated distillation 271
Simulation 1–24, 26
- azeotrope system 422
- bug 407
- characterization of feed components 1, 2, 11–14, 402, 406
- component accumulation 46–47
- convergence 8, 11, 46
- correct chemistry 1, 402–404
- diagnose unexplained mysteries 70–71, 199, 404, 405, 413, 420–423, 596
- efficiency estimate (*see also* Efficiency, measurement) 1, 14–17, 37, 407–408
- entrainment in vapor draw 407
- equilibrium in condensers 18–20
- feed entry 2, **406**, 413
- graphical troubleshooting techniques 1, 2, 6–11, **15–18**, **28–30**, **35–37**, 405, **407**, 411, **421–422**
- hydraulic predictions 2, **20–24**, 409
- leak, external 284
- leak, internal 221, 170, 406
- matching plant data 1, 12, 14–18, **85–90**, 283–284, 347, 399, 404–407, 420, 474
- misleading 494
- pumparound 409
- two liquid phases 2, 5–11, 60, 406, **420–421**
- unable to diagnose problem 488, 494
- validation using temperature-reflux dependence 87–89
- vapor/liquid loadings 2, 22, 26, 61, 409
- VLE 1, 3, 5–12, 17–18, 60, **398–402**, 404–405

Siphon breaker 308
Siphoning 55–57, 95–96, 180, 191, 305–309, 335–339, 456, **509–510**, 534, 609
Skimming 545–546, 552, 618
Skirt 511, 579
Slip plates (*see* Blinding)
Slop 225, 512, 577, 629
Sloped downcomer 85, 492
Slot area (*see* Valve trays, open slot area)
Sludge 521, 528
Slug flow (*also see* Thermosiphon) 127–128, 456, 611
Slugging 38, 42, 145, 147, 151–152, 155, 270, 378, 472
Slurry
- flash point 447
- line rupture 527
- pumparound 170–171, 272, **276–279**, 426, 436, 449, 460, 463, 569, **573**, 587
- recycle 436

Smell (*see also* Odor) 51–52
Soapy water/polyalcohol oligomers 241, 438, 459, 557
Soave-Redlich-Kwong (equation of state) 398
Soda ash 495
Sodium chloride 543
Solidification (*see* Plugging)
Solvent (*see also* Extractive distillation)
- deasphalting 241, 550, 557
- recovery 43–46, 313–314, 388–389, 400, 404, 408, 410, 422–423, 511, 520, 566, 586, 624
- residue batch still 241, 557
- wash column 550

Sounds
- banging 588
- hissing 499
- rumbling ("domino effect") 307
- scream 297

Sour water stripper 155, 259, 309–313, 474, 559, 566, 600
Sparger (*see* Distributor, vapor)
Spark 579
Specialty chemicals 90, 157, 254, 258, 300, 448, 502
Specific gravity 347, 355–356, 618–619
Spin (heat integration) 61, **68–69**, 426
Spiral-wound pipe 353–354, 616
Splash decks (*see* Shed decks)
Splitter
- aromatic isomer 122
- C_2 28–30, 52–55, 61–63, 339–340, 358–360, 395–396, 399, 411, 418, 575, 606, 609, 621, 623, 633
- C_3 2–4, 212–213, 343–344, 351–354, 405, 418, 445, 460, 494, 498, 502, 524, 563, 611, 616, 621, 623
- C_3-C_4 71, 134–138, 469, 478, 596
- iC_4-nC_4 (*also see* Alkylation DIB) 581, 601–602, 612, 630, 635, 639, 647
- C_5 diolefins 635
- ethylbenzene-styrene 152, 446, 503, 624
- hydrocarbon 441
- isomer separation 155–157, 308–309, 629
- naphtha 432, 450, 491, 565, 576, 601, 635
- petrochemical 261, 325, 433, 447
- raffinate 512
- xylene 362–363, 446

Sponge absorber 500, 553
Sponge oil 413, 455, 545
Spray condenser 340–343
Spray height 187

Spray nozzles (*also see* Distributor, spray)
coking 275
damage 154, 272, 451, 457, 472
entrainment 571
flashing 455
high pressure drop (*see also* Sprays plugging) 571
homogenous 110, 122
hot-tapping 440, 588
interference with supports 460
internals missing 497
oversized 447, 457
plugging 110, 122, 235, 430, **449–451**, 497
poor performance 447
poor spray pattern 447, 457
single- 121–122
spray angle collapse 430
testing 110, 122, 272, 497
Spray tower 95–96, **440**
Squeezing column 237–239
SRK equation of state 398
Stabilizer 14–18, 26, 405, 485, 576, 592
C_5-C_6 isomerization 218–219, 467
cat polymerization 525
crude 101–104, 482
diesel 26
naphtha 431, 493, 535, 607, 629
natural gas 425, 630, 645
reformer 220–221
Stability test, VLE calculations 6
Stacking rings on supports 264–265
Standby person 510–511
Startup (*see also* Base level, high; Total reflux) 81–83, 215–223, 291–292, 295–296, 300–302, 310–311, 440, 450, 468, 499–528, 535, 537, 575, 579, 583, 590–591, 597, 607
bringing feed in 295–296, 310–313, 468, 579
inert gas addition 305, 307, 311, 346
instrument problems 351, 354, 468, 499
inventory 170–171, 295–296, 502
level control, at start-up 222–223, 310–311, 347–348, 356, 468
pressure control, at start-up 313, 385, **510**, 583
procedure 311, 313, 499, 501–504, 506–508, 512–528, 534, 537, 579
reboiler control at start-up (*see also*, Control, reboiler) 388–393
relief at- 537
stability diagram (for downcomer sealing) 81–83, 434
total reflux (*see* Total reflux)
vacuum columns 297–298
without proper instrumentation 216–218, 351, 468, 499, 576
Static electricity 579
Static head (*see also* Gas lifting) 301, 308, 354, 592, 603
boiling point suppression 328, 589, 603
damage 592
kettle reboiler (*see* Kettle)
measurement 176, 481, 613
to overcome friction 56, 338, 342
over weir 599
pulling vacuum 342
in reboiler circuits 328, 334
Static mixer 245, 467
Steam (*see also* Ejector; Stripper)
cleaning of packing 255, 529–531
desuperheat 393, 507
emergency 154, 266, 304–308
generator 583, 607, 640
hammer (*see* Hammering)
inerts in (*see* Inerts)
pressure fluctuations 597, 625, 628, 644
system fouling 285
tracing 523
trap 412
-water operation 216, 266, 292, 295, 309–313, 429, **506**, 578, **591**
wet 184, 226, 349–350
Steaming 216, 309, 506, 508, 516, 521, 578, 586
side-draw line 285–286
Steamout 322–324, 594
Stepped trays 435
Sticking (valve trays) 257–258, **261–262**, 300, 322–324, 564
Still (*see also* Lean oil still) 63–66, 209, 495
Storage 244, 471, 475, 492, 536, 540–541, 544
Strainer (*also see* Filter) 279, 438–439, 449, 495, 502, 576
Stress corrosion 473, 578
Stress relieved 160
Stripper (*see also* Amine regenerator; Control assembly; Crude fractionator; Deethanizer; Hot pot regenerator; Methanol; Selexol)
air 407, 559, 566–567
ammonia 47, 241, 310–311, 407, 557, 581
aromatics unit 490
asphaltene 550
BTX 628
butadiene 632, 645
condensate 241, 247–250, 457, 469, 555
deaerator 429
diesel 26–27, 412, 470
extractive (*see also* Extractive distillation) 547
gasoline 318–319, 597

gas oil 226, 412
glycol 580
ground water 567
HCN 462
HF (*see* Alkylation)
hydrogen sulfide 27, 456, 602, 644
inert gas 152, 412
iso- (*see* Alkylation)
jet fuel 412, 489
kerosene 440
lights 112, 128, 411, 439, 474, 536
LCO 413, 627
LNG 462
lube oil 474
naphtha 27, 534, 647
NRU LP 562
olefins 468
organics from water 57, 63–66, 83–85, 236, 425, 437, 492
overhead line undersized 488–489
re-absorption 566
refinery 319–321, 486, 492
side 488
solvent 505
solvent-water 400, 408
sour water 155, 259, 309–313, 474, 559, 566, 600
steam 26–27, 155, 473, 507, 566, 574, 587, 634, 641
temperatures 412, 470
urea 561
vacuum 314
VCM 574
wastewater 295, 310–312, 454, 463

Stripping
gas 418
heat 411
insufficient 25–30, **410–411**, 574, 644
no stripping 412
partial pressure 412
restricted by steam inlet line 475
steam (*also see* Water-induced pressure surges) 26–27, 149–152, 412, 427, 471, 503, 574, 617
trays 410

Structured packing (*see* Packing, structured)
Styrene (*also see* Splitter) 398, 532, 605
Subcooling 61, 428–430
of absorption solvent 209–211, 428
causing plugging 301
damage, aqueous systems 292, 295, 305, 309–310, 591
effect on control 487, 637–639
effect on distribution 428, 430
effect on impurities 33–35
effect on simulation 2, **409**
effect on stripping 61, 412
effect on vapor and liquid loads 61, 72, **428–429**
of entrainer (azeotropic/extractive distillation) 305, 423
of feed 107, 301, 426, 428–430, 444, 484
of internal reflux, refinery fractionators 487–488
of reboiler condensate 332, 334
of reflux 19, 33, 60, 61, 428–429, 637–639
quenching at inlet zone 61, 133, 295, **429–430**, 464
trapping lights 417, 428

Suffocation 503, 510–511
Sulfinol (*see* Amine)
Sulfolane 547
Sulfur compounds (*see also* Hydrogen sulfide, Sulfuric acid) 3, 399
Sulfur plant 243
Sulfuric acid (*see* Alkylation)
Superfractionator (*see* Splitter)
Superheat
of feed 152, 272, 276–279, 409, 430
of reboiler steam 393

Support, grid 588
Support, packing 73
damage 95, 155, 457, 472
I-beam interference 73, 133, 460, 466
mesh screens cover 439
migration through 73, 439, 550
open area 85, 134, **439**
ring 85
strength 439

Support, pipe 457, 506
Support ring
bolt holes 175, 493
removal 85, 193, 449, **495–496**

Support, trays 294–295, 301, 303–305, 310–311, 480, 483, 486, 591
damage to 591, 593–594
heavy-duty 311, 587
I-beam 480, 591, 593
ledges 278, 304, 312
splitting trays into compartments 445, 467
stabilizer bars 311
tie-rods 593
trusses 312, 433, 466, 581, 587

Surfactant 438
Surface
temperature survey 112–113, 131–132, 404–405, 441, 444, 451–452, 458, 472, 484, 488, 515, 602
tension, high, in packed tower 73, 438, 459, 557

Surge
compressor (*see* Compressor)
pressure (*see* Downward; Pressure surges; Rapid; Vaporization, rapid; Water-induced pressure surges)
reboiler (*see* Thermosiphon)
Surge drum
absorber solution 244
base of tower 158–159, 485
cooling 429
feed 429, 644–645
reflux drum 295–296
volume 146, 157–159, 429
Swings (*see also* Base level; Boilup; Feed; Hiccups; Pressure, swings; Pressure, surges; Reflux, instability; Reflux drum; Thermosiphon; Vibrations):
Synthesis gas 453

Tails tower 95–96
Tall oil 285
TAME 492
Tangent pinch (see Pinch)
Tangential feed (*see* Distributor, vapor horn)
Tar 522, 529, 539, 566
TBP 11–12, 402
TCE 400
TEG 50, 438, 462
Temperature (*see also* Thermowell)
approach 108–110, 122–124, 141–143, 440, 449
bottom, excessive 25, 27, 237, 410, 518–522, 538, **539**
bottom, too low 27, 197, 322–324, 332, 359, 596
coil outlet 26, 32, 217–218, 272, 513
control (*see* Control, assembly; Control, temperature)
cooling water return 344–346, 386–387, 640
feed 40, 42–43, 47–49, 63, 81, 284, 313, 415, **417**, 424, 426, **428–429**, 507, 578, 625, 640, 643
measurement 88–89, 217–218, 405, 538, 573, 614
measurement for level indication 614
multiplicity 30–33, 412
monitoring for hot spots 236, 530–532
overhead, too high 427
overhead, variation 30–33
pinch 156, 596
profile 2, 15, 88–89, 197, 438
radial spreads 463
reboiler inlet 596
reboiler outlet 155, 217–218, 630
-reflux dependence 30–33, 87–89, 404
rise at turnaround 236, 521
survey (*see* Surface temperature survey)
Test (*see also* Foam test)
bench scale 60, 219, 240
column 289, 348–350, 447, 495, 543, 545, 547, 550, 557
at commissioning 212
control response 376
efficiency (*see* Efficiency measurement)
ejectors 348–350
exchanger leaks 239, 282–284
flood 90–95
higher loads 80
overflow 169
plugging 93
reboiler troubleshooting 322–325, 392, 474, 596, 617
rigorous 14–17
for solids 115–116
for simulation validation 14–17, **85–90**, 404–405
for troubleshooting 213, 284, 351–354, 411, 471–472, 589, 596, 608, 617
Tetra solvent 490
Thermal expansion
chimney tray 166, 175–176, 458, **479–480**, 482
at draw pan 184
fired heater tubes 576
pipes 506, 513, 602
spray header 457
stresses 526
trays 493
Thermal stress, shock 579
Thermocouple (*see* Thermowell)
Thermodynamically inconsistent 400
Thermosiphon reboilers
driving head/base level 319–322, 324–328, 518, 596, 598
dryout near top of tubes 518
erratic action 328–330, 604
excessive circulation 315, 596
failure to thermosiphon 316, 318–321, 326–329, 368, 597–598, 603
gas lifting to start thermosiphon 320–321, 326–329, 598, 603
horizontal, circulating 155, 315, 642
insufficient circulation 518, 524, 596, 642
-kettle in series 602–603
liquid supply to 155–157, 196–198, 319–325, 597–598
once-through 315, 319–322, 472, 492, 575, 597–599

oversized 388–393
pinching **315–317**, 326–328, 332, 597, 606
puffing 597
pulsation 322, 605
rods in tubes 598
short tubes 319
slug flow in outlet line 324, 599
surging 223, 315–317, 322–325, **596–597**
venting distribution baffles 597
vertical, circulating 223, 315–319, 596–597, 605–606
water accumulation (*see* Water)

Thermowell
cutting nozzle 529
fouling 347, 522, 614
not contacting fluid 347, 521–522, 573
supporting trays 593

Threaded connection 527
Three pass trays 134
Through-bolting (grid packing) 588
Tie rods (cartridge trays; grid packing) 588, 593
Time lag 324, 359, 373–**376**, 389, 621, 623, 628, **633–635**
Time studies (gamma scans) 78, 127–128, 326, 468, 599, 617
Toluene (*see also* BTX) 398, 454, 463, 466, 475, 532, 538, 596, 605, 625
Toluene azeotrope 625
Total reflux
base baffle problem 222–223
concentrates unstable component 521
start-up 222–223, 414
testing for leaks 239
for testing separation 400
undesirable reaction 541
vaporization of lights 509, 521
for water removal 414

Touchy (*see* Instability)
Tower skirt 511, 579
Tracer 129–131, 281–283, 449, 453, 462, 465, 474, 575–577, 619
Transfer line 225, 513
Transition tray 444, 566
Trap (*see* Liquid trap; Oil-water separation trap; Steam trap)
Trapped chemicals released 234, 501
Trapout pan (*see* Draw-off)
Trapping of intermediate component (*see* Accumulation, Hiccups)
Tray (*see* Assembly; Bubble-cap trays; Chimney trays; Downcomers; Dual flow trays; Feed/reflux entry to tray towers; Multi-pass trays; Plugging; Screen trays; Shed decks; Sieve trays; Stepped trays; Support, trays; Three-pass trays; Transition trays; Tunnel trough trays; Two-pass trays; Valve trays; Vibration)
deflection 594
downsizing 432–433
dry 73, 428, **435–436**, 487, 489, 491, 562, 624, 626–627
fatigue failure 594
high capacity 81, 106–107, 354–355, 434, 437, 441, 466, 490
hydraulic gradient 133, 433
hydraulic predictions 74–77, 80–81, 176, 340–342, 354
layout 73
levelness 352
natural frequency 594
replacement by grid 228
replacement by packing 90, 99, 112, 122, 127, 134, 183, 193, 241, 262, 427–428, 447, 450–451, 455, 458, 462, 464, 477–478, 480, 485, 495–497, 516, 571–572, 574, 591–592, 626–627
replacement by other trays 139, 178, 258–259, 261, 284, 314, 427, 431–435, 441, 466, 487, 489–491, 498, 552, 556, 566, 574, 599
spacing, low 251, 313, 431, 434, 441, 443, 467, 478, 489, 545, 553, 556
uniform liquid-flow devices 566

Trichloroethane 400
Triethylamine 578
Trips 279, 287, 425, 469, 525, **581–585**, 586
Troubleshooting procedure 353
Trough (*see* Distributor)
Trousers downcomer 140, 266–270
True boiling point 11–12, 402
Truncated downcomers 434
Trusses, trays 312, 433, 466, 581, 587
Tube leak (*see also* Rupture)
condenser 239, 281, 517, 576
diagnosing 239, 282–284, 575–577
preheater 281, 283–284, 526, **576–577**, 582, 605
pumparound heat exchanger 228, 281, 284–285, 576–577
reboiler 239, 281–283, **575–576**, 610

Tunnel trough trays 437
Turndown (*also see* Weeping) 79–80, 107, 194, 332, 387, 432, 444, 462, 466, 541, 593–594, 597–598, 641
Two columns in series (one separation) 308–309

Two liquid phases (*see also* Water, free)
in bottom sump 152, 517, 589, 618
in batch drum 520
in condenser 287, 335, 609
in decanter only 51, 55–57, 229–231, **419**, 524, 538
delayed boiling 517, 589
in reboiler base 316, 319, 597
in tower 2, 25, 40, 46–47, 57–60, 101,108, 229–231, 406, 413, 417, **419–423**, 516, 545–546, 630–631
Two-pass tray 138–140, 200–201, 248, 266, 303, 433, 467, 594, 615

UDEX 490
Unblinding (*see* Blinding)
Underground pipe 212–213, 349
Upflow, rapid (*see also* Vaporization, rapid) 291, 294, **589**
Uplift of trays or packing (*see also* Base level, uplift; Depressuring; Relief; Upflow; Vaporization; Water-induced pressure surges) 322, 499, 587–589, 591–594
Upset trays (*see* Damage)
Upward tray damage (*see also* Uplift) 310
Urea 561

V-baffle 113–114, 462–463
V-notch (*see* Distributor, liquid)
Vacuum
barometric pressure effects 348–350, 613
breaking 305, 311, 521, 580, 586, 590
chemical tower 33–37, 90–95, 128, 236–239, 285–286, 297–298, 300–303, 335–343, 386, 404, 438, 450, 458, 501–502, 517, 520–523, 533, 538, 542, 557, 579, 586, 590, 593, 607, 609–610, 614, 628, 633–634
cooling water side of condenser 387, 509
difficulty to achieve 286, 340–343, 348–350
drawing 297–298, 301–303
ejector (*see* Ejector)
implosion 216, 586
instrument purge 340–343, 348–350
liquid leg 342, 509
local 295, 305, 310, 312, 471
during maintenance 511
pump 297–298, 335–337
reboiler steam chest 292
reclaiming 555
relief 305, 308, 502, 586
sudden loss (*see* Failure, vacuum generation)
startup 297–298, 301–302

Vacuum refinery tower
asphalt 172, 515
asphaltenes 167, 271, 457
asphaltenes balance 571
assembly of internals 480, 493–494
base level, high 471–472
black gas oil 276, 402, 481
blinding 536
cavitation, bottom pump 485
chimney tray coking 402, 477, 572
chimney tray leakage 164–171, 175–176, 477–479
chimney tray overflow 164–170, 172–173, 477–478, 480–482, 497
chimney tray refractory 479
chimney tray/wash bed supports interaction 466
condenser waxing 608
cracking 199–200, 472, 608
cut point 275, 446, 477, 571–572, 608
damage 225, 227–228, 457, 460, 471–472, 477, 479, 513–517, 587–588, 591
damp 477
deep-cut 371, 402, 451, 475, 571–572
distributor plugging 497
draw nozzle plugging 402
draw-off (non-chimney-tray) 184, 487, 572
dry (no steam) 165, 167, 477
entrainment from flash zone 2, 272–274, 276, 464, 571–572
entrainment from sprays 571, 608
entrainment from tower top 165
feed simulation 2, 271, 275, 406
fire 506, 529, 531, 534
flash zone 2, 183–184, 271–276, 429, 464, 479, 573
flash zone pressure 272–276, 608, 612
flash zone temperature 429, 573, 608
fractionation bed distribution 446–447, 452
fuel oil to heater flow measurement 573
gravity distributor 275, 451
grid packing 168, 272–276, 464, 477, 571–572, 587–588
heat balance 164–171, 273, 477
heat transfer limiting 164–166, 408, 460, 477–478
heater pass water accumulation 513
high coil outlet temperature 199, 272, 573, 608
hot overhead 477
HVGO bleed to LVGO 166–167
HVGO flush system 504
HVGO/LVGO fractionation 446, 452
HVGO PA 20–21, 164–170, 228, 440, 477–478, 504

HVGO product tail 457, 464
hydrocracking 591
inlet too close to wash bed 464, 483
inlet velocity 168, 274, 464
instrument purge gas plugging 570
insufficient vacuum 164–167, 228, 412, 477, 577
leak 506, 517
level bridle plugging 402
level measurement/control problems 163, 169–170, 175–176, 272, 276, 477, 481–483
loss of vacuum, tray tower 591
low gas oil yield 166–170, 172–173, 273, 406, 429, 471–472, 477, 479–480, 571–572, 608
low lube cut yield 183–184, 199–200, 487
low metals crude 272–276
low tray efficiency 408
lube cut separation 433, 472
lube tower 183–184, 199–200, 408–409, 433, 464, 472, 487, 516, 588
LVGO PA 164–166, 477–478
materials of construction 494
metals balance 273
metals content of gas oil 271–276, 406, 451, 457, 483, 571
optimistic capacity prediction 409
overflash 167, 571–572
overflash pump 481, 572
overflow (slop wax chimney tray) 276
packing corrosion damage 438
packing plugging 504, 572
penetration 172–173, 515
petroleum fraction characterization 11–12, 271, 402
polymerization 464
pressure measurement 613, 615
pumparound exchanger leak 228, 577
quench 200, 429, 444, 479, 485
random packing 572
residence time 200, 272, 572, 608
residue 11–12, 167–170, 172–173, 175, 199–200, 271–272, 406, 608
short runs 275, 464, 504, 571–573
side-stream accumulator uplifted 514
simulation 1, 2, 11–12, 271, 275, 402, **406**, 409
sleeve (smaller diameter section) 200
slop in ejector condensate 165, 577
slop wax production, high 272
slop wax PA 572
spray distributor 154, 168, 274–275, 447, 450–451, 497, 571
spray (empty) PA section 440
start-up procedure 276, 504
stripping steam sparger 587
stripping steam line undersized 475
stripping trays 184, 279, 471–472, 493–494, 587
structured packings 272–276, 460, 477, 571–572
supports 460, 466, 480, 493
thermocouple, heater outlet 573
transfer line pressure drop 168, 199, 274
tray weep, turndown 184, 432–433, 467
vacuum depth 608, 612–613
vapor horn 168, 272–274, **464**, 615
vapor load 274, 467
vapor maldistribution 272, 464, 466–467
wash oil vaporization 271–276, 571–572
wash section 20–21, **271–276**, **406**, 571–573
wash section coking, dryout 73, 122, 168, **271–276**, 402, 406, 450–451, 457, 460, 464, 466, 482, **571–573**
wash section, cooler temperatures 169
wash section, grid cleaning 571–572
wash section, insufficient wash rate 271, 402, 406, **571–572**
wash section, high dP 272–276, 406, **571–572**, 615
wash section, packing inspection 276, 572
wash section, too tall/efficient 271–276, 571–572
wash section, sprays damage 457, 497
wash section, sprays issues 272, 457, 460, 571
wash section, sprays plugging 122, 168, **450–451**, 497, 529
water freeze in unused line 526
water pocket at pump 514–515
waxing 608
wet (using stripping steam) 515–516
wet stripping steam 515–516
Validation of plant data 1, 14–18, 90–95, 404–407
Valve
check 219, 297, 387–388, 535
control (*see* Control valve)
leak (*see* Leak)
removal (*see also* Blinding) 499
Valve trays 13, 79, 80, 97, 101, 138–139, 176, 229–231, 247–250, 266–270, 305, 308–310, 314, 363, 466–467, 475, 486–487, 491, 512, 556, 563–564, 581, 590–594, 625
blanking 73, 143, **194–195**, 251, **432–433**, 467, 481, 489
caged 262, 300
channeling 138–139, 141–143, 466, 481

Valve trays (*Continued*)
  directional valves 195–196
  downward damage (*see* Downward)
  fixed 141–143, 259–261, 300, 311–313, 431, 467, 481, 512, 552, 566, 581, 587
  flush with tray floor 432, 597
  fouling resistant 257, 566
  home-made valves 299, 590
  interrupter bars 258
  leak-resistant 73, 432
  legs bent 299
  leg corrosion 299–300
  long-legged 139
  low efficiency (*see also* Valve tray weeping) 408, 432
  nibs 432, 597
  open slot area, too large 139, 143, 467, 481, 489
  popping out 258, **298–300**, 587, **590**, 593
  removing valve floats 261
  reverse flow (*see* Reverse flow)
  seat corrosion 298–300
  spin 300
  sticking closed 257–258, **261–262**, 322–324, 564
  sticking open 262, 300
  valves beneath downcomer 491
  VCFC 138–139
  venturi 319–321, 466, 487
  weeping 73, 184, 319–321, **432–433**, 486–487, 489–490, 597–598
Van Laar method 401
Vanadium (*see* Vacuum refinery tower, metals)
Vane distributor 106, 277–278
Vapor
  cloud 216, 233, 281, 521, **524–528**, 534, 537
  collapse (*see also* Implosion; Steam-water operation) 107–108, 384, 389, 444, 586, 609, 637
  cross flow channeling 133, 138–139, **466**
  entrainment in outlet liquid 146, 179–181, 191, 335–340, 342, 484–486, 492
  gap damage 300–302, 591
  gap in flooded reflux drum 382
  horn, in crude tower 149–151
  horn, in vacuum tower 168, 272–274, **464**, 615
  inlet (*see also* Vapor maldistribution; Distributor, vapor; Vapor horn) 139, 145
  -liquid equilibrium (*see* VLE)
  loadings (*see* Hydraulic loadings)
  phase association (*see* Association)
  phase treatment for packing fires prevention 531
  pressure (*see also* Reid vapor pressure) 399–400
  static head 134
  side draw (*see* Draw-off, vapor)
  surge (*see* Depressuring, rapid; Upflow, rapid; Vaporization, rapid; Water-induced pressure surges)
Vapor maldistribution (*see also* Distributor, vapor)
  CFD modeling 464
  chimney tray 141–143, 464, 467, 478, 481
  condenser 335, 607, 610–611
  downflow in packing 462
  draw tray 489
  height between nozzle (or source of maldistribution) & bed/bottom tray 139, 277, 464, 483
  inlet 21, 22, 113, 133, 139–140, 272, 277, 449, **462–465**, 467, **472–475**, 489
  inlet velocities 110, 133, 151, 277, **462–464**
  internals damage, plugging 119–120, 463
  manhole 277
  obstruction by downcomer, support 140, 467
  packing 21, 22, 113, 119–120, 133, 264–265, 277, 449, **462–465**, 478, 568
  quenching at feed 464–465
  shed decks 110, 445, 464
  split to tray passes 133
  trays 97, 133, 139–143, 433, **466–467**, 481
  zig-zag flow path 133, 138–139, 433
Vaporization
  of desuperheating liquid 507
  due to superheat 430, 460, 507, 571–574
  of hazardous materials 4, 334, 389, 392, 538
  rapid (*see also* Depressuring) 152, 289, 295, 297–298, 313, 452, 516–517, 584, 589, 603, 636
VCFC (Vapor cross flow channeling) 133, 138–139, **466**
VCM 489, 527, 566, 574
Vent condenser 68–69, 335–339, **343–346**, 421, 579, **611**
  cooling water outlet temperature, high 344–346
  decanting 421
  entrainment/flooding 68–69, 335, 343–346, 611
  flow metering 579
  liquid removal from 335–339
Venting (*see also* Inerts)
  to atmosphere 234, 511, 528
  commissioning 212
  condensate pot 390–392, 642

condensation in vent line stack 410
condenser 297, 335, 340–343, 382, 386, 581, 597, 607–609
cooling water line 212–213
high point 180, 501
light ends 5, 33–35, 342, 386, 417, 581, 597, 607–609
line- 4, 180–181, 191, 211–212, 484, 503, 511, 538, 579, 586, 607–608
liquid 522, 538
low leg in vent pipe 211–212, 342, 586
off-gas 18–20, 521, 610
reboiler 316, 518, 524, 597, **605**, 642
reboiler horizontal baffle 597
reflux drum 381–382, 607, 636–637
from relief valve 610
scrubbing 95–96, 117–120, 494
sewer 4
startup/shutdown 534
storage tank 211–212, 399
stripping steam lines 515–516
wastewater tank 399
Venturi valve 319–321, 466, 487
Vertacoke 567, 577
Vibrations, flow-induced
column 291, 313–314, 593–594
condenser 288
line 456, 509, 527
monitoring 314
pumparound exchanger 285
reboiler 599
Viewing ports 120, 189, 313–314, 467
Vinyl acetate 540
Vinyl acetylene 521, 645
Vinyl chloride 489, 527, 566, 574
Visbreaker fractionator 241, 272, 488, 546, 572
Viscometer 634
Viscosity, cycling 439
Viscosity, high (*see also* Packing, high viscosity) 74, 438, 507, 528
Viscosity, too low 184
Viscosity runaway 262, 507, 509
Vitamin A 523
VLE (*see also* chemical systems VLE, Equations of state) 1–12, 17–18, 86, **398–402**, 538
association of molecules 9–11, **401–402**
characterization of components in petroleum fractions 1, 2, 11–12, **402**
close-boiling systems 1, **398**
data extrapolation 6–12, **400–402**
inaccuracies 17–18, 37, 398, 404–405
non-idealities 1–11, 38–42, 45–47, **399–402**, 538
VLLE 5–11, 60, 400
VOC 411
Vortex 160–161, **476**, 486
Vortex breaker 152, **161**

Wall temperature (*see* Surface temperature)
Warped 458
Wash section (*see* Vacuum refinery tower)
Wash tower 538
Washers 594
Washing (*see also* Flushing) 216, 502–505, 562
absorber-regenerator system 502–503
acid 503, 552, 559, 563
boiling hydrocarbon 503
boiling water 503
caustic 219–220, 438, 523, 536
chemical 262, 503, 505, 510, 530
detergent 438
dissolving deposits 115, 219–221, 258, 261, 446, 504–505, 559–560
drying after water-wash 261, 503
hydrocarbon 505
insufficient 501–502
inventory 502
mist 437
on-line 220–221, 258, 261, **504–505**
for packing fires prevention 234–235, 530–532
for packing wetting 438
surfactant 438
water 220, 236, 322–324, 502–503
Waste gas 511
Wastewater 2, 4, 295, 399, 414, 418, 454, 463, 538, 554, 566
Water
accumulation, induced to stabilize boil-up 316
accumulation in tower 4, 35, **37–42**, 49–50, 52–55, 414–419, 427, 620
accumulation in tower base 316–319, 596–597
balance 620
chlorides content 506
condensation near top of HC tower 427, 430
deoxygenator 458
depletion from tower base 596, 620
dissolved (in hydrocarbons) 38
draw (tower side draw) 39–40, 42, 101–104, 415
free (in hydrocarbons or in water-insoluble organics) 38, 40, 230–231, 249–250, 316, 414–417, 516, 609, 630
freezing (*see* Freezing, Hydrates)

Water (*Continued*)
ground- 559, 567
hammer (*see* Hammering)
impurity 33–35, 37–42, 49–50, 52–55, 414–419
in HC condenser 609
in HC stripper 248–250, 412
makeup 120, 474, 544, 552, 620
marks 119, 161, 189, 438, 459, 461
milky appearance 120
quench tower (*see also* Aftercooler tower) 108–110, 122–124, 140–143, 200–201, 229–231, 295, 399, 405, 440, 445, 449, 481, 495, 588
reducing residue thermal stability 523
refluxing (into HC or organic tower) 42, 225, **229–231**, 416, 420, 427, **516–517**, 558, 636
removal (*see also* Pressure surges, water-induced) 215, 504–505, **508**
scaling (*see* Plugging)
settling in tower base 316–319
soapy (*see* Soapy water)
-steam operation (*see* Steam-water operation)
Step-up of cold- 292, **295**, 305, **308–313**
suspended solids 120, 543
test (*see also* Distributor, water test) 166, 169, 171, **184–185**, 193, 205, 312
wastewater 2, 4, 295, 399, 414, 418, 454, 463, 538, 554, 566
Water-induced pressure surges 215–216, **225–231**, 291–292, **512–517**
dead pockets 225, 513–514
during abnormal operation 216, 512–517
heat exchanger leak 228, 517
heater passes/outlet piping 225–226, **513**
hot oil entry into water region 225, **517**
leaking valve, pump seal 228, 512
no liquid circulation at startup 514
pump/spare pump circuits 225–226, 228, 512, **514–515**
refluxed water/condensate 225, 229–231, **516–517**, 636
tank pumpout 226, 512
transfer lines accumulation 225, **513**
undrained stripping steam lines 225, 228, **515–516**
water in feed/slop to tower 225–226, 289, **512**
wet stripping steam 184, 226, 228, **515–516**
Waterfall pool effect 141, 189, 481
Wax fractionator 487
Waxing 608
Weep holes 128, 189, 456–457, 478, 514, 583
Weeping (*also see* Turndown; Valve trays) 73, 79, 588
at bubble caps 435
at drawoff 184, 427, 486–487, 490
effect of hole area 431
at intermediate weir 435
link to vibrations 291, 314, 593–594
poor assembly 588
promoting quench 305, 312–313
at reboiler trapout pan 315–316, 319–321, 597–598
at tray inlet 258, 433, 466
Weir
assembly 196, 198, 435, 491
bathtub overflow 602
decanter overflow- 55–57, 421
fouling-resistant design 566
height 433, 556
inlet 73, 259, 433, 435, 574
inlet, chimney tray 141–142, 481
inlet, at feed 104, 443
intermediate 435
interrupter bars 258
kettle reboiler overflow 599–601
length 134, 464
outlet 81, 574
overflow, chimney tray 101, 248
overflow, base (*see* Baffle, reboiler)
overflow, draw-off sump 181–182, 598
overflow, reboiler draw 321–322, 598
picket fence 196, 428, 433, 435
Wet packing
for efficiency improvement 438
for fire prevention 234–235, 530–532
loading into water 264
Wetted-wall column 338
What-if analysis 215, 347
Whiskey 435, 437
White smoke 235, 532
Wide boiling mixture 18–20, 25, 295–296, 335, 412, 609
Wilson equation (chemical systems VLE) 6, 8, 401
Wire mesh packing 22, 35–37, 204, 236, 254, 450, 458, 501, 532, 625

X-ray
debris in draw box 269
debris in pipe 269
Xylene (*see also* BTX; Splitter) 349, 622
impurity in feed 47

About the Author

Henry Z. Kister is a Senior Fellow and director of fractionation technology at Fluor Corporation. He has 30 years of experience in troubleshooting, revamping, field consulting, design, control, and startup of fractionation processes and equipment. Previously, he was Brown & Root's staff consultant on fractionation and also worked for ICI Australia and Fractionation Research, Inc. (FRI). He is the author of the textbooks *Distillation Design* and *Distillation Operation*, as well as 80 published technical articles, and has taught the IChemE-sponsored "Practical Distillation Technology" course more than 260 times. A recipient of *Chemical Engineering* magazine 2002 award for personal achievement in chemical engineering, and of the AIChE's 2003 Gerhold Award for outstanding contributions to chemical separation technology, Kister obtained his BE and ME degrees from the University of NSW in Australia. He is a Fellow of IChemE, a Member of the AIChE, and serves on the FRI Technical Advisory and Design Practices Committees.